Reinforced Concrete Design

Tenth Edition

Reinforced Concrete Design

Tenth Edition

Abi O. Aghayere, Ph.D., P.Eng.
Professor of Structural Engineering
Drexel University

Jason Vigil, P.E., S.E.
Consulting Engineer, Rochester, N.Y.

Mercury Learning and Information
Boston, Massachusetts

Publisher: David Pallai
Mercury Learning and Information
121 High Street
Boston, MA 02110
info@merclearning.com
www.merclearning.com
(800) 232-0223

A. Aghayere and J. Vigil. *Reinforced Concrete Design, Tenth Edition.*
ISBN: 978-1-68392-907-9

Library of Congress Control Number: 2023948768

242526321 This book is printed on acid-free paper in the United States of America.

To our families who have provided us
with much love and support that has made this project possible.

CONTENTS

PREFACE

The primary objective of *Reinforced Concrete Design, Tenth Edition*, remains the same as that of the previous editions that were co-authored with George F. Limbrunner who is now retired: to provide a basic and thorough understanding of the strength and behavior of reinforced concrete members and simple reinforced concrete structural systems.

With the recent changes to the ACI 318 Code, and relevant reinforced concrete research and literature continuing to become available at a rapid rate, it is the intent of this book to translate this vast amount of information and data into an integrated source that reflects the latest information available. This book is intended to help the reader understand the fundamentals of reinforced concrete design and behavior, and good practices in the industry. It will be useful to students in undergraduate civil and architectural engineering programs, engineering technology and architecture programs as well as construction engineering programs. In addition, practicing structural engineers and engineers preparing for the licensure exams will find this text to be a helpful and practical resource.

This tenth edition has been prepared with the primary objective of updating its contents to conform to the latest *Building Code Requirements for Structural Concrete* (ACI 318-19) of the American Concrete Institute. Throughout the text, frequent references are made to the pertinent sections of the ACI Code. Because the ACI Code serves as the design standard for concrete design and construction in the United States, it is strongly recommended that the code be used as a companion publication to this book.

In addition to the necessary changes to conform to the new code, some sections have been edited and new sections have been added, and the student design project exercises have been enhanced and the drawings updated. Working on the design project will help the student appreciate how and where the individual reinforced concrete structural elements covered in the different chapters fit within the context of a real-world building project. Thus, they can see how what they have learned is readily applicable to, and usable in, the real world of engineering and construction. Answers to selected end-of-chapter exercises are presented at the back of the text.

This text is suitable for any undergraduate course which would typically include topics from Chapters 1, 2, 3, 4, 5, 6, 7, and parts of Chapters 9 and 10. The remaining topics and chapters could make up a significant portion of a second undergraduate reinforced concrete design course. This text covers more topics than those required in the "Concrete I" course, and many of the topics required in the "Foundation Design" course, of the "Basic Education for Structural Engineers Curriculum" published by the Structural Engineering Institute (SEI) of the American Society of Civil Engineers (ASCE), the National Council of Structural Engineering Associations (NCSEA), and the Council of American Structural Engineers (CASE).

Throughout the ten editions, the text content has remained primarily a fundamental and practice-oriented approach to the design and analysis of reinforced concrete structural members using numerous examples and a step-by-step solution format. In addition, there are chapters that provide a conceptual approach on such topics as prestressed concrete and detailing of reinforced concrete structures. The metric system (SI) is introduced in Appendix C with several example problems.

Form design is an important consideration in most structural design problems involving concrete members, and Chapter 12 illustrates procedures for the design of job-built forms for slabs, beams, and columns. Appropriate tables are included that will expedite the design process. In Chapter 14, we introduce the reader to several practical considerations and rules of thumb for the design of reinforced concrete beams, girders, columns, and one-way slabs, and methods for strengthening, existing reinforced concrete structures.

What's New in the Tenth Edition

- The entire text has been revised to conform to the latest ACI Code: ACI 318-19, and many chapters have been enhanced with new drawings and photos. The following new or enhanced features have also been added:
- In Chapter 1, the different types of concrete floor systems have been added with their typical thickness and span ranges, and concrete durability and tolerances are discussed.
- Addition of pole footings and Sonutube foundations – commonly used in practice for lightly loaded footings – in Chapter 10, and an example presented.

This book has been thoroughly tested over the years in civil and architectural engineering, and engineering technology programs and should serve as a valuable design guide and resource for engineering and architectural students, technologists, technicians, and design engineers. In addition, it will aid engineers and architects preparing for state licensing examinations for professional registration.

Instructor's Resources

Upon adoption, these files are also available to instructors by contacting the publisher at *info@merclearning.com*:

- Solutions to most of the exercises at the ends of the chapters
- MS PowerPoint Slides
- Sample syllabus for a course based on this book

Notice to the Reader

Acknowledgments

Thanks are due to George Limbrunner, the founding co-author of this text since 1977 and my co-author for the sixth through the eight editions, for the rich legacy and practical imprints he left on this text. I would like to welcome Jason Vigil, my new co-author for this tenth edition. We are excited about continuing the practice-oriented and easy-to-understand emphasis of this textbook that effectively connects the theory of reinforced concrete to the practical design of reinforced concrete structures, while introducing some new topics of interest based on the feedback we have received on the previous editions.

As in the past, appreciation is extended to students, past and present, and colleagues who, with their constructive comments, criticisms, and enthusiasm, have provided input and encouragement for this new edition.

We would like to express our gratitude to our families for their unyielding support and encouragement, and in particular to our wives - Josie Aghayere and Michele Vigil - for their endearing support and encouragement.

Abi O. Aghayere and Jason Vigil
Avondale, PA and Rochester, NY
December 2023

CHAPTER 1

Materials and Mechanics of Bending and Concrete Slab Systems

1.1 CONCRETE

Concrete, the most commonly used construction material worldwide, is a composite material that consists primarily of a mixture of cement and fine and coarse aggregates (sand, gravel, crushed rock, and/or other materials) to which water has been added as a necessary ingredient for the chemical reaction during the curing process. The bulk of the mixture consists of fine and coarse aggregates. The resulting concrete strength and durability are a function of the proportions of the mix as well as other factors, such as the concrete placing, finishing, and curing history. A typical cross-section of concrete is shown in Figure 1.1a. Note the presence of the coarse aggregates, the fine aggregates, and the cement paste that binds them together. The two main types of concrete used in building construction are normal-weight and light-weight concrete:

- Normal-weight concrete has a density of 145 pounds per cubic foot (pcf) and is more commonly used in construction.
- Lightweight concrete has a density that varies between 90 and 115 pcf. They are used to reduce the self-weight of the structure and therefore the load on multi-story buildings, resulting in thinner members and smaller floor-to-floor heights and a reduced total load on the foundation system. Lightweight concrete can provide the same fire rating or protection as normal-weight concrete at a smaller thickness than normal-weight concrete. Though it has a higher per cubic foot cost than normal-weight concrete, structural lightweight concrete is competitive in high-rise building construction and for structures with long floor spans for the reasons cited previously.

The compressive strength of concrete is relatively high, with typical values varying from 3000 psi to higher than 10,000 psi, and concrete strengths of up to 16,0000 psi have been used at lower levels of high-rise building columns and shear walls in the United States. Yet, concrete is a relatively brittle material, with a small tensile strength compared with its compressive strength. Hence, steel reinforcing rods (which have high tensile and compressive strength) are combined with the concrete, with the steel resisting the tension and the concrete resisting the compression. *Reinforced concrete* is the result of this combination of steel and concrete. In many instances, steel and concrete are positioned in members so that they both resist compression.

FIGURE 1.1a Typical concrete cross-section. (Abi Aghayere)

1.2 DESIGN CONSTRAINTS AND THE ACI BUILDING CODE

The design of structures typically involve satisfying the following constraints:

- Strength and serviceability requirements
- Economy
- Constructability, durability, and sustainability

The design and construction of reinforced concrete buildings in the United States are controlled by the *Building Code Requirements for Structural Concrete* (ACI 318-19) of the American Concrete Institute (ACI) [1]. Concrete construction is carried out per ACI 301-20, Specifications for Concrete Construction, which contains the means and method for placing and transporting the concrete. It contains information on formwork, reinforcement, concrete materials, and specialty sections such as architectural concrete, posttensioning and mass concrete, lightweight concrete, shrinkage compensating concrete (SCC), industrial floors, precast concrete, and tilt-up walls. ACI 301 complements the ACI 318 Code requirements and is a mandatory document meant to be used by contractors. Thus, ACI 318 provides directions to the structural designer, while ACI 301 provides directions to the contractor on how to build the concrete structure. The term *code* in this text refers to the ACI 318 Code unless otherwise stipulated. The code, which prescribes the minimum requirements to ensure life safety for the public, is revised, updated, and reissued on a 3 to 6-year cycle. It has been incorporated into the building codes of almost all states and municipalities throughout the United States. When so incorporated by a municipality, it has an official sanction, becomes a legal document, and is part of the law controlling reinforced concrete design and construction in that jurisdiction.

1.3 CEMENT AND WATER

Structural concrete uses, almost exclusively, hydraulic cement. With this cement, water is necessary for the chemical reaction of *hydration*. The cement sets and bonds the fresh concrete into one mass in the hydration process. *Portland cement*, which originated in England in 1824, is undoubtedly the most common form of cement. Portland cement consists chiefly of calcium and aluminum silicates. The raw materials are limestones, which provide calcium oxide (CaO), and clays or shales, which furnish silicon dioxide (SiO_2) and aluminum oxide (Al_2O_3). The raw materials are ground, blended, and then baked in an oven or kiln at high temperatures of up to 2900°F until the CO_2 in the limestone is dissipated. The resulting material is ground to a fine powder to form the Portland cement. The manufacturing process for Portland cement utilizes a lot of energy because of the high temperatures required, and cement is the most expensive component material in concrete. Admixtures are sometimes added to replace portions of the cement in concrete. However, they can also be used to replace a portion of the fine aggregate or used in addition to the cement to achieve a desired characteristic or property [13]. Following processing, cement is marketed in bulk or in 94-lb (1-ft^3) bags.

There are eight types of Portland cement recognized by ASTM: types I, IA, II, IIA, III, IIIA, IV, and V, with type I cement being the most commonly used in reinforced concrete construction (see ASTM C150/C150M – Standard Specification for Portland Cement).

Types of Cement per ASTM C150/C150M

ASTM Types of Portland Cement	When to Use
Type I	General construction that does not require any special properties. This is the most used type of cement.
Type IA	Same as type I, but where air entrainment is required to ensure durability against freeze-thaw action and to ensure better workability.
Type II	General construction, but where moderate sulfate resistance or moderate heat of hydration is required.
Type IIA	Same as type II, but where air entrainment is required to ensure durability against freeze-thaw action and to ensure better workability.
Type III	When high early strength is required.
Type IIIA	Same as type III, but where air entrainment is required to ensure durability against freeze-thaw action and to ensure better workability.
Type IV	When low heat of hydration is required.
Type V	When high sulfate resistance is required.

To reduce the embodied carbon in concrete, ASTM C150/C150 types I and II cement have been replaced by Portland limestone cement or PLC that conforms to ASTM C595/C595M type IL in many parts of the United States. The use of PLC can "reduce the carbon footprint of concrete by up to 10%" [16].

In fresh concrete, the ratio of the amount of water to the amount of cement, by weight, is termed the *water-cementitious material ratio*. This ratio can also be expressed in terms of gallons of water per bag of cement. For complete hydration of the cement in a mix, a water-cementitious material ratio of 0.35 to 0.40 (4 to 4½ gal of water/bag of cement) is required. To increase the *workability* of the concrete (i.e., the ease with which it can be mixed, handled, and placed), higher water-cementitious material ratios are normally used, or superplasticizers are added to the mix.

1.4 AGGREGATES

In ordinary structural concretes, the aggregates occupy approximately 70% to 75% of the volume of the hardened mass. Gradation of aggregate size to produce close packing is desirable because, in general, the more densely the aggregate can be packed, the better the strength and durability.

Aggregates are typically classified as fine or coarse. *Fine aggregate* is generally sand and may be categorized as consisting of particles that will pass a No. 4 sieve (four openings per linear inch). *Coarse aggregate* consists of particles that would be retained on a No. 4 sieve. The maximum size of coarse aggregate in reinforced concrete is governed by various ACI Code requirements. These requirements are established primarily to ensure that the concrete can be placed with ease into the forms without any danger of a jam-up between adjacent bars or between bars and the sides of the forms. Section 26.4.2.1 of ACI 318 states that the maximum size of coarse aggregate should not exceed the least of the following:

i. one-fifth the narrowest dimension between sides of forms

ii. one-third the depth of slabs

iii. three-fourths the minimum specified clear spacing between individual reinforcing bars or wires, bundles of bars, prestressed reinforcement, individual tendons, bundled tendons, or ducts.

Note that smaller aggregate sizes have a relatively larger total surface area and thus require more cement paste to coat the surfaces of the aggregate, which results in higher drying shrinkage. In typical building construction, the nominal maximum coarse aggregate size of ¾″ is commonly specified, though a maximum aggregate size of up to 1½″ is also sometimes specified to achieve reduced shrinkage.

1.5 CONCRETE MIXES AND DURABILITY

There are numerous factors to consider when determining all the components of a concrete mix. The most basic mix consists of water, cement, coarse aggregate, and fine aggregate (sand). Beyond these basic components, one or more admixtures can be used depending on the required performance of the concrete. Since the admixtures are a proprietary product, the dosage is based on the manufacturer's recommendations. Admixtures are often dosed in units of fluid ounces per 100 pounds of cementitious material. The strength of the concrete could be adversely impacted by certain admixtures, and there could also be compatibility issues with multiple admixtures. To control these issues and ensure the performance of the concrete, concrete suppliers will typically go through the process of creating batches of concrete and testing them. Once a certain mix is validated, it can then be used in the future using the same components and proportions. A sample mix design data is shown in the table below. The slump

Sample Mix Design Data

The following provides sample mix design data. Note that the code requires that the contractor provide a verified mix design.

Mix Design Number: 4001-A (this number is specific to the concrete supplier)
Minimum compressive strength at 28 days, f_c' = 4000 psi
Used for: **Footings, piers, and foundation walls**
Slump: 4 in.; 8 in. with HRWR
Air content: 6.0% +/– 1.5%
water-cementitious material ratio, w/cm: 0.40

MATERIALS	WEIGHT PER CUBIC YARD
Cement, Type I/II, ASTM C150**	540 lbs
Fly Ash, ASTM C618	95 lbs
Natural Sand, ASTM C33	1230 lbs
#1 and #2 crushed stone, ASTM C33	1780 lbs
Potable Water, ASTM C1602	255 lbs
Air Entrainment, ASTM C260	1.5 oz./cwt
Water Reducer, ASTM C494 Type A	3.0 oz./cwt
High-Range Water Reducer (HRWR), ASTM C494 Type F*	7 oz. to 10 oz./cwt
Nonchloride Accelerator, ASTM C494*	0 oz. to 26 oz./cwt

*Added at the site upon request

**In some parts of the country, Portland Limestone Cement (PLC), which is produced with marginally lower greenhouse gas emissions (~10%–)—of which type 1L (ASTM C-595) cement is the most common—has replaced types I/II C-150 cement. According to the New York State Construction Materials Association, "PLC is substituted for C-150 cement at a one-to-one ratio. Ultimate strength, rate of strength gain, durability, and performance characteristics are equal. Supplementary Cementitious Materials (SCMs) like fly ash, slag, and other pozzolans are compatible with PLC. Mixtures utilizing SCMs in combination with PLC can further reduce concrete embodied carbon. Existing concrete mixture designs can be used as-is, substituting the PLC for C-150 cement." C-595 allows a higher lime content (up to 15%) in the cement during the process of manufacturing compared to up to 5% for C-150 cement.

is a measure of the workability of the concrete mix that would allow for adequate placement of the concrete in the forms (see ASTM C143/C143M-20, "Standard Test Method for Slump of Hydraulic-Cement Concrete"). The higher the slump, the more flowable or workable the mix is and the easier it is to place the concrete. Typical slump ranges from 2 inches to about 5 inches.

The most critical parameter in the design and selection of a concrete mix is the design strength, f_c', which is defined as the compressive strength of the concrete at 28 days after placement (ACI 318-19 Section 19.2.1.3). In addition, Chapter 26 of the ACI 318 Code requires consideration for concrete durability. For each group of structural members, the exposure classes must first be assigned within each of the four categories in ACI 318 Table 19.3.1.1 :

- Exposure to Freeze-thaw cycles (**F**),
- Exposure to water-soluble sulfates (**S**),

- Contact with water (**W**), and
- Protection of reinforcement from corrosion (C)

Once the exposure classes are known, ACI 318 Table 19.3.2.1 is used to determine the required minimum concrete requirements: compressive strength, the maximum water-cementitious materials ratio, and other additional requirements such as air content, concrete cover, types of cementitious materials, maximum water-soluble chloride ion content in the concrete as a percentage by mass of cementitious materials, and limits on the quantity of supplementary cementitious materials (SCM) such as fly ash, silica fume, and slag cement. As an example, concrete exposed to seawater is Exposure **C2,** and this requires a minimum concrete compressive strength of 5000 psi, and a maximum water-to-cementitious materials ratio of 0.40. Where a member is assigned more than one exposure class, the most stringent durability requirements should be specified. ACI 318 Table 19.3.2.1 is also used to determine whether air entrainment is required. If air entrainment is required, then ACI 318 Table 19.3.3.1 is used. The amount of air entrainment needed is a function of the exposure class and the aggregate size. Air entraining typically ranges between 3.5% and 7.5%.

The most common admixtures used in concrete are outlined as follows:

Air Entrainment

Air entrainment provides small air bubbles in the concrete to act as a buffer against volumetric changes during freeze/thaw cycles and thus makes the concrete more durable. It also makes the concrete more workable. A rule of thumb is that the compressive strength of concrete will decrease by 5% for every 1% increase in entrained air.

Water Reducers

When it is desired to make the concrete more workable, a common, cost-effective solution is to simply add water, but this will have a negative impact on the strength of the concrete. Water-reducing admixtures will increase the slump and, thus, the workability of concrete while not adversely impacting the water-cementitious material ratio.

When a much higher degree of workability is needed, a superplasticizer (high-range water reducer) can be used, which will yield slumps around 8 in. and higher. There are also products available that are called mid-range water reducers that provide a more moderate amount of slump increase.

Accelerators and Retarders

It is sometimes desired to decrease or increase the setting time of the concrete, and so an accelerator or retarder product could be used. Accelerators can be used to increase the rate at which the concrete reaches its design strength. This may be desired due to weather conditions or to meeting a certain construction schedule. Calcium chloride is a common accelerator component, but it cannot be used in reinforced concrete since the chloride will promote corrosion in the steel reinforcement.

Set retarders are used to delay the chemical reaction with the cement that allows the concrete to set. This might be used to offset the effect of high temperatures since high temperatures can produce a faster setting time. They might also be used, for example, during a paving operation to allow more time between concrete batches and reduce the possibility of a cold joint in the *pavement.*

Corrosion-Inhibitor

Corrosion-inhibiting admixtures are used to deter corrosion of reinforcing steel in concrete that is exposed to water or salt. This will increase the durability and life span of the concrete structure.

Other chemical admixtures include shrinkage compensating and permeability reducing admixtures. Refer to ACI 212.3R-16 Report on Chemical Admixtures for Concrete.

Concrete that is placed between 50°F and 85°F do not require any special mix design or temperature considerations. When concrete is placed during extreme temperatures outside of these ranges, special considerations for cold and hot weather concrete placement must be accounted for [13, 14].

When concrete is poured in cold temperatures without any precautionary measures being taken, the concrete will freeze, preventing the hydration process from taking place, and this will destroy the bond between the concrete and the steel reinforcement. Conversely, pouring concrete at excessily high temperatures causes the concrete to set faster than anticipated, leading to formation of cold joints between the different concrete pours.

High temperatures also leads to uncontrolled and significant cracking of the concrete which negatively impacts strength and durability. Detailed information on the temperature requirements for concrete can be found in ACI 305-Guide to Hot Weather Concreting, and ACI 306 - Guide to Cold Weather Concreting. Special mix considerations also apply to mass concrete pours, such as thick mat foundations, where the high temperatures generated from the hydration process, if not adequately controlled, can lead to cracking of the concrete.

1.6 VOLUMETRIC CHANGES IN CONCRETE DUE TO TEMPERATURE, SHRINKAGE, AND CREEP

Volumetric changes occur in concrete due to temperature change, shrinkage, and creep. When there is a change in temperature, the concrete can increase in volume with an increase in temperature or decrease in volume with a decrease in temperature. Volumetric changes can also occur due to applied sustained stress and change in moisture content. The coefficient of linear expansion for concrete is approximately 6×10^{-6} in./in./°F.

Shrinkage

Shrinkage in concrete begins while the concrete is still in its plastic state and continues even after the concrete hardens. Shrinkage continues as the hardened concrete loses water through evaporation over time and undergoes drying (i.e., drying shrinkage), causing a reduction in the volume of the concrete. Water can also be lost to the formwork through suction or to the subgrade for slab-on grades. In the plastic state, restraint to shrinkage (RTS) is provided by the rebar, resulting in hairline surface cracks—which in slabs will be more uniformly distributed because of the temperature and shrinkage rebar provided. External restraints to shrinkage prevent the concrete from reducing in volume, and this would lead to full-depth cracks. To prevent these cracks from forming, pour gaps or pour strips are often used for large concrete pours. Factors affecting shrinkage include water-to-cementitious materials ratio, temperature, moisture content, aggregate size, rate of drying, relative humidity, and thickness of the concrete member [13].

To minimize drying shrinkage in concrete,

1. Avoid adding water to the concrete mix than is necessary. Do not allow water to be added to the mix on site.
2. Use shrinkage compensating cement (SCC) to avoid the use of pour gaps or strips. There are also proprietary "mechanical reinforcement splice systems" that can eliminate pour strips or gaps while still allowing volumetric changes in the slab and providing structural integrity at the joint.
3. Provide adequate curing of the concrete using wet blankets or spraying with water.
4. Limit concrete pours to 60 ft x 60 ft without construction joints.
5. Movement joints such as construction joints, expansion/contraction joints, control joints (such as used in slab on grades), expansion joints, and pour gaps can be used to mitigate the effects of shrinkage in concrete structures.
6. Provide temperature and shrinkage rebar near the top and bottom surfaces of concrete slabs and provide skin rebar in deep beams.

Creep in Concrete

When a concrete member is loaded, there is an instantaneous elastic deformation that occurs. Over time, under the same constant load, the concrete member will continue to deflect, and the final long-term deflection due to creep may be up to three times the instantaneous deflection. Creep depends on the sustained applied stress and the amount of free moisture in the concrete, the water-cementitious materials ratio, the curing conditions, temperature, cement type, size and shape of the concrete member, and composition and the presence of admixtures. As the ratio of the water-to-cementitious materials increases, so does the creep of concrete. Thus, creep is the continuous long-term deformation of the concrete member under constant load. The effect of creep on deflections is discussed in Chapter 7.

About 70% of the additional long-term deflections due to creep occur within 12 months of the concrete pour, and 100% of the creep deflection is assumed to occur at the 5-year mark and beyond. The author once worked on a project with a specification for very little shrinkage and creep where the pour strips had to be left open for at least 24 months to allow most of the shrinkage to occur before the pour strips were closed.

1.7 CONCRETE IN COMPRESSION

The theory and techniques relative to the design and proportioning of concrete mixes, as well as the placing, finishing, and curing of concrete, are outside the scope of this book text and are adequately discussed in many other publications [2–5]. Field testing, quality control, and inspection are also adequately covered elsewhere. This is not to imply that these are of less importance in overall concrete construction technology but only to reiterate that the objective of this book is to deal with the design and analysis of reinforced concrete members.

We are concerned primarily with how a reinforced concrete member behaves when subjected to load. It is generally accepted that the behavior of a reinforced concrete member under load depends on the stress–strain relationship of the materials, as well as the type of stress to which it is subjected. With concrete used principally in compression, the compressive stress–strain curve is of primary interest.

The compressive strength of concrete is denoted f_c' and is assigned the unit of *pounds per square inch* (psi). This is the unit of f_c' used in the ACI Code equations. For calculations, f_c' is frequently used with the unit *kips per square inch* (ksi).

A test that has been standardized by the American Society for Testing and Materials (ASTM C39—Standardized Test Method for Compressive Strength of Cylindrical Concrete Specimens) [6] is used to determine the compressive strength (f_c') of concrete. The test specimens are obtained at the point of delivery of the concrete per ASTM C172—Standard Practice for Sampling Freshly Mixed Concrete, and the specimens are prepared and cured in accordance with ASTM C31—Standard Practice for Making and Curing Concrete Test Specimens in the Field. The test involves compression loading to failure of several 6-in diameter x 12-in. long concrete cylinders in a compression testing machine. The compressive strength so determined is the highest compressive stress to which the specimen is subjected. Note in Figure 1.1b that f_c' is not the stress that exists in the specimen at failure but that which occurs at a strain of about 0.002 (though the concrete strain at f_c' may vary between approximately 0.0015 and 0.0025). Currently, 28-day concrete strengths (f_c') range from 2500 to upwards of 10,000 psi, with 3000 to 4000 psi being common for reinforced concrete structures and 5000 to 6000 psi being common for prestressed concrete members, and higher strengths used for columns in high-rise buildings. High-strength concrete of up to 16,000 psi at 56 days has been used for columns and shear walls of a building in New York City. For normal-weight concrete (density = 145 pcf), the compressive strength at 28 days is specified in the ACI Code as the design strength. However, for higher-strength concrete such as concrete used in thick mat foundations, the compressive strength at 56 or 90 days is commonly used and allowed in ACI 318-19 Section 19.2.1.3 to cope with the higher temperatures generated from the cement hydration process and the desired slow rate of hydration [15]. The curves shown in Figure 1.1b represent the result of compression tests on 28-day standard cylinders for varying design mixes. The higher the concrete strength, the lower the ductility, that is, the steeper and shorter the descending portion of the stress–strain curve. The strain at rupture of concrete varies, as indicated by the plots in Figure 1.1b, but in developing the ACI 318 Code equations for flexure, a limiting crushing strain of 0.003 is assumed (ACI 318-19, Section 22.2.2.1). The acceptance of the concrete used in a structure is determined from the concrete cylinder tests.

According to ACI Code, Section 26.12.3, the acceptance or compliance criteria for standard-cured concrete cylinder specimens is as follows (ACI 318 Section 26.12.3.1):

1. Every average of three consecutive concrete cylinder tests is not less than the specified concrete compressive strength, f_c', and
2. No concrete cylinder tests have a compressive strength lower than $f_c' - 500$ psi for $f_c' \leq 5000$ psi.
3. No concrete cylinder tests have a compressive strength lower than $0.90f_c'$ for $f_c' > 5000$ psi.

If the above compressive strength compliance criteria are not met, then an investigation of the in-place concrete strength has to be carried out. Per ACI 562-16, Section 6.4.1.1 [14], the investigation can be carried out using core tests only or by a combination of concrete core tests and nondestructive testing (NDT) methods such as rebound hammer, pulse velocity measurements, etc. Note that ACI 562-16, Section 6.4.3.2, does not permit the use of NDT methods alone. The reader should refer to ACI 228.1R-19, "Report on Methods for Estimating In-Place Concrete Strength," for detailed information on the implementation and interpretation of the results of in-place concrete testing.

Concrete coring must be carried out in accordance with ASTM C42/C42M, "Standard Test Method for Obtaining and Testing Drilled Cores and Sawed Beams of Concrete." In accordance with ACI 318-19 Section 26.12.6.1e, concrete in an area of a structure evaluated by testing of concrete cores is deemed to be adequate if the average compressive strength of three cores is at least $85\%f_c'$ and no concrete core compressive strength is lower than $75\%f_c'$. If these conditions are not satisfied, "additional testing of cores extracted from locations represented by erratic strength results shall be permitted" per ACI 318-19 Section 26.12.6.1(f). The reader should refer to ACI 318 Section 26.12.6.1 for other requirements for drilled concrete cores.

A review of the stress–strain curves for different-strength concretes reveals that the maximum compressive strength, f_c', is generally achieved at a unit strain of approximately 0.002 in./in. The stress then decreases, accompanied by additional strain. Higher-strength concretes are more brittle and will fracture at a lower maximum strain than will the lower-strength concretes. The initial slope of the curve varies, unlike that of steel, and only approximates a straight line. For steel, where stresses are below the yield point and the material behaves elastically, the stress–strain plot will be a straight line. The slope of the straight line is the modulus of elasticity. For concrete, however, we observe that the straight line portion of the plot is very short, if it exists at all. Therefore, there exists no constant value of modulus of elasticity for a given concrete because the stress–strain ratio is not constant. It may also be observed that the slope of the initial portion of the curve (if it approximates a straight line) varies with concretes of different strengths. Even if we assume a straight line portion, the modulus of elasticity is different for concretes of different strengths. At low and moderate stresses (up to about $0.5f_c'$), concrete is commonly assumed to behave elastically.

The ACI 318 Code, Section 19.2.2, provides the accepted empirical expression for *modulus of elasticity*:

$$E_c = w_c^{1.5} 33\sqrt{f_c'}$$

where

E_c = modulus of elasticity of concrete in compression (psi)

w_c = unit weight of concrete (lb/ft^3)

f_c' = compressive strength of concrete (psi)

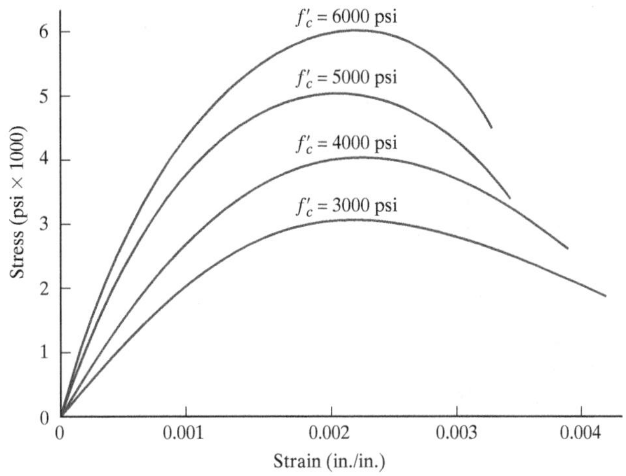

FIGURE 1.1b Typical stress–strain curves for concrete.

This expression is valid for concretes having unit weights or density, w_c, between 90 and 160 lb/ft^3. For normal-weight concrete, the unit weight w_c will vary with the mix proportions and with the character and size of the aggregates. If the unit weight is taken as 144 lb/ft^3, the resulting expression for the modulus of elasticity is

$$E_c = 57{,}000\sqrt{f_c'} \qquad (\textit{where}, f_c' \text{ is in psi; see Table A.6 for values of } E_c)$$

It should also be noted that the stress–strain curve for the same-strength concrete may be of different shapes if the condition of loading varies appreciably. With different *rates of strain* (loading), we will have different shape curves. Generally, the maximum strength of a given concrete is smaller at slower rates of strain.

Concrete strength varies with time, and the specified concrete strength is usually that strength that occurs 28 days after the placing of concrete. A typical strength–time curve for normal stone concrete is shown in Figure 1.2. Generally, concrete attains approximately 70% of its 28-day strength in 7 days and approximately 85% to 90% in 14 days. The compressive strength of concrete can be increased by one or more of the following ways:

- Increasing the cementitious material content.
- Reducing the water-to-cementitious materials ratio.
- Use water-reducing admixtures.

As previously discussed, concrete under constant load exhibits a phenomenon termed *creep*. This is the property by which concrete continues to deform (or strain) over long periods of time while under constant load. Creep occurs at a decreasing rate over a period of time and may cease after several years. Generally, high-strength concretes exhibit less creep than do lower-strength concretes. The magnitude of the creep deformations is proportional to the magnitude of the applied load as well as to the length of time of load application.

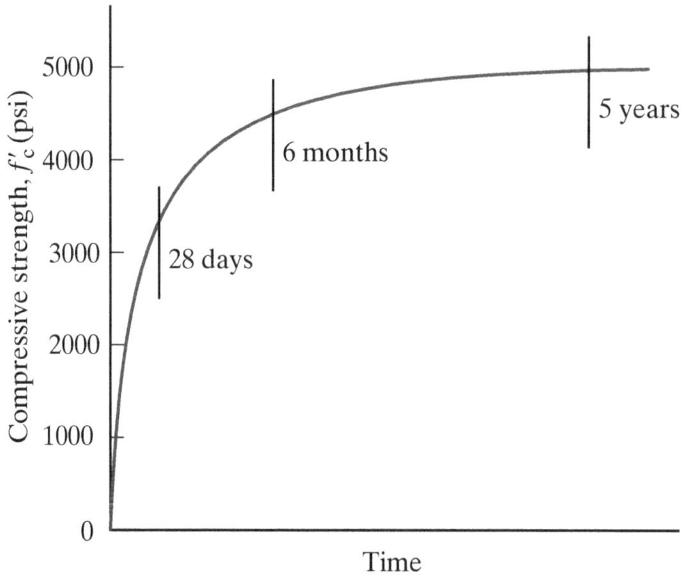

FIGURE 1.2 Strength–time relationship for concrete.

1.8 CONCRETE IN TENSION

The tensile and compressive strengths of concrete are not proportional, and an increase in compressive strength is accompanied by an appreciably smaller percentage increase in tensile strength. According to the ACI 318 Commentary, the tensile strength of normal-weight concrete in flexure is about 10% to 15% of the compressive strength.

The true tensile strength of concrete is difficult to determine. The *split-cylinder test* (ASTM C496) [6] has been used to determine the tensile strength of lightweight aggregate concrete and is generally accepted as a good measure of the true tensile strength. The split-cylinder test uses a standard 6-in.-diameter, 12-in.-long cylinder placed on its side in a testing machine. A compressive line load is applied uniformly along the length of the cylinder, with support furnished along the full length of the bottom of the cylinder. The compressive load produces a transverse tensile stress, and the cylinder will split in half along a diameter when its tensile strength is reached.

The tensile stress at which splitting occurs is referred to as the *splitting tensile strength*, f_{ct}, and may be calculated by the following expression derived from the theory of elasticity:

$$f_{ct} = \frac{2P}{\pi LD}$$

where

f_{ct} = splitting tensile strength of lightweight aggregate concrete (psi)

P = applied load at splitting (lb)

L = length of cylinder (in.)

D = diameter of cylinder (in.)

Another common approach has been to use the *modulus of rupture*, f_r (which is the maximum tensile bending stress in a plain concrete test beam at failure), as a measure of tensile strength (ASTM C78) [6]. The moment that produces a tensile stress just equal to the modulus of rupture is termed the *cracking moment*, M_{cr}, and may be calculated using methods discussed in Section 1.11. The ACI Code recommends that the modulus of rupture f_r be taken as $7.5\lambda\sqrt{f_c'}$, where f_c' is in psi (ACI 318 Equation 19.2.3.1). Greek lowercase lambda (λ) is a modification factor reflecting the lower tensile strength of lightweight concrete relative to normal-weight concrete. Normal-weight concrete has a unit weight or density of approximately 145 pounds per cubic feet (pcf), while lightweight concrete has densities of between 90 and 135 pcf. The values for λ are as follows:

Normal-weight concrete: 1.0

Sand-lightweight concrete: 0.85

All-lightweight concrete: 0.75

Interpolation between these values is permitted. See ACI 318 Tables 19.2.4.1a, b for details.

1.9 REINFORCING STEEL

Concrete cannot withstand very much tensile stress without cracking; therefore, tensile reinforcement must be embedded in the concrete to overcome this deficiency. In the United States, this reinforcement is in the form of steel reinforcing bars or welded wire reinforcing composed of steel wire. In addition, reinforcing in the form of structural steel shapes, steel pipe, steel tubing, and high-strength steel tendons is permitted by the ACI Code. Many other approaches have been taken in the search for an economical concrete reinforcement. Among these are the fiber-reinforced concretes, where the reinforcement is obtained through the use of short steel fibers or other materials, such as fiberglass. In this book, our discussion will primarily include steel reinforcing bars and welded wire reinforcing. High-strength steel tendons are used mainly in prestressed concrete construction (see Chapter 11).

The specifications for steel reinforcement published by the ASTM are generally accepted for the steel used in reinforced concrete construction in the United States and are identified in ACI 318 Section 20.2.

Though the ACI Code does not expressly prohibit the wet-setting or wet-sticking of rebar in freshly poured concrete, ACI Code, Section R26.6.2.2(a) does state that: "Reinforcement, including bundled bars, should be adequately supported in the forms to prevent displacement by concrete placement or workers." While it is not uncommon in residential construction to see some contractors place or stick rebar in wet concrete, this practice should not be encouraged because the rebar displaces the aggregate, and proper bond between the concrete and rebar cannot easily be ensured.

The steel bars used for reinforcing are, almost exclusively, round, deformed bars with some form of patterned ribbed projections rolled onto their surfaces. The patterns vary depending on the producer, but all patterns should conform to ASTM specifications. Steel reinforcing bars are readily available in straight lengths of 60 ft. Smaller sizes are also available in coil stock for use in automatic bending machines. The bars vary in designation from No. 3 through No. 11, with two additional bars, No. 14 and No. 18. See Figure 1.3 for rebar marks or designation.

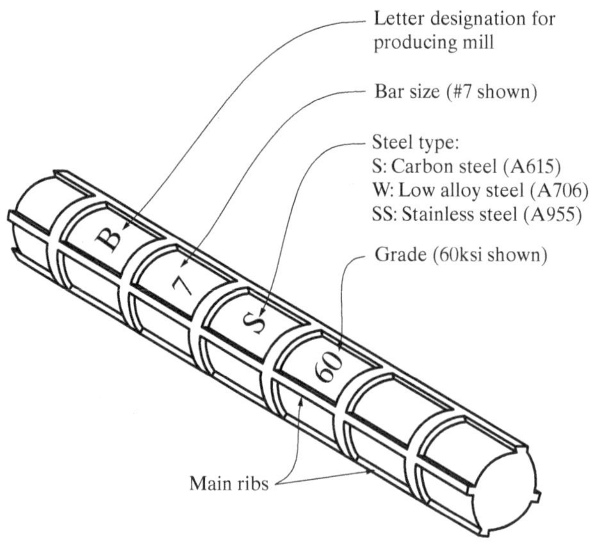

FIGURE 1.3 Rebar marks and designation.

For bars No. 3 through No. 8, the designation represents the bar diameter in eighths of an inch. The No. 9, No. 10, and No. 11 bars have diameters that provide areas equal to 1-in.-square bars, 1⅛-in.-square bars, and 1¼-in.-square bars, respectively. The No. 14 and No. 18 bars correspond to or have the same cross-sectional areas as 1½-in.-square bars and 2-in.-square bars, respectively, and are commonly available only by special order. Round, plain reinforcing bars are permitted for spirals (lateral reinforcing) in concrete compression members. The largest rebar size permitted in the ACI Code is No.18 bars because there is lack of available information on the performance, bond behavior, and bar bending characteristics of larger size bars (ACI Code, Section R20.2.1.3).

ASTM specifications require that identification marks be rolled onto the bar to provide the following information: a letter or symbol indicating the producer's mill, a number indicating the size of the bar, a symbol or letter indicating the type of steel from which the bar was rolled, and for grade 60 bars, either the number 60 or a single continuous longitudinal line (called a *grade line*) through at least five deformation spaces. The *grade* indicates the minimum specified yield stress in ksi. For instance, a grade 60 steel bar has a minimum specified yield stress of 60 ksi. Reference [7] is an excellent resource covering the various aspects of bar identification.

Reinforcing bars are usually made from newly manufactured steel (billet steel). Steel types and ASTM specification numbers for bars are tabulated in Table A.1. Note that ASTM A615, which is billet steel, is available in grades 40, 60, 80, and 100 (see ACI 318, Table 20.2.1.3(a)). Grade 80 steel is allowed for nonseismic applications per ASTM 615 and ASTM 706 [8]. ASTM A706, low-alloy steel, was developed to satisfy the requirement for reinforcing bars with controlled tensile properties and controlled chemical composition to ensure weldability for grades 60 and 80. They are available in grades 60, 80, and 100 (f_y = 100 ksi); see ACI 318 Table 20.2.1.3(b). ACI Table 20.2.2.4(a) gives the maximum allowed value of f_y that can be used in design calculations depending on the usage of the member, such as flexure, lateral ties, shear, torsion, etc. For example, the maximum f_y for flexure is 100,000 psi, whereas for lateral ties and stirrups, the maximum allowed f_y is 80,000 psi, and 60,000 psi for longitudinal and transverse torsion reinforcement. Tables A.2 and A.3 contain useful information on cross-sectional areas of bars. Note that Table A.2 is used for slabs and walls, while Table A.3 is used for beams, girders, and columns.

The most useful physical properties of reinforcing steel for reinforced concrete design calculations are yield strength (f_y), tensile strength, and modulus of elasticity. Note that for reinforcing steel in compression, the spacing of the confining ties along the length of the reinforcement is also important in preventing buckling of the longitudinal rebar in compression. A typical stress–strain diagram for reinforcing steel is shown in Figure 1.4a. The idealized stress–strain diagram of Figure 1.4b is used in Chapter 2.

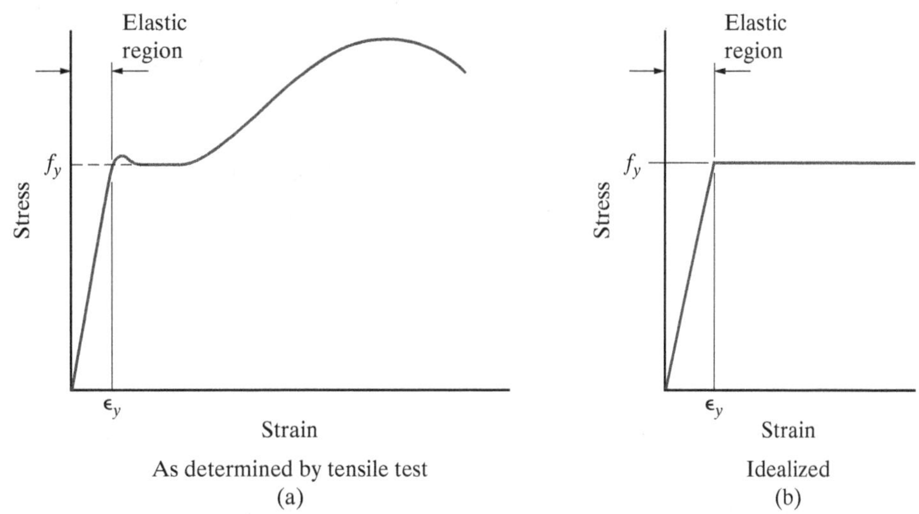

FIGURE 1.4 Stress–strain diagram for reinforcing steel.

The yield stress (or yield point) of steel is determined through procedures governed by ASTM standards. For practical purposes, the yield stress may be thought of as the stress at which the steel exhibits increasing strain with no increase in stress. For reinforcement without a sharply defined yield point, ACI 318-19, like the ASTM standards, defines the yield strength as the 0.2% proof stress (that is, the offset stress at a 0.2% strain). The yield stress of the steel will usually be one of the known (or given) quantities in the design or analysis of reinforced concrete members. See Table A.1 for the range of values of the yield strength, f_y.

The modulus of elasticity of carbon reinforcing steel (the slope of the stress–strain curve in the elastic region) varies over a very small range and has been adopted as 29,000,000 psi (ACI 318 Section 20.2.2.2).

Unhindered corrosion of reinforcing steel will lead to cracking and spalling of the concrete in which it is embedded. Under normal conditions, quality concrete provides good protection against corrosion for steel embedded in the concrete with adequate cover (minimum requirements are discussed in Chapter 2). This protection is attributed to, among other factors, the high alkalinity of the concrete. Where reinforced concrete structures (or parts of structures) are subjected to corrosive conditions, however, some type of corrosion protection system should be used to prevent deterioration. Such structures include bridge decks, parking garage decks, wastewater treatment plants, and industrial and chemical processing facilities.

One method to minimize the reinforcing steel's corrosion is to coat the bars with a suitable protective coating. The protective coating can be a nonmetallic material such as epoxy or a metallic material such as zinc (galvanizing). The ACI Code requires epoxy-coated reinforcing bars to comply with ASTM A775 or ASTM A934 and galvanized (zinc-coated) bars to comply with ASTM A767 (ACI 318 Section 20.5.2). The bars to be epoxy-coated or zinc-coated (galvanized) must meet the code requirements for uncoated bars as tabulated in Table A.1.

Welded wire reinforcing (WWR) (commonly called *mesh*) is another type of reinforcement and is mostly used in slabs-on-grade. It consists of cold-drawn wire in orthogonal patterns, square or rectangular, welded at all intersections. It may be supplied in either rolls or sheets, depending on the wire size. WWR with wire diameters larger than about ¼ in. is usually available only in sheets.

Both plain and deformed WWR products are available. According to ACI 318 Section 20.2.1.7, deformed wire, plain wire, welded deformed wire reinforcement, and welded plain wire reinforcement shall conform to ASTM A1064 for carbon steel and ASTM A1022 for stainless steel. Both materials have a minimum yield strength of 60,000 psi but can be as high as 100,000 psi (see ACI 318 Tables 20.2.2.4a and 20.2.2.4b). The deformed wire is usually more expensive, but it can be expected to have an improved bond with the concrete.

A rational method of designating wire sizes to replace the formerly used gauge system has been adopted by the wire industry. Plain wires are described by the letter W followed by a number equal to 100 times the cross-sectional area of the wire in square inches. Deformed wire sizes are described similarly, but the letter D is used. Thus, a W9 wire has an area of 0.090 in.^{2}, and a D8 wire has an area of 0.080 in.^{2} A W8 wire has the same cross-sectional area as the D8 but is plain rather than deformed. Sizes between full numbers are given by decimals, such as W9.5.

Generally, the material is indicated by the symbol WWR, followed by spacings first of longitudinal wires, then of transverse wires, and last by the sizes of longitudinal and transverse wires. Thus, WWR 6 × 12-W16 × W8 indicates a plain WWR with 6-in. longitudinal spacing, 12-in. transverse spacing, and a cross-sectional area equal to 0.16 in.^{2} (or 0.16 in.^{2}/6-in. = 0.32

in.^{2}/ft width) for the longitudinal wires and 0.08 in.^{2} (or 0.08 in.^{2}/12-in. = 0.08 in.^{2}/ft width) for the transverse wires.

Additional information about WWR, as well as tables relating size number with wire diameter, area, and weight, may be obtained through the Wire Reinforcement Institute [9] or the Concrete Reinforcing Steel Institute [9 and 10]. Table 1.1 contains common WWR sizes with the area of steel in in.^{2}/ft.

TABLE 1.1 Welded Wire Reinforcement (Meets ASTM A185, f_y = 65 ksi).

Wire gauge	W-number	Area of steel in.^{2}/ft (each direction)	Wire diameter (in.)
6× 6 – 10 × 10	**6 × 6-W1.4 × W1.4**	0.029	0.135 (10ga)
6 × 6 – 8 × 8	**6 × 6-W2.0 × W2.0**	0.041	0.162 (8ga)
6 × 6 – 6 × 6	**6 × 6-W2.9 × W2.9**	0.058	0.192 (6ga)
6×6 – 4 × 4	**6 × 6-W4.0 × W4.0**	0.080	0.225 (4ga)
4 × 4 – 10 × 10	**4 × 4-W1.4 × W1.4**	0.043	0.135 (10ga)
4 × 4 – 8 × 8	**4 × 4-W2.0 × W2.0**	0.062	0.162 (8ga)
4 × 4 – 6 × 6	**4 × 4-W2.9 × W2.9**	0.087	0.192 (6ga)
4 × 4 – 4 × 4	**4 × 4-W4.0 × W4.0**	0.120	0.225 (4ga)

Most concrete is reinforced in some way to resist tensile forces. In reinforced concrete, the reinforcing steel resists the tension stresses after the concrete has cracked. Figures 1.5a through 1.5e show reinforced concrete structures under different stages of construction. Some structural elements, particularly strip footings, are sometimes made of *plain concrete*, however. Plain concrete is defined as structural concrete with no reinforcement or with less reinforcement than the minimum amount specified for reinforced concrete. Plain concrete is discussed further in Chapter 10.

FIGURE 1.5a Concrete construction in progress. Note formwork, reinforcing bars, and pumping of concrete. (George Limbrunner)

FIGURE 1.5b High-rise reinforced concrete building construction in progress.(Abi Aghayere)

FIGURES 1.5c, d, and e Reinforced concrete buildings under construction. (Abi Aghayere)

1.10 CONCRETE COVER

The clear distance between the concrete surface and the face of the rebar is called the cover or clear cover. The reinforcement in concrete needs to have this cover for several reasons:

- To protect the reinforcement against corrosion
- To provide adequate surface area for bond between the concrete and the reinforcement
- To protect the reinforcement against loss of strength in a fire (i.e., provide fire resistance to the rebar)
- To provide for construction tolerance for concrete cover and rebar placement (See section 2.2.2 of ACI 117—Specifications for Tolerances for Concrete Construction for construction tolerances), and to account for other constructability issues.

According to ACI 117-10(15), "Tolerances are a means to establish permissible variation in dimension and location, giving both the designer and the contractor limits within which the work is to be performed," because "no structure is exactly level, plumb, straight, and true." Below are some examples of the tolerances for concrete cover and rebar placement allowed by ACI 117:

- **Concrete <u>Cover</u> Tolerances (ACI 117-10 Section 2.2.)**—The reduction in the specified concrete cover (or tolerances) should not be greater than the following (note that the cover and tolerance are measured relative to the rebar face closest to the concrete surface):
  - ✓ Members up to 12 in. thick = –3/8 in. ≤ 1/3 of the specified concrete cover
  - ✓ Members thicker than 12 in.= –½ in. ≤ 1/3 of the specified concrete cover
- **Rebar <u>Placement</u> Tolerances (ACI 117-10 Section 2.2.2)** – Limits minimum and maximum distances between the surface of the rebar and the nearest concrete surface as follows:
  - ✓ Members up to 4 in. thick = ± ¼ in. (i.e., Actual cover = specified cover ± ¼ in.)
  - ✓ Members up to 12 in. thick = ± 3/8 in. (i.e., Actual cover = specified cover ± 3/8 in.)
  - ✓ Members thicker than 12 in.= ± ½ in. (i.e., Actual cover = specified cover ± ½ in.)

Note that the cover tolerances "limits only the minimum distance from the bar to the surface of the concrete, while the tolerance on placement limits both the minimum and maximum distances."

For protection against corrosion and to ensure adequate bond between the reinforcement and the surrounding concrete, ACI 318-19 Table 20.5.1.3.1 specifies the concrete cover requirements for cast-in-place reinforced concrete members. The cover requirements apply equally to both noncoated rebar and epoxy-coated rebar. For slabs and walls not exposed to weather or in contact with the ground, the minimum clear cover to the outermost reinforcement surface is ¾″ for No. 11 and smaller bars, and 1.5″ for No. 14 and larger bars. For beams and columns not exposed to weather or in contact with the ground, the minimum clear cover to the outermost reinforcement surface is 1.5″. For all structural elements cast against and permanently in contact with soil or ground (e.g., footings, caissons, grade beams, foundation, basement walls, etc.), the specified concrete cover is 3″. These members are typically placed without the use of forms. For concrete elements not cast against soil or ground but exposed to the weather or in contact with ground, the specified concrete cover is 2″ for No. 6 through No. 18 bars, and 1.5″ for No. 5 and smaller bars. The surfaces of these members would typically be formed. The presence of forms allows for greater accuracy in establishing the proper clear cover distance.

The fire protection requirements of the Code may sometimes necessitate a higher concrete cover than the above-specified concrete covers, depending on the required fire rating and whether the flexural member is "*restrained*" or "*unrestrained*." An example of an unrestrained system is the end span of a continuous slab or continuous beam or girder. Interior spans of continuous slabs and beams or girders are considered "restrained." Examples of minimum clear concrete covers required by the Code for different fire ratings are as follows:

- 1 hour fire rating: clear concrete cover = ¾″
- 3 hours fire rating: clear concrete cover = 1¼″
- 4 hours fire rating: clear concrete cover = 1-5/8″

1.11 BEAMS: MECHANICS OF BENDING - REVIEW

The concept of bending stresses in homogeneous elastic beams is generally discussed at great length in all strength of materials textbooks and courses. Beams composed of a material such as steel or timber are categorized as homogeneous, with each exhibiting elastic behavior up to some limiting point. Within the limits of elastic behavior, the internal bending stress distribution developed at any cross-section is linear (straight line), varying from zero at the neutral axis to a maximum at the outer fibers.

The accepted expression for the maximum bending stress in a beam is termed the *flexure formula*, and is given as:

$$f_b = \frac{Mc}{I}$$

where

f_b = calculated bending stress at the outer fiber of the cross-section

M = the applied moment

c = distance from the neutral axis to the outside tension or compression fiber of the beam

I = moment of inertia of the cross-section about the neutral axis

The flexure formula represents the relationship between bending stress, bending moment, and the geometric properties of the beam cross-section. By rearranging the flexure formula, the maximum moment that may be applied to the beam cross-section, called the *resisting moment*, M_R, may be found:

$$M_R = \frac{F_b I}{c}$$

where F_b = the allowable bending stress

This procedure is straightforward for a beam of known cross-section for which the moment of inertia can easily be found. For a reinforced concrete beam, however, the use of the flexure formula presents some complications, because the beam is not homogeneous, and concrete does not behave elastically over its full range of strength. As a result, a somewhat different approach that uses the beam's internal bending stress distribution is recommended. This approach is termed the *internal couple method*.

Recall from strength of materials that a couple is a pure moment composed of two equal, opposite, and parallel forces separated by a distance called the *moment arm*, which is commonly denoted Z. In the internal couple method, the couple represents an internal resisting moment and is composed of a compressive force C above the neutral axis (assuming a single-span, simply supported beam that develops compressive stress above the neutral axis) and a parallel internal tensile force T below the neutral axis.

As with all couples, and because the forces acting on any cross-section of the beam must be in equilibrium, C must equal T. The internal couple must be equal and opposite to the bending moment at the same location, which is computed from the external loads. It represents a couple developed by the bending action of the beam.

The internal couple method of determining beam strength is more general and may be applied to homogeneous or nonhomogeneous beams having linear (straight line) or nonlinear stress distributions. For reinforced concrete beams, it has the advantage of using the basic resistance pattern found in the beam.

The following three analysis examples dealing with plain (unreinforced) concrete beams introduce the internal couple method. Note that the unreinforced beams are considered homogeneous and elastic. This is valid if the moment is small and tensile bending stresses in the concrete are low (less than the tensile bending strength of the concrete) with no cracking of the concrete developing. For this condition, the entire beam cross-section carries bending stresses. Therefore, the analysis for bending stresses in the uncracked beam can be based on the properties of the gross cross-sectional area using the elastic-based flexure formula. The use of the flexure formula is valid if the maximum tensile stress in the concrete does not exceed the modulus of rupture f_r. If a moment is applied that causes the maximum tensile stress just to reach the modulus of rupture, the cross-section will be on the verge of cracking. This moment is called the *cracking moment*, M_{cr}.

These examples use both the internal couple approach and the flexure formula approach so that the results may be compared.

EXAMPLE 1.1

A normal-weight plain concrete beam is 6 in. × 12 in. in cross-section, as shown in Figure 1.6. The beam is simply supported on a span of 4 ft and is subjected to a midspan concentrated load of 4500 lb. Assume $f'_c = 3000$ psi.

a. Calculate the maximum concrete tensile stress using the internal couple method.

b. Repeat part (a) using the flexure formula approach.

c. Compare the maximum concrete tensile stress with the value for modulus of rupture f_r using the ACI-recommended value based on f'_c.

Solution:

Calculate the weight of the beam (weight per unit length):

$$\begin{aligned}\text{weight of beam} &= \text{volume per unit length} \times \text{unit weight}\\ &= \frac{6\text{ in.}(12\text{ in.})}{144\text{ in.}^2/\text{ft}^2}(150\text{ lb/ft}^3)\\ &= 75\text{ lb/ft}\end{aligned}$$

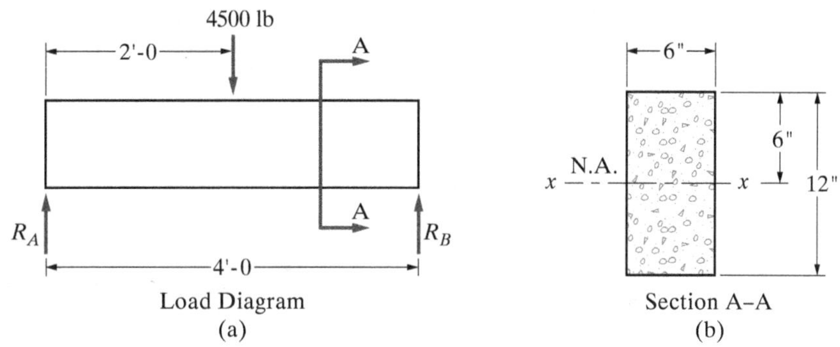

FIGURE 1.6 Loading diagram and section for Example 1.1.

Calculate the maximum applied moment:

$$\begin{aligned}M_{\max} &= \frac{PL}{4} + \frac{wL^2}{8}\\ &= \frac{4500\text{ lb}(4\text{ ft})}{4} + \frac{75\text{ lb/ft}(4\text{ ft})^2}{8}\\ &= 4650\text{ ft-lb}\end{aligned}$$

a. Internal couple method:

1. Because the beam is homogeneous, elastic, and symmetrical with respect to both the *X–X* and *Y–Y* axes, the neutral axis (N.A.) is at mid-height. Stresses and strains vary linearly from zero at the N.A. (which is also the centroidal axis) to a maximum at the outer fiber. As the member is subjected to a positive moment, the area above the N.A. is stressed in compression, and the area below the N.A. is stressed in tension. These stresses result from the bending behavior of the member and are shown in Figure 1.7.
2. *C* represents the resultant compressive force above the N.A. *T* represents the resultant tensile force below the N.A. *C* and *T* each act at the centroid of their respective triangles of stress distribution. Therefore, *Z* = 8 in. *C* and *T* must be equal (since $\Sigma H_F = 0$). The two forces act together to form the internal couple (or internal resisting moment) of magnitude *CZ* or *TZ*.
3. The internal resisting moment must equal the bending moment due to external loads at any section. Therefore,

$$M = CZ = TZ$$
$$4650 \text{ ft-lb } (12 \text{ in./ft}) = C\ (8 \text{ in.})$$

from which

$$C = 6975 \text{ lb} = T$$

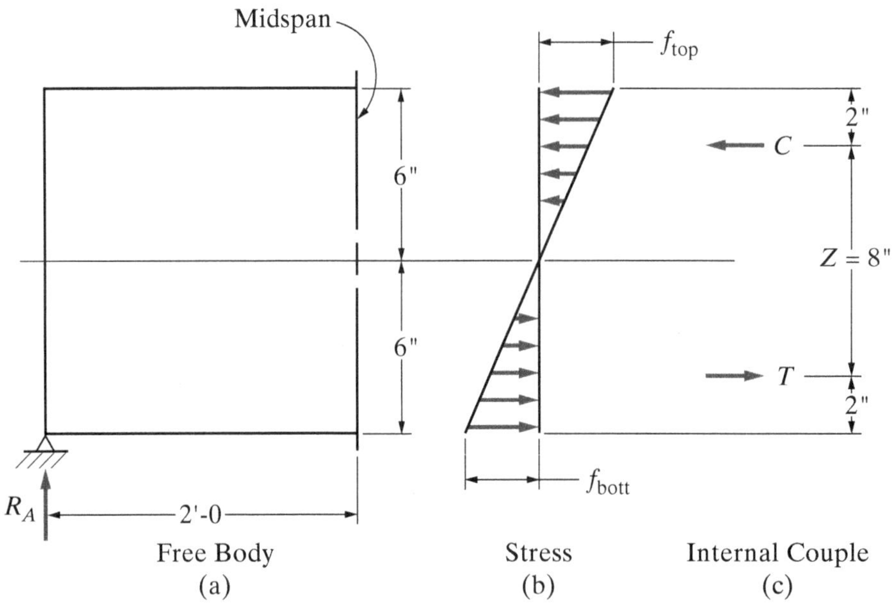

FIGURE 1.7 Stress and internal couple diagram for Example 1.1.

4. C = average stress × area of beam on which stress acts

$$C = \tfrac{1}{2} f_{top} \left(y_{top}\right)(b) = \tfrac{1}{2} f_{top} (6 \text{ in.})(6 \text{ in.}) = 6975 \text{ lb}$$

Solving for f_{top} yields

$$f_{top} = 388 \text{ psi} = f_{bott}$$

The modulus of elasticity of the concrete,

$$E_c = 57{,}000\sqrt{f_c'} = 57{,}000\sqrt{3000 \text{ psi}} = 3{,}122{,}019 \text{ psi}$$

The concrete strain at the top of the beam is $\varepsilon_{top} = \dfrac{f_{top}}{E_c} = \dfrac{388 \text{ psi}}{3{,}122{,}019 \text{ psi}} = 0.000124 \text{ in./in.}$

The curvature, $\varphi = \dfrac{\varepsilon_{top}}{y_{top}} = \dfrac{0.000124 \text{ in./in.}}{6 \text{ in.}} .= 0.000021 / \text{in.}$

b. Flexure formula approach:

$$I = \frac{bh^3}{12} = \frac{6(12^3)}{12} = 864 \text{ in.}^4$$

$$f_{top} = f_{bott} = \frac{Mc}{I} = \frac{4650(12)(6)}{864} = 388 \text{ psi}$$

c. The ACI-recommended value for the modulus of rupture (based on f'_c) is

$$f_r = 7.5\lambda\sqrt{f'_c} = 7.5(1.0)\sqrt{3000}$$
$$f_r = 411 \text{ psi}$$

The calculated tensile stress (f_{bott}) of 388 psi is about 6% below the modulus of rupture, the stress at which flexural cracking would be expected.

Example 1.1 is based on elastic theory and assumes the following: (1) a plane section before bending remains a plane section after bending (the variation in strain throughout the depth of the member is linear from zero at the N.A.), and (2) the modulus of elasticity is constant; therefore, stress is proportional to strain, and the stress distribution throughout the depth of the beam is also linear from zero at the N.A. to a maximum at the outer fibers.

The internal couple approach may also be used to find the moment strength (resisting moment) of a beam.

EXAMPLE 1.2

Calculate the cracking moment M_{cr} for the plain concrete beam shown in Figure 1.8. Assume normal-weight concrete and $f'_c = 4000$ psi.

a. Use the internal couple method.

b. Check using the flexure formula.

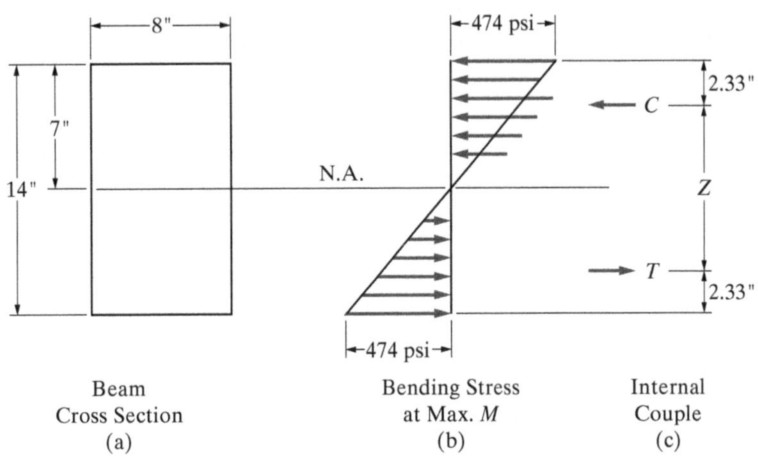

FIGURE 1.8 Sketch for Example 1.2.

Solution:

The moment that produces a tensile stress just equal to the modulus of rupture f_r is called the cracking moment, M_{cr}. The modulus of rupture for normal-weight concrete is calculated from ACI Equation 19.2.3.1:

$$f_r = 7.5\sqrt{f_c'} = 7.5\sqrt{4000} = 474 \text{ psi}$$

For convenience, we will use force units of kip (1 kip = 1000 lb). Therefore, $f_r = 0.474$ ksi.

a. Using the internal couple method:

$$Z = 14 - 2(2.33) = 9.34 \text{ in.}$$

$$C = T = \tfrac{1}{2}(0.474)(8)(7) = 13.27 \text{ kip}$$

$$M_{cr} = CZ = TZ = \frac{13.27(9.34)}{12} = 10.33 \text{ ft.-kip}$$

b. Check using the flexure formula:

$$f = \frac{Mc}{I}$$

$$M_R = M_{cr} = \frac{f_r I}{c}$$

$$I = \frac{bh^3}{12} = \frac{8(14)^3}{12} = 1829 \text{ in.}^4$$

$$M_{cr} = \frac{f_r I}{c} = \frac{0.474(1829)}{7(12)} = 10.32 \text{ ft.-kip}$$

The internal couple method may also be used to analyze irregularly shaped cross sections, although for homogeneous beams, it is more cumbersome than the use of the flexure formula.

EXAMPLE 1.3

Calculate the cracking moment (resisting moment) for the T-shaped unreinforced concrete beam shown in Figure 1.9. Use $f_c' = 4000$ psi. Assume positive moment (compression in the top). Use the internal couple method and check using the flexure formula.

Solution:

The N.A. must be located so that the strain and stress diagrams may be defined. The location of the N.A. with respect to the noted reference axis is calculated from

$$\bar{y} = \frac{\Sigma(Ay)}{\Sigma A}$$
$$= \frac{4(20)(22) + 5(20)(10)}{4(20) + 5(20)}$$
$$= 15.33 \text{ in.}$$

The bottom of the cross-section is stressed in tension. Note that the stress at the bottom will be numerically larger than at the top because of the relative distances from the N.A.. The stress at the bottom of the cross-section will be set equal to the modulus of rupture ($\lambda = 1.0$ for normal-weight concrete):

$$f_{\text{bott}} = f_r = 7.5\lambda\sqrt{f'_c} = 7.5(1.0)\sqrt{4000} = 474 \text{ psi} = 0.474 \text{ ksi}$$

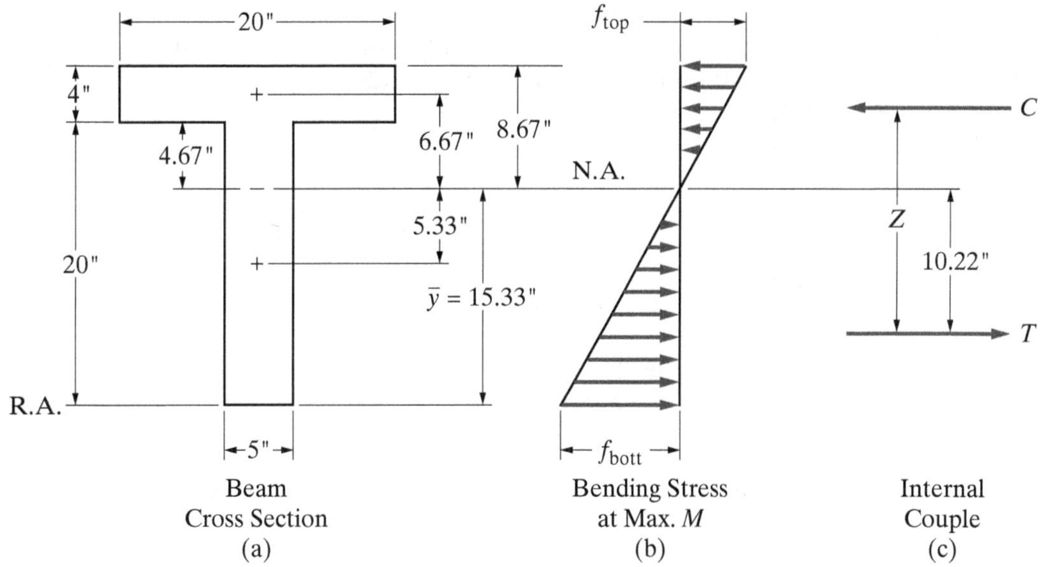

FIGURE 1.9 Sketch for Example 1.3.

Using similar triangles in Figure 1.9b, the stress at the top of the flange is

$$f_{\text{top}} = \frac{8.67}{15.33}(0.474) = 0.268 \text{ ksi}$$

The modulus of elasticity of the concrete,

$$E_c = 57{,}000\sqrt{f'_c} = 57{,}000\sqrt{4000 \text{ psi}} = 3{,}605{,}000 \text{ psi}$$

The concrete strain at the top of the beam is $\varepsilon_{\text{top}} = \dfrac{f_{\text{top}}}{E_c} = \dfrac{268 \text{ psi}}{3{,}605{,}000 \text{ psi}} = 0.000074 \text{ in./in.}$

Similarly, the stress at the bottom of the flange is

$$f_{\text{bott of flange}} = \frac{4.67}{15.33}(0.474) = 0.1444 \text{ ksi}$$

The total tensile force can be evaluated as follows:

$$\begin{aligned} T &= \text{average stress} \times \text{area} \\ &= \tfrac{1}{2}(0.474)(15.33)(5) = 18.17 \text{ kip} \end{aligned}$$

and its location below the N.A. is calculated from

$$\tfrac{2}{3}(15.33) = 10.22 \text{ in. (below the N.A.)}$$

The compressive force will be broken up into components because of the irregular area, as shown in Figure 1.10. Referring to both Figures 1.9 and 1.10, the component internal compressive forces, component internal couples, and M_R may now be evaluated. The component forces are first calculated:

$$\begin{aligned} C_1 &= 0.1444(20)(4) = 11.55 \text{ kip} \\ C_2 &= \tfrac{1}{2}(0.1236)(20)(4) = 4.94 \text{ kip} \\ C_3 &= \tfrac{1}{2}(0.1444)(5)(4.67) = 1.686 \text{ kip} \\ \text{total } C &= C_1 + C_2 + C_3 = 18.18 \text{ kip} \\ C &\approx T \quad \text{(O.K.)} \end{aligned}$$

Next, we calculate the moment arm distance from each component compressive force to the tensile force T:

$$\begin{aligned} Z_1 &= 10.22 + 4.67 + \tfrac{1}{2}(4.00) = 16.89 \text{ in.} \\ Z_2 &= 10.22 + 4.67 + \tfrac{2}{3}(4.00) = 17.56 \text{ in.} \\ Z_3 &= 10.22 + \tfrac{2}{3}(4.67) = 13.33 \text{ in.} \end{aligned}$$

The magnitudes of the component internal couples are then calculated from force × moment arm as follows:

$$\begin{aligned} M_{R_1} &= 11.55(16.89) = 195.1 \text{ in.-kip} \\ M_{R_2} &= 4.94(17.56) = 86.7 \text{ in.-kip} \\ M_{R_3} &= 1.686(13.33) = 22.5 \text{ in.-kip} \\ M_{cr} &= M_R = M_{R_1} + M_{R_2} + M_{R_3} = 304 \text{ in.-kip} \end{aligned}$$

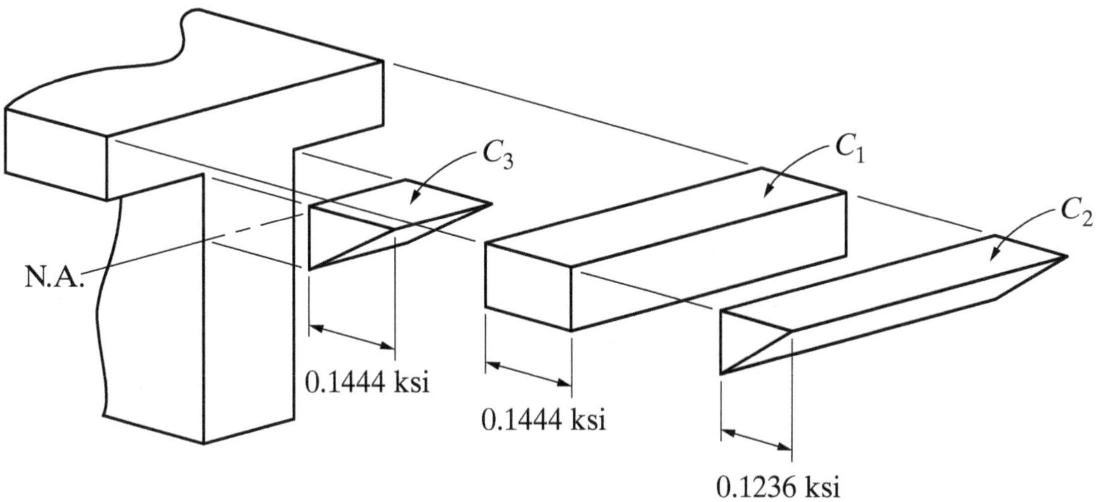

FIGURE 1.10 Component compression forces for Example 1.3.

Check using the flexure formula. The moment of inertia is calculated using the transfer formula for moment of inertia from strength of materials:

$$I = \sum I_o + \sum Ad^2$$

$$I = \tfrac{1}{12}(20)(4^3) + \tfrac{1}{12}(5)(20^3) + 4(20)(6.67^2) + 5(20)(5.33^2)$$

$$= 9840 \text{ in.}^4$$

$$M_{cr} = M_R = \frac{f_r I}{c} = \frac{0.474(9840)}{15.33} = 304 \text{ in.-kip}$$

(Checks O.K.)

As mentioned previously, the three examples are for plain, unreinforced, and uncracked concrete beams that are considered homogeneous and elastic within the bending stress limit of the modulus of rupture. The internal couple method is also applicable to nonhomogeneous beams with nonlinear stress distributions of any shape, however. Because reinforced concrete beams are nonhomogeneous, the flexure formula is not directly applicable. Therefore, the basic approach used for reinforced concrete beams is the internal couple method. (See Chapters 2 and 3).

1.12 CONCRETE SLAB SYSTEMS

Reinforced concrete roof and floor framing systems include the following [10]:

- **Beam-supported slab systems**
    - One-way slab supported on beams, with the beams supported on girders and on columns along the column lines.
        - ✓ Span range: 15 ft
        - ✓ Slab thickness: 4 in. to 10 in.

- Two-way slabs supported on beams and girders.
  - ✓ Span range: 15-40 ft
  - ✓ Slab thickness: 5 in. to 10 in.
- **Flat plate** – That is, slab directly supported on columns with no beams.
  - ✓ Span range: 15-30 ft.
  - ✓ Slab thickness: 5 in. to 10 in. (may also include column capitals).
- **Flat slab** – That is, slab supported on columns with drop panels or slab thickening at the column locations and with no beams.
  - ✓ Span range: 15-30 ft.
  - ✓ Slab thickness: 5 in. to 12 in.
  - ✓ Drop panel thickness = 2¼ in. to 8 in. (may also include column capitals).
- **Other reinforced concrete slab systems**
  - One-way rib joist – That is, thin slab (3 in. to 3.5 in.) supported on one-way rib joists that are supported on girders. The one-way joist is formed with metal pan formwork.
  - Two-way rib joist – That is, slab supported on two-way rib joists (this is also called a waffle slab).
  - Proprietary voided slab systems such as the filigree composite slab system, which is compressed of filigree planks (typically 2¼ in. thick) with polystyrene voids and embedded steel trusses spaced at regular intervals and spanning between filigree beams, thus obviating the need for slab formwork and leading to a reduction in the number of shores per floor. This voided slab system, which uses a reduced volume of concrete in the slab results in smaller loads on the beams, columns, and foundations in addition to a smaller structural height, which leads to reduced overall cost of the building [15].

One-Way Slab Systems

One-way concrete floor systems are usually supported on all four sides by stiff beams or walls, and the slab spans or bends predominantly in one direction (usually in the shorter direction), and the load transfer to the members supporting the slab occurs predominantly in the shorter direction (see Figure 1.11). This is the case when the clear span of the longer side of the rectangular slab panel (L_2) is greater than or equal to twice the clear span of the shorter side of the slab (L_1); that is, $L_2/L_1 \geq 2$. In a reinforced or prestressed concrete one-way slab system, the curvature of the slab is predominantly in *the shorter* direction (i.e., parallel to the L_1 dimension). The design of one-way reinforced concrete slab systems is covered in Chapter 2.

Two-Way Slab Systems

Two-way concrete slab systems are supported on columns and span in two orthogonal directions, or the slab is supported on all four sides by beams or walls, and the curvature of the slab and the load transfer occurs in both orthogonal directions (see Figure 1.12). A two-way slab system occurs when the clear span of the longer side of the rectangular slab panel (L_2) is less than twice the clear span of the shorter side of the slab panel (L_1); that is, $L_2/L_1 \leq 2$. Examples of two-way slabs include flat plates, flat slabs, flat slabs with beams, and slabs supported on stiff beams or walls on all four sides of a rectangular slab panel. Note that L_2 is the larger dimension of the rectangular slab panel bounded on all four sides by columns or beams or walls, and L_1 is the smaller dimension of the rectangular slab panel. Further treatment of two-way slab systems is covered in Chapter 6.

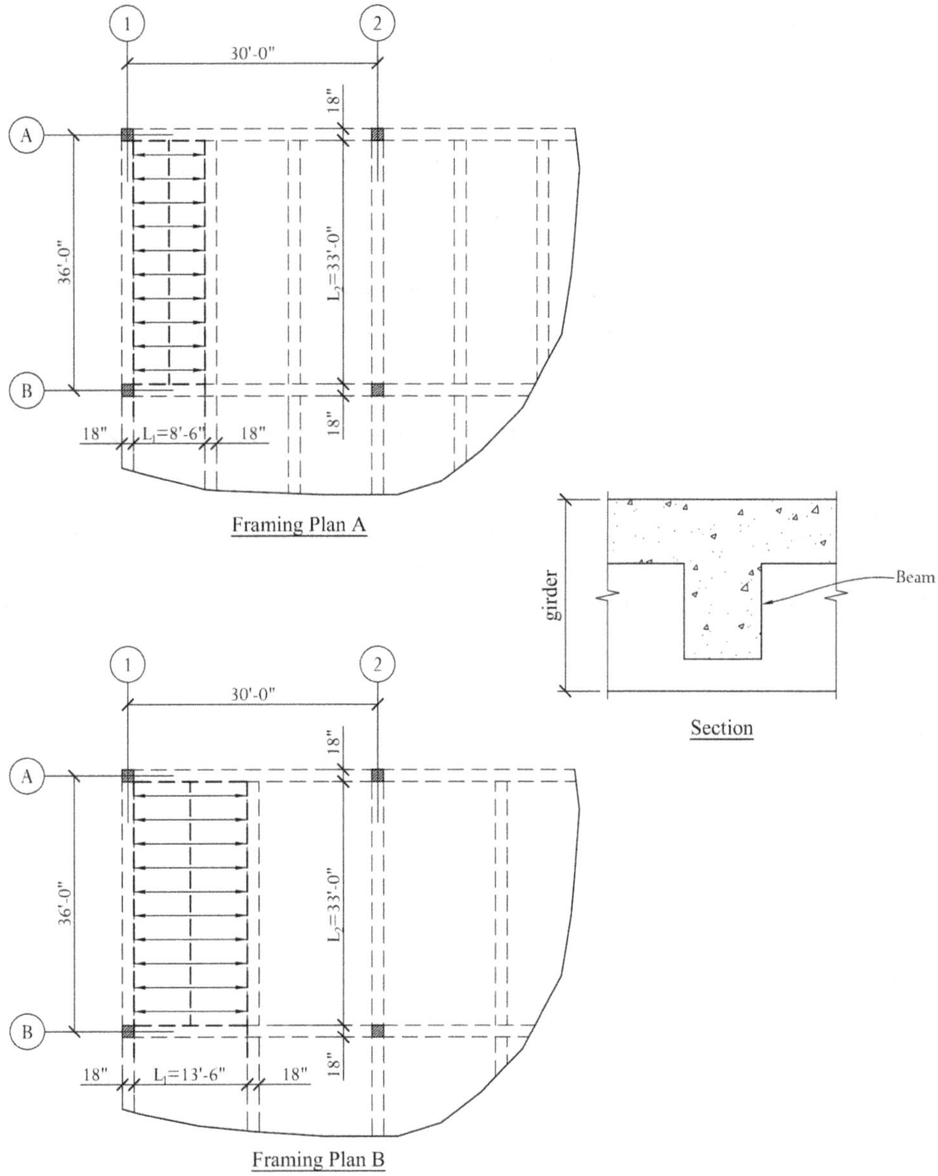

FIGURE 1.11 Load distribution in one-way slab systems.

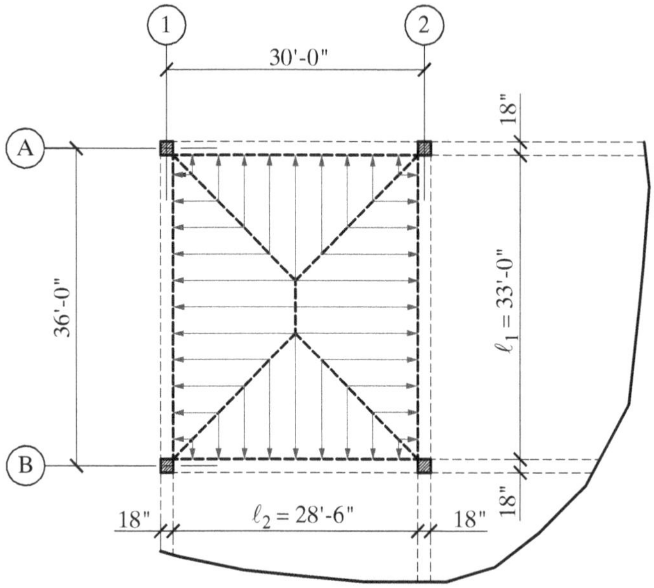

FIGURE 1.12 Load distribution in two-way slab systems.

1.13 GRAVITY LOAD DISTRIBUTION IN CONCRETE SLAB SYSTEMS

Reinforced concrete structures are subjected to a variety of loads, which include dead loads, floor live loads, roof live loads, snow loads, hydrostatic pressures, lateral soil pressures, wind loads, and seismic or earthquake loads. In this section, we introduce the reader to the distribution of gravity loads in concrete slab systems.

Gravity loads that act on structures include dead loads, floor live loads, snow loads, and roof live loads. In concrete floor and roof systems, the pertinent load combinations that usually govern ultimate strength design are as follows:

Floor slab:

$1.4D$

$1.2D + 1.6L$

Roof slab:

$1.4D$

$1.2D + 1.6(L_r \text{ or } S \text{ or } R) + 0.5W$

$1.2D + 1.0W + 0.5(L_r \text{ or } S \text{ or } R)$

where,

D = dead load

L = floor live load

L_r = roof live load

S = snow load

R = rain load

W = wind load

When designing for serviceability limit state (e.g., deflections and vibrations), use the allowable stress design (ASD) load combinations from the ASCE 7 Load Standard. For floors with dead and floor live loads, and roofs with dead, snow, rain or roof live loads, all the load factors in the load combinations given above (i.e., the "1.2" and "1.6") default to 1.0 since serviceability is checked at working or service loads. For a detailed treatment of the different structural loads and the load combinations prescribed in the Code, the reader should refer to References [11] and [12].

Dead Loads

The dead load is the weight of anything permanently attached to the structure, including the self-weight of the structure. Dead loads include the weight of the following:

- Floor finishes
- Partitions
- Mechanical and electrical equipment and conduits (M & E)
- Glazing
- Cladding

The weights of common building materials are given in Tables C3.1-1a, C3.1-1b, and C3.1-2 in Chapter C3 of the Appendix of ASCE 7-22. The density of unreinforced concrete is approximately 145 lb/ft^3 and the density of reinforced concrete is approximately 150 lb/ft^3.

A sample dead load calculation for a building is as follows:

Components of Roof Dead Load in a Reinforced Concrete Structure

5 ply + gravel = 6.5 psf

Insulation and membrane = 3.5 psf

4″ concrete slab ≅ 50 psf
(slab weight = 4″/12 × 150 lb/ft^3 = 50 psf)

Mechanical/Electrical (M & E) ≅ 10 psf (20 psf for industrial buildings)

Suspended ceiling ≅ 2 psf

Total roof dead load, D ≅ 72 psf (unfactored)

Components of Floor Dead Load in a Reinforced Concrete Structure

6-in. concrete slab[1] = 75 psf

Floor finishes ≅ 8 psf (estimate the weight of actual finishes specified by the architect)

M & E ≅ 10 psf (20 psf for industrial building)

Suspended ceiling ≅ 2 psf

Partitions[2, 3] ≅ 15 psf

Total floor dead load, D = 115 psf (unfactored)

[1]Weight of 6-in. slab = 6"/12 x 150 lb/ft^3 = 75 psf

[2]The minimum value per ASCE 7-22, Section 4.3.2, is 15 psf. Actual weight of partitions may be higher.

[3]The Code allows partition loads to be neglected whenever the floor live load, L, is greater than or equal to 80 psf (ASCE 7-22, Section 4.3.2)

Tributary Width and Areas

In this section, we introduce the concept of tributary area and tributary width. Beams in reinforced concrete floor or roof systems share the uniformly distributed loads in proportion to their distance from adjacent parallel beams. The tributary width (TW) of a beam is a measure of the total width of the floor or roof supported by the beam. The tributary area (A_T) is the area of the floor supported by a structural element. For a beam, it is the product of the TW and the span of the beam. For a column, it is the floor or roof area bounded by lines mid-way to adjacent columns above, below, to the right, and to the left of the column under consideration. In a roof or floor system, the load path for gravity loads involves the beams supporting uniform loads from the floor and the beams transferring their reactions to the girders, which then in turn transfer their reactions to the columns.

Tributary width (TW) = Width of floor or roof supported by a **beam** or **girder**

= (½ × the distance to the adjacent beam on the right + ½ × the distance to the adjacent beam on the left)

Tributary area (A_T) = Area of roof or floor supported by a **beam**, **girder**, or **column**

The tributary area of a beam, $A_{T\,(\text{beam})}$ = TW of beam × span of beam

The tributary area of a column,

$A_{T\,(\text{column})}$ = Floor or roof area supported by the column

= the rectangular area bounded by one-half the distance to adjacent columns on all sides of the column under consideration

EXAMPLE 1.4 Load Calculations for One-Way Slabs and Supporting Beams

The elevation and section through a reinforced concrete building is shown in Figure 1.13. Determine the factored design loads for the floor slab, interior beam (A), and spandrel beam (B). Assume the following loads:

- Office building live load (i.e., floor live load = 50 psf)
- Block wall weighs 60 psf of vertical surface area
- Brick veneer or cladding weighs 40 psf of vertical surface area

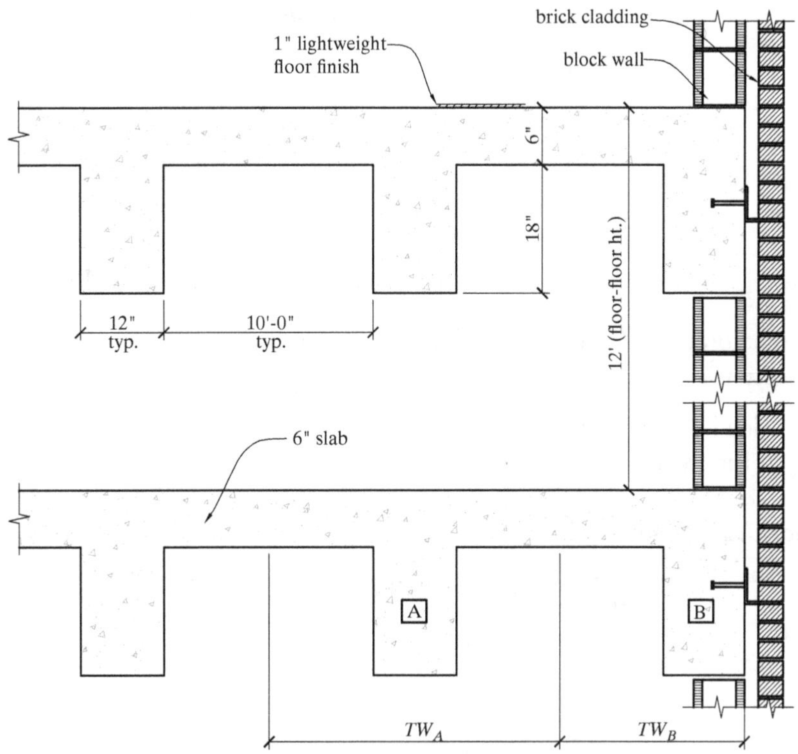

FIGURE 1.13 Building Section for Example 1.4.

Solution:

Floor Dead Load:

Finishes	8 psf
6″ concrete slab	75 psf
(6″/12 × 150 pcf)	
Suspended ceiling	2 psf
M & E	10 psf
Partitions	<u>20 psf</u>

Total dead load, $D \cong 115$ psf

Live load, L = 50 psf (for office buildings)

Slab Loads:

Factored load, $w_u = 1.2\,D + 1.6\,L$

$= 1.2\,(115) + 1.6\,(50)$

$= \underline{218 \text{ psf}}$

In the design of slabs, a 12-inch-wide strip is usually considered. Therefore, the load on the slab in pounds per linear foot (Ib/ft) per foot width of slab is calculated as

$$w_u = (218 \text{ psf})(1 \text{ ft}) = 218 \text{ Ib/ft per foot width of slab}$$

This load is used in the ultimate strength design of the slab.

$$\begin{aligned} \text{Service or unfactored load, } w_s &= 1.0\,D + 1.0\,L \\ &= 115 + 50 \\ &= 165 \text{ psf} \end{aligned}$$

$$w_s = (165 \text{ psf})(1 \text{ ft width}) = 165 \text{ Ib/ft per foot width of slab}$$

This unfactored load will be used for the design of the slab for serviceability limit state such as deflections.

The loads to the interior and spandrel beams are based on the TWs of each beam, as shown in Figure 1.14.

Typical Interior Beam Loads

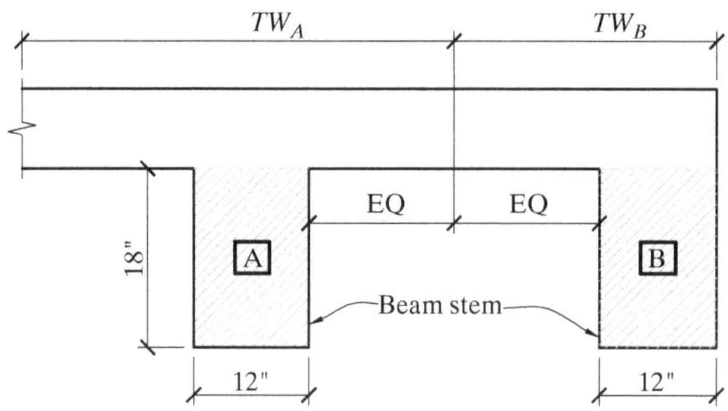

FIGURE 1.14 Tributary widths of beams for Example 1.4.

$$\text{Tributary width of beam A, } TW_A = \frac{10'-0''}{2} + 1'-0'' + \frac{10'-0''}{2} = 11'-0''$$

$$\text{Unfactored weight of beam stem} = \left(\frac{12''}{12}\right)\left(\frac{18''}{12}\right)(150 \text{ pcf}) = 225 \text{ lb/ft}$$

$$\begin{aligned} \text{Factored load on beam A, } w_u &\equiv w_u(\text{slab}) \times TW_A + \text{factored dead weight of beam A stem} \\ &\equiv (218 \text{ psf})(11') + (1.2)(225 \text{ lb/ft}) \\ &\equiv 2668 \text{ lb/ft} \\ &= \mathbf{2.67 \text{ kip/ft}} \end{aligned}$$

Service or unfactored load on beam, $w_s = w_s(\text{slab}) \times TW_A$ + unfactored dead weight of beam A stem

$$\begin{aligned} &= (165 \text{ psf})(11\text{-ft.}) + 225 \text{ lb/ft} \\ &= 2040 \text{ lb/ft} \\ &\equiv \mathbf{2.04 \text{ k/ft}} \end{aligned}$$

Spandrel Beam Loads

Tributary width of spandrel beam B, $TW_B = \dfrac{10'-0"}{2} + 1'-0" = 6'-0"$

Unfactored weight of spandrel beam stem = $\left(\dfrac{12"}{12}\right)\left(\dfrac{18"}{12}\right)(150 \text{ pcf}) = 225 \text{ lb/ft}$

Height of block backup wall = 12 ft – 2 ft = 10 ft

Height of exterior brick veneer/cladding = 12 ft

Unfactored weight of exterior block and brick wall = (60 psf)(10-ft.) + (40 psf)(12-ft.)
= 1080 lb/ft

Factored load on beam,

$w_u = w_{u,\text{slab}} \times TW_B$ + factored weight of beam stem + factored weight of wall

= (218 psf) (6′) + (1.2) (225 lb/ft) + (1.2) (1080 lb/ft)

= 2874 lb/ft

= **2.87 kip/ft**

Service load on beam,

$w_s = w_{s,\text{slab}} \times TW_B$ + unfactored weight of beam stem + unfactored weight of wall

= (165 psf)× (6-ft.) + (225 lb/ft) + (1080 lb/ft)

= 2295 lb/ft

≡ **2.30 kip/ft**

REFERENCES

[1] *Building Code Requirements for Structural Concrete* (ACI 318-19). American Concrete Institute, Farmington Hills, MI, 2019, 623 pp. (Reapproved 2022)

[2] ACI Committee 211. *Standard Practice for Selecting Proportions for Normal, Heavyweight, and Mass Concrete* (ACI 211.1-91). American Concrete Institute, P.O. Box 9094, Farmington Hills, MI, 1991. (Reapproved 2002)

[3] George E. Troxell, Harmer E. Davis, and Joe W. Kelly. *Composition and Properties of Concrete*, 2nd ed. New York: McGraw-Hill Book Company, 1968.

[4] Steven H. Kosmatka, Beatrix Kerkhoff, and William C. Panarese. *Design and Control of Concrete Mixtures*, 14th ed. Engineering Bulletin of the Portland Cement Association, 5420 Old Orchard Road, Skokie, IL 60077, 2002.

[5] Joseph J. Waddell, ed. *Concrete Construction Handbook*, 3rd ed. New York: McGraw-Hill Book Company, 1993.

[6] *ASTM Standards*. American Society for Testing and Materials, West Conshohocken, PA.

[7] *Manual of Standard Practice,* 28th ed. Concrete Reinforcing Steel Institute, Schaumburg, IL.

[8] Clifford W. Schwinger. "ASTM A615 Grade 75 Reinforcing Steel–When, Why & How to Use It," *STRUCTURE Magazine*, pp. 34–35, August 2011.
[9] *Manual of Standard Practice*, (WWR-500-R-16) Wire Reinforcement Institute, Hartford, CT, 2016.
[10] CRSI Design Handbook 2002, Concrete Reinforcing Steel Institute, 933 North Plum Grove Road, Schaumburg, IL 60173.
[11] Abi Aghayere and Jason Vigil. *Structural Steel Design*, 3rd ed. Boston: Mercury Learning, 2020.
[12] David A. Fanella. *Structural Loads – 2012 IBC and ASCE/SEI 7-10.*, International Code Council, 2012.
[13] Sidney Mindess, J. Francis Young, and David Darwin. *Concrete*, 2nd ed. Prentice Hall, 2003.
[14] *ACI Committee 562, "Code Requirements for Assessment, Repair, and Rehabilitation of Existing Concrete Structures and Commentary (ACI 562-16),"* American Concrete Institute, Farmington Hills, MI, 2016, 86 pp.
[15] Adam Beckman, Chris Christoforou, and Michael Squartzini, "Hudson Yards – Filigree in NYC," STRUCTURE Magazine, June 2020, pp. 26-27.
[16] "Portland-limestone-cement, "Portland Cement Association, Skokie, IL. *www.greenercement.com/* (Accessed July 24, 2023)

EXERCISES

NOTE *In the following exercises, assume plain concrete to have a unit weight of 145 pcf (conservative) unless otherwise noted.*

1.1 The unit weight of normal-weight reinforced concrete is commonly assumed to be 150 lb/ft^3 to account for the added weight of the reinforcement. Find the weight per lineal foot (lb/ft) for a normal-weight reinforced concrete beam that

a. Has a rectangular cross-section 16 in. wide and 28 in. deep.

b. Has a cross-section as shown in the accompanying diagram.

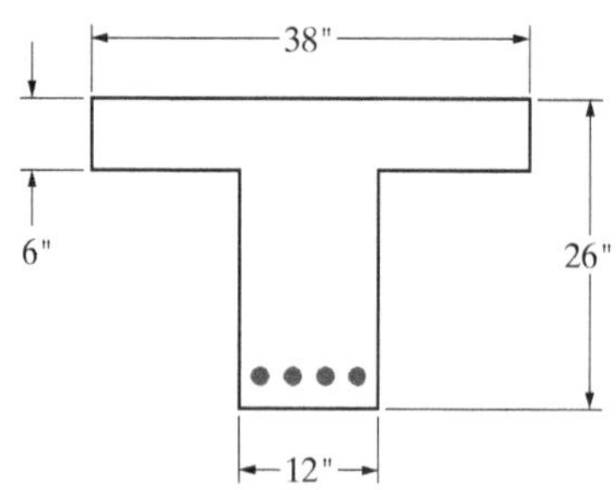

FIGURE FOR EXERCISE 1.1

1.2 Develop a spreadsheet application that will display in a table the values of modulus of elasticity E_c for concrete having unit weight ranging from 95 pcf to 155 pcf (in steps of 5 pcf) and compressive strength ranging from 3500 to 7000 psi (in steps of 500 psi). Display the modulus of elasticity rounded to the nearest 1000 psi.

1.3 A normal-weight concrete test beam 6 in. by 6 in. in cross-section and supported on a simple span of 24 in. was loaded with a point load at midspan. The beam failed at a load of 2100 lb. Using this information, determine the modulus of rupture f_r of the concrete and compare it with the ACI-recommended value based on an assumed concrete strength f'_c of 3000 psi.

1.4 A plain concrete beam has cross-sectional dimensions of 10 in. by 10 in. The concrete is known to have a modulus of rupture f_r of 350 psi. The beam spans between simple supports. Determine the span length at which this beam will fail due to its own weight. Assume a unit weight of 145 pcf.

1.5 The normal-weight plain concrete beam shown is on a simple span of 10 ft. It carries a dead load (which includes the weight of the beam) of 0.5 kip/ft. There is a concentrated load of 2 kip located at midspan. Use $f'_c = 4000$ psi. Compute the maximum bending stress. Use the internal couple method and check with the flexure formula.

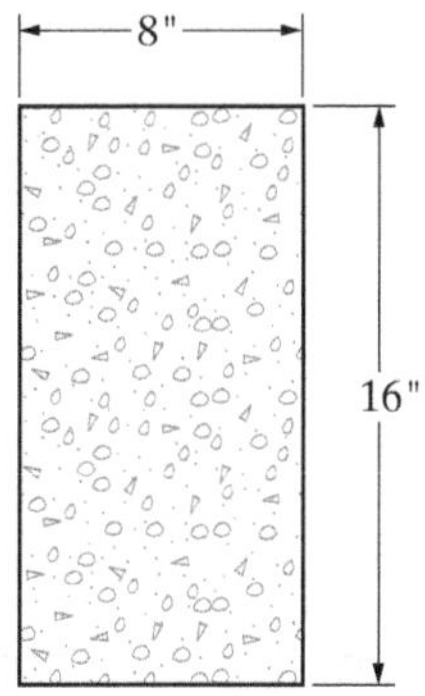

FIGURE FOR EXERCISE 1.5

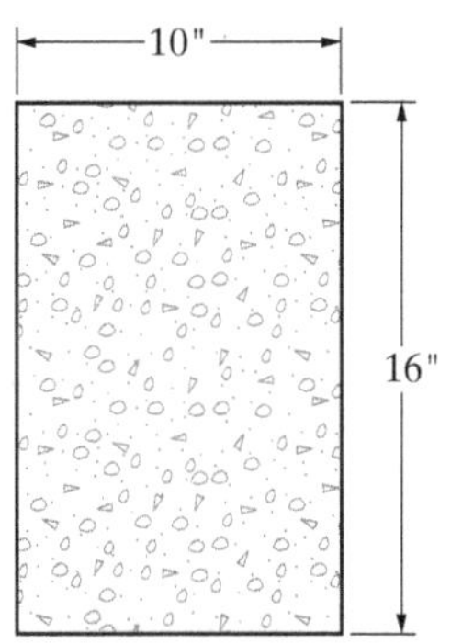

FIGURE FOR EXERCISE 1.6

1.6 Calculate the cracking moment (resisting moment) for the unreinforced concrete beam shown. Assume normal-weight concrete with $f'_c = 3000$ psi. Use the internal couple method and check with the flexure formula.

1.7 Develop a spreadsheet application to solve Exercise 1.6. Set up the spreadsheet so a table will be generated in which the width of the beam varies from 8 in. to 16 in. (1-in. increments) and the depth varies from 12 in. to 24 in. (1-in. increments.) The spreadsheet should allow the user to input any value for f'_c between 3000 psi and 8000 psi.

1.8 Rework Example 1.3 but invert the beam so that the flange is on the bottom and the web extends vertically upward. Calculate the cracking moment using the internal couple method and check using the flexure formula. Assume positive moment.

1.9 Calculate the cracking moment (resisting moment) for the U-shaped unreinforced concrete beam shown. Assume normal-weight concrete with $f_c' = 3500$ psi. Use the internal couple method and check with the flexure formula. Assume positive moment.

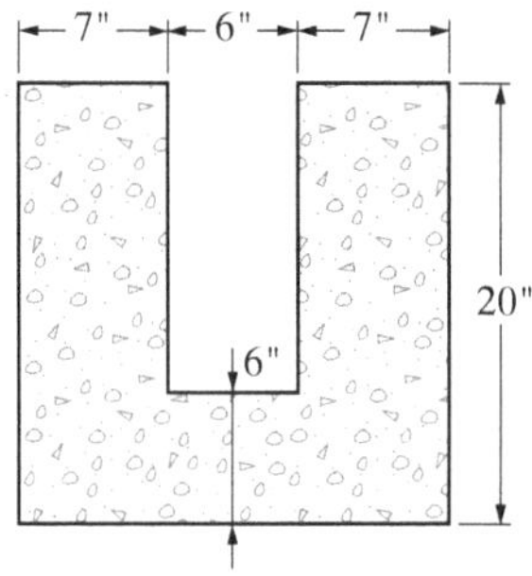

FIGURE FOR EXERCISE 1.9

1.10 The plain concrete beam shown is used on a 12-ft simple span. The concrete is normal weight with $f_c' = 3000$ psi. Assume positive moment.

a. Calculate the cracking moment.

b. Calculate the value of the concentrated load P at midspan that would cause the concrete beam to crack. (Be sure to include the weight of the beam.)

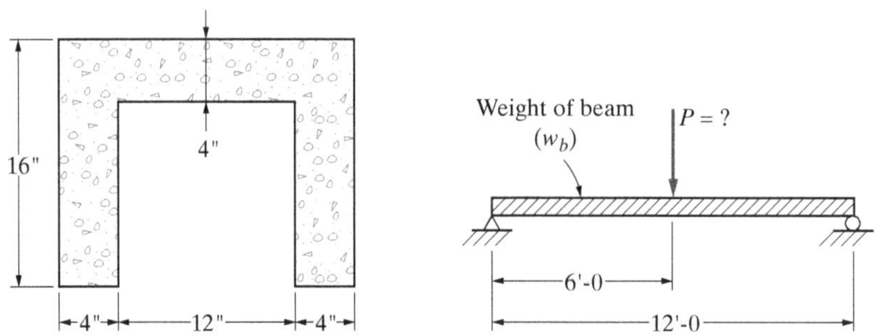

FIGURE FOR EXERCISE 1.10

1.11 Using the section below, determine the following:

a. The service dead load in psf supported by the slab.

b. The factored total uniform load in psf supported by the slab. Assume a floor live load of 100 psf.

c. The service dead and service live load in kip/ft supported by the spandrel or edge beam. Assume the block backup wall weighs 55 psf and the brick cladding weighs 40 psf of the vertical surface are of the wall.

d. The factored total uniform load in kip/ft supported by the spandrel beam.

e. The factored total uniform load in kip/ft supported by a typical interior beam.

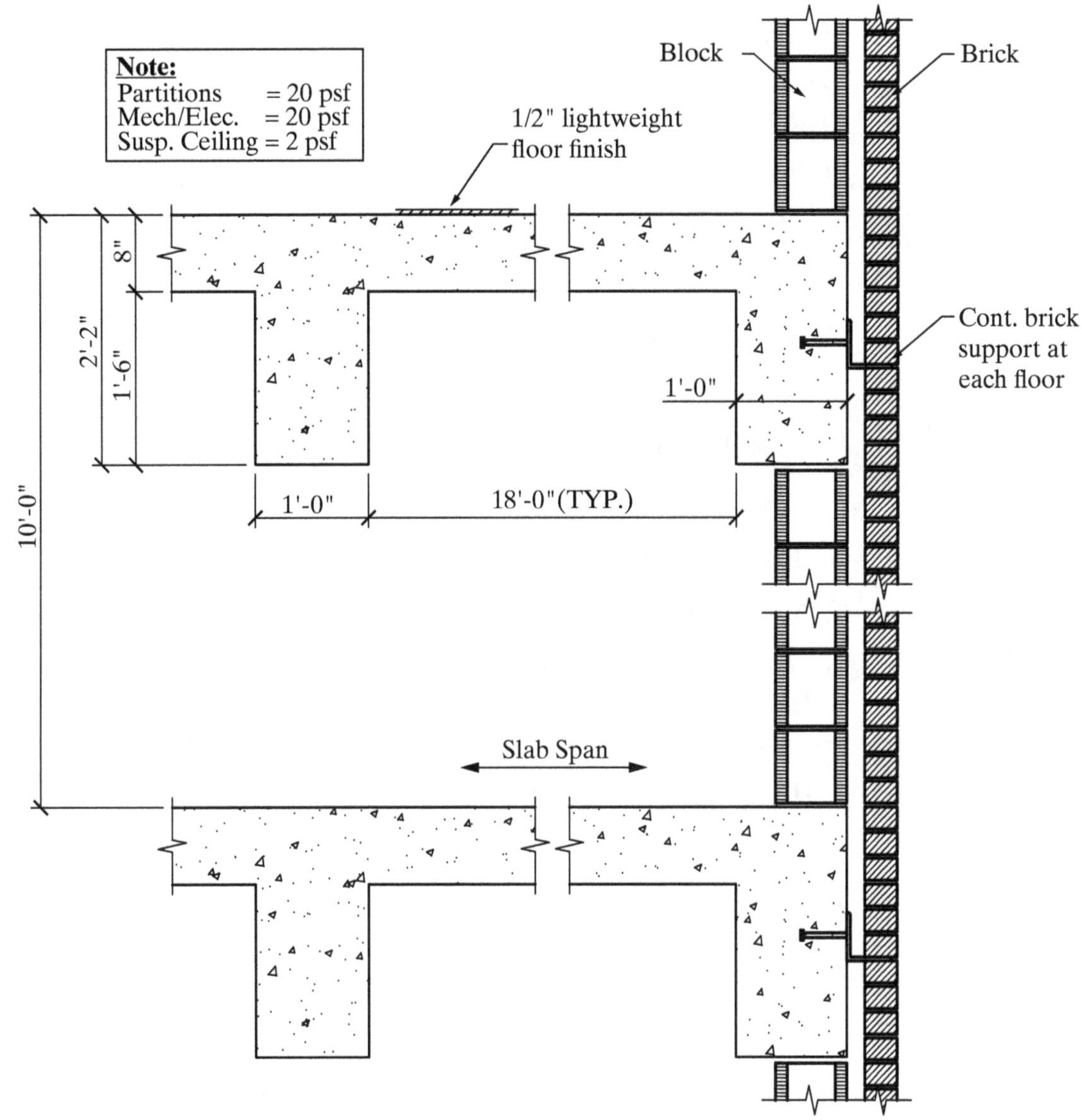

FIGURE FOR EXERCISE 1.11

1.12 A partial floor plan for a three-story reinforced concrete office building with 10 ft floor-to-floor heights, located in Rochester, New York, is shown in the figure. Assume the following:

- 7″ thick slab on the roof and floors
- 5-ply roofing plus gravel on the roof slab plus 3.5 psf for insulation
- 0.5 in lightweight finish on the floor slabs

- 20 psf for the partitions
- 10 psf for mechanical and electrical equipment
- 2 psf for the suspended ceiling
- Exterior cladding:
- 55 psf backup block wall—7 ft 10 in. high
- 40 psf brick veneer—10 ft high
- Beams: 16″ wide × 26″ deep; Girders: 18″ wide × 26″ deep
- Ignore live load reduction.

Determine the following:

a. The factored total uniform load in psf on the roof and floor slabs

b. The factored total uniform load in kip/ft supported by the

  i. Typical interior roof beam

  ii. Typical interior floor beam

  iii. Typical roof spandrel beam

  iv. Typical floor spandrel beam

c. The factored concentrated load from the beam reactions in kip on the following girders and the factored self-weight of the girder stem in kip/ft. For the purposes of calculating the beam reactions, assume the beams are simply supported:

  i. Typical interior roof girder

  ii. Typical interior floor girder

  iii. Typical roof spandrel girder

  iv. Typical floor spandrel girder

d. The total factored load and the total service load in kip on Column B-2. Assume 16″ × 16″× columns. For the purposes of calculating the beam and girder reactions, assume the beams and girders are simply supported. Present your results in a tabular format.

e. The total factored load and the total service load in kip on Column A.1. Assume 16″ × 16″ columns. For the purposes of calculating the beam and girder reactions, assume the beams and girders are simply supported. Present your results in a tabular format.

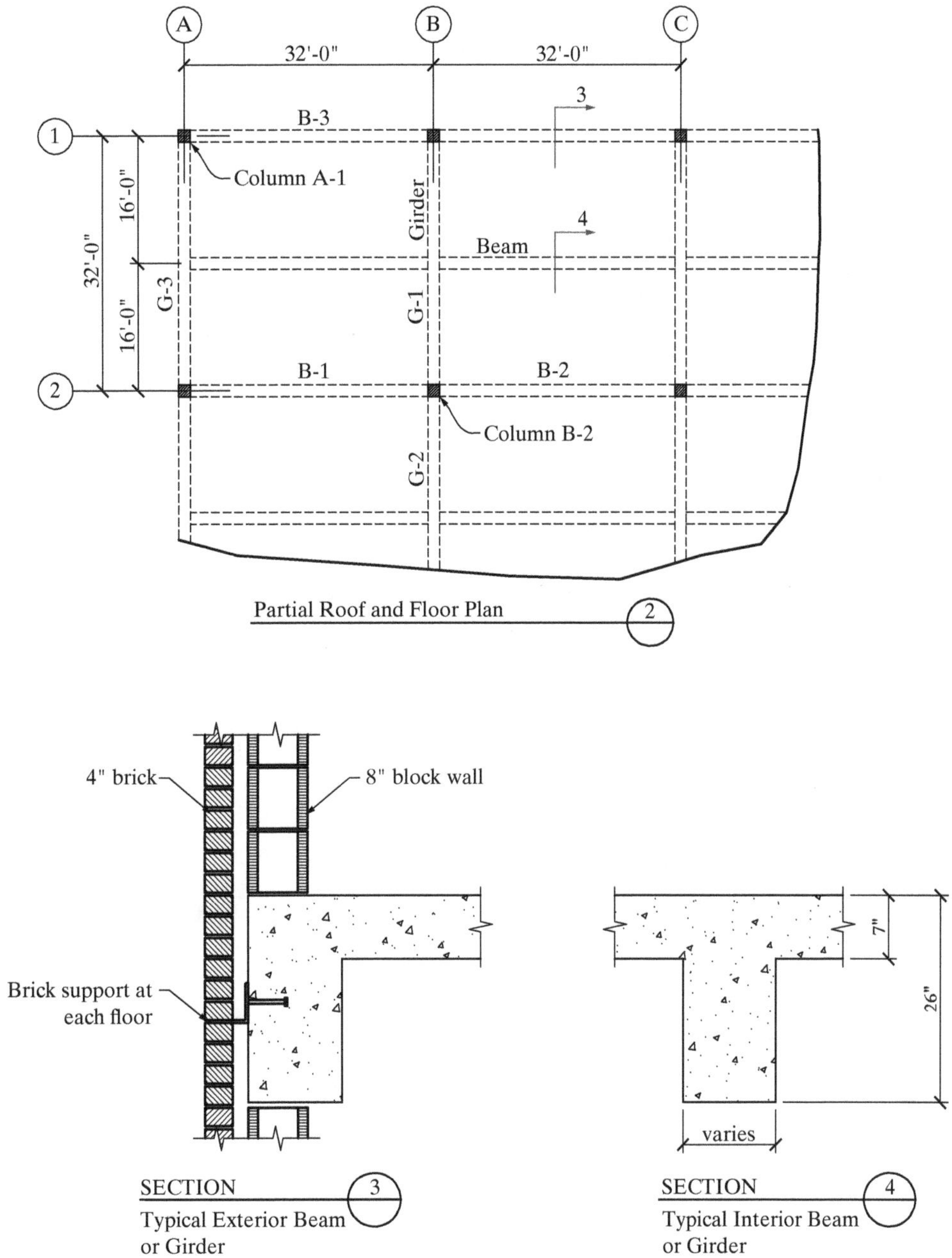

FIGURE FOR EXERCISE 1.12

CHAPTER 2

RECTANGULAR REINFORCED CONCRETE BEAMS AND SLABS: TENSION STEEL ONLY

2.1 INTRODUCTION

In the ACI Code, two types of limit states are considered: the ultimate limit state (i.e., strength design), where the failure of the structure or structural member due to bending, shear, crushing or buckling, etc., is imminent, and this occurs at the factored loads; and the serviceability limit state which results from excessive deflection, cracking, and vibrations under service loads.

When a beam is subjected to bending moments (also termed *flexure*), bending strains are produced. Under positive moment (as normally defined), compressive strains are produced in the top of the beam and tensile strains are produced in the bottom. These *strains* produce *stresses* in the beam, compression in the top, and tension in the bottom. Bending members must therefore be able to resist both tensile and compressive stresses.

Recalling that concrete is weak in tension and strong in compression, therefore, for a concrete flexural member (beam, wall, slab, and so on) to have any significant load-carrying capacity, its basic inability to resist tensile stresses must be overcome. By embedding reinforcement (usually deformed steel bars) in the tension zones, a *reinforced concrete* member that can resist both tensile and compressive stresses is created. When properly designed and constructed, members composed of these materials perform very adequately when subjected to flexure.

Initially, we will consider simply supported single-span beams that, as they carry only positive moment (tension in the bottom), will be reinforced with steel bars placed near the bottom of the beam. However, note that most reinforced concrete flexural members in practice are continuous over several spans, and simply supported single-span concrete flexural members are the exception rather than the rule.

2.2 ANALYSIS AND DESIGN METHOD

In the beam examples in Chapter 1, we assumed both a straight-line strain distribution and straight-line stress distribution from the neutral axis to the outer fibers. This, in effect, stated that stress was proportional to strain. This analysis is sometimes called *elastic design*.

As stated in Chapter 1, elastic design is considered valid for the homogeneous plain concrete beam as long as the tensile stress does not exceed the modulus of rupture, that is, the stress at which tensile cracking commences. With homogeneous materials used in construction, such as structural steel and timber, the limit of stress–strain proportionality is generally termed the *proportional limit.* Note that the modulus of rupture for the plain concrete beam may be considered analogous to the proportional limit for structural steel and timber with respect to the limit of stress–strain proportionality.

With structural steel, the proportional limit and yield stress have nearly the same value, and when using the allowable stress design (ASD) method, an allowable bending stress is determined by applying a factor of safety to the yield stress.

With timber, the determination of allowable bending stress is less straightforward, but it may be thought of as some fraction of the bending stress that causes failure. Using the allowable bending stress and the assumed linear stress–strain relationship, both the analysis and design of timber members and structural steel members (using the ASD method) can be carried out using a simplified method that is similar to that used in Chapter 1 examples.

Even though a reinforced concrete beam was known to be a nonhomogeneous member, for many years, the elastic behavior approach was considered valid for concrete design, and it was known as the *working stress design* (WSD) method. The basic assumptions for the WSD method were as follows: (1) A plane section before bending remains a plane section after bending; (2) Hooke's law (stress is proportional to strain) applies to both the steel and the concrete; (3) the tensile strength of concrete is zero and the reinforcing steel carries all the tension; and (4) the bond between the concrete and the steel is perfect, so no slip occurs.

Based on these assumptions, the flexure formula was still used even though the beam was nonhomogeneous. This was accomplished by theoretically transforming one material into another based on the ratio of the concrete and steel moduli of elasticity.

Although the WSD method was convenient and was used for many years, it has been replaced with a more modern and realistic approach for the analysis and design of reinforced concrete. One basis for this approach is that at some point in the loading, the proportional stress–strain relationship for the compressive concrete ceases to exist. When first developed, this method was called the *ultimate strength design* (USD) method. Since then, the name has been changed to the *strength design method.*

The assumptions for the strength design method are similar to those itemized for the WSD method, with one notable exception. Research has indicated that the compressive concrete stress is approximately proportional to strain up to only moderate loads. With an increase in load, the approximate proportionality ceases to exist, and the compressive stress diagram takes shape similar to the concrete compressive stress–strain curve of Figure 1.1b. Additional assumptions for strength design are discussed in Section 2.4.

A major difference between the ASD and strength design methods lies in the way the applied loads (i.e., *service loads*—the loads that are specified in the general building code) are handled and in the determination of the capacity (strength) of the reinforced concrete members. In the strength design method, service loads are amplified using load factors. Members are then designed so that their practical or design strength at failure, which is somewhat less than the true strength at failure, is sufficient to resist the amplified or factored loads. The strength at failure is still commonly called the *ultimate strength*, and the load at or near failure is commonly called the *ultimate or factored load*. The stress pattern assumed for strength design is such that predicted strengths are in substantial agreement with test results.

2.3 BEHAVIOR UNDER LOAD

Before discussing the strength design method, let us review the behavior of a long-span, rectangular reinforced concrete beam as the load on the beam increases from zero to the magnitude that would cause failure. The reinforced concrete simple beam of Figure 2.1 is assumed to be subjected to downward loading, which will cause a positive moment in the beam (i.e., a bending moment that causes tension stresses in the bottom fiber of the beam and compression stresses at the top fiber of the beam). The steel reinforcing, three bars in this example, is located near the bottom of the beam, which is the tension side. Note that the overall depth of the beam is designated h, whereas the location of the centroid of the reinforcing steel, relative to the compression face, is defined as the *effective depth, d*. In this example, the centroid is at the center of a single layer of bars. If there are multiple layers of bars, then the effective depth is measured from the compression face to the centroid of the bar group.

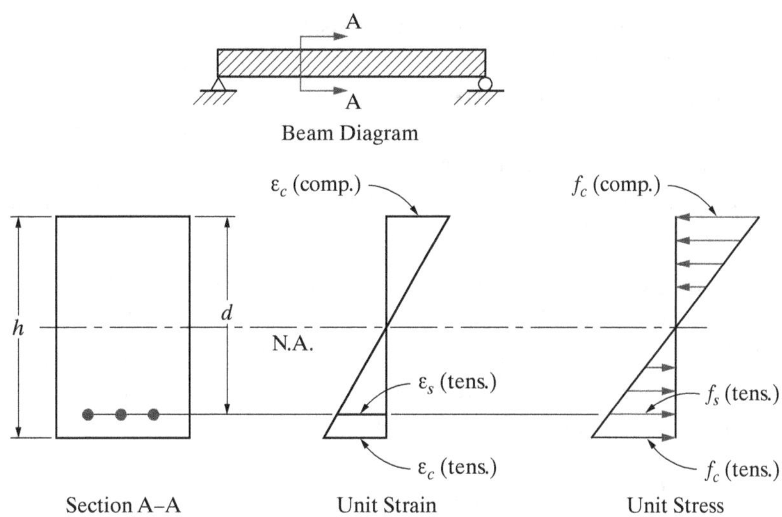

FIGURE 2.1 Flexural behavior at very small loads.

At very small loads, assuming that the concrete has not cracked, both concrete and steel will resist the tension, and concrete alone will resist the compression. The stress distribution is shown in Figure 2.1. The strain variation will be linear from the neutral axis to the outer fiber. Note that stresses also vary linearly from zero at the neutral axis and are, for all practical purposes, proportional to strains. This will be the case when stresses are low (below the modulus of rupture).

At moderate loads, the tensile strength of the concrete will be exceeded, and the concrete will crack (hairline cracks) in the manner shown in Figure 2.2. Because the concrete cannot transmit any tension across a crack, the steel bars will then resist the entire tension. The stress distribution at or near a cracked section then becomes as shown in Figure 2.2. This stress pattern exists up to approximately a concrete stress f_c of about $0.5f_c'$. The concrete compressive stress is still assumed to be proportional to the concrete strain.

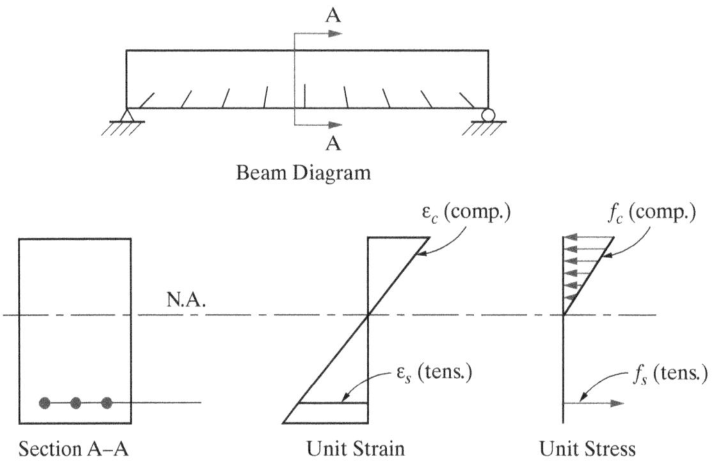

FIGURE 2.2 Flexural behavior at moderate loads.

With further load increase, the compressive strains and stresses will increase; they will cease to be proportional, however, and some nonlinear stress curves will result on the compression side of the beam. This stress curve above the neutral axis will be essentially the same shape as the concrete stress–strain curve (see Figure 1.1b). The stress and strain distribution that exists at or near the ultimate load is shown in Figure 2.3. Eventually, the ultimate capacity of the beam will be reached, and the beam will fail. The actual mechanism of the failure is discussed later in this chapter.

At this point the reader may well recognize that the process of attaining the ultimate capacity of a member is irreversible. The member has cracked and deflected significantly; the steel has yielded and will not return to its original length. If other members in the structure have similarly reached their ultimate capacities, the structure itself is probably crumbling and in a state of distress or partial ruin, even though it may not have completely collapsed. Naturally, although we cannot ensure that this state will never be reached, factors are introduced to create the commonly accepted margins of safety. Nevertheless, the ultimate capacities of members are, at present, the basis for reinforced concrete analysis and design. In this text, it is in such a context that we will speak of the failures of members.

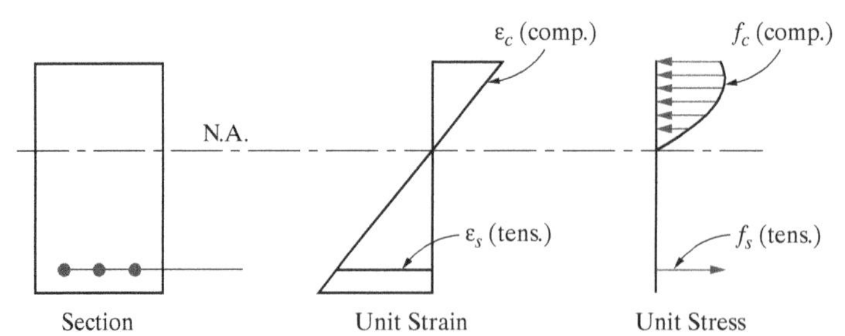

FIGURE 2.3 Flexural behavior near ultimate load.

2.4 STRENGTH DESIGN METHOD ASSUMPTIONS

The development of the strength design approach depends on the following basic assumptions:

1. A plane section before bending remains a plane section after bending. That is, the strain throughout the depth of the member varies linearly from zero at the neutral axis. Tests have shown this assumption to be essentially correct.
2. Stresses and strains are approximately proportional only up to moderate loads (assuming that the concrete stress does not exceed approximately $\frac{f'_c}{2}$). When the load is increased and approaches an ultimate load, stresses and strains are no longer proportional. Hence, the variation in concrete stress is no longer linear.
3. In calculating the ultimate moment capacity of a beam, the tensile strength of the concrete is neglected.
4. The maximum usable concrete compressive strain at the extreme fiber is assumed to be equal to 0.003. This value is based on extensive testing, which indicated that the flexural concrete strain at failure for rectangular beams generally ranges from 0.003 to 0.004 in./in. Hence, the assumption that the concrete is about to crush when the maximum strain reaches 0.003 is slightly conservative.
5. The steel is assumed to be uniformly strained to the strain that exists at the level of the centroid of the steel. Also, if the strain in the steel (ε_s) is less than the yield strain of the steel (ε_y), the stress in the steel is $E_s\varepsilon_s$. This assumes that for stresses less than f_y, the steel stress is proportional to strain. For strains equal to or greater than ε_y, the stress in the reinforcement will be considered independent of strain and equal to f_y. See the idealized stress–strain diagram for steel shown in Figure 1.4b.
6. The bond between the steel and concrete is perfect and no slip occurs.

Assumptions 4 and 5 constitute what may be termed *code criteria* with respect to failure. The true ultimate strength of a member will be somewhat greater than that computed using these assumptions. The strength method of design and analysis of the ACI Code is based on these criteria; however, and consequently, so is our basis for bending member design and analysis. Note that "cracking in reinforced concrete is not a defect. The very basis of reinforced concrete design is that concrete has no tensile strength and that sufficient reinforcement is provided to control crack widths." Therefore, "cracking is an inherent part of reinforced concrete, and if controlled, will not be detrimental to the performance of the structure." [10]

2.5 FLEXURAL STRENGTH OF RECTANGULAR BEAMS

Based on the assumptions previously stated, we can now examine the strains, stresses, and forces that exist in a reinforced concrete beam subjected to its *ultimate moment*, that is, the moment that exists just prior to the failure of the beam. In Figure 2.4, the assumed beam has a width b and an effective depth d, and it is reinforced with a steel area of A_s. (A_s is the total *cross-sectional area* of tension steel present.)

Based on the preceding assumptions, it is possible that a beam may be loaded to the point where the maximum tensile steel unit stress equals its yield stress (as a limit), and the concrete compressive strain is less than 0.003 in./in. It is also possible that in another beam, the maximum concrete compressive strain will equal 0.003 in./in. and the tensile steel unit stress will be less than its yield stress f_y. When either condition occurs, it implies a specific mode of failure, which will be discussed later.

As stated previously, the compressive stress distribution above the neutral axis for a flexural member is similar to the concrete compressive stress–strain curve as depicted in Figure 1.1b. As may be observed in Figure 2.4, the ultimate compressive stress f_c' does not occur at the outer fiber, neither is the shape of the curve the same for different-strength concretes. The magnitudes of the compressive concrete stresses are defined by some irregular curve, which could vary not only from concrete to concrete but also from beam to beam. Present theories accept that, at ultimate moment, compressive stresses and strains in concrete are not proportional. Although strains are assumed linear, with maximum strain of 0.003 in./in. at the extreme outer compressive fiber, the maximum concrete compressive stress f_c' develops at some intermediate level near, but not at, the extreme outer fiber.

The flexural strength or resisting moment of a rectangular beam is created by the development of these internal stresses that, in turn, may be represented as internal forces. As observed in Figure 2.4, N_C represents a theoretical internal resultant compressive force that in effect constitutes the total internal compression above the neutral axis. N_T represents a theoretical internal resultant tensile force that in effect constitutes the total internal tension below the neutral axis.

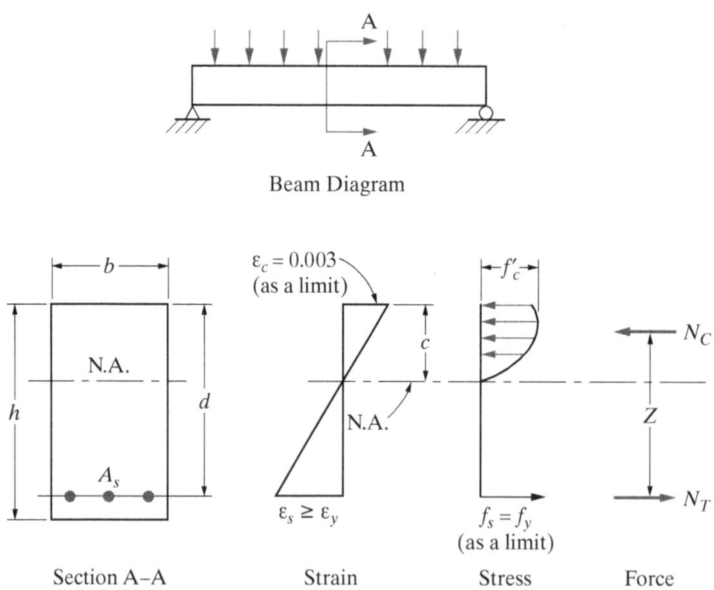

FIGURE 2.4 Beam subjected to ultimate moment.

These two forces, which are parallel, equal, and opposite and separated by a distance Z, constitute an internal resisting couple whose maximum value may be termed the *nominal moment strength* of the bending member. As a limit, this nominal moment strength must be capable of resisting the design bending moment induced by the applied loads. Consequently, if we wish to design a beam for a prescribed loading condition, we must arrange its concrete dimensions and the steel reinforcements so that it can develop a moment strength at least equal to the maximum bending moment induced by the loads.

The determination of the moment strength is complex because of the shape of the compressive stress diagram above the neutral axis. Not only is N_C difficult to evaluate but its location relative to the tensile steel is difficult to establish. Because the moment strength is a function of the magnitude of N_C and Z, however, it is not really necessary to know the exact shape of the compressive stress distribution above the neutral axis. To determine the moment strength, it is necessary to know only (1) the total resultant compressive force N_C in the concrete and (2) its location from the outer compressive fiber (from which the distance Z may be established). These two values may easily be established by replacing the unknown complex compressive stress distribution with a fictitious one of simple geometrical shape, provided the fictitious distribution results in the same total compressive force N_C applied at the same location as in the actual distribution when it is at the point of failure.

2.6 EQUIVALENT STRESS DISTRIBUTION

For purposes of simplification and practical application, a fictitious but equivalent rectangular concrete stress distribution was proposed by Whitney [1] and subsequently adopted by the ACI Code (Sections 22.2.2.3 and 22.2.2.4). The ACI Code also stipulates that other compressive stress distribution shapes may be used, provided results are in substantial agreement with comprehensive test results. Because of the simplicity of the rectangular shape, however, it has become the most widely used fictitious stress distribution for design purposes. Though fictitious, the Whitney rectangular stress block yields values of moment and axial load capacities that are consistent with test results.

With respect to this equivalent stress distribution, as shown in Figure 2.5, the average compressive stress intensity is taken as $0.85f'_c$ and is assumed to act over the upper area of the beam cross-section defined by the width b and a depth of a. The magnitude of the depth of the rectangular stress block, a may be determined by

$$a = \beta_1 c$$

where

c = distance from the outer compressive fiber to the neutral axis

β_1 = a factor that is a function of the strength of the concrete as follows and as shown in Figure 2.6:

For 2500 psi $\leq f'_c \leq$ 4000 psi: β_1 = 0.85.

For f'_c > 4000 psi: $\beta_1 = 0.85 - \dfrac{0.05(f'_c - 4000)}{1000} \geq 0.65$

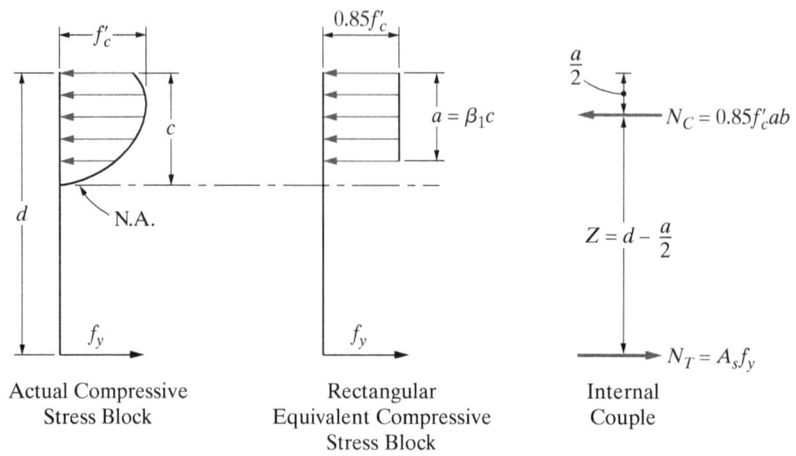

FIGURE 2.5 Equivalent stress block for strength design and analysis.

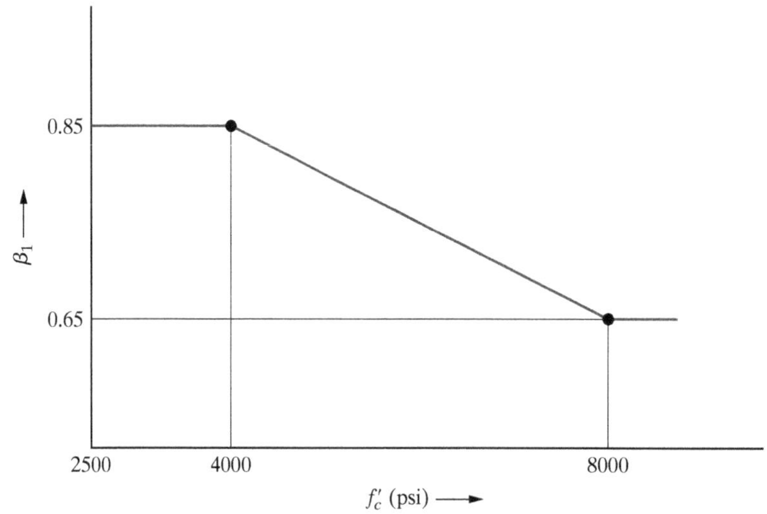

FIGURE 2.6 β_1 versus f'_c.

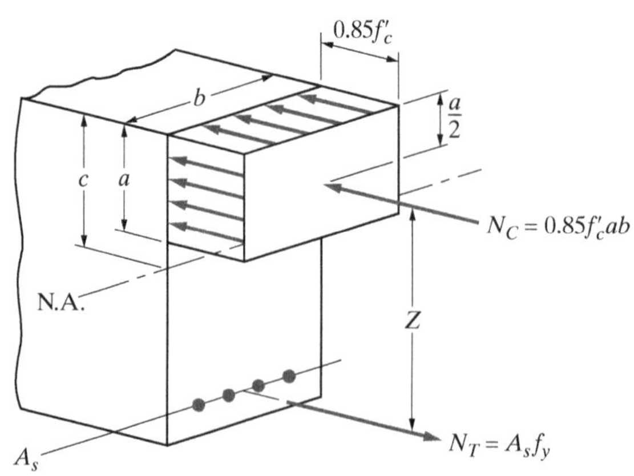

FIGURE 2.7 Equivalent stress block for strength design and analysis - isometric view.

It is in no way implied that the concrete compressive stresses are actually distributed in this most unlikely manner as shown in Figure 2.5, but this equivalent rectangular stress distribution produces strength values that are close to those obtained from the complex actual stress distribution. An isometric view of the accepted internal relationships is shown in Figure 2.7.

Using the equivalent stress distribution in combination with the strength design assumptions, we may now determine the nominal moment strength M_n of rectangular reinforced concrete beams that are reinforced for tension only.

The nominal moment strength determination is based on the assumption that the member will have the exact dimensions and material properties used in the design computations. As discussed in Section 2.9, the nominal moment strength will be further reduced when it is used in practical analysis and design work.

EXAMPLE 2.1

Determine the nominal or theoretical moment capacity, M_n for a beam with the cross-section shown in Figure 2.8, where $f_c' = 4000$ psi. Assume A615 grade 60 steel.

Solution:

1. We will assume that f_y exists in the steel, subject to later check. By $\Sigma H = 0$,

$$N_C = N_T$$
$$(0.85\ f_c')ab = A_s f_y$$
$$a = \frac{A_s f_y}{0.85\ f_c' b} = \frac{2.37(60)}{0.85(4)(10)} = 4.18 \text{ in.}$$

This is the depth of the stress block that must exist if there is to be horizontal equilibrium.

2. Calculate the length of the lever or moment arm, Z:

$$Z = d - \frac{a}{2} = 23 - \frac{4.18}{2} = 20.9 \text{ in.}$$

3. Calculate the nominal or theoretical moment capacity, M_n:

$$M_n = N_C Z = N_T Z$$

Based on the concrete,

$$\begin{aligned} M_n &= N_C Z = (0.85 f_c')ab\,(20.9) \\ &= 0.85(4.0)(4.18)(10)(20.9) \\ &= 2970 \text{ in.-kip} \end{aligned}$$

$$\frac{2970 \text{ in.-kip}}{12 \text{ in./ft}} = 248 \text{ ft.-kip}$$

or, based on the steel,

$$\begin{aligned} M_n &= A_s f_y Z \\ &= 2.37(60)(20.9) \\ &= 2970 \text{ in.-kip} \end{aligned}$$

4. In the foregoing computations, the assumption was made that the steel reached its yield strain (and therefore its yield stress) before the concrete reached its "ultimate" (by definition) strain of 0.003. This assumption will now be checked by calculating the strain ε_t in the steel when the concrete strain reaches 0.003. ε_t is defined as the net tensile strain at the centroid of the extreme tension steel at nominal strength.

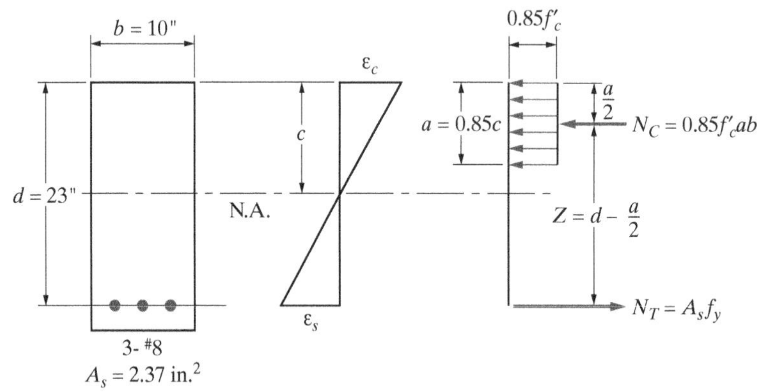

FIGURE 2.8 Sketch for Example 2.1.

Referring to Figure 2.9, we may locate the neutral axis as follows:

$$\begin{aligned} a &= \beta_1 c \quad \text{(ACI Code, Section 22.2.2.4.1)} \\ \beta_1 &= 0.85 \quad \text{because } f'_c = 4000 \text{ psi} \end{aligned}$$

Therefore

$$c = \frac{a}{0.85} = \frac{4.18}{0.85} = 4.92 \text{ in.}$$

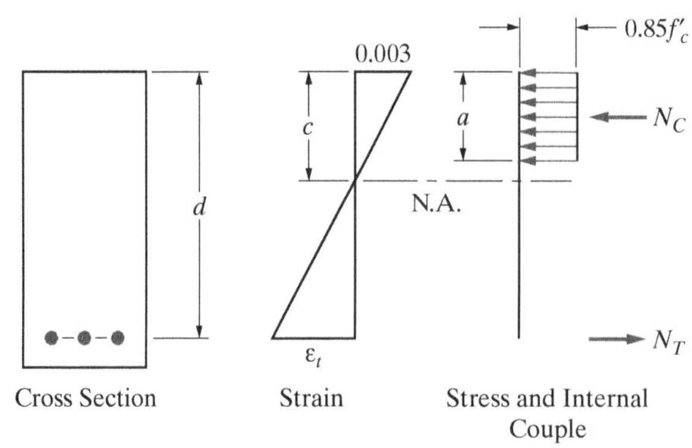

FIGURE 2.9 Steel strain check.

By similar triangles in the strain diagram, we may find the strain in the steel when the concrete strain is 0.003:

$$\frac{0.003}{c} = \frac{\varepsilon_t}{d-c}$$

Then

$$\varepsilon_t = \frac{d-c}{c}(0.003) = \frac{23-4.92}{4.92}(0.003)$$
$$= 0.011 \text{ in./in.}$$

The strain at which the steel yields (ε_y) may be determined from the basic definition of the modulus of elasticity, E = stress/strain:

$$\varepsilon_y = \frac{f_y}{E_s} = \frac{60{,}000}{29{,}000{,}000} = 0.00207 \text{ in./in.} \text{ (\textit{see} Table A.1)}$$

This represents the strain in the steel when the stress first reaches 60,000 psi.

Because the computed strain in the steel (0.011) is greater than the yield strain (0.00207), the steel reaches its yield stress before the concrete reaches its crushing strain of 0.003, and the assumption that the stress in the steel is equal to the yield stress was correct. (See assumption 5 in Section 2.4 and the idealized stress–strain diagram for steel shown in Figure 1.3b.)

2.7 BALANCED, BRITTLE, AND DUCTILE FAILURE MODES

In Figure 2.9, the *ratio* between the steel strain and the maximum concrete strain is fixed once the neutral axis is established. The location of the neutral axis will vary based on the amount of tension steel in the cross-section, as the stress block is just deep enough to ensure that the resultant compressive force is equal to the resultant tensile force ($\Sigma H = 0$). If more tension bars are added to the bottom of a reinforced concrete cross-section, the depth of the compressive stress block will be greater, and therefore, the neutral axis will be lower. Referring again to Figure 2.9, if there were just enough steel to put the neutral axis at a location where the yield strain in the steel and the maximum concrete strain of 0.003 existed at the same time, the cross-section would be said to be *balanced*. The amount of steel required to create this condition is relatively large. However, the balanced condition is the dividing line between two distinct types of reinforced concrete beams that are characterized by their failure modes. If a beam has more steel than is required to create the balanced condition, the beam will fail in a *brittle mode*. The additional steel will cause the neutral axis to be low (see Figure 2.10). This will in turn cause the concrete to reach a strain of 0.003 *before* the steel yields. Should more moment (and therefore strain) be applied to the beam cross-section, failure will be initiated by a sudden crushing of the concrete. The brittle failure mode is undesirable.

If a beam has less steel than is required to create the balanced condition, the beam will fail in a *ductile mode*. The neutral axis will be higher than the balanced neutral axis, and the steel will reach its yield strain (and therefore its yield stress) before the concrete reaches a strain of 0.003. Figure 2.10 shows these variations in neutral axis location for beams that are on the verge of failure. In each case, the concrete has been strained to 0.003. Following the ductile failure mode case through to failure, we see that a slight additional load will cause the steel to stretch a considerable amount. The strains in the concrete and the steel continue to increase. The tensile force is not increasing since the steel stress has reached the maximum tensile stress f_y. Because the compressive force cannot increase ($\Sigma H = 0$), and because the concrete strain and therefore its stress are increasing, the area under compression must be decreasing, and the neutral axis must rise. This process continues until the reduced area fails in compression as a secondary effect. This failure due to yielding is a gradual one, with the beam showing greatly increased deflection after the steel reaches the yield point; hence, there is adequate warning of impending failure. The steel, being ductile, will not actually be pulled apart even at the failure of the beam. The ductile failure mode is desirable and is required by the ACI Code, as discussed in the following section.

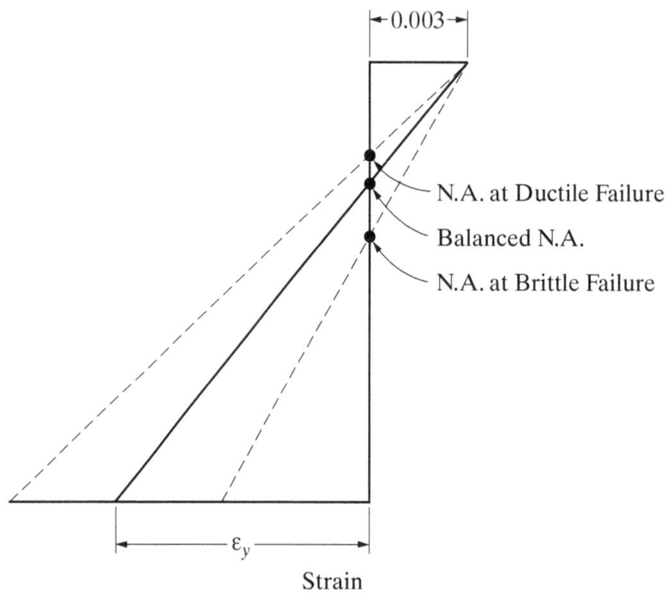

FIGURE 2.10 Strain distribution and failure modes in flexural members.

2.8 DUCTILITY REQUIREMENTS

Although failure due to yielding of steel is gradual, with adequate warning of collapse, failure due to crushing of concrete is sudden and without warning. We have seen that flexural members with less steel than is required to produce the balanced condition will fail by yielding of the steel due to the strain in the steel exceeding the yield strain. In previous editions of the ACI Code prior to 2005, in order to assure a ductile failure mode, the maximum tension reinforcement was limited to 75% of the steel reinforcement required to achieve the balanced condition, and a uniform strength reduction factor, $\phi = 0.9$ was prescribed for calculating the flexural capacity. In the current version of the ACI Code, the strength reduction factor is a function of the net tensile strain in the extreme tension reinforcement, ε_t. The three regions of the net tensile strains are specified as:

- Compression-controlled region,
- Transition region, and
- Tension-controlled region.

The ACI Code (Section R21.2.2) defines a section as *tension-controlled* when the net tensile strain ε_t in the extreme tension steel is equal to or greater than ε_{ty} + 0.003 when the concrete has reached its assumed crushing strain limit of 0.003. The ACI Code requires reinforced concrete flexural members to be *tension-controlled* to prevent brittle failure (ACI 318-19 Sections 7.3.3.1, 8.3.3.1, 9.3.3.1, and Table 21.2.2). With reference to Figure 2.11, note that the centroid of the extreme tension steel is located a distance of d_t from the extreme compression face. For a single layer of tension steel in the flexural member, $d_t = d$, the effective depth. For multiple layers of steel, $d_t > d$.

Further, ACI Code (Section R21.2.2) defines a section as *compression-controlled* when the net tensile strain ε_t in the extreme tension steel is equal to or less than the yield strain ε_{ty} of the steel just as the concrete in compression reaches its assumed strain limit of 0.003. See Table A.1 for values of ε_{ty}. (The code permits ε_{ty} to be taken as 0.002 for grade 60 steel (see ACI 318, Section 21.2.2.1).)

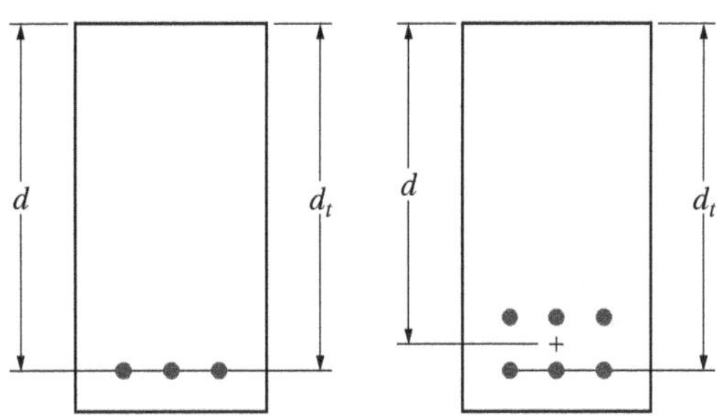

FIGURE 2.11 Definition of extreme tension steel.

For non-prestressed flexural members subjected to little or no axial load, the net tensile strain ε_t at nominal strength shall not be less than ε_{ty} + 0.003, which ensures a tension-controlled flexural member, one that will exhibit ductility and fail by yielding of the steel.

Sections in which ε_t falls between the *compression-controlled* strain limit ε_{ty} and the *tension-controlled* limit of ε_{ty} + 0.003 constitute a *transition* region between compression-controlled and tension-controlled sections. Whether a section is tension-controlled, compression-controlled, or in the transition region has implications on the value of the strength reduction factor, which will be discussed shortly.

To account for high-strength reinforcement, ACI 318 now permits the use of Grade 100 reinforcement (i.e., f_y = 100 ksi).

For the **tension-controlled zone**, the net tensile strain in the extreme tension reinforcement,$\varepsilon_t \geq 0.003 + \varepsilon_{ty}$.

For 60 ksi reinforcement, $\varepsilon_{ty} = 60/29000 \approx 0.002$, therefore $\varepsilon_t \geq 0.003 + 0.002 = 0.005$.

For the **compression-controlled zone**, the net tensile strain in the extreme tension reinforcement, $\varepsilon_t \leq \varepsilon_{ty}$.

For 60 ksi reinforcement, $\varepsilon_{ty} = 60/29000 \approx .002$, therefore $\varepsilon_t \leq 0.002$.

For the **transition zone**, the net tensile strain in the extreme tension reinforcement should be such that, $\varepsilon_{ty} \leq \varepsilon_t < 0.003 + \varepsilon_{ty}$.

For 60 ksi reinforcement, $\varepsilon_{ty} = 60/29000 \approx 0.002$, therefore $0.002 \leq \varepsilon_t < 0.005$ indicates a concrete member in the transition zone. See Figure 2.12 for a graphical representation of the strain limit definitions.

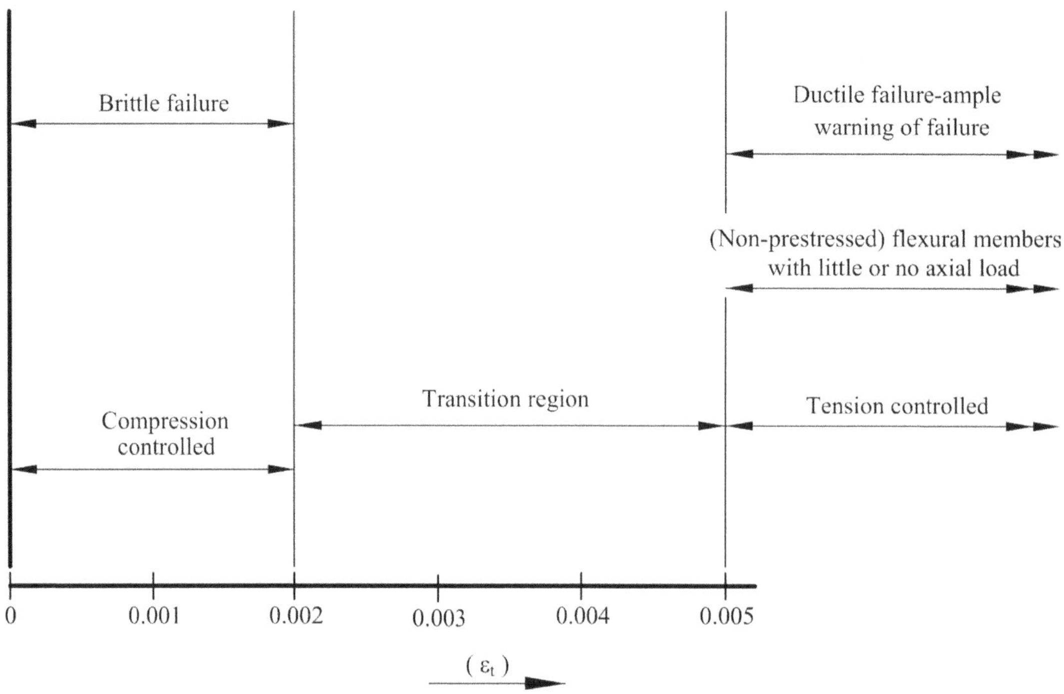

FIGURE 2.12 Strain limit definitions (for f_y = 60 ksi).

EXAMPLE 2.2

For the beam cross-section of Example 2.1, determine the amount of steel A_s required to cause the strain in the tension steel ε_t to be 0.005 just as the maximum strain in the concrete reaches 0.003. $f_c' = 3000$ psi and $f_y = 60{,}000$ psi.

Solution:

The cross-section, strain, stress, and internal couple are shown in Figure 2.13. Note that $d_t = d$.

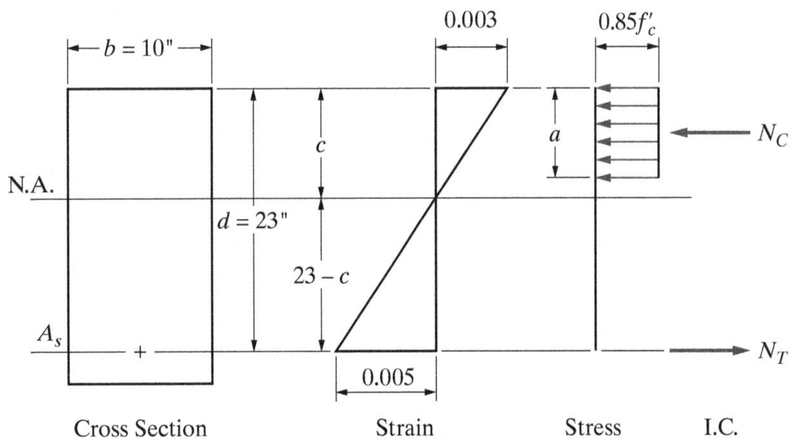

FIGURE 2.13 Sketch for Example 2.2.

1. Determine the location of the neutral axis. By similar triangles:

$$\frac{c}{0.003} = \frac{23 - c}{0.005}$$
$$0.005\,c = 0.003(23 - c)$$
$$0.005\,c + 0.003\,c = 0.003(23) = 0.0690$$

from which

$$c = \frac{0.0690}{0.008} = 8.63 \text{ in.}$$

2. Determine N_C:

$$a = \beta_1 c = 0.85(8.63) = 7.34 \text{ in.}$$
$$N_C = 0.85 f_c' ab$$
$$= 0.85(3 \text{ kip/in.}^2)(7.34 \text{ in.})(10 \text{ in.}) = 187.2 \text{ kip}$$

3. Determine A_s:

$$N_C = N_T = 187.2 \text{ kip}$$
$$N_T = A_s f_y$$
$$A_s = \frac{N_T}{f_y} = \frac{187.2 \text{ kip}}{60 \text{ kip/in.}^2} = 3.12 \text{ in.}^2$$

For ε_t to reach 0.005 just as the maximum concrete strain reaches 0.003, A_s must be 3.12 in.^{2} This is the maximum amount of steel for the section to be a tension-controlled section.

Any steel in excess of 3.12 in.^{2} will cause ε_t to be less than 0.005. Recall that the ACI Code does not allow the net tensile strain in the steel reinforcement, ε_t, to be less than 0.005 in a non-prestressed concrete flexural member with little or no axial load.

At this point, note again that heavily reinforced beams are less efficient than their more lightly reinforced counterparts. One structural reason for this is that, for a given beam size, an increase in A_s is accompanied by a decrease in the lever arm of the internal couple ($Z = d - a/2$). This may be illustrated by doubling the tension steel for the beam of Example 2.1 and recalculating M_n:

$$A_s = 2(2.37) = 4.74 \text{ in.}^2 \quad (100\% \text{ increase})$$

$$a = \frac{A_s f_y}{0.85 f'_c b} = \frac{4.74(60)}{0.85(4)(10)} = 8.36 \text{ in.}$$

$$M_n = 0.85 f'_c ab\left(d - \frac{a}{2}\right) = 0.85(4)(8.36)(10)\left(23 - \frac{8.36}{2}\right)$$

$$= 446 \text{ ft.-kip}$$

For the beam of Example 2.1, M_n was 248 ft.-kip; therefore,

$$\frac{446 - 248}{248} \times 100\% = 80\% \text{ increase}$$

Sections 9.6.1.1 and 9.6.1.2 of the ACI Code also establish a lower limit on the amount of tension reinforcement for flexural members. The code states that where tensile reinforcement is required by analysis, the steel area A_s shall not be less than that given by

$$A_{s,\min} = \frac{3\sqrt{f'_c}}{f_y} b_w d \geq \frac{200}{f_y} b_w d$$

(Note that for rectangular beams, $b_w = b$.)

$A_{s,\min}$ is conveniently calculated using Table A.5, where the larger of $\frac{3\sqrt{f'_c}}{f_y}$ and $200/f_y$ is tabulated. Note that $A_{s,\min}$ is the product of the tabulated value and $b_w d$.

The lower limit guards against sudden failure essentially by ensuring that a beam with a very small amount of tensile reinforcement has a greater moment strength as a reinforced concrete section than that of the corresponding plain concrete section computed from its modulus of rupture. Alternatively, it is satisfactory to provide an area of tensile reinforcement that is one-third greater than that required by analysis (ACI Code, Section 9.6.1.3). This requirement applies especially to grade beams, wall beams, and other deep flexural members where the minimum reinforcement requirement specified in ACI 9.6.1.2 would result in an excessively large amount of steel.

Minimum required reinforcement in structural slabs (see Section 2.13) is governed by the required shrinkage and temperature steel of 0.0018 x the gross cross-sectional area of the concrete slab as outlined in the ACI Code (Sections 24.4.1 and 24.4.3).

2.9 STRENGTH REQUIREMENTS

The basic criterion for strength design may be expressed as

Strength furnished or design capacity ≥ strength required or the factored load effect.

All members and all sections of members must be proportioned to meet this criterion.

The required strength may be expressed in terms of design loads or their related moments, shears, and forces. Design loads may be defined as service loads multiplied by the appropriate *load factors*. (When the word *design* is used as an adjective [e.g., design load] throughout the ACI Code, it indicates that load factors are included.) The subscript u is used to indicate design loads, moments, shears, and forces.

The ACI Code, Section 5.3, specifies load factors to be used and load combinations to be investigated. Loads to be considered are dead loads, live loads, fluid loads, loads due to weight and pressure of soil, and snow loads, among others. For applications in this text, we will consider only dead load, live load, and loads due to weight and pressure of soil. For the combination of dead load and live load, the general representation is as follows:

$$U = 1.2D + 1.6L \geq 1.4D$$

where U is defined as the required strength to resist factored loads or related factored internal moments and forces, D is the service dead load, and L is the service live load. (The term *service load* generally refers to the load specified in applicable building codes as representing minimum requirements.) The factors 1.2 and 1.6 (and 1.4) represent load factors. The load factors are part of the overall safety provision in reinforced concrete structures and are meant to reflect the variability in load effects. Dead loads can be more accurately estimated than live loads, and hence a lower load factor is used for dead loads. Live loads have a greater variability than dead loads, and hence a higher load factor is used for live loads.

A second part of the overall safety provision, provided in the ACI Code, Section 21.2, is the reduction of the theoretical capacity of a structural element by a *strength-reduction factor,* ϕ. This provides for the possibility that small adverse variations in material strengths, workmanship, and dimensions, although within acceptable tolerances and limits of good practice, may combine to result in under-capacity. In effect, the nominal strength of a member, when multiplied by the ϕ factor, will furnish us with a practical or design strength that is obviously less than the nominal strength.

The ACI Code, Section 21.2, provides for these variables by using the following strength reduction or ϕ factors:

Tension-controlled sections	0.90
Compression-controlled sections	
– spirally reinforced concrete members	0.75
– other reinforced concrete members	0.65
Shear and torsion	0.75
Bearing on concrete	0.65
Bearing on concrete (post-tension anchorage zones)	0.85

Additionally, for sections in which ε_t is between the limits for tension-controlled sections and compression-controlled sections, ϕ shall be permitted to be linearly increased from that for compression-controlled sections to 0.90 as ε_t increases from the compression-controlled strain limit to a strain of 0.005 (for Grade 60 rebar).

In each case, the practical strength of a reinforced concrete member will be the product of the nominal strength and the ϕ factor. Therefore, in terms of moment, we can say that:

Practical or design moment strength = ϕM_n

Therefore, the general design equation for the ultimate limit state is:

$\phi R_n \geq U$

Where,

ϕR_n = *the design strength or capacity*

U = *the required strength or the factored load effects or the demand*

The individual values assigned to ϕ factors are determined using statistical studies of structural resistances and loading variabilities tempered by subjective factors reflecting the type and consequences of a structural failure.

The variation in ϕ for tension-controlled sections, transition sections, and compression-controlled (nonspirally reinforced) sections is shown in Figure 2.14. For the transition section, an expression for ϕ as a function of ε_t can be developed:

$$\phi = 0.65 + \left(\frac{0.90-0.65}{0.005-\varepsilon_y}\right)(\varepsilon_t - \varepsilon_y) \leq 0.9$$

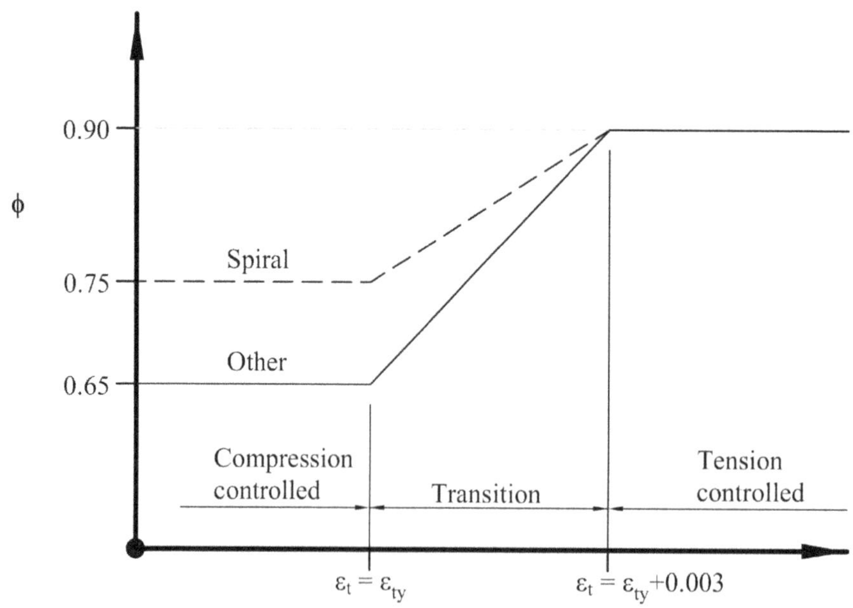

FIGURE 2.14 ϕ versus ε_t (Adapted from ACI Code Figure R21.2.2b)

This is valid for the range of $\varepsilon_y \le \varepsilon_t \le 0.005$. Assuming $\varepsilon_y = 0.002$ (i.e., for Grade 60 rebar), this expression becomes:

$$\phi = 0.65 + (\varepsilon_t - 0.002)\left(\frac{250}{3}\right) \le 0.9$$

The calculation of ε_t can also be simplified using the depth of the stress block a, which is commonly calculated in analysis exercises. Refer to Figure 2.9 and set $d_t = d$.

$$\varepsilon_t = \frac{d_t - c}{c}(0.003)$$

Substitute $c = a/\beta_1$:

$$\varepsilon_t = \left(\frac{d_t - \frac{a}{\beta_1}}{\frac{a}{\beta_1}}\right)(0.003) = \frac{0.003\beta_1 d_t}{a} - 0.003 \quad (2.1)$$

Summary of Strength Reduction Factor for the Three Zones:

- **Tension-controlled zone (Required for flexural members to ensure ductile behavior)**
  - $\varepsilon_t \ge 0.003 + \varepsilon_y$ (for Grade 60 rebar, $\varepsilon_t \ge 0.005$)
  - => $c/d_t \le 0.375$
  - => $\phi = 0.9$
- **Compression-controlled zone (has less ductility)**
  - $\varepsilon_t \le 0.002$ or
  - => $c/d_t \ge 0.6$
  - => $\phi = 0.65$ for tied columns
  - => $\phi = 0.75$ for spiral columns (spiral columns are more ductile than tied columns)
- **Transition zone (see ACI 318 Table 21.2.2)**
  - $0.002 \le \varepsilon_t \le 0.005$
  - $\phi = 0.65 + 0.25[(1/(c/d_t)) - 5/3]$ for tied columns
  - $\phi = 0.65 + (250/3)[\varepsilon_t - \varepsilon_y]$ for tied columns
  - $\phi = 0.75 + 0.15[(1/(c/d_t)) - 5/3]$ for spiral columns
  - $\phi = 0.75 + 50[\varepsilon_t - \varepsilon_y]$ for spiral columns

Note

- For Grade 60 rebar, the yield strain, $\varepsilon_y = 60\text{ksi}/29000\text{ ksi} = 0.002$
- The net tensile strain is determined using the strain compatibility equation as follows:

$$\frac{\varepsilon_t}{d-c} = \frac{\varepsilon_{cu}}{c} = \frac{0.003}{c}$$

2.10 RECTANGULAR BEAM ANALYSIS FOR MOMENT (TENSION REINFORCEMENT ONLY)

The *flexural analysis problem* is characterized by knowing precisely what comprises the cross-section of a beam. That is, the following data are *known*: tension bar size and number (or A_s), beam width (b), effective depth (d) or total depth (h), f'_c, and f_y. To be found, basically, is the beam strength, although this may be manifested in various ways: Find ϕM_n, check the adequacy of the given beam, or find an allowable load that the beam can carry. The *flexural design problem*, on the other hand, requires the determination of one or more of the dimensions of the cross-section or the determination of the main tension steel to use. It will be important to recognize the differences between these two types of problems because the methods of solving are different.

To expedite reinforced concrete analysis and design calculations, use is frequently made of tables of pertinent quantities. Tables find their greatest use in the design process, but because they are also useful for analysis, they are developed here.

In Example 2.1, we calculated the nominal moment strength M_n for a rectangular reinforced concrete section (tension steel only). Based on concrete:

$$M_n = 0.85 f'_c ab\left(d - \frac{a}{2}\right)$$

and

$$a = \frac{A_s f_y}{0.85 f'_c b}$$

We will now develop a modified expression for M_n (and ϕM_n), which will then be adapted for table use. It is convenient to use the concept of *reinforcement ratio* ρ (lowercase Greek "rho"), the ratio of tension steel area to effective concrete area:

$$\rho = \frac{A_s}{bd}$$

from which

$$A_s = \rho bd$$

Substituting this into the expression for a:

$$a = \frac{A_s f_y}{(0.85 f'_c)b} = \frac{\rho d f_y}{0.85 f'_c}$$

Arbitrarily define ω (omega):

$$\omega = \rho \frac{f_y}{f'_c}$$

Then

$$a = \frac{\omega d}{0.85}$$

Substitute into the ϕM_n expression:

$$\phi M_n = \phi(0.85 f_c')(b)\frac{\omega d}{0.85}\left[d - \frac{\omega d}{2(0.85)}\right]$$

Simplify and rearrange:

$$\phi M_n = \phi b d^2 f_c' \omega (1 - 0.59\omega)$$

Arbitrarily define $\bar{k}$:

$$\bar{k} = f_c' \omega (1 - 0.59\omega)$$

The term $\bar{k}$ is sometimes called the *coefficient of resistance*. It varies with ρ, f_c', and f_y. The general expression for ϕM_n can now be written as

$$\phi M_n = \phi b d^2 \bar{k}$$

Before introducing the tables, we discuss another important quantity that depends on these same three factors and that is included in the tables.

In Example 2.1, we calculated the strain ε_t in the extreme tension steel at nominal strength (when the maximum concrete strain reaches 0.003). Assuming one layer of steel ($d = d_t$), it can be shown that ε_t is fixed if ρ, f_c', and f_y are known. Referring to Figure 2.8 and Example 2.1:

$$\varepsilon_t = \frac{d_t - c}{c}(0.003)$$

Also,

$$a = \beta_1 c$$

From which

$$c = \frac{a}{\beta_1}$$

We previously developed the following relationship:

$$a = \frac{\rho d f_y}{0.85 f_c'}$$

By substitution,

$$c = \frac{\rho d f_y}{0.85 f_c' \beta_1}$$

Again, by substitution,

$$\varepsilon_t = \frac{d-c}{c}(0.003) = \frac{\left(d - \dfrac{\rho d f_y}{0.85 f_c' \beta_1}\right)}{\dfrac{\rho d f_y}{0.85 f_c' \beta_1}}(0.003)$$

From which

$$\varepsilon_t = \frac{0.00255 f_c' \beta_1}{\rho f_y} - 0.003 \tag{2.2}$$

Note, again, that this is based on the assumption that $d = d_t$. For multiple layers of steel, use basic principles or Equation (2.1).

Tables A.7 through A.11 give the values of the coefficient of resistance $\bar{k}$ for values of ρ and various combinations of f_c' and f_y. The maximum tabulated value of ρ is that which corresponds to a net tensile strain, ε_t, of 0.004. Note that the minimum value allowed by code for non-prestressed flexural members with little or no axial load is a net tensile strain, ε_t, of 0.005 (ACI Code, Sections 7.3.3.1, 8.3.3.1, and 9.3.3.1). Values of ε_t are tabulated between 0.004 and 0.005 because in this transition region, ϕ must be determined (ACI Code, Section 21.2.2). When $\varepsilon_t \geq 0.005$, sections are tension-controlled and $\phi = 0.90$. Note that for the tables, it is assumed that $d = d_t$, as it would be for one layer of steel. If there are multiple layers of steel (see Figure 2.11), then $d_t > d$ and ε_t will be larger than the tabulated ε_t. Therefore, the tables are conservative with regard to the determination of ϕ based on ε_t.

EXAMPLE 2.3

Determine if the beam shown in Figure 2.15 is adequate as governed by the ACI Code (318-19). The loads shown are service loads. The uniformly distributed load is DL = 0.65 kip/ft, LL = 0.80 kip/ft. The dead load excludes the beam weight. The point load is a live load; $f_c' = 4000$ psi and $f_y = 60{,}000$ psi.

Solution:

A logical approach to this type of problem is to compare the practical moment strength (ϕM_n) with the applied design moment resulting from the *factored* loads. This latter moment will be noted as M_u. If the beam is adequate for the moment, $\phi M_n \geq M_u$. The procedure outlined here for ϕM_n is summarized in Section 2.11.

Determination of ϕM_n

1. Given:

$$\begin{aligned} f_y &= 60{,}000 \text{ psi} \\ b &= 12 \text{ in.} \\ d &= 17.5 \text{ in.} \\ A_s &= 3.16 \text{ in.}^2 \quad \text{(Table A.2)} \\ f_c' &= 4000 \text{ psi} \end{aligned}$$

2. To be found: ϕM_n and M_u.

3. $\rho = \dfrac{A_s}{bd} = \dfrac{3.16}{12(17.5)} = 0.01505$

4. From Table A.5:

$$A_{s,\min} = 0.0033\, b_w d$$
$$= 0.0033(12)(17.5) = 0.69 \text{ in.}^2 \qquad \text{(O.K.)}$$
$$3.16 \text{ in.}^2 > 0.69 \text{ in.}^2$$

5. From Table A.10, ε_t is not tabulated; therefore $\varepsilon_t > 0.005$, the section is tension-controlled, and $\phi = 0.90$. Also, from Table A.10, $\bar{k} = 0.7809$. (It is conservative to use the lower tabulated value. An interpolation between 0.7809 and 0.7853 could be done, but it is not warranted.)

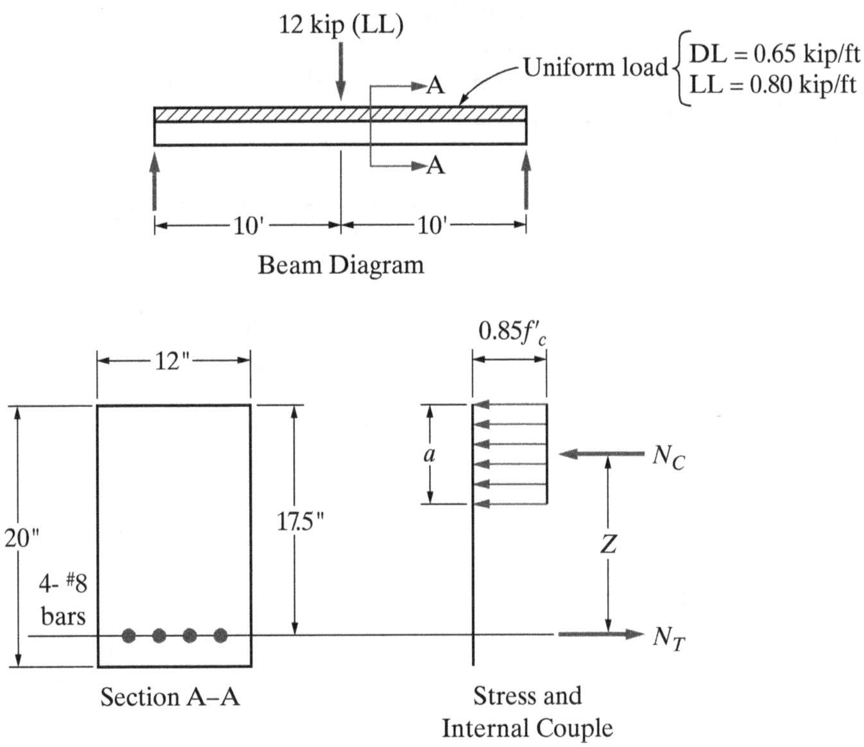

FIGURE 2.15 Sketch for Example 2.3.

6. Calculate ϕM_n from $\phi N_C Z$ or $\phi N_T Z$:

$$a = \frac{A_s f_y}{0.85 f'_c b} = \frac{3.16(60)}{0.85(4)(12)} = 4.65 \text{ in.}$$

$$Z = d - \frac{a}{2} = 17.5 - \frac{4.65}{2} = 15.18 \text{ in.}$$

Based on steel:

$$M_n = A_s f_y Z$$
$$= 3.16(60)(15.18) = 2880 \text{ in.-kip}$$
$$\frac{2880}{12} = 240 \text{ ft.-kip}$$
$$\phi M_n = 0.90(240) = 216 \text{ ft.-kip}$$

or

Calculate ϕM_n using the coefficient of resistance $\bar{k}$:

$$\phi M_n = \phi b d^2 \bar{k}$$
$$= \frac{0.90(12 \text{ in.})(17.5 \text{ in.})^2 (0.7809 \text{ kip/in.}^2)}{12 \text{ in./ft}}$$
$$= 215 \text{ ft.-kip}$$

Find M_u: Because the given service dead load excluded the beam weight, it will now be calculated. As the beam weight is a uniformly distributed load, it will be found in terms of weight per linear foot (kip/ft):

$$\text{beam weight} = \text{beam volume per foot of length} \times 0.150 \text{ kip/ft}^3$$
$$= \frac{20 \text{ in.}(12 \text{ in.})}{144 \frac{\text{in.}^2}{\text{ft}^2}} \times 1 \text{ ft} \times 0.150 \text{ kip/ft}^3$$
$$= 0.250 \text{ kip (per linear foot)}$$
$$= 0.250 \text{ kip/ft}$$

Then summarizing the loads,

Superimposed service uniform dead load = 0.65 kip/ft

Total service uniform dead load = 0.250 + 0.65 = 0.90 kip/ft = w_{DL}

Total service uniform live load = 0.80 kip/ft = w_{LL}

Total factored uniform load:

$$w_u = 1.2w_{DL} + 1.6w_{LL}$$
$$= 1.2(0.90) + 1.6(0.80) = 2.36 \text{ kip/ft}$$

Superimposed concentrated live load = 12 kip = P_{LL}.

Total factored concentrated load:

$$P_u = 1.6P_{LL} = 1.6(12) = 19.2 \text{ kip}$$

$$\begin{aligned} M_u &= \frac{w_u \ell^2}{8} + \frac{P_u \ell}{4} \\ &= \frac{2.36(20)^2}{8} + \frac{19.2(20)}{4} \\ &= 214 \text{ ft.-kip} < 215 \text{ ft.-kip} \end{aligned}$$

Therefore, the beam is adequate.

2.11 SUMMARY OF PROCEDURE FOR RECTANGULAR BEAM ANALYSIS FOR ϕM_n (TENSION REINFORCEMENT ONLY)

1. List the known quantities. Use a sketch.
2. Determine what is to be found. (An "analysis" may require any of the following to be found: ϕM_n, allowable service live load or dead load, maximum allowable span.)
3. Calculate the reinforcement ratio:

$$\rho = \frac{A_s}{bd}$$

4. Calculate $A_{s,\min}$ and compare with A_s (use Table A.5).
5. Determine ε_t by calculation or table ($\varepsilon_t \geq 0.005$). Determine ϕ. Select $\bar{k}$ if it is to be used.
6. Calculate ϕM_n from $\phi N_C Z$ or $\phi N_T Z$, where

$$\begin{aligned} N_C &= 0.85 f_c' ab \\ N_T &= A_s f_y \\ a &= \frac{A_s f_y}{0.85 f_c' b} \\ Z &= d - \frac{a}{2} \end{aligned}$$

or

calculate ϕM_n from

$$\phi M_n = \phi b d^2 \bar{k}$$

A flow diagram of this procedure is presented in Appendix B.

2.12 SLABS: INTRODUCTION

Slabs constitute a specialized category of bending members and are used in both structural steel and reinforced concrete structures. As discussed in Chapter 1, slabs can be classified as one-way or two-way slabs with regards to the direction of their curvature under gravity loads. Probably the most basic and common type of slab is the *one-way slab*. A one-way slab may be described as a structural reinforced concrete slab supported on two opposite sides so that the bending or curvature occurs in one direction only—that is, perpendicular to the supported edges. One-way slabs can be found in floors and roofs, and in concrete stairs.

If a slab panel is supported along all four edges, it may be designated as a *two-way slab*, with bending or curvature occurring in the two orthogonal directions of the slab panel. If the ratio of the lengths of the sides of the slab panel in the two orthogonal directions is greater than or equal to 2.0, however, the slab may be assumed to act as a one-way slab with bending primarily occurring in the short direction of the slab panel.

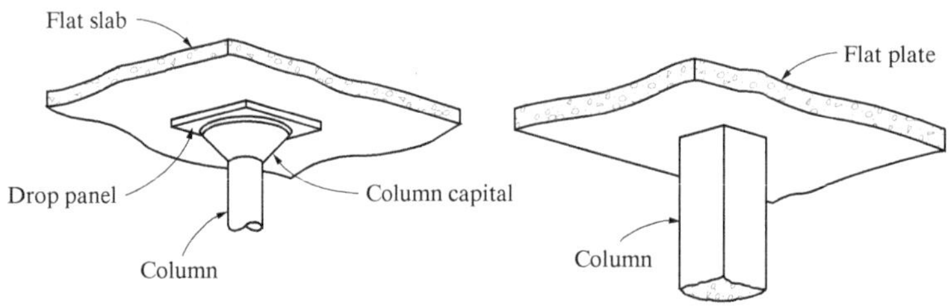

FIGURE 2.16a Reinforced concrete slabs.

FIGURE 2.16b Flat plate with circular columns (Abi Aghayere).

FIGURE 2.16c Flat slab in a basement of a reinforced concrete building under construction (Abi Aghayere).

FIGURE 2.16d Flat plate with square columns, and one-way slabs supported on concrete walls (Abi Aghayere).

FIGURE 2.16e Flat plate with rectangular columns (Abi Aghayere).

Two-way slabs that are supported directly on columns are categorized as flat plates or *flat slabs*. A flat plate may be defined as a concrete slab reinforced in two or more directions and supported directly on columns, generally without beams or girders to transfer the loads to the supporting members. The slab, then, could be considered to be supported on a grid of shallow beams, which are themselves integral to and have the same depth as the slab. The columns tend to punch upward through the slab, resulting in high shearing stress along with inclined slab cracking. (This "punching shear" is considered later in the discussion on column footings in Chapter 10 and two-way slabs in Chapter 6). For a flat slab, the slab is thickened in the vicinity of the column and this thickened portion of the slab is called a *drop panel*; in addition to drop panels, flat slabs can also have *column capitals* where the top of the column is enlarged in the shape of an inverted frustum (see Figure 2.16a). Flat plates, unlike flat slabs, do not have drop panels and column capitals; hence, they are slabs of constant thickness supported directly on columns. Generally, the flat plate is used where spans are smaller and loads lighter than those requiring a flat slab design (see Figure 2.16a). See Figures 2.16b through 2.16e for photos of different types of slabs in concrete buildings under construction.

2.13 ONE-WAY SLABS: ANALYSIS FOR MOMENT

In our discussion of slabs in this chapter, we will be primarily concerned with one-way slabs. Examples of such slabs are shown in Figures 2.17a and 2.17b. Such a slab is assumed to be a rectangular beam with a width b = 12 in. When loaded with a uniformly distributed load, the slab deflects so that it has curvature, and therefore bending moment, in only one direction. Hence, the slab is analyzed and designed as though it were composed of 12.in.-wide "beam" segments placed side by side with a total depth equal to the slab thickness. With a width of 12 in., the uniformly distributed load, generally specified in pounds per square foot (psf) for buildings, automatically becomes the load per linear foot (lb/ft) for the design of the slab (see Figure 2.17a).

In a one-way slab, the main reinforcement for bending is placed perpendicular to the supports. Because analysis and design will be done for a typical 12-in.-wide segment, it will be necessary to specify the amount of steel in that 12-in. wide segment (i.e., in in.^{2} per foot width of slab). Reinforcing steel in slabs is normally specified by bar size and center-to-center spacing, and the amount of steel considered to exist in the 12-in.-wide typical segment is an *average amount*. Table A.4 is provided to facilitate this determination. For example, if a slab is reinforced with No. 7 bars spaced 15 in. apart (center-to-center), a typical 12-in.-wide segment of the slab would contain an average steel area of 0.48 in.^{2} The area of reinforcing steel is denoted as 0.48 in.^{2}/ft width of slab.

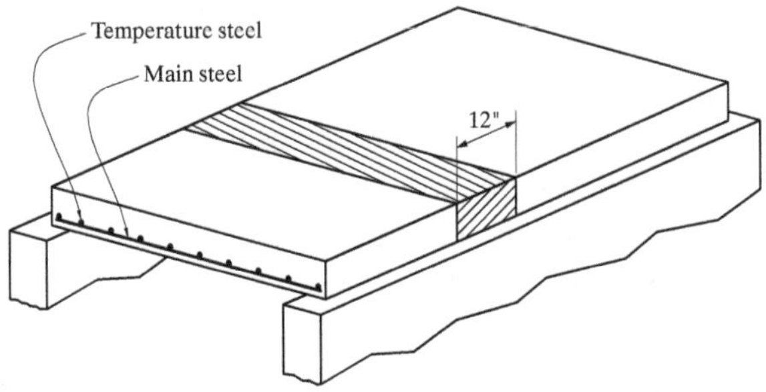

FIGURE 2.17a Floor or roof slab.

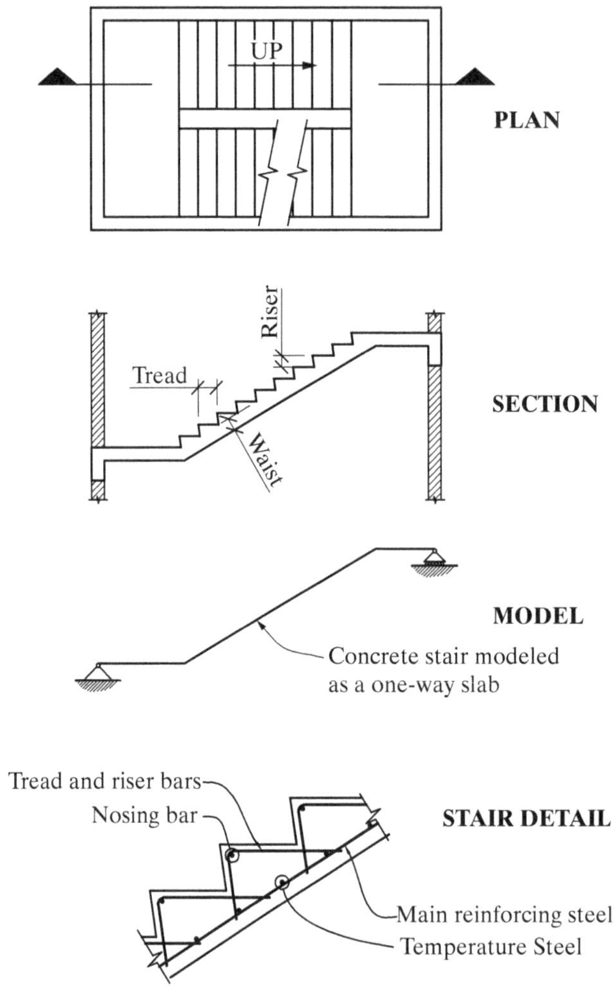

FIGURE 2.17b Reinforced concrete stair.

In addition, the ACI Code stipulates that reinforcement for shrinkage and temperature stresses perpendicular to the principal reinforcement must be provided in structural floor and roof slabs, where the principal reinforcement extends in one direction only. The required area of temperature and shrinkage reinforcement is 0.0018bh, and the spacing should not exceed five times the slab thickness (5h) nor 18 inches (see ACI Code, Section 24.4.3.3). The minimum area of the principal or main reinforcement is $A_s = 0.0018bh$, where b = width of member (12 in. for slabs) and h = total slab thickness. The ACI Code also stipulates that in structural slabs of uniform thickness, the minimum amount of reinforcement in the direction of the span (i.e., the principal reinforcement) must not be less than that required for shrinkage and temperature reinforcement (ACI Code, Section 7.6.1). Further, according to the ACI Code, Section 7.7.2.3, the principal reinforcement shall not be spaced farther apart than three times the slab thickness (3h) nor more than 18 in.

The required thickness of a one-way slab may depend on the bending, deflection, or shear strength requirements. The ACI Code imposes span/depth criteria in an effort to prevent excessive deflection, which might adversely affect the strength or performance of the structure at service loads. Table 7.3.1.1 (see also Chapter 6) of the ACI Code establishes minimum thicknesses for beams and one-way slabs in terms of fractions of the span length. According to the code, these may be used for members *not* supporting or attached to construction or finishes or equipment likely to be damaged by large deflections. For a structure supporting deflection-sensitive elements, it is advisable to always check deflections, and use the values given in ACI 318 Table 7.3.1.1 as the absolute minimum slab thickness or beam depth. For members of lesser thickness than that indicated in the table, deflection should be computed, and if satisfactory, the member may be used. The tabulated values are for use with non-prestressed reinforced concrete members made with normal-weight concrete and grade 60 reinforcement. If a different grade of reinforcement is used, the tabulated values must be multiplied by a factor equal to

$$0.4+\frac{f_y}{100{,}000}$$

where f_y is in units of psi. As an example, for simply supported solid one-way slabs of normal-weight concrete and grade 60 steel, the minimum thickness required when deflections are not computed equals 1/20 of the span of the slab (i.e., $\frac{\ell}{20}$), where ℓ is the span length of the slab. Deflections are discussed further in Chapter 7.

The ACI Code, Section 20.6.1.3.1, discusses concrete *cover*, which is measured from the surface of the steel to the nearest surface of the concrete and provides protection for the reinforcement against weather and other effects, as previously discussed in Chapter 1. The clear cover for reinforcement in concrete slabs must not be less than $\frac{3}{4}$ in. for surfaces not exposed directly to the weather or in contact with the ground. This is applicable for No. 11 and smaller bars. For surfaces exposed to the weather or in contact with the ground, the minimum cover for reinforcement is 2 in. for No. 6 through No. 18 bars and 1½ in. for No. 5 and smaller bars. If a slab is cast against and permanently exposed to the ground, the minimum clear concrete cover for all reinforcement is 3 in. Note that the concrete cover requirements for fire protection may be more stringent for slabs, beams, and columns with higher fire ratings than the clear cover specified in the ACI Code, Section 20.6.1.3.1.

EXAMPLE 2.4

A one-way structural interior slab having the cross-section shown spans 12 ft. The steel is A615 grade 40. The concrete strength is 3000 psi, and the cover is ¾ in. Determine the service live load (psf) that the slab can support.

Solution:

From Figure 2.18, the bars in this slab are No. 5 bars spaced 7 in. apart. This is sometimes denoted as "7 in. o.c." (7 in. on center), meaning 7 in. center-to-center distance. These bars are perpendicular to the supports.

1. Given:

$$A_s = 0.53 \text{ in.}^2/\text{ft} \quad (\text{Table A.4})$$
$$f'_c = 3000 \text{ psi}, \quad f_y = 40{,}000 \text{ psi}$$
$$b = 12 \text{ in.}$$
$$d = 6.5 - 0.75 - \frac{0.625}{2} = 5.44 \text{ in.}$$

2. Find ϕM_n and the permissible service live load.

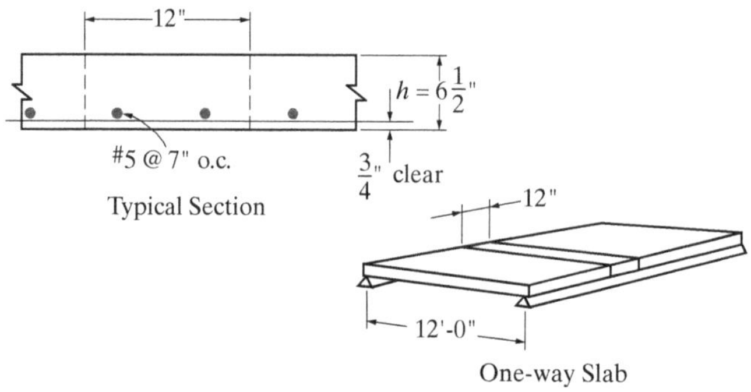

FIGURE 2.18 Sketch for Example 2.4.

3. $\rho = \dfrac{A_s}{bd} = \dfrac{0.53}{12(5.44)} = 0.0081$

4. The minimum flexural reinforcement for slabs is equal to what is required for shrinkage and temperature steel:

$$A_{s,\min} = 0.0018bh = 0.0018(12)(6.5) = 0.14 \text{ in.}^2/\text{ft}$$
$$0.53 \text{ in.}^2/\text{ft} > 0.14 \text{ in.}^2/\text{ft} \qquad (O.K.)$$

5. From Table A.7, for ρ = 0.0081, ε_t is not tabulated. Therefore, $\varepsilon_t > 0.005$, this section is tension-controlled, and ϕ = 0.90. Also, from Table A.7, $\bar{k} = 0.3034$ *ksi*

6. Calculate a, Z, and ϕM_n:

$$a = \frac{A_s f_y}{0.85 f'_c b} = \frac{0.53(40)}{0.85(3)(12)} = 0.693 \text{ in.}$$
$$Z = d - \frac{a}{2} = 5.44 - \frac{0.693}{2} = 5.09 \text{ in.}$$
$$\phi M_n = \phi A_s f_y Z = 0.9(0.53)(40)(5.09)$$
$$= 97.1 \text{ in.-kip (per foot of slab width)}$$
$$\phi M_n = \frac{97.1 \text{ in.-kip}}{12 \text{ in./ft}} = 8.09 \text{ ft.-kip}$$

or

from Table A.7, $\bar{k} = 0.3034\,ksi$, from which

$$\phi M_n = \phi b d^2 \bar{k}$$

$$= \frac{0.90(12 \text{ in.})(5.44 \text{ in.})^2 (0.3034 \text{ kip/in.}^2)}{12 \text{ in./ft}} = 8.08 \text{ ft.-kip}$$

The service live load that the slab can support will be found next. (The notation M_u is used to denote the moment resulting from *factored* applied loads.) As

$$M_u = \frac{w_u \ell^2}{8}$$

the total factored design load that can be supported by the slab is

$$w_u = \frac{8M_u}{\ell^2}$$

Because, as a limit, $M_u = \phi M_n$,

$$w_u = \frac{8\phi M_n}{\ell^2} = \frac{8(8.09)}{12^2} = 0.449 \text{ kip/ft}$$

The slab self-weight is

$$w_{DL} = \frac{6.5(12)}{144}(0.150) = 0.0813 \text{ kip/ft}$$

As the total factored design load is

$$w_u = 1.2w_{DL} + 1.6w_{LL}$$

w_{LL} may be found from

$$w_{LL} = \frac{w_u - 1.2w_{\text{DL}}}{1.6} = \frac{0.449 - 1.2(0.0813)}{1.6}$$
$$= 0.220 \text{ kip/ft}$$
$$= 220 \text{ 1b/ft}$$

Because the slab segment is 12 in. wide or 1 foot, this load is equivalent to 220 Ib/ft per ft width of slab, or 220 psf. It will be noted that the procedure for finding ϕM_n for a one-way slab is almost identical to that for a beam. See Section 2.11 for a summary of this procedure.

2.14 RECTANGULAR BEAM DESIGN FOR MOMENT (TENSION REINFORCEMENT ONLY)

In the design of rectangular sections for moment, with f_c' and f_y usually prescribed, *three basic quantities* are to be determined: beam width, beam depth, and steel area. It should be recognized that there is a large multitude of combinations of these three quantities that will satisfy the moment strength required in a particular application. Theoretically, a wide, shallow beam may have the same ϕM_n as a narrow, deep beam. It must also be recognized that practical considerations and code constraints will affect the final choices of these quantities. There is no easy way to determine the *best* cross-section because economy depends on much more than simply the volume of concrete and amount of steel in a beam.

We have previously developed the analysis expression for the resisting moment of a rectangular beam with tension reinforcement only:

$$\phi M_n = \phi N_C Z = \phi N_T Z$$

We subsequently modified the equation for ϕM_n for the use of tables (Tables A.7 through A.11):

$$\phi M_n = \phi b d^2 \bar{k}$$

This equation will now be used for the design of rectangular reinforced concrete sections. The first example is one where the cross-section width b and overall depth h are known by either practical or architectural considerations, leaving the selection of the reinforcing bars as the only unknown.

Material properties, sizes, and availability of reinforcing steel in the form of bars and welded wire fabric were discussed in Section 1.9. In our discussion, we will be concerned with bars only and the ACI Code recommendations as to details governing minimum clearance and cover requirements for steel reinforcing bars. Clearance details are governed by the requirement for concrete to pass through a layer of bars without undue segregation of the aggregates. Cover details are governed by the necessity that the concrete must provide protection against corrosion for the bars. Required minimums of both spacing and cover also play a role in preventing the splitting of the concrete in the proximity of highly stressed tension bars.

Spacing requirements in the ACI Code indicate that the clear space between bars in a single layer shall be not less than the following:

1. The bar diameter, but not less than 1 in. (ACI Code, Section 25.2.1).
2. 4/3 × maximum aggregate size (ACI Code, Sections 25.2.1 and 26.4.2.1).

Also, should multiple layers of bars be necessary, a 1-in. minimum clear distance is required between layers (ACI Code, Section 25.2.2) and bars in the upper layers shall be placed directly above bars in the bottom layer. When multiple layers of steel are required, short transverse *spacer bars* may be used to separate the layers and support the upper layers. A No. 8 spacer bar will provide the minimum 1-in. separation between layers. The detail of a beam with multiple layers of tension reinforcement is illustrated in Figure 3.6. In general, in this text, the clear distance between layers of steel is noted with a dimension.

Cover requirements for cast-in-place concrete are stated in the ACI Code, Section 20.5.1.3.1. This listing is extensive. For beams, girders, and columns not exposed to the weather or in contact with the ground, the minimum concrete cover on any steel is 1½ in. Cover requirements for slabs were discussed in Section 2.13 of this text.

Table A.3 combines spacing and cover requirements into a tabulation of minimum beam widths for multiples of various bars. The assumptions are stated. It should be noted that No. 3 stirrups are assumed in developing this table. Stirrups are a special form of reinforcement that primarily resists shear forces and will be discussed in Chapter 4. Stirrups are common in rectangular beams and will be assumed to exist in all further rectangular beam examples and exercises in this text. One type of stirrup, called a *loop or closed stirrup*, can be observed in Figure 2.19, in Example 2.5.

EXAMPLE 2.5

Design a rectangular reinforced concrete beam to support a service dead load moment of 100 ft.-kip (which includes the moment due to the weight of the beam) and a service live load moment of 75 ft.-kip. Architectural considerations require the beam width to be 10 in. and the total depth (h) to be 25 in. Use $f_c' = 3000$ psi and $f_y = 60{,}000$ psi.

Solution:

Of the three basic quantities to be found, two are specified in this example, and the solution for the required steel area is direct. The procedure outlined here is summarized in Section 2.15.

1. The total design moment is

$$\begin{aligned} M_u &= 1.2M_{\text{DL}} + 1.6M_{\text{LL}} \\ &= 1.2(100) + 1.6(75) \\ &= 240 \text{ ft.-kip} \end{aligned}$$

2. Estimate d to be equal to $h - 3$ in. This is conservative for a single layer of bars. The effective depth d in the resulting section will be a bit larger.

$$d = 25 - 3 = 22 \text{ in.}$$

Because $\phi M_n = \phi bd^2\bar{k}$ and because ϕM_n must equal M_u as a lower limit, the expression may be written as $M_u = \phi bd^2\bar{k}$ and, assuming $\phi = 0.90$, subject to later check,

$$\text{Required } \bar{k} = \frac{M_u}{\phi bd^2} = \frac{240(12)}{0.9(10)(22)^2} = 0.6612 \text{ ksi}$$

3. From Table A.8, $\bar{k}$ of 0.6649 ksi will be provided if the steel ratio $\rho = 0.0131$. Therefore, required $\rho = 0.0131$. Also, from Table A.8, because ε_t is not tabulated, $\varepsilon_t > 0.005$, and therefore, this is a tension-controlled section, and $\phi = 0.90$.

4. required $A_s = \rho bd$

$$= 0.0131(10)(22) = 2.88 \text{ in.}^2$$

Check $A_{s,\min}$. From Table A.5,

$$A_{s,\min} = 0.0033 b_w d$$
$$= 0.0033(10)(22) = 0.73 \text{ in.}^2$$

5. Select the bars. Theoretically, any bar or combination of bars that provides at least 2.88 in.^{2} of steel area will satisfy the design requirements. Preferably, no fewer than two bars should be used. The bars should be of the same diameter and placed in one layer whenever possible. We next consider the bar selection based on the foregoing required A_s (2.88 in.^{2}). The following combinations may be considered:

$$2 \text{ No. } 11 \text{ bars: } A_s = 3.12 \text{ in.}^2$$
$$3 \text{ No. } 9 \text{ bars: } A_s = 3.00 \text{ in.}^2$$
$$4 \text{ No. } 8 \text{ bars: } A_s = 3.16 \text{ in.}^2$$
$$5 \text{ No. } 7 \text{ bars: } A_s = 3.00 \text{ in.}^2$$

A review of Table A.3 indicates that the most acceptable combination is three No. 9 bars. The minimum width of beam required for three No. 9 bars is 9½ in. which is satisfactory since it is less than the 10-in. beam width.

At this point we should check the actual effective depth d and compare it with the estimated d:

$$\text{actual } d = h - \text{cover} - \text{stirrup} - d_b/2$$
$$= 25 - 1.5 - 0.38 - \frac{1.128}{2} = 22.6 \text{ in.}$$

This is slightly in excess of the estimated d and is therefore conservative (on the safe side). Because of the small difference, no revision is either suggested or required.

6. Generally, concrete dimensions should be to an increment no smaller than ½ in. In this example, whole inch dimensions are given. The final design sketch should show the following (see Figure 2.19):

a. Beam width

b. Total beam depth

c. Main reinforcement size and number of bars

d. Cover on reinforcement

e. Stirrup size

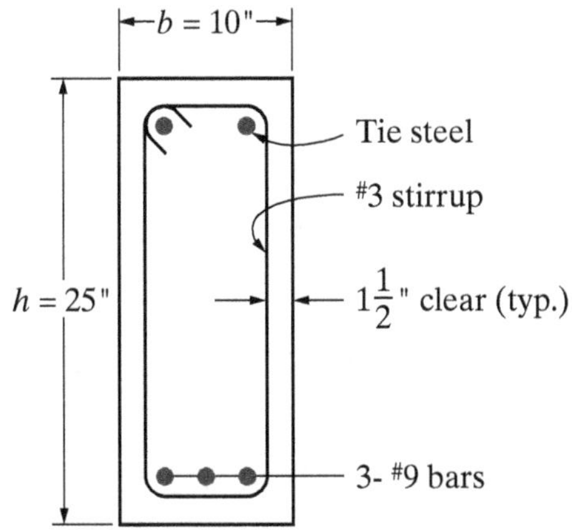

FIGURE 2.19 Design sketch for Example 2.5.

As a final note for Example 2.5, the reader has probably recognized that there is a direct solution that avoids the use of Tables A.7 through A.11. A quadratic equation results when the design expression is written as

$$M_u = \phi M_n = \phi A_s f_y \left(d - \frac{a}{2} \right)$$

where

$$a = \frac{A_s f_y}{0.85 f'_c b}$$

With known quantities of M_u, f'_c, f_y, b, and d, the quadratic equation can be solved for the required steel area A_s. Generally, the use of the tables results in a much faster solution when performing hand calculations.

Rearranging the equation above yields,

$$M_u = \phi M_n = \phi A_s f_y \left(d - \frac{A_s f_y}{1.7 f'_c b} \right),$$

from which we obtain the following quadratic equation:

$$A_s^2 - \left(\frac{1.7 f'_c bd}{f_y} \right) A_s + \left(\frac{1.7 f'_c b}{\phi f_y^2} \right) M_u = 0$$

The quadratic equation can be represented as $ax^2 + bx + c = 0$, for which the two solutions are:

$$x = \frac{-b \pm \sqrt{b^2 - 4ac}}{2a}$$

Where

$$x = A_s \qquad (\text{in.}^2)$$

$$a = 1$$

$$b = -\left(\frac{1.7f_c' bd}{f_y}\right)$$

$$c = \left(\frac{1.7f_c' b}{\phi f_y^2}\right) M_u$$

f_c' and f_y are in ksi and M_u is in units of in.-kip.

Where there are two positive solutions to the quadratic equation, the smaller of the two solutions is the correct area of steel. In the case where one solution is positive and the second solution is negative, the positive solution is the correct area of steel.

A rule of thumb or approximate equation commonly used in design practice to determine the approximate area of flexural reinforcement is [2]

$$A_s = \frac{M_u}{4d}$$

where

M_u = the factored moment in ft.-kip

d = the effective depth in inches

A_s = reinforcement area, in.^{2}

This equation is derived from

$$M_u = \phi M_n = \phi A_s f_y \left(d - \frac{a}{2}\right)$$

by substituting f_y of 60 ksi, assuming tension-controlled section or ϕ = 0.9, and assuming the internal moment arm

$$\left(d - \frac{a}{2}\right) \approx 0.9d$$

Since the factored moment, M_u in the equation above is in ft.-kip, the right hand side of the equation must be divided by 12 to convert it to ft.-kip.

Both the quadratic equation and the rule of thumb methods can be used to determine the required area of flexural reinforcement for rectangular beams, T-beams, L-beams, and one-way slabs. Both methods will now be used to determine the required area of steel for the rectangular beam in Example 2.5.

Quadratic Equation Method:

$$a = 1$$

$$b = -\left(\frac{1.7 f'_c bd}{f_y^2}\right) = -\frac{1.7(3\text{ ksi})(10\text{ in.})(22\text{ in.})}{60\text{ ksi}}$$

$$= -18.7\text{ in.}^2$$

$$c = \left(\frac{1.7 f'_c b}{\phi f_y^2}\right) M_u = \frac{1.7(3\text{ ksi})(10\text{ in.})}{0.9(60\text{ ksi})^2}(240\text{ ft.-kip})(12\text{ in./ft})$$

$$= 45.33\text{ in.}^4$$

Therefore, $A_s = 2.86\text{ in.}^2$ (the smaller of the two positive solutions). This required area of flexural steel compares well with 2.88 in.^2 obtained using the Tables.

Rule of Thumb Method:

$$A_s = \frac{M_u\ (\text{ft.-kip})}{4d\ \ (\text{in.})} = \frac{240\text{ ft.-kip}}{4(22\text{ in.})} = 2.73\text{ in.}^2$$

This required area of flexural steel compares well with 2.88 in.^2 obtained using the Tables.

A second type of design problem is presented in Example 2.6. This may be categorized as a *free design* because of the *three* unknown variables: beam width, beam depth, and area of reinforcing steel. There are, therefore, a large number of combinations of these variables that will theoretically solve the problem. We do have some idea as to the required or desired relationships between these unknowns, however.

Whenever possible, flexural members should be proportioned so they are tension-controlled sections ($\varepsilon_t \geq 0.005$) This allows the strength-reduction factor ϕ to be taken at its maximum value of 0.90. The reinforcement ratio ρ corresponding to $\varepsilon_t = 0.005$ may be determined from Tables A.7 through A.11. Additionally, we know that the steel area must not be less than $A_{s,\min}$ (from Table A.5). This in effect establishes the range of the acceptable amount of steel. Table A.5 contains recommended values of ρ and associated values of $\bar{k}$ to use for design purposes. These are recommended maximum values. Try not to use larger ρ values. If larger values are used, a smaller concrete section will result, with potential deflection problems.

These ρ values are based on a previous ACI Code, which stipulated that where

$$\rho > \frac{0.18 f'_c}{f_y}, \quad \textit{deflection must be checked}$$

$$\rho < \frac{0.18 f'_c}{f_y}, \quad \textit{deflection need not be checked}$$

This stipulation was deleted in more current ACI Codes. Nevertheless, it remains a valid guide for selecting a preliminary value for the reinforcement ratio. The assumption of a value for reinforcement ratio will reduce from three to two the number of unknown quantities to be determined (yet to be found are b and d).

Experience and judgment developed over the years have also established a range of acceptable and economical depth/width ratios for rectangular beams. Although there is no code requirement for the d/b ratio to be within a given range, rectangular beams commonly have d/b ratios between 1 and 3. *Desirable* d/b ratios lie between 1.5 and 2.2. There are situations in which d/b ratios outside this range have applications, however; therefore, it is the designer's choice.

EXAMPLE 2.6

Design a simply supported rectangular reinforced concrete beam with tension steel only to support a service dead load of 1.35 kip/ft and a service live load of 1.90 kip/ft. (The dead load does not include the weight of the beam.) The span is 18 ft. Assume No. 3 stirrups. Use f_c' = 4000 psi and f_y = 60,000 psi. See Figure 2.20.

Solution:

1. Find the applied design moment M_u, temporarily neglecting the beam weight, which will be included at a later step:

$$\begin{aligned} w_u &= 1.2w_{DL} + 1.6w_{LL} \\ &= 1.2(1.35) + 1.6(1.90) = 4.66 \text{ kip/ft} \end{aligned}$$

Note that this is the factored design load,

$$M_u = \frac{w_u \ell^2}{8} = \frac{4.66(18)^2}{8} = 189 \text{ ft.-kip}$$

2. Assume a value for ρ. Use ρ = 0.0120 (see Table A.5).
3. From Table A.5, the associated $\bar{k}$ value is 0.6438 ksi (alternatively, Table A.10 can be used).

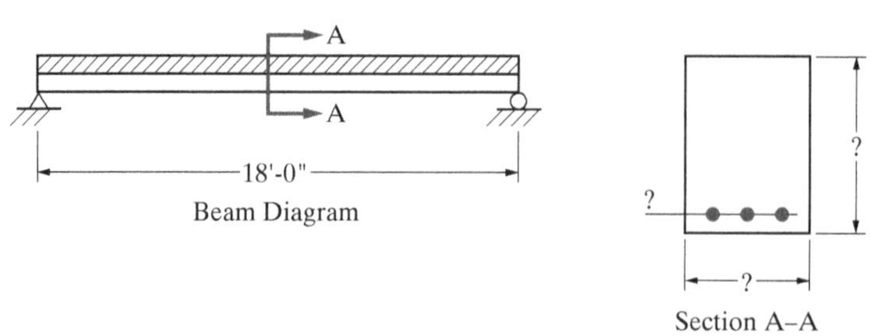

FIGURE 2.20 Sketch for Example 2.6.

4. At this point, we have two unknowns, b and d, which can be established using two different approaches. One approach is to assume b and then solve for d. This is logical as practical and architectural considerations often establish b within narrow limits. Assume $\phi = 0.90$, subject to later check.

   Assuming that $b = 11$ in. and utilizing the relationship

$$M_u = \phi b d^2 \bar{k}$$

$$\text{required } d = \sqrt{\frac{M_u}{\phi b \bar{k}}} = \sqrt{\frac{189(12)}{0.9(11)(0.6438)}} = 18.9 \text{ in.}$$

   A check of the d/b ratio gives 18.9/11 = 1.72, which is a reasonable value.

5. At this point, the beam weight may be estimated. Realizing that the total design moment will increase, the final beam size may be estimated to be about 11 in. × 23 in. for purposes of calculating its weight. Thus

$$\textit{beam self-weight (i.e., dead load)} = \frac{11(23)}{144}(0.150) = 0.264 \text{ kip/ft}$$

6. The additional M_u due to beam self-weight is

$$M_u = 1.2\left(\frac{0.264(18)^2}{8}\right) = 12.8 \text{ ft.-kip}$$

$$\text{total } M_u = 189 + 12.8 = 202 \text{ ft.-kip}$$

7. Using the same ρ, $\bar{k}$ and b as previously, compute the new required d:

$$\text{required } d = \sqrt{\frac{M_u}{\phi b \bar{k}}} = \sqrt{\frac{202(12)}{0.9(11)(0.6438)}} = 19.5 \text{ in.}$$

   A check of the d/b ratio gives 19.50/11 = 1.77, which is reasonable.

8. $$\text{required } A_s = \rho b d$$
$$= 0.0120(11)(19.50) = 2.57 \text{ in.}^2$$

   Check $A_{s,\min}$. From Table A.5,

$$A_{s,\min} = 0.0033 b_w d$$
$$= 0.0033(11)(19.50) = 0.71 \text{ in.}^2 \quad \text{(O.K.)}$$

9. From Table A.2, select three No. 9 bars. Therefore, A_s = 3.00 in.^{2}. From Table A.3, Minimum required width of beam, b = 9.5 in. (O.K.)

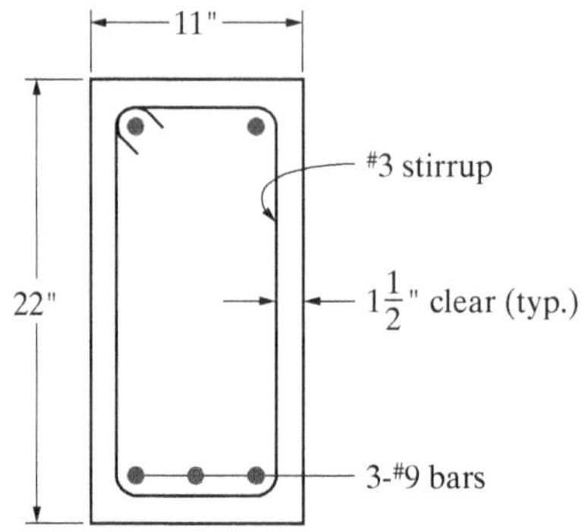

FIGURE 2.21 Design sketch for Example 2.6.

10. Determine the total beam depth h:

$$\text{required } h = 19.50 + \frac{1.13}{2} + 0.38 + 1.5 = 21.95 \text{ in.}$$

Rounding up by a ½ in. increment, we will use $h = 22.0$ in. The actual effective depth d may now be checked:

$$d = 22.0 - 1.5 - 0.38 - \frac{1.13}{2} = 19.55 \text{ in.} > 19.50 \text{ in.} \quad \text{(O.K.)}$$

11. A design sketch is shown in Figure 2.21. This design is not necessarily the best. The final ρ is

$$\frac{3.0}{11(19.55)} = 0.01395$$

Check ε_t: from Table A.10, $\varepsilon_t > 0.005$. Therefore, the assumed ϕ is O.K. The final d/b ratio is

$$\frac{19.55}{11} = 1.78 \qquad \text{(O.K.)}$$

The design moment was based on a section 11 in. × 23 in., whereas the final designed beam size turned out to be 11 in. × 22 in. Small modifications in proportions could very well make more efficient use of the steel and concrete provided.

The percentage of overdesign could be determined (not necessary, however) by analyzing the designed cross-section and comparing moment strength with applied moment.

There are several alternative approaches to this problem. At step 4, we could establish a desired d/b ratio and then mathematically solve for b and d. For example, if we establish a desirable d/b ratio of 2.0, then $d = 2b$. Again, using the relationship $M_u = \phi b d^2 \bar{k}$,

$$\text{required } bd^2 = \frac{M_u}{\phi \bar{k}} = \frac{189(12)}{0.9(0.6438)} = 3914 \text{ in.}^3$$

We can substitute for d as follows:

$$\text{required bd}^2 = 3914 \text{ in.}^3$$
$$b(2b)^2 = 3914 \text{ in.}^3$$
$$b^3 = \frac{3914}{4} = 979 \text{ in.}^3$$
$$\text{required } b = \sqrt[3]{979} = 9.93 \text{ in.}$$

Hence, assuming that b = 10 in., the required effective depth d can then be calculated in a manner similar to that previously used. Also, at step 7, with a width b and a depth h established, along with the related design moment M_u, we could revert to the procedure for design of a rectangular section where the area is known (see Example 2.5, step 2).

2.15 SUMMARY OF PROCEDURE FOR RECTANGULAR REINFORCED CONCRETE BEAM DESIGN FOR MOMENT (TENSION REINFORCEMENT ONLY)

A. Cross-Section (b and h) are known; find the required A_s:

1. Convert the service loads or moments to design M_u (include the beam weight which must first be assumed, and then checked later, since the beam size is not yet determined).
2. Based on knowing h, estimate d by using the relationship $d = h - 3$ in. (conservative for bars in a single layer). Calculate the required $\bar{k}$ using an assumed ϕ value of 0.90, subject to later check.

$$\text{required } \bar{k} = \frac{M_u}{\phi b d^2}$$

3. From Tables A.7 through A.11, find the required steel ratio ρ and ensure that the net tensile strain in the extreme tension rebar, $\varepsilon_t \geq 0.005$. If ε_t is within the range $0.004 \leq \varepsilon_t \leq 0.005$, then ϕ will be less than 0.9.
4. Compute the required A_s:

$$\text{Required } A_s = \rho bd$$

Check $A_{s,\min}$. Use Table A.5.

5. Select the bars. Check to see if the bars can fit within the beam width in one layer (preferable). Check the actual effective depth and compare it with the assumed effective depth. If the actual effective depth is slightly in excess of the assumed effective depth, the design will be slightly conservative (on the safe side). If the actual effective depth is less than the assumed effective depth, the design is on the unconservative side and should be revised.
6. Sketch the design. (See Example 2.5, step 6, for a discussion.)

B. Design for Cross-Section and Required A_s (i.e., b, h, and A_s are unknown):

1. Convert the service loads or moments to design moment M_u. An estimated beam weight may be included in the dead load if desired. Be sure to apply the load factor to this additional dead load.
2. Select a desired steel ratio ρ. (See Table A.5 for recommended values. Use the ρ values from Table A.5 unless a small cross-section or decreased steel is desired.)
3. From Table A.5 (or from Tables A.7 through A.11), find $\bar{k}$
4. Assume b and compute the d required:

$$\text{required } d = \sqrt{\frac{M_u}{\phi b \bar{k}}}$$

   If the d/b ratio is reasonable (1.5 to 2.2), use these values for the beam. If the d/b ratio is not reasonable, increase or decrease b and compute the new required d.

5. Estimate h and compute the beam weight. Compare this with the initially assumed estimated beam weight if an estimated beam weight was included.
6. Revise the design factored moment, M_u, to include the moment due to the beam's own weight using the latest weight determined (remember to apply the 1.2 load factor for dead load to this self-weight). Note that at this point, one could go to step 2 in design procedure A, where the cross-section is known.
7. Using b and $\bar{k}$ previously determined along with the new total factored design moment, M_u, find the new required d:

$$\text{required } d = \sqrt{\frac{M_u}{\phi b \bar{k}}}$$

   Check to see if the d/b ratio is reasonable.

8. Find the required A_s:

$$\text{Required } A_s = \rho b d$$

   Check $A_{s,\min}$. Use Table A.5.

9. Select the bars and check to see if the bars can fit within the beam width, b in one layer (preferable).
10. Establish the final depth, h, rounding this upward to the next ½ in. This will make the actual effective depth greater than the design effective depth, and the design will be slightly conservative (on the safe side).
11. Check ε_t. Check the previously assumed strength reduction factor, ϕ. Sketch the design. (See Example 2.5, step 6, for a discussion.)

   A flow diagram of this procedure is presented in Appendix B.

2.16 DESIGN OF ONE-WAY SLABS FOR MOMENT (TENSION REINFORCEMENT ONLY)

As higher-strength steel and concrete have become available for use in reinforced concrete members, the sizes of the members have decreased. *Deflections* of members are affected very little by material strength but are affected greatly by the size of a cross-section and its related moment of inertia. Therefore, deflections will be larger for a member of high-strength materials than would be the deflections for the same member fabricated from lower-strength materials, because the latter member will be essentially larger in cross-sectional area. Deflections are discussed in detail in Chapter 7. As discussed previously, one method that the ACI Code allows for limiting adverse deflections is the use of a minimum slab thickness (see Section 2.13). A slab that meets the minimum thickness requirement from ACI 318 Table 7.3.1.1. must still be designed for flexure. Deflections need not be calculated or checked unless the slab supports or is attached to nonstructural elements likely to be damaged by large deflections.

Example 2.7 illustrates the use of the ACI minimum thickness for one-way slabs. With respect to the span length to be used for design, a rational approach found in previous versions of ACI 318 [9] recommends the following for beams and slabs not integral with supports:

$$\text{Span length} = \text{clear span} + \text{depth of member}$$

but not to exceed the distance between centers of support. For design purposes, we will use the distance between the centers of supports, as the slab thickness is not yet determined.

EXAMPLE 2.7

Design a simple-span one-way slab to carry a uniformly distributed live load of 400 psf. The span is 10 ft (center-to-center of supports). Use f'_c = 4000 psi and f_y = 60,000 psi. Select the thickness to be not less than the ACI minimum thickness requirement.

Solution:

Determine the required minimum h and use this to estimate the slab dead weight.

1. From ACI Table 7.3.1.1, for a simply supported, solid, one-way slab,

$$\text{minimum } h = \frac{\ell}{20} = \frac{10(12)}{20} = 6.0 \text{ in.}$$

   Try h = 6 in. and design a 12-in.-wide slab segment.

2. Determine the slab weight dead load:

$$\frac{6(12)}{144}(0.150) = 0.075 \text{ kip/ft}$$

The total design load is

$$w_u = 1.2w_{DL} + 1.6w_{LL} \\ = 1.2(0.075) + 1.6(0.400) \\ = 0.730 \text{ kip/ft}$$

3. Determine the design moment:

$$M_u = \frac{w_u \ell^2}{8} = \frac{0.730(10)^2}{8} = 9.13 \text{ ft.-kip/ft width}$$

4. Establish the approximate d. Assuming No. 6 bars and minimum concrete cover on the bars of ¾ in.,

$$\text{assumed d} = 6.0 - 0.75 - 0.375 = 4.88 \text{ in.}$$

5. Determine the required $\bar{k}$ assuming $\phi = 0.90$:

$$\text{required } \bar{k} = \frac{M_u}{\phi b d^2} \\ = \frac{9.13(12)}{0.90(12)(4.88)^2} = 0.4260 \text{ ksi}$$

6. From Table A.10, for a required $\bar{k}$=0.4260,the required ρ = 0.0077. (Note that the required ρ selected is the next *higher* value from Table A.10). Also note that ε_t is not tabulated. Therefore, $\varepsilon_t > 0.005$, this is a tension-controlled section, and $\phi = 0.90$.

   Use $\rho = 0.0077$.

7. required $A_s = \rho bd = 0.0077(12)(4.88) = 0.45$ in.^{2}/ft width of slab

8. Select the main reinforcing steel (from Table A.4). Select No. 5 bars at 8 in. o.c. (A_s = 0.46 in.^{2}). The assumption on bar size was satisfactory (actual d > assumed d). The code requirements for maximum spacing have been discussed in Section 2.13. The minimum spacing of bars in slabs, practically, should not be less than 4 in., although the ACI Code allows bars to be placed closer together, as discussed in Example 2.5. Check the maximum spacing (ACI Code, Section 7.7.2.3):

$$\text{maximum spacing} = 3h \text{ or } 18 \text{ in.} \\ 3h = 3(6) = 18 \text{ in.} \quad \text{(O.K.)} \\ 8 \text{ in.} < 18 \text{ in.}$$

   Therefore, use No. 5 bars at 8 in. o.c.

9. Select shrinkage and temperature reinforcement (ACI Code, Table 24.4.3.2):

$$\text{required } A_s = 0.0018bh \\ = 0.0018(12)(6) = 0.13 \text{ in.}^2\text{/ft}$$

Select No. 3 bars at 10 in. o.c. (A_s = 0.13 in.^{2}/ft) or No. 4 bars at 18 in. o.c. (A_s = 0.13 in.^{2}/ft):

$$\text{Maximum spacing} = 5h \text{ or } 18 \text{ in.}$$

Use No. 3 bars at 10 in. o.c.

10. The main steel area must exceed the area required for shrinkage and temperature steel (ACI Code, Section 7.6.1.1):

$$0.46 \text{ in.}^2 > 0.13 \text{ in.}^2 \text{ (O.K.)}$$

11. A design sketch is shown in Figure 2.22.

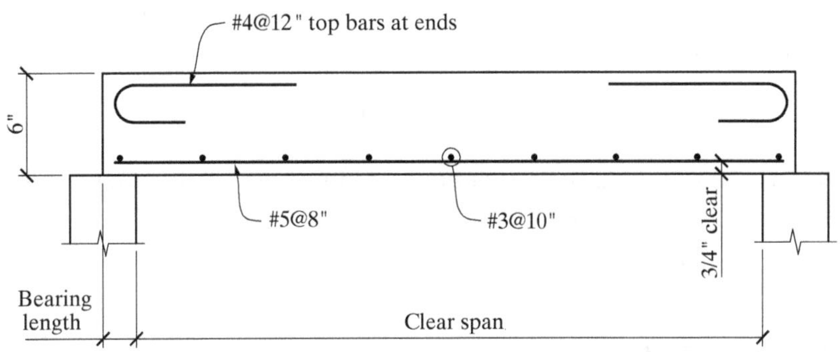

FIGURE 2.22 Design sketch for Example 2.7.

2.17 SUMMARY OF PROCEDURE FOR DESIGN OF ONE-WAY SLABS FOR MOMENT (TO SATISFY ACI MINIMUM THICKNESS, *h*)

1. Compute the minimum h based on the ACI Code, Table 7.3.1.1. For practical purposes, the slab thickness h can be rounded to the next higher ½ in.
2. Compute the slab weight and compute w_u (total design load).
3. Compute the factored design moment M_u.
4. Calculate an assumed effective depth d (assuming No. 6 bars and ¾ in. cover for interior slabs) by using the relationship

$$d = h - 1.12 \text{ in.}$$

For slabs exposed to the weather, use a cover of 1½ in. for No. 5 and smaller bars, and 2 in. for No. 6 and larger bars; use 3 in. concrete cover for slabs cast against and permanently in contact with the ground).

5. Calculate the required $\bar{k}$ assuming ϕ = 0.90:

$$\text{required } \bar{k} = \frac{M_u}{\phi b d^2}$$

6. From Tables A.7 through A.11, find the required steel ratio ρ. Check ε_t to verify the ϕ assumption. If $\varepsilon_t < 0.005$, the slab must be made thicker. Note that the ACI 318 Code (Sections 7.3.3.1, 8.3.3.1, and 9.3.3.1) does not allow flexural members with axial load, $P_u < 0.10f_c'A_g$ (e.g., slabs, beams, and girders) to have extreme tensile strain, $\varepsilon_t < 0.005$; these members must be tension-controlled in accordance with ACI 318 Table 21.2.2.

7. Compute the required A_s:

$$\text{required } A_s = \rho bd$$

8. Select the main reinforcing steel (Table A.4) with a maximum rebar spacing of 3*h* or 18 in. Check the assumption of step 4.

9. Select shrinkage and temperature steel as per the ACI Code, with a maximum rebar spacing of 5*h* or 18 in.:

$$\text{required } A_s = 0.0018bh$$

10. The main reinforcing steel area provided must *not* be less than the area of steel required for shrinkage and temperature.

11. Sketch the design.

2.18 SLABS-ON-GRADE

The previous discussion primarily focused on framed structural slabs. Another category of slabs, generally used at the lowest floor, are called *slab-on-grade*. As the name implies, it is a slab that is supported throughout its entire area by some form of subgrade.

Concrete placed directly on soil is used as a floor surface in almost every structure and in other areas outside of a building such as sidewalks, equipment pads, and roads. The design of slabs-on-grade is significantly different from the design of a framed structural slab; there are numerous factors to consider in the design of slabs-on-grade, but we will cover the most basic and most common aspects of their design here.

The subgrade is the natural ground, graded and compacted, on which the lowest floor slab is built. The pressure on the subgrade is generally low due to the rigidity of the concrete floor slab. The floors do not necessarily require strong support from the subgrade. It is important that the subgrade support be reasonably uniform horizontally, without abrupt changes from hard to soft subgrade, however. The upper portion of the subgrade should be of uniform material and density.

The subbase is usually a thinner layer of material placed on top of the prepared subgrade. It is generally used when a uniform subgrade cannot be developed by grading and compaction. It will serve to equalize minor surface defects as well as provide a capillary break and a working platform for construction activities. A 4-in. thickness of the sub-base, usually of compacted granular material, is sometimes used.

A typical slab-on grade detail is shown in Figure 2.23. Each of the components shown play a role in the performance and load-carrying capacity of the slab-on-grade. In a typical project, the existing native soil at the ground level is removed down to a suitable bearing surface. The slab-on-grade could be placed directly on native soil, but the amount removed during the construction process is typically greater than the proposed concrete slab thickness. It is also usually desirable to have a free-draining and well-compacted base directly under the slab-on-grade, and once the existing soil is removed down to the suitable bearing layer of native soil, a layer of compacted granular fill is placed next. This layer can vary from a few inches up to about 24 inches.

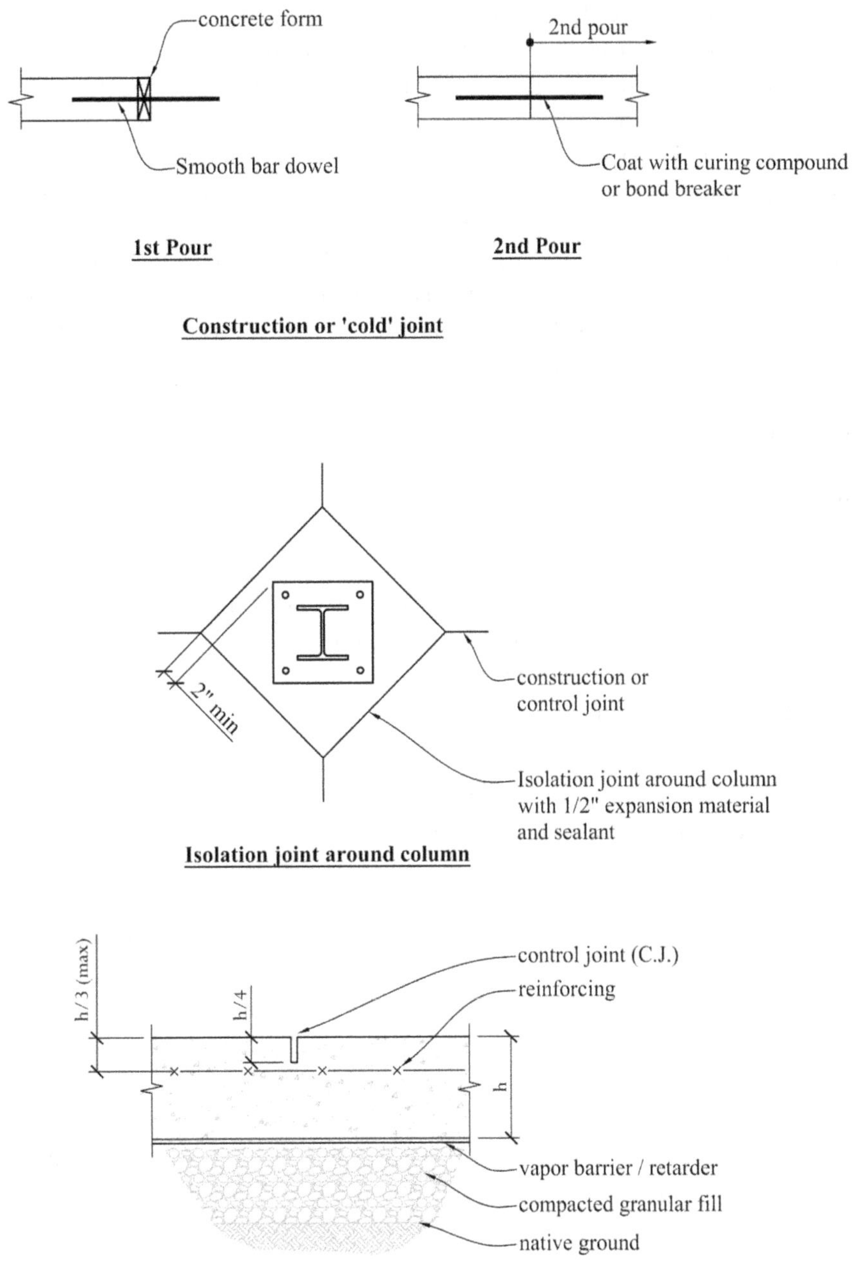

FIGURE 2.23 Slab-on-grade detail.

Another item typically used in a slab-on-grade is a vapor barrier or vapor retarder placed directly beneath the concrete slab. ACI 302 identifies three possible cases for a slab-on-grade construction regarding the use of the vapor barrier as indicated in Table 2.1. A vapor barrier would often be used where upward moisture migration through the slab would be detrimental to floor finishes (such as carpet, tiles, or hardwood flooring) or other items such as products that are sensitive to moisture or water infiltration, or moisture-sensitive equipment that will be placed on the floor. The IBC (Ref. 11) states that a vapor retarder is not required "for buildings of other occupancies where migration of moisture through the slab from below will not be detrimental to the intended occupancy of the building." It is often very difficult to establish the exact occupancy and conditions that may be present in a structure throughout its life cycle, so the use of a vapor barrier is typically recommended as good construction practice. Regarding the location of the vapor barrier, it is required to be placed directly underneath the slab when moisture-sensitive finishes are used. Furthermore, the only condition where a layer of dry granular material (i.e., sand) can be placed on the vapor barrier is when a watertight roof is in place. This would be required to prevent moisture from accumulating in this layer of sand which could be detrimental to the slab curing and the occupancy. In the experience of the authors, it is therefore best to place the vapor barrier directly below the slab, which allows for greater latitude in the construction sequence and for all possible occupancy conditions. The location of the vapor barrier needs to be considered in the layout of the control joints, and that will be discussed later. Table 2.1 is provided to summarize the possible locations for a vapor barrier. Note that per ACI 318 Section 1.4.8, slabs-on-grade "that do not transmit vertical or lateral forces from other portions of the structure to the soil" are not designed using ACI 318; the reader should refer to ACI 360R, "Guide to Design of Slabs-on-Ground" [4]. Two options are presented in ACI 360R for slabs-on-grade to avoid random cracks outside of the sawcut joints, and to ensure closely spaced cracks for slabs without saw-cut joints. If control joints are not provided or desirable, then the minimum slab-on-grade reinforcement required according to ACI 360R-10 is 0.5% of the gross area of the slab.

TABLE 2.1 Vapor Barrier Location.

Vapor Barrier Location	Occupancy	Construction Conditions
Vapor barrier not used	Moisture infiltration through the slab is not detrimental to the occupancy	N/A
Vapor barrier directly below the concrete slab	Moisture infiltration through the slab is detrimental to the occupancy and any occupancy with a moisture-sensitive floor covering	Any
Vapor barrier is above the subbase and below a layer of dry granular material (e.g., sand)	Moisture infiltration through the slab is detrimental to the occupancy	Watertight roof must be in place

For slabs-on-grade in expansive soils, the vapor barrier is attached to the underside of the slab-on-grade on top of void forms on top of the granular layer, and the slab-on-grade is reinforced with steel reinforcement top and bottom each way to resist flexure in the slab.

The thickness of a concrete slab-on-grade is a function of several factors:

1. Strength of subgrade and subbase
2. Strength of concrete
3. Magnitude and type of load (including contact area of applied loads)

To determine the slab thickness, several methods can be used, most of which are empirical or based on rules of thumb. It is also possible to analyze the slab by isolating a typical section. The slab-on-grade can be modeled as a beam element supported by an infinite number of supports, each with a spring constant. Under a perfectly uniform load, the soil under the slab would compress uniformly, and the slab would only see direct compression stress. The typical loading is often quite varied, and thus, differential displacement will occur across the slab, and flexural stresses will occur in the tension zones of the slab. Figure 2.24 shows the loading and displacement that a slab could experience. In this case, the unloaded zone between applied loads w_1 and w_2 would experience flexural tension stress, and if the slab were not sufficiently thick or if this tension zone was not sufficiently reinforced, a flexural crack would occur. In most typical structures, the slab-on-grade design does not involve any significant structural analysis, rather rules of thumb and historical practices are used. For most structures, the slab-on-grade thickness varies from 4 to 8 inches. For slabs-on-grade with heavy loading, and specifically heavy point loads, the reader should refer to ACI 360R, "Guide to Design of Slabs-on-Ground" [4], for several empirical methods for determining the slab thickness and reinforcing.

The concrete strength for industrial and commercial slabs-on-grade should not be less than 4000 psi at 28 days. This furnishes satisfactory wear resistance in addition to strength. In residential construction, the slabs-on-ground will have concrete strengths of 2500 or 3000 psi. A minimum of 3000 psi is recommended.

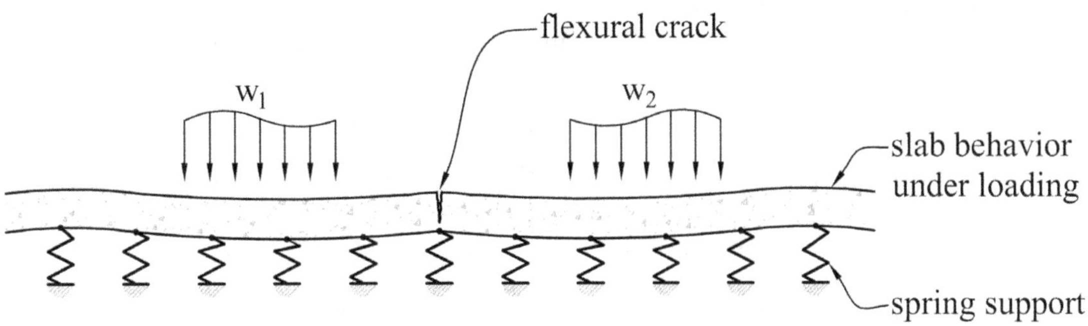

FIGURE 2.24 Behavior of slab-on-grade under loading.

The reinforcement in the slab-on-grade often consists of welded wire fabric (WWF) but reinforcing bars in both directions could also be used for more heavily loaded slabs. The reinforcing is mainly used to mitigate flexural cracking from loading when the slab spans over soft spots in the subgrade, and to control and minimize shrinkage cracks. Despite several other relatively minor advantages that result from the use of reinforcing steel, questions have been raised as to whether reinforcement is always necessary, particularly with uniform support of the slab and short joint spacings. The reinforcing steel does not prevent cracking, nor does it add significantly to the load-carrying capacity of the slab. It is also usually more economical to obtain increased strength in concrete slabs-on-grade by increasing the thickness of the slab.

While there are methods for determining the exact amount of reinforcing to use in certain slabs, the reinforcing typically used is based on rules of thumb and historical practices (see Table 2.2). The reinforcement should be placed within the *upper one-third of the slab* to be effective in limiting the crack widths at the top of the slab and it also should be secured and in place prior to pouring the concrete. The reinforcing should typically *not* be higher than 1-in. below the finished concrete surface to allow for proper clear cover and construction tolerances for finishing. Infact, 1½ to 2 inches of concrete cover to the reinforcement is recommended to minimize the risk of cracking [15].

Control joints are critical to the performance of the slab-on-grade. The control joints (also called contraction joints) in a slab-on-grade are intentionally created "planes of weakness" in the slab that causes the shrinkage cracks to occur along these straight preplanned joints which are "formed by tooling, sawcuts, or mechanical devices" [14]. The timing of the joint placement and the location of the joints need to be properly addressed to avoid undesirable random cracking and curling. Joints are typically placed within the first 12 to 24 hours of the concrete pour, and they typically must be placed within 24 hours in order to be effective. As the concrete sets, moisture is released, and the concrete undergoes volumetric changes such that it shrinks. This shrinkage will cause the concrete to crack along the control joints – the deliberately weakened section or "planes of weakness" of the concrete. The concrete slab will also tend to curl because one face of the concrete slab dries (and therefore shrinks) faster than the other side. Concrete slabs are often cured by placing a vapor barrier on top of the slab, and water is regularly applied to this concrete surface for a few weeks after the slab is poured to allow the slab to maintain a constant moisture content through the entire slab thickness. Also note that with the presence of a vapor barrier directly under the slab, this constant moisture content is more easily achieved than with the vapor barrier under a layer of sand or if the vapor barrier is not present at all.

The required control joint dimensions are as follows:

- The depth of the joint from the top of the slab should be *one-quarter (¼)* of the slab thickness.
- The maximum joint spacings should be between *24 and 36 times the slab thickness*, and they should be placed such that each slab panel between the control joints has an aspect ratio not greater than *1.5* (i.e., the ratio of the long side to the short side of the slab panel).
- Slabs should also be allowed to float away from other building elements to avoid cracking. Where a slab abuts a wall, there should be a joint between the slab and the building element filled with an expansion material to prevent the concrete slab-on-grade from bonding to the wall and to allow for differential movement.
- In addition to the control joints, *isolation joints* are also provided in the slab-on-grade at column locations and at perimeter walls to mitigate the effect of the restraint to movement of the slab due to temperature and shrinkage from these stiff elements. These formed

joints are usually provided with a keyway and a bond separator, and the concrete around the column within the diamond pattern is typically cast at least 7 days after casting the slab-on-grade. At building columns, control joints in the slab-on-grade are typically placed in a diamond pattern to isolate the concrete slab and prevent it from bonding to the column. The diamond pattern of the isolation joint around the columns in conjunction with the control joints in the slab helps to avoid reentrant corners in the slab where diagonal cracking can occur. Additionally, care should be taken to ensure a separation between the footing and the slab-on-grade. Where the slab-on-grade sits directly on top of the footing without any bond breaker, circular cracks in the slab-on-grade around the columns can occur because of the mechanical bond that develops between the underside of the slab and the top of the footing [15].

- Another type of joint that occurs in slabs-on-grade and framed structural concrete elements are *construction joints*; the location of these joints determines the extent of concrete slab that can be cast in one continuous pour. Thus, construction joints delineate the extent of multiple concrete pours. In a slab-on-grade, construction joints would either have a keyway or dowels between the two adjacent slab segments, with one end of the dowel greased or inserted in a sleeve to allow relative movement of the two portions of the slab-on-grade, and thus prevent restraint of the slab segments. When dowels are used, the construction joint between the slab segments is a vertical joint without a keyway. The slab is also sawcut along the construction joint, and the sawcut joint is filled with a semi-rigid epoxy joint filler. The use of dowels is preferred to keyways because the construction of keyways is complex especially because of the smaller thickness of slabs-on-grade, which necessitates an increase in slab thickness to accommodate the keyway; keyways can also be damaged due to movement of the adjacent slab segments and the use of keyways can result in restraint to slab movement which can cause cracking as one slab segment slides on the concrete keyway of the adjacent slab segment. Note that in a framed structural element, both keyways and dowel reinforcement that is continuous through the construction joint are typically provided [14, 15]. Note that ACI places no limit on the amount of concrete that can be poured at one time, but "the size of a pour is limited by what the contractor has resources and technology to accomplish." A continuous single-day concrete pour of up to 75,000 square feet or more has been reported. [15]

The table below summarizes typical slab types and uses

TABLE 2.2 Typical Slab-on-Grade Types.

Slab thickness	Common occupancies	Typical reinforcing	Load capacity (Ref [12])	Recommended control joint spacing
4″	Residential slab-on-grade; lightly loaded sidewalks; Residential garages	None or 6 × 6-W1.4 × W1.4 WWF	100 psf	8ft to 12ft
5″	Commercial, institutional slab-on-grade; public garages	6 × 6-W2.1 × W2.1 WWF or 6 × 6-W2.9 × W2.9 WWF	100 psf-200 psf	10ft to 15ft
6″	Industrial slab-on-grade with moderate vehicle traffic (passenger vehicles, forklifts)	6 × 6-W2.9 × W2.9 WWF or 6x6-W4.0xW4.0	400 psf-500 psf	12ft to18ft

Slab thickness	Common occupancies	Typical reinforcing	Load capacity (Ref [12])	Recommended control joint spacing
7"	Industrial slab-on-grade with heavier vehicle traffic	Two layers of 6 × 6-W2.9 × W2.9 WWF or 6x6-W4.0xW4.0	600psf-800psf	14ft to 21ft
8″	Heavy vehicle loading for loading docks around a building	#4@12″ or #4@16″ each way	1500psf	16ft to 24ft
10"	Heavy Industrial	Two layers of #5@12" each way	3000psf – 3500psf	20ft – 30ft

A typical plan of a slab-on-grade showing control joint and isolation joint locations at columns is shown in Figure 2.25. In laying out the control and isolation joints in the slab-on-grade, reentrant corners should be avoided.

Design aids and procedures (based primarily on research done for highway and airport pavements) have been developed and are available in specialized publications. (See references [3] through [8]). Figure 2.26 shows slab-on-grade control joints and isolation joints at a concrete column location.

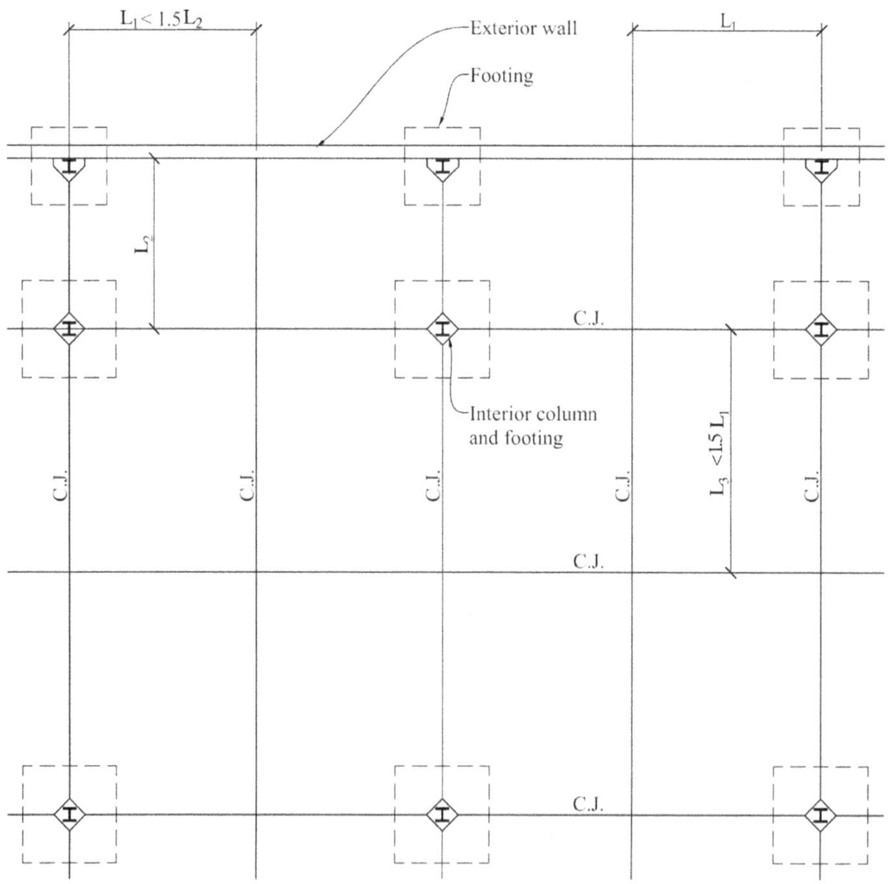

FIGURE 2.25 Foundation plan showing locations of control joints in a slab-on-grade.

FIGURE 2.26 Slab-on-grade control joints and isolation joints at a concrete column. (Abi Aghayere)

REFERENCES

[1] Charles S. Whitney, "Plastic Theory of Reinforced Concrete Design." *Trans. ASCE*, V. 68, 1942.

[2] JoAnn Browning, "Rule-of-Thumb for Flexural Steel Area," *Concrete Q & A, Concrete International*, October 2007, p. 91.

[3] *Guide for Concrete Floor and Slab Construction.* ACI 302.1R-15, Farmington Hills, MI: American Concrete Institute, P.O. Box 9094, 48333-9094, 2004.

[4] *Design of Slabs on Grade.* ACI 360R-10, Farmington Hills, MI: American Concrete Institute, P.O. Box 9094, 48333-9094, 2010.

[5] *Slabs on Grade.* Concrete Craftsman Series 1, Farmington Hills, MI: American Concrete Institute, P.O. Box 9094, 48333-9094, 1994.

[6] Robert G. Packard, *Slab Thickness Design for Industrial Concrete Floors on Grade.* Skokie, IL: Portland Cement Association, 5420 Old Orchard Road, 60077, 1996.

[7] J. A. Farny and S. M. Tarr, *Concrete Floors on Ground*. Skokie, IL: Portland Cement Association, 5420 Old Orchard Road, 60077, 2008.
[8] Robert B. Anderson, *Innovative Ways to Reinforce Slabs-on-Ground*. Hartford, CT: Tech Facts, Wire Reinforcement Institute, Inc., 942 Main Street, Suite 300, 06103, 1996.
[9] *Building Code Requirements for Structural Concrete* (ACI 318-19(22)). Farmington Hills, MI: American Concrete Institute, P.O. Box 9094, 2019 (Reapproved 2022).
[10] *Ronald L. Kozikowski and Bruce A. Suprenant*, "Controlling Early-Age Cracking in Mass Concrete," *Concrete International*, March 2015, pp. 59–62.
[11] International Codes Council, International Building Code (IBC)—2021, Falls Church, VA: ICC, 2021.
[12] *CRSI Handbook 1980*, 4 ed., Concrete Reinforcing Steel Institute, 1980.
[13] American Society of Civil Engineers, "Design Loads on Structures During Construction," (ASCE/SEI 37-14), Reston VA, 2014.
[14] *Alexander Newman*, "Avoiding Problematic Uses of Slabs on Ground," *STRUCTURE Magazine*, January 2021, pp. 12–15.
[15] *T. J. Cate*, "Concrete Slabs-on-Ground, Crying for Attention," *STRUCTURE Magazine*, April 2021, pp. 14–16.

EXERCISES

NOTE *In the following exercises, consider moment only and tension reinforcing only. For beams, assume 1 ½ in. cover and No. 3 stirrups. For slabs, assume ¾ in. cover. Unless noted otherwise, given loads are superimposed service loads and do not include weights of the members.*

2.1 **a.** The beam of the cross-section shown has $f_c' = 3000$ psi and $f_y = 60{,}000$ psi. Neglect the tensile strain check and $A_{s,\min}$ check. Calculate M_n.+

b. Same cross-section as part (a), but the steel is changed to four No. 10 bars. Calculate M_n. Calculate the percent increase in M_n and A_s.

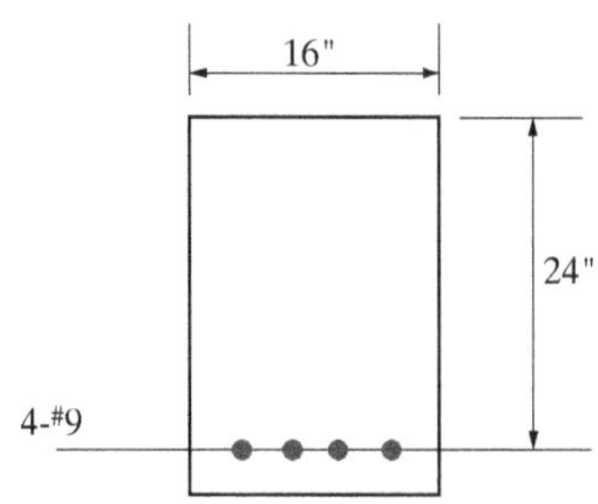

FIGURE FOR EXERCISE 2.1

c. Same cross-section as part (a), but the depth d is increased to 28 in. Calculate M_n. Calculate the percent increase in M_n and d.

d. Same cross-section as part (a), but f_c' is increased to 4000 psi. Calculate M_n. Calculate the percent increase in M_n and f_c'.

2.2 For the cross-section of Exercise 2.1(a), verify that the tension steel yields.

2.3 **a.** Calculate the practical moment strength ϕM_n for a rectangular reinforced concrete cross-section having a width b of 13 in. and an effective depth d of 24 in. The tension reinforcing steel is four No. 8 bars. $f_c' = 4000$ psi and $f_y = 40{,}000$ psi.

b. Same as part (a) but $f_y = 60{,}000$ psi. Also calculate the percent increase in ϕM_n and f_y.

2.4 Determine ϕM_n for a reinforced concrete beam 16 in. wide by 32 in. deep reinforced with seven No. 10 bars (placed in two layers: five in the bottom layer, two in the top layer, with 1 in. clear between layers). Use $f_c' = 4000$ psi and $f_y = 60{,}000$ psi.

2.5 A reinforced concrete beam having the cross-section shown is on a simple span of 28 ft. It carries uniform service loads of 3.60 kip/ft live load and 2.20 kip/ft dead load. Check the adequacy of the beam with respect to moment. Use $f_c' = 4000$ psi and $f_y =$ 40,000 psi.

a. Reinforcing is six No. 10 bars.

b. Reinforcing is six No. 11 bars.

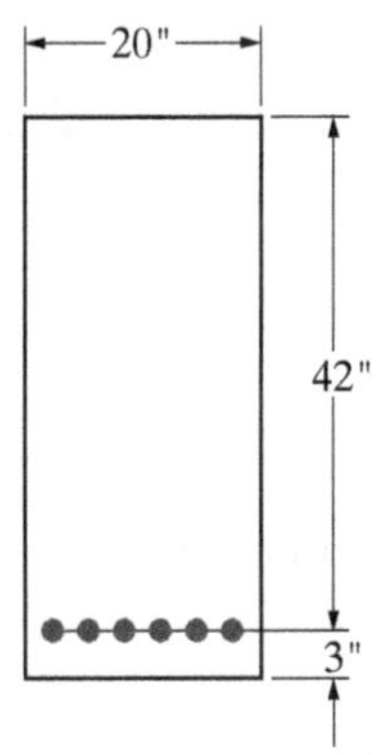

FIGURE FOR EXERCISE 2.5

2.6 Develop a spreadsheet application that will allow a user to input the basic information for the analysis of a rectangular reinforced concrete cross-section and that will then calculate the design moment strength or capacity, ϕM_n. Set up the spreadsheet to be "user friendly" and fully label the output.

2.7 A 12-in.-wide by 20-in.-deep concrete beam is reinforced with three No. 8 bars. The beam supports a service live load of 2.5 kip/ft and a service dead load of 0.7 kip/ft on a simple span of 16 ft. Use $f_c' = 4000$ psi and $f_y = 60{,}000$ psi. Check the adequacy of the beam with respect to moment.

2.8 A reinforced concrete beam having the cross-section shown is on a simple span of 26.5 ft. It supports uniformly distributed service loads of 3.20 kip/ft live load and 1.80 kip/ft dead load (excluding the beam weight). Reinforcing is as shown. Check the adequacy of the beam with respect to moment. Use $f_c' = 3000$ psi and $f_y =$ 60,000 psi.

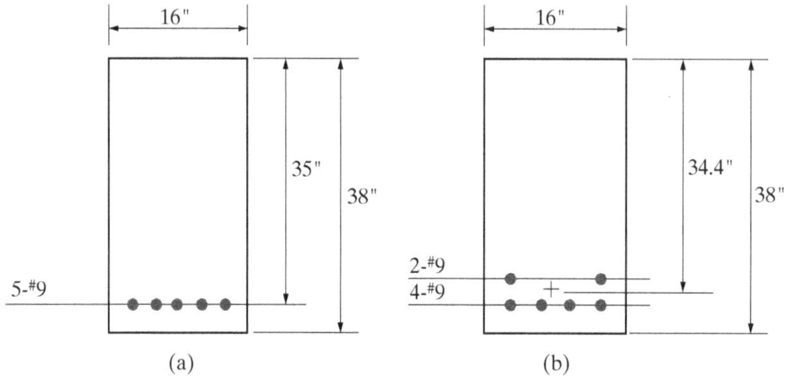

FIGURE FOR EXERCISE 2.8

2.9 A rectangular reinforced concrete beam carries *service loads* on a span of 20 ft as shown. Use f'_c = 3000 psi and f_y = 60,000 psi; b = 14.5 in., h = 26 in., and reinforcing is three No. 10 bars. Determine whether the beam is adequate with respect to moment.

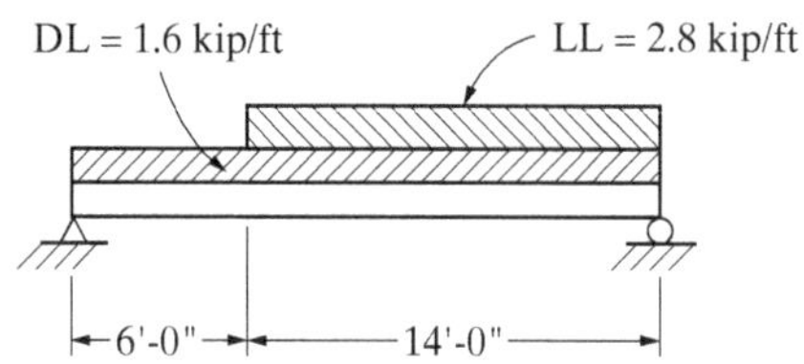

FIGURE FOR EXERCISE 2.9

2.10 A rectangular reinforced concrete beam 14 in. wide by 24 in. deep is to support a service dead load of 0.6 kip/ft and a service live load of 1.4 kip/ft. Reinforcing is four A615 grade 60 No. 9 bars. Use f'_c = 4000 psi. Determine the maximum simple span length for which this beam may be utilized.

2.11 A 10-in.-thick one-way slab supports a superimposed service live load of 600 psf on a simple span of 16 ft. Reinforcement is No. 7 at 6 in. on center. Check the adequacy of the slab with respect to moment. Use f'_c = 3000 psi and f_y = 60,000 psi.

2.12 The one-way slab shown spans 12 ft from center of support to center of support. Calculate ϕM_n and determine the *service live load* (psf) that the slab may carry. (Assume that the only dead load is the weight of the slab.) Use f'_c = 3000 psi and f_y = 40,000 psi.

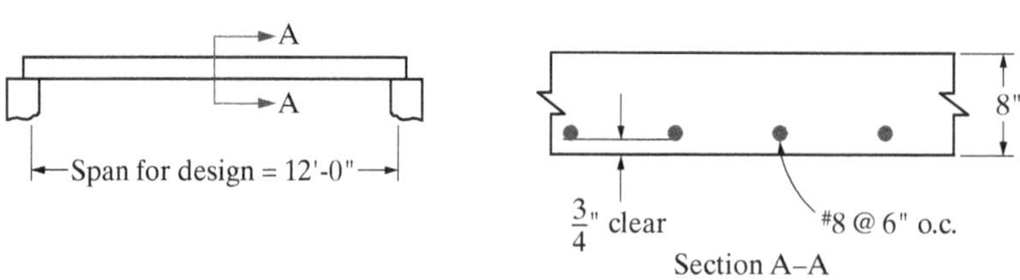

FIGURE FOR EXERCISE 2.12

2.13 An 8½ in.-thick one-way reinforced concrete slab overhangs a simple support. The span of the overhang is 8 ft. Drawings called for the reinforcement to be placed with top cover of 1 in. The steel was misplaced, however, and later was found to be as much as 3½ in. below the top of the concrete slab. Find the ϕM_n for the slab as-designed and as-built, and the percent of reduction in flexural strength. Use $f_c' = 4000$ psi and $f_y = 60{,}000$ psi. Bars are No. 7 at 11 in. o.c.

2.14 Design a rectangular reinforced concrete beam to resist a total design moment M_u of 133 ft.-kip. (This includes the moment due to the beam self-weight.) Architectural considerations require that the width (b) be 11½ in. and the overall depth (h) be 23 in. Use $f_c' = 4000$ psi and $f_y = 60{,}000$ psi. Sketch your design.

2.15 Rework Exercise 2.14 with $M_u = 400$ ft.-kip, $b = 16$ in., $h = 28$ in., $f_c' = 4000$ psi, and $f_y = 60{,}000$ psi.

2.16 For the beam designed in Exercise 2.15, if the main reinforcement were incorrectly placed so that the actual effective depth was 24 in., would the beam be adequate? Check by comparing M_u with the ϕM_n resulting from the beam using actual steel, actual b, and $d = 24$ in.

2.17 Design a rectangular reinforced concrete beam (tension steel only) for a simple span of 32 ft. Uniform service loads are 0.85 kip/ft dead load and 1.0 kip/ft live load. The beam is to be 11½ in. wide and 26 in. deep overall (from reuse consideration). Use $f_c' = 4000$ psi and $f_y = 60{,}000$ psi. Calculate ϕM_n for the beam designed.

2.18 Design a rectangular reinforced concrete beam (tension steel only) for a simple span of 30 ft. There is no superimposed dead load (other than the weight of the beam), and the superimposed live load is 1.35 kip/ft. The beam is to be 12 in. wide and 27 in. deep overall. Use $f_c' = 5000$ psi and A615 grade 60 steel. As a check, calculate ϕM_n for the beam *designed*.

2.19 Rework Exercise 2.18 assuming that the superimposed live load has increased to 1.75 kip/ft and there is now a 1.0 kip/ft superimposed dead load (in addition to the beam weight). As before, check ϕM_n for the beam designed.

2.20 Design a simply supported rectangular reinforced concrete beam to span 22 ft and to carry uniform service loads of 1.6 kip/ft dead load and 1.4 kip/ft live load. The assumed dead load includes an estimated beam weight. Use A615 grade 60 steel and $f_c' = 3000$ psi. Use the recommended ρ from Table A.5. Make the beam width 15 in. and keep the overall depth (h) to full inches. Assume No. 3 stirrups. Check the adequacy of the beam you design by comparing M_u with ϕM_n. Sketch your design.

2.21 Design a rectangular reinforced concrete beam for a simple span of 30 ft. The beam is to carry uniform *service* loads of 1.0 kip/ft dead load and 2.0 kip/ft live load. Because of column sizes, the beam width should not exceed 16 in. Use $f_c' = 3000$ psi and $f_y = 60{,}000$ psi. Sketch your design.

2.22 Rework Exercise 2.21 assuming that the total depth h is not to exceed 30 in. and that there is no limitation on the width b.

2.23 Design a rectangular reinforced concrete beam for a simple span of 32 ft. Uniform service loads are 1.5 kip/ft dead load and 2.0 kip/ft live load. The width of the beam is limited to 18 in. Use $f_c' = 3000$ psi and $f_y = 60{,}000$ psi. Sketch your design.

2.24 Rework Exercise 2.23 assuming that the total depth h is not to exceed 32 in. and that there is no limitation on the width b.

2.25 Design a rectangular reinforced concrete beam for a simple span of 40 ft. Uniform service loads are 0.8 kip/ft dead load and 1.4 kip/ft live load. Use $f_c' = 4000$ psi and $f_y = 60{,}000$ psi. Sketch your design.

2.26 Design a rectangular reinforced concrete beam (tension steel only) for the span and *superimposed* service loads shown. Use $f_c' = 5000$ psi and A615 grade 60 steel.

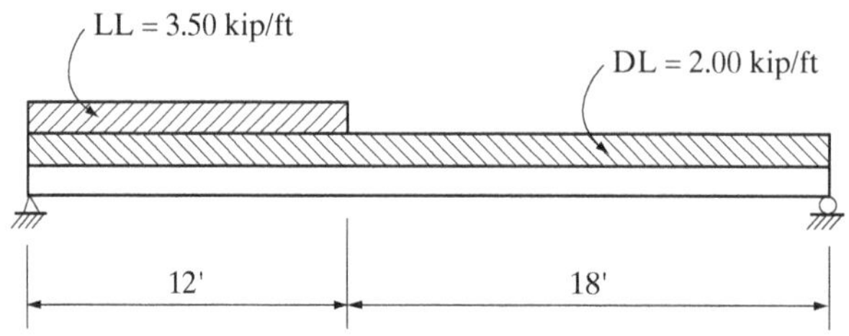

FIGURE FOR EXERCISE 2.26

2.27 Design a 28-ft, simple-span, rectangular reinforced concrete beam to support a uniform live load of 0.8 kip/ft and concentrated loads at midspan of 10-kip dead load and 14-kip live load. Use $f_c' = 5000$ psi and $f_y = 60{,}000$ psi.

2.28 Design a simply supported rectangular reinforced concrete beam for the span and *service loads* shown. Use $f_c' = 3000$ psi and $f_y = 60{,}000$ psi.

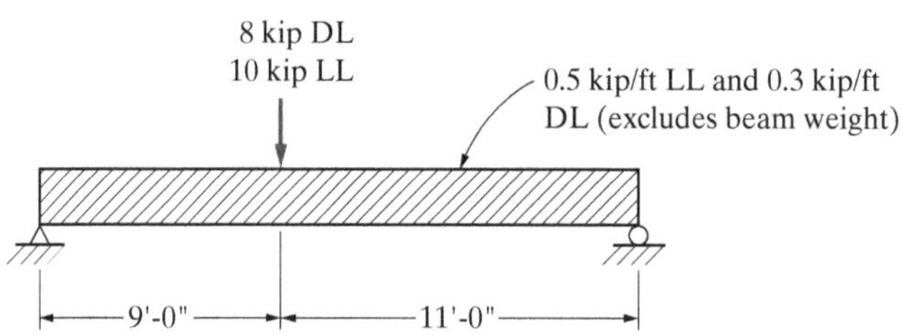

FIGURE FOR EXERCISE 2.28

2.29 Rework Exercise 2.28 assuming that the beam is to be extended to overhang the right support by 10 ft. The uniform load also extends to the end of the overhang.

2.30 Design a simply supported one-way reinforced concrete floor slab to span 8 ft and carry a service live load of 300 psf. Use $f_c' = 3000$ psi and $f_y = 60{,}000$ psi. Sketch your design.

2.31 Design a simply supported one-way reinforced concrete floor slab to span 10 ft and carry a service live load of 175 psf and a service dead load of 25 psf. Use $f_c' = 3000$ psi and $f_y = 60{,}000$ psi. Make the slab thickness to ½ -in. increments.

a. Design the slab for the ACI Code minimum thickness.

b. Design the thinnest possible slab allowed by the ACI Code.

2.32 Design the simply supported one-way reinforced concrete slab as shown. The service live load is 200 psf. Use $f_c' = 3000$ psi and $f_y = 60{,}000$ psi. Sketch your design.

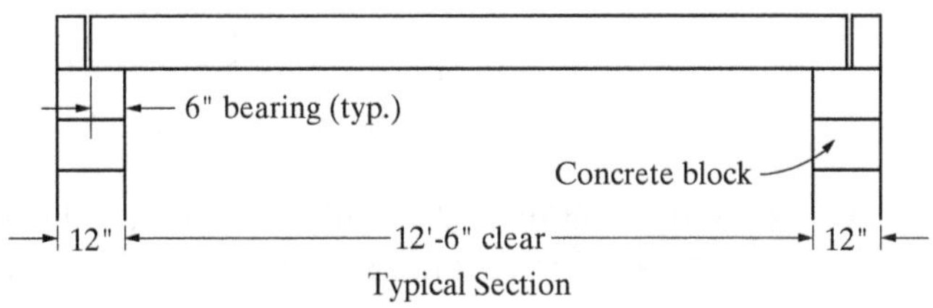

FIGURE FOR EXERCISE 2.32

2.33 a. Determine the service live load in psf that the slab can support. Ignore the design checks for $A_{s,\min}$ and spacing for the main reinforcing,

b. Determine the required temperature reinforcing (select bar size and spacing)

- Assume a simple span of 14 ft. for the slab.
- $f_c' = 3000$ psi, $g_c = 150$ pcf
- $f_y = 60{,}000$ psi
- Service dead load is 50 psf plus the self-weight of the slab.
- Use the load combination for dead plus live loads only (ignore live load reduction)

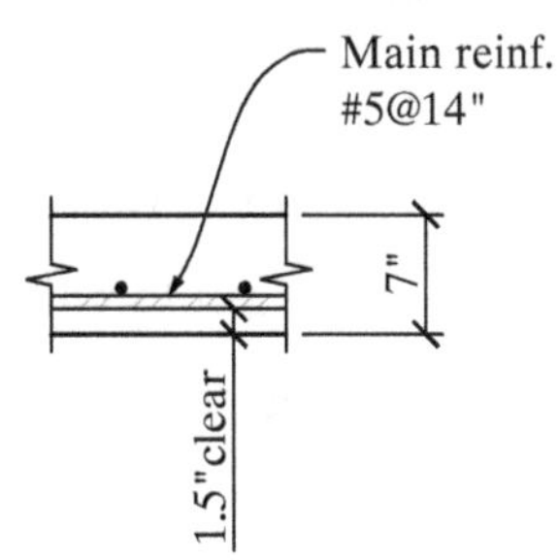

FIGURE FOR EXERCISE 2.33

2.34 Determine the required area of steel for bending (bar size and number).

- Self-weight is included as shown.
- $f_c' = 4000$ psi
- $f_y = 60{,}000$ psi
- Use the load combination for dead plus live loads only (ignore live load reduction)

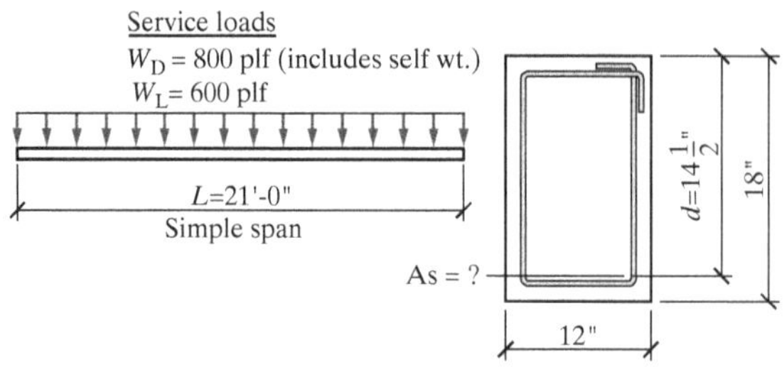

FIGURE FOR EXERCISE 2.34

2.35 The detail is a cross-section of two precast concrete lintels used to support gravity loads. Each lintel is 4″ × 8″ and has a single #4 for flexural reinforcing. The total factored uniform load on the assembly, including self-weight, is w_u = 500 plf. Use f'_c = 5000 psi and f_y = 60,000 psi. Use the load combination for dead plus live loads only (ignore live load reduction). Determine the maximum span in feet that this assembly could span based on bending strength only.

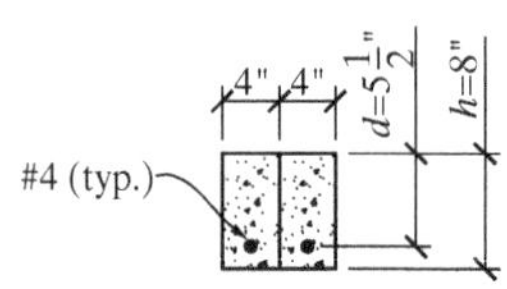

FIGURE FOR EXERCISE 2.35

2.36 For the floor plan and details shown below, assume simple spans for the slab. Use f'_c = 3500 psi and f_y = 60,000 psi. Use a service dead load of 110 psf (includes self-weight) The service live load is L = 175 psf. Use the load combination for dead plus live loads only (ignore live load reduction). Consider the design of a single 32″ wide slab section.

Determine the following:

a. The required reinforcing for flexural loads only

b. The required temperature reinforcing (bar size and spacing)

c. Sketch the design showing a cross-section through the 32″ slab width and the actual number of bars

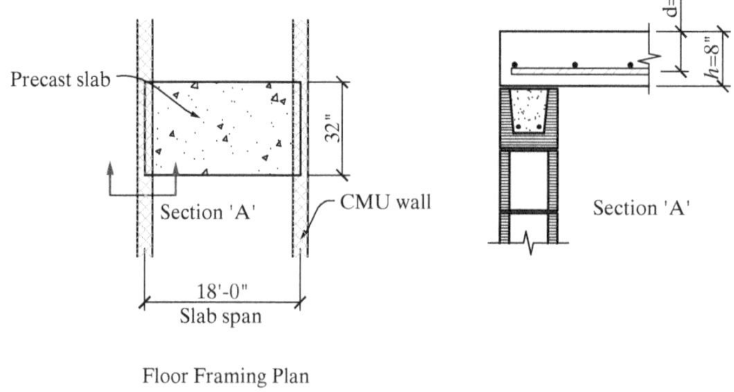

FIGURE FOR EXERCISE 2.36

2.37 Assume a rectangular concrete beam with b = 11″ and h = 20″. Use f_c' = 4000 psi and f_y = 60,000 psi. The factored moment is M_u = 175 ft-kip (includes beam self-weight)

Determine the following:

a. The required reinforcing for flexural loads; check minimum steel provisions.

b. Sketch the design showing the reinforcing and all dimensions.

2.38 Assume a 9″ thick concrete slab with d = 7″; simple span of 18 ft. between supporting beams. Use f_c' = 3000 psi and f_y = 60,000 psi. The service live load is 125 psf and the service dead load is 20 psf plus the self-weight of the slab. Use the load combination for dead plus live loads only (ignore live load reduction).

Determine the following:

a. The required reinforcing for flexural loads; check minimum steel provisions.

b. Sketch the design showing the reinforcing and all dimensions through the slab section and sketch a floor plan showing the reinforcing.

CHAPTER 3

REINFORCED CONCRETE BEAMS: T-BEAMS AND DOUBLY REINFORCED BEAMS

3.1 T-BEAMS: INTRODUCTION

Floors and roofs in reinforced concrete buildings may be composed of slabs that are supported directly on the columns so that the roof or floor loads are carried directly to the columns, and then to the building foundation. As previously discussed, these are termed *flat slabs* or *flat plates*. The span of such a slab cannot become very large before its own dead weight causes it to become uneconomical. Many types of systems have been devised to allow greater spans without the problem of excessive weight.

One such system, called a *beam and girder system*, is composed of a slab on supporting reinforced concrete beams and girders. The beam and girder framework, in turn, is supported by columns. In such a system, the beams and girders are commonly cast monolithically with the slab. Systems other than the monolithic system do exist, and these may make use of some precast (e.g., hollow core slabs) and some cast-in-place concrete. These are generally of a proprietary nature. The typical monolithic system is shown in Figure 3.1. The beams are commonly spaced so that they intersect the girders at the midpoint, third points, or quarter points, as shown in Figure 3.2. To minimize the cost of concrete formwork, the same overall depth is often used for the beams and girders.

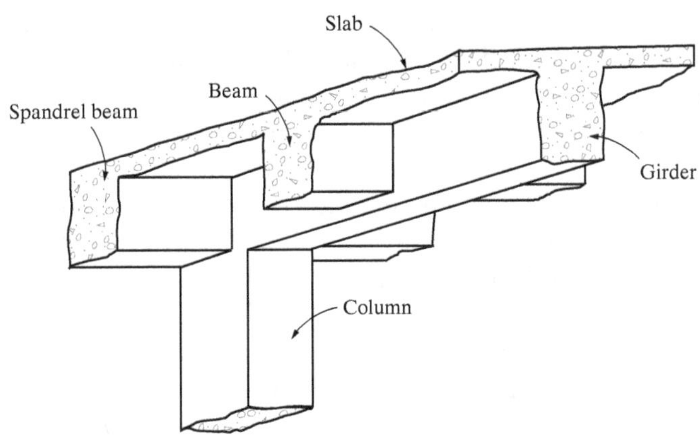

FIGURE 3.1 Beam and girder floor system.

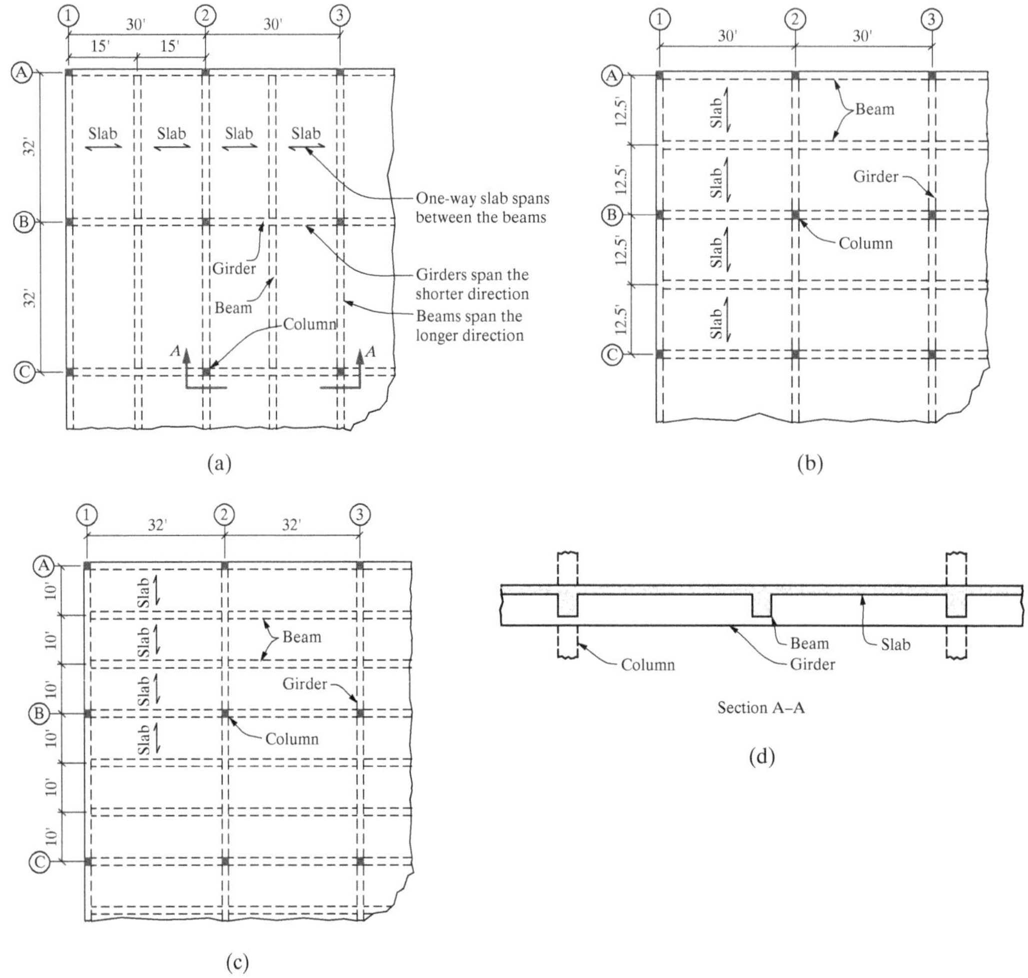

FIGURE 3.2 Common beam and girder layouts.

In the analysis and design of such floor and roof systems, it is common practice to assume that the monolithically placed slab and supporting beam interact as a unit in resisting *positive bending moment*. As shown in Figure 3.3, the slab becomes the compression flange, and the supporting beam becomes the web or stem. For the typical interior beams or girders (see Figure 3.2), the interacting flange and stem produce the cross-section having the typical T-shape from which the T-beam gets its name (see Figure 3.3). Similarly, for a typical edge or spandrel beam or girder (e.g., the beams along grid line 1 in Figure 3.2a and grid line A in Figures 3.2b and 3.2c, and the girders along grid line A in Figure 3.2a and grid line 1 in Figures 3.2b and 3.2c), the interacting flange and stem produce the cross-section having an inverted L-shape as shown in Figure 3.3. It should be noted that the slab, which comprises the T-beam flange, must itself be designed to span across the supporting beams. Therefore, the slab behaves as a bending member supporting the floor loads in two orthogonal directions. It should also be noted that should the T-beam cross-section be subjected to *negative bending moment*, the slab at the top of the stem will be in tension while the bottom of the stem is in compression. It will be seen that this situation will occur at interior supports of continuous beams, which will be discussed later.

To simplify the complex two-way behavior of the flange, the ACI Code, for design and analysis purposes, has established criteria whereby the flange, when acting together with the web, will have a limited width that may be considered effective in resisting the compression stress from the applied positive moment within the span of the beam. The fact that this effective width is smaller than the spacing between adjacent beams is caused by the nonuniform compression stress distribution in the flange of the beam. This phenomenon is called *shear lag*. The effective flange width for symmetrical shapes will always be equal to or less than the beam spacing (see Figure 3.3).

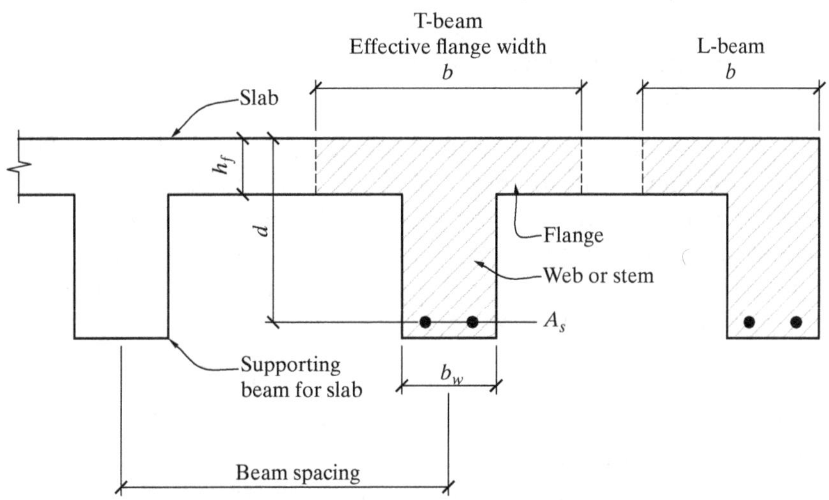

FIGURE 3.3 T-beam and L-beam as part of a floor system.

3.2 T-BEAM AND L-BEAM ANALYSIS

For purposes of analysis and design, the ACI Code, Section 6.3.2.1, has established limits on the effective flange width in the compression zone for T-beams (and inverted L-beams) in the positive moment regions as follows:

1. For T-beams (i.e., interior beams or girders), the **effective** flange width must not exceed the width of the beam web plus one-fourth of the clear span length of the beam, and the effective overhanging flange width on each side of the web must not exceed eight times the thickness of the slab nor one-half of the clear distance between adjacent beam webs. In other words, the effective flange width, b, must not exceed

   a. b_w + one-fourth of the clear span length.

   b. $b_w + 16h_f$.

   c. Center-to-center spacing of beams.

   The smallest of the three values will control.

2. For beams having a flange on one side only (i.e., spandrel or perimeter beams, which are also referred to as inverted L-beams), the effective overhanging flange width must not exceed one-twelfth of the clear span length of the beam, nor six times the slab thickness, nor one-half of the clear distance to the next beam web. Thus, the effective width, b, for inverted L-beams or girders is the smallest of the following:

   a. b_w + one-twelfth of the clear span length of the beam.

   b. $b_w + 6h_f$.

   c. b_w + ½ (clear spacing between the webs of the L-beam and the adjacent interior beam.)

3. For isolated beams in which the T-shape is used only for the purpose of providing additional compressive area, the flange thickness must not be less than one-half of the width of the web, and the total flange width must not be more than four times the width of the beam web (i.e., $4b_w$).

   In the positive moment regions of a beam (i.e., where the bottom of the beam is in tension and the top flange is in compression), such as the mid-span of the beam, the width of the compression zone is the effective width calculated above (i.e., b). In the negative moment regions of a beam (i.e., where the bottom of the beam is in compression and the top flange is in tension), such as at the beam support locations, the width of the compression zone is the width of the beam web or stem (i.e., b_w).

The ductility requirements for T-beams are similar to those for rectangular beams. To ensure ductile behavior, the ACI Code, Sections 9.3.3.1 and 21.2.2 require a net tensile strain $\varepsilon_t \geq 0.003+\varepsilon_{ty}$ for flexural members with little or no axial load, which for Grade 60 steel means that $\varepsilon_t \geq 0.005$. Such a section is tension-controlled, that is, it is completely ductile since the net tensile strain $\epsilon_t \geq 0.005$. Consequently, the strength reduction factor, ϕ, for such a section is 0.90. It is always desirable and more efficient in the design of flexural members to strive for a tension-controlled section. The T-shape can be a factor in the determination of net tensile strain for a T-beam.

The procedure for determining the minimum steel for a T-beam is the same as for a rectangular beam when the T-beam flange is in compression (positive moment). Where tensile reinforcement is required by analysis, the steel area, A_s, shall not be less than that given by

$$A_{s,\min} = \frac{3\sqrt{f'_c}}{f_y} b_w d \geq \frac{200}{f_y} b_w d$$

Note that for T-beams, b_w represents the width of the web. Also, note that the first expression controls only if $f'_c > 4440$ psi. The above expressions for minimum steel also apply to continuous T-beams.

For negative moment (flange in tension) in statically determinate members,

$$A_{s,\min} = \text{the smaller of } \frac{6\sqrt{f'_c}}{f_y} b_w d \quad or \quad \frac{3\sqrt{f'_c}}{f_y} bd$$

The minimum flexural reinforcement provisions are needed to prevent undesirable brittle failure immediately after the concrete member cracks, and to help limit the crack widths in the member. The minimum A_s in the ACI Code "provides a reasonable margin of safety between first cracking and flexural failure" [1, 2], thus ensuring a ductile failure mode. Per ACI 318 Section 9.6.1.3, the minimum steel requirements need not be applied if, at every section along the member, at least 33% more steel than required by analysis is provided (i.e., minimum steel provisions do not apply if $A_{s,\text{provided}} \geq 1.33A_{s,\text{required}}$). Where multiple layers of reinforcement are used in a beam, the clear spacing between horizontal layers of reinforcement should be a minimum of 1 inch (ACI 318-Code, Section 25.2.2). This spacing can be provided by using at least a #8 spacer bar.

Because of the relatively large compression area available in the flange of the T-beam, the moment strength is usually limited by the yielding of the tensile steel. Therefore, it is usual to assume that the tensile steel will yield before the concrete reaches its ultimate strain and crushes. The total tensile force, N_T, at the ultimate condition, may then be found by

$$N_T = A_s f_y$$

To proceed with the analysis, the shape of the compressive stress block must be defined. As in our previous analyses, the total compressive force N_C must be equal to the total tensile force N_T. The shape of the stress block must be compatible with the area in compression. Two conditions may exist: The stress block may be completely within the flange, or it may cover the flange and extend into the web. These two conditions will result in what we will term, respectively, a *rectangular T-beam* and a *true T-beam*. In addition to the shape of the stress block, the basic difference between the two is that the *rectangular* T-beam has a rectangular compression zone with effective flange width, b and is analyzed in the same way as a rectangular beam of width, b; on the other hand, the *true* T-beam has a T-shaped compression zone and its analysis must consider the T-shaped compression stress block.

EXAMPLE 3.1

The T-beam shown in Figure 3.4 is part of a floor system. Determine the practical moment strength or design moment capacity, ϕM_n if $f_y = 60{,}000$ psi (A615 Grade 60) and $f_c' = 3000$ psi.

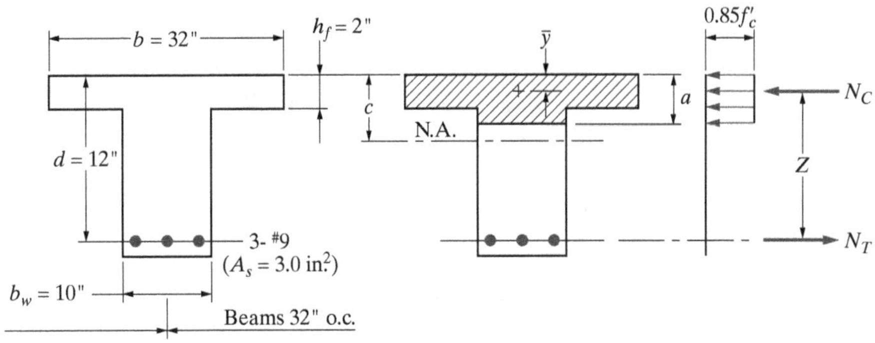

FIGURE 3.4 Sketch for Example 3.1.

Solution:

1. Because the span length is not given, determine the effective flange width in terms of the flange thickness and beam spacing:

$$b_w + 16h_f = 10 + 16(2) = 42 \text{ in.}$$
$$\text{beam spacing} = 32 \text{ in. o.c.}$$
$$\text{Use } b = 32 \text{ in.}$$

2. Check $A_{s,\min}$. We know that from Table A.5:

$$A_{s,\min} = 0.0033 b_w d$$
$$= 0.0033(10)(12) = 0.40 \text{ in.}^2$$
$$0.40 \text{ in.}^2 < 3.0 \text{ in.}^2 \qquad \text{(O.K)}$$

3. Assume that the steel yields and find N_T:

$$N_T = A_s f_y = 3.00(60{,}000) = 180{,}000 \text{ lb}$$

4. The flange alone, if fully stressed to $0.85 f_c'$, would produce a total compressive force of

$$N_{Cf} = (0.85 f_c')h_f b$$
$$= 0.85(3000)(2)(32) = 163{,}200 \text{ lb}$$

5. Because 180,000 > 163,200, the stress block must extend below the flange far enough to provide the remaining compression required for equilibrium:

$$180{,}000 - 163{,}200 = 16{,}800 \text{ lb}$$

Hence, the stress block extends below the flange and the analysis is one for a true T-beam.

6. The remaining compression $(N_T - N_{Cf})$ may be obtained by the additional web area:

$$N_T - N_{Cf} = (0.85 f_c')b_w(a - h_f)$$

Solving for a, we obtain

$$a = \frac{N_T - N_{Cf}}{(0.85 f_c')b_w} + h_f = \frac{16{,}800}{0.85(3000)(10)} + 2$$
$$= 2.66 \text{ in.}$$

7. Determine net tensile strain ε_t (check ductility). Using the relationship, $a = \beta_1 c$, which is approximate for T-beams:

$$c = \frac{a}{\beta_1} = \frac{2.66 \text{ in.}}{0.85} = 3.13 \text{ in. (see Figure 3-4)}$$

The distance d_t of the extreme tensile reinforcement from the compression face is 12 in. Therefore, the net tensile strain in the extreme tensile reinforcement is

$$\varepsilon_t = 0.003\frac{(d_t - c)}{c} = 0.003\frac{(12 - 3.13)}{3.13} = 0.0085$$

8. Determine the strength-reduction factor ϕ: because 0.0085 > 0.005, this is a tension-controlled section and $\phi = 0.90$ (see Section 2.9).

9. To calculate the magnitude of the internal couple, it is necessary to know the lever-arm distance between N_C and N_T. The location of N_T is assumed at the centroid of the steel area, and we will locate N_C at the centroid of the T-shaped compression area (see Figure 3.5a). Using a reference axis at the top of the section, the centroid may be located a distance $\bar{y}$ below the reference axis, as follows:

$$\bar{y} = \frac{\Sigma(Ay)}{\Sigma A}$$

$$A_1 = 32(2) = 64 \text{ in.}^2, \quad A_2 = 10(0.66) = 6.6 \text{ in.}^2$$

$$\bar{y} = \frac{64(1) + 6.6(2 + 0.33)}{64 + 6.6} = 1.12 \text{ in.}$$

This locates N_C. Therefore,

$$Z = d - \bar{y}$$
$$= 12 - 1.12 = 10.88 \text{ in.}$$

The nominal (or ideal or theoretical) moment strength may be found:

$$M_n = N_T Z = \frac{180{,}000(10.88)}{12{,}000} = 163 \text{ ft.-kip}$$

from which the practical moment strength or design moment capacity is

$$\phi M_n = 0.9(163) = 147 \text{ ft.-kip}$$

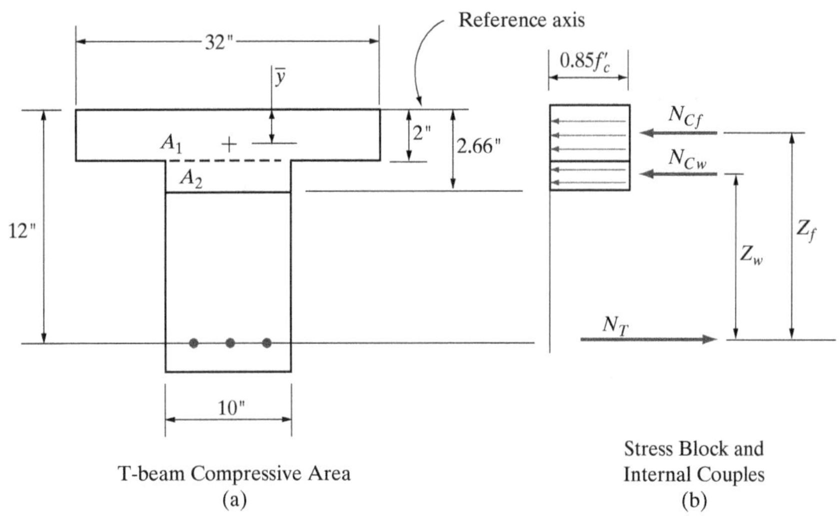

FIGURE 3.5 Sketch for Example 3.1.

In the solution of Example 3.1, step 9 may be accomplished in a slightly different way. If the total internal couple M_n is assumed to be composed of two component couples, a flange couple (using compressive force N_{Cf} in Figure 3.5b) and a web couple (using compressive force N_{Cw} in Figure 3.5b), then its magnitude can be calculated from

$$\begin{aligned} M_n &= \text{flange couple} + \text{web couple} \\ &= N_{Cf}Z_f + N_{Cw}Z_w \\ &= N_{Cf}\left(d - \frac{h_f}{2}\right) + \left(N_T - N_{Cf}\right)\left[d - h_f - \left(\frac{a - h_f}{2}\right)\right] \end{aligned}$$

This avoids the calculation of the centroid location and results in the same moment strength. The concept of two component couples is used again in the design of true T-beams.

EXAMPLE 3.2

For the T-beam shown in Figure 3.6, determine the practical moment strength or design moment capacity, ϕM_n if $f_c' = 3000$ psi and $f_y = 60{,}000$ psi. The beam span length is 24 ft.

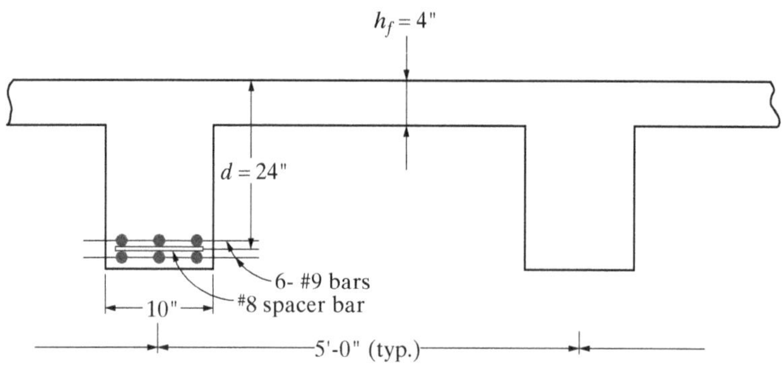

FIGURE 3.6 Sketch for Example 3.2.

Solution:

1. Find the effective flange width, b:

$$b_w + \tfrac{1}{4}\text{ span length} = 10 + \frac{24(12)}{4} = 82 \text{ in.}$$

$$b_w + 16h_f = 10 + 16(4) = 74 \text{ in.}$$

$$\text{beam spacing} = 60 \text{ in.}$$

$$\text{Use } b = 60 \text{ in.}$$

2. Check $A_{s,\min}$. From Table A.5:

$$\begin{aligned} A_{s,\min} &= 0.0033 b_w d \\ &= 0.0033(10)(24) = 0.79 \text{ in.}^2 \end{aligned}$$

$$0.79 \text{ in.}^2 < 6.00 \text{ in.}^2 \qquad \text{(O.K.)}$$

3. Assume the steel yields and find N_T.

$$N_T = A_s f_y = 6.0(60{,}000) = 360{,}000 \text{ lb}$$

4. The flange itself is capable of furnishing a compression force of

$$N_{Cf} = (0.85 f_c')bh_f = 0.85(3000)(60.0)(4) = 612{,}000 \text{ lb}$$

5. Because 612,000 > 360,000, the flange furnishes sufficient compression area and the stress block lies entirely in the flange. Therefore, analyze the T-beam as a rectangular T-beam of width $b = 60$ in.

6. Solve for the depth of the stress block.

$$a = \frac{A_s f_y}{0.85 f_c' b} = \frac{6.00(60{,}000)}{0.85(3000)(60)} = 2.35 \text{ in.}$$

7. Determine the net tensile strain (i.e., check ductility). The depth to the extreme tension steel d_t is determined using *a* #8 spacer bar between the two layers of #9 bars, as shown in Figure 3.6.

$$d_t = 24 + \frac{1.00}{2} + \frac{1.125}{2} = 25.1 \text{ in.}$$

$$c = \frac{a}{\beta_1} = \frac{2.35}{0.85} = 2.76 \text{ in.}$$

$$\varepsilon_t = \frac{0.003(d_t - c)}{c} = \frac{0.003(25.1 - 2.76)}{2.76} = 0.0243$$

(Or use Equation [2.1] from Section 2.9.)

Because 0.0243 > 0.005, ductility is ensured, and this is a tension-controlled section.

8. Find ϕ. Because $\varepsilon_t > 0.005$, $\phi = 0.90$ (see Figure 2.14).

9. Find ϕM_n.

$$\phi M_n = \phi A_s f_y \left(d - \frac{a}{2}\right)$$

$$= \frac{0.90(6.00)(60)\left(24 - \frac{2.35}{2}\right)}{12} = 616 \text{ ft.-kip}$$

This T-beam behaves like a rectangular beam having a width b of 60 in., so the expression for ϕM_n developed in Section 2.10 could be used at step 9:

$$\phi M_n = \phi b d^2 \bar{k}$$

$$\rho = \frac{A_s}{bd} = \frac{6.00}{60(24)} = 0.0042$$

From Table A.8, $\bar{k} = 0.2396$ ksi and $\varepsilon_t > 0.005$. Therefore, $\phi = 0.90$.

Then,

$$\phi M_n = \frac{0.90(60)(24)^2(0.2396)}{12} = 621 \text{ ft.-kip}$$

3.3 ANALYSIS OF BEAMS HAVING IRREGULAR CROSS-SECTIONS

Beams with cross-sections that are not rectangular or L- or T-shaped are commonly used in structures with precast elements. Examples include box beams that are used in bridge structures. The approach for the analysis of such beams is to use the internal couple in the normal way, taking into account any variation in the shape of the compressive stress block. The method is similar to that used for *true* T-beam analysis.

EXAMPLE 3.3

The cross-section shown in Figure 3.7 is sometimes referred to as an inverted T-girder. Find the practical moment strength ϕM_n. (The ledges in the beam cross-section will possibly be used for support of precast slabs.) Use $f_y = 60{,}000$ psi (A615 Grade 60) and $f'_c = 3000$ psi.

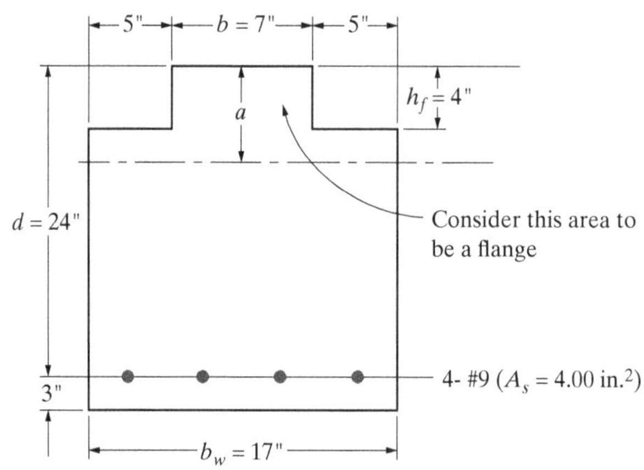

FIGURE 3.7 Sketch for Example 3.3.

Solution:

1. The effective flange width may be considered to be 7 in.
2. Check $A_{s,\min}$. From Table A.5:

$$A_{s,\min} = 0.0033 b_w d$$
$$= 0.0033(17)(24) = 1.35 \text{ in.}^2$$
$$1.35 \text{ in.}^2 < 4.00 \text{ in.}^2 \qquad \text{(O.K.)}$$

3. Find N_T:

$$N_T = A_s f_y = 4.00(60{,}000) = 240{,}000 \text{ lb}$$

4. Determine the amount of compression that the 7 in. × 4 in. area is capable of furnishing. (This is the area we are considering to be the flange.)

$$\begin{aligned} N_{Cf} &= (0.85 f_c')h_f b \\ &= 0.85(3000)(4)(7) = 71{,}400 \text{ lb} \end{aligned}$$

5. Because 240,000 lb > 71,400 lb, the compressive stress block must extend below the ledges to provide the remaining compression ($a > h_f$):

$$N_T - N_{Cf} = 240{,}000 - 71{,}400 = 168{,}600 \text{ lb}$$

6. The remaining compression will be furnished by an additional beam area below the ledges. Referring to Figure 3.7,

$$\begin{aligned} a &= \frac{N_T - N_{Cf}}{(0.8 f_c')b_w} + h_f \\ &= \frac{168{,}600}{0.85(3000)(17)} + 4 \\ &= 7.89 \text{ in. from top of beam} \end{aligned}$$

7. Check ductility.

$$c = \frac{a}{\beta_1} = \frac{7.89 \text{ in.}}{0.85} = 9.28 \text{ in.}$$

The distance of the extreme tensile reinforcement from the compression face d_t is 24 in. Therefore, the net tensile strain in the extreme tensile reinforcement is

$$\varepsilon_t = 0.003\frac{(d_t - c)}{c} = 0.003\frac{(24 - 9.28)}{9.28} = 0.00476$$

8. Therefore, this is a transition section ($0.004 < \varepsilon_t < 0.005$), and the corresponding strength-reduction factor is

$$\begin{aligned} \phi &= 0.65 + (\varepsilon_t - 0.002)\left(\frac{250}{3}\right) \\ &= 0.65 + (0.00476 - 0.002)\left(\frac{250}{3}\right) = 0.875 \end{aligned}$$

Note that $0.65 < 0.875 < 0.90$.

9. ϕM_n will be calculated considering two component internal couples, a flange couple and a web couple. Refer to Figure 3.8.

$$\phi M_n = \phi\left(N_{Cf}Z_f + N_{Cw}Z_w\right)$$

$$= \phi\left\{N_{Cf}\left(d - \frac{h_f}{2}\right) + \left(N_T - N_{Cf}\right)\left[d - h_f - \left(\frac{a - h_f}{2}\right)\right]\right\}$$

$$= \frac{0.875\left\{71.4\left(24 - \frac{4}{2}\right) + 168.6\left[24 - 4 - \left(\frac{3.89}{2}\right)\right]\right\}}{12}$$

$$= 337 \text{ ft.-kip}$$

Section 9.3.3.1 of the ACI 318 Code requires that flexural members be tension-controlled, therefore, the maximum area of steel in the beam permitted by the ACI Code is that area of steel that would result in a net tensile strain of 0.005 (i.e., for Grade 60 rebar); any excess reinforcement is neglected.

Note that the tensile reinforcement corresponding to ε_t = 0.005 results in an economical amount of reinforcement. Using a greater amount of reinforcement than this results in a section that is not fully ductile, which requires a strength reduction factor that is lower than 0.9, and therefore results in an uneconomical section and a member that is not fully ductile. If the example is repeated with 6#7 tension bars (i.e., tension steel area = 3.6 in.^{2}) instead of 4#9, the reader should verify that the section will be in the tension-controlled zone with the depth of the compressive stress block, a = 7.34", and thus fully ductile, with a strength reduction factor of 0.9 and a design moment capacity of 317 ft.-kip.

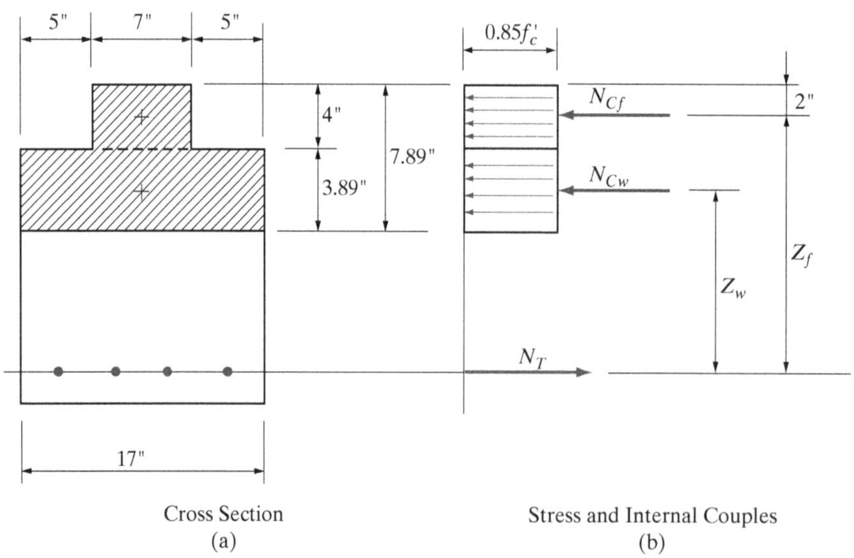

FIGURE 3.8 Inverted T-girder, Example 3.3.

3.4 T-BEAM DESIGN (FOR MOMENT)

The design of the T-sections (and inverted L-sections) involves the dimensions of the flange and web and the area of the tension steel, a total of five unknowns. In the normal progression of a design, the flange thickness is determined by the design of the slab, and the web size is determined by the shear and moment requirements at the end supports of a beam in continuous construction. Practical considerations, such as column sizes and forming, may also dictate web width. Therefore, when the T-section is designed for positive moment, most of the five unknowns would have been previously determined.

As indicated previously, the ACI Code dictates permissible effective flange width, b, when the flange is in compression (e.g., at the midspan). The flange itself generally provides more than sufficient compression area; therefore, the compressive stress block usually lies completely within the depth of the flange in the positive moment region of the beam. Thus, most T-beams (and inverted L-beams) are wide rectangular beams with respect to flexural behavior when the top of the beam is in compression (e.g., at the midspan). Note that when the flange of the T-beam or inverted L-beam is in tension as would occur in the negative moment regions (e.g., at the beam or girder supports), the web of the beam or girder will be in compression and the width of the beam or girder in the compression zone will be b_w (i.e., $b = b_w$)

The recommended method for the design of T-beams will depend on whether the T-beam behaves as a rectangular T-beam or a true T-beam. The first step will be to answer this question. If the T-beam is determined to be a rectangular T-beam, the design procedure is the same as for the tensile reinforced rectangular beam, where the size of the cross-section is known (see Section 2.15). If the T-beam is determined to be a true T-beam, the design proceeds by designing a flange component and a web component and combining the two. The design of inverted L-beams follows a similar approach.

EXAMPLE 3.4

Design the T-beam for the floor system shown in Figure 3.9. The floor has a 4-in. slab supported by 22-ft-span-length beams cast monolithically with the slab. Beams are 8 ft-0 in. on center and have a web width of 12 in. and a total depth = 22 in.; $f'_c = 3000$ psi and $f_y = 60{,}000$ psi (A615 Grade 60). Service loads are 0.125-ksf live load and 0.200-ksf dead load. The given dead load does not include the weight of the floor system.

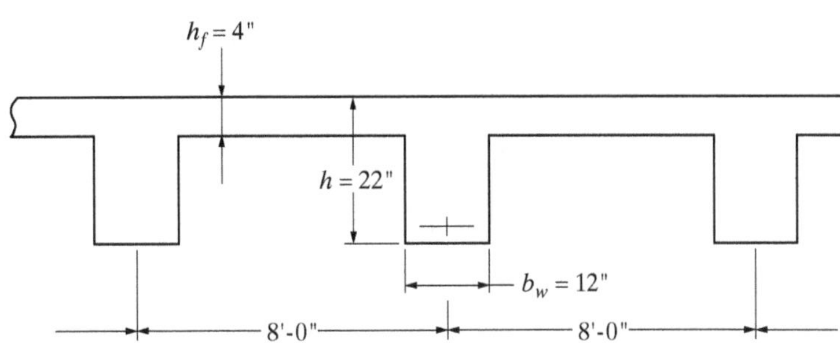

FIGURE 3.9 Typical floor section, Example 3.4.

Solution:

1. Establish the factored design moment:

$$\text{slab weight} = \frac{96(4)}{144}(0.150) = 0.400 \text{ kip/ft}$$

$$\text{stem weight} = \frac{12(18)}{144}(0.150) = 0.225 \text{ kip/ft}$$

$$\text{total weight} = 0.625 \text{ kip / ft}$$

$$\text{service DL} = 8(0.200) = 1.60 \text{ kip/ft}$$

$$\text{service LL} = 8(0.125) = 1.00 \text{ kip/ft}$$

Calculate the factored load and moment:

$$w_u = 1.2(0.625 + 1.60) + 1.6(1.00) = 4.27 \text{ kip/ft}$$

$$M_u = \frac{w_u \ell^2}{8} = \frac{4.27(22)^2}{8} = 258 \text{ ft.-kip}$$

2. Assume an effective depth $d = h - 3$ in.:

$$d = 22 - 3 = 19 \text{ in.}$$

3. Determine the effective flange width:

$$b_w + \tfrac{1}{4}\text{span length} = 12 + 0.25(22)(12) = 78 \text{ in.}$$

$$b_w + 16h_f = 12 + 16(4) = 76 \text{ in.}$$

$$\text{beam spacing} = 96 \text{ in.}$$

Using the smallest value, the effective flange width b = 76 in.

4. Assume a tension-controlled section—that is, the net tensile strain $\varepsilon_t \geq 0.005$; this assumption will be checked later. The net tensile strain value of 0.005 gives a strength-reduction factor ϕ = 0.90.

5. Determine whether the beam behaves as a true T-beam or as a rectangular beam by computing the practical moment strength ϕM_{nf} with the full effective flange assumed to be in compression. This assumes that the bottom of the compressive stress block coincides with the bottom of the flange, as shown in Figure 3.10. Thus,

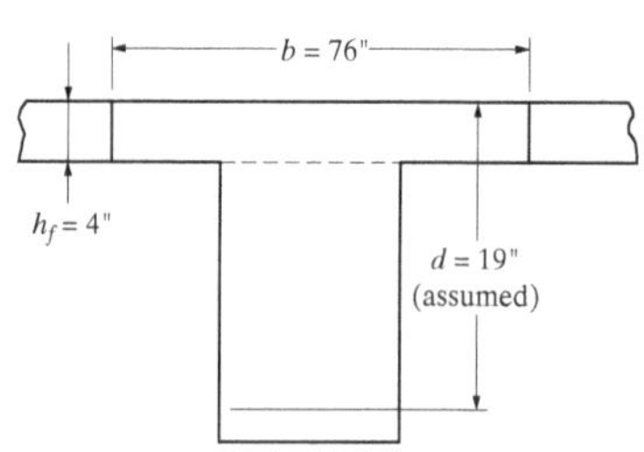

FIGURE 3.10 T-beam for Example 3.4.

$$\phi M_{nf} = \phi(0.85 f_c')bh_f\left(d - \frac{h_f}{2}\right)$$
$$= \frac{0.9(0.85)(3)(76)(4)(19 - 4/2)}{12} = 988 \text{ ft.-kip}$$

6. Because 988 ft.-kip > 258 ft.-kip, the total effective flange need not be completely utilized in compression (i.e., $a < h_f$), and the T-beam behaves as a wide rectangular beam with a width b of 76 in.

7. Design a rectangular beam with b and d as known values (see Section 2.15):

$$\text{required } \bar{k} = \frac{M_u}{\phi bd^2} = \frac{258(12)}{0.9(76)(19)^2} = 0.1254 \text{ ksi}$$

8. From Table A.8, select the required steel ratio to provide a $\bar{k}$ of 0.1254 ksi:

$$\text{required } \rho = 0.0021$$

9. Calculate the required steel area:

$$\text{required } A_s = \rho bd$$
$$= 0.0021(76)(19) = 3.03 \text{ in.}^2$$

10. Select the steel bars. Use four No. 8 bars ($A_s = 3.16$ in.^{2}):

From Table A.3:

$$\text{minimum } b_w = 11 \text{ in.} \qquad \text{(O.K.)}$$

Check the effective depth d. Assume a No. 3 stirrup and $1\frac{1}{2}$-in. cover, as shown in Figure 3.11.

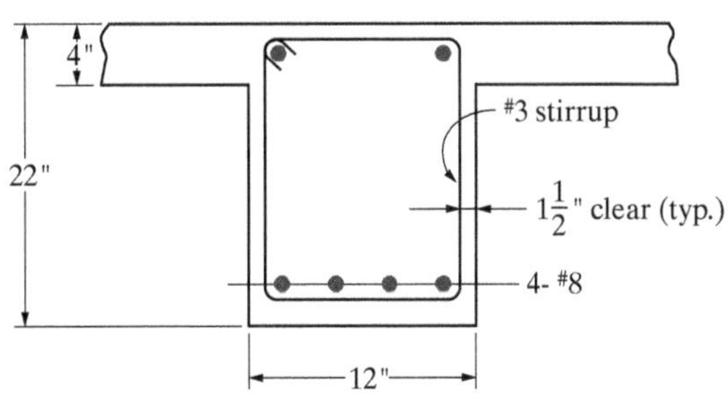

FIGURE 3.11 Design sketch for Example 3.4.

$$d = 22 - 1.5 - 0.38 - \frac{1.00}{2} = 19.62 \text{ in.}$$

$$19.62 \text{ in.} > 19 \text{ in.} \qquad \text{(O.K.)}$$

11. Check $A_{s,\min}$. From Table A.5:

$$A_{s,\min} = 0.0033 b_w d$$
$$= 0.0033(12)(19.62) = 0.78 \text{ in.}^2$$
$$0.78 \text{ in.}^2 < 3.16 \text{ in.}^2 \qquad \text{(O.K.)}$$

12. Check ε_t to ensure a tension-controlled section ($\varepsilon_t \geq 0.005$). From Section 2.10, for a rectangular section:

$$\varepsilon_t = \frac{0.00255 f_c' \beta_1}{\rho f_y} - 0.003$$
$$= \frac{0.00255(3)(0.85)}{0.0025(60)} - 0.003 = 0.0404$$

Therefore, the net tensile strain is much larger than 0.005; this is a tension-controlled section, and $\phi = 0.90$, as assumed. (Note that $\varepsilon_t > 0.005$ could also be confirmed from Table A.8.)

13. Sketch the design (see Figure 3.11).

EXAMPLE 3.5

Design a T-beam having a cross-section, as shown in Figure 3.12. Assume that the effective flange width given is acceptable. The T-beam will carry a total design moment M_u of 340 ft.-kip. Use $f_c' = 3000$ psi and $f_y = 60{,}000$ psi. Use $1\frac{1}{2}$-in. cover and No. 3 stirrups.

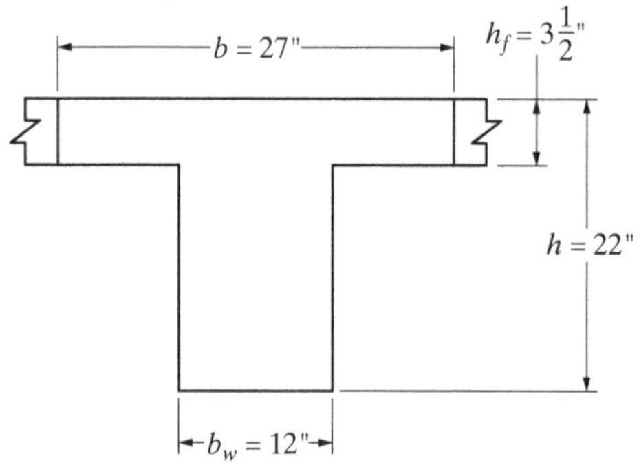

FIGURE 3.12 Sketch for Example 3.5.

Solution:

1. The factored design moment $M_u = 340$ ft.-kip (given).
2. Assume an effective depth of

$$d = 22 - 3 = 19 \text{ in.}$$

3. The effective flange width = 27 in. (given).
4. Assume a tension-controlled section—that is, the net tensile strain $\varepsilon_t \geq 0.005$; this assumption will be checked later. The net tensile strain value of 0.005 gives a strength-reduction factor $\phi = 0.90$.
5. Determine ϕM_{nf} assuming the effective flange to be in compression over its full depth:

$$\phi M_{nf} = \phi(0.85 f_c')bh_f\left(d - \frac{h_f}{2}\right)$$

$$= \frac{0.9(0.85)(3)(27)(3.5)(19 - 3.5/2)}{12} = 312 \text{ ft.-kip}$$

6. $\phi M_{nf} < M_u$; therefore, the beam must behave as a true T-beam.
7. Two component couples will be designed, a flange couple (subscript *f*) and a web couple (subscript *w*). Refer to Figure 3.13.

   Calculate the required steel area A_{sf} for the flange couple:

   estimated $d_f = h - 3 \text{ in.} = 22 - 3 = 19 \text{ in.}$

   estimated $Z_f = d_f - h_f/2 = 19 - 3.5/2 = 17.25 \text{ in.}$

$$\text{required } A_{sf} = \frac{\phi M_{nf}}{\phi f_y Z_f}$$

$$= \frac{312(12)}{0.9(60)(17.25)}$$

$$= 4.02 \text{ in.}^2$$

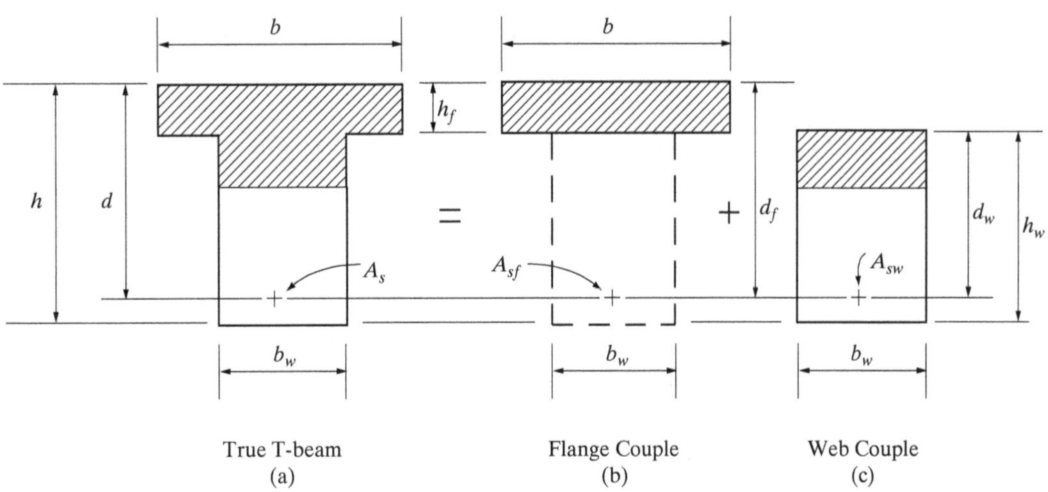

FIGURE 3.13 True T-beam design, Example 3.5.

8. The web couple will be designed for the remaining applied moment $(M_u - \phi M_{nf})$. The design is for a rectangular reinforced concrete beam having a depth $h_w = h - h_f$ and a width of b_w.

$$h_w = 22 - 3.5 = 18.5 \text{ in.}$$

$$\text{estimated } d_w = h_w - 3 \text{ in.} = 18.5 - 3 = 15.5 \text{ in.}$$

$$\text{required } \bar{k} = \frac{M_u - M_{nf}}{\phi b_w d_w^2} = \frac{(340 - 312)(12)}{0.9(12)(15.5)^2}$$

$$= 0.1295 \text{ ksi}$$

From Table A.8, the required $\rho = 0.0023$, from which we calculate

$$\text{required } A_{sw} = \rho b_w d_w = 0.0023(12)(15.5) = 0.43 \text{ in.}^2$$

9. Total area of reinforcing steel required

$$A_s = A_{sf} + A_{sw} = 4.02 + 0.43 = 4.45 \text{ in.}^2$$

10. From Table A.2, select three No. 11 bars. $A_s = 4.68$ in.^{2} and minimum b_w is 11.0 in. Check d assuming No. 3 stirrups and $1\frac{1}{2}$ in. cover:

$$d = 22 - 1.5 - 0.38 - \frac{1.41}{2} = 19.42 \text{ in.}$$

$$19.42 \text{ in.} > \text{estimated } d_f = 19.0 \text{ in.} \qquad \text{(O.K.)}$$

11. Check $A_{s,\min}$. From Table A.5:

$$A_{s,\min} = 0.0033 b_w d$$

$$= 0.0033(12)(19.42) = 0.77 \text{ in.}^2$$

$$0.77 \text{ in.}^2 < 4.68 \text{ in.}^2 \qquad \text{(O.K.)}$$

12. Check ε_t to ensure a tension-controlled section ($\varepsilon_t \geq 0.005$):

$$d_t = d = 19.42 \text{ in.}$$

$$N_T = A_s f_y = 4.68(60) = 281 \text{ kip}$$

$$N_{Cf} = 0.85 f_c' b h_f = 0.85(3)(27)(3.5) = 241 \text{ kip}$$

$$a = \frac{N_T - N_{Cf}}{0.85 f_c' b_w} + h_f = \frac{281 - 241}{0.85(3)(12)} + 3.5 = 4.81 \text{ in.}$$

From Section 2.9, Equation (2.1):

$$\varepsilon_t = \frac{0.003\beta_1 d_t}{a} - 0.003$$

$$= \frac{0.003(0.85)(19.42)}{4.81} - 0.003 = 0.00730$$

0.00730 > 0.005; therefore, this is a tension-controlled section and $\phi = 0.90$, as assumed. (O.K.)

13. Sketch the design. See Figure 3.14.

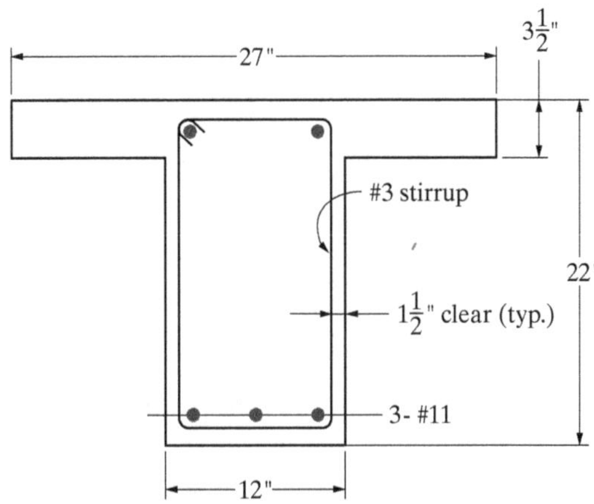

FIGURE 3.14 Design sketch for Example 3.5.

3.5 SUMMARY OF PROCEDURE FOR ANALYSIS OF T-BEAMS (FOR MOMENT)

1. Establish the effective flange width (i.e., the width of the beam in the compression zone) in the positive moment regions based on ACI criteria.In the negative moment regions, the width of the beam in the compression zone is equal to the width of the beam web, b_w.

2. Check $A_{s,\min}$. Use Table A.5.

3. To ensure ductility, assume a net tensile strain, ε_t, greater than or equal to 0.005 (for Grade 60 rebar) in accordance with ACI 318 Section 9.3.3.1; this assumption will be checked later. Compute the total tension in the steel:

$$N_T = A_s f_y$$

4. Compute the magnitude of the compression that the flange itself is capable of furnishing:

$$N_{Cf} = 0.85 f_c' b h_f$$

5. If $N_T > N_{Cf}$, the beam will behave as a true T-beam and the remaining compression, which equals $N_T - N_{Cf}$, will be furnished by additional web area. If $N_T < N_{Cf}$, the beam will behave as a rectangular beam of width b.

Rectangular T-Beam

1. Solve for the depth of the stress block:

$$a = \frac{A_s f_y}{0.85 f'_c b}$$

2. Check ductility; find ε_t.
3. Find ϕ $(0.65 \le \phi \le 0.90)$.
4. Calculate ϕM_n.

$$\phi M_n = \phi A_s f_y \left(d - \frac{a}{2} \right)$$

5. Or, in place of steps 6 through 9, calculate ρ, obtain $\bar{k}$, check ε_t, determine ϕ, and use

$$\phi M_n = \phi b d^2 \bar{k}$$

A flow diagram of this procedure is presented in Appendix B.

True T-Beam

1. Determine the depth of the compressive stress block:

$$a = \frac{N_T - N_{Cf}}{0.85 f'_c b_w} + h_f$$

2. Check ductility; find ε_t.
3. Find ϕ $(0.65 \le \phi \le 0.90)$.
4. **a.** Locate the centroid of the total compressive area referenced to the top of the flange using the relationship

$$\bar{y} = \frac{\Sigma(Ay)}{\Sigma A}$$

from which

$$Z = d - \bar{y}$$

Compute the practical moment strength or design moment capacity, ϕM_n:

$$\phi M_n = \phi N_C Z \quad \text{or} \quad \phi N_T Z$$

Or alternatively,

b. Calculate ϕM_n using a summation of internal couples contributed by the flange and the web:

$$\phi M_n = \phi\left\{N_{cf}\left(d - \frac{h_f}{2}\right) + \left(N_T - N_{Cf}\right)\left[d - h_f - \left(\frac{a - h_f}{2}\right)\right]\right\}$$

The procedures presented above apply also to spandrel or edge beams and girders which are shaped like inverted L-beams (see Figure 3.3). The main difference between a T-beam/girder and an L-beam/girder in the positive moment region is in the value of the effective flange width, b (see Section 3.2). In the negative moment region, the width of the T-beam/girder and L-beam/girder in the compression zone is equal to the width of the beam or girder stem, b_w.

3.6 SUMMARY OF PROCEDURE FOR DESIGN OF T-BEAMS AND L-BEAMS (FOR MOMENT)

1. Compute the design moment M_u. In many cases, the size of the beam is unknown and an estimate of the beam depth and width of the beam stem has to be made before proceeding with the calculation of the factored load on the beam and the factored design moment, M_u. The reader should refer to Chapter 6 for the calculation of the factored moments at the positive and negative moment regions of continuous beams.
2. Assume that the effective depth $d = h - 3$ in.

   The "3-in." in the effective depth equation above is an average value. For beams with one layer of tension reinforcement, an effective depth of $h - 2.5$ in. can be used, while for beams with two layers of tension reinforcement, an effective depth of $h - 3.5$ in. can be used.
3. Establish the effective flange width based on ACI 318 criteria.
4. Assume a net tensile strain $\varepsilon_t \geq 0.005$ (i.e., for Grade 60 rebar); this will give a strength-reduction factor $\phi = 0.90$. This assumption will be checked later.
5. Compute the practical moment strength ϕM_{nf} assuming that the total effective flange is in compression:

$$\phi M_{nf} = \phi(0.85 f_c')bh_f\left(d - \frac{h_f}{2}\right)$$

6. If $\phi M_{nf} > M_u$, the beam will behave as a rectangular T-beam of width b. If $\phi M_{nf} < M_u$, the beam will behave as a true T-beam.

Rectangular T-Beam

1. Design a rectangular beam with b and d as known values. Compute the required $\bar{k}$:

$$\text{required } \bar{k} = \frac{M_u}{\phi b d^2}$$

2. From the tables in Appendix A, determine the required ρ for the required $\bar{k}$ from step 7.
3. Compute the required A_s:

$$\text{required } A_s = \rho b d$$

4. Select the bars and check the beam width. Check the actual d and compare it with the assumed d. If the actual d is slightly in excess of the assumed d, the design will be slightly conservative (i.e., on the safe side). If the actual d is less than the assumed d, the design may be on the nonconservative side (depending on the steel provided) and should be more closely investigated for possible revision.
5. Check $A_{s,\min}$. Use Table A.5.
6. Check ductility. Find ε_t to ensure that $\varepsilon_t \geq 0.005$. If $d = d_t$, Tables A.7 through A.11 can be used or use Equation (2.2) from Section 2.10. If $d \neq d_t$, use basic principles or Equation (2.1) from Section 2.9. Check the assumed value of ϕ.

   For $d = d_t$:

$$\varepsilon_t = \frac{0.00255 f'_c \beta_1}{\rho f_y} - 0.003 \tag{2.2}$$

   For $d \neq d_t$:

$$\varepsilon_t = \left(\frac{d_t - \frac{a}{\beta_1}}{\frac{a}{\beta_1}} \right)(0.003) = \frac{0.003\beta_1 d_t}{a} - 0.003 \tag{2.1}$$

7. Sketch the design.

True T-Beam

1. Using an estimated $d_f = h - 3''$ and $Z_f = d_f - h_f/2$, determine the steel area A_{sf} required for the flange couple:

$$\text{required } A_{sf} = \frac{\phi M_{nf}}{\phi f_y Z_f}$$

2. Design the web couple as a rectangular reinforced concrete beam having a total depth $h_w = h - h_f$, using an estimated $d_w = h_w - 3''$ and a beam width of b_w. Design for an

applied factored moment of $M_u - \phi M_{nf}$. Determine the required $\overline{k}$, the required ρ, and the required A_{sw}.

3. Total required area of reinforcing steel, $A_s = A_{sf} + A_{sw}$.
4. Select the bars. (Bars must fit within the beam width, b_w.) Check d as in step 10 of the rectangular T-beam design.
5. Check $A_{s,\min}$. Use Table A.5.
6. Check ductility. Find ε_t. Determine the stress block depth, a, and use the expression for ε_t at the end of Section 2.9.
7. Sketch the design.

Flowcharts of these procedures are presented in Appendix B.

The procedures presented above for T-beams or girders (i.e., interior beams or girders) also apply to inverted L-beams or girders (i.e., spandrel beams or girders)

3.7 DOUBLY REINFORCED BEAMS: INTRODUCTION

As indicated in Chapter 2, the practical moment strength or design moment capacity of a rectangular reinforced concrete beam reinforced with only tensile steel may be determined by the expression $\phi M_n = \phi b d^2 \overline{k}$, where $\overline{k}$ is a function of the steel ratio ρ, f'_c, and f_y.

If we assume a given rectangular section with tension-only reinforcing that is tension-controlled, the upper limit of tension steel area can be established using the reinforcement ratio ρ associated with net tensile strain $\varepsilon_t = 0.005$. The maximum practical moment strength or design moment capacity, ϕM_n for the section may then be calculated using the associated value of $\overline{k}$.

Occasionally, practical and architectural considerations may dictate and limit beam sizes, whereby it becomes necessary to develop more moment strength from a given cross-section. When this situation occurs, the ACI Code, Section 22.2.3.1, permits the addition of tensile steel over and above the code maximum provided that compression steel is also added in the compression zone of the cross-section. The result constitutes a combined tensile and compressive reinforced beam commonly called a *doubly reinforced beam*.

Where beams span more than two supports (continuous construction), practical considerations are sometimes the reason for the existence of main steel in compression zones. In Figure 3.15, positive moments exist at A and C; therefore, the main tensile reinforcement would be placed in the bottom of the beam. At B, however, a negative moment exists and the bottom of the beam is in compression. The tensile reinforcement must be placed near the top of the beam. It is general practice that at least some of the tension steel in each of these cases will be extended beyond the tension regions of the beam to the compression zones. In this case the compression steel may sometimes be used for additional strength. In Chapter 7, we will see that compression reinforcement aids significantly in reducing long-term deflections. In fact, the use of compression steel to increase the bending strength of a reinforced concrete beam is an inefficient way to utilize steel; however, in shallow concrete beams, the extra moment capacity might be needed. More commonly, deflection control will be the reason for adding compression steel to a beam or girder.

In general, the advantages of compression steel include reduction of long-term deflections due to creep and increase in beam ductility which is advantageous in seismic regions; they can also serve as stirrup support bars which eases fabrication of the reinforcement cage. When compression steel is not needed for design, two nominal bars (one at each top corner) are provided to help with the fabrication of the stirrups.

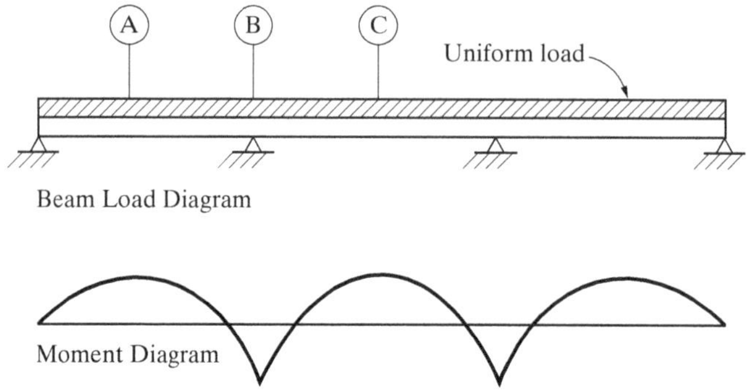

FIGURE 3.15 Continuous beam.

3.8 DOUBLY REINFORCED BEAM ANALYSIS FOR MOMENT (CONDITION I)

The basic assumptions for the analysis of doubly reinforced beams are similar to those for tensile reinforced beams. One additional significant assumption is that the *compression steel stress* (f_s') is a function of the strain at the level of the centroid of the compression steel. As discussed previously, the steel will behave elastically up to the point where the strain exceeds the yield strain ε_y. In other words, as a limit, $f_s' = f_y$ when the compression steel strain $\varepsilon_s' \geq \varepsilon_y$. If $\varepsilon_s' < \varepsilon_y$, the stress in the compression steel, $f_s' = \varepsilon_s' E_s$, where E_s is the modulus of elasticity of the steel.

With two different materials, concrete and steel, resisting the compressive force N_C, the total compression will now consist of two forces: N_{C1}, the compression resisted by the concrete, and N_{C2}, the compression resisted by the compressive steel. For analysis, the total resisting moment of the beam will be assumed to consist of two parts or two internal couples: the part due to the resistance of the compressive concrete and tensile steel, and the part due to the compressive steel and additional tensile steel. The two internal couples are illustrated in Figure 3.16.

Notation for doubly reinforced beams is as follows:

A_s' = total compression steel cross-sectional area

d = effective depth of tension steel

d' = depth to centroid of compression steel from compression face of beam

A_{s1} = amount of tension steel used by the concrete–steel couple

A_{s2} = amount of tension steel used by the steel–steel couple

A_s = total tension steel cross-sectional area $(A_s = A_{s1} + A_{s2})$

M_{n1} = nominal moment strength of the concrete–steel couple

M_{n2} = nominal moment strength of the steel–steel couple

M_n = nominal moment strength of the (doubly reinforced) beam

ε_s = unit strain at the centroid of the tension steel

ε_s' = unit strain at the centroid of the compression steel

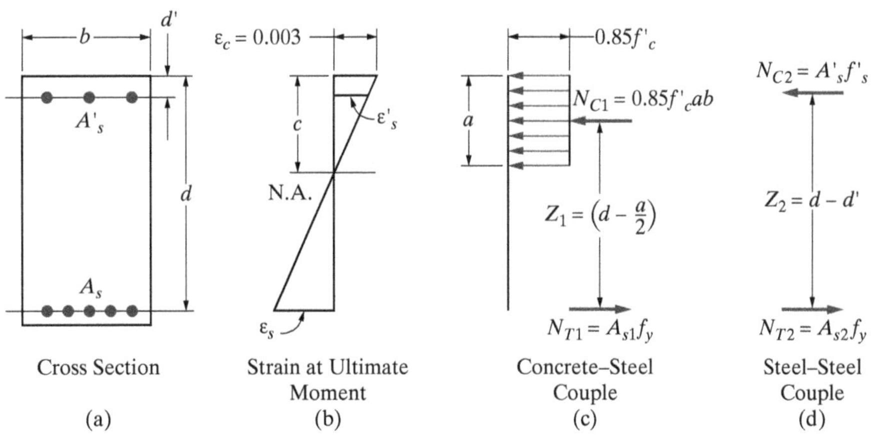

FIGURE 3.16 Doubly reinforced beam analysis.

The total nominal moment strength may be developed as the sum of the two internal couples, neglecting the concrete displaced by the compression steel.

The strength of the steel–steel couple is evaluated as follows:

$$M_{n2} = N_{T2} Z_2$$

Assuming that $f_s = f_y$ (tensile steel yields),

$$M_{n2} = A_{s2} f_y (d - d')$$

Also, because $\sum H_F = 0$ and $N_{C2} = N_{T2}$,

$$A_s' f_s' = A_{s2} f_y$$

If we assume that the compression steel yields and that $f_s' = f_y$, then

$$A_s' f_y = A_{s2} f_y$$

from which

$$A_s' = A_{s2}$$

Therefore,

$$M_{n2} = A_s' f_y (d - d')$$

The strength of the concrete–steel couple is evaluated as follows:

$$M_{n1} = N_{T1} Z_1$$

Assuming that the tensile steel yields and $f_s = f_y$,

$$M_{n1} = A_{s1} f_y \left(d - \frac{a}{2} \right)$$

Also, because $A_s = A_{s1} + A_{s2}$, then

$$A_{s1} = A_s - A_{s2}$$

and because $A_{s2} = A_s'$, then

$$A_{s1} = A_s - A_s'$$

Therefore,

$$M_{n1} = (A_s - A_s') f_y \left(d - \frac{a}{2} \right)$$

Summing the two couples, we arrive at the nominal moment strength of a doubly reinforced beam:

$$\begin{aligned} M_n &= M_{n1} + M_{n2} \\ &= (A_s - A_s') f_y \left(d - \frac{a}{2} \right) + A_s' f_y (d - d') \end{aligned}$$

The practical moment strength or design moment capacity, ϕM_n may then be calculated.

The foregoing expressions are based on the assumption that both tension and compression steels yield prior to concrete strain reaching 0.003. This may be checked by determining the strains that exist at the nominal moment, which depend on the location of the neutral axis. The neutral axis may be located, as done previously, by using the relationship between the depth of the compressive stress block and the neutral axis depth, that is, $a = \beta_1 c$. Thus,

$$N_T = N_{C1} + N_{C2}$$

$$A_s f_y = (0.85 f_c')ab + A_s' f_y$$

$$a = \frac{(A_s - A_s') f_y}{(0.85 f_c')b}$$

which may also be expressed as

$$a = \frac{A_{s1} f_y}{(0.85 f_c')b}$$

With the calculation of the depth of the compression stress block, a, the neutral axis location c may be determined and the assumptions checked.

Check the net tensile strain ε_t in the extreme tensile reinforcement and ensure that $\varepsilon_t \geq 0.005$ to satisfy the ACI Code, Section 9.3.3.1 and ensure a fully ductile section

$$\varepsilon_t = 0.003 \frac{(d_t - c)}{c}$$

where d_t = the distance of the extreme tensile reinforcement from the compression face. The corresponding strength-reduction factor is then calculated as described in Section 2.9.

The case where both tensile and compressive steel yield prior to the concrete strain reaching 0.003 will be categorized as *condition I* (see Example 3.6). The case where the tensile steel yields but the compressive steel does not yield prior to the concrete strain reaching 0.003 is categorized as *condition II* (see Example 3.7). The maximum area of steel in the beam permitted by the ACI Code for flexural members with little or no axial load` in order to ensure a fully ductile section is the area of steel that would result in a net tensile strain, ε_t of 0.005 (i.e., for Grade 60 rebar).

EXAMPLE 3.6

Compute the practical moment strength or design moment capacity, ϕM_n for a beam having a cross-section as shown in Figure 3.17. Use $f_c' = 3000$ psi and $f_y = 60{,}000$ psi.

Solution:

1. Assume that all the steel (i.e., both tension and compression steels) yields:

$$f_s' = f_y \quad \text{and} \quad f_s = f_y$$

Compression steel: 2 #8, A_s' = 1.58 in.^{2}
Tension steel: 6 #8, A_s = 4.74 in.^{2}

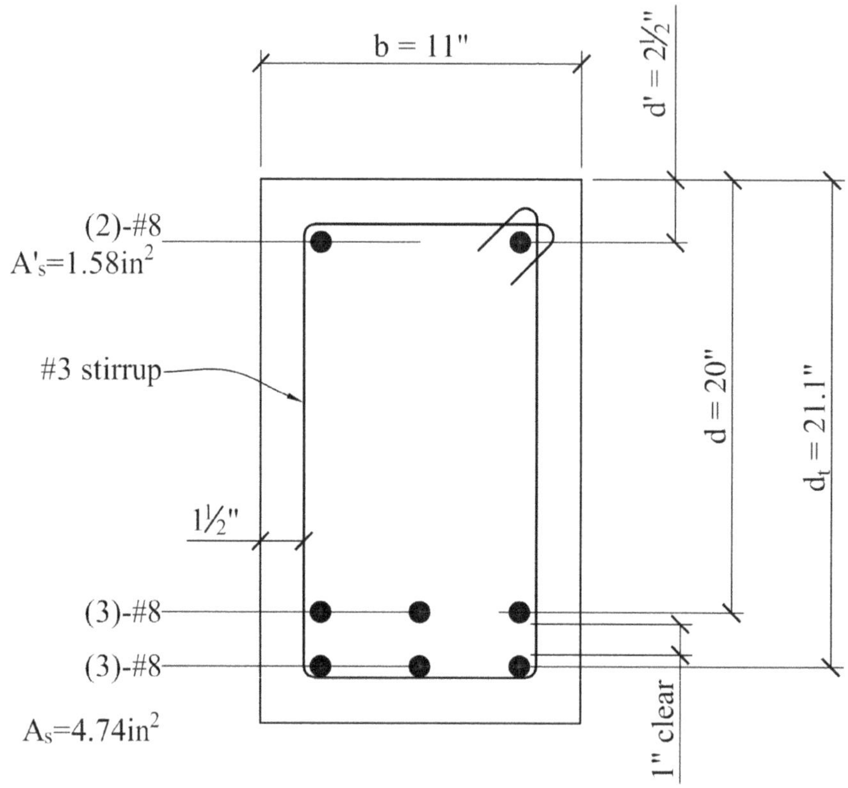

FIGURE 3.17 Sketch for Example 3.6.

Therefore, $A_{s2} = A'_s$ (see Figure 3.16d).

2. $A_s = A_{s1} + A_{s2}$

$$A_{s1} = A_s - A'_s$$
$$= 4.74 - 1.58 = 3.16 \text{ in.}^2$$

From the concrete–steel couple, the stress block depth can be found:

$$a = \frac{A_{s1} f_y}{(0.85 f'_c) b} = \frac{3.16(60)}{0.85(3.0)(11)} = 6.76 \text{ in.}$$

3. Assuming that the same relationship ($a = \beta_1 c$) exists between the depth of the stress block and the beam's neutral axis as existed in singly reinforced beams, the neutral axis may now be located for purposes of checking the steel strains. From Figure 3.16 at the nominal moment,

$$c = \frac{a}{\beta_1} = \frac{a}{0.85} = \frac{6.76}{0.85} = 7.95 \text{ in.}$$

This value of c is based on the assumption in step 1 and will be verified in step 4.

4. Check the strains to determine whether the assumptions are valid and that both steels yield before the concrete crushes (see Figure 3.16b). The strains calculated exist at the nominal moment:

$$\varepsilon'_s = \frac{c - d'}{c}(0.003)$$

$$= \frac{7.95 - 2.5}{7.95}(0.003) = 0.00206 \approx 0.00207 = \varepsilon_y$$

Therefore, $\varepsilon'_s \approx \varepsilon_y = 0.00207$ (from Table A.1). That is, the compression steel yields.

Check the ductility of the beam by calculating the net tensile strain in the extreme or farthest tensile reinforcement:

$$\varepsilon_t = 0.003\frac{(d_t - c)}{c} = 0.003\frac{(21.2 - 7.95)}{7.95} = 0.005 \qquad \text{(O.K.)}$$

Therefore, the tensile steel yields and the section is tension-controlled, and the strength reduction factor $\phi = 0.9$.

Because $\varepsilon'_s \approx \varepsilon_y$, the compression steel will yield before the concrete strain reaches 0.003, and $f'_s = f_y$. Therefore, the assumption concerning the compression steel stress is O.K.

5. From the concrete–steel couple:

$$M_{n1} = A_{s1} f_y \left(d - \frac{a}{2}\right)$$

$$= 3.16(60)\left(20 - \frac{6.76}{2}\right) = 3151 \text{ in.-kip}$$

$$\frac{3151}{12} = 263 \text{ ft.-kip}$$

From the steel–steel couple:

$$M_{n2} = A'_s f_y (d - d') = 1.58(60)(20 - 2.5)$$

$$= 1659 \text{ in.-kip}$$

$$\frac{1659}{12} = 138 \text{ ft.-kip}$$

$$M_n = M_{n1} + M_{n2}$$

$$= 263 + 138 = 401 \text{ ft.-kip}$$

6. $\phi M_n = 0.90(401)$
$= 361$ ft.-kip

In this example, because both the compression steel and tension steel yield prior to the concrete reaching a compressive strain of 0.003 and because the net tensile train in the tension steel is at least 0.005, a ductile failure mode is assured.

3.9 DOUBLY REINFORCED BEAM ANALYSIS FOR MOMENT (CONDITION II)

As has been pointed out, usually the compression steel (A_s') will reach its yield stress before the concrete reaches a strain of 0.003. This may not occur in shallow beams reinforced with the higher-strength steels, however. Referring to Figure 3.16b, if the neutral axis is located relatively high in the cross-section, it is possible that $\varepsilon_s' < \varepsilon_y$ at the nominal moment. The magnitude of ε_s' (and therefore f_s') depends on the location of the neutral axis. The depth of the compressive stress blocks a also depends on c, because $a = \beta_1 c$.

The total compressive force must be equal to the total tensile force $A_s f_y$, and an equilibrium equation can be written to solve for the exact required value of c. This turns out to be a quadratic equation. This situation and its solution may be observed in the following example.

EXAMPLE 3.7

Compute the practical moment strength ϕM_n for a beam having a cross-section as shown in Figure 3.18. Use $f_c' = 5000$ psi and $f_y = 60{,}000$ psi.

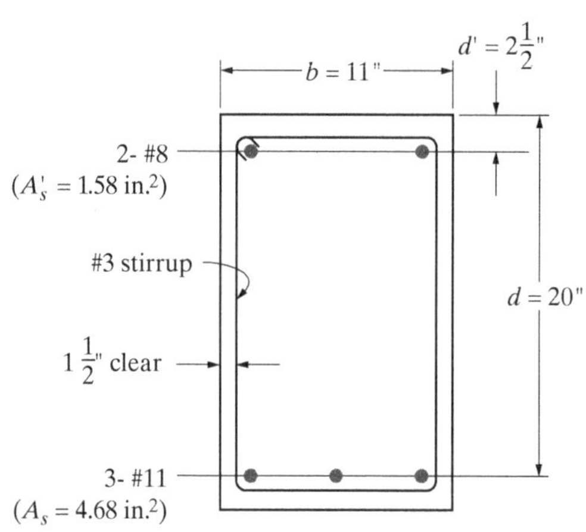

FIGURE 3.18 Sketch for Example 3.7.

Solution:

1. Assume that all the steel yields. This results in

$$A_{s2} = A_s'$$

2. With reference to Figure 3.19b,

$$a = \frac{(A_s - A_s')f_y}{(0.85f_c')b}$$

$$= \frac{(4.68 - 1.58)60}{0.85(5)(11)} = \frac{3.10(60)}{0.85(5)(11)}$$

$$= 3.98 \text{ in.}$$

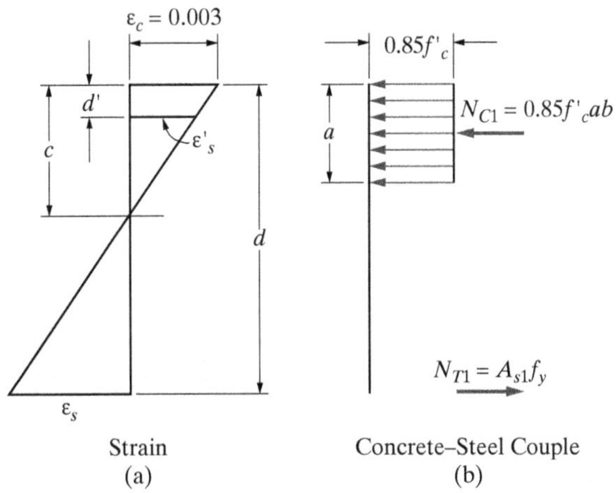

FIGURE 3.19 Compatibility check, Example 3.7.

3. Locate the neutral axis:

$$a = \beta_1 c, \quad \beta_1 = 0.80 \quad \text{(reference, ACI Code, Section 22.2.2.4.3)}$$

$$c = \frac{a}{\beta_1} = \frac{3.98}{0.80} = 4.98 \text{ in.}$$

This value of c is based on the assumption in step 1 and will be verified in step 4.

4. By similar triangles of Figure 3.19a, check the steel strains:

$$\text{compressive steel: } \varepsilon_s' = \frac{0.003(c - d')}{c}$$

$$\varepsilon_s' = \frac{2.48}{4.98}(0.003) = 0.0015$$

Calculate the net tensile strain in the extreme reinforcement based on the depth of the neutral axis obtained in step 3. Note that $d_t = d$ and, therefore, $\varepsilon_s = \varepsilon_t$.

$$\varepsilon_t = 0.003\frac{(d_t - c)}{c} = 0.003\frac{(20 - 4.98)}{4.98} = 0.009$$

$$0.009 > 0.004 \qquad \text{(O.K.)}$$

For Grade 60 steel, $\varepsilon_y = 0.00207$ (from Table A.1). Because $\varepsilon_s > \varepsilon_y > \varepsilon_s'$, the tensile steel *has* yielded and the compression steel *has not* yielded. Therefore, the assumptions of step 1 are incorrect.

5. With the original assumptions incorrect, a solution for the location of the neutral axis must be established. With reference to Figure 3.16, c will be determined by using the condition that horizontal equilibrium exists. That is, $\Sigma H_F = 0$. Thus,

$$N_T = N_{C1} + N_{C2}$$
$$A_s f_y = (0.85 f_c')ba + f_s' A_s'$$

But

$$a = \beta_1 c$$

and

$$f_s' = \varepsilon_s' E_s = \left[\frac{c - d'}{c}(0.003)\right] E_s$$

Then, by substitution,

$$A_s f_y = (0.85 f_c') b \beta_1 c + \left[\frac{c - d'}{c}(0.003)\right] E_s A_s'$$

Multiplying by c and expanding, we obtain

$$A_s f_y c = (0.85 f_c') b \beta_1 c^2 + c(0.003) E_s A_s' - d'(0.003) E_s A_s'$$

Rearranging yields

$$(0.85 f_c' b \beta_1) c^2 + (0.003 E_s A_s' - A_s f_y) c - d'(0.003) E_s A_s' = 0$$

With E_s = 29,000 ksi, the expression becomes

$$(0.85 f_c' b \beta_1) c^2 + (87 A_s' - A_s f_y) c - 87 d' A_s' = 0$$

where

A_s = 4.68 in.^{2}

f_y = 60 ksi

$f_c' = 5$ ksi

b = 11 in.

β_1 = 0.80

$A_s' = 1.58$ in.^{2}

$d' = 2.5$ in.

Substitution yields

$$
\begin{aligned}
\left[0.85(5)(11)(0.80)\right]c^2 + \left[87(1.58) - 4.68(60)\right]c& \\
-87(2.5)(1.58) &= 0 \\
37.4c^2 - 143.34c - 343.65 &= 0 \\
c^2 - 3.83c - 9.19 &= 0
\end{aligned}
$$

This may be solved using the usual formula for the roots of a quadratic equation:

$$\frac{-b \pm \sqrt{b^2 - 4ac}}{2a}$$

where the coefficients are

$$
\begin{aligned}
a &= 1.0 \\
b &= -3.83 \\
c &= -9.19
\end{aligned}
$$

Or the square may be completed as follows:

$$
\begin{aligned}
c^2 - 3.83c &= 9.19 \\
c^2 - 3.83c + \left(\frac{-3.83}{2}\right)^2 &= 9.19 + \left(\frac{-3.83}{2}\right)^2 \\
c^2 - 3.83c + 3.67 &= 9.19 + 3.67 = 12.86 \\
(c - 1.92)^2 &= 12.86 \\
c - 1.92 &= \sqrt{12.86} = 3.59 \\
c &= 5.51 \text{ in.}
\end{aligned}
$$

The solution of the quadratic equation for c may be simplified as follows:

$$c = \pm\sqrt{Q + R^2} - R$$

where

$$
\begin{aligned}
R &= \frac{87A_s' - A_s f_y}{1.7 f_c' b \beta_1} \\
Q &= \frac{87d'A_s'}{0.85 f_c' b \beta_1}
\end{aligned}
$$

Note that basic units are kips and inches, so the value of f_y, for instance, must be in ksi, not in psi.

6. With this value of c, all the remaining unknowns may be found:

$$f'_s = \frac{c-d'}{c}(87) = \frac{5.51-2.50}{5.51}(87)$$
$$= 47.5 \text{ ksi} < 60 \text{ ksi} \quad \text{(the compression steel has not yielded, as expected)}$$

7. $a = \beta_1 c = 0.80(5.51) = 4.41$ in.

The actual net tensile strain is calculated as

$$\varepsilon_t = 0.003\frac{(d_t - c)}{c} = 0.003\frac{(20-5.51)}{5.51} = 0.0079$$
$$0.0079 > 0.005 \qquad \text{(O.K.)}$$

Thus, the beam is ductile as assumed and because $\varepsilon_t \geq 0.005$, the strength-reduction factor $\phi = 0.90$, as discussed in Section 2.9.

8. $N_{C1} = (0.85f'_c)ab = 0.85(5)(4.41)(11.0) = 206.2$ kip (O.K.)

$$N_{C2} = A'_s f'_s = 47.5(1.58) = 75.1 \text{ kip}$$
$$N_C = 281.3 \text{ kip}$$
$$\text{Check: } N_T = A_s f_y = 4.68(60) = 281 \text{ kip}$$
$$N_T \approx N_C$$

9. $M_{n1} = N_{C1}Z_1 = N_{C1}\left(d - \frac{a}{2}\right) = 206.2\left(20 - \frac{4.41}{2}\right)$

$$= 3670 \text{ in.-kip} = 306 \text{ ft.-kip}$$

$$M_{n2} = N_{C2}Z_2 = N_{C2}(d - d') = 75.1(20 - 2.5)$$
$$= 1314 \text{ in.-kip} = 110 \text{ ft.-kip}$$

$$M_n = 306 + 110 = 416 \text{ ft.-kip}$$

10. $\phi M_n = 0.9(416)$

$$= 374 \text{ ft.-kip}$$

3.10 SUMMARY OF PROCEDURE FOR ANALYSIS OF DOUBLY REINFORCED BEAMS (FOR MOMENT)

1. Assume that all the steel yields, $f_s = f'_s = f_y$. Therefore,

$$A_{s2} = A'_s$$

2. Using the concrete–steel couple and $A_{s1} = A_s - A_s'$, compute the depth of the compression stress block:

$$a = \frac{A_{s1} f_y}{(0.85 f_c')b} = \frac{(A_s - A_s') f_y}{(0.85 f_c')b}$$

3. Compute the location of the neutral axis:

$$c = \frac{a}{\beta_1}$$

This value of c is based on the assumption in step 1 and will be verified in step 4.

4. Using the strain diagram, check the strain in the compression reinforcement and the net tensile strain in the extreme tensile reinforcement to determine whether the assumption in step 1 is valid:

$$\varepsilon_s' = \frac{0.003(c - d')}{c}$$

$$\varepsilon_t = \frac{0.003(d_t - c)}{c}$$

It is required that $\varepsilon_t \geq 0.005$. Therefore, the tensile steel has yielded. The following two conditions may exist with regard to the compression steel. In both cases, the strength-reduction factor ϕ is determined as discussed previously in Section 2.9:

a. Condition I: $\varepsilon_s' \geq \varepsilon_y$. This indicates that the assumption of step 1 is correct and the compression steel has yielded.

b. Condition II: $\varepsilon_s' < \varepsilon_y$. This indicates that the assumption of step 1 is incorrect and the compression steel has not yielded.

Condition I

1. If ε_s' and ε_s both exceed ε_y, compute the nominal moment strengths M_{n1} and M_{n2}. For a steel–steel couple:

$$M_{n2} = A_s' f_y (d - d')$$

For a concrete–steel couple:

$$M_{n1} = A_{s1} f_y \left(d - \frac{a}{2} \right)$$

and

$$M_n = M_{n1} + M_{n2}$$

2. Practical moment strength or design moment capacity $= \phi M_n$.

Condition II

1. If ε'_s is less than ε_y and $\varepsilon_s \geq \varepsilon_y$, compute c using the following formula:

$$(0.85 f'_c b \beta_1)c^2 + (87A'_s - A_s f_y)c - 87d'A'_s = 0$$

   and solve the quadratic equation for c, or use the simplified formula approach from Example 3.7, step 5. Note that the basic units are kips and inches.

2. Compute the compressive steel stress (to be less than f_y):

$$f'_s = \frac{c - d'}{c}(87)$$

3. Solve for a using

$$a = \beta_1 c$$

   To check ductility, recalculate the net tensile strain,

$$\varepsilon_t = 0.003\frac{(d_t - c)}{c}$$

   The strength-reduction factor ϕ is determined as discussed in Section 2.9.

4. Compute the compressive forces:

$$N_{C1} = (0.85 f'_c)ba$$
$$N_{C2} = A'_s f'_s$$

   Check these by computing the tensile force:

$$N_T = A_s f_y$$

   Note that N_T should equal $N_{C1} + N_{C2}$.

5. Compute the ideal resisting moment strengths of the individual couples:

$$M_{n1} = N_{C1}\left(d - \frac{a}{2}\right)$$

   and

$$M_{n2} = N_{C2}(d - d')$$
$$M_n = M_{n1} + M_{n2}$$

6. Practical moment strength or design moment capacity $= \phi M_n$.

3.11 DOUBLY REINFORCED BEAM DESIGN FOR MOMENT

If a check shows that a singly reinforced rectangular section is inadequate and the size of the beam cannot be increased, a doubly reinforced section may be designed using a procedure that consists of the separate design of the two-component couples such that their summation will result in a beam of the required strength.

EXAMPLE 3.8

Design a rectangular reinforced concrete beam to carry a factored design moment M_u of 697 ft.-kip. Physical limitations require that b = 14 in. and h = 30 in. If compression steel is needed, d' = 3 in. Use $f_c' = 3000$ psi and f_y = 60,000 psi. The beam cross-section is shown in Figure 3.20.

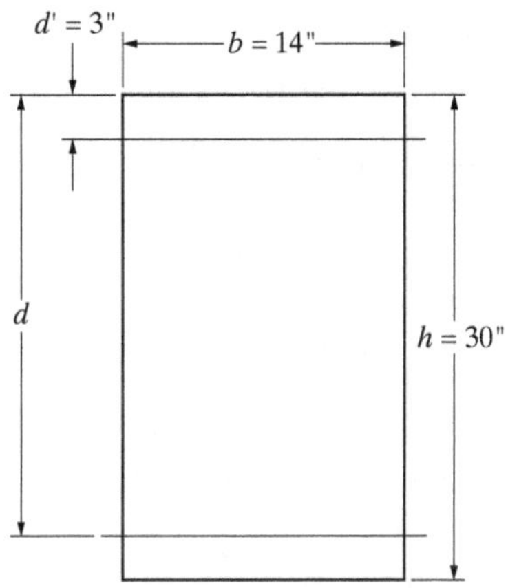

FIGURE 3.20 Sketch for Example 3.8.

Solution:

1. Assume that $d = h - 4 = 26$ in. (because of the probability of two rows of steel).
2. The factored design moment M_u is given; M_u = 697 ft.-kip.
3. Determine if a singly reinforced beam will work. From Table A.8, maximum ρ for ε_t = 0.005 is 0.01355. This is if $d = d_t$. For $d_t > d$, the maximum ρ for moment-strength calculation can be found by proportion. Assuming a #8 bar, #3 stirrups, and $1\frac{1}{2}$-in. cover,

$$d_t = 30 - 1.5 - 0.38 - \frac{1.00}{2} = 27.6 \text{ in.}$$

Then

$$\rho_{max} = \frac{d_t}{d}(\rho_{max} \text{ from Table A-8})$$
$$= \frac{27.6}{26}(0.01355) = 0.01438$$

The associated $\bar{k}$ (Table A.8) is 0.7177 ksi. Assuming a tension-controlled section ($\varepsilon_t \geq 0.005$), ϕ will be 0.90. This assumption will be checked later.

$$\text{maximum } \phi M_n = \phi b d^2 \bar{k}$$
$$= \frac{0.90(14)(26^2)(0.7177)}{12} = 509 \text{ ft.-kip}$$

4. 509 ft.-kip < 697 ft.-kip. Therefore, a doubly reinforced beam is required.

5. Provide a concrete–steel couple having ϕM_{n1} of 509 ft.-kip. Therefore, $\rho = 0.01438$ and $\bar{k} = 0.7177$ ksi:

$$\text{required } A_{s1} = \rho b d = 0.01438(14)(26) = 5.23 \text{ in.}^2$$

6. The steel–steel couple must be proportioned to have moment strength equal to the remainder of the design moment:

$$\text{required } \phi M_{n2} = M_u - \phi M_{n1} = 697 - 509 = 188 \text{ ft.-kip}$$

7. Considering the steel–steel couple, we have

$$\phi M_{n2} = \phi N_{C2}(d - d')$$
$$N_{C2} = \frac{\phi M_{n2}}{\phi(d - d')} = \frac{188(12)}{0.90(26-3)} = 109.0 \text{ kip}$$

8. Because $N_{C2} = A'_s f'_s$, compute f'_s using the neutral axis location of the concrete–steel couple and check the strain ϵ'_s in the compression steel (see Figure 3.21).

$$a = \frac{A_{s1} f_y}{0.85 f'_c b} = \frac{5.23(60)}{0.85(3)(14)} = 8.79 \text{ in.}$$
$$c = \frac{a}{\beta_1} = \frac{8.79}{0.85} = 10.34 \text{ in.}$$
$$\varepsilon'_s = \frac{(c - d')}{c}(0.003) = \frac{(10.34 - 3.00)}{10.34}(0.003)$$
$$= 0.00213$$

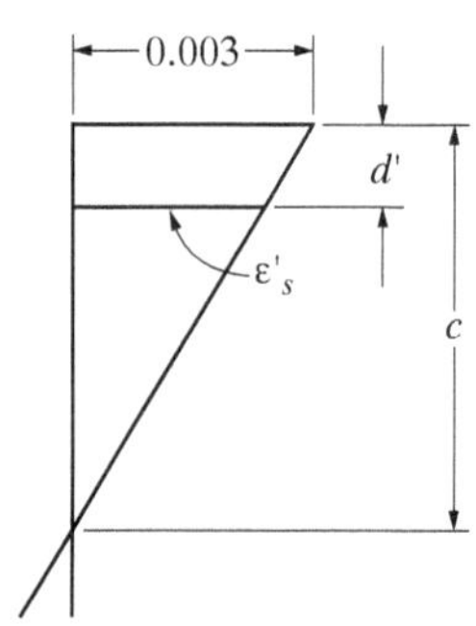

FIGURE 3.21 Concrete strain diagram for Example 3.8.

$\varepsilon_y = 0.00207$ from Table A.1. Because $\varepsilon_s' > \varepsilon_y$, the compression steel will yield before the concrete strain reaches 0.003, and $f_s' = f_y$.

9. Determine the required compression steel:

$$\text{required } A_s' = \frac{N_{C2}}{f_s'} = \frac{N_{C2}}{f_y} = \frac{109.0}{60} = 1.82 \text{ in.}^2$$

10. Because $f_s' = f_y$, required A_{S2} = required $A_s' = 1.82$ in.^{2}

11. The total required tension steel is

$$A_s = A_{s1} + A_{s2} = 5.23 + 1.82 = 7.05 \text{ in.}^2$$

12. Select the compression steel. To ensure yielded compression steel and a fully ductile section (i.e., section in the tension-controlled zone), try Two No. 10 bars which will provide $A_s' = 2.54$ in.^{2}

13. Select the tension steel. Try Six No. 10 bars which will provide $A_s = 7.62$ in.^{2} Place the tension steel in two layers of three bars each with 1 in. clear between layers. Minimum beam width for three #10 bars is 10.5 in., from Table A.3.

14. Assume No. 3 stirrups and determine the actual depth to the centroid of the bar group by considering the total depth (30 in.), required cover ($1\frac{1}{2}$ in.), stirrup size (No. 3), tension bar size (No. 10), and required 1-in. minimum clear space between layers:

$$\text{actual } d = 30 - 1.5 - 0.38 - 1.27 - 0.5 = 26.35 \text{ in.}$$

The assumed d was 26 in.

$$26.35 \text{ in.} > 26 \text{ in.} \qquad \text{(O.K.)}$$

15. Check d_t.

$$d_t = 30 - 1.5 - 0.38 - \frac{1.27}{2} = 27.5 \text{ in.} \approx 27.6 \text{ in.} \qquad \text{(Say O.K.)}$$

Recalculating the design is an option. The differences will be very small.

16. Check the assumption of step 3 ($\phi = 0.90$). We will use the cross-section designed (see the design sketch: Figure 3.22).

$$A_s = 7.62 \text{ in.}^2,\ A_s' = 2.54 \text{ in.}^2$$
$$d = 26.4 \text{ in.} \text{ (step 14)},\ d_t = 27.5 \text{ in.} \text{ (step 15)}$$
$$d' = 1.5 + 0.38 + 1.27/2 = 2.52 \text{ in.}$$

$$\text{Assume } f_s = f_s' = f_y; \quad A_{s2} = A_s'$$

$$a = \frac{(A_s - A_s')f_y}{0.85 f_c' b} = \frac{(7.62 - 2.54)(60)}{0.85(3)(14)} = 8.54 \text{ in.}$$

$$c = \frac{a}{\beta_1} = \frac{8.54}{0.85} = 10.04 \text{ in.}$$

$$\varepsilon_s' = \frac{0.003(c - d')}{c} = \frac{0.003(10.04 - 2.52)}{10.04}$$
$$= 0.00225 > 0.00207 \qquad \text{(O.K.)}$$

$$\varepsilon_t = \frac{0.003(d_t - c)}{c} = \frac{0.003(27.5 - 10.04)}{10.04}$$
$$= 0.0052 > 0.005$$

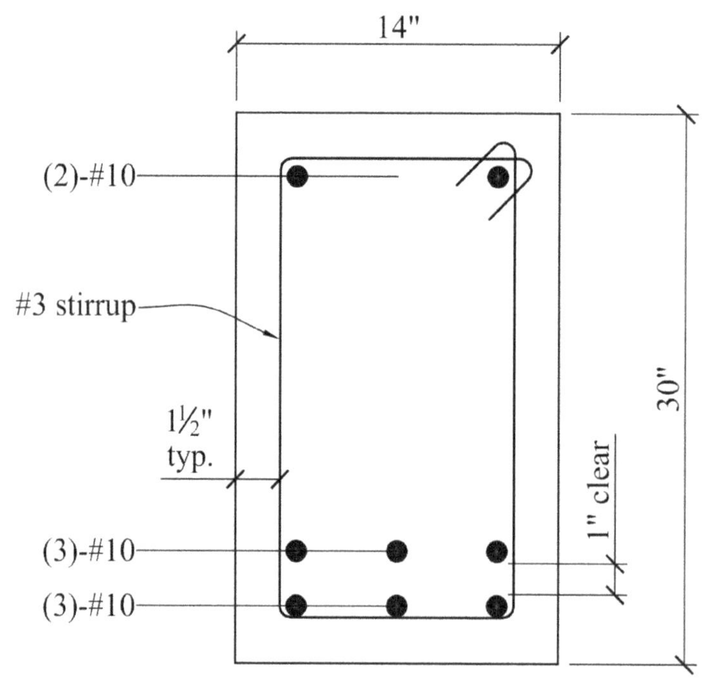

FIGURE 3.22 Design sketch for Example 3.8.

Therefore, this is a tension-controlled section, which implies a ductile failure mode; therefore, the strength reduction factor, ϕ = 0.9.:

$$M_{n1} = A_{s1} f_y \left(d - \frac{a}{2} \right)$$

$$= \frac{(7.62 - 2.54)(60)\left(26.4 - \frac{8.54}{2}\right)}{12} = 562 \text{ ft.-kip}$$

$$M_{n2} = A'_s f_y (d - d') = \frac{2.54(60)(26.4 - 2.52)}{12}$$

$$= 303 \text{ ft.-kip}$$

$$\phi M_n = \phi(M_{n1} + M_{n2}) = 0.90(562 + 303)$$

$$= 779 \text{ ft.-kip}$$

$$779 \text{ ft.-kip} > 697 \text{ ft.-kip} \qquad \text{(O.K.)}$$

17. Figure 3.22 is a sketch of the design.

3.12 SUMMARY OF PROCEDURE FOR DESIGN OF DOUBLY REINFORCED BEAMS (FOR MOMENT)

The size of the beam cross-section is fixed.

1. Assume that $d = h - 4$ in. Estimate d_t.
2. Establish the total design moment M_u.
3. Check to see if a doubly reinforced beam is necessary. Compute maximum ϕM_n for a **singly reinforced beam**. Assume a tension-controlled section (net tensile strain $\varepsilon_t \geq 0.005$) and use the corresponding strength-reduction factor, $\phi = 0.90$; use maximum $\bar{k}$ from the tables in Appendix A and the corresponding maximum steel ratio ρ_{max} for $\varepsilon_t = 0.005$. If $d_t > d$, adjust ρ_{max} for $\varepsilon_t = 0.005$ and select maximum $\bar{k}$ accordingly.

$$\text{maximum } \phi M_n = \phi b d^2 \bar{k}$$

4. If $\phi M_n < M_u$, design the beam as a doubly reinforced beam. If $\phi M_n \geq M_u$, the beam can be designed as a beam reinforced with tension steel only.

For a Doubly Reinforced Beam

1. Provide a concrete–steel couple having the maximum ϕM_n from step 3. This is ϕM_{n1}.

   Using ρ from step 3, find the steel required for the concrete–steel couple:

$$\text{required } A_{s1} = \rho b d$$

2. Find the remaining moment that must be resisted by the steel–steel couple:

$$\text{required } \phi M_{n2} = M_u - \phi M_{n1}$$

3. Considering the steel–steel couple, find the required compressive force in the steel (assume that $d' = 3$ in.):

$$N_{C2} = \frac{\phi M_{n2}}{\phi(d - d')}$$

4. Because $N_{C2} = A'_s f'_s$, compute f'_s so that A'_s may eventually be determined. This can be accomplished by using the neutral-axis location of the concrete–steel couple and checking the strain ε'_s in the compression steel with ε_y from Table A.1. Thus,

$$a = \frac{A_{s1} f_y}{(0.85 f'_c) b}$$

$$c = \frac{a}{\beta_1}$$

$$\varepsilon'_s = \frac{0.003(c - d')}{c}$$

If $\varepsilon'_s \geq \varepsilon_y$, the compressive steel has yielded at the nominal moment and $f'_s = f_y$. If $\varepsilon'_s < \varepsilon_y$, then calculate $f'_s = \varepsilon'_s E_s$ and use this stress in the following steps.

5. Because $N_{C2} = A'_s f'_s$,

$$\text{required } A'_s = \frac{N_{C2}}{f'_s}$$

6. Determine the required A_{s2}:

$$A_{s2} = \frac{f'_s A'_s}{f_y}$$

7. Find the total tension steel required:

$$A_s = A_{s1} + A_{s2}$$

8. Select the compressive steel (A'_s), ensuring that the compression steel yields, and the section is in the tension-controlled or fully ductile zone.

9. Select the tensile steel (A_s), ensuring that the section is in the tension-controlled or fully ductile zone. Check the required beam width. Preferably, place the bars in one layer.

10. Check the actual d and compare it with the assumed d. If the actual d is slightly in excess of the assumed d, the design will be slightly conservative (i.e., on the safe side). If the actual d is less than the assumed d, the design may be on the unconservative side and an analysis and possible revision should be considered.

11. Check d_t and compare with the assumed d_t.

12. Check the ϕ value assumption.

13. Sketch the design.

3.13 ADDITIONAL CODE REQUIREMENTS FOR DOUBLY REINFORCED BEAMS

The compression steel in beams or girders, whether it is in place to increase flexural strength or to control deflections, will act similarly to all typical compression members in that it will tend to buckle, as shown in Figure 3.23. Should this buckling occur, it will naturally be accompanied by spalling of the concrete cover. To help guard against this type of failure, the ACI Code requires that the compression bars be tied within the beam in a manner similar to that used for reinforced concrete columns (discussed in Chapter 9). Compression reinforcement in beams or girders must be enclosed by closed ties or stirrups. The size of the ties or stirrups is to be at least No. 3 for No. 10 longitudinal bars or smaller in the beam, and No. 4 for No. 11 longitudinal bars or larger in the beam. The center-to-center spacing of the closed ties or stirrups, s, for the compression reinforcement, is not to exceed the least of the following:

- 16 longitudinal bar diameters,
- 48 tie (or stirrup) bar diameters, or
- the least dimension of the beam.

The ties or stirrups are to be used throughout the area where compression reinforcement is required (see ACI Code, Section 9.7.6.4).

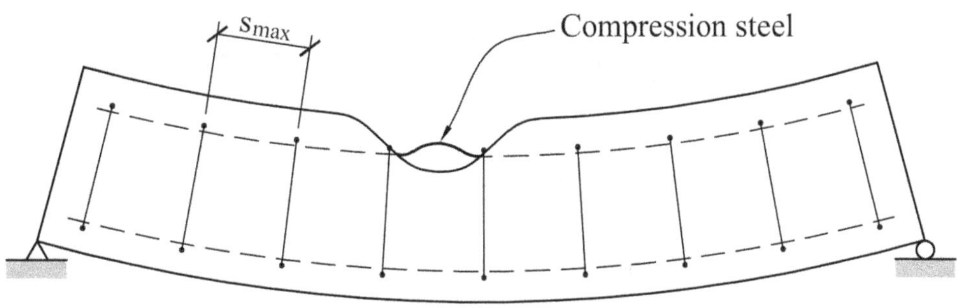

FIGURE 3.23 Possible failure mode for compression steel.

REFERENCES

1. Stephen J. Seguirant, Richard Brice, and Bijan Khaleghi, "Making sense of minimum flexural reinforcement requirements for reinforced concrete members," PCI Journal, Summer 2010.
2. William L. Gamble, "Minimum Flexural Reinforcement, Revisited," SP-353: Design of Slabs for Serviceability and Punching Shear Strength – Honoring Professor Amin Ghali, July 1, 2022, pp. 17-28.

EXERCISES

NOTE **The following note applies to all the exercises in this chapter**. *Unless otherwise noted, assume No. 3 stirrups, $1\frac{1}{2}$-in. cover for beams, and 1-in. clear space between layers of bars. In all exercises, check the net tensile strain e_t in the extreme tension steel to ensure that it is within the allowable limits. Unless otherwise noted or shown, $d_t = d$.*

3.1 Find ϕM_n for the following T-beam: b = 36 in., b_w = 12 in., h_f = 4 in., d = 22 in., $f'_c = 4000$ psi, and $f_y = 60{,}000$ psi. The reinforcement is four No. 8 bars.

3.2 Rework Exercise 3.1 with b = 48 in.

3.3 Rework Exercise 3.1 with reinforcement of three No. 11 bars.

3.4 Rework Exercise 3.1 with $f'_c = 5000$ psi.

3.5 The simple-span T-beam shown is part of a floor system of span length 20 ft-0 in. and beam spacing 45 in. o.c. Use b = 45 in., $f'_c = 3000$ psi and $f_y = 60{,}000$ psi. The bars are placed with 1-in. clear space between layers.

a. Find the practical moment strength ϕM_n.

b. How much steel would be required in this beam cross-section to make the compressive stress block completely cover the flange?

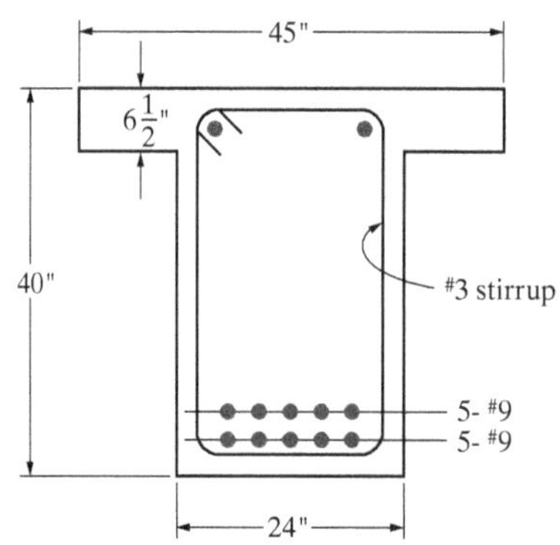

3.6 Find ϕM_n for a typical T-beam in the floor system shown. The beams span 24 ft and are spaced 6′-0″ on center. Use $f'_c = 3000$ psi and $f_y = 60{,}000$ psi.

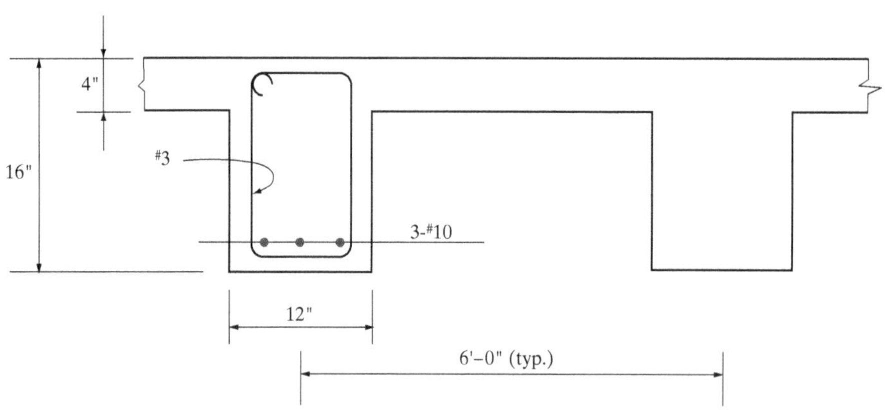

3.7 Find the practical moment strength ϕM_n for the T-beam in the floor system shown. The beam span is 31 ft-6 in. Use $f_c' = 4000$ psi and $f_y = 60{,}000$ psi.

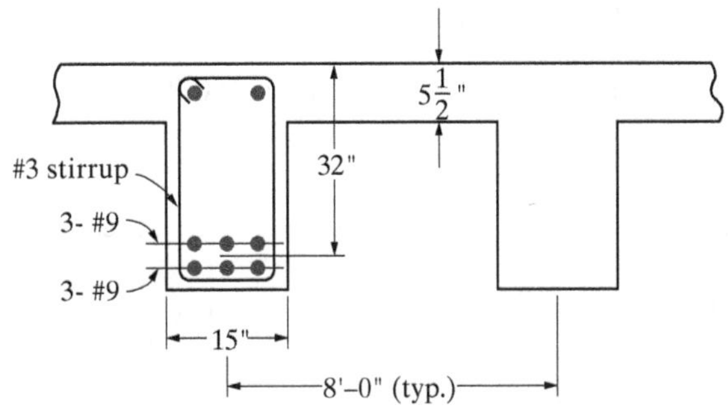

3.8 The T-beam shown is on a simple span of 30 ft. Use $f_c' = 3000$ psi and $f_y = 60{,}000$ psi. No dead load exists other than the weight of the floor system.

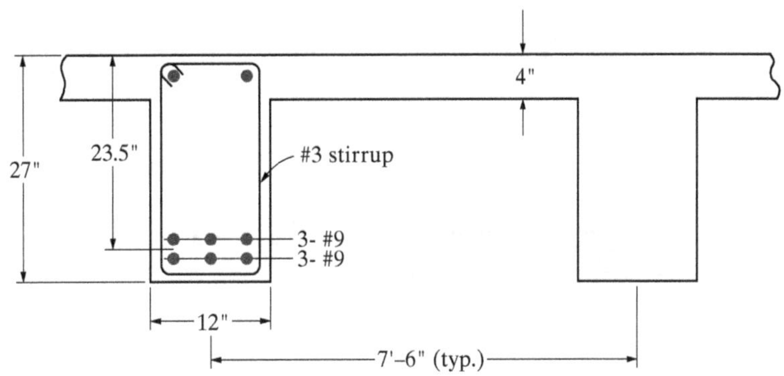

Assume that the slab design is adequate.

a. Find the practical moment strength ϕM_n.

b. Compute the permissible service live load that can be placed on the floor (psf).

3.9 The simple-span T-beam shown is part of a floor system of span length 20 ft-0 in. and beam spacing of 8 ft-0 in. o.c. Find the practical moment strength ϕM_n. Use $f_c' = 3000$ psi and $f_y = 60{,}000$ psi. Assume $d_t = 27$ in.

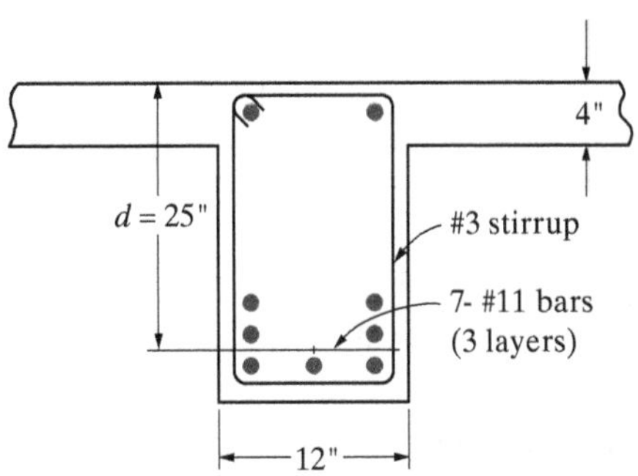

3.10 Find ϕM_n for the beams of cross-section shown. Assume that the physical dimensions are acceptable.

a. Spandrel beam with a flange on one side only. Use $f_c' = 4000$ psi and $f_y = 60{,}000$ psi.

b. Box beam: $f_c' = 3000$ psi and $f_y = 60{,}000$ psi.

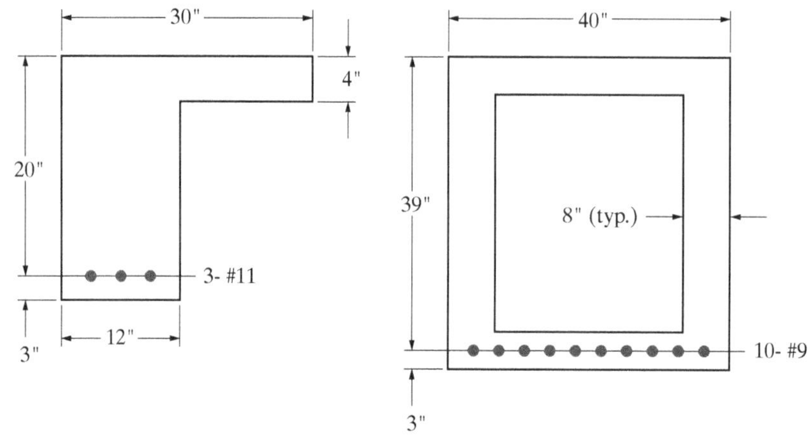

(a) (b)

3.11 Determine ϕM_n for the cross-section that has a rectangular duct cast in it, as shown. Use $f_c' = 3000$ psi, $f_y = 60{,}000$ psi, 1-in. clear space between bar layers, $1\frac{1}{2}$-in. cover, and eight No. 7 bars, as shown.

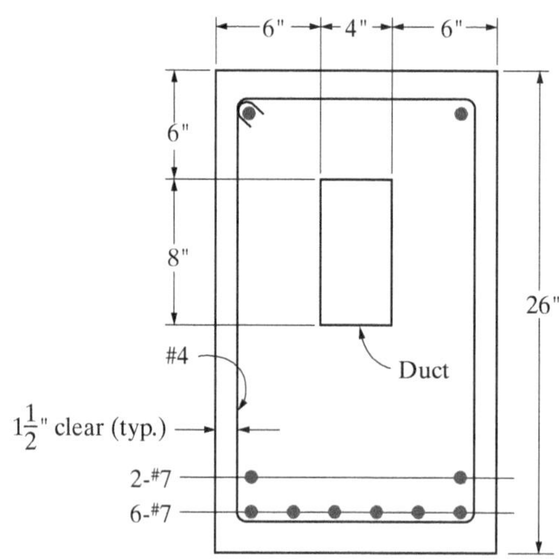

3.12 Design a typical interior tension-reinforced T-beam to resist positive moments. A cross-section of the floor system is shown. The service loads are 50 psf dead load (this does not include the weight of the beam and slab) and 325 psf live load. The beam is on a simple span of 18 ft. Use $f_c' = 4000$ psi and $f_y = 60{,}000$ psi.

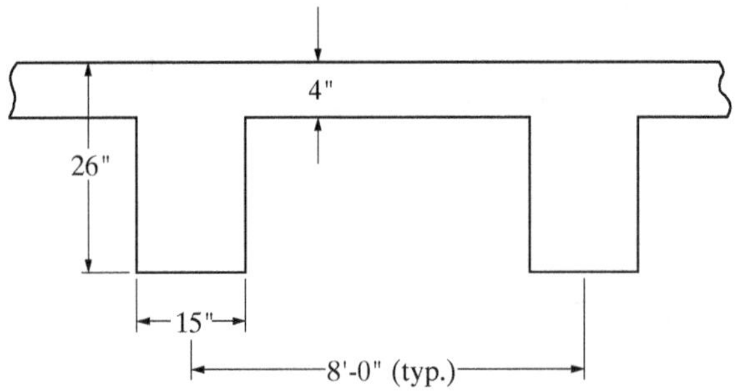

3.13 A reinforced concrete floor system consists of a 3-in. concrete slab supported by continuous-span T-beams of 24-ft spans. The T-beams are spaced 4 ft-8 in. o.c. The web dimensions, determined by negative moment and shear requirements at the supports, are shown. Select the steel required at midspan to resist a total positive design moment M_u of 575 ft.-kip (this includes the weight of the floor system). Use $f'_c = 3000$ psi and $f_y = 60{,}000$ psi.

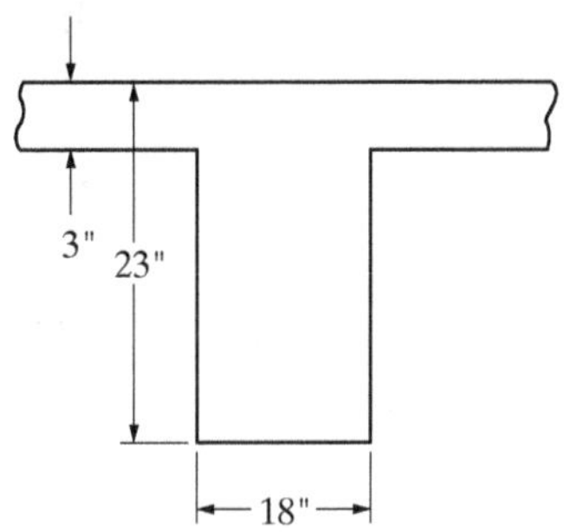

3.14 A reinforced concrete floor system is to have a 4-in.-thick slab supported on 16-in.-wide beams, as shown. At one location, penetrations through the floor slab limit the effective flange width for the supporting beam to 20 in. (Use b = 20 in.) The positive factored moment M_u at this section is 320 ft.-kip. Design the T-beam using $f'_c = 3000$ psi and $f_y = 60{,}000$ psi.

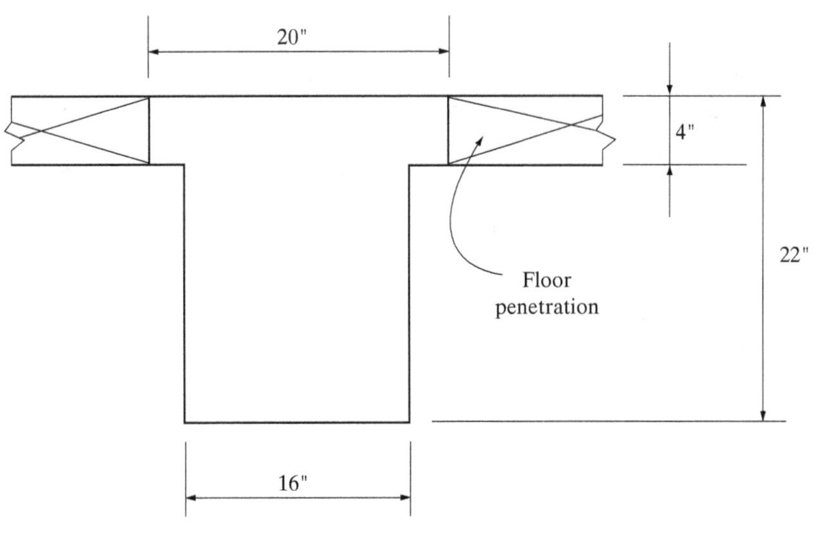

3.15 Select steel for the beams of cross-section shown. The positive service moments are 160 ft.-kip live load and 100 ft.-kip dead load (this includes beam weight). Use $f_c' = 3000$ psi and $f_y = 60{,}000$ psi.

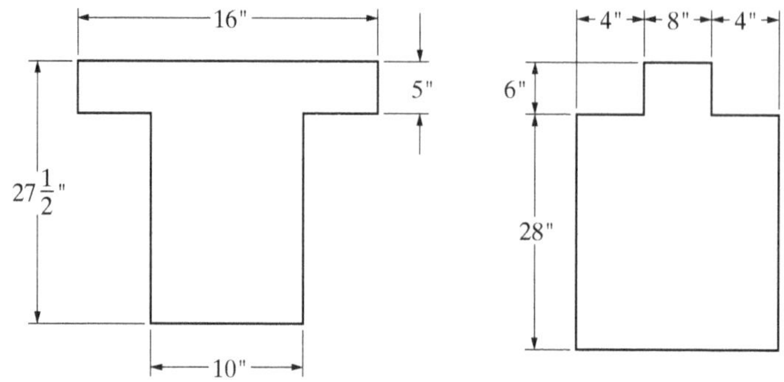

3.16 Find ϕM_n for the beam of cross-section shown. Use $f_c' = 4000$ psi and $f_y = 60{,}000$ psi.

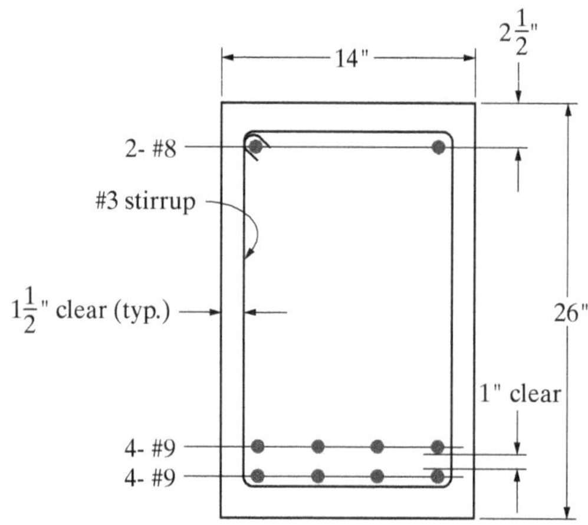

3.17 The beam of cross-section shown is to span 28 ft on simple supports. The uniform load on the beam (in addition to its own weight) will be composed of equal service dead load and live load. Use $f_c' = 4000$ psi and $f_y = 60{,}000$ psi.

a. Find the service loads that the beam can carry (in addition to its own weight).

b. Compare the practical moment strength ϕM_n of the beam as shown with ϕM_n of a beam of similar size reinforced with the area of steel associated with (1) $\varepsilon_t = 0.005$ and (2) $\varepsilon_t = 0.004$.

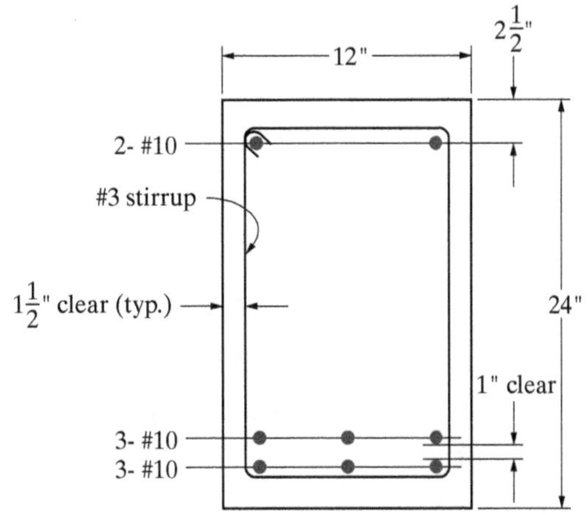

3.18 Find ϕM_n for the beam cross-section shown. Use $f_c' = 4000$ psi and $f_y = 60{,}000$ psi.

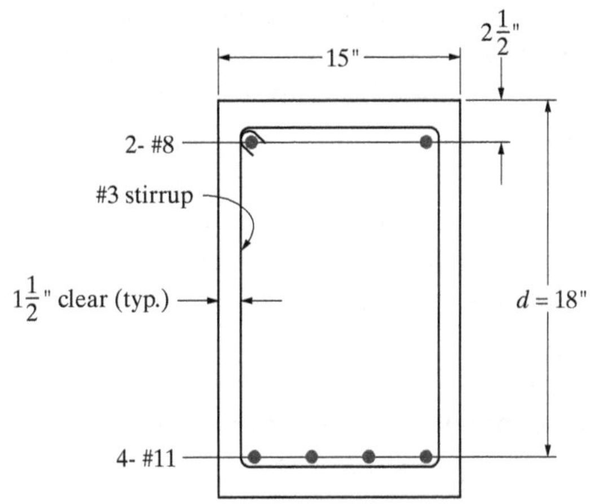

3.19 Compute the practical moment strength ϕM_n for the beam of cross-section shown. How much can ϕM_n be increased if four No. 8 bars are added to the top of the beam? Use $d' = 2\frac{1}{2}$ in., $f_c' = 3000$ psi, and $f_y = 60{,}000$ psi.

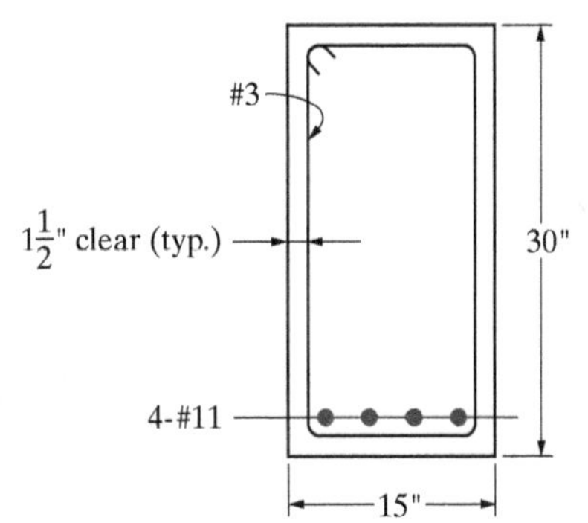

3.20 Compute the practical moment strength ϕM_n for the simply supported precast inverted T-girder shown. Use $f_c' = 3000$ psi and $f_y = 60{,}000$ psi.

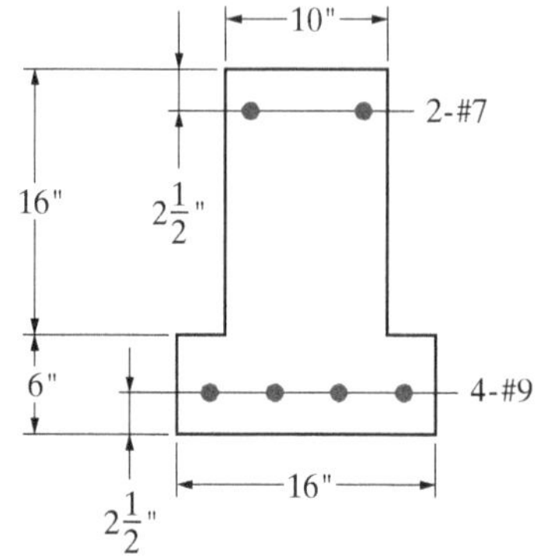

3.21 Design a rectangular reinforced concrete beam to resist a total design moment M_u of 765 ft.-kip (this includes the moment due to the weight of the beam). The beam size is limited to 15 in. maximum width and 30 in. maximum overall depth. Use $f_c' = 3000$ psi and $f_y = 60{,}000$ psi. If compression steel is required, make $d' = 2\frac{1}{2}$ in.

3.22 Design a rectangular reinforced concrete beam to resist service moments of 150 ft.-kip dead load (includes moment due to weight of beam) and 160 ft.-kip due to live load. Architectural considerations require that width be limited to 11 in. and overall depth be limited to 23 in. Use $f_c' = 3000$ psi and $f_y = 60{,}000$ psi.

3.23 Design a rectangular reinforced concrete beam to carry service loads of 1.25 kip/ft dead load (includes beam weight) and 2.60 kip/ft live load. The beam is a simple span and has a span length of 18 ft. The overall dimensions are limited to a width of 10 in. and overall depth of 20 in. Use $f_c' = 3000$ psi and $f_y = 60{,}000$ psi.

3.24 Redesign the beam of Exercise 3.23 for tension reinforcing only and increased width. Keep an overall depth of 20 in.

3.25 A simply supported precast inverted T-girder having the cross-section shown is subjected to a total positive design moment M_u of 280 ft.-kip. Select the required reinforcement (both tensile and compressive). Assume a No. 3 stirrup and $1\frac{1}{2}$ in. cover on the tension side of the beam. Use $f_c' = 4000$ psi and $f_y = 60{,}000$ psi.

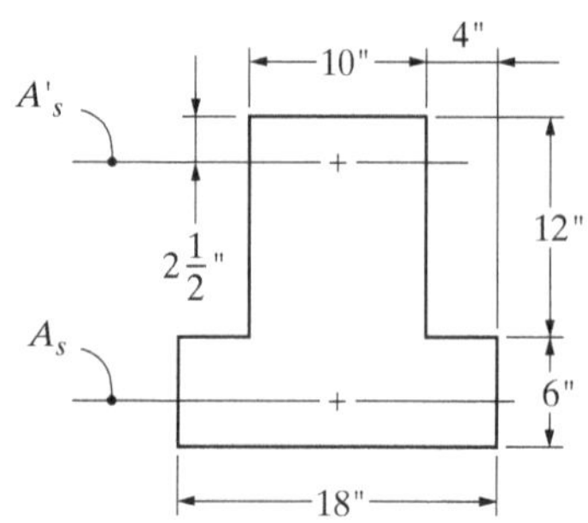

CHAPTER 4

SHEAR AND TORSION

4.1 INTRODUCTION

In prior chapters, we have been concerned primarily with the bending strength of reinforced concrete beams and slabs. The shear forces create additional tensile stresses that must be considered. In these members, steel reinforcing must be added specifically to provide additional shear strength if the shear is in excess of the shear strength of the concrete itself.

The concepts of bending stresses and shearing stresses in homogeneous elastic beams are generally discussed at great length in most strength-of-materials texts. The accepted expressions are

$$f = \frac{Mc}{I} \quad \text{and} \quad v = \frac{VQ}{Ib} \tag{4.1}$$

where f, M, c, and I are as defined in Chapter 1; v is the shear stress; V is the external shear; Q is the statical moment of area about the neutral axis, and; b is the width of the cross-section.

All points in the length of the beam, where the shear and bending moment are not equal to zero, and at locations other than the extreme fiber or neutral axis, are subject to both shearing stresses and bending stresses. The combination of these stresses is of such a nature that maximum normal and shearing stresses at a point in a beam exist on planes that are inclined with respect to the axis of the beam. It can be shown that maximum and minimum normal stresses exist on two perpendicular planes. These planes are commonly called the *principal planes*, and the stresses that act on them are called *principal stresses*. The principal stresses in a beam subjected to shear and bending may be calculated using the following formula:

$$f_{pr} = \frac{f}{2} \pm \sqrt{\frac{f^2}{4} + v^2} \tag{4.2}$$

where f_{pr} is the principal stress and f and v are the bending and shear stresses, respectively, calculated from Equation (4.1).

The *orientation* of the principal planes may be calculated using the following formula:

$$\tan 2\alpha = \frac{2v}{f} \tag{4.3}$$

where α is the angle measured from the horizontal.

The magnitudes of the shearing stresses and bending stresses vary along the length of the beam and with the distance from the neutral axis. It then follows that the inclination of the principal planes as well as the magnitude of the principal stresses will also vary. At the neutral axis, the principal stresses will occur at a 45° angle. This may be verified by Equation (4.3), substituting $f = 0$, from which $\tan 2\alpha = \infty$ and $\alpha = 45°$.

In Figures 4.1a and 4.1b, we isolate a small, square unit element from the neutral axis of a beam (where $f = 0$). The vertical shear stresses are equal and opposite on the two vertical faces by reason of equilibrium. If these were the only two stresses present, the element would rotate. Therefore, there must exist equal and opposite horizontal shear stresses on the horizontal faces and of the same magnitude as the vertical shear stresses. (The concept of horizontal shear stresses equal in magnitude to the vertical shear stresses at any point in a beam can also be found in almost any strength-of-materials text.)

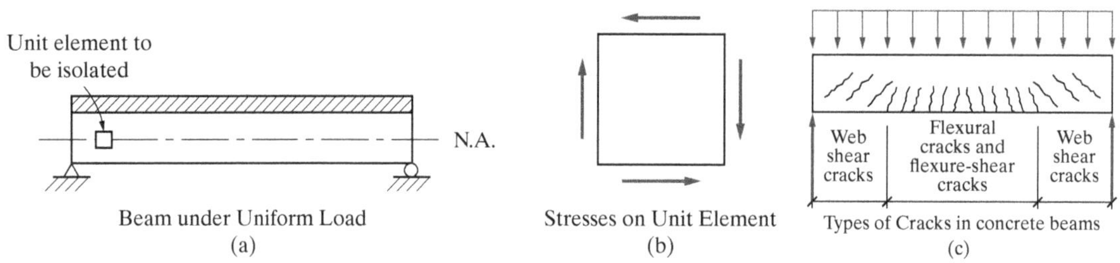

FIGURE 4.1 Shear stress relationship.

If we consider a set of orthogonal planes that are inclined at 45° with respect to the original element and resolve the shear stresses into components that are parallel and perpendicular to these planes, the effect will be as shown in Figure 4.2. Note that the components combine so that one of the inclined planes is in compression while the other is in tension. Concrete is strong in compression but weak in tension, and there is a tendency for the concrete to crack on the plane subject to tension should the stress become large enough. The tensile force resulting from the tensile stress acting on a diagonal plane has historically been designated as *diagonal tension*. When it becomes large enough, it will necessitate that shear reinforcing be provided.

As stated previously, tensile stresses of various inclinations and magnitudes, resulting from either shear alone or the combined action of shear and bending, exist in all parts of a beam and must be taken into consideration in both analysis and design. There are two types of inclined cracks in a concrete flexural member:

- Diagonal web-shear cracks which occur near the zone of maximum shear when the principal tensile stress exceeds the concrete tensile strength, and
- Flexural-shear cracks which start out as flexural cracks and then becomes an inclined crack; they occur when the tension stress due to combined shear and bending exceeds

the concrete tensile strength (ACI 318-19, Section R22.5.8.3). See Figure 4.1c. The web-shear crack originates in the web of the beam, while the flexural-shear crack starts in the flexural tension zone.

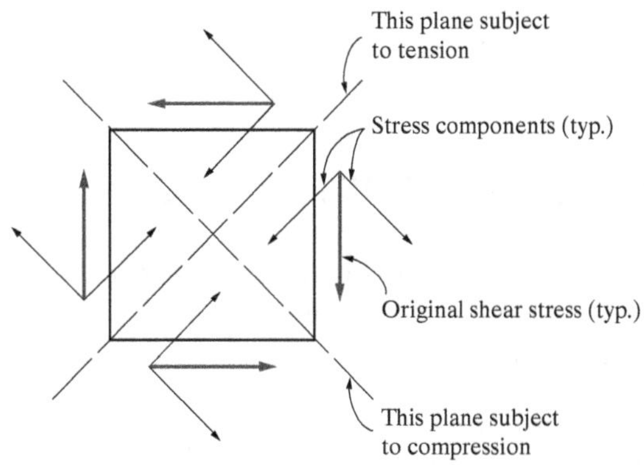

FIGURE 4.2 Effect of shear stresses on inclined planes.

The preceding discussion is a fairly accurate conceptual description of what occurs in a plain concrete beam. In the beams with which we are concerned, where the length over which a shear failure could occur (the *shear span*) is in excess of approximately three times the effective depth, the diagonal tension failure would be the mode of failure in shear. Such a failure is shown in Figure 4.3. For shorter spans, the failure mode would actually be some combination of shear, crushing, and splitting. The design of deep members such as pile caps and deep beams which have relatively short shear spans (a_v/d less than or equal to 2.0) is covered in Chapter 10. For the longer shear spans in plain concrete beams, cracks due to flexural tensile stresses would occur long before cracks due to diagonal tension. The earlier flexural cracks would initiate the failure, and shear would be of little consequence. In a concrete beam reinforced for flexure (moment) where tensile strength is furnished by steel, however, tensile stresses due to flexure and shear will continue to increase with increasing load. The steel that is placed in the beam to resist the moment is not located where the large diagonal tension stresses (due to shear) occur. The problem then becomes one of furnishing additional reinforcing steel to resist the diagonal tension stresses.

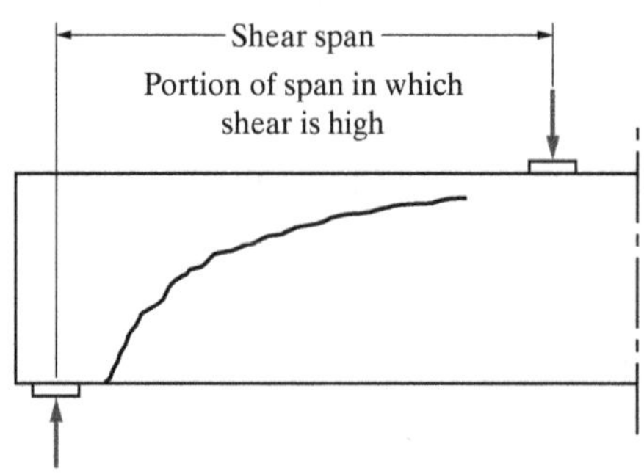

FIGURE 4.3 Typical diagonal tension failure.

Considerable research over the years has attempted to establish the exact distribution of the shear stresses over the depth of the beam cross-section. Despite extensive studies and ongoing research, the precise shear-failure mechanism is still not fully understood.

As with several previous codes, ACI 318-19 furnishes design guidelines for shear reinforcement based on the vertical shear force V_u that develops at any given cross-section of a member. Although it is really the diagonal tension for which shear reinforcing must be provided, diagonal tensile forces (or stresses) are not calculated. Historically, vertical shear force (and in older codes, vertical shear stress) has been taken to be a good indicator of the diagonal tension present.

4.2 SHEAR REINFORCEMENT DESIGN REQUIREMENTS

ACI 318-19, Chapters 9 and 22, addresses shear and torsion design provisions for both non-prestressed and prestressed concrete members. Our discussion in this chapter is limited to non-prestressed concrete members.

The design of bending members for shear is based on the assumption that the concrete resists part of the shear, and any excess over and above what the concrete is capable of resisting has to be resisted by shear reinforcement. The concrete resistance to shear is provided by the uncracked concrete in the compression zone, the aggregate interlock along the diagonal crack, the dowel action of the longitudinal reinforcement, and through direct truss action of compression struts for deep beams. The basic rationale for the design of the shear reinforcement, or *web reinforcement* as it is usually called in beams, is to provide steel to cross the diagonal tension cracks and subsequently keep them from opening. Visualizing this basic rationale with reference to Figure 4.3, it is seen that the web reinforcement may take several forms.

The code allows vertical stirrups and welded wire reinforcement with wires located perpendicular to the axis of the member as well as spirals, circular ties, or hoops.

Additionally, for non-prestressed members, the code allows shear reinforcement to be composed of inclined or diagonal stirrups and main reinforcement bent to act as inclined stirrups.

The most common form of web reinforcement used is the vertical stirrup. The web reinforcement contributes very little to the shear resistance prior to the formation of the inclined cracks but appreciably increases the ultimate shear strength of a bending member. After the concrete cracks, the shear strength provided by the concrete alone is due to the friction from the interlocking aggregates within the concrete, the dowel action of the longitudinal reinforcement crossing the shear crack, and the shear resistance of the uncracked concrete in the compression zone.

From ACI 318-19 Section 22.5.5.1, the concrete member shear strength, V_c, is a function of the depth of the member and the percentage of flexural reinforcement in the member as follows:

$$V_c = \left[2\lambda\sqrt{f'_c} + \frac{N_u}{6A_g}\right]b_w d \qquad \text{[ACI Eq. 22.5.5.1a]}$$

$$V_c = \left[8\lambda(\rho_w)^{1/3}\sqrt{f'_c} + \frac{N_u}{6A_g}\right]b_w d \qquad \text{[ACI Eq. 22.5.5.1b]}$$

$$V_c = \left[8\lambda_s\lambda(\rho_w)^{1/3}\sqrt{f'_c} + \frac{N_u}{6A_g}\right]b_w d \qquad \text{[ACI Eq. 22.5.5.1c]}$$

Where,

ρ_w = flexural reinforcing ratio = $A_s/b_w d$

N_u = axial load on the member (+ve for compression)

A_g = gross cross-sectional area of the member

b_w = beam width in inches; d = effective depth of the member in inches

λ = 1.0 for normal-weight concrete (see ACI 318 Table 19.2.4.1(a))

λ_s = size-effect factor (ACI 318-19 Section 22.5.5.1.3);

$$\lambda_s = \sqrt{\frac{2}{1+\frac{d}{10}}} \leq 1$$

The conditions and the applicable V_c equations are as follows:

Condition	Applicable V_c equation
$A_v < A_{v,min}$	Use ACI Eq. 22.5.5.1c
$A_v \geq A_{v,min}$	Use either ACI Eq. 22.5.5.1a or ACI Eq. 22.5.5.1b
Normal weight concrete with no axial load	Use modified ACI Eq. 22.5.5.1a: $V_c = 2\sqrt{f'_c}\, b_w d$
Flexural members with ρ_w > 1.56%	Use ACI Eq. 22.5.5.1b because it will give a higher V_c value than $2\sqrt{f'_c}\, b_w d$

Where,

ρ_w = flexural reinforcing ratio = $A_s/b_w d$

A_v = shear reinforcement (i.e. stirrups)

$A_{v,min}$ = ACI Code-required minimum shear reinforcement (to be defined later)

b_w = beam width, inches;

d = effective depth, inches

ACI Code, Sections 22.5.5.1.1 and 22.5.5.1.2 places limits on V_c of $5\lambda\sqrt{f'_c}\, b_w d$ and $N_u/6A_g$ of $0.05 f'_c$. This V_c limit hardly ever controls in regular beam design. Only in the case of a member with compressive axial load may the V_c exceed this limit.

The columns in moment frames, and concrete piles and drilled shafts or caissons, are subject to shear forces. For rectangular columns, the effective depth, d, can be taken as 0.8h, where h is the depth of the rectangular column; this simplified approach is "in lieu of calculating d as a function of column section geometry, reinforcement layout, and applied moment and axial load." [1]. For solid circular columns, ACI 318-19, Section 22.5.2.2 prescribes the effective depth as 0.8D, where D is the diameter of the circular column and b_w = D. In calculating the shear strength of columns, the compressive axial load, N_u, should be included.

For members of normal-weight concrete that are subject to shear and flexure only with no axial load, the amount of shear force that the concrete alone, unreinforced for shear, can resist is V_c:

$$V_c = 2\sqrt{f'_c}\, b_w d \qquad \textbf{[mod ACI Eq. (22.5.5.1a)]}$$

In the expression for V_c, the terms are as previously defined, with units for f'_c in psi and units for b_w and d in inches. Lamda (λ) is described in Chapter 1, and for normal-weight concrete, λ = 1.0. In this chapter, we will consider only normal-weight concrete. V_c will be in units of pounds. For rectangular beams, b_w is equivalent to b. The nominal shear strength of the concrete will be reduced to a dependable shear strength by applying a strength-reduction factor ϕ of 0.75 (ACI Code, Section 21.2.2). For most designs, it is convenient and conservative to use the modified ACI Equation (22.5.5.1a) above.

The design shear force is denoted V_u and results from the application of factored loads. Values of V_u are most conveniently determined using a typical shear force diagram. Theoretically, no web reinforcement should be required if $V_u \le \phi\, V_c$. The code, however, requires that a minimum area of shear reinforcement be provided in all reinforced concrete flexural members where V_u exceeds $\frac{1}{2}\phi V_c$ except for the following cases:

1. In slabs and footings.
2. In one-way concrete joist construction as defined by the ACI Code, Section 9.8.
3. In beams with a total depth of not greater than 10 in., or $2\frac{1}{2}$ times the flange thickness ($2.5t_f$), or one-half the width of the web ($0.5b_w$), whichever is greater. Research and tests of large wide beams [2] suggests that size effect, which is not modeled by ACI Eq. (22.5.5.1a), and the size of shear cracks play a major role in determining the shear capacity of these beams. The researchers found that "beams without stirrups fail in shear at lower values of shear stress as the members become deeper and as the maximum aggregate becomes smaller." Therefore, they recommend that beams with an overall depth of 30 inches or more "contain at least minimum shear reinforcement" throughout the length of the beam [2].
4. In beams built integral with the slab and with a total depth not exceeding 24 in.

This provision of the code is primarily to guard against those cases where an unforeseen overload would cause failure of the member due to shear. Tests have shown the shear failure of a flexural member to be sudden and without warning. In cases where it is not practical to provide shear reinforcement (e.g., spread footings and one-way slabs) and sufficient thickness is provided to resist V_u, the minimum area of web reinforcement is not required. In cases where shear reinforcement is required for strength or because $V_u > \frac{1}{2}\phi V_c$, the minimum area of shear reinforcement shall be calculated from ACI Code Table 9.6.3.3 as

$$A_v = 0.75\sqrt{f'_c}\,\frac{b_w s}{f_{yt}} \ge \frac{50 b_w s}{f_{yt}}$$

In the preceding equation and with reference to Figure 4.4,

A_v = total cross-sectional area of web reinforcement within a distance s; for single-loop or 2-leg stirrups, $A_v = 2A_s$, where A_s is the cross-sectional area of one leg of the stirrup bar (in.^{2}). For double-loop or 4-leg stirrups, $A_v = 4A_s$, where A_s is the cross-sectional area of one leg of the stirrup bar (in.^{2}). For triple-loop or 6-leg stirrups, $A_v = 6A_s$, where A_s is the cross-sectional area of one leg of the stirrup bar (in.^{2}). Note that closed loop stirrups are required for spandrel or rectangular beams and girders, and for beams and girders subject to torsion.

b_w = web width = b for rectangular sections (in.)

s = center-to-center spacing of shear reinforcement in a direction parallel to the longitudinal reinforcement (in.)

f_{yt} = yield strength of web reinforcement steel (psi)

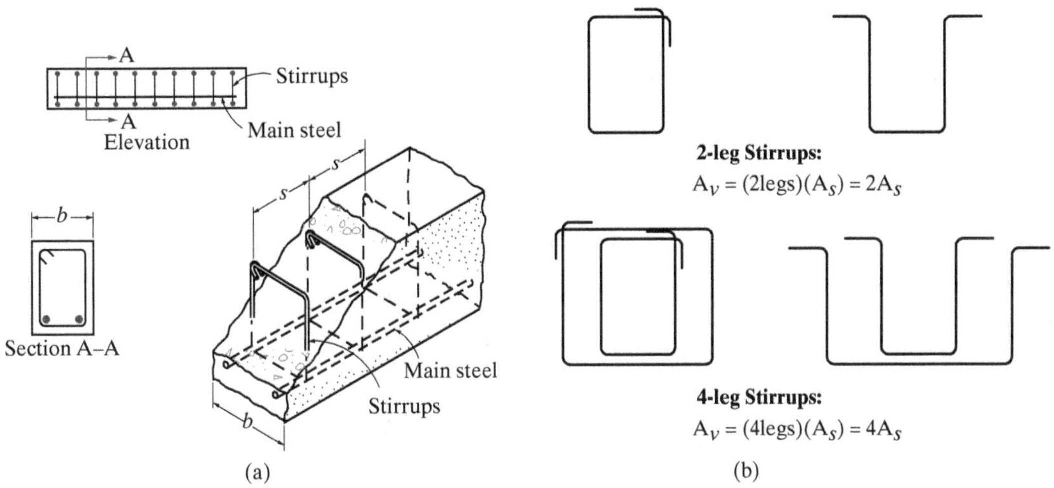

FIGURE 4.4 Isometric section showing stirrups partially exposed.

Note that for $f'_c \leq 4444$ psi, the minimum area of shear reinforcement will be controlled by

$$\frac{50 b_w s}{f_{yt}}$$

When determining the shear strength, V_c, of reinforced or prestressed concrete beams and concrete joist construction, the value of $\sqrt{f'_c}$ is limited to 100 psi unless minimum web reinforcement is provided.

In any span, the portion in which web reinforcement is theoretically necessary can be determined by using the shear (V_u) diagram. When the applied shear, V_u, exceeds the capacity of the concrete web, ϕV_c, web reinforcement is required. In addition, according to the code, web reinforcement at least equal to the minimum required must be provided elsewhere in the span, where the applied shear is greater than one-half of ϕV_c. The ACI Code, Section 9.5.1.1(b), states that the basis for shear design at the ultimate limit state must be

$$\phi V_n \geq V_u$$

where

$$V_n = V_c + V_s \qquad \textbf{[ACI Eq. (22.5.1.1)]}$$

from which

$$\phi V_c + \phi V_s \geq V_u$$

where V_u, ϕ, and V_c are as previously defined; V_n is the total nominal shear strength; and V_s is the nominal shear strength provided by shear reinforcement. In the design process, the design of the stirrups usually follows the selection of the beam size. Therefore, V_c can be determined, as can the complete shear (V_u) diagram. The stirrups to be designed will provide the shear strength, V_s. Therefore, it is convenient to write the preceding expression as

$$\text{required } \phi V_s = V_u - \phi V_c$$

For vertical stirrups, V_s may be calculated from

$$V_s = \frac{A_v f_{yt} d}{s} \qquad \textbf{[ACI Eq. (22.5.8.5.3)]}$$

where all terms are as previously defined. Also, for inclined stirrups at 45°, V_s may be calculated using

$$V_s = \frac{1.414 A_v f_{yt} d}{s}$$

from ACI Equation (22.5.8.5.4), where s is the horizontal center-to-center distance of stirrups parallel to the main longitudinal steel. Note that for inclined stirrups, the orientation of the stirrups must be such that the stirrups interrupt or cross the diagonal tension cracks.

It will be more practical if ACI Equations (22.5.8.5.3) and (22.5.8.5.4) are rearranged as expressions for spacing because the stirrup bar size, strength, and beam effective depth are usually predetermined. The design is then for stirrup spacing. For vertical stirrups,

$$\text{required } s = \frac{A_v f_{yt} d}{\text{required } V_s}$$

Because it is the required ϕV_s that will be conveniently determined, the preceding expression is rewritten as

$$\text{required } s = \frac{\phi A_v f_{yt} d}{\text{required } \phi V_s}$$

or

$$\text{required } s = \frac{\phi A_v f_{yt} d}{V_u - \phi V_c}$$

Similarly, for 45° inclined stirrups,

$$\text{required } s = \frac{1.414\phi A_v f_{yt} d}{V_u - \phi V_c}$$

Note that these equations give the *maximum* spacing of stirrups based on *required strength.*

4.3 SHEAR ANALYSIS PROCEDURE

The shear analysis procedure involves checking the shear strength in an existing member and verifying that the various code requirements have been satisfied. The members may be reinforced or plain concrete.

EXAMPLE 4.1

A reinforced concrete beam of rectangular cross-section shown in Figure 4.5 is reinforced for moment only (no shear reinforcement). Beam width b = 18 in., d = 10.25 in., and the reinforcing is five No. 4 bars. Calculate the maximum factored shear force V_u permitted on the member by the ACI Code. Use $f'_c = 4000$ psi and f_y = 60,000 psi.

Solution:

Because no shear reinforcement is provided, the ACI Code, Section 9.6.3.1, requires that V_u not exceed $\phi\sqrt{f'_c}\, b_w d$ or $0.5\,\phi V_c$:

$$\begin{aligned}\text{maximum } V_u &= 0.5\,\phi V_c \\ &= 0.5\,\phi(2\sqrt{f'_c}\, b_w d) \\ &= 0.5(0.75)(2)(\sqrt{4000})(18)(10.25) \\ &= 8750 \text{ lb}\end{aligned}$$

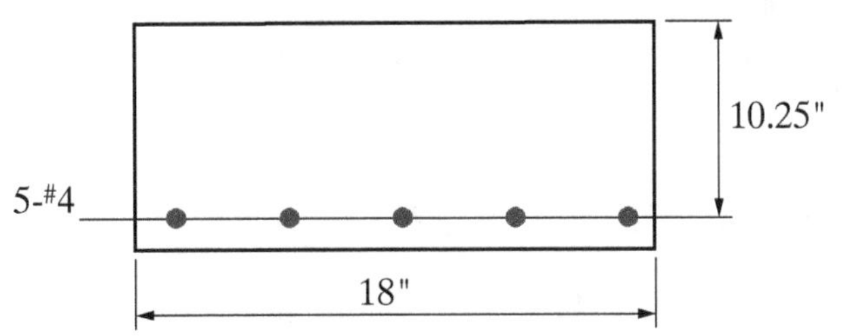

FIGURE 4.5 Cross-section, Example 4.1.

EXAMPLE 4.2

A reinforced concrete beam of the rectangular cross-section shown in Figure 4.6 is reinforced with seven No. 6 bars in a single layer. Beam width b = 18 in., d = 33 in., single-loop No. 3 stirrups are placed 12 in. on center, and typical cover is $1\frac{1}{2}$ in. Find V_c, V_s, and the maximum factored shear force V_u permitted on this member. Use $f'_c = 4000$ psi and f_y = 60,000 psi.

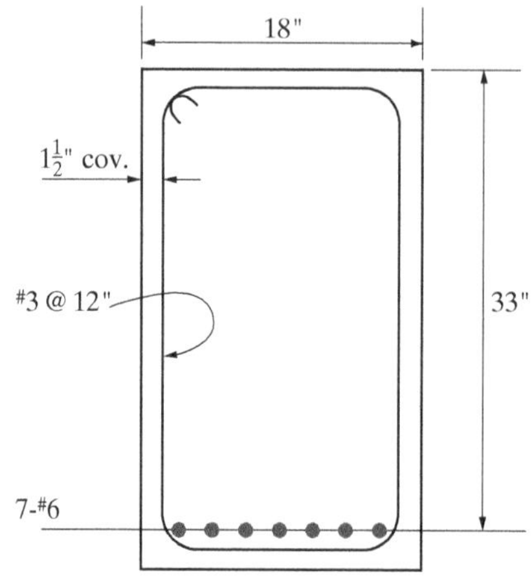

FIGURE 4.6 Cross-section, Example 4.2.

Solution:

V_c and V_s will be expressed in units of kip.

$$V_c = 2\sqrt{f'_c}b_w d = \frac{2\sqrt{4000}(18)(33)}{1000} = 75.1 \text{ kip}$$

$$V_s = \frac{A_v f_{yt} d}{s} = \frac{2(0.11)(60)(33)}{12} = 36.3 \text{ kip}$$

$$\text{maximum } V_u = \phi V_n = \phi V_c + \phi V_s = 0.75(75.1 + 36.3) = 83.6 \text{ kip}$$

In the general case of shear analysis, one must ensure that at all locations in the member, $\phi V_n = \phi V_c + \phi V_s \geq V_u$. In addition, all other details of the reinforcement pattern must be checked to ensure that they comply with code provisions. (Refer to "Notes on Stirrup Design" in Section 4.4.)

4.4 STIRRUP DESIGN PROCEDURE

In the design of stirrups for shear reinforcement, the end result is a determination of stirrup size and spacing pattern. A *general procedure* may be adopted as follows:

1. Determine the shear values based on clear span and draw the factored shear diagram and determine the factored shear (V_u) at the critical section. The location of the critical section is dependent on whether confining action on the concrete is present at the beam supports and is determined according to ACI 318-19, Section 9.4.3.2 (see Figure 4.7). When a uniform load is applied at the top of a flexural member with the supports as shown in Figures 4.7a and 4.7b, the concrete in the member at the supports will be

subject to confinement, which increases the shear strength. If the uniform load on the beam is applied at or near the bottom of the beam rather than at the top of the beam, as shown in Figures 4.7c and 4.7d, the location of the critical section will occur at the face of the support.

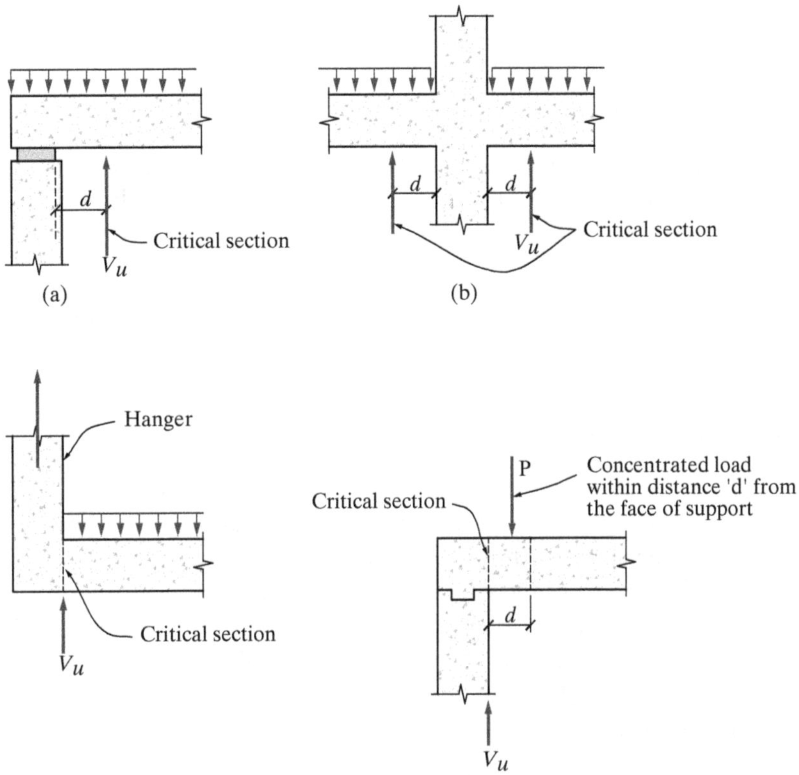

FIGURE 4.7 **Location of critical section for shear.**

2. Determine if stirrups are required.
3. Determine the length of span over which stirrups are required (assuming that stirrups *are* required). Note that no beam or girder should be without stirrups for its entire length. Even if the design calculations indicate that a beam or girder does not require stirrups for its entire length, the minimum area of stirrups with the required maximum spacing must still be provided in the beam to avoid shear failure, which would be brittle and sudden if no stirrup is present to interrupt a diagonal tension crack.
4. On the V_u diagram, determine the area representing "required ϕV_s." This will display the required strength of the stirrups to be provided.
5. Select the size of the stirrup. See item 2a in "Notes on Stirrup Design." Find the spacing required at the critical section (a distance *d* from the face of the support). See "Notes on Stirrup Design" item 3b.
6. Establish the ACI Code maximum spacing requirements.

7. Determine the spacing requirements based on shear strength to be furnished by web reinforcing.
8. Establish the spacing pattern and show sketches or indicate the stirrups required in a beam schedule.

Notes on Stirrup Design

1. Materials and maximum stresses
   a. To reduce excessive crack widths in beam webs subject to diagonal tension, the ACI Code, Table 20.2.2.4a, limits the design yield strength of deformed bar shear reinforcement to 60,000 psi. This increases to 80,000 psi for deformed welded wire reinforcing.
   b. The value of V_s must not exceed $8\sqrt{f'_c}b_w d$ irrespective of the amount of web reinforcement (ACI Code, Section 22.5.1.2) to ensure ductile failure by yielding of the stirrups.
2. Bar sizes for stirrups
   a. The most common stirrup size used is a No. 3 bar. Under span and loading conditions where the shear values are relatively large, it may be necessary to use a No. 4 bar. Rarely is anything larger than a No. 4 bar stirrup ever required, however. In large beams, multiple stirrup sets are sometimes provided in which a diagonal crack would be crossed by four or more vertical bars at one location of a beam. Single-loop stirrups, as shown in Figure 4.4, are generally satisfactory for $b \leq 24$ in.; double-loop stirrups are satisfactory for 24 in. $< b \leq 48$ in.; and triple-loop stirrups are satisfactory for $b > 48$ in.
   b. When conventional single-loop stirrups are used, the web area A_v provided by each stirrup is twice the cross-sectional area of the bar (No. 3 bars, $A_v = 0.22$ in.^{2}; No. 4 bars, $A_v = 0.40$ in.^{2}) because each single-loop stirrup crosses a diagonal crack twice.
   c. If possible, do not vary the stirrup bar sizes; use the same bar sizes unless all other alternatives are not reasonable. Spacing should generally be varied and size held constant.
3. Stirrup spacings *along* the length of the beam or girder
   a. When stirrups are required, the maximum spacing of vertical stirrups along the length of the beam or girder, s_{max}, must not exceed $d/2$ or 24 in., whichever is smaller (ACI Code, Section 9.7.6.2.2). If V_s exceeds $4\sqrt{f'_c}b_w d$, the maximum spacing of the stirrups must not exceed $d/4$ or 12 in., whichever is smaller (ACI Code, Section 9.7.6.2.2). The maximum spacing may also be governed by the relationship for the minimum area of shear reinforcement given in ACI Code Table 9.6.3.3, which gives

$$s_{\max} = \frac{A_v f_{yt}}{0.75\sqrt{f'_c}b_w} \leq \frac{A_v f_{yt}}{50 b_w}$$

The maximum stirrup spacing is to ensure that a diagonal tension crack will be interrupted by at least one stirrup.

b. It is usually undesirable to space vertical stirrups closer than 4 inches. This is to provide adequate space between adjacent stirrups for the aggregates and to enable adequate vibration of the concrete between the stirrups.

c. It is generally economical and practical to compute the spacing required at several sections and to place stirrups accordingly in groups of varying spacing. Spacing values should be made to not less than 1-in. increments.

d. The ACI code (Section 9.4.3.2) permits the maximum shear to be considered to be at the section a distance d from the face of the support (except for brackets, short cantilevers, and special isolated conditions), when the support reaction introduces a vertical compression into the end region of a member, and no concentrated load occurs between the face of the support and distance d from the face of the support, and the beam is loaded at or near the top. For stirrup design, the section located a distance d from the face of the support will be called the *critical section*. Sections located less than a distance d from the face of the support may be designed for the same V_u as that at the critical section. Therefore, stirrup spacing should be constant from the critical section back to the face of the support based on the spacing requirements *at* the critical section. The first stirrup should be placed at a maximum distance of $s/2$ from the face of the support, where s equals the immediately adjacent required stirrup spacing (a distance of 2 inches is commonly used). For the balance of the span, the stirrup spacing is a function of the shear strength required to be provided by the stirrups or the maximum spacing limitations.

e. The actual stirrup pattern used in the beam is the designer's choice. The choice will be governed by strength requirements and economy. Many patterns will satisfy the strength requirements. In most cases, the shear decreases from the support to the center of the span, indicating that the stirrup spacing could be continually increased from the critical section up to the maximum spacing allowed by the code. This would create tedious design, detailing, and bar placing operations but would nevertheless result in the least amount of steel being used. This is not warranted economically or within the framework of the philosophy outlined in Appendix B. In the usual uniformly loaded beams, no more than two or three different spacings should be used within a span. Longer spans or concentrated loads may warrant more detailed spacing patterns.

4. Stirrup spacings *across* the width of the beam or girder (i.e., transverse spacing of the stirrup legs)

The maximum transverse spacing of the stirrup legs across the width of the beam, $s_{t,max}$, is d or 24 inches, whichever is smaller. If V_s exceeds $4\sqrt{f'_c}b_wd$, the maximum transverse spacing of the stirrups must not exceed $d/2$ or 12 in. (see ACI Code, Table 9.7.6.2.2).

EXAMPLE 4.3

A simply supported rectangular concrete beam shown in Figure 4.8 is 16 in. wide and has an effective depth of 25 in. The beam supports a total factored load (w_u) of 11.5 kip/ft on a clear span of 20 ft. The given load includes the weight of the beam. Design the shear reinforcement. Use $f_c' = 4000$ psi and f_y = 60,000 psi.

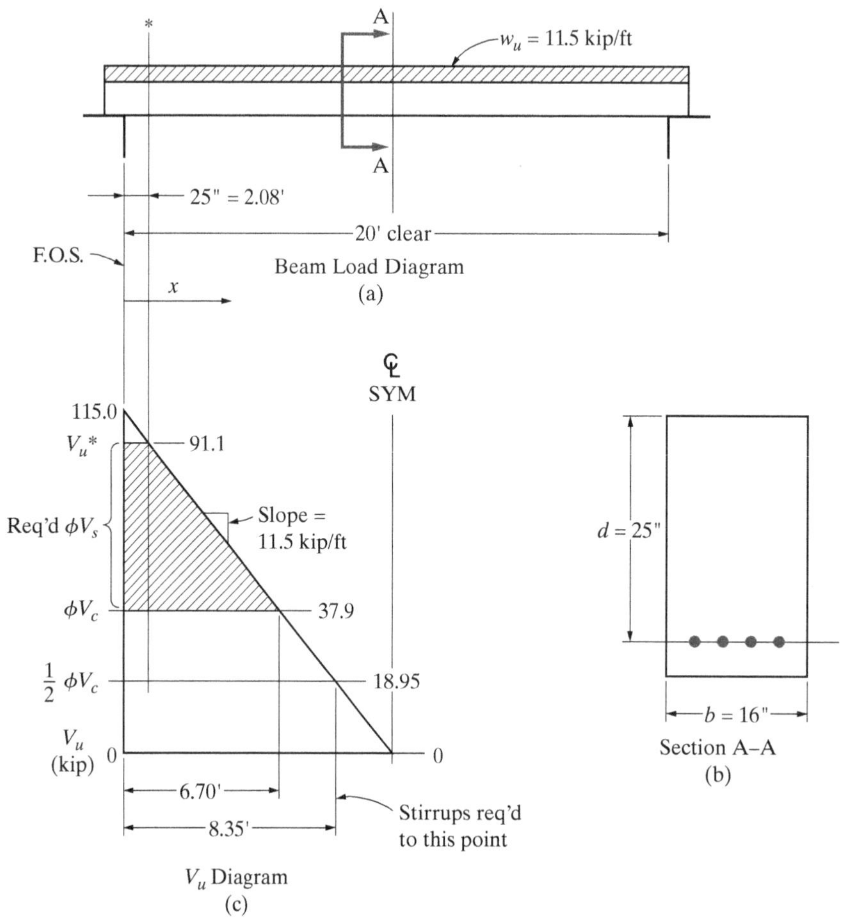

FIGURE 4.8 Sketch for Example 4.3.

Solution:

1. Draw the shear force (V_u) diagram:

$$\text{maximum } V_u = \frac{w_u \ell}{2} = \frac{11.5(20)}{2} = 115.0 \text{ kip}$$

The quantities at the critical section (25 in. or 2.08 ft from the face of the support) are designated with an asterisk. Therefore,

$$V_u^* = 115.0 - 2.08(11.5) = 91.1 \text{ kip}$$

2. Determine if stirrups are required. The ACI Code, Section 9.6.3.1, requires that stirrups be provided if $V_u > 0.5\phi V_c$. Thus,

$$\phi V_c = \phi(2\sqrt{f'_c}b_w d) = 0.75\left[\frac{2\sqrt{4000}(16)(25)}{1000}\right] = 37.9 \text{ kip}$$

$$0.5\phi V_c = 0.5(37.9) = 18.95 \text{ kip}$$

Therefore, because 91.1 kip > 18.95 kip, stirrups are required.

3. Find the length of span over which stirrups are required. As stirrups must be provided to the point where $V_u = 0.5\phi V_c = 18.95$ kip, find where this shear exists on the V_u diagram of Figure 4.8c. From the face of the support,

$$\frac{115.0 - 18.95}{11.5} = 8.35 \text{ ft}$$

Note this location on the V_u diagram as well as the location at 6.70 ft from the face of the support, where $V_u = \phi V_c = 37.9$ kip.

4. Designate as "required ϕV_s" the area enclosed by the ϕV_c line, the V_u^* line, and the sloping V_u line. This shows the required strength of the shear reinforcing at any point along the span and graphically depicts the relationship

$$\phi V_c + \phi V_s \geq V_u$$

At any location, the required ϕV_s can be determined from the diagram as the distance between the V_u^* (or V_u) line and the ϕV_c line (the height of the crosshatched area). For this particular V_u diagram, designating the slope (kip/ft) as m, taking x (ft) from the face of the support, and considering the range 2.08 ft $\leq x \leq$ 6.70 ft, we have

$$\begin{aligned} \text{required } \phi V_s &= \text{maximum } V_u - \phi V_c - mx \\ &= 115.0 - 37.9 - 11.5x \\ &= 77.1 - 11.5x \end{aligned}$$

5. Assume a No. 3 vertical stirrup ($A_v = 0.22$ in.^{2}) and establish the spacing requirement at the critical section based on the required ϕV_s. At this location, the stirrups will be most closely spaced. Substituting ACI Equation (22.5.8.5.3) into ACI Equation (22.5.8.1), and multiplying both the numerator and denominator by the strength-reduction factor, ϕ, we obtain the required stirrup spacing as

$$\text{required } s^* = \frac{\phi A_v f_{yt} d}{\text{required } \phi V_s^*} = \frac{0.75(0.22)(60)(25)}{91.1 - 37.9} = 4.65 \text{ in.}$$

(Note that the denominator in the preceding expression, required ϕV_s^*, is equal to $V_u^* - \phi V_c$).

We will use a 4-in. spacing. This is the spacing used in the portion of the beam between the face of the support and the critical section, which lies a distance d from the face of the support, and it is based on the amount of shear strength that must be provided by the shear reinforcing. Had the required spacing in this case turned out to be less than 4 in. (see item 3b in "Notes on Stirrup Design" in Section 4.4), a larger bar (i.e., #4 bar) would have been selected for the stirrups.

6. Establish ACI Code maximum spacing requirements. Recall that if V_s is less than $4\sqrt{f'_c}b_wd$, the maximum spacing is $d/2$ or 24 in., whichever is smaller. Therefore, compare V_s^* at the critical section with $4\sqrt{f'_c}b_wd$:

$$4\sqrt{f'_c}b_wd = \frac{4\sqrt{4000}(16)(25)}{1000} = 101.2 \text{ kip}$$

$$V_s^* = \frac{\phi V_s^*}{\phi} = \frac{91.1 - 37.9}{0.75} = 70.9 \text{ kip}$$

Because 70.9 kip < 101.2 kip, the maximum spacing should be the smaller of $d/2$ or 24 in.:

$$\frac{d}{2} = \frac{25}{2} = 12.5 \text{ in.}$$

Use 12 in.

A second criterion is based on the code minimum area requirement (ACI Code, Section 9.6.3.3). The equation for minimum shear reinforcement from ACI Code Table 9.6.3.3 may be rewritten in the form

$$s_{\max} = \frac{A_v f_{yt}}{0.75\sqrt{f'_c}b_w} \le \frac{A_v f_{yt}}{50b_w}$$

where the units of f_{yt} should be carefully noted as psi. Evaluate these two expressions:

$$\frac{A_v f_{yt}}{0.75\sqrt{f'_c}b_w} = \frac{0.22(60{,}000)}{0.75\sqrt{4000}(16)} = 17.39 \text{ in.}$$

$$\frac{A_v f_{yt}}{50b_w} = \frac{0.22(60{,}000)}{50(16)} = 16.50 \text{ in.}$$

Therefore, of these two, 16.50 in. controls.

Of the foregoing two maximum spacing criteria, the smaller value will control. Therefore, the 12-in. maximum spacing (determined previously above) controls throughout the length of the beam wherever stirrups are required.

7. Determine the spacing requirements based on shear strength to be furnished. At this point, we know that the spacing required at the critical section is 4.65 in., that the maximum spacing allowed in this beam is 12 in. where stirrups are required, and that stirrups are required to 8.35 ft from the face of the support (F.O.S.)

To establish a spacing pattern for the rest of the beam, the spacing required should be established at various distances from the face of support. This will permit the placing of stirrups in groups, with each group having a different spacing. The number of locations at which the required spacing should be determined is based on judgment and should be a function of the shape of the required ϕV_s portion of the V_u diagram.

To aid in the determination of an acceptable spacing pattern, a plot is developed using the formula

$$\text{required } s = \frac{\phi A_v f_{yt} d}{\text{required } \phi V_s}$$

where the denominator can be determined from the expression given in step 4.

For plotting purposes, the required spacing will arbitrarily be found at 1-ft intervals beyond the critical section. At 3 ft from the face of the support,

$$\text{required s} = \frac{\phi A_v f_{yt} d}{\text{required } \phi V_s} = \frac{0.75(0.22)(60)(25)}{77.1 - 11.5(3)} = 5.81 \text{ in.}$$

Similarly, required spacing may be found at other points along the beam. The results of these calculations are tabulated and plotted in Figure 4.9. Note that required spacings need not be determined beyond the point where s_{max} (12 in.) has been exceeded.

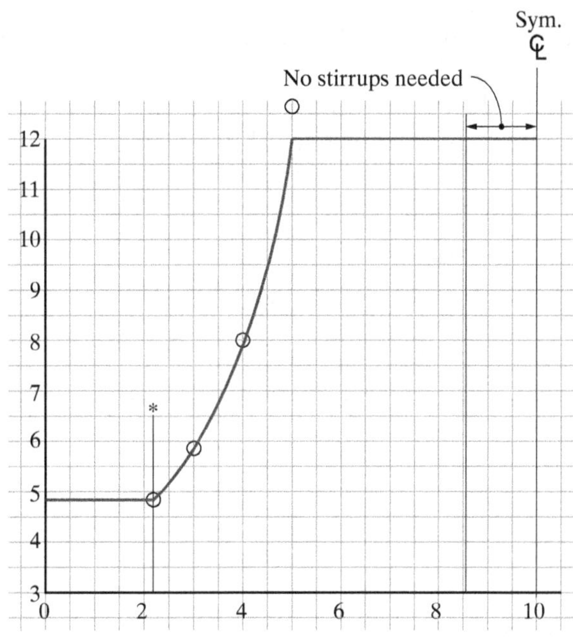

x (ft)	Req'd s (in.)
3	5.81
4	7.96
5	12.63

FIGURE 4.9 Stirrup spacing requirements for Example 4.3.

The maximum transverse spacing of the stirrup legs, $s_{t,max}$, is 24 inches (i.e., the smaller of d or 24"). The actual transverse spacing of the stirrup legs across the width of the beam is

$s_t = b - 2(1.5\text{" cover}) - ½\,(3/8\text{" stirrups}) - ½\,(3/8\text{" stirrups})$

$= 16 - 3 - 0.75 = 12.25 \text{ in.} < 24 \text{ in.}$ OK.

8. The plot of Figure 4.9 may be readily used to aid in establishing a final pattern. For example, the 5-in. spacing could be started about 2.4 ft from the face of the support (F.O.S.), and the maximum spacing of 12 in. could be started at about 4.9 ft from the face of the support. The first stirrup will be placed away from the face of the support a distance equal to one-half the required spacing at the critical section. The rest of the stirrup pattern may be selected using an approach as shown in the following table.

Spacing(in.)	Theoretical stopping point (from F.O.S.)	Length required to cover (in.)	Number of spaces to use	Actual beam length covered (in.)	Actual stopping point (inches from F.O.S.)
2	—	—	1	2	2
4	2.4′ = 29″	27	7	28	30
5	4.9′ = 59″	29	6	30	60
12	120″	60	5	60	120

The spacings and theoretical stopping points for those spacings are determined from Figure 4.9. The 4-in. spacing must run from the end of the 2-in. spacing (i.e., 2 in. from the face of the support) to the theoretical stopping for the 4-in. spacing (29 in.). Therefore, the 4-in. spacings must cover 27 in. For the 4-in. spacing, the required number of spaces is then calculated from 27/4 = 6.75 spaces; use 7 spaces. This places the last stirrup of the 4-in. spacing group at 30 in. from the face of the support, where the 5-in. spacing will begin. The final pattern for No. 3 single-loop stirrups is: one at 2 in., seven at 4 in., six at 5 in., and five at 12 in. This places the last stirrup in the 12-in. spacing group on the beam centerline—thus, remembering symmetry about the centerline, providing stirrups across the full length of the beam. This is common practice and is conservative. The final stirrup pattern is shown in the design sketch of Figure 4.10.

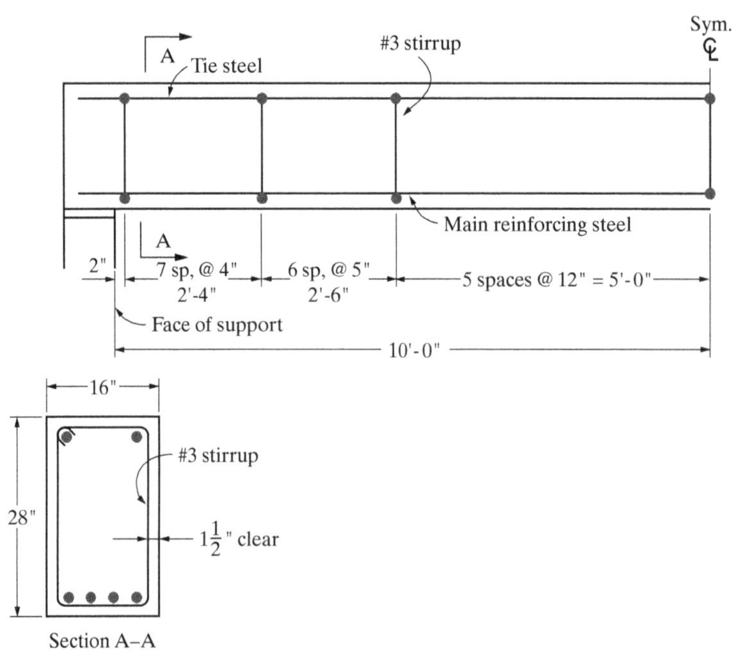

FIGURE 4.10 Design sketch for Example 4.3.

An alternative approach to stirrup longitudinal spacing is to select a desired spacing (in this case between 4 in. and 12 in.) and to compute the distance from the face of the support at which that spacing may begin. This can be accomplished by rewriting the expression

$$\text{required } s = \frac{\phi A_v f_{yt} d}{\text{required } \phi V_s} = \frac{\phi A_v f_{yt} d}{(\max V_u - \phi V_c - mx)}$$

in the form

$$x = \frac{\max V_u - \phi V_c - \dfrac{\phi A_v f_{yt} d}{s}}{m}$$

This expression will furnish the distance x at which any desired stirrup spacing s may commence. Returning to Example 4.3, we can compute where a 5-in. spacing may begin relative to the face of the support,

$$x = \frac{115.0 - 37.9 - \dfrac{0.75(0.22)(60)(25)}{5}}{11.5} = 2.40 \text{ ft}$$

and a 12-in. spacing may begin at

$$x = \frac{115.0 - 37.9 - \dfrac{0.75(0.22)(60)(25)}{12}}{11.5} = 4.91 \text{ ft}$$

Note that Figure 4.9 may be generated equally well with these data.

EXAMPLE 4.4

A continuous reinforced concrete beam shown in Figure 4.11 is 15-in. wide and has an effective depth of 31 in. The factored loads are shown. (The factored uniform load includes the weight of the beam.) Design the web or shear reinforcement using the V_u diagram shown in Figure 4.12. Use $f'_c = 4000$ psi and $f_y = 60{,}000$ psi.

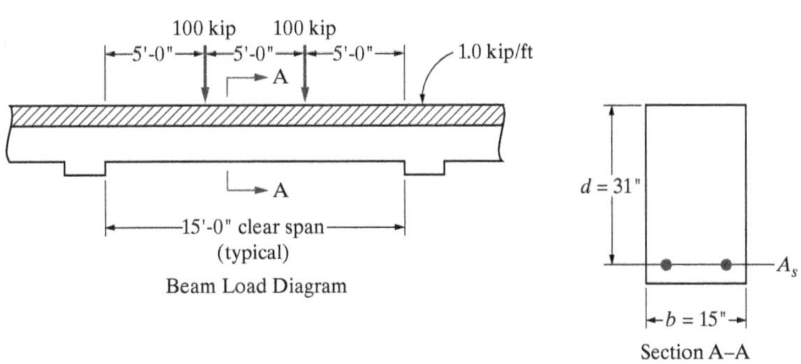

FIGURE 4.11 Sketch for Example 4.4.

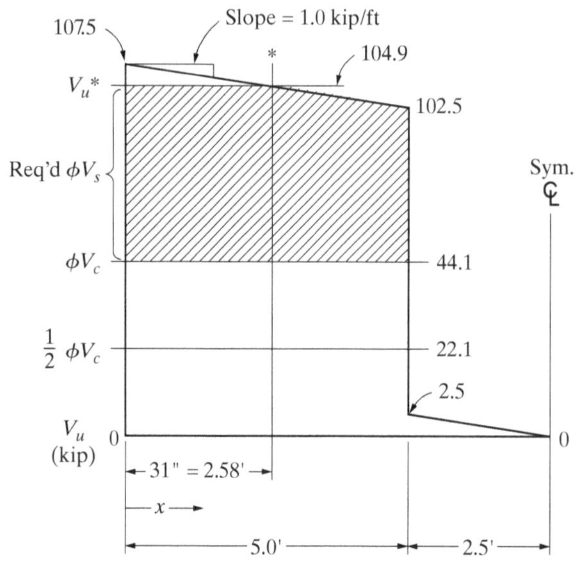

FIGURE 4.12 V_u diagram for Example 4.4.

Solution:

1. Note that, by reason of symmetry, it is necessary to show only half of the V_u diagram.
2. Determine if stirrups are required:

$$\phi V_c = \phi 2\sqrt{f'_c}b_w d = \frac{0.75\left(2\sqrt{4000}\right)(15)(31)}{1000} = 44.1 \text{ kip}$$

$$0.5\phi V_c = 0.5(44.1) = 22.1 \text{ kip}$$

Therefore, because V_u^{*} = 104.9 kip > 22.1 kip, stirrups are required.

3. Find the length along the beam span over which stirrups are required. Stirrups are required to the point where

$$V_u = 0.5\phi V_c = 22.1 \text{ kip}$$

From Figure 4.12, the point where V_u is equal to 22.1 kip may be determined by inspection to be at the concentrated load, 5 ft-0 in. from the face of the support. No stirrups are required between the two concentrated loads.

4. Designate "required ϕV_s" on the V_u diagram:

$$\begin{aligned} \text{required } \phi V_s &= \text{maximum } V_u - \phi V_c - mx \\ &= 107.5 - 44.1 - 1.0x \\ &= 63.4 - 1.0x \quad (\text{applies for the range } 2.58 \text{ ft} \le x \le 5.0 \text{ ft}) \end{aligned}$$

Note that 2.58 ft. is the distance d from the face of the support.

5. Assume a No. 3 vertical stirrup (A_v = 0.22 in.^{2}):

$$\text{required } s^* = \frac{\phi A_v f_{yt} d}{\text{required } \phi V_s^*} = \frac{0.75(0.22)(60)(31)}{104.9 - 44.1}$$
$$= 5.05 \text{ in.}$$

Use 5 in.

6. Establish ACI Code maximum spacing requirements:

$$4\sqrt{f_c'}b_w d = \frac{4\sqrt{4000}(15)(31)}{1000} = 117.6 \text{ kip}$$

$$V_s^* = \frac{\phi V_s^*}{\phi} = \frac{V_u^* - \phi V_c}{\phi} = \frac{104.9 - 44.1}{0.75} = 81.1 \text{ kip}$$

Because 81.1 kip < 117.6 kip, the maximum spacing should be the smaller of $d/2$ or 24 in.:

$$\frac{d}{2} = \frac{31}{2} = 15.5 \text{ in.}$$

Also check s_{max}. Because $f_c' < 4444$ psi,

$$s_{max} = \frac{A_v f_{yt}}{50 b_w} = \frac{0.22(60{,}000)}{50(15)} = 17.6 \text{ in.}$$

Therefore, use a maximum longitudinal spacing, s, of 15 in.

The maximum transverse spacing of the stirrup, s_t,max, is 24 inches (i.e., the smaller of d or 24"). The actual transverse spacing of the stirrup legs across the width of the beam is

$s_t = b$ – 2(1.5" cover) – ½ (3/8" stirrups) – ½ (3/8" stirrups)
= 15 – 3 – 0.375 = 11.63 in. < 24 in. OK.

7. Determine the spacing requirements between the critical section and the concentrated load based on shear strength to be furnished. The denominator of the following formula for required spacing uses the expression for required ϕV_s from step 4:

$$\text{required } s = \frac{\phi A_v f_{yt} d}{\text{required } \phi V_s} = \frac{0.75(0.22)(60)(31)}{63.4 - 1.0x}$$

The results of these calculations are tabulated and plotted in Figure 4.13.

8. With reference to Figure 4.13, no stirrups are required in the portion of the beam between the point load and the center of the beam. A spacing of 5 in. will be used between the face of the support and the point load. In the center portion of the beam, between the point loads, stirrups will be placed at a spacing slightly less than the maximum spacing as a conservative measure to create a convenient spacing pattern. The design sketches are shown in Figure 4.14. Note the symmetry about the span centerline.

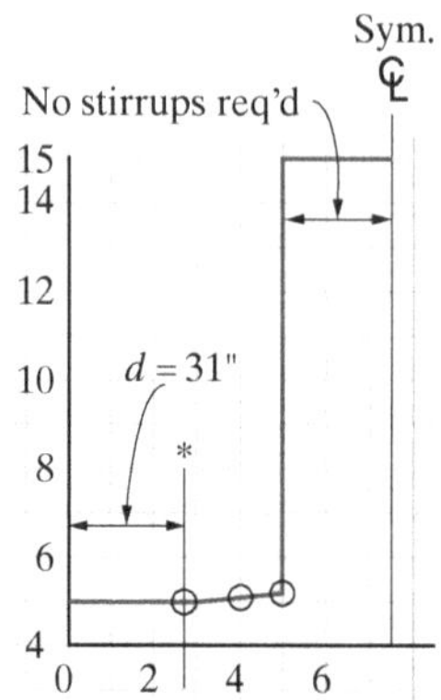

x (ft)	Req'd s (in.)
3	5.08
4	5.17
5	5.26

FIGURE 4.13 Stirrup spacing requirements for Example 4.4.

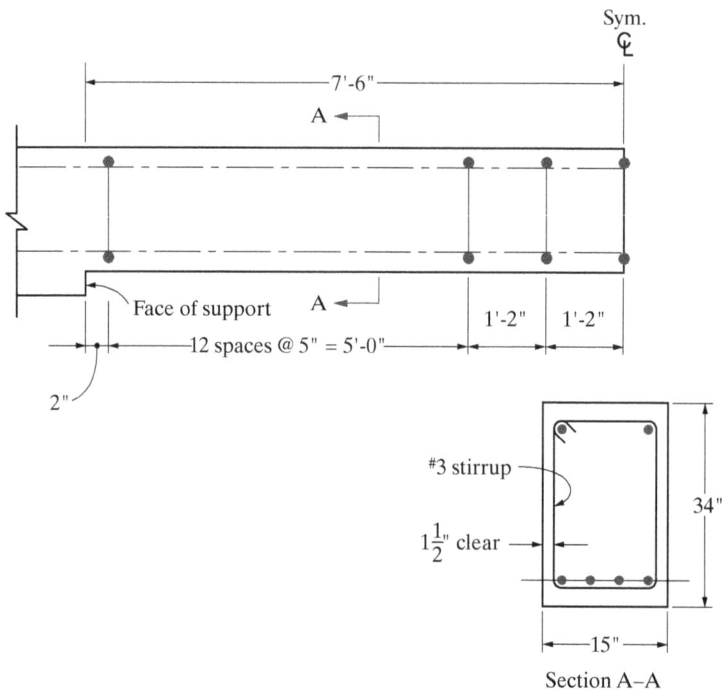

FIGURE 4.14 Design sketches for Example 4.4.

4.5 TORSION OF REINFORCED CONCRETE MEMBERS

The torsion or twisting of reinforced concrete members is caused by a torsional moment that acts about the longitudinal axis of the member due to unbalanced loads applied to the member. The torsional moment usually acts in combination with a bending moment and shear force, as shown in Figure 4.15.

A typical example of torsion in concrete members occurs in a rectangular beam supporting precast hollow-core slabs (or planks). The torsion may be due to unequal live loads on adjacent spans of the hollow-core planks or due to unequal adjacent spans of the hollow-core planks supported on the beam. Torsion in such beams could also be due to the construction sequence that has the hollow-core planks fully installed on one side of the beam before being installed on the adjacent side of the beam. Rectangular and L-beams are more susceptible to torsion than T-beams. In the ACI Code, the design for torsion in solid and hollow concrete beams is based on a thin-walled tube space truss model (see Figures 4.16 and 4.17). In the thin-walled space truss model, the outer concrete cross-section that is centered on the stirrups is assumed to resist the torsion, while the concrete in the core is neglected because, after cracking, this core is relatively ineffective in resisting torsion. Torsional moments cause additional shear stresses that result in diagonal tension stresses in the concrete member. These diagonal tension stresses cause inclined cracks that form a spiral around the surfaces of the concrete member, as shown in Figure 4.17. After cracking, the torsional resistance of a concrete member is provided by the outermost closed stirrups and the longitudinal reinforcement located near the surface of the beam, and this is modeled by the space truss shown in Figure 4.17, where the longitudinal reinforcement acts as the truss tension members, the stirrups act as the tension web members, and the inclined concrete struts between the diagonal cracks act as the compression web members of the space truss.

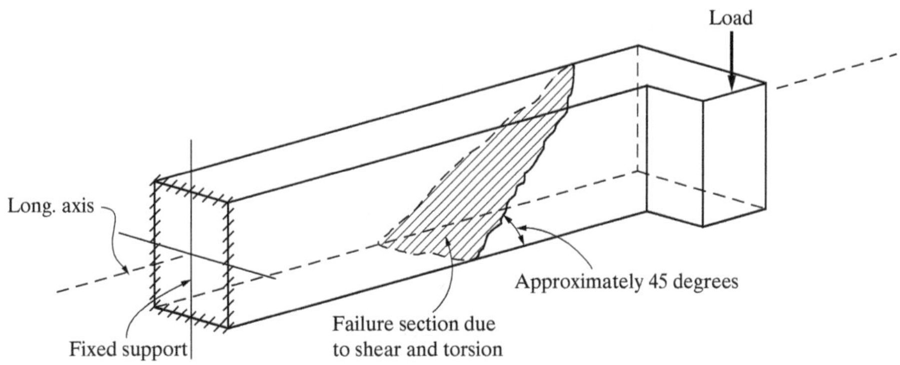

FIGURE 4.15 Cantilever beam subject to combined shear, moment, and torsion.

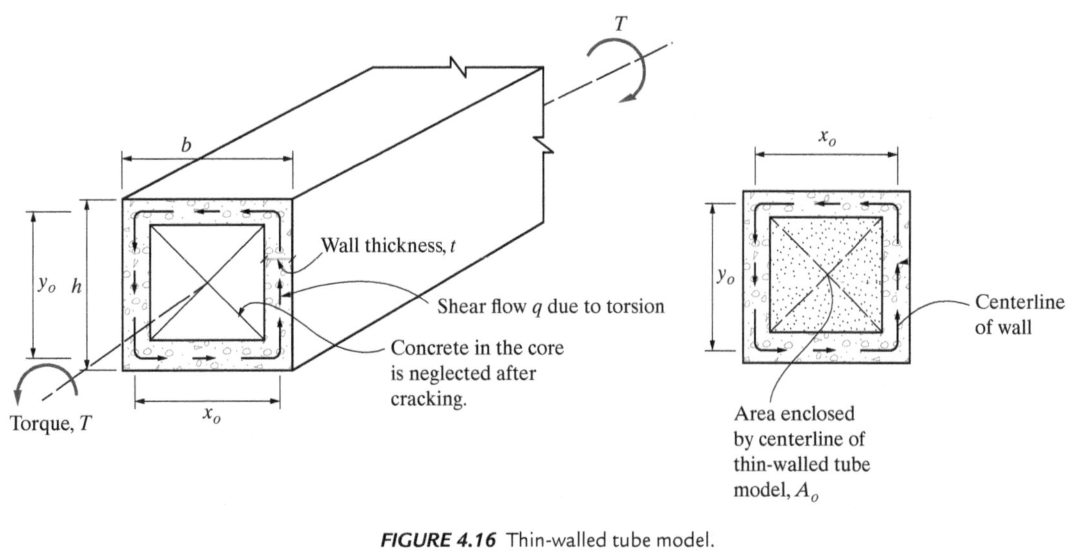

FIGURE 4.16 Thin-walled tube model.

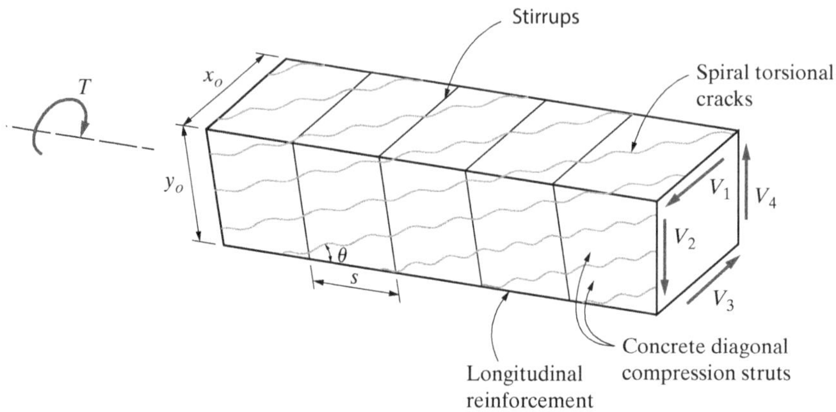

FIGURE 4.17 Space truss model for torsion.

In the thin-walled tube model, the shear flow q, which is assumed to be constant around the perimeter of the beam, is equal to the product of the shear stress τ and the wall thickness, t. Using the thin-walled tube model, and summing the torques, the equilibrium of torsional moments yields

$$T-(q\,x_o)y_o-(q\,y_o)x_o=0$$

or

$$T=2q\,x_o y_o=2q\,A_o$$

Therefore,

$$q=T/2A_o \tag{4.4}$$

where

A_o = area enclosed by centerline of the shear flow path = $x_o y_o$

q = shear flow (i.e., force per unit length)

x_o and y_o are the width and height of the space truss model measured between the centerlines of the tube walls—that is, the centerlines of the longitudinal corner bars.

There are two types of torsion that may occur in reinforced concrete members:

- Primary or equilibrium torsion, and
- Secondary or compatibility torsion.

Compatibility Torsion

Compatibility torsion occurs in statically indeterminate structures, and the design torque, which cannot be obtained from statics alone, may be reduced due to redistribution of internal forces to maintain compatibility of deformations. Members subjected to compatibility torsion may be designed for the cracking torque multiplied by the resistance factor (i.e., ϕT_{cr}), but the redistribution of internal forces due to the reduction of the torque to ϕT_{cr} must be taken into account in the design of all the adjoining structural members. One example of compatibility torsion occurs in spandrel beams (see Figure 4.18), where the rotation of the slab is restrained by the spandrel beam. For compatibility, the restraining moment at the exterior end of the slab is equal to the uniform torsional moment per unit length on the spandrel beam. As the slab rotates and cracks, and the slab moments are redistributed, the torsional moment on the spandrel beam is reduced until it reaches the cracking torque of the spandrel beam, at which point a hinge is formed at the exterior end of the slab.

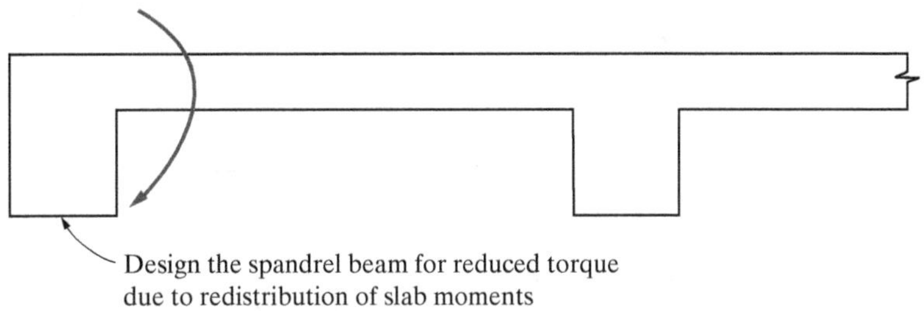

FIGURE 4.18 Compatibility torsion in spandrel beams.

Equilibrium Torsion

For statically determinate structures, the design torque, which can be obtained from statics considerations alone, cannot be reduced because redistribution of internal moments and forces is not possible in such structures, and in order to maintain equilibrium, the full design torque has to be resisted by the beam. This is called *equilibrium torsion.* Examples of concrete members in equilibrium torsion are shown in Figures 4.19 and 4.20.

FIGURE 4.19 Equilibrium torsion. (George Limbrunner)

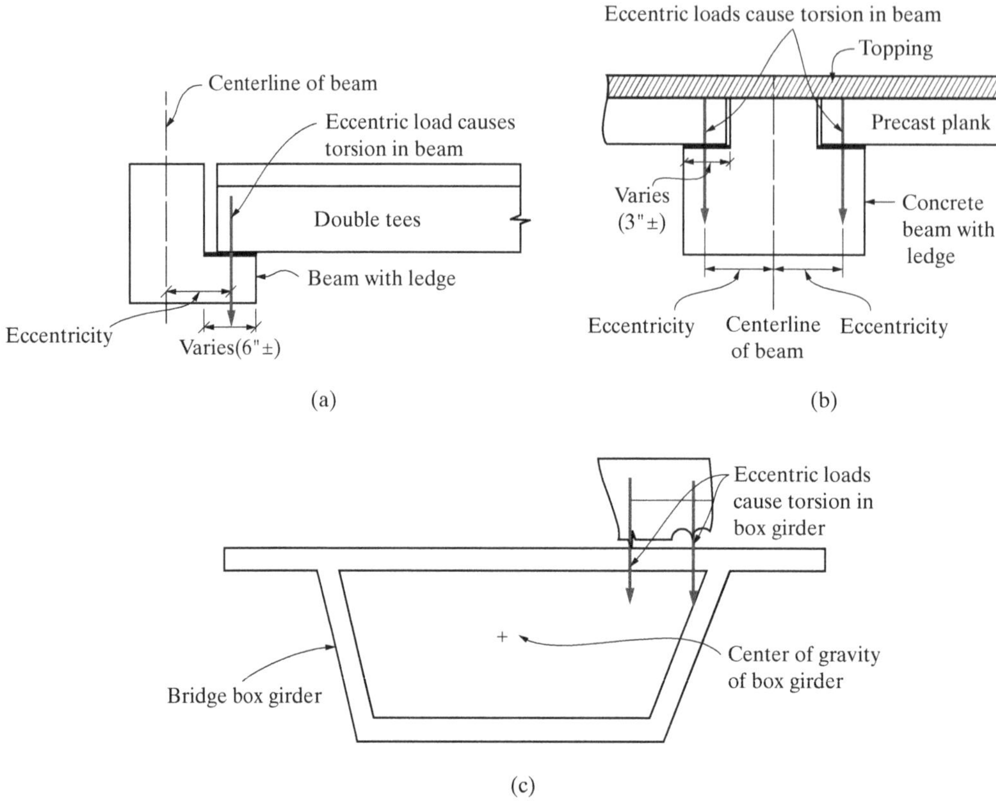

FIGURE 4.20 Equilibrium torsion.

Torsion Design of Reinforced Concrete Members

The ACI Code design approach for torsion follows a similar approach to the design for shear (ACI Code, Sections 9.2.4.4 and 22.7). Like shear, the critical section for torsion is located at a distance d from the face of a support, but where a concentrated torque occurs within a distance of d from the face of a support, the critical section for torsion shall be at the face of the support. Torsion can be neglected if the factored torque is less than or equal to one-quarter of the cracking torque of the beam section. That is, if

$$T_u \le 0.25\phi T_{cr} \tag{4.5}$$

where the cracking torque for non-prestressed members not subject to axial tension or compression force is

$$T_{cr} = 4\lambda\sqrt{f_c'}\frac{(A_{cp})^2}{p_{cp}} \tag{4.6}$$

Thus, torsion may be neglected when

$$T_u \le 0.25\phi\left(4\lambda\sqrt{f_c'}\frac{(A_{cp})^2}{p_{cp}}\right) = \lambda\phi\sqrt{f_c'}\,\frac{(A_{cp})^2}{p_{cp}} \tag{4.7}$$

where

$\phi = 0.75$ (ACI Code, Section 21.2.1)

A_{cp} = area of outside perimeter of the cross-section = bh (for rectangular beams not cast monolithic with a slab)

p_{cp} = outside perimeter of the cross-section = $2(b + h)$

b and h = cross-section width and depth (see Figure 4.16)

The term, $\lambda\sqrt{f_c'}\,\dfrac{(A_{cp})^2}{p_{cp}}$, called the threshold torsion, T_{th}, in ACI Code, Table 22.7.4.1(a) is 25% of the cracking torsion, T_{cr}.

For beams cast monolithic with a slab, the area of the outside perimeter of the section in torsion, A_{cp}, can be determined from ACI Section R9.2.4.4.

λ is the lightweight aggregate factor discussed in Chapter 1, and for normal weight concrete, $\lambda = 1.0$. When the torque is small enough such that Equation (4.5) or (4.7) is satisfied, closed stirrups are not required.

Torsional reinforcement is required to resist the full applied torsional moment as specified in ACI Code, Section 22.7.3.1, when

$$T_u > \phi T_{cr} = \lambda\phi\sqrt{f_c'}\,\frac{(A_{cp})^2}{p_{cp}} \qquad \textbf{(4.8)}$$

To reduce unsightly cracks on the surface of the beam and to prevent crushing of the surface concrete from stresses in the inclined concrete compression struts (see Figure 4.17) due to combined shear and torsion, the ACI Code requires that *solid and hollow concrete beam* cross-sectional dimensions be such that ACI Code Equations (22.7.7.1(a)) and (22.7.7.1(b)), respectively, are satisfied. The size of the beam should be increased if these relationships are not satisfied.

Solid section: $$\sqrt{\left(\frac{V_u}{b_w d}\right)^2 + \left(\frac{T_u p_h}{1.7(A_{oh})^2}\right)^2} \le \phi\left(\frac{V_c}{b_w d} + 8\sqrt{f_c'}\right) \qquad \textbf{[ACI Eq. } (22.7.7.1(a))\textbf{]}$$

Similarly, the cross-sectional dimensions of *hollow concrete beams* are limited as follows:

Hollow section: $$\frac{V_u}{b_w d} + \frac{T_u p_h}{1.7(A_{oh})^2} \le \phi\left(\frac{V_c}{b_w d} + 8\sqrt{f_c'}\right) \qquad \textbf{[ACI Eq. } (22.7.7.1(b))$$

where

$V_c = 2\lambda\sqrt{f_c'}\,b_w d$

A_{oh} = area enclosed by the centerline of the outermost closed stirrup = $x_1 y_1$

$p_h = 2(x_1 + y_1)$

x_1 and y_1 are the width and height of the space truss model measured to the centerline of the outermost closed stirrup, as shown in Figure 4.22a. Note that for hollow sections, the shear stress due to direct shear force is directly additive to the shear stress due to torsion as indicated in ACI Eq. 22.7.7.1(b) because, for hollow sections, the shear stresses due to direct shear and torsion are concentrated in the two webs of the hollow section (see Figures 4.16, 4.17, and 4.21a). In contrast, for solid sections, the shear stress due to direct shear force is distributed across the full width of the cross-section, while the shear stress due to torsion is concentrated on the outermost fibers of the cross-section, with very low torsional shear stress near the center of the solid cross-section (see Figure 4.21b). Consequently, the shear stress components from direct shear force and torsion are not directly additive for solid sections; hence, for combined torsion and shear, ACI Eq. 22.7.7.1(a) uses the square root of the sum of the squares expression (ACI Code, Section R22.7.7.1). It should be noted that the nominal concrete shear strength, V_c, is assumed to be unaffected by torsion.

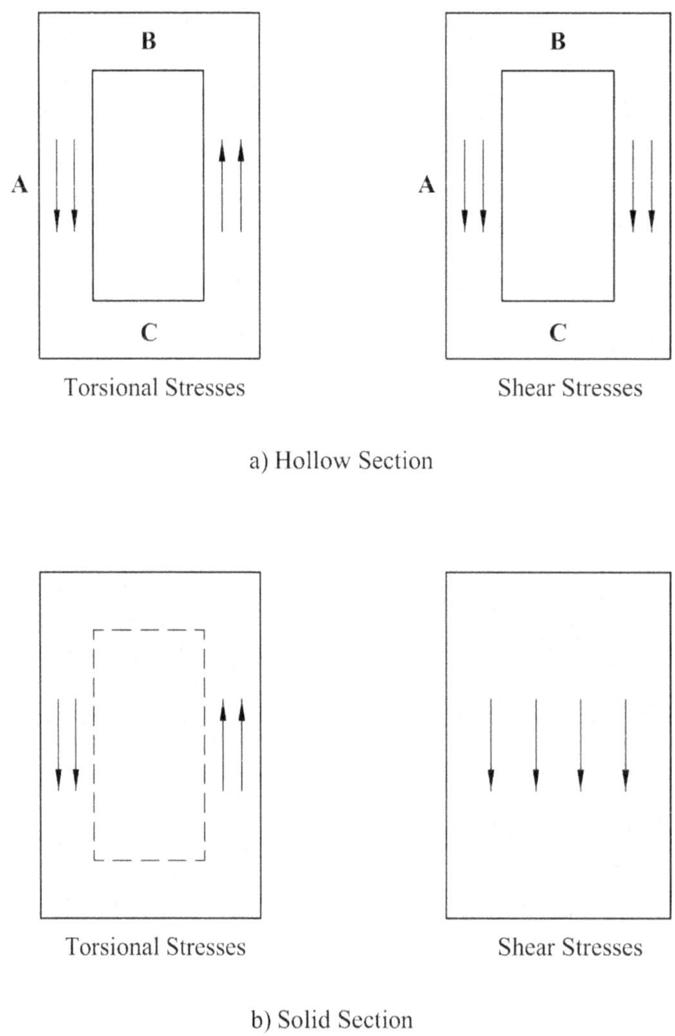

FIGURE 4.21 Torsional and shear stresses in solid and hollow sections (Adapted from ACI 318-19, Figure R22.7.7.1)

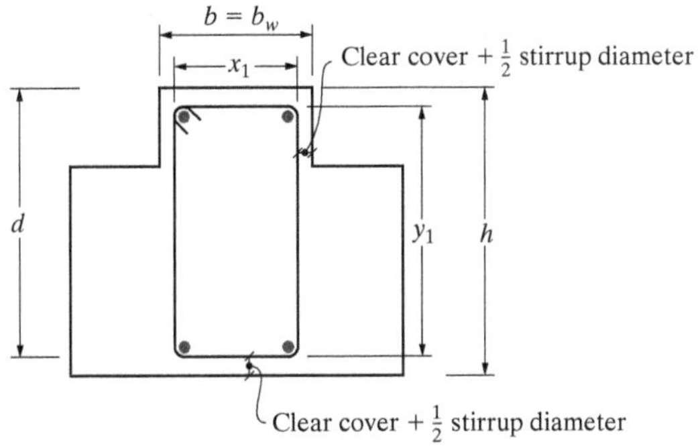

FIGURE 4.22a Definition of x_1 and y_1.

For hollow sections with varying wall thickness, ACI Equation (22.7.7.1(b)) shall be evaluated at a location where the left-hand side of the equation is at its maximum value. If the wall thickness of the hollow concrete beam, t, is less than A_{oh}/p_h at the location where the torsional stresses are being determined, the second term in ACI Equation (22.7.7.1(b)) shall be taken as $T_u/(1.7A_{oh}\, t)$.

Torsion Reinforcement

The reinforcement required for torsion shall be added to that required for other load effects that act in combination with the torsional moment, and the most restrictive spacing requirements must be satisfied (ACI Code, Section 9.5.4.3). The torsional reinforcement is determined using the space truss model shown in Figure 4.17. The yield strength of the torsional reinforcement shall not be greater than 60,000 psi per ACI Code Section 20.2.2.4. The torsional reinforcement consists of closed stirrups and longitudinal reinforcement at the corners of the beam.

Vertical Equilibrium of Forces

Considering a free body diagram of the vertical forces acting on the front wall of the space truss model (Figure 4.17), as shown in Figure 4.22b, the equilibrium of the vertical forces yields

$$V_2 = A_t f_{yt}\left(\frac{y_o \cot\theta}{s}\right) \tag{4.9}$$

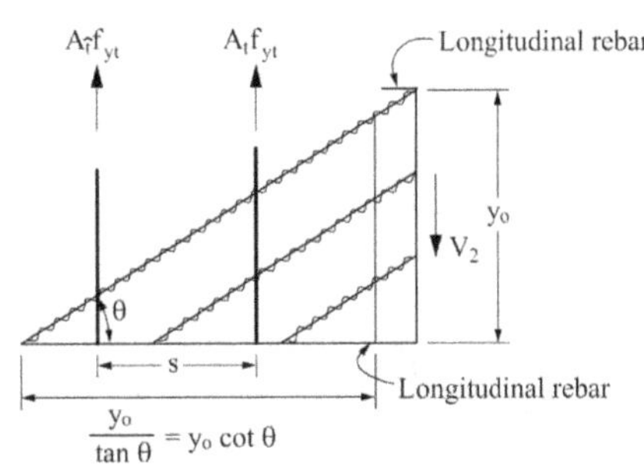

FIGURE 4.22b Free body diagram for vertical equilibrium of forces in the space truss model.

From Equation (4.4) and noting that the shear flow is constant, V_2 can be calculated as

$$V_2 = q\,y_o = \left(\frac{T}{2A_o}\right) y_o \tag{4.10}$$

Substituting Equation (4.10) into Equation (4.9) and rearranging, the ratio of the area of the torsional stirrup to the spacing of the stirrup is

$$\frac{A_t}{s} = \frac{T}{2A_o f_{yt} \cot\theta} \tag{4.11}$$

Horizontal Equilibrium of Forces

Considering a free-body diagram of the horizontal forces acting on the front wall of the space truss model (Figure 4.17), as shown in Figure 4.23, the equilibrium of the horizontal forces yields

$$A_\ell f_{y\ell} = \Sigma V_i \cot\theta = \Sigma (q y_i) \cot\theta = q \cot\theta\, \Sigma y_i \tag{4.12}$$

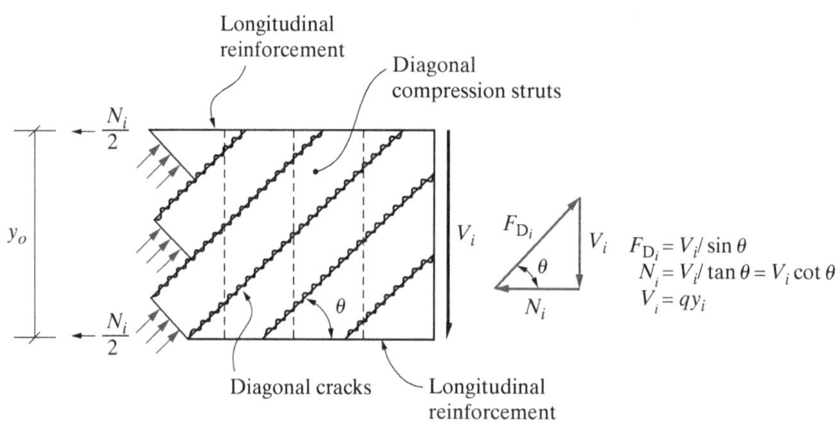

FIGURE 4.23 Free body diagram for horizontal equilibrium of forces in the space truss model.

Substituting Equation (4.4) into Equation (4.12) yields

$$\begin{aligned} A_\ell f_{y\ell} &= \left(\frac{T}{2A_o}\right) \cot\theta\, \Sigma y_i \\ &= \left(\frac{T}{2A_o}\right) \cot\theta\, (x_o + x_o + y_o + y_o) \end{aligned}$$

Therefore,

$$A_\ell f_{y\ell} = \left(\frac{T}{2A_o}\right) \cot\theta\, [2(x_o + y_o)] \tag{4.13}$$

Substituting for the torque, T, in Equation (4.13) using Equation (4.11) gives

$$A_\ell f_{y\ell} = \left[\left(\frac{A_t}{s}\right) f_{yt} \cot\theta\right] \cot\theta \left[2(x_o + y_o)\right]$$

or

$$A_\ell = \left(\frac{A_t}{s}\right)\frac{f_{yt}}{f_{y\ell}} \cot^2\theta \left[2(x_o + y_o)\right] \tag{4.14}$$

It should be recalled that x_o and y_o are the width and height measured to the centerlines of the tube wall in the thin-walled tube model, while x_1 and y_1 are the width and height measured to the centerlines of the outermost closed stirrup. After cracking, the shear flow path is defined more by the center-to-center dimensions between the outermost closed stirrup. Therefore, in subsequent equations, assuming that $x_o = x_1$ and $y_o = y_1$ yields the equation for the longitudinal torsion reinforcement as

$$A_\ell = \left(\frac{A_t}{s}\right)\frac{f_{yt}}{f_{y\ell}} \cot^2\theta \left[2(x_1 + y_1)\right] \tag{4.15}$$

That is,

$$A_\ell = \left(\frac{A_t}{s}\right)\frac{f_{yt}}{f_{y\ell}} \cot^2\theta \, (p_h)$$

Transverse Reinforcement Required for Torsion (Stirrups)

Using Equation (4.11) and the limit states design requirement that the design torsional strength ϕT_n be greater than or equal to the factored torsional moment, T_u, yields the required torsional stirrup area,

$$\frac{A_t}{s} = \frac{T_u}{2\phi A_o f_{yt} \cot\theta} \tag{4.16}$$

where

A_t = area of *one* leg of the torsional or outermost closed stirrups, in.^{2}

s = stirrup spacing

θ = 30° to 60°; *use* θ = 45° for non-prestressed members (ACI Code, Sections 22.7.6.1.2)

f_{yt} = yield strength of stirrups = 60,000 psi

T_u = factored torque at the critical section

The *critical section* for torsion is permitted to be at a distance d from the face of the beam support, provided no concentrated torque occurs within a distance of d from the face of the support. If a concentrated torque occurs within a distance d from the face of support, the critical section will be at the face of the support (ACI Code, Sections 9.4.4.2 and 9.4.4.3).

d = effective depth of the beam

$A_o \approx 0.85\, A_{oh} = 0.85\, x_1 y_1$

A_{oh} = area enclosed by the centerline of outermost closed stirrup

The nominal torsional strength, T_n, is obtained from Equation (4.11) as

$$T_n = \frac{2A_o A_t f_{yt}}{s} \cot\theta \qquad \textbf{[ACI Eq. (22.7.6.1(a))]}$$

$$T_n \text{ should be} \leq \frac{2A_o A_\ell f_{yt}}{p_h} \tan\theta \qquad \textbf{[ACI Eq. (22.7.6.1(b))]}$$

The *total* equivalent transverse reinforcement or stirrups required for *combined shear plus torsion* is obtained from the ACI Code, Section 9.6.4.2, as

$$\frac{A_{vt}}{s} = \frac{A_v}{s} + \frac{2A_t}{s} \geq \frac{50b_w}{f_{yt}}$$

$$\geq 0.75\sqrt{f'_c}\,\frac{b_w}{f_{yt}} \qquad \textbf{(4.17)}$$

where

A_{vt} = area of *two* legs of closed stirrups required for combined shear plus torsion

A_t = area of *one* leg of closed stirrups required for torsion

A_v = area of *two* legs of closed stirrups required for shear

f_{yt} = yield strength of stirrups = 60,000 psi

b_w = width of beam stem

$p_h = 2(x_1 + y_1)$; see Figure 4.22a.

s = spacing of stirrups $\leq p_h/8$ and 12 in. (ACI Code, Section 9.7.6.3.3)

Note that A_v/s is a slightly modified form for the stirrup requirement for shear from Section 4.2.

Additional Longitudinal Reinforcement Required for Torsion

The additional longitudinal reinforcement required to resist torsion, and to be added to the reinforcement required for bending, is obtained from Equation 4.15 and ACI Equations (9.6.4.3(a)) and (9.6.4.3(b)) as

$$A_\ell = \left(\frac{A_t}{s}\right) p_h \frac{f_{yt}}{f_y} \cot^2\theta \qquad \textbf{(4.18)}$$

A_ℓ should have a minimum value of the lesser of ACI Equations (9.6.4.3(a)) and (9.6.4.3(b)) reproduced below:

$$5\sqrt{f'_c}\,\frac{A_{cp}}{f_y} - \left(\frac{A_t}{s}\right) p_h \frac{f_{yt}}{f_y} \qquad \textbf{[ACI Eq. (9.6.4.3(a))]}$$

$$5\sqrt{f'_c}\frac{A_{cp}}{f_y} - \left(\frac{25b_w}{f_{yt}}\right)p_h\frac{f_{yt}}{f_y} \qquad \textbf{[ACI Eq. (9.6.4.3(b))]}$$

where

$\frac{A_t}{s}$ = torsion stirrup area from Equation (4.11)

$\geq 25b_w / f_{yt}$ (as indicated in ACI Eq. (9.6.4.3(b))for calculating the minimum additional longitudinal reinforcement.)

$p_h = 2(x_1 + y_1)$

$A_{cp} = bh$

b_w = width of beam stem

f_{yt} = stirrup yield strength, psi

$f_{y\ell}$ = longitudinal steel yield strength, psi

The following should be noted regarding the required torsional reinforcement:

- The additional longitudinal reinforcement must be distributed around the surface of the beam with a maximum spacing of 12 in., and there should be at least one longitudinal bar in each corner of the closed stirrups to help transfer the compression strut forces into the stirrups.
- The additional longitudinal reinforcement diameter should be at least 0.042 times the stirrup spacing (i.e., 0.042*s*), but not less than a $\frac{3}{8}$ in. diameter bar (ACI Code, Sections 9.7.5.1, 9.7.5.2, and 25.7.1.2) to prevent buckling of the longitudinal reinforcement due to the horizontal component of the diagonal compression strut force.
- The additional longitudinal rebar area should be added to the longitudinal rebar area required for bending.
- The closed torsional stirrups should be enclosed with 135° hooks (ACI Code, Sections 9.7.6.3.1, 25.7.1.6, and 25.7.2.5) and there should be at least one longitudinal bar enclosed by and at each corner of the stirrup. Note that 90° hooks are ineffective after the corners of the beam spall off due to torsion failure unless the 90° hook of the stirrup in the beam is restrained against spalling by a slab or a flange (see Figure 5.16a and ACI Code, Figure R25.7.1.6).

Torsion Design Procedure

The design procedure for torsion is as follows:

1. Determine the maximum factored concentrated or uniformly distributed torsional load and the corresponding factored gravity load that occurs simultaneously.

   Note that pattern or checkered live loading may need to be considered to maximize the torsional load and moment.

2. Determine the factored torsional moment, T_u, the factored shear, V_u, and the factored bending moment, M_u.
3. Determine the reinforcement required to resist the factored bending moment, M_u.
4. Calculate the concrete shear strength, ϕV_c.
5. Determine the cracking torque, T_{cr} and if torsion can be neglected (check if $T_u \leq 0.25T_{cr}$).
6. Determine if the torsion in the member is caused by compatibility torsion or by equilibrium torsion:
   a. For compatibility torsion, redistribution of internal forces is possible because the torsional moment is not required to maintain equilibrium; therefore, design the member for a reduced torque of ϕT_{cr}.
   b. For equilibrium torsion, redistribution of internal forces is not possible because the torsional moment is required to maintain equilibrium; therefore, the member must be designed for the full torsional moment, T_u, calculated in step 1.
7. Check the limits of the member cross-section using ACI Code Equation (22.7.7.1(a)) to prevent crushing of the diagonal concrete compression struts in solid cross-sections.
8. Determine the required torsional stirrup area, A_t/s, the stirrup area required for shear, A_v/s, and the total stirrup area required for combined shear and torsion, A_{vt}/s. Check that maximum stirrup spacing is not exceeded and check minimum stirrup area. Using the torsional moment diagram and shear force diagram, the required stirrup spacing can be laid out to match the variation in shear and torsional moment.
9. Determine the additional longitudinal reinforcement required for torsion.
10. Draw the details of the torsional reinforcement and/or show the reinforcement in a beam schedule.

EXAMPLE 4.5 Design of Beams for Torsion

The floor framing in the operating rooms in a hospital building consists of reinforced concrete beams 18″ × 24″ deep that support precast concrete planks, as shown in Figure 4.24. The clear span of the beam is 27 ft between columns. The planks are 10 in. deep with 2-in. topping and supports stud wall partitions that weigh 10 psf and mechanical/electrical equipment that weigh 5 psf. The weight of the precast planks is 70 psf. The centerline-to-centerline spans of the planks is 30 ft on the left-hand side of the supporting beam and 24 ft on the right-hand side of the beam. Design the beam for torsion and shear assuming normal weight concrete (i.e., $\lambda = 1.0$) and $f'_c = 4000$ psi. Assume the beam has already been designed for bending. The live load for hospital operating rooms is 60 psf.

Solution:

1. Determine the maximum factored concentrated or uniformly distributed torsional load and the corresponding factored gravity load that occurs simultaneously.

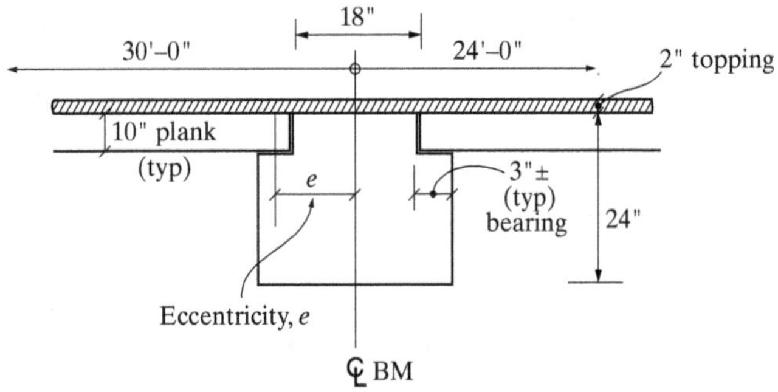

FIGURE 4.24 Beam section for Example 4.5.

Dead Load

$$
\begin{aligned}
\text{10 in. plank} + \text{2 in. topping} &= 95\ \text{psf} \\
\text{Mechanical and electrical} &= 5\ \text{psf} \\
\text{Stud wall partitions} &= 10\ \text{psf} \\
\text{Total dead load, } D &= 110\ \text{psf} \\
\text{Floor live load (operating room), } L &= 60\ \text{psf}
\end{aligned}
$$

Tributary Widths (TW) of Beam

$$TW\ (\text{due to the 30-ft-span hollow-core plank}) = \frac{30\ \text{ft}}{2} = 15\ \text{ft}$$

$$TW\ (\text{due to the 24-ft-span hollow-core plank}) = \frac{24\ \text{ft}}{2} = 12\ \text{ft}$$

A review of Figure 4.24 shows that the torsion in this beam is equilibrium torsion caused by the eccentricity of the plank loads.

Eccentricity of the Hollow-Core Plank Load

$$\text{Eccentricity, } e = \frac{18\ \text{in.}}{2} + \frac{3\ \text{in.}}{2} = 10.5\ \text{in.} = 0.88\ \text{ft}$$

The maximum uniform torsional loading will occur due to checkerboard or partial loading on the hollow-core slabs in which the full design live load is assumed on the 30-ft-span hollow-core slab and *one-half of the design live load is assumed* on the 24-ft-span hollow-core slab. This is common practice among some designers and will generally result in a slightly more conservative design. This partial loading is similar to what is prescribed in Section 7.5 of the ASCE 7 Load Standard. The maximum torsion will be considered together with the corresponding maximum uniform vertical load that occurs at the same time.

The maximum factored uniform torsional load is

$$
\begin{aligned}
w_{tu} = \{&[1.2(110\ \text{psf}) + 1.6(60\ \text{psf})](15\ \text{ft}) \\
&- [1.2(110\ \text{psf}) + 1.6(\tfrac{1}{2})(60\ \text{psf})](12\ \text{ft})\} \\
&\times 0.88\ \text{ft} = 1109\ \text{ft-lb/ft} = 1.11\ \text{ft.-kip/ft}
\end{aligned}
$$

The corresponding maximum factored uniform vertical load is

$$\begin{aligned} w_u &= [1.2(110\text{ psf}) + 1.6(60\text{ psf})](15\text{ ft}) + [1.2(110\text{ psf}) \\ &\quad + 1.6(\tfrac{1}{2})(60\text{ psf})](12\text{ ft}) \\ &= 5580\text{ lb/ft} = 5.58\text{ kip/ft} \end{aligned}$$

2. Determine the factored torsional moment, T_u, and the factored shear, V_u, at d from the face of the support, and the factored bending moment, M_u:

   Assuming two layers of reinforcement, d = 24 in. – 3.5 in. = 20.5 in. For torsion and shear, use a reduced span commencing at d from the face of the beam supports. This reduced span is

$$\ell = 27\text{ ft} - \frac{2(20.5\text{ in.})}{12} = 23.6\text{ ft}$$

   Maximum design torsional moment,

$$T_u = \frac{w_{tu}\ell}{2} = \frac{1.11\text{ ft.-kip/ft}(23.6\text{ ft})}{2} = 13.1\text{ ft.-kip}$$

   Maximum design shear that occurs at the same time as the maximum torsion is

$$V_u = \frac{w_u\ell}{2} = \frac{5.58\text{ kip/ft}(23.6\text{ ft})}{2} = 65.8\text{ kip}$$

3. The reinforcement required to resist the bending moment is assumed to have previously been designed and is not calculated here.
4. The concrete shear strength for this beam is

$$\begin{aligned} \phi V_c &= 0.75(2)\sqrt{4000}(18\text{ in.})(20.5\text{ in.}) = 35{,}000\text{ lb} \\ &= 35\text{ kip} \end{aligned}$$

5. Torsion can be neglected if the factored torsional moment is less than or equal to the concrete torsional strength, that is, if $T_u \le 0.25\phi T_{cr}$, where the concrete torsional strength is

$$\begin{aligned} 0.25\phi T_{cr} &= \lambda\phi\sqrt{f'_c}\,\frac{(A_{cp})^2}{p_{cp}} \\ &= (1.0)(0.75)(\sqrt{4000})\frac{(18\text{ in.}\times 24\text{ in.})^2}{2(18\text{ in.} + 24\text{ in.})} \\ &= 105.4\text{ in.-kip} = 8.8\text{ ft.-kip} \end{aligned}$$

   Because T_u = 13.1 ft.-kip > 8.8 ft.-kip, this beam therefore, must be designed for torsion.

6. This is equilibrium torsion as redistribution of internal forces is *not* possible because the torsional moment is required to maintain equilibrium. The member thus must be designed for the full torsional moment, T_u, calculated in step 2.

7. Check the limits of the member cross-section using ACI Code Equation (22.7.7.1(a)) to prevent crushing of the diagonal concrete compression struts:

$$b = 18 \text{ in.}$$
$$h = 24 \text{ in.}$$

Effective depth, $d = 24 - 3.5$ in. (assuming 2 layers of rebar) = 20.5 in.

$$\theta = 45°$$
$$\phi = 0.75$$
$$x_1 = 18 \text{ in.} - (2 \text{ sides})\left(1.5\text{-in. cover} + \frac{0.5\text{-in. stirrup}}{2}\right)$$
$$= 14.5 \text{ in.}$$
$$y_1 = 24 \text{ in.} - (2 \text{ sides})\left(1.5\text{-in. cover} + \frac{0.5\text{-in. stirrup}}{2}\right)$$
$$= 20.5 \text{ in.}$$
$$f_{yt} = f_{yv} = 60{,}000 \text{ psi}$$
$$A_{oh} = x_1 y_1 = (14.5 \text{ in.})(20.5 \text{ in.}) = 297.3 \text{ in.}^2$$
$$A_o = 0.85 A_{oh} = 0.85 x_1 y_1 = 0.85(14.5 \text{ in.})(20.5 \text{ in.})$$
$$= 252.7 \text{ in.}^2$$
$$p_h = 2(x_1 + y_1) = 2(14.5 + 20.5) = 70 \text{ in.}$$
$$A_{cp} = bh = (18 \text{ in.})(24 \text{ in.}) = 432 \text{ in.}^2$$

The limits on the solid beam cross-sectional dimensions will now be checked using ACI Equation (22.7.7.1(a)):

$$\sqrt{\left(\frac{V_u}{b_w d}\right)^2 + \left(\frac{T_u p_h}{1.7(A_{oh})^2}\right)^2} \le \phi\left(\frac{V_c}{b_w d} + 8\sqrt{f'_c}\right)$$

That is,

$$\sqrt{\left(\frac{65.8 \text{ kip}(1000)}{(18 \text{ in.})(20.5 \text{ in.})}\right)^2 + \left(\frac{(13.1 \text{ ft.-kip})(12{,}000)(70 \text{ in.})}{1.7(297.3 \text{ in.}^2)^2}\right)^2}$$
$$\le \frac{35 \text{ kip}(1000)}{(18 \text{ in.})(20.5 \text{ in.})} + 8(0.75)\sqrt{4000}$$

$$192.8 \text{ psi} < 474.3 \text{ psi} \qquad \text{OK}$$

Thus, the diagonal concrete compression struts are not crushed, and the size of the beam is adequate to resist the torsional moments.

8. Determine the required torsional stirrup area, A_t/s, the stirrup area required for shear, A_v/s, and the total stirrup area required for combined shear and torsion, A_{vt}/s. Check that maximum stirrup spacing is not exceeded and check minimum stirrup area.

From Equation (4.16), the torsional stirrup required is

$$\frac{A_t}{s} = \frac{T_u}{2\phi A_o f_{yt} \cot\theta} = \frac{13.1 \text{ ft.-kip}(12{,}000)}{2(0.75)(252.7)(60{,}000)\cot 45^\circ}$$
$$= 0.0069$$

The stirrup area required to resist the maximum factored shear acting together with the maximum torsion is

$$\frac{A_v}{s} = \frac{V_u - \phi V_c}{\phi f_{yv} d} = \frac{(65.8 \text{ kip} - 35 \text{ kip})(1000)}{(0.75)(60{,}000)(20.5 \text{ in.})} = 0.033$$

The total stirrup area required (2-leg stirrups) is calculated from

$$\frac{A_{vt}}{s} = \frac{A_v}{s} + \frac{2A_t}{s} = 0.033 + 2(0.0069) = \mathbf{0.047}$$
$$\geq \frac{50 b_w}{f_{yt}} = \frac{50(18 \text{ in.})}{60{,}000} = 0.015$$
$$\geq 0.75\sqrt{f_c'}\frac{b_w}{f_{yt}} = 0.75\sqrt{4000}\,\frac{18 \text{ in.}}{60{,}000} = 0.014 \quad \text{(O.K.)}$$

Using No. 4 stirrups, A_{vt} (2 legs) = 2 (0.2 in.^{2}) = 0.4 in.^{2}, the spacing of the stirrups required to resist the maximum combined shear and torsion is calculated as

$$s = \frac{0.4 \text{ in.}^2}{0.047} = 8.5 \text{ in. (controls)}$$
$$\leq p_h/8 = 70 \text{ in.}/8 = 8.75 \text{ in.} \qquad \text{(O.K.)}$$
$$\leq 12 \text{ in.} \qquad \text{(O.K.)}$$

Therefore, use No. 4 closed stirrups at 8-in. on center.

The shear and torsion are at their maximum values at the face of the beam support and decrease linearly to zero at the midspan of the beam; the stirrup spacing thus can be varied accordingly, as done previously in the shear design examples.

9. Additional Longitudinal Reinforcement

$$b_w = \text{width of beam stem} = 18 \text{ in.}$$
$$f_{yt} = \text{stirrup yield strength} = 60{,}000 \text{ psi}$$
$$f_y = \text{longitudinal steel yield strength} = 60{,}000 \text{ psi}$$
$$\frac{A_t}{s} = 0.0069 \,(\text{as previously calculated}), \geq \frac{25b_w}{f_{yt}}$$
$$= \frac{25(18 \text{ in.})}{60{,}000} = 0.0075$$

Therefore, use 0.0075. (as required by ACI Eq. 9.6.4.3(b))

From Equation (4.18), the additional longitudinal reinforcement is calculated as

$$A_\ell = \left(\frac{A_t}{s}\right) p_h \frac{f_{yt}}{f_y} \cot^2\theta = (0.0075 \text{ in.})(70 \text{ in.}) \frac{60{,}000}{60{,}000} \cot^2 45°$$
$$= 0.53 \text{ in.}^2$$
$$\geq \textit{the lesser of } 5\sqrt{f_c'}\frac{A_{cp}}{f_y} - \left(\frac{A_t}{s}\right) p_h \frac{f_{yt}}{f_y} \quad \textit{and} \quad 5\sqrt{f_c'}\frac{A_{cp}}{f_y} - \left(\frac{25b_w}{f_{yt}}\right) p_h \frac{f_{yt}}{f_y}$$
$$\geq \textit{the lesser of } 5\sqrt{4000}\frac{432 \text{ in.}^2}{60{,}000} - (0.0075 \text{ in.})(70 \text{ in.})\frac{60{,}000}{60{,}000} \quad \textit{and}$$
$$5\sqrt{4000}\frac{432 \text{ in.}^2}{60{,}000} - \left(\frac{25(18 \textit{ in.})}{60{,}000}\right)(70 \text{ in.})\frac{60{,}000}{60{,}000}$$
$$\geq \textit{the lesser of } 1.75 \text{ in.}^2 \quad \textit{and} \quad 1.75 \text{ in.}^2$$

Therefore, the required additional longitudinal steel area is $A_\ell = 1.75 \text{ in.}^2$

This additional longitudinal reinforcement should be distributed at the corners of the beam, but the spacing between these bars should be no greater than 12 in. Where the spacing exceeds 12 in., provide additional longitudinal bars at the midwidth or middepth of the beam as required. This longitudinal reinforcement is in addition to the reinforcement required to resist the bending moments on the beam.

If the additional longitudinal reinforcement due to torsion is concentrated on the top and bottom layers (i.e., 50% each to the top and bottom layers), therefore, the total areas of the top and bottom longitudinal reinforcement in the beam are calculated as

$$A_{s,\text{top}} = A_{s,\text{top}}\,(\text{due to bending}) + 0.5(1.75 \text{ in.}^2)$$
$$= A_{s,\text{top}}\,(\text{due to bending}) + 0.88 \text{ in.}^2$$
$$A_{s,\text{bottom}} = A_{s,\text{bottom}}\,(\text{due to bending}) + 0.5(1.75 \text{ in.}^2)$$
$$= A_{s,\text{bottom}}\,(\text{due to bending}) + 0.88 \text{ in.}^2$$

However, for the beam in this example, the spacing of the longitudinal reinforcement will exceed the maximum 12 in. because the center-to-center distance between the top and bottom rebars is approximately 18 in. The additional longitudinal reinforcement should thus be distributed in three layers as follows (see Figure 4.25):

$$\begin{aligned} A_{s,top} &= A_{s,top}\,(\text{due to bending}) + \left(\frac{1}{3}\right)(1.75\text{ in.}^2) \\ &= A_{s,top}\,(\text{due to bending}) + 0.58\text{ in.}^2 \end{aligned}$$

$$A_{s,midheight} = \left(\frac{1}{3}\right)(1.75\text{ in.}^2) = 0.58\text{ in.}^2$$

$$\begin{aligned} A_{s,bottom} &= A_{s,bottom}\,(\text{due to bending}) + \left(\frac{1}{3}\right)(1.75\text{ in.}^2) \\ &= A_{s,bottom}\,(\text{due to bending}) + 0.58\text{ in.}^2 \end{aligned}$$

The *minimum* diameter of the longitudinal reinforcement is the largest of the following:

$$0.042\,s = 0.042(8\text{ in.}) = 0.34\text{ in.}$$

or $\frac{3}{8}$ in. (controls)

Note that on the top and bottom faces of the beam, the distance between the corner top layer and bottom layer bars will also be greater than 12 inches. An additional bar should be added at the midwidth of the beam in the top and bottom layers.

10. Torsional reinforcement detail is shown in Figure 4.25.

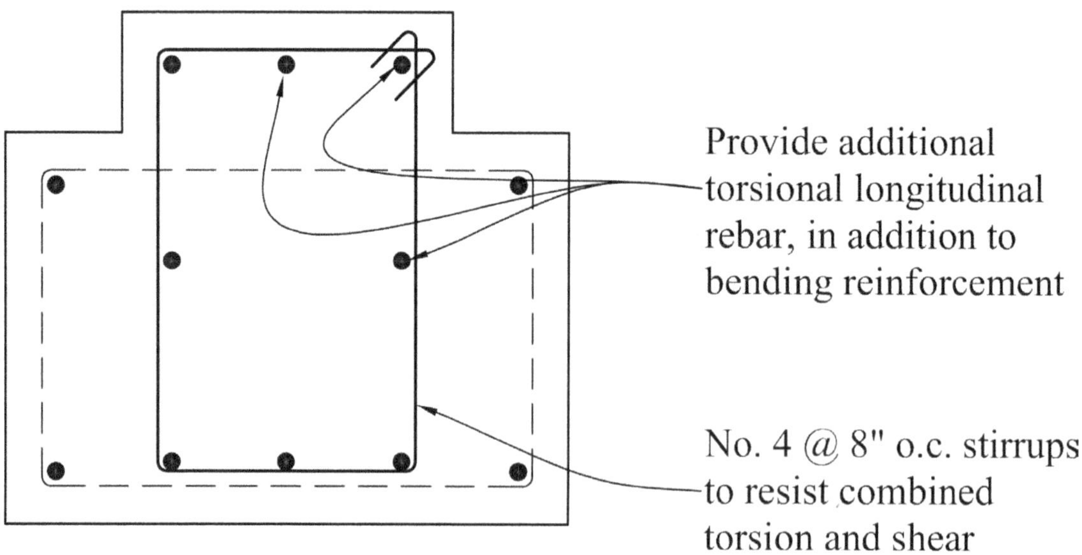

FIGURE 4.25 Beam torsional reinforcement detail.

4.6 CORBELS AND BRACKETS

Corbels and brackets are short cantilever deep beams that are supported off the face of columns where the distance from the face of the supporting column to the line of action of the concentrated load that is being supported (i.e., a_v) is no greater than the effective depth, d, of the corbel. Corbels and brackets are used to support loads at expansion joints, in precast construction to support precast beams, at slightly offset columns, and in retrofit situations where the corbel is cast against an existing column to support a new beam [2] (see Figures 4.26, 4.27, and 4.28). Corbels are also used where steel floor framing members connect to concrete core walls.

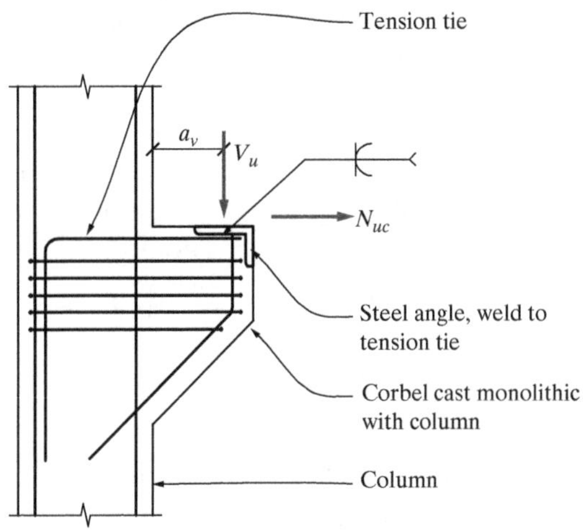

FIGURE 4.26 Corbels and brackets at expansion joints or in precast construction.

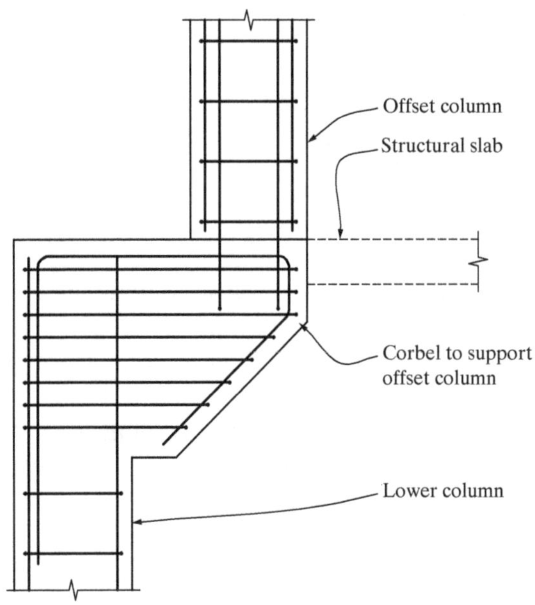

FIGURE 4.27 Corbels at offset columns.

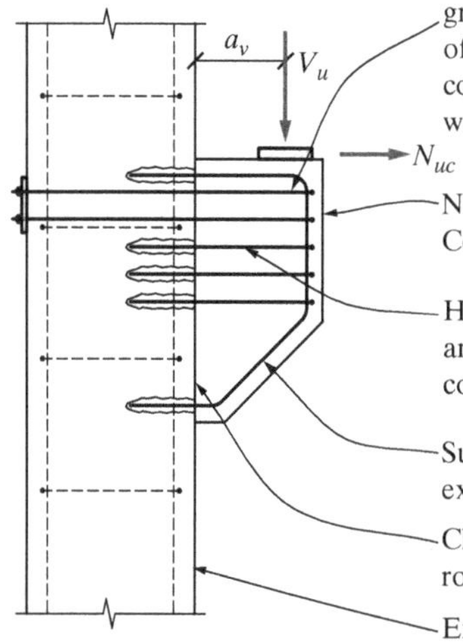

FIGURE 4.28 Corbels at existing columns (see Ref. [3]).

The following possible failure modes must be considered in the design of corbels [ACI 318-19, Section R16.5.1]:

- Shearing along the vertical interface between the face of the column and the corbel
- Tension failure of the primary tension rebar
- Crushing or splitting of the compression strut
- Local failure of the concrete under the bearing plate as a result of excessive bearing stresses or shear stresses

Two methods are presented in the ACI Code for the design of corbels: the shear friction method presented in ACI 318-19 Code, Section 22.9, and the strut-and-tie method presented in ACI 318-19 Code, Section 23. The dimensional limits for corbels are presented in ACI 318-19 Code, Section 16.5.

In the shear friction method, the relative vertical displacement at the interface between the corbel and the column is resisted by dowel action of the main horizontal reinforcement crossing the corbel-column interface, and by friction between the vertical faces of the column and the corbel at the interface. The shear friction method can be used for corbels with $a_v/d \leq 1.0$ and $N_{uc} \leq V_u$ because the method has only been validated experimentally for this range of a_v/d and N_{uc} (ACI 318-19 Code, Section R16.5.1). The strut-and-tie method, introduced in Chapter 10, can be used for all corbels, but in this chapter, only the shear friction method will be covered. To prevent cracking that starts below the bearing plate and propagates to the sloped surface of the corbel, the ACI Code (Section R16.5.2.2 and Figure R16.5.1a) requires the depth of the vertical face of the corbel to be at least 50% of the effective depth of the corbel at the face of the column (see Figure 4.29). The primary tension tie reinforcement must be fully developed on both sides of the shear planer at the column-corbel interface. This usually requires a 90° hook in the column; in order to develop the tension tie reinforcement within the corbel, the rebar can be welded to a horizontal anchor bar near the outside face of the corbel. Alternatively, the tension tie reinforcement can be welded to an embedded steel angle located at the top outside edge of the corbel; this steel angle also serves as a bearing plate for the supported load, V_u (see Figures 4.26, 4.29 and 4.30).

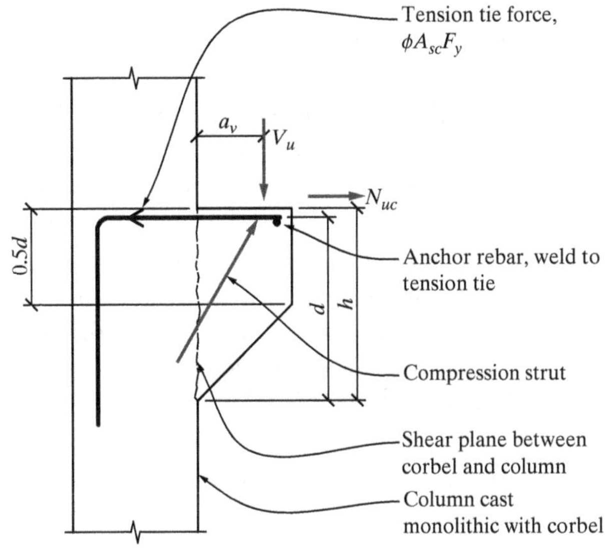

FIGURE 4.29 Corbels and brackets with anchor bar.

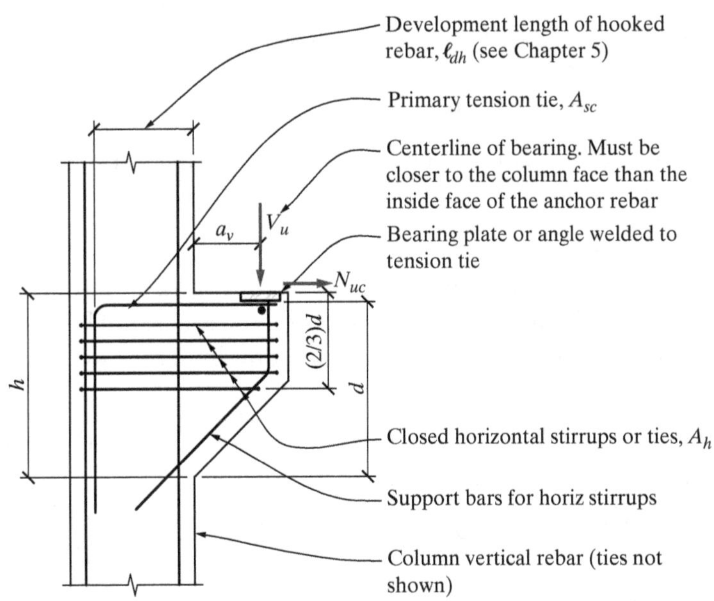

FIGURE 4.30 Corbels and brackets with steel angle anchor and bearing plate.

In accordance with ACI 318-19 Code, Section 16.5, the requirements and procedure for the design of corbels and brackets using the shear friction method are as follows:

- In order to use the shear friction method, the following limits (which have been validated experimentally; see ACI Code, Section 16.5.1.1) must be satisfied: $a_v/d \leq 1.0$ and $N_{uc} \leq V_u$
- The clear concrete cover to the reinforcement in the corbels is ¾" for corbels not exposed to the weather or in contact with the ground; for corbels exposed to the weather or in contact with ground, the clear concrete cover is 1½-in. for #5 rebar or smaller, and 2-in. for #6 rebar or larger (ACI 318-19, Sections 16.5.6.1 and 20.5.1.3)
- The depth of the vertical outside face of the corbel should be at least $0.5d$, where d is the effective depth of the corbel at the supporting column face. This prevents premature failure from a vertical shear crack propagating from the outermost edge of the bearing plate down to the sloped surface of the corbel (ACI 318-19 Code, Section 16.5.2.2).
- No part of the bearing area of the corbel should project beyond the end of the straight portion of the primary tension tie reinforcement within the corbel or the interior face of the horizontal anchor bar where an anchor bar is used. This will prevent the formation of a diagonal shear crack projecting from the outermost edge of the bearing plate down to the vertical face of the corbel (ACI 318-19, Section 16.5.2.3).
- For normal-weight concrete, the corbel should be such that the factored concentrated load on the corbel, V_u, satisfies the following relationships (ACI 318-19 Code, Section 16.5.2.4 and Table 22.9.4.4):

$$\frac{V_u}{\phi} \leq 0.2 f_c' b_w d$$

$$\leq \left(480 + 0.08 f_c'\right) b_w d$$

$$\leq 1600 b_w d$$

  Where, b_w = width of the corbel.

  In practice, the width of the corbel is usually selected to be the same as the column width. For the corresponding equations for lightweight concrete, see ACI Code, Section 16.5.2.5.

- $N_{uc} \geq 0.2\, V_u$ (ACI 318-19 Code, Section 16.5.3.5)
  The code mandates a minimum horizontal tensile force of $0.2V_u$ unless tension forces are prevented from being applied to the corbel or bracket. This factored horizontal tension force could occur from the effects of temperature, creep, and shrinkage.
- Since shear is the predominant mode of failure for corbels, a single strength-reduction factor of 0.75 is used for all potential failure modes. Strength-reduction factor for corbels, $\phi = 0.75$ (ACI 318-19 Code, Table 21.2.1, row (f)).
- At the face of the supporting column (i.e., at the interface of the column and the corbel), the corbel will be subjected to the following forces (ACI 318-19 Code, Section 16.5.3.1):
  - Factored vertical shear force, V_u
  - Factored horizontal tension force, N_{uc} (At a minimum, $N_{uc} = 0.2\, V_u$)
  - Factored moment, $M_u = V_u a_v + N_{uc}$ (h - d)

- The limits for the total area of reinforcement required to resist flexure and direct tension at the top of the corbel, A_{sc}, are given as (ACI 318-19 Code, Section 16.5.5.1):

$$A_{sc} \geq A_f + A_n$$
$$\geq \left(\frac{2}{3}\right)A_{vf} + A_n$$
$$\geq 0.04\left(\frac{f_c'}{f_y}\right)(b_w d)$$

The third equation above represents the minimum amount of reinforcement that is required to cross the corbel-column interface in order to prevent sudden failure of the corbel should the corbel crack vertically at the corbel-column interface under the flexural and tensile forces (ACI 318-19, Section R16.5.5.1). The area of the direct tension tie reinforcement, A_n, is given as $A_n = \dfrac{N_{uc}}{\phi f_y}$ (ACI 318-19 Code, Section 16.5.4.3) and the area of the flexural reinforcement, A_f, required to resist the factored moment, $M_u = V_u a_v + N_{uc}(\mathrm{h} - \mathrm{d})$, is determined from the flexure theory previously presented in Chapters 2 and 3 as follows:

$$A_f = \frac{M_u}{\phi f_y (d - 0.5a)}$$
$$a = \frac{A_f f_y}{0.85 f_c' b_w}$$

Some iteration is required to calculate A_f since the equation for A_f is a function of the depth of the stress block, a, which in turn is a function of A_f. To use the above equations, first assume a value (e.g., initially assume $a \approx 0.1\mathrm{d}$) for the depth of the compression stress block, a, and then calculate A_f. Recalculate the stress block depth using the A_f obtained and repeat the iterative process until the calculated depth of the stress block is close enough to the initially assumed depth of the stress block.

- Using ACI 318-19 Section 22.9.4.2, calculate the area of the shear friction reinforcement, A_{vf}, as follows using ACI Code Equation (22.9.4.2).

$$A_{vf} = \frac{V_u}{\phi \mu f_y}$$

where μ is the coefficient of friction with values obtained from ACI 318-19 Code, Table 22.9.4.2. For normal-weight concrete, $\mu = 1.4$ for concrete cast monolithically, which applies to concrete corbels. For lightweight concrete, $\mu = (1.4)(0.75) = 1.05$. See ACI Code Table 22.9.4.2 for values of μ for other conditions. The strength-reduction factor, ϕ, for brackets and corbels is 0.75 (ACI 318-19, Table 21.2.1).

- To prevent diagonal tension failure of the corbel, closed horizontal stirrups or ties are required and these horizontal stirrups must be located with a depth of $\frac{2}{3}d$ from the primary tension tie reinforcement at the top of the corbel (ACI 318-19 Code, Section 16.5.6.6). The minimum area of the closed horizontal stirrups or ties, A_h, is given in ACI 318-19 Code Sections R16.5.5.1 and 16.5.5.2 as follows:

$$A_h = \frac{A_{vf}}{3}$$
$$\geq 0.5\left(A_{sc} - A_n\right)$$

- For the reinforcement supporting the horizontal stirrups, select nominal sizes since these are framing rebars that only provide support to the horizontal stirrups, A_h.
- The area of the bearing plate can be determined from ACI Code, Section 22.8 as follows:

$$A_{bearing\ plate} = \frac{V_u}{\phi_{bearing} 0.85 f'_c}$$

where $\phi_{bearing}$ is 0.65 (see ACI Table 21.2.1, row (d)).

Select the bearing plate dimensions assuming a width of b_w – 4 in. (i.e., assuming the bearing plate is set in 2 inches on both sides of the corbel). The bearing plate dimension in the orthogonal direction perpendicular to the column-corbel interface = $\dfrac{A_{bearing\ plate}}{(b_w - 4\text{ in.})}$

Where, b_w = width of corbel.

- Check the required development length, $\ell_{dh,\ required}$, of the main tension tie reinforcement with a 90° standard hook in the column. Compare this to the actual development length provided, $\ell_{dh,\ \text{provided}}$ = column width – 1.5 in concrete cover – 3/8" ties – diameter of column vertical rebar. Note that the hooked bar will be within the column vertical rebar cage.
- The main tension tie reinforcement must also be developed in the corbel either using an anchor bar or a steel angle to which the tension tie is welded.

EXAMPLE 4.6 Corbels and Brackets

The column-corbel shown in Figure 4.31 supports a service dead load of 35 kip and a service live load of 75 kip. The concrete strength, f'_c, is 4500 psi and the steel yield strength, f_y, is 60,000 psi. Design the corbel, selecting the dimensions of the corbel and all the required reinforcement per the ACI Code using the shear friction method.

- For the corbel, the distance from the face of the supporting column to the line of action of the concentrated load that is being supported, a_v = 9 in. – ½(length of leg of bearing angle) = 9 in. – ½(6 in.) = 6 in.
- The factored vertical load, V_u = 1.2 (35 kip) + 1.6 (75 kip) = 162 kip.
- The factored tension force, $N_{uc} = 0.2\ V_u$ = (0.2)(162 kip) = 32.4 kip.
- Strength-reduction factor for corbels, ϕ = 0.75 (ACI Code, Table 21.2.1).
- In order to be able to use the shear friction method, the following limits must be satisfied: $a_v/d \leq 1.0$ and $N_{uc} \leq V_u$.

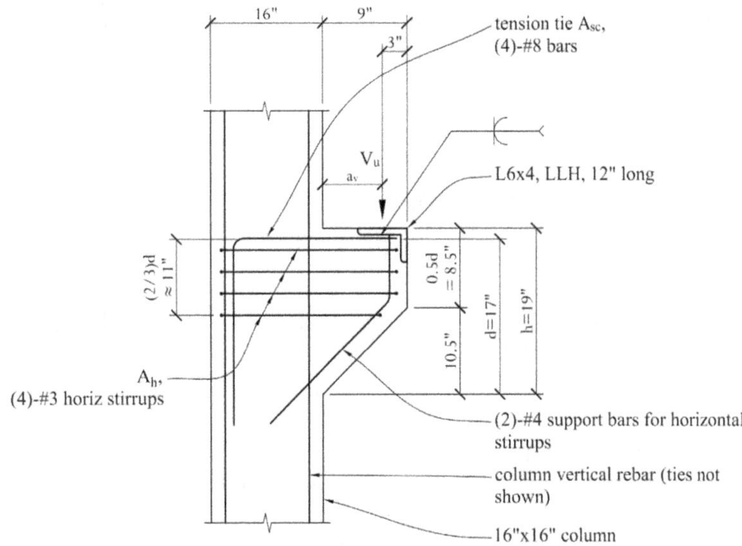

FIGURE 4.31 Corbel for Example 4.6

- The depth of the vertical outside face of the corbel should be at least $0.5d$, where d is the effective depth of the corbel at the supporting column face (ACI Code, Section 16.5.2.2).
- No part of the bearing area of the corbel should project beyond the end of the straight portion of the primary tension tie reinforcement within the corbel or the interior face of the horizontal anchor bar where an anchor bar is used.
- The width of the column is 16 inches; assuming the width of the corbel is equal to the column width, therefore, b_w = 16 inches.
- For normal-weight concrete, the corbel should be such that the factored concentrated load on the corbel, V_u satisfies the following relationships (ACI Code, Section 16.5.2.4 and Table 22.9.4.4):

$$\frac{V_u}{\phi} = \frac{162\,\text{kip}}{0.75} \leq 0.2 f_c' b_w d = \frac{\left[0.2(4500\,\text{psi})(16\text{ in.})d\right]}{1000}$$

$$\leq (480 + 0.08 f_c') b_w d = \frac{\left[480 + 0.08(4500\,\text{psi})\right](16\text{ in.})d}{1000}$$

$$\leq 1600 b_w d = \frac{1600(16\text{ in.})d}{1000}$$

Solving the above three equations yields the following values for d: 15 in.; 16.1 in.; 8.4 in.

The largest value governs, therefore, assume d = 16.1 inches ≈ 17 inches.

The total depth of the bracket, $h = d$ + concrete cover to primary tie rebar + ½ (diameter of tie rebar)

Assuming #8 tie rebar and 1.5 in. cover, h = 17 + 1.5 + ½ (1″) = 19 in.

a_v/d = 6 in./17 in. = 0.35 < 1, and N_{uc} = 32.4 kip ≤ V_u = 162 kip

Since both conditions are satisfied, therefore, the shear friction method can be used.

- At the face of the supporting column (i.e., at the interface of the column and the corbel), the corbel will be subjected to the following forces (ACI Code, Section 16.5.3.1):
  - Factored vertical shear force, V_u = 162 kip
  - Factored horizontal tension force, N_{uc} = 0.2 V_u = 32.4 kip
  - Factored moment, $M_u = V_u a_v + N_{uc}(h\text{-}d)$
    $= (162 \text{ kip})(6 \text{ in.}) + (32.4 \text{ kip})(19 \text{ in.} - 17 \text{ in.})$
    $= 1037$ kip-in.
- $A_n = \dfrac{N_{uc}}{\phi f_y} = \dfrac{32.4 \text{ kip}}{(0.75)(60 \text{ ksi})} = 0.72 \text{ in}^2$ (ACI Code, Section 16.5.4.3) and the flexural reinforcement, A_f, required to resist the factored moment,
  M_u = 1037 kip-inch is calculated as follows:
  Assume initial depth of the stress block, a = 1.5 in.

$$A_f = \frac{M_u}{\phi f_y(d-0.5a)} = \frac{1037 \text{ kip-in.}}{(0.75)(60 \text{ ksi})[17"-0.5(1.5")]} = 1.42 \text{ in}^2$$

$$a = \frac{A_f f_y}{0.85 f_c^{'} b_w} = \frac{(1.42 \text{ in}^2)(60 \text{ ksi})}{0.85(4.5 \text{ ksi})(16 \text{ in.})} = 1.4 \text{ in.} \approx 1.5 \text{ in. assumed. O.K.}$$

  Use $A_f = 1.42 \text{ in.}^2$

- The coefficient of friction for a corbel cast monolithic with the column and using normal-weight concrete, μ = 1.4. Using ACI Code, Section 22.9.4.2, calculate the area of the shear friction reinforcement, A_{vf}, as follows:

$$A_{vf} = \frac{V_u}{\phi \mu f_y} = \frac{162 \text{ kip}}{(0.75)(1.4)(60 \text{ ksi})} = 2.57 \text{ in.}^2$$

- The limits for the total area of reinforcement required to resist flexure and direct tension at the top of the corbel, A_{sc}, are given as (ACI Code, Section 16.5.5.1):

$$A_{sc} \geq A_f + A_n = 1.42 \text{ in.}^2 + 0.72 \text{ in.}^2 = 2.14 \text{ in.}^2$$

$$\geq \left(\frac{2}{3}\right)A_{vf} + A_n = \left(\frac{2}{3}\right)(2.57 \text{ in.}^2) + 0.72 \text{ in.}^2 = 2.43 \text{ in.}^2$$

$$\geq 0.04\left(\frac{f_c^{'}}{f_y}\right)(b_w d) = 0.04\left(\frac{4.5 \text{ ksi}}{60 \text{ ksi}}\right)(16 \text{ in.})(17 \text{ in.}) = 0.82 \text{ in.}^2$$

- Therefore, A_{sc} = 2.43 in^2. Use 4 #8 bars (A_{sc} provided = 3.14 in^2).
- To prevent diagonal tension failure of the corbel, closed horizontal stirrups or ties are required and these horizontal stirrups must be located with a depth of $\frac{2}{3}d$ from the primary tension tie reinforcement at the top of the corbel (ACI Code, Section 16.5.6.6). The minimum area of the closed horizontal stirrups or ties, A_h, is given in ACI 318 Code, Sections R16.5.5.1 and 16.5.5.2, as follows:

$$A_h = \frac{A_{vf}}{3} = \frac{2.57\,\text{in.}^2}{3} = 0.86\ \text{in.}^2$$

$$\geq 0.5\left(A_{sc} - A_n\right) = 0.5\left(2.43\ \text{in.}^2 - 0.72\ \text{in.}^2\right) = 0.86\ \text{in.}^2$$

Therefore, $A_h = 0.86\ \text{in.}^2$

Use 4 #3 closed horizontal stirrups placed within (2/3)d or (2/3)(17 in.) or 11 inches below the main tension tie rebar.

A_h provided = (4)(2 legs)(0.11 in.^{2} per leg) = 0.88 in.^{2} > A_h required. O.K.

- For the framing rebars supporting the horizontal stirrups, use 2 #4 bars.
- The area of the bearing plate can be determined from

$$A_{bearing\ plate} = \frac{V_u}{\phi_{bearing} 0.85 f_c'} = \frac{162\ \text{kip}}{(0.65)(0.85)(4.5\,\text{ksi})} = 65\ \text{in.}^2$$

$\phi_{bearing} = 0.65$. Given that the width of the corbel b_w = 16 inches, and assuming a 12-in.-wide bearing plate which provides at least a 2 in edge distance on each side of the plate, therefore, the required length of the bearing plate = 65 in^2/ 12 in = 5.5 in. < 6 inch angle leg provided, O.K.

- Check the required development length, $\ell_{dh,\ required}$, of the main tension tie reinforcement with a 90° standard hook in the column. Compare this to the actual development length provided, $\ell_{dh,\ \text{provided}}$ = 16 in. – 1.5 in. concrete cover – 3/8″ ties – 1" (assumed column rebar diameter) = 13.1 in. Note that the hooked bar will be within the column vertical rebar cage. Reviewing what is discussed in Chapter 5 (see also Table A.13), the reader can verify that the development provided (i.e., 16.4 in. × 2.43 in.^{2}/3.14 in.^{2} = 12.7 in.) is greater than the required development length, $\ell_{dh,\ required}$ for the 4#8 tension tie rebar.
- The main tension tie reinforcement must also be developed in the corbel by welding the tension ties to the underside of the L 6 × 4 angle.
- The final reinforcement detail for the corbel is shown in Figure 4.31.

4.7 HANGER REINFORCEMENT FOR BEAMS FRAMING INTO GIRDERS

When a beam is cast monolithically with, and frames into the side face of, a supporting girder, the soffit of the girder may be prone to premature failure unless additional reinforcement known as hanger reinforcement are provided in the girder (see Figure 4.32 and ACI Code, Section 9.7.6.2.1). This hanger reinforcement, which will be in the form of stirrups, will be in addition to the reinforcement required for shear and torsion.

The required area of one leg of the hanger reinforcement, $A_{s(hanger)}$, in the girder, which is placed in the form of stirrups in the girder over the width of the supported beam, b_w, is

$$A_{s(hanger)} = \frac{V_{u(supported\ beam)}}{\phi f_y}$$

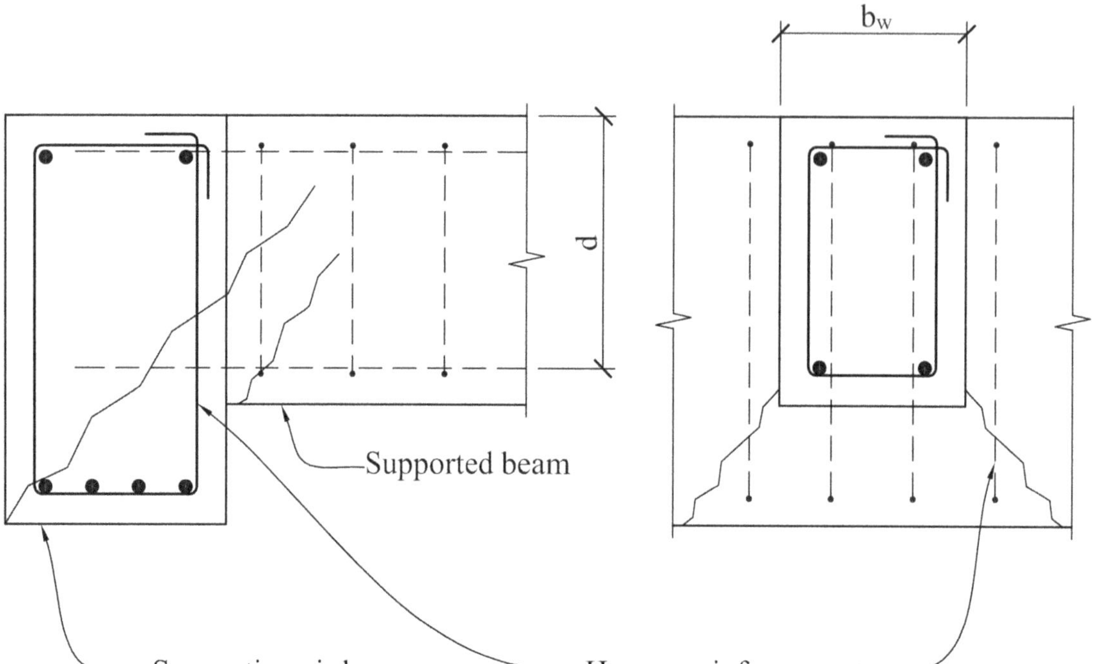

FIGURE 4.32 Hanger reinforcement in Girders at Supported Beams (Adapted from ACI 318-19 Figure R9.7.6.2.1)

The hanger reinforcement can be neglected if the following conditions are satisfied:

- $h_{beam} \leq 0.5h_{girder}$
- $V_{u,\,supported\ beam} < 3\sqrt{f_c'}b_w d$

  Where,

  $\phi = 0.75$

  b_w = width of the supported beam

  d = effective depth of the supported beam

REFERENCES

1. Halil Sezen, Sergio M. Alcocer, and Jack P. Moehle, "Effective Depth of Rectangular and Circular Columns for Shear Strength Calculations," Concrete International, July 2021, pp. 37-41.
2. Adam Lubell, Ted Sherwood, Evan Bentz, and Michael P. Collins. "Safe Shear Design of Large Wide Beams," Concrete International, January 2004.
3. Terry McDonald and Homa Ghaemi. "From Office-to-Condominium, The Park Monroe - Adaptive Reuse of the 55/65 E. Monroe near Chicago's Millennium Park, Structure Magazine, April 2011.

EXERCISES

4.1 A reinforced concrete beam of rectangular cross-section is reinforced for moment only and subjected to a shear V_u of 9000 lb. Beam width b = 12 in., d = 7.25 in., f'_c = 3000 psi, and f_y = 60,000 psi. Is the beam satisfactory for shear?

4.2 An 8-in.-thick one-way slab is reinforced for a positive moment with No. 6 bars at 6 in. on center. Cover is 1 in. Determine the maximum shear V_u permitted. Use f'_c = 4000 psi and f_y = 60,000 psi.

4.3 Assume that the beam of Exercise 4.1 has an effective depth of 18 in. and is reinforced with No. 3 single-loop stirrups spaced at 10 in. on center. Determine the maximum shear V_u permissible.

4.4 The simply supported beam shown is on a clear span of 30 ft. The beam carries uniformly distributed service loads of 1.9 kip/ft live load and 0.7 kip/ft dead load (excluding the weight of the beam). Additionally, the beam carries two concentrated service loads of 8 kip dead load each, one load being placed 5 ft in. from the face of each support. The beam is reinforced with No. 3 single-loop stirrups placed in the following pattern, starting at the face of the support (symmetry about midspan): one space at 2 in., five spaces at 8 in., four spaces at 10 in., and eight spaces at 12 in. Beam width b = 15 in., h = 28 in., and d = 25.4 in. Use f'_c = 4000 psi and f_y = 60,000 psi.

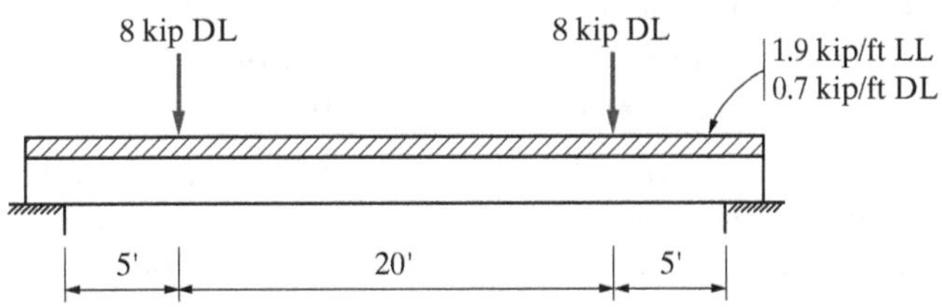

EXERCISE 4.4

a. Draw an elevation view of the beam showing the stirrup layout.

b. Draw the V_u diagram.

c. Calculate $\phi(V_c + V_s)$ for each spacing group (omit the 2-in. spacing).

d. Superimpose the results from part (c) on the V_u diagram and comment on the results.

4.5 A uniformly loaded beam is subjected to a shear V_u of 60 kip at the face of the support. Clear span is 32 ft. Beam width b = 12 in., d = 22 in., f'_c = 4000 psi, and f_y = 60,000 psi. Determine the maximum spacing allowed for No. 3 single-loop stirrups at the critical section.

4.6 A simply supported, rectangular, reinforced concrete beam having d = 24 in. and b = 15 in. supports a uniformly distributed load w_u of 7.50 kip/ft as shown. The given load includes the beam weight. Assume No. 3 single-loop stirrups. The concrete strength is 4000 psi and the steel is grade 60.

a. Select the stirrup spacing to use at the critical section (d distance from the face of support).

b. Determine the maximum stirrup spacing allowed for this beam.

c. Using the two spacings determined in parts (a) and (b), devise an appropriate stirrup spacing layout for this beam.

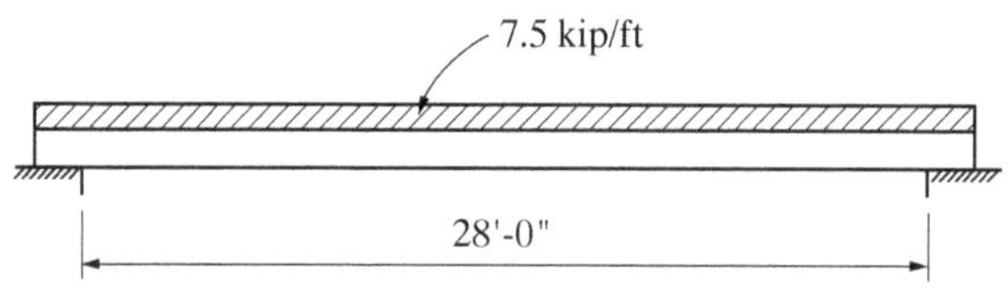

EXERCISE 4.6

4.7 A simply supported beam carries a total factored load w_u of 5.0 kip/ft. The span length is 28 ft center–to-center of supports, and the supports are 12-in. wide. Beam width b = 14 in., d = 20 in., $f'_c = 4000$ psi, and f_y = 60,000 psi. Determine spacings required for No. 3 stirrups and show the pattern with a sketch. (Recall that clear span is used for determining shears.)

4.8 A simply supported rectangular reinforced concrete beam, 13-in. wide and having an effective depth of 20 in., supports a total factored load (w_u) of 4.5 kip/ft on a 30-ft clear span. (The given load includes the weight of the beam.) Design the web reinforcement if $f'_c = 3000$ psi and f_y = 40,000 psi.

4.9 A rectangular reinforced concrete beam supports a total factored load (w_u) of 10 kip/ft on a simple clear span of 40 ft. (The given load includes the weight of the beam.) The effective depth d = 40 in., b = 24 in., $f'_c = 3000$ psi, and f_y = 60,000 psi. Design double-loop No. 3 stirrups.

4.10 Design stirrups for the beam shown. The supports are 12-in. wide, and the loads shown are service loads. The dead load includes the weight of the beam. Beam width b = 16 in., d = 20 in., $f'_c = 4000$ psi, and f_y = 60,000 psi. Sketch the stirrup pattern.

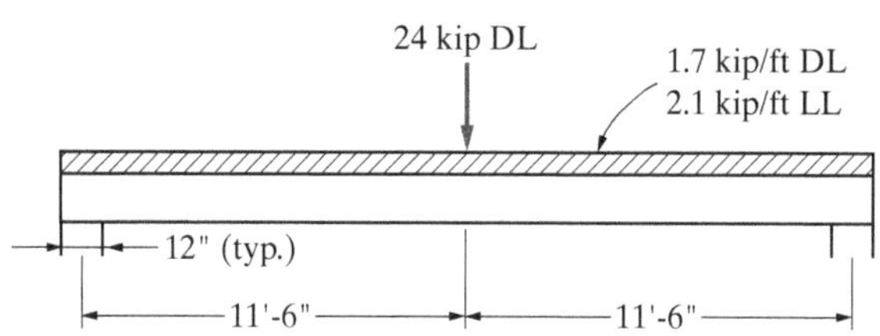

EXERCISE 4.10

4.11 For the beam shown, check moment and design single-loop stirrups. The loads shown are factored loads. Assume the supports to be 12-in. wide. Use $f'_c = 3000$ psi and f_y = 60,000 psi. The uniformly distributed load includes the beam weight.

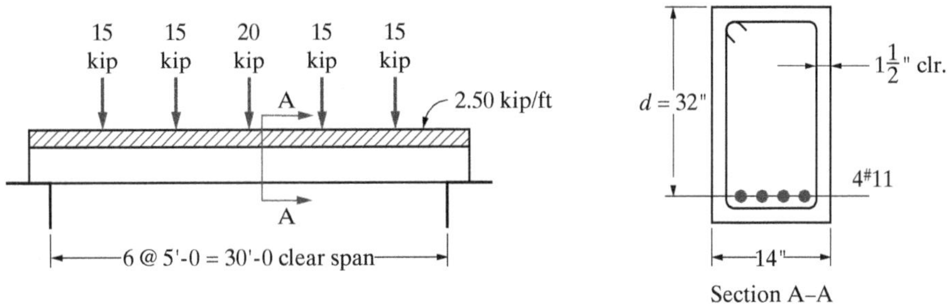

EXERCISE 4.11

4.12 Design stirrups for the beam shown. Service loads are 1.5 kip/ft dead load (includes beam weight) and 1.9 kip/ft live load. The supports are 12-in. wide. Beam width b = 13 in. and d = 24 in. for both top and bottom steel. Use $f_c' = 3000$ psi and f_y = 60,000 psi. Sketch the stirrup arrangement.

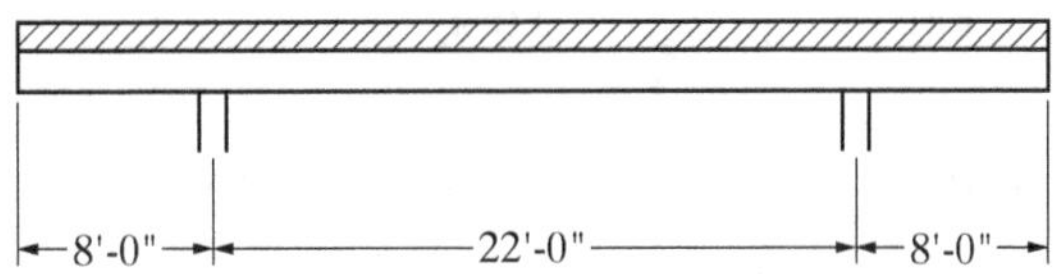

EXERCISE 4.12

4.13 Design stirrups for the beam shown. The supports are 12-in. wide, and the loads shown are factored design loads. The dead load includes the weight of the beam. Beam width b = 14 in., d = 24 in., $f_c' = 3000$ psi, and f_y = 60,000 psi. Sketch the stirrup pattern.

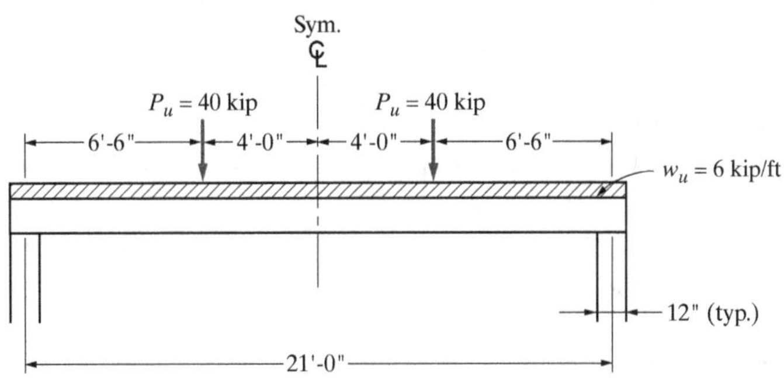

EXERCISE 4.13

4.14 The beam shown is supported on pedestals 20-in. wide. The loads are service loads, and the distributed load includes the beam weight. The clear span is 24′-4″. Use $f_c' = 4000$ psi; the reinforcing steel is A615 Grade 60. Design shear reinforcing for the beam:

a. between the point load and the left reaction

b. between the point load and the right reaction

Show design sketches, including the stirrup pattern.

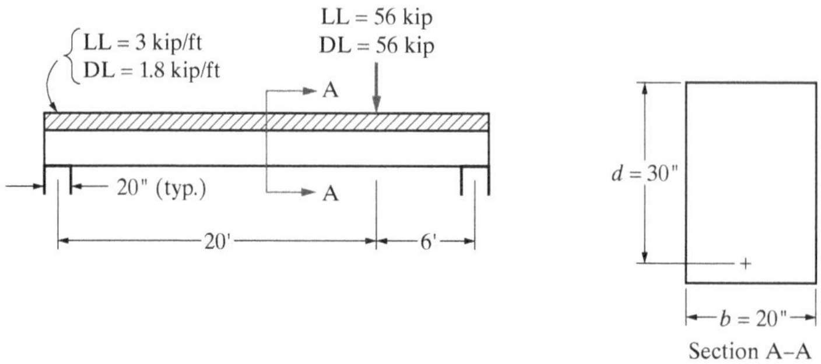

EXERCISE 4.14

4.15 Design the rectangular reinforced concrete beam for *moment and shear.* Use only tension steel for flexure. The loads shown are service loads. The uniform load is composed of 1 kip/ft dead load (does not include beam weight) and 1 kip/ft live load. The concentrated loads are dead load. Assume the supports to be 12-in. wide. Use $f_c' = 4000$ psi and $f_y = 60{,}000$ psi. Show design sketches, including the stirrup pattern.

4.16 Explain, with examples, the difference between equilibrium and compatibility torsion.

4.17 A rectangular reinforced concrete beam 14 in. × 20 in. deep is subject to a maximum factored torque of 24 ft-k. Calculate the cracking torque, T_{cr}, and determine if torsion can be neglected in the design of this beam. Assume normal-weight concrete (i.e., $\lambda = 1.0$) and $f_c' = 4000$ psi.

4.18 The floor framing in an office building consists of reinforced concrete beams 18 in. × 24 in. deep that support precast concrete planks on a 3-in. ledge similar to that shown in Figure 4.23. The clear span of the beam is 27 ft between columns. The planks are 10-in. deep with 2-in. topping and supports stud wall partitions that weigh 10 psf and mechanical/electrical equipment that weigh 5 psf. The weight of the precast planks is 70 psf. The centerline-to-centerline spans of the planks is 30 ft on the left-hand side of the beam and 10 ft on the right-hand side of the beam. The live load on the 30-ft span is 50 psf while the live load on the 10-ft corridor span is 100 psf. Design the beam for torsion and shear assuming normal-weight concrete (i.e., $\lambda = 1.0$), $f_c' = 4000$ psi, and $f_y = 60{,}000$ psi. Assume the beam has already been designed for bending.

4.19 A reinforced concrete beam was designed with vertical stirrups (see the top figure below), but the stirrups were inadvertently fabricated with its height larger than the beam depth. To avoid having to discard all the incorrectly fabricated stirrups, the contractor's engineer has suggested using the fabricated stirrups in an inclined orientation, as shown in the bottom figure below. You have been asked to review this suggested stirrup layout.

a. Is the beam stirrup arrangement suggested by the contractor's engineer adequate?

b. If the stirrup arrangement is adequate, why is it adequate? If it's not adequate, why is it not adequate? And, if it's not adequate, what is the solution?

(Note: No new stirrups will be fabricated, and no calculation is needed)

c. Research the Harmon Tower in Las Vegas, an unfinished multi-story building, that had to be demolished.

(i) What were the reason(s) why the building was demolished?

(ii) What happened to the reinforcement in link or coupling beams between the shear walls?

(iii) Discuss the lessons learned from this building?

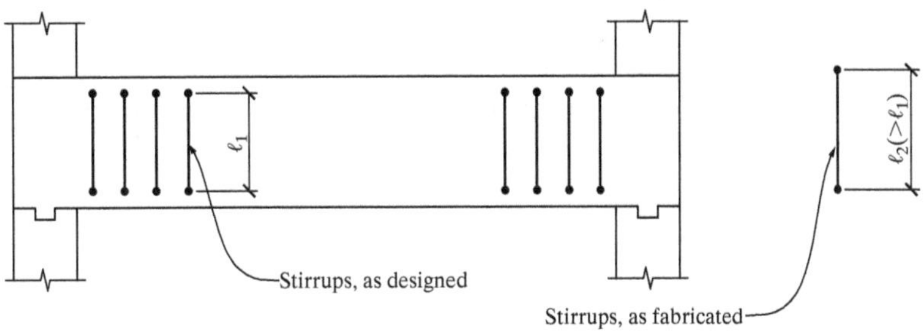

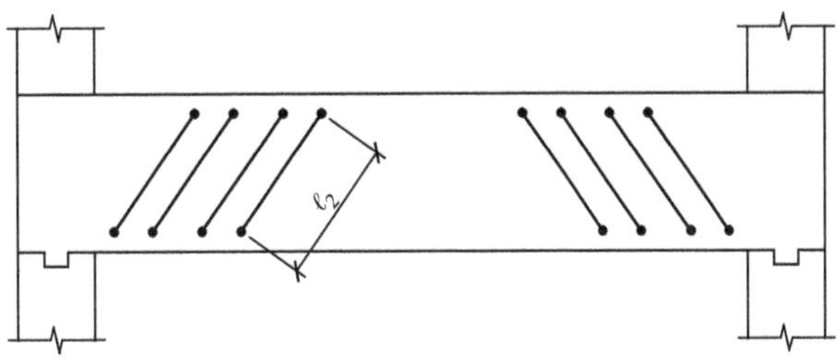

Contractor's Suggestion

EXERCISE 4.19

4.20 Refer to the given detail and shear diagram for a simple span beam. Use $f_c' =$ 6,000 psi, f_y = 60,000 psi

a. Determine the required spacing of the shear reinforcement in zones A and B (select uniform spacing in each zone).

b. Sketch the loading that goes along with this shear diagram.

4.21 Refer to the given detail and shear diagram. Use f_c' = 3,000 psi, f_y = 60,000 psi. Determine the required spacing of the shear reinforcement along the beam.

4.22 Assume the clear span of a beam is 36′-0″ and has dimensions of 12″ × 24″ beam, with d = 21.5″. There are (3) concentrated loads of 23 kip (factored) that are applied at 1/4 points. Add the self-weight as a uniform load (density = 150 pcf). Draw the loading diagram and shear diagram and determine the shear reinforcement for the beam. Sketch the complete design.

4.23 Design the corbel for a 20″ × 20″ column that supports a service dead load of 50 kip and a service live load of 110 kip. The concrete strength, f_c', is 4500 psi and the steel yield strength, f_y, is 60,000 psi. Select the dimensions of the corbel and all the required reinforcement per the ACI Code using the shear friction method.

4.24 Refer to the given detail and shear diagram for a simple span beam. Using f_c' = 3,000 psi and f_y = 60,000 psi:

a. Determine the required spacing of the shear reinforcement in zones A, B, & C (select uniform spacing in each zone).

b. Sketch the factored loading that goes along with this shear diagram.

CHAPTER 5

Development, Splices, and Simple-Span Bar Cutoffs

5.1 BOND STRESS AND DEVELOPMENT LENGTH: INTRODUCTION

One of the fundamental assumptions of reinforced concrete design is that at the interface of the concrete and the steel bars, perfect bonding exists, and no slippage occurs. That means the tension in the concrete is transferred from the concrete through bond to the steel reinforcement. Based on this assumption, it follows that some form of bond stress exists at the contact surface between the concrete and the steel bars. The bond action between the concrete and the deformed bar is primarily due to the wedge action of the protruding ribs of the reinforcement against the surrounding concrete (see Figure 5.1), creating equal and opposite radial forces that act on the bar ribs and the surrounding concrete. The effect of friction between the reinforcement surface and the concrete is small compared to the wedge action of the ribs. In beams, this bond stress is caused by the change in bending moment along the length of the beam and the accompanying change in the tensile stress in the bars and has historically been termed *flexural bond*. The actual distribution of bond stresses along the reinforcing steel is highly complex, due primarily to the presence of concrete cracks. Research has indicated that large local variations in bond stress are caused by flexural and diagonal cracks and very high bond stresses have been measured adjacent to these cracks [1, 2]. These high bond stresses may result in small local slips adjacent to the cracks, with a resultant widening of cracks as well as increased deflections. This will be harmless as long as failure does not propagate all along the bar with the resulting complete loss of bond. It is possible, if *end anchorage* is reliable, that the bond can be severed along the entire length of bar, excluding the anchorage, without endangering the carrying capacity of the beam. The resulting behavior is like that of a tied arch with the beam tension reinforcement acting as a tension tie anchored at the end supports.

End anchorage may be considered reliable if the bar is embedded into concrete a prescribed distance known as the *development length*, ℓ_d, of the straight bar or the reinforcement is hooked (see Figure 5.2), in which case the required development length of the hooked bar would be ℓ_{dh}. If in the beam the actual extended length of a bar is equal to or greater than this required development length, no premature bond failure will occur—that is, the predicted strength of the beam will not be controlled by bond but rather by some other factor.

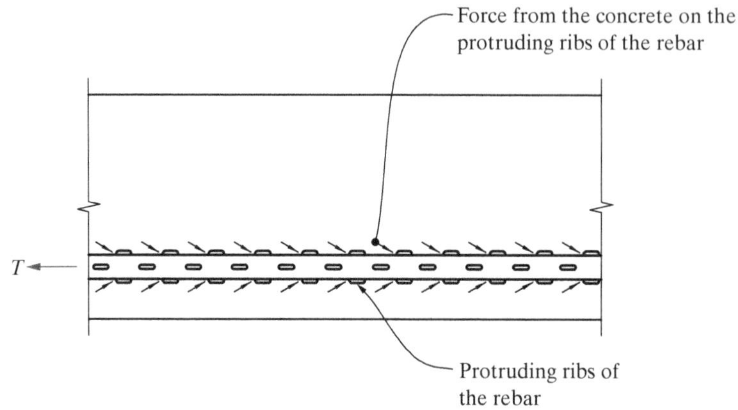

FIGURE 5.1 Bond transfer in concrete with deformed bars.

FIGURE 5.2 90° bends for development of No. 18 bars. Seabrook station, New Hampshire (George Limbrunner).

Hence, the main requirement for safety against bond failure is that the length of the bar from any point of given steel stress f_s (or, as a maximum, f_y) to its nearby free end must be at least equal to its development length. If this requirement is satisfied, the magnitude of the flexural bond stress along the beam is of only secondary importance because the integrity of the member is assured even in the face of possible minor local failures. If the actual available length is inadequate for full development of a straight bar, however, special anchorages, such as hooks, must be provided to ensure adequate strength.

The average bond stress, f_{bd}, between two cracks in a beam is the tension force in the steel reinforcement divided by the product of the sum of the circumference of the reinforcement and the length of the beam between the cracks and can be derived as follows using:

- Equilibrium of forces between the tension force in the reinforcement at the crack and the bond force that exists between the concrete and the steel between the two cracks.
- The principles of flexural behavior from Chapter 2.
- The relationship between shear and moment ($V = dM/dx$).

Average bond stress, $$f_{bd} = \frac{T}{\left(\sum \pi d_b\right) dx} = \frac{dM/jd}{\left(\sum \pi d_b\right) dx}$$

Since V = dM/dx.

The average bond stress can be rewritten as

$$f_{bd} = \frac{dM/jd}{\left(\sum \pi d_b\right) dx} = \frac{V}{\left(\sum \pi d_b\right) jd}$$

where

jd = the flexural lever arm

dM = the change in moment in the beam between the two cracks

dx = length of the beam between the two cracks

$\sum \pi d_b$ = sum of the circumference of all the tension bars in the beam between the two cracks

V = shear at the section.

Note that f_{bd} is an average bond stress, but the actual bond stress varies considerably between the cracks [1, 2]. The smaller the diameter of the reinforcement, the higher the bond stress (strength) as is evident from the preceding equation, and in regions of zero shear, the bond stress is zero. The strain in the concrete is zero at the cracks and at a maximum near the mid-distance between the cracks. The corresponding strain in the steel reinforcement is at a maximum at the crack locations and at a minimum midway between the cracks, where the stress is fully transferred from the steel to the concrete. The stress distribution in the steel and concrete between the cracks follows a similar pattern [1].

Current design methods based on the ACI Code (i.e., ACI 318-19) do not require calculation of the bond stress and disregard high localized bond stress even though it may result in localized slip between steel and concrete adjacent to the cracks. Instead, attention is directed toward providing an adequate length of embedment past the location at which the bar is fully stressed, which will ensure the development of the full strength of the bar (see Figure 5.3). The development length is the shortest length of reinforcement for which the rebar stress can increase from zero to the yield strength, f_y, without pulling out of the concrete. If the actual length of the straight rebar within the concrete cantilever beam in Figure 5.3 is less than ℓ_d, the rebar will not develop its maximum strength ($A_s f_y$). Similarly, if the length of the straight portion of the hooked bar within the column in Figure 5.3 is less than ℓ_{dh}, the hooked bar will not develop its maximum strength ($A_s f_y$). Consequently, the rebar will pull out of the concrete before its yield strength is reached.

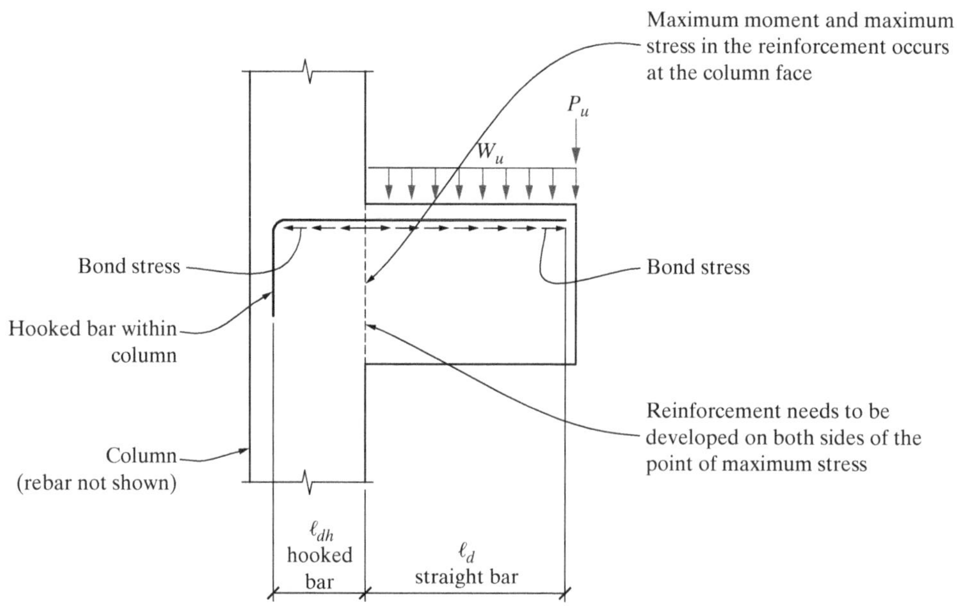

FIGURE 5.3 Development of tension reinforcement in a cantilever beam.

5.2 DEVELOPMENT LENGTH: TENSION BARS

Referring to Figure 5.3, the maximum tension force, T, in the steel reinforcement is

$$T = A_s f_y = \pi \frac{(d_b)^2}{4} f_y$$

The concrete transfers this tension force into the steel reinforcement through bond stress. The force transferred through bond action (i.e., the bond force) is the product of the average bond stress, f_{bd}, and the surface area of the rebar in contact with the concrete. This surface area is the product of the sum of the circumference of the steel reinforcement and the development length, ℓ_d.

Thus, the force transferred by bond = $f_{bd}\,\ell_d\,(\pi d_b)$

Equilibrium requires that the tension force be equal to the total bond force, thus

$$f_{bd}\,\ell_d\,(\pi d_b) = \pi\frac{(d_b)^2}{4}f_y$$

Therefore, the expression for development length, ℓ_d is given as

$$\ell_d = \left(\frac{f_y}{4f_{bd}}\right)d_b$$
$$\equiv K\left(\frac{f_y}{\sqrt{f_c'}}\right)d_b$$

Where K is an empirical factor determined from analysis of test results (ACI 318-19, Sections 25.4.2.2 and 25.4.2.3). The equations given in the ACI Code for calculating rebar development length take the same form as the preceding equation. The factors that affect the bond strength of a concrete flexural member, and hence the tension development length, are summarized below [1]:

1. Concrete compressive strength, f_c': The higher the concrete strength, the higher the tensile strength of the concrete, and therefore, the higher the bond strength, and the smaller the required development length.
2. Steel reinforcement yield strength, f_y: The higher the steel yield strength, the higher the required development length.
3. Steel reinforcement diameter: The smaller the rebar diameter, the smaller the required development length. With regards to bond strength, it is advantageous to use more smaller diameter bars than fewer larger diameter bars for the same total area of steel because of larger surface area that is obtained with the many smaller bars compared to fewer larger size rebar.
4. Concrete Cover: The smaller the side and bottom cover to the rebar, the greater the likelihood of concrete spalling and cracking due to the lack of adequate confinement to contain the radial wedging forces. Inadequate bottom covers lead to the development of vertical splitting cracks propagating from the rebar to the bottom surface. This leads to a lower bond strength and, therefore, a higher required development length.
5. Rebar spacing: The wider the spacing between the steel reinforcement, the better the bond between each rebar and the surrounding concrete, leading to a higher bond strength and a smaller development length. Inadequate center-to-center spacing between the tension bars leads to the development of horizontal splitting cracks between the bars that extend to the vertical surfaces of the beam. This leads to a reduced bond strength and a higher required development length.

6. Type of stress (tension or compression) in the reinforcement: The development length required for tension reinforcement is higher than that required for compression reinforcement because of the presence of flexural tension cracking.
7. Vertical location of the reinforcement within the depth of a beam or girder: When steel reinforcement is located such that there is more than 12 inches of concrete depth below the reinforcement, there is a tendency for entrapped air to gravitate to the underside of the steel rebar and for excess water to accumulate below the reinforcement due to settlement of the concrete during the concrete vibration process. This leads to a reduced bond between the rebar and the surrounding concrete and thus a higher required development length. The higher the concrete slump, the greater the tendency for the concrete to settle, and therefore, the lower the bond strength.
8. Transverse reinforcement (i.e., stirrups): Confinement of the concrete around the tension rebar by transverse reinforcement helps to prevent the occurrence of horizontal and vertical splitting cracks at the tension rebar locations. This leads to higher bond strength.
9. Lightweight concrete: The tensile strength of concrete is proportional to $\sqrt{f_c'}$, and from the equation developed earlier for development length, it can be seen that the greater the tensile strength, the lower the required development length, thus, the higher the bond strength. Conversely, the lower the tensile strength, as is the case with lightweight concrete compared to normal-weight concrete, the lower the bond strength and the higher the required development length.
10. Epoxy-coated rebar: The coating on epoxy-coated reinforcement protects it from corrosion, but studies of the anchorage of epoxy-coated bars show that bond strength is reduced because the coating prevents adhesion and friction between the bar and the concrete. This leads to a higher required development length.

The ACI Code, Section 25.4.2.1, specifies that the development length ℓ_d for deformed bars and deformed wires in tension shall be determined using either the tabular criteria of ACI Code, Section 25.4.2.3 or the general equation of ACI Code, Section 25.4.2.4, but in either case, ℓ_d shall not be less than 12 in.

The general equation of Section 25.4.2.3 (ACI Equation [25.4.2.4a]) offers a simple approach that allows the user to see the effect of all variables controlling the development length. The tabular criteria of Section 25.4.2.3 also offer a simple, conservative approach that recognizes commonly used practical construction techniques. Based on a sampling of numerous cases, the authors have found that significantly shorter development lengths are computed using ACI Equation (25.4.2.4a) of Section 25.4.2.4. Therefore, the development length ℓ_d computation used in this text will be based on ACI Equation (25.4.2.4a):

$$\ell_d = \frac{3}{40}\left(\frac{f_y}{\lambda\sqrt{f_c'}}\right)\left[\frac{\psi_t\psi_e\psi_s\psi_g}{\left(\dfrac{c_b + K_{tr}}{d_b}\right)}\right]d_b$$

in which the confinement term, $(c_b + K_{tr})/d_b$, shall not be taken greater than 2.5, the range in which splitting failure is likely to control the rebar embedment length needed to develop the yield strength of the rebar, and not bond failure or concrete breakout, and where

ℓ_d = development length (in.)

f_y = specified yield strength of non-prestressed reinforcement (psi)

f_c' = specified compressive strength of concrete (psi); the value of $\sqrt{f_c'}$ shall not exceed 100 psi (ACI Code, Section 25.4.1.4)

d_b = nominal diameter of bar or wire (in.).

The other factors used in ACI Equation (25.4.2.4a) are defined as follows:

1. ψ_t is a casting position factor that accounts for the position of the reinforcement in freshly placed concrete:

Where horizontal reinforcement is located such that more than 12 in. of fresh concrete is cast in the member below the reinforcement, use ψ_t = 1.3 (ACI Code, Table 25.4.2.5). This is because this condition lends itself to the formation of entrapped air and moisture on the underside of the horizontal reinforcement, resulting in partial loss of bond between the concrete and steel. Typically, ψ_t = 1.3 will apply to the top bars in beams and girders. For all other conditions, ψ_t = 1.0.

For other reinforcement, use ψ_t = 1.0.

2. ψ_e is a coating factor reflecting the effects of epoxy coating (ACI Code, Table 25.4.2.5). Studies of the anchorage of epoxy-coated bars show that bond strength is reduced because the coating prevents adhesion and friction between the bar and the concrete.

For epoxy-coated or zinc and epoxy dual-coated reinforcement with clear cover less than $3d_b$ or the clear spacing between bars less than $6d_b$, use ψ_e = 1.5.

For epoxy-coated or zinc and epoxy dual-coated reinforcement for all other conditions, use ψ_e = 1.2.

For uncoated or zinc-coated (galvanized) reinforcement, use ψ_e = 1.0.

The product of ψ_t and ψ_e need not be taken greater than 1.7 (ACI Code, Table 25.4.2.5).

3. ψ_s is a reinforcement size factor (ACI Code, Table 25.4.2.5).

For No. 7 and larger bars, use ψ_s = 1.0

For No. 6 and smaller bars, and deformed wires, use ψ_s = 0.8

4. ψ_g is a reinforcement grade factor that accounts for the yield strength of the reinforcement (ACI Code, Table 25.4.2.5).

For Grade 60 reinforcement, use ψ_g = 1.0

For Grade 80 reinforcement, use $\psi_g = 1.15$

For Grade 100 reinforcement, use $\psi_g = 1.3$

5. Lambda (λ) is the lightweight-aggregate concrete factor and has been discussed in Section 1.8 of Chapter 1. For purposes of development length calculation where lightweight concrete is used, λ shall not exceed 0.75 unless average f_{ct} is specified, in which case it can be calculated (see Section 1.8). For normal-weight concrete, $\lambda = 1.0$.

6. The factor c_b represents a spacing or cover dimension (in.).

   The value of c_b will be smaller of either the distance from the center of the bar to the nearest concrete surface or one-half the center-to-center spacing of the bars being developed.

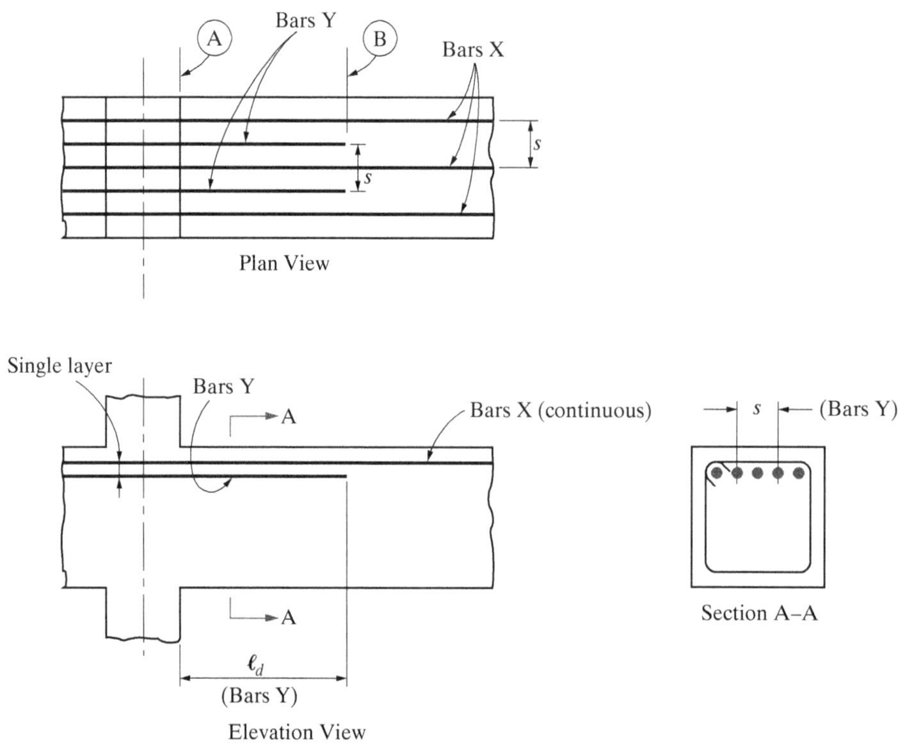

FIGURE 5.4 Spacing criteria for bars being developed.

The bar spacing will be the actual center-to-center spacing between the bars if adjacent bars are all being developed at the same location. If, however, an adjacent bar has been developed at another location, the spacing to be used will be greater than the actual spacing to the adjacent bar. Note in Figure 5.4 that the spacing for bars Y may be taken the same as for bars X because bars Y are developed in length AB, whereas bars X are developed at a location other than AB. When the bottom cover of the rebar is smaller than one-half the spacing between the longitudinal bars, vertical splitting cracks propagating from the rebar to the bottom surface occur. When the bottom concrete cover to the rebar is greater than one-half the horizontal spacing between longitudinal bars, horizontal splitting cracks occur between the bars, and the cracks extend to the exterior vertical surfaces of the beam [1].

7. The transverse reinforcement index, K_{tr}, is computed from

$$\frac{40A_{tr}}{sn}$$

where

A_{tr} = total cross-sectional area of all transverse reinforcement that is within the spacing s and that crosses the potential plane of splitting through the reinforcement being developed (in.^{2})

s = maximum center-to-center spacing of transverse reinforcement within ℓ_d (in.)

n = number of bars or wires being developed along the plane of splitting

To simplify the design, the ACI Code permits the use of $K_{tr} = 0$. This is conservative and may be used even if transverse reinforcement is present. To further simplify development length computations, we designate a portion of ACI Equation (25.4.2.4a) as K_D, where

$$K_D = \frac{3}{40}\frac{f_y}{\sqrt{f_c'}}$$

Values of K_D as a function of various combinations of f_y and f_c' are tabulated in Table 5.1.

A reduction in the required straight bar development length ℓ_d is permitted (see ACI Code, Section 25.4.10.1) where reinforcement is in excess of that required by analysis (except where anchorage or development for f_y is specifically required or where the design includes provisions for seismic considerations). We designate this reduction factor as K_{ER}. Excess reinforcement factor K_{ER} does *not* apply to the following conditions (ACI Code, Section 25.4.10.2):

- At discontinuous support.
- At locations where anchorage or development of the yield strength, f_y, is required.
- Development of shrinkage and temperature reinforcement.
- At locations where the reinforcement is required to be continuous.
- For hooked, headed, and mechanically anchored deformed reinforcement.
- For anchorage of concrete piles and concrete-filled pipe piles to pile caps in structures in Seismic Design Category (SDC) C, D, E, and F.
- For anchorage of concrete piles and concrete-filled pipe piles to pile caps in structures in Seismic Design Category (SDC) C, D, E, and F.

Although K_{ER} is not reflected in ACI Equation (25.4.2.4a), it may be calculated from

$$K_{ER} = \frac{A_s \text{ required}}{A_s \text{ provided}}$$

and subsequently applied to (i.e., multiplied by) the ℓ_d computed from ACI Equation (25.4.2.4a).

When bundled bars are used, the ACI Code, Section 25.6.1.5, stipulates that calculated development lengths are to be made for individual bars within a bundle (either in tension or in compression) and then increased by 20% for three-bar bundles and by 33% for four-bar bundles. Bundled bars consist of a group of not more than four parallel reinforcing bars in contact with each other and are assumed to act as a unit. For determining the appropriate factors, as discussed previously, a unit of bundled bars shall be treated as a single bar of a diameter derived from the equivalent total area.

TABLE 5.1 Coefficient K_D for ACI Code Equation (25.4.2.4a).

$K_D = \frac{3}{40}\frac{f_y}{\sqrt{f_c}}$				
f'_c (psi)	f_y = 40,000 psi	f_y = 50,000 psi	f_y = 60,000 psi	f_y = 75,000 psi
3000	54.8	68.5	82.2	102.7
4000	47.4	59.3	71.2	88.9
5000	42.4	53.0	63.6	79.5
6000	38.7	48.4	58.1	72.6

Summary of Procedures for Calculation of ℓ_d (Using ACI Equation (25.4.2.4a))

1. Determine K_D from Table 5.1.
2. Determine applicable factors (use 1.0 unless otherwise determined).

    a. Use ψ_t = 1.3 for top reinforcement, when applicable.

    b. Coating factor ψ_e applies to epoxy-coated bars. Use ψ_e = 1.5 if cover < $3d_b$ or clear space < $6d_b$; Use ψ_e = 1.2 otherwise. For uncoated bars, use ψ_e = 1.0.

    c. Use ψ_s = 0.8 for No. 6 bars and smaller (ψ_s = 1.0 otherwise).

    d. ψ_g = 1.0 for Grade 60 steel, 1.15 for Grade 80 rebar, and 1.3 for Grade 100 rebar.

    e. Use λ = 0.75 for lightweight concrete with f_{ct} not specified and λ = 1.0 for normal-weight concrete. Use

    $$\lambda = f_{ct}/\left(6.7\sqrt{f'_c}\right) \le 1.0$$

    for lightweight concrete with f_{ct} specified.
3. Check that $\psi_t\,\psi_e \le 1.7$
4. Determine c_b, the smaller of the concrete cover, measured to the center of the bar, or one-half the center-to-center spacing between the bars.
5. Calculate $K_{tr} = 40A_{tr}/(sn)$ or use $K_{tr} = 0$ (conservative).
6. Check $(c_b + K_{tr})/d_b \le 2.5$.

7. Calculate K_{ER} if applicable:

$$K_{ER} = \frac{A_s \text{ required}}{A_s \text{ provided}}$$

8. Calculate ℓ_d:

$$\ell_d = \frac{K_D}{\lambda}\left[\frac{\psi_t \psi_e \psi_s \psi_g}{\left(\frac{c_b + K_{tr}}{d_b}\right)}\right] K_{ER}\, d_b \geq 12 \text{ in.}$$

EXAMPLE 5.1

Calculate the tensile development length ℓ_d required for No. 8 top bars (more than 12 in. of fresh concrete to be cast below the bars) in a lightweight-aggregate concrete beam as shown in Figure 5.5. The clear cover is 2 in., and the clear space between bars is 3 in. Use f_y = 60,000 psi and f_c' = 4000 psi. Stirrups are No. 4 bars. All bars are uncoated. f_{ct} is not specified.

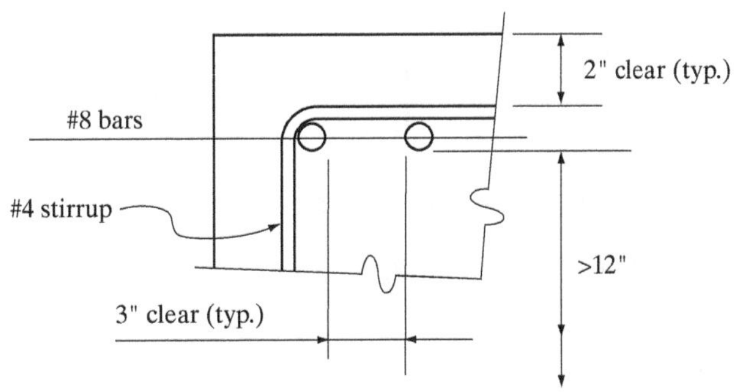

FIGURE 5.5 Partial cross-section.

Solution:

Use ACI Equation (25.4.2.4a) and follow the procedural outline that precedes this example.

1. From Table 5.1, K_D = 71.2.
2. Establish values for the factors ψ_t, ψ_e, ψ_s, ψ_g and λ.
   - **a.** ψ_t = 1.3 (the bars are top bars).
   - **b.** The bars are uncoated; ψ_e = 1.0.
   - **c.** The bars are No. 8; ψ_s = 1.0.
   - **d.** ψ_g = 1.0 for Grade 60 rebar
   - **e.** Lightweight-aggregate concrete is used; λ = 0.75.
3. The product $\psi_t \times \psi_e$ = 1.3 < 1.7. (O.K.)

4. Determine c_b. Based on cover (center of bar to nearest concrete surface), consider the clear cover, the No. 4 stirrup diameter, and one-half the diameter of the No. 8 bar:

$$c_b = 2 + 0.5 + 0.5 = 3.0 \text{ in.}$$

Based on bar spacing (one-half the center-to-center distance),

$$c_b = 0.5[3 + 2(0.5)] = 2.0 \text{ in.}$$

Therefore, use c_b = 2.0 in.

5. In the absence of data needed for a calculation, K_{tr} may be conservatively taken as zero.
6. Check $(c_b + K_{tr})/d_b \le 2.5$:

$$\frac{c_b + K_{tr}}{d_b} = \frac{2.0 + 0}{1.0} = 2.0 < 2.5 \qquad \text{(O.K.)}$$

7. The excess reinforcement factor is assumed not applicable and is omitted.
8. Calculate ℓ_d:

$$\ell_d = \frac{K_D}{\lambda}\left[\frac{\psi_t \psi_e \psi_s \psi_g}{\left(\frac{c_b + K_{tr}}{d_b}\right)}\right] d_b$$

$$= \frac{71.2}{0.75}\left[\frac{1.3(1.0)(1.0)(1.0)}{2.0}\right](1.0) = 61.7 \text{ in.} > 12 \text{ in.} \quad \text{(O.K.)}$$

EXAMPLE 5.2

Calculate the development length required for the No. 9 bars in the top of a 15-in.-thick reinforced concrete slab (see Figure 5.6). Note that these bars are the tension reinforcement for negative moments in the slab at the supporting beam. As this is a slab, no stirrups are used. Use f_y = 60,000 psi and f_c' = 4000 psi (normal-weight concrete). The bars are epoxy-coated.

Solution:

1. From Table 5.1, K_D = 71.2.
2. Establish values for the factors ψ_t, ψ_e, ψ_s, ψ_g and λ.
    a. ψ_t = 1.3 (the bars are top bars).
    b. The bars are epoxy-coated. Compare the cover (3/4 in.) with $3d_b$. If necessary, calculate the clear spacing between bars and compare with $6d_b$.

$$3d_b = 3(1.13) = 3.39 \text{ in.}$$

$$0.75 \text{ in.} < 3.39 \text{ in.}$$

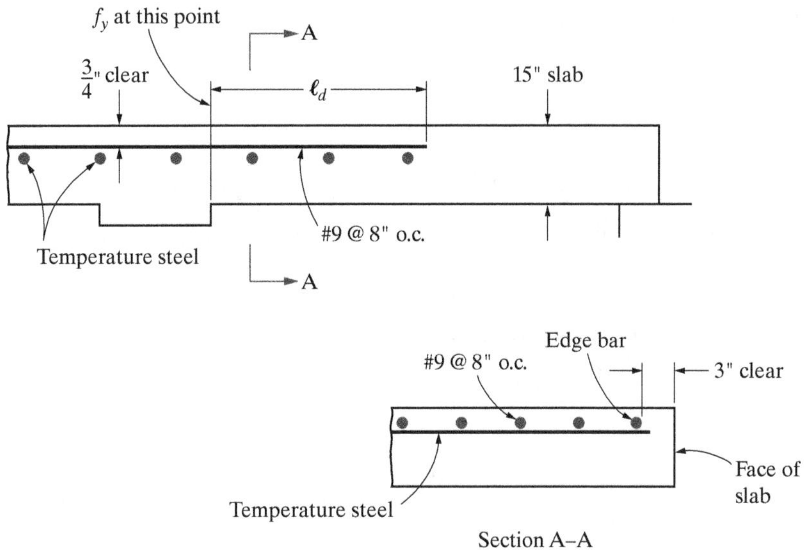

FIGURE 5.6 Sketch for Example 5.2.

Therefore, use $\psi_e = 1.5$

c. The bars are No. 9. Use $\psi_s = 1.0$

d. For Grade 60 rebar, $\psi_g = 1.0$

e. Normal-weight concrete is used. $\lambda = 1.0$

3. Check the product of ψ_t and ψ_e:

$$\psi_t \times \psi_e = 1.3(1.5) = 1.95 > 1.7$$

Therefore, use $\psi_t \times \psi_e = 1.7$.

4. Determine c_b. Based on cover (center of bar to nearest concrete surface), consider the clear cover and one-half the diameter of the No. 9 bar:

$$c_b = 0.75 + \frac{1.128}{2} = 1.314 \text{ in.}$$

Based on bar spacing (one-half the center-to-center distance),

$$c_b = 0.5(8.0) = 4.0 \text{ in.}$$

Therefore, use $c_b = 1.314$ in.

5. K_{tr} is taken as zero. (There is no transverse reinforcement crossing the plane of splitting.)

6. Check $(c_b + K_{tr})/d_b \le 2.5$:

$$\frac{c_b + K_{tr}}{d_b} = \frac{1.314 + 0}{1.128} = 1.165 < 2.5 \qquad \text{(O.K.)}$$

7. The excess reinforcement factor is assumed not applicable and is omitted.

8. Calculate ℓ_d (recall that $\psi_t \times \psi_e$ will be taken as 1.7 from step 3):

$$\ell_d = \frac{K_D}{\lambda}\left[\frac{\psi_t \psi_e \psi_s \psi_g}{\left(\dfrac{c + K_{tr}}{d_b}\right)}\right] d_b$$

$$= \frac{71.2}{1.0}\left[\frac{1.7(1.0)(1.0)(1.0)}{1.165}\right](1.128) = 117.2 \text{ in.} > 12 \text{ in.} \quad \text{(O.K.)}$$

EXAMPLE 5.3

Calculate the development length required for the interior two No. 7 bars in the beam shown in Figure 5.7. The two No. 7 outside bars are continuous for the full length of the beam. Use f_y = 60,000 psi and $f_c' = 4000$ psi (normal-weight concrete). The bars are uncoated. Assume that, from the design of this member, the required tension steel area was 2.28 in.^{2}.

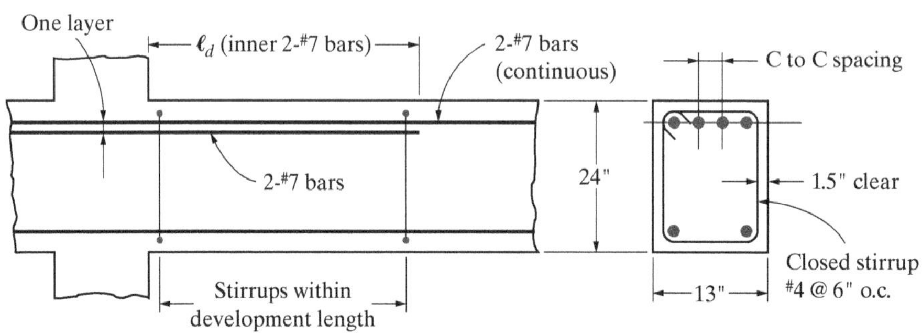

FIGURE 5.7 Sketch for Example 5.3.

Solution:

1. From Table 5.1, K_D = 71.2.

2. Establish values for the factors ψ_t, ψ_e, ψ_s, ψ_g and λ.

   a. ψ_t = 1.3 (the bars are top bars).

   b. The bars are uncoated; ψ_e = 1.0.

c. The bars are No. 7; $\psi_s = 1.0$.

d. For Grade 60 rebar, $\psi_g = 1.0$

e. Normal-weight concrete is used; $\lambda = 1.0$.

3. The product $\psi_t \times \psi_e = 1.3 < 1.7$. (O.K.)

4. Determine c_b. Based on cover (center of bar to nearest concrete surface), consider the clear cover, the No. 4 stirrup diameter, and one-half the diameter of the No. 7 bar:

$$c_b = 1.5 + 0.5 + \frac{0.875}{2} = 2.44 \text{ in.}$$

Based on bar spacing (one-half the center-to-center distance),

$$c_b = \frac{13 - 2(1.5) - 2(0.5) - 2\left(\frac{0.875}{2}\right)}{3(2)} = 1.354 \text{ in.}$$

Therefore, use $c_b = 1.354$ in.

5. Using data on stirrups from Figure 5.7:

$$K_{tr} = \frac{40A_{tr}}{(sn)} = \frac{40(0.40)}{(6)(2)} = 1.333$$

6. Check $(c_b + K_{tr})/d_b \leq 2.5$:

$$\frac{c_b + K_{tr}}{d_b} = \frac{1.354 + 1.333}{0.875} = 3.07 > 2.5$$

Therefore, use 2.5.

7. Calculate the excess reinforcement factor: (tension steel provided is 4 No. 7)

$$K_{ER} = \frac{A_s \text{ required}}{A_s \text{ provided}} = \frac{2.28 \text{ in.}^2}{2.40 \text{ in.}^2} = 0.95$$

8. Calculate ℓ_d:

$$\ell_d = \frac{K_D}{\lambda}\left[\frac{\psi_t\psi_e\psi_s\psi_g}{\left(\frac{c_b + K_{tr}}{d_b}\right)}\right]K_{ER}d_b$$

$$= \frac{71.2}{1.0}\left[\frac{1.3(1.0)(1.0)(1.0)}{2.5}\right](0.95)(0.875) = 30.8 \text{ in.} > 12 \text{ in.} \quad \text{(O.K.)}$$

5.3 DEVELOPMENT LENGTH: COMPRESSION BARS

Whereas tension bars produce flexural tension cracking in the concrete, this effect is not present with compression bars, and thus shorter development lengths are allowed. For deformed bars in compression, the development length ℓ_{dc} is calculated from (see ACI Code, Section 25.4.9):

$$\ell_{dc} = \left(\frac{0.02 f_y \psi_r}{\lambda \sqrt{f'_c}} \right) d_b \geq 0.0003 f_y \psi_r d_b$$

in inches, where all quantities are as described previously for tension development length. This value of ℓ_{dc} is tabulated in Table A.12 for $\lambda = 1.0$, grade 60 bars, and various values of f'_c. The required compression development length may be further reduced by multiplying ℓ_{dc} by the following excess reinforcement modification factor:

$$\frac{A_s \text{ required}}{A_s \text{ provided}}$$

NOTE *ψ_r accounts for the effect of confining reinforcement. Bars enclosed within a spiral that is not less than $\frac{1}{4}$ in. in diameter and not more than 4 in. in pitch or within No. 4 ties in conformance with ACI Code, Section 25.7.2, and spaced at not more than 4 in. on center, or hoops in accordance with ACI Code, Section 25.7.4 and spaced at not more than 4 in. on center: Use ψ_r = 0.75 (For all other cases, use ψ_r = 1.0)*

The value of ℓ_{dc} shall not be less than 8 in.

5.4 DEVELOPMENT LENGTH: STANDARD HOOKS IN TENSION

If the desired development length in tension cannot be furnished, it is necessary to provide mechanical anchorage at the end of the bars. Although the ACI Code (Section 25.4) allows any mechanical device to serve as anchorage if its adequacy is verified by testing, anchorage for reinforcement is usually accomplished by means of a 90° or 180° hook. The dimensions and bend radii for these hooks have been standardized by the ACI Code. Additionally, 90° and 135° hooks have been standardized for stirrups and tie reinforcement. Hooks in compression bars are ineffective and cannot be used as anchorage. Standard reinforcement hooks are shown in Figure 5.8. The bend diameters are measured on the inside of the bar.

The anchorage strength, and therefore, the required development length for hooked bars is dependent on the spacing of the bars and the degree of confinement of the hooked bars by closed stirrups. If the hooked bars are closely spaced, the anchorage strength decreases and therefore the required development length increases. If, however, the hooked bars are closely spaced, but are confined with closed stirrups, the anchorage strength reduction effect of the closely spaced hooked bars is much less [7]. Therefore, higher anchorage strength can be achieved either by using widely spaced hooked bars *or* by providing adequate confinement of the closely spaced hooked bars using closed stirrups with sufficient total area within the hooked bar development length.

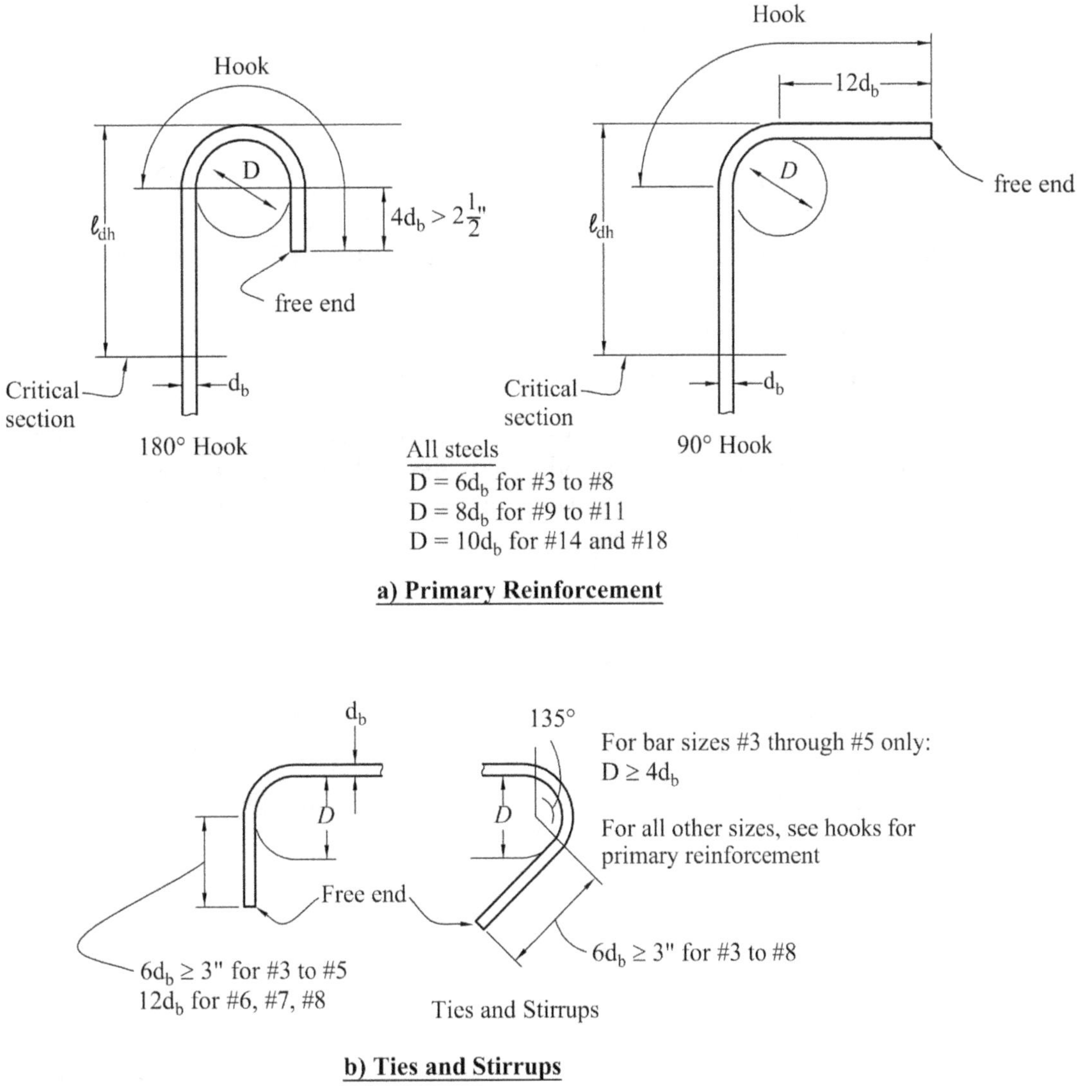

FIGURE 5.8 ACI Standard Hook Dimensions: (a) Primary Reinforcement, (b) Ties and Stirrups.

The ACI Code, Section 25.4.3, specifies that the development length ℓ_{dh} (see Figure 5.9) for deformed bars in tension that terminates in a standard hook, be computed using the following expression:

$$\ell_{dh} = \left[\frac{f_y \Psi_e \Psi_r \Psi_o \Psi_c}{55\lambda\sqrt{f_c'}}\right](d_b)^{1.5}$$

$$\geq 8d_b$$

$$\geq 6 \ in.$$

Where,

ψ_e = 1.2 for epoxy-coated or zinc and epoxy dual-coated reinforcing (1.0 otherwise)

ψ_r = (effect of confining reinforcement) = 1.0 for No. 11 and smaller bars with $A_{th} \geq 0.4A_{hs}$, or center-to-center spacing of the hooked bars, $s \geq 6d_b$. For all other conditions, ψ_r = 1.6.

Where

A_{th} = total cross-sectional area of both legs of ties or stirrups confining the hooked bars, in.^{2}, located within the hooked bar development length, ℓ_{dh}.

A_{hs} = total cross-sectional area of hooked or headed bars being developed at a critical section, in.^{2}

d_b = diameter of the hooked bars

s = minimum center-to-center spacing of the hooked bars

The confining ties or stirrups must consist of 135° hooks that can be placed parallel to the straight portion of the tension bars (i.e., perpendicular to the hook tail extension) or perpendicular to the straight portion of the tension bars, which terminate with a 90° or 180° hook and the distribution and spacing of the confining reinforcement to the hooked bars are as shown in Figure 5.9 (see ACI 318-19, Figures 25.4.3.3a and 25.4.3.3b).

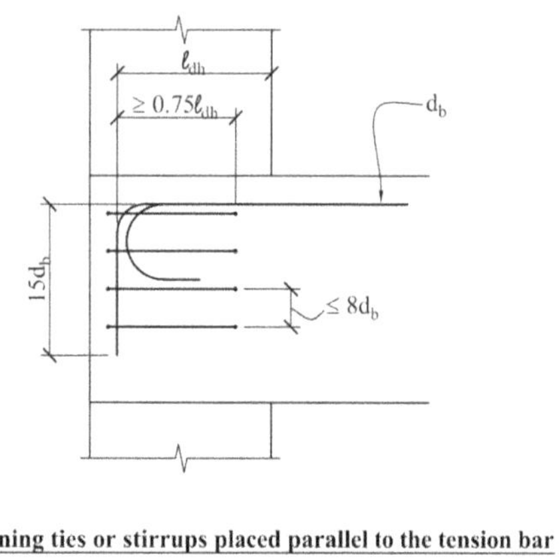

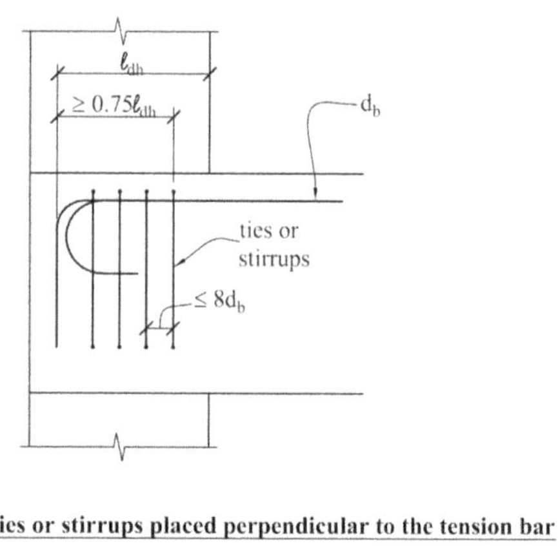

FIGURE 5.9 Distribution of Confining Ties and Stirrups for 90° and 180° Hooked Bars at a Beam-Column Joint. (Adapted from ACI 318-19 Figures 25.4.3.3a and 25.4.3.3b)

When the closed ties or stirrups confining the hook are placed *parallel* to the straight portion of the top bars (that terminate with a standard 90° or 180° hook), the closed ties or stirrups (whose horizontal dimension parallel to the beam should be no less than length $0.75\ell_{dh}$) should be located within a depth of $15d_b$ from the centroid of the straight portion of the hooked top bars of the beam, and spaced vertically apart no greater than $8d_b$, where d_b is the diameter of the hooked bar (see Figure 5.9a).

When the closed ties or stirrups are placed *perpendicular* to the straight portion of the top bars (that terminates with a standard 90° or 180° hook), the closed ties or stirrups should be located at the end of the beam within a horizontal distance of at least $0.75\ell_{dh}$ from the hook, and spaced horizontally apart no greater than $8d_b$, where d_b is the diameter of the hooked bar (see Figure 5.9b).

Recent research and tests [7] have shown that closely spaced hooked bars or hooked bars that are not well confined with closed ties or stirrups will fail by concrete breakout at a lower load than previously indicated in the ACI 318-2014 Code, hence the new equation for hooked bar development length in the ACI 318-2019 Code as presented above. For No. 14 and 18 bars, use a confining reinforcement factor, $\psi_r = 1.6$ unless the bars are widely spaced such that the spacing of the top bars, $s \geq 6d_b$.

$\psi_o = 1.0$ for No. 11 and smaller diameter hooked bars satisfying the following conditions (see Figure 5.10a for the definition of side cover to hooked bars):

(1) Terminating inside column core with *side cover* normal to plane of hook ≥ 2.5 in., or

(2) With *side cover* normal to plane of hook $\geq 6d_b$

For all other cases, the location or side cover factor, $\psi_o = 1.25$.

The concrete strength factor, $\psi_c = 1.0$ for $f'_c \geq 6000$ psi.

For $f'_c < 6000$ psi, $\psi_c = (f'_c/15{,}000) + 0.6$

Where f'_c is in psi.

$\lambda = 0.75$ for lightweight concrete (1.0 for normal-weight concrete). See ACI 318 Section 25.4.3.2.

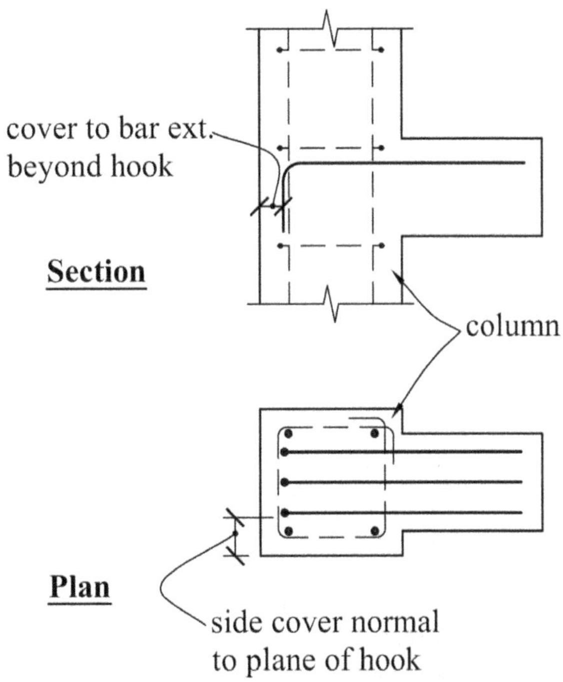

FIGURE 5.10a Covers to Hooked Bars.

Table A.13 gives values of ℓ_{dh} for various values of f_c'; and $f_y = 60{,}000$ psi; $\psi_e = \psi_r = \psi_o = \psi_c = 1.0$. The final ℓ_{dh} shall not be less than $8d_b$ nor less than 6 in. Note that the effect of excess reinforcement, K_{ER}, and the effect of casting position of the reinforcement, ψ_t, that were considered previously for straight tension bar development, ℓ_d, are not applicable for hooked bars (see ACI 318-19, Sections 25.4.10.2 and R25.4.3.2).

For the development of headed deformed bars in tension, the reader should refer to ACI 318, Section 25.4.4.

In addition, the ACI Code, Section 25.4.3.4, establishes criteria for hooked bars that terminate at the discontinuous ends of members such as:

- simply supported beams, and
- free ends of cantilevers.

If the full strength (f_y) of the hooked bar must be developed and if both the side cover and the top (or bottom) cover over the hook are less than $2\frac{1}{2}$ in., closed ties or stirrups perpendicular to the bar being developed and spaced at $3d_b$ maximum are required over the development length ℓ_{dh}. The first tie or stirrup shall enclose the bent portion of the hook, within $2d_b$ of the bend (see Figure 5.10b). This requirement does not apply to the discontinuous ends of slabs with concrete confinement provided by the slab continuous on both sides perpendicular to the plane of the hook.

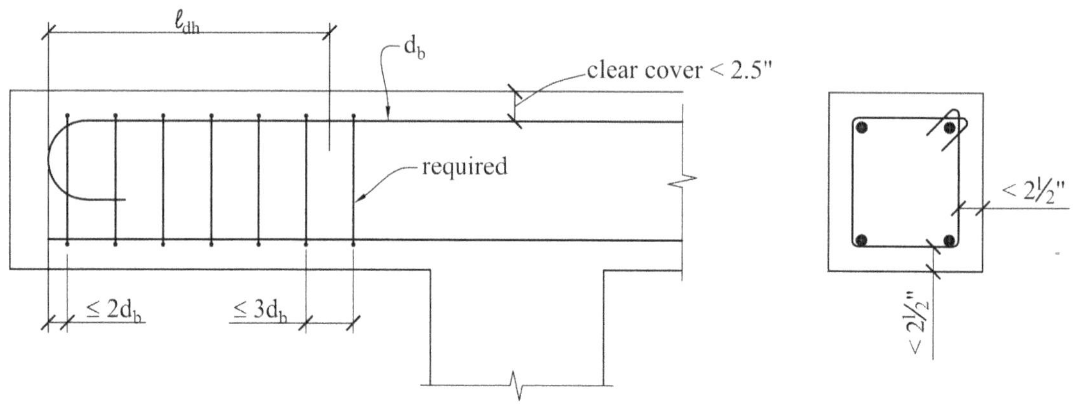

FIGURE 5.10b Hooked Tension Bars at Discontinuous Ends.

Development of Tension Bars at Wide Beam-to-Column Connections

One condition in which a beam top reinforcement may not be fully developed within a column, occurs when a wide beam or girder connects to a column that has a smaller width such that some of the beam top reinforcement fall outside the width or footprint of the column. If the analysis assumes that a fully rigid connection exists between the wide beam and the column, then enough top rebar in the beam must be placed within the column width (i.e., within the column vertical reinforcement cage) to ensure adequate development of the rebar in the column and a realization of the moment capacity of the beam-column connection that was assumed in the structural analysis. The dimension of the column parallel to the beam top rebar must also be at least equal to the required development length of the hooked top reinforcement in the beam plus the concrete cover to the vertical reinforcement in the column (see Figures 5.11 and 5.14).

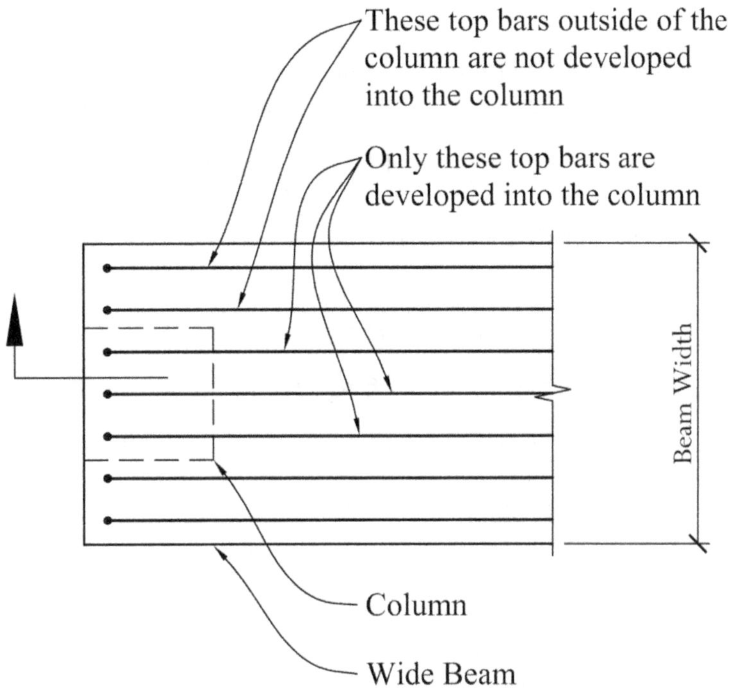

FIGURE 5.11 Development of a wide beam top rebar in a column (plan view) at a Wide Beam-to-Column Joint.

EXAMPLE 5.4

Determine the anchorage or development length required for the tension (top) bars for the conditions shown in Figure 5.12. Use $f_c' = 3000$ psi (normal-weight concrete) and $f_y = 60{,}000$ psi. The No. 8 bars may be categorized as top bars. Assume a side cover on the main bars of $2\frac{1}{2}$ in. minimum. Bars are uncoated.

Solution:

a. Anchorage of straight top bars into the exterior column

  1. From Table 5.1, $K_D = 82.2$.

  2. Establish values for the factors ψ_t, ψ_e, ψ_s, ψ_g, and λ.

    a. The bars are top bars with at least 12 inches of concrete cast below them: $\psi_t = 1.3$

    b. The bars are uncoated: $\psi_e = 1.0$.

c. The bars are No. 8 bars: $\psi_s = 1.0$.

d. For Grade 60 reinforcement: $\psi_g = 1.0$.

e. Normal-weight concrete is used: $\lambda = 1.0$.

3. The product $\psi_t \times \psi_e = 1.3 < 1.7$ (O.K.).

4. Determine c_b. Based on cover (center of bar to nearest concrete surface), consider the clear cover, the No. 4 stirrup diameter, and one-half the diameter of the No. 8 bar:

$$c_b = 2.0 + 0.5 + \frac{1.0}{2} = 3.0 \text{ in.}$$

Based on bar spacing of one-half the center-to-center distance between adjacent No. 8 bars,

$$c_b = \left(\frac{1}{2}\right)\left(\frac{14 - 2(2.0) - 2(0.5) - 2(0.5)}{(2\ spaces)}\right) = 2.0 \text{ in.}$$

Therefore, using the smaller of the two values: $c_b = 2.0$ in.

5. Figure 5.12 shows stirrups in the beam. However, there are no stirrups in the column and K_{tr} is taken as 0.

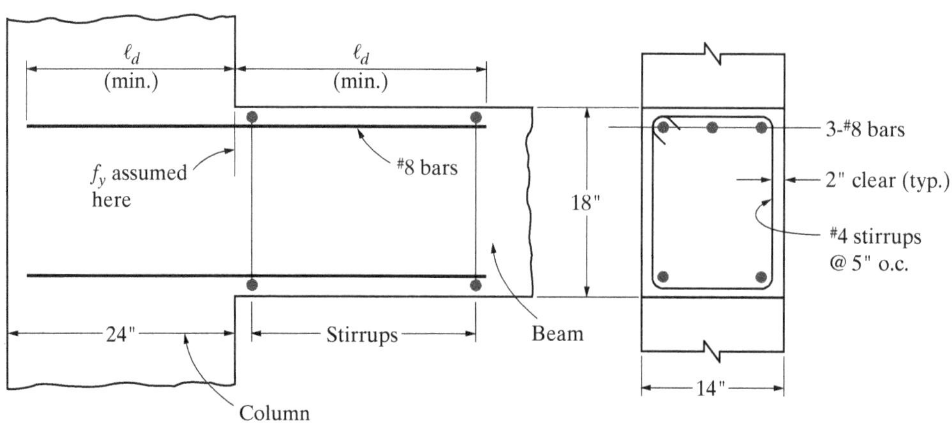

FIGURE 5.12 Sketch for Example 5.4.

6. Check $(c_b + K_{tr})/d_b \leq 2.5$:

$$\frac{c_b + K_{tr}}{d_b} = \frac{2.0 + 0}{1.0} = 2.0 < 2.5 \qquad \text{(O.K.)}$$

7. The excess reinforcement factor is assumed not applicable and is omitted.

8. Calculate the straight rebar tension development length, ℓ_d:

$$\ell_d = \frac{K_D}{\lambda}\left[\frac{\psi_t\psi_e\psi_s\psi_g}{\left(\dfrac{c_b + K_{tr}}{d_b}\right)}\right]d_b$$

$$= \frac{82.2}{1.0}\left[\frac{(1.3)(1.0)(1.0)(1.0)}{2.0}\right](1.0) = 53.4 \text{ in.} > 12 \text{ in.} \quad \text{(O.K.)}$$

Because 53.4 in. > 24 in. column width, use a standard hook, either a 90° hook or a 180° hook.

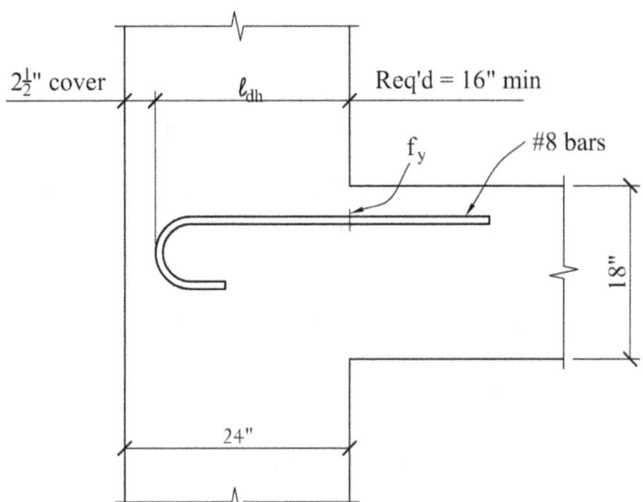

FIGURE 5.13 Sketch for Example 5.4: Hooked Tension Bars at Beam-Column Joint.

b. Anchorage using a standard 180° hook

1. The development length ℓ_{dh} for the hooked bar shown in Figure 5.13 is calculated from

$$\ell_{dh} = \left[\frac{f_y\psi_e\psi_r\psi_o\psi_c}{55\lambda\sqrt{f'_c}}\right](d_b)^{1.5}$$
$$\geq 8d_b$$
$$\geq 6 \; in.$$

2. Establish values for the factors ψ_e, ψ_r, ψ_o, ψ_c, and λ.

a. The bars are uncoated rebar: $\psi_e = 1.0$

b. Minimum center-to-center spacing of hooked No. 8 top bars is
s = ½ [14 in. – 2(2 in. cover) –(0.5 in.) – (0.5 in.) – ½ (1 in.) – ½ (1 in.)] = 4.0 in. < $6d_b$ = 6(1 in.) = 6 in.

or

A_{hs} = total cross-sectional area of hooked bars being developed at the critical section = area of 4 No. 8 top bars = 3.16 in.^{2}

0.4 A_{hs} = 0.4(3.16) = 1.26 in.^{2}

A_{th} = total cross-sectional area of both legs of ties or stirrups confining the hooked bars, in.^{2}, located within the hooked bar development length, ℓ_{dh}.

= area of both legs of No. 4 stirrups = 2(0.2)(16 in./5 in.) = 1.28 in.^{2} > 0.4 A_{hs}

Therefore, these hooked bars are closely spaced, but there is confining reinforcement, and therefore the confining reinforcement factor, ψ_r = 1.0

c. The hooked bars are No. 8 bars (i.e., No. 11 and smaller) terminate inside a column core with a side cover normal to the top bars = 2 in. clear cover + 0.5 in. stirrup = 2.5 in. ≥ 2.5 in. Therefore, the location factor, ψ_o = 1.0.

d. For f_c' = 3000 psi: ψ_c = (3000/15000) + 0.6 = 0.8

e. Normal-weight concrete: λ = 1.0.

The hooked bar development length is

$$\ell_{dh} = \left[\frac{f_y \Psi_e \Psi_r \Psi_o \Psi_c}{55\lambda\sqrt{f_c'}}\right](d_b)^{1.5} = \left[\frac{(60{,}000)(1.0)(1.0)(1.0)(0.8)}{55(1.0)\sqrt{(3000)}}\right](1.0)^{1.5} = 16\ in.$$

$$\geq 8d_b = 8(1.0) = 8\ in.$$

$$\geq 6\ in.$$

(Check this with Table A.13: (1.0)(1.0)(1.0)(0.8)(19.9 in.) = 16 in.)

2. Therefore, the required development length of the No. 8 hooked top bars is,

$$\ell_{dh} = 16\text{ in.}$$

The minimum width of column required is

$$16\text{ in.} + 2.5\text{ in.} = 18.5\text{ in.} < 24\text{ in.} \qquad \text{(O.K.)}$$

The hook therefore will be adequately developed within the column and the detail is satisfactory.

If the hooked rebar in the beam had not been adequately developed within the column, the beam width could be increased to achieve a spacing between the top bars of at least $6d_b$, or adequate confining closed stirrups or ties provided within ℓ_{dh} so that the confinement factor, ψ_r, can be 1.0 instead of 1.6. The beam top bars must fit within the column footprint and column rebar cage.

Anchorage into beam: The development length required if bars are straight is (conservatively) 53.4 in., as determined previously. Therefore, the bars must extend at least this distance into the span. The ACI Code has additional requirements for the extension of tension bars in areas of negative moments. These requirements are covered in the discussion on continuous construction in Chapter 6.

Adequate Development of Hooked Beam Rebar in Columns

It is very important that reinforcement be adequately developed into the supporting structural elements (e.g., columns), especially for statically determinate structures such as cantilevers (e.g., balcony slabs), which lack redundancy and therefore have no alternate load paths. The collapse of a garage at an Atlantic City Casino Resort in 2003 [3, 4] was believed to have been caused by the lack of adequate anchorage of the top reinforcement of a cantilever slab/beam. For a cantilever beam framing into a column, the column size parallel to the beam must be at least equal to the required development length of the hooked top reinforcement in the beam, plus the clear concrete cover to the bar extension beyond the hook within the column. Without adequate development, the beam top reinforcement will not be fully developed, and hence, the moment capacity of the beam at the beam-column joint will not be fully realized (see Figure 5.14).

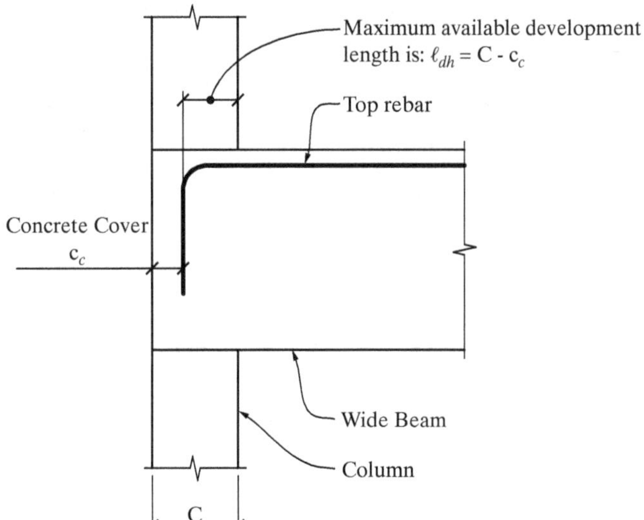

FIGURE 5.14 Development of a beam top rebar in a column.

For cantilevered balcony slabs, improper placement of the top reinforcement—resulting in a reduced effective depth—is the leading cause of failures for these nonredundant structures. It also leads to excessive deflections of the balcony slab which could induce unanticipated forces in the balcony railing system [5].

Note that the development lengths of hooked bars required in ACI 318-19 can be as much as 50-percent longer than required in previous ACI Codes (ACI 318-19, Section R25.4.3.2). In addition, to avoid potential concrete breakout failure, the excess reinforcement factor that applies to straight tension bars is not applicable for hooked bars (ACI 318-19, Section R25.4.10.2). Similarly, the casting position factor is also not applicable to hooked top bars.

5.5 DEVELOPMENT OF WEB OR TRANSVERSE REINFORCEMENT (STIRRUPS)

Anchorage of web reinforcement must be furnished in accordance with the ACI Code, Section 25.7. Stirrups must be carried as close to the compression and tension surfaces as possible. Close proximity to the compression face is necessary as flexural tension cracks penetrate deeply as the ultimate load is approached.

The ACI Code stipulates that ends of single-leg stirrups, simple U-stirrups, or multiple U-stirrups shall be anchored by one of the following means (see Figure 5.15):

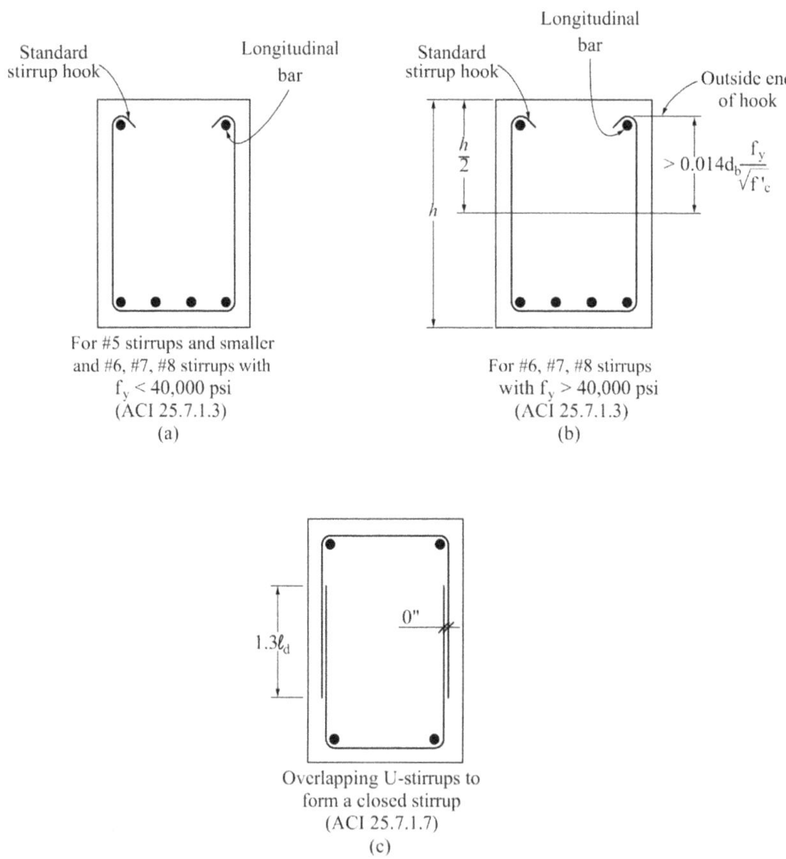

FIGURE 5.15 Web or Transverse reinforcement anchorage.

1. For a No. 5 bar or smaller, and for Nos. 6, 7, and 8 bars of f_y = 40,000 psi or less, anchorage is provided by a standard stirrup hook, bent around a longitudinal bar (ACI Code, Section 25.7.1.3a).

2. For Nos. 6, 7, and 8 stirrups with f_y > 40,000 psi, anchorage is provided by a standard stirrup hook bent around a longitudinal bar *plus* an embedment between midheight of the member and the outside end of the hook equal to or greater than

$$0.014d_b \frac{f_y}{\left(\lambda\sqrt{f'_c}\right)}$$

   (See ACI Code, Section 25.7.1.3(b)). A 135° or a 180° hook is preferred, but a 90° hook is acceptable provided that the free end of the hook is extended the full $12d_b$ beyond the center of the bend as required in ACI Code, Table 25.3.2.

It should be noted that the ACI standard hooks for ties and stirrups, as shown in Figure 5.15, include 90° and 135° hooks only. This does not imply that the 180° hook is unacceptable, but from a constructability consideration, they should be avoided when closed stirrups are used. The 135° hook is easier to fabricate. Many stirrup bending machines now in use are not designed to fabricate a 180° hook. The anchorage strength of either a 135° or 180° hook is approximately the same. Also, the typical sizes of stirrups used in most buildings are No. 3 to No. 5 bars.

In addition, the ACI Code, Section 25.7.1.7, permits closed stirrups to be made of lapped pairs of U-stirrups. The legs of the pair of U-stirrups shall be considered properly spliced when the length of lap splice is 1.3 ℓ_d, as depicted in Figure 5.15c. Each bend of each simple U-stirrup must enclose a longitudinal bar. If the vertical lap length of 1.3 ℓ_d of the U-stirrups cannot fit within the depth of a shallow beam or girder that is at least 18 in. deep, double U-stirrups may still be used and considered adequate provided each U-stirrup leg extends the full available depth of the member and the force in each leg of the stirrup does not exceed 9000 lb. (i.e., $A_b f_y \leq 9000$ lb).

Where torsional reinforcing is required or desired, a commonly used alternative stirrup is the one-piece closed stirrup. Use of the one-piece closed stirrup is disadvantageous, however, in that the entire beam reinforcing (longitudinal steel and stirrups) may have to be prefabricated as a cage and then placed as a unit. This may not be practical if the longitudinal bars must be passed between the column vertical bars. Alternatively, and at a greater cost, the longitudinal bars could be threaded through the closed stirrups and column bars. Two commonly used types of one-piece closed stirrups are shown in Figure 5.16. If the spalling of the member at the transverse torsional reinforcement anchorage (hooks) is restrained by a flange or similar member, 90° standard hooks around a longitudinal bar are allowed, as shown in Figure 5.16a. Otherwise, 135° standard hooks around a longitudinal bar are required (Figure 5.16b). Numbers 6, 7, and 8 stirrups with f_y > 40,000 psi require additional anchorage, as previously described. See ACI Code, Section 25.7.1.3, for full details.

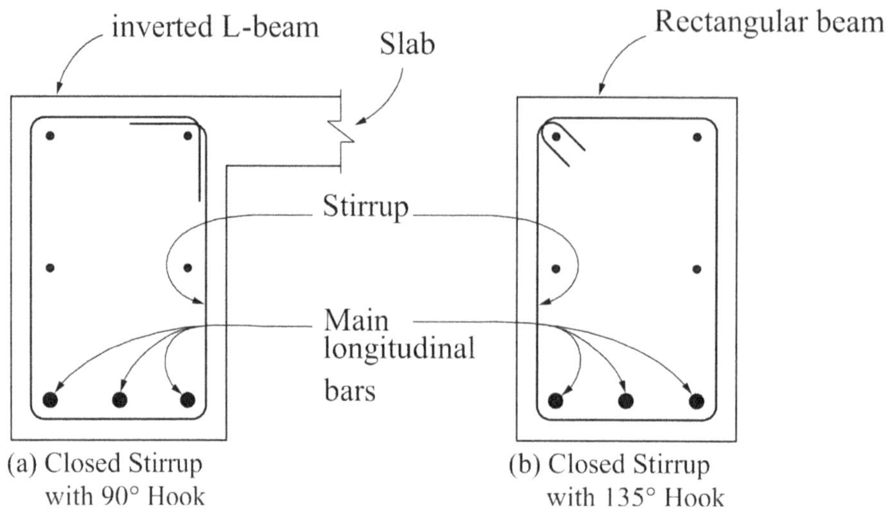

FIGURE 5.16 Closed stirrups.

5.6 SPLICES

The need to splice reinforcing steel is a reality due to the limited lengths of steel available. All bars are readily available in lengths up to 60 ft; No. 3 and No. 4 bars will tend to bend in handling when longer than 40 ft, however. Typical stock straight lengths are as follows:

No. 3 bar: 20, 40, 60 ft

No. 4 bar: 30, 40, 60 ft

Nos. 5 to 18 bars: 60 ft

Splicing may be accomplished by welding, by mechanical means, or, most commonly for No. 11 bars and smaller, by lapping bars, as shown in Figure 5.17. As shown in Figure 5.17, the stress in lap spliced bars vary from zero at the end of the bars to f_y at the end of the lap length, as the stress is transferred from one bar to the other. Lap splices may not be used for bars larger than No. 11 except for compression splices at footings, as provided in the ACI Code, Sections 16.3.5.4 and 25.5.5.3, and for compression splices of bars of different sizes, as provided in the ACI Code, Section 25.5.5.4. The splice composed of lapped bars is usually more economical than mechanical or welded splices. The lapped bars are commonly tied in contact with each other (i.e., contact lap splices). They may, however, be spaced transversely apart center-to-center up to one-fifth of the required lap splice length, with an upper limit of 6 in. resulting in a noncontact lap splice (ACI Code, Section 25.5.1.3). Transverse spacings between lapped rebars larger than these will create an unreinforced concrete section between the noncontact lapped bars resulting in cracking in the concrete between the bars. Lap splices in regions of maximum moment or maximum tensile stress should, if possible, be avoided.

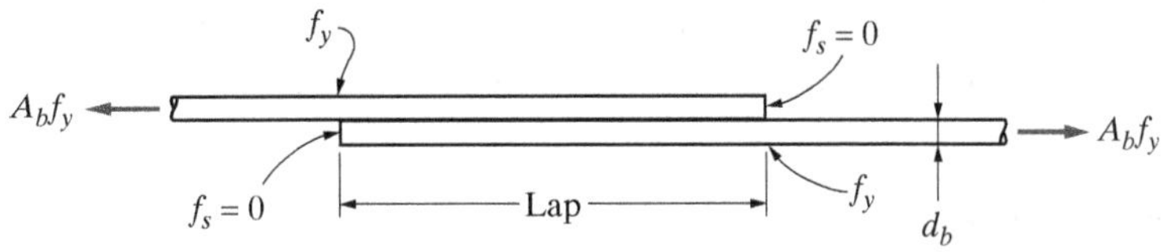

FIGURE 5.17 Stress transfer in tension lap splice.

The ACI Code (Section 1.8.1 and 26.6.1.1) requires that the design drawings show the location and length of lap splices and the type and location of mechanical and welded splices of reinforcement.

5.7 TENSION SPLICES

The required length of tension lap splices is based on the class in which the splice is categorized. The required lap length increases with increased stress and increased amount of steel spliced in close proximity. Lap length is expressed in terms of the tensile development length ℓ_d for the bar, as shown in Section 5.2. The 12-in. minimum for the ℓ_d calculation is not considered, nor is the excess reinforcement factor considered for tension splices, as the splice classification already reflects any excess reinforcement at the splice location.

Two classes of tension lap splices are prescribed in the ACI Code: **Class A** and **Class B** tension lap splices (ACI 318-19, Section 25.5.2.1). The length of the tension lap splice is a function of the percentage of bars spliced at the same location and the level of stress in the bars.

- **Class A tension lap splice**: Lap length, $\ell_{st} = 1.0\ell_d \geq 12$ in.

  Class A lap splices are permitted if the following two conditions are satisfied:

  1. Area of reinforcement provided for the entire length of the splice is twice that required (i.e., $A_{s,provided} / A_{s,required} \geq 2.0$), and
  2. Not more than 50% of the total reinforcement is spliced within the required lap length.

  If the two conditions above are not satisfied, then a Class B lap splice is required.

- **Class B tension lap splice**: Lap length, $\ell_{st} = 1.3\ell_d \geq 12$ in.

  The minimum length of lap for tension lap splices is 12 in. (ACI Code, Table 25.5.2.1). Requirements for mechanical and welded splices for tension and compression reinforcement are contained in the ACI Code, Section 25.5.7. Mechanical or welded splices must have a yield strength in tension or compression of at least 125% of the yield strength of the spliced reinforcing bars (ACI Code, Section 25.5.7.1), and mechanical or welded splices need not be staggered except in tension tie members (ACI Code Section 25.5.7.3). In addition, splices in tension tie members must be made with a welded splice or mechanical splice, and their locations must be staggered a longitudinal distance of at least 30 in., in accordance with the ACI Code, Section 25.5.7.4. Staggering of all nonmechanical and nonwelded tension splices is encouraged. Use tension lap splices for #11 and smaller bars and mechanical or welded splices for #14 and #18 bars.

5.8 COMPRESSION SPLICES

The ACI Code, Section 25.5.5, contains requirements for lap splices for compression bars. For $f'_c \geq 3000$ psi, the required compression lap lengths, in multiples of bar diameters d_b, are as follows:

$$f_y \leq 60{,}000 \text{ psi:} \qquad \ell_{sc} = 0.0005 f_y d_b \geq 12 \text{ in.}$$

$$60{,}000 \text{ psi} < f_y \leq 80{,}000 \text{ psi:} \qquad \ell_{sc} = (0.0009 f_y - 24) d_b \geq 12 \text{ in.}$$

$$f_y > 80{,}000 \text{ psi:} \qquad \ell_{sc} \geq (0.0009 f_y - 24) d_b \geq \ell_{st}$$

(from Section 5.7 and ACI Code, Section 25.5.2.1)

Consequently, the required compression lap lengths, in multiples of bar diameters d_b, for three specific rebar yield strengths are:

$$For\ f_y = 40{,}000 \text{ psi:} \quad \ell_{sc} = 20d_b$$
$$For\ f_y = 60{,}000 \text{ psi:} \quad \ell_{sc} = 30d_b$$
$$For\ f_y = 80{,}000 \text{ psi:} \quad \ell_{sc} = 48d_b$$

but not less than 12 in. For $f'_c < 3000$ psi, the length of lap should be increased by one-third. For compression splices within tied columns with ties within the lap splice length having an effective area of at least $0.0015hs$ where h is the column dimension and s is the tie spacing, the compression lap splice length calculated above can be multiplied by a factor of 0.83 (note that the legs of the ties perpendicular to the column dimension, h, shall be used in calculating the effective area). For compression splices within spirally reinforced columns within the lap splice length, the compression lap splice length calculated above can be multiplied by a factor of 0.75 (see ACI Code, Section 10.7.5.2.1). In no case should the compression lap splice length be less than 12 in.

Compression splices may also be of the end-bearing type, where bars are cut square, then butted together and held in concentric contact by a suitable device. End-bearing splices must not be used except in members containing closed ties, closed stirrups, or spirals. Welded splices and mechanical connections are also acceptable and are subject to the requirements of the ACI Code, Section 25.5.5. Special splice requirements for columns are furnished by the ACI Code, Section 10.7.5. Compression lap splices shall not be used for bars larger than #11 except that compression lap splices of #14 or #18 bars to #11 or smaller bars are permitted (ACI Code, Section 25.5.5.3). Where bars of different sizes are lap-spliced in compression, it is permitted for the compression lap splice length, ℓ_{sc}, to be the longer of the compression development length, ℓ_{dc}, of the *larger* bar and the compression lap splice length ℓ_{sc}, of the *smaller* bar (ACI 318-19, Section 25.5.5.4).

5.9 SIMPLE-SPAN BAR CUTOFFS AND BENDS

The maximum required A_s for a beam is needed only where the moment is maximum. This maximum steel may be reduced at points along a bending member where the bending moment is smaller. This is usually accomplished by either stopping or bending the bars in a manner consistent with the theoretical requirements for the strength of the member, as well as the requirements of the ACI Code.

Bars can theoretically be stopped or bent in flexural members whenever they are no longer needed to resist moment. The ACI Code, Sections 7.7.3.3 and 9.7.3.3, however, requires that each bar be extended beyond the point at which it is no longer required for flexure for a distance equal to the effective depth of the member, d, or $12d_b$, whichever is greater, except at supports of simple spans and at free ends of cantilevers. In effect, this prohibits the cutting of a bar at the theoretical cutoff point but can be interpreted as permitting bars to be bent at the theoretical cutoff point. If bars are to be bent, a general practice that has evolved is to commence the bend at a distance equal to one-half the effective depth beyond the theoretical cutoff point. The bent bar should be anchored or made continuous with reinforcement on the opposite face of the member. It should be pointed out, however, that the bending of reinforcing bars in slabs and beams to create *truss bars* (see types 3 through 7, Figure 13.5) has fallen into disfavor over the years because of placement problems and the labor involved. It is more common to use straight bars and place them in accordance with the strength requirements.

With simple-span flexural members of constant dimensions, we can assume that the required A_s varies directly with the bending moment and that the shape of a *required A_s curve* is identical to that of the moment diagram. The moment diagram (or curve) may then be used as the required A_s curve by merely changing the vertical scale. Steel areas may be used as ordinates, but it is more convenient to use the required number of bars (assuming that all bars are of the same diameter). The necessary correlation is established by using the maximum ordinate of the curve as the maximum required A_s in terms of the required number of bars. Because of possible variations in the shape of the moment diagram, either a graphical or a mathematical approach may be more appropriate. A graphical approach requires that the moment diagram be plotted to scale. Example 5.5 lends itself to a mathematical solution.

In determining bar cutoffs, it should be remembered that the stopping of bars should be accomplished by using a symmetrical pattern so that the remaining bars will also be in a symmetrical pattern. In addition, the ACI Code, Sections 7.7.3.8 and 9.7.3.8, requires that for simple spans, at least one-third of the positive moment steel extends into the support a distance of 6 in. In practice, this requirement is generally exceeded. Normally, recommended bar details for single-span solid concrete slabs indicate that all bottom bars should extend into the support. See Figure 5.18 for recommended bar details.

Economy sometimes dictates that reinforcing steel should be cutoff in a simple span. Example 5.5 shows one approach to this problem.

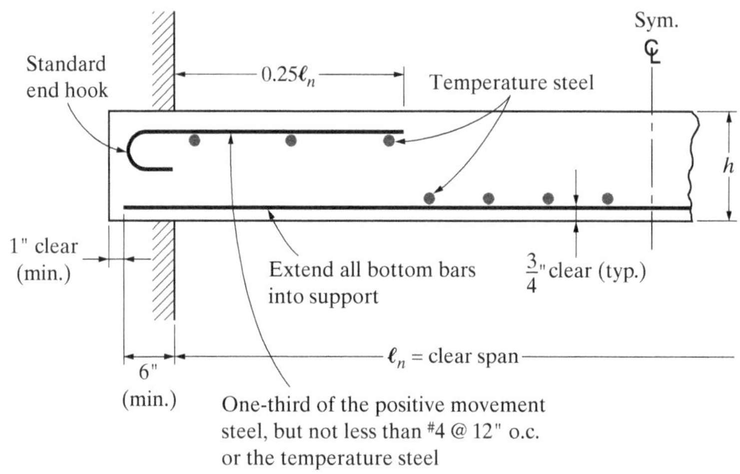

FIGURE 5.18 Recommended bar details: one-way simple-span slabs; tensile-reinforced simple-span beams similar.

EXAMPLE 5.5

A simple-span, uniformly loaded beam, shown in Figures 5.19a and 5.19b, requires six No. 7 bars for tensile reinforcement. If the effective depth d is 18 in., determine where the bars may be stopped. Use $f'_c = 4000$ psi and $f_y = 60{,}000$ psi. Assume that there is no excess steel (required A_s = furnished A_s = 3.60 in.^{2}). Assume normal-weight concrete and that the stirrups extend to the end of the beam at the support. The bars are uncoated.

Solution:

We will try to establish a bar cutoff scheme whereby the two center bars are cut first, followed by the cutting of the other two inside bars. The two corner bars are to run the full length of the beam. First, we check the minimum steel requirement. From Table A.5:

$$A_{s,\min} = 0.0033\,bd = 0.0033(16.5)(18) = 0.98 \text{ in.}^2$$

The two corner bars provide a steel area of 1.20 in.^{2} Therefore, the minimum steel area requirement is met.

1. Determine where the first two bars may be cutoff. The first two bars may be stopped where only four are required. This distance from the centerline of the span is designated x_1 in Figure 5.19d.

   As the moment diagram is a second-degree curve (i.e., a parabola), offsets to the line tangent to the curve at the point of maximum moment vary as the squares of the distances from the centerline of the span. The solution for the distance x_1 may be formulated as follows:

$$\frac{(x_1)^2}{\left(\dfrac{\ell}{2}\right)^2} = \frac{y_1}{Y}$$

$$\frac{(x_1)^2}{12^2} = \frac{2 \text{ bars}}{6 \text{ bars}}$$

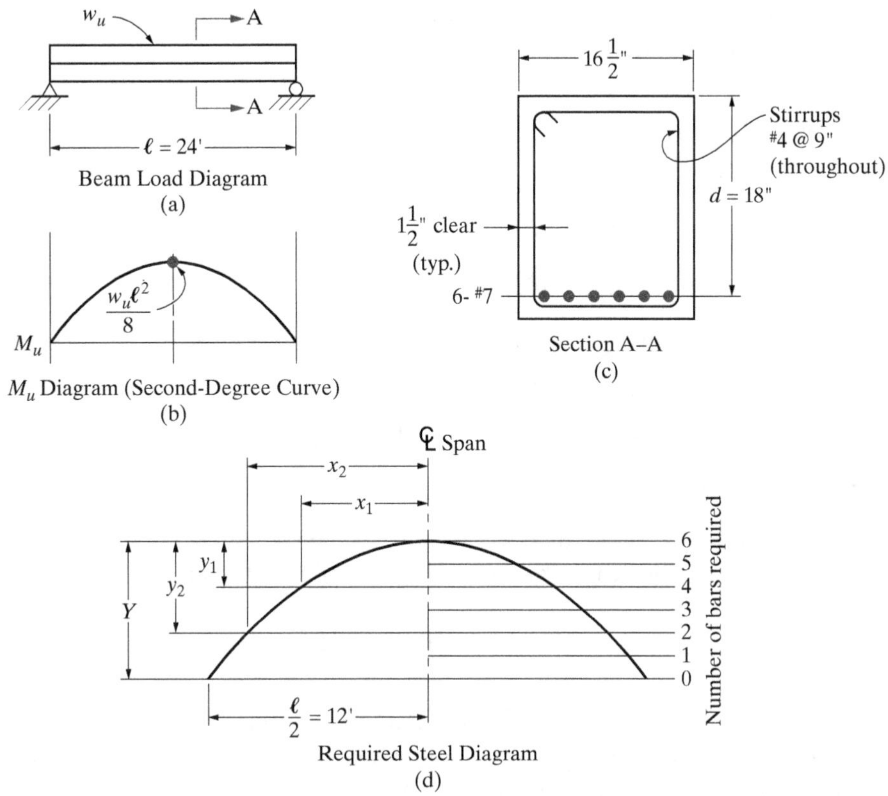

FIGURE 5.19 Sketches for Example 5.5.

from which x_1 = 6.93 ft. This locates the *theoretical* point where two bars may be terminated. Additionally, bars must be extended past this point a distance of d or $12d_b$, whichever is larger. Thus,

$$\begin{aligned}12 \text{ bar diameters} &= 12(0.875)\\ &= 10.5 \text{ in.}\end{aligned}$$

Because d = 18 in., the bars should be extended 18 in. (or 1.50 ft). Then, the minimum distance from the centerline of the span to the cutoff of the first two bars is

$$6.93 + 1.50 = 8.43 \text{ ft}$$

2. Determine where the next two bars may be cutoff. The two remaining bars, which are the corner bars, will continue into the support. With reference to Figure 5.19d:

$$\frac{(x_2)^2}{12^2} = \frac{4}{6}$$

$$x_2 = 9.80 \text{ ft}$$

Therefore, the minimum distance from the centerline of the span to the cutoff of the next two bars is

$$9.80 + 1.50 = 11.30 \text{ ft}$$

3. A check must be made of the required development length ℓ_d for the first *two* bars. Because the bar cutoff occurs 8.43 ft from the centerline of the span, ℓ_d must be less than 8.43 ft for the stress f_y to be developed at the centerline.

   a. From Table 5.1, $K_D = 71.2$.

   b. Establish values for the factors ψ_t, ψ_e, ψ_s, ψ_g, and λ.

      1. $\psi_t = 1.0$ (the bars are not top bars).
      2. The bars are uncoated; $\psi_e = 1.0$.
      3. The bars are No. 7; $\psi_s = 1.0$.
      4. The bars are Grade 60 reinforcement: $\psi_g = 1.0$.
      5. Normal-weight concrete is used; $\lambda = 1.0$.

   c. The product $\psi_t \times \psi_e = 1.0 < 1.7$. (O.K.)

   d. Determine c_b: Based on cover (center of bar to nearest concrete surface), consider the clear cover, the No. 4 stirrup, and one-half the diameter of the No. 7 bar:

$$c_b = 1.5 + 0.5 + \frac{0.875}{2} = 2.44 \text{ in.}$$

Based on bar spacing (one-half the center-to-center distance) and considering the concrete cover and the stirrup legs on both sides of the beam, one-half of the center-to-center spacing between the No. 7 bars is

$$c_b = \frac{1}{2}\left(\frac{16.5 - 2(1.5) - 2(0.5) - 2\left(\frac{0.875}{2}\right)}{5}\right) = 1.163 \text{ in.}$$

Therefore, use the smaller of the two values (see Section 5.2), use $c_b = 1.163$ in.

   e. Using data on stirrups from Figure 5.19 where s = spacing of the stirrups in inches, n = number of rebars being developed, and A_{tr} is the cross-section area of the legs of the stirrups in.^{2} (see Section 5.2):

$$K_{tr} = \frac{40A_{tr}}{sn} = \frac{40(0.40)}{(9)(2)} = 0.889$$

   f. Check $(c_b + K_{tr})/d_b \le 2.5$:

$$\frac{c_b + K_{tr}}{d_b} = \frac{1.163 + 0.889}{0.875} = 2.35 < 2.5 \quad \text{(O.K.)}$$

   g. The excess reinforcement factor is neglected (conservative).

   h. Calculate the straight bar tension development length, ℓ_d:

$$\ell_d = \frac{K_D}{\lambda}\left[\frac{\psi_t\psi_e\psi_s\psi_g}{\left(\frac{c_b + K_{tr}}{d_b}\right)}\right]d_b$$

$$= \frac{71.2}{1.0}\left[\frac{(1.0)(1.0)(1.0)(1.0)}{2.35}\right](0.875)$$

$$= 26.5 \text{ in.} > 12 \text{ in.} \qquad \text{(O.K.)}$$

$$26.5 \text{ in.} = 2.21 \text{ ft}$$

As the required tensile development length of 2.21 ft is much less than the distance from the centerline to the actual cutoff point (8.43 ft), the cutoff for the first two bars is satisfactory. Similarly, if we consider the second two bars, for which ℓ_d can conservatively be taken as 2.21 ft (conservative, as the c_b distance, based on bar spacing, is larger for these two bars than for the first two cut bars), the point at which they are stressed to maximum (the theoretical cutoff point for the first two bars), measured from the centerline, is 6.93 ft. The second two bars must extend at least ℓ_d past this point. Again, measured from the centerline, the bars must extend at least

$$6.93 + 2.21 = 9.14 \text{ ft}$$

Because this is less than the actual distance to the cutoff point (11.30 ft), the cutoff for the second pair is satisfactory (see Figure 5.20).

4. The remaining pair of corner bars, representing one-third of the positive moment steel, must continue into the support. As the corner bars are stressed to f_y at a distance of 9.80 ft from the centerline of the span (where the second pair of bars may be theoretically cut), however, the development length of these bars must be satisfied from the end of the bar back to the point of maximum stress f_y. The development length required is again conservatively taken as 2.21 ft. The straight development length furnished (or available) is 2.58 ft (see Figure 5.20). Because 2.58 ft > 2.21 ft, there is adequate development length available. If the development length available were insufficient, either of two solutions could be used: (a) Provide a standard 180° hook at the end of each of the two corner bars or (b) extend all four of the outside bars to the end of the beam (thereby cutting only the two center bars).

Example 5.5 could also be solved by plotting the M_u diagram to scale and superimposing on the M_u diagram the values of ϕM_n for four bars and two bars. The theoretical cutoff points are established where the ϕM_n lines intersect the M_u curve. For instance, the theoretical cutoff point for the two center bars is established where the ϕM_n line for four bars intersects the M_u curve. This method has several advantages: (1) Any excess flexural steel in the design will result in the two center bars being cut closer to the midpoint of the beam, thus saving steel; (2) the calculation of ϕM_n better reflects the available moment strength than does the division of the M_u diagram ordinate into six equal parts (in Example 5.5); and (3) it can be used for any shape M_u diagram.

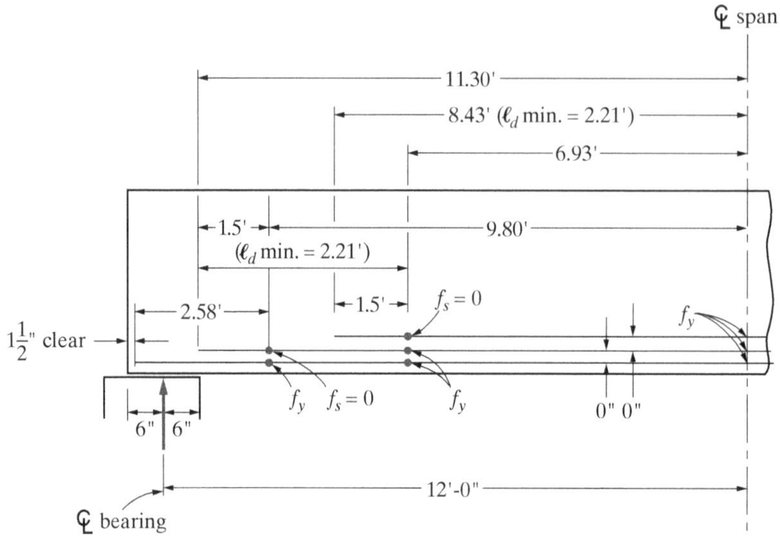

FIGURE 5.20 Sketch for Example 5.5.

The bars in Example 5.5 are terminating in a tension zone. When this occurs, the member undergoes a reduction in shear capacity. Therefore, special shear and stirrup calculations must be performed in accordance with the ACI Code, Section 9.7.3.5, which stipulates that no flexural reinforcement shall be terminated in a tension zone unless *one* of the following conditions is satisfied:

(a) **(ACI Code, Section 9.7.3.5(a)):** Factored shear at the cutoff point does not exceed two-thirds of the design shear strength ϕV_n. This may be written as

$$V_u \le \tfrac{2}{3}\phi V_n = \tfrac{2}{3}\phi(V_c + V_s)$$

(b) **(ACI Code, Section 9.7.3.5(b)):** For No. 11 bars and smaller, the continuing reinforcement provides double the area required for flexure at the cutoff point, and the factored shear does not exceed three-quarters of the design shear strength, that is:

$$V_u \le \tfrac{3}{4}\phi V_n = \tfrac{3}{4}\phi(V_c + V_s)$$

(c) **(ACI Code, Section 9.7.3.5(c)):** Stirrup area *in excess* of that required for shear is provided along each bar terminated over a distance equal to three-fourths the effective depth of the member (i.e., $0.75d$) from the point of bar cutoff. The excess stirrup area shall not be less than $60b_w s/f_y$. The stirrup spacing, s, shall not exceed $d/(8\beta_b)$, where β_b is the ratio of area of reinforcement cutoff to the total area of tension reinforcement at the section.

With respect to the preceding conditions, the first is interpreted as a check only. Although the V_s term could be varied, the code is not specific concerning the length of beam over which this V_s must exist. The second condition involves moving the cutoff point to another location. The third condition is a design method whereby additional stirrups may be introduced if the shear strength is inadequate as determined by the ACI Code, Section 9.7.3.5(a). This condition will be developed further.

It is convenient to determine the *number* of additional stirrups to be added in the length $0.75d$ along the end of the bar from the cutoff point. The required excess stirrup area is

$$A_v = \frac{60 b_w s}{f_{yt}}$$

If the size of the stirrup (and therefore A_v) is known, the maximum spacing is

$$s = \frac{A_v f_{yt}}{60 b_w}$$

Therefore, the number of stirrups N_s to be added over the length $0.75d$ is

$$N_s = \frac{0.75d}{s} + 1 = \frac{0.75d}{\left(\dfrac{A_v f_{yt}}{60 b_w}\right)} + 1 = \left(\frac{45 b_w d}{A_v f_{yt}}\right) + 1$$

Also, because maximum $s = d/(8\beta_b)$,

$$N_{s(\text{min.})} = \frac{0.75d}{\left(\dfrac{d}{8\beta_b}\right)} + 1 = 6\beta_b + 1$$

The larger resulting N_s value controls.

EXAMPLE 5.6

In the beam shown in Figure 5.21, the location at which the top layer of two No. 9 bars is to be terminated is in a tension zone. At the cutoff point, V_u = 52 kip and the No. 3 stirrups are spaced at 11 in. on center. Check shear in accordance with the ACI Code, Section 9.7.3.5, and redesign the stirrup spacing if necessary. Assume that $f'_c = 3000$ psi and f_y = 60,000 psi.

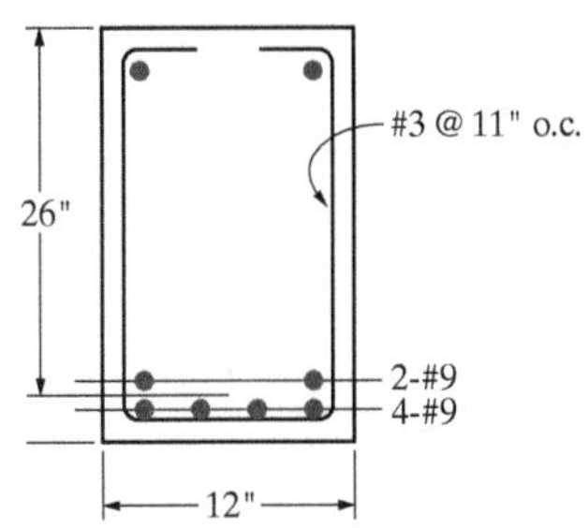

FIGURE 5.21 Sketch for Example 5.6.

Solution:

In accordance with the ACI Code, Section 9.7.3.5(a), check $V_u \le \frac{2}{3}\phi(V_c + V_s)$:

$$\tfrac{2}{3}\phi(V_c + V_s) = \tfrac{2}{3}(0.75)\left(2\sqrt{f'_c}bd + \frac{A_v f_{yt} d}{s}\right)$$

$$= \tfrac{2}{3}(0.75)\left[2\sqrt{3000}(12)(26) + \frac{(0.22)(60{,}000)(26)}{11}\right]$$

$$= 32{,}700 \text{ lb}$$

$$= 32.7 \text{ kip} < 52 \text{ kip} \qquad \text{(N.G.)}$$

Therefore, add excess stirrups over a length of 0.75d along the terminated bars from the cut end in accordance with ACI Code 9.7.3.5(c):

$$N_s = \left(\frac{45b_w d}{A_v f_{yt}} + 1\right) \text{or } (6\beta_b + 1)$$

$$\frac{45(12)(26)}{0.22(60{,}000)} + 1 = 2.06$$

$$6\left(\tfrac{1}{3}\right) + 1 = 3$$

Add three stirrups over a length of 0.75d = 19.5 in. Find the new spacing required in the 19.5-in. length:

$$\frac{19.5}{11} = 1.77 \text{ stirrups at 11-in. spacing}$$

$$\underline{+\,3.00} \text{ additional stirrups}$$

$$4.77 \text{ stirrups in 19.5-in. length}$$

The new stirrup spacing is

$$\frac{19.5}{4.77} = 4.09 \text{ in.}$$

Use 4 in.

The stirrup pattern originally designed should now be altered to include the 4-in. spacing along the last 19.5 in. of the cut bars.

The problems associated with terminating bars in a tension zone may be avoided by extending the bars in accordance with the ACI Code, Section 9.7.3.5(c), or by extending them into the support.

In summary, a general representation of the bar cutoff requirements for positive moment steel in a simple span may be observed in Figure 5.22. If bars A are to be cutoff, they must (1) project ℓ_d past the point of maximum positive moment and (2) project beyond their theoretical cutoff point a distance equal to the effective depth of the member, d, or 12 bar diameters

($12d_b$), whichever is greater. The remaining positive moment bars B must extend ℓ_d past the theoretical cutoff point of bars A and extend at least 6 in. into the support.

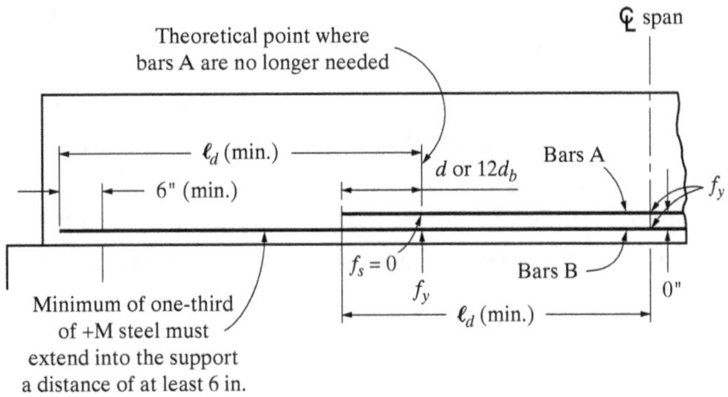

FIGURE 5.22 Bar cutoff requirements for simple spans (positive moment steel).

5.10 CODE REQUIREMENTS FOR DEVELOPMENT OF POSITIVE MOMENT STEEL AT SIMPLE SUPPORTS

The ACI Code, Sections 7.7.3.8 and 9.7.3.8, contains requirements concerning the development of straight, positive moment bars *at simple supports* and at points of inflection. The intent of this code section is the same as the check on the development of the two bars extending into the support in Example 5.5. The method of Example 5.5 may be used if we are working with the actual moment diagram. The code approach does not require the use of a moment diagram.

The ACI Code requirement (ACI Code, Sections 7.7.3.8.3(b) and 9.7.3.8.3(b)) places a restriction on the size of the bar that may be used such that the tension development length

$$\ell_d \le \frac{M_n}{V_u} + \ell_a$$

Where,

M_n = nominal moment strength, $\left[A_s f_y\left(d - \frac{a}{2}\right)\right]$, assuming all reinforcement at the section to be stressed to f_y

V_u = total applied design shear force at the section

ℓ_a (at a support) = the embedment length beyond the center of the support

ℓ_a (at a point of inflection) = the effective depth of the member, d, or $12d_b$, whichever is greater

This requirement need not be satisfied for reinforcement terminating beyond the centerline of simple supports by a standard hook or a mechanical anchorage at least equivalent to a standard hook.

The effect of this code restriction is to require that bars be small enough so that they can become fully developed before the applied moment has increased to the magnitude where they *must* be capable of carrying f_y. The M_n/V_u term approximates the distance from the section in question to the location where applied moment M_u exists that is equal to ϕM_n (and where f_y must exist in the bars).

Therefore, the distance from the end of the bar to the point where the bar must be fully developed is $(M_n/V_u) + \ell_a$, and the bars must be chosen so that their ℓ_d is less than this distance. The code allows M_n/V_u to be increased by 30% (ACI Code, Sections 7.7.3.8.3(a) and 9.7.3.8.3(a)) when the ends of the reinforcement are confined by a compressive reaction such as is found in a simply supported beam (a beam supported by a wall).

EXAMPLE 5.7

At the support of a simply supported beam, a cross-section exists as shown in Figure 5.23. Check the bar diameters in accordance with the ACI Code, Section 9.7.3.8.3. Assume a support width of 12 in., $1\frac{1}{2}$-in. cover, $f'_c = 4000$ psi, and $f_y = 60{,}000$ psi. Assume that V_u at the support is 80 kip. Normal-weight concrete is used. The stirrups begin at 3 in. from the face of the support. The bars are uncoated.

Solution:

Because this beam has its ends confined by a compressive reaction, we will check ACI Section 9.7.3.8.3(a):

$$\ell_d \le 1.3\left(\frac{M_n}{V_u}\right) + \ell_a$$

We next calculate the required tensile development length ℓ_d:

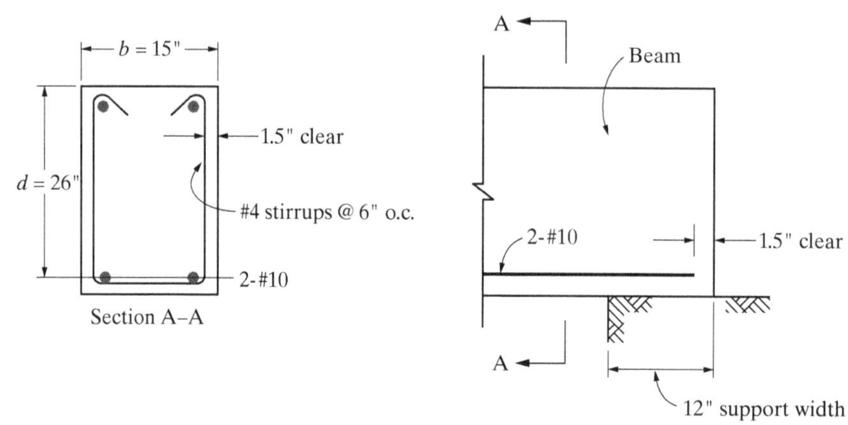

FIGURE 5.23 Sketch for Example 5.7.

1. From Table 5.1, $K_D = 71.2$.
2. Establish values for the factors ψ_t, ψ_e, ψ_s, ψ_g, and λ.

a. $\psi_t = 1.0$ (the bars are not top bars).

b. The bars are uncoated; $\psi_e = 1.0$.

c. The bars are No. 10; $\psi_s = 1.0$.

d. The bars are Grade 60 reinforcement: $\psi_g = 1.0$.

e. Normal-weight concrete is used; $\lambda = 1.0$.

3. The product $\psi_t \times \psi_e = 1.0 < 1.7$ (O.K.).

4. Determine c_b: Based on cover (center of bar to the nearest concrete surface), consider the clear cover, the No. 4 stirrup, and one-half the diameter of the No. 10 bar:

$$c_b = 1.5 + 0.5 + \frac{1.27}{2} = 2.64 \text{ in.}$$

Based on bar spacing, one-half the center-to-center spacing between the bars being developed is,

$$c_b = \frac{1}{2}\left[15 - 2(1.5) - 2(0.5) - 2\left(\frac{1.27}{2}\right)\right] = 4.87 \text{ in.}$$

Therefore, based on the smaller of the two values, use $c_b = 2.64$ in.

5. K_{tr} is taken as zero, as stirrups do not extend to the ends of the bars.

6. Check $(c_b + K_{tr})/d_b \le 2.5$:

$$\frac{c_b + K_{tr}}{d_b} - \frac{2.64 + 0}{1.27} = 2.08 < 2.5 \qquad \text{(O.K.)}$$

7. The excess reinforcement factor is not applicable and is omitted.

8. Calculate ℓ_d:

$$\ell_d = \frac{K_D}{\lambda}\left[\frac{\psi_t \psi_e \psi_s \psi_g}{\left(\dfrac{c_b + K_{tr}}{d_b}\right)}\right] d_b$$

$$= \frac{71.2}{1.0}\left[\frac{(1.0)(1.0)(1.0)}{2.08}(1.0)\right](1.27)$$

$$= 43.5 \text{ in.} > 12 \text{ in.} \qquad \text{(O.K.)}$$

Now check ACI 318-19, Section 9.7.3.8.3(a):

ℓ_a = the embedment length of the rebar beyond the center of the support

= 12 in. - 1.5 in. clear cover – ½(12 in.) = 4.5 in.

$$\rho = \frac{A_s}{bd} = \frac{2.54}{15(26)} = 0.0065$$

$$\bar{k} = 0.3676 \text{ ksi (see Table A.10)}$$

$$M_n = bd^2\bar{k} = 15(26)^2(0.3676) = 3728 \text{ in.-kip}$$

The maximum permissible required ℓ_d is calculated using ACI 318-19, Section 9.7.3.8.3(a) as

$$1.3\frac{M_n}{V_u} + \ell_a = 1.3\left(\frac{3728}{80}\right) + 4.5 = 65.1 \text{ in.}$$

$$65.1 \text{ in.} > 43.5 \text{ in.} \qquad \text{(O.K.)}$$

Therefore, the bar diameter is adequately small, and the bar can be developed as required. (If the required ℓ_d were more than 65.1 in., the use of a standard hook (or a mechanical anchorage equivalent to a standard hook) beyond the centerline of support would satisfy the development requirement and this code section would not apply. Also, the use of smaller bars would result in a smaller required ℓ_d.)

EXAMPLE 5.8

Tension and compression development lengths in a column footing

For the footing shown in Figure 5.24:

(a) Determine the tension development length, ℓ_d, for the No. 6 bottom rebars in the footing.

(b) Determine the compression development length, ℓ_{dc}, for the No. 7 column rebars.

(c) Determine the minimum footing thickness from bond and anchorage considerations.

(d) Determine the compression lap splice length, ℓ_{sc}, for the No. 7 column rebars.

Use $f_c' = 4000$ psi and $f_y = 60{,}000$ psi.

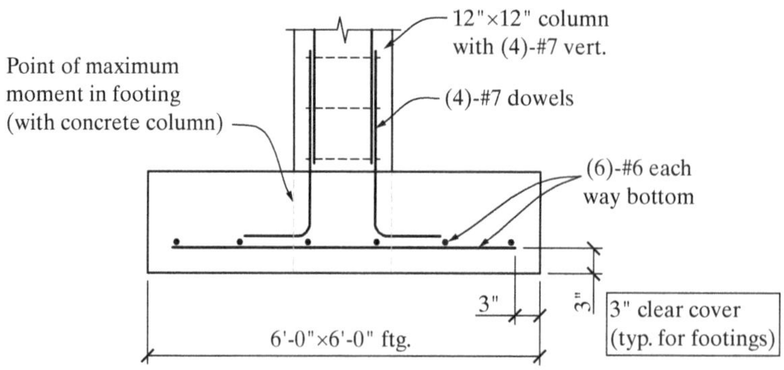

FIGURE 5.24 Detail for Example 5.8.

Solution:

(a) Tension development length:

Actual development length available within the spread footing,

$$\ell_{d,\text{provided}} = \frac{(6\,\text{ft.})(12\,\text{in./ft}) - 12\,\text{in. column}}{2} - 3\,\text{in. cover} = 27\text{ in.}$$

#6 bars in the footing (d_b = 0.75")
f_c' = 4000 psi
f_y = 60000 psi
$\psi_t = \psi_e = 1.0$
$\psi_t\,\psi_e = 1.0 < 1.7$ (O.K.)
For No. 6 bars, $\psi_s = 0.8$
For Grade 60 reinforcement, $\psi_g = 1.0$
K_{ER}=1.0 (assumed)
K_{tr}=0

Determine c_b, which is the smaller of:

$$c_b = \text{cover} + \frac{d_b}{2} = 3\text{"} + \frac{0.75\text{"}}{2} = 3.38\text{"}$$

$$c_b = \left(\frac{1}{2}\right)\left(\frac{W - (2)(\text{cover}) - d_b}{N_{bars} - 1}\right) = \left(\frac{1}{2}\right)\left(\frac{72\text{"} - (2)(3\text{"}) - 0.75\text{"}}{6-1}\right) = 6.5\text{"}$$

Using the smaller of the two values, c_b = 3.38"

$$\frac{c_b + K_{tr}}{d_b} \le 2.5$$

$$= \frac{3.38+0}{0.75} = 4.5 \le 2.5$$

Therefore, use 2.5 for this term.

$$\ell_d = \frac{3}{40}\left(\frac{f_y}{\lambda\sqrt{f_c'}}\right)\left[\frac{\psi_t\psi_e\psi_s\psi_g}{\left(\frac{c_b + K_{tr}}{d_b}\right)}\right]K_{ER}d_b \ge 12\text{ in.}$$

$$= \frac{3}{40}\left(\frac{60{,}000}{(1.0)\sqrt{4000}}\right)\left[\frac{(1.0)(1.0)(0.8)(1.0)}{2.5}\right](1.0)(0.75) \ge 12\text{ in.}$$

$$= 17.1\text{ in.} \ge 12\text{ in.}$$

ℓ_d = 17.1" < 27", OK for tension development length

(b) Compression development length of column dowels:

$\psi_r = 1.0$

$K_{ER} = 1.0$

$$\ell_{dc} = \left(\frac{0.02 f_y \psi_r}{\lambda\sqrt{f_c'}}\right) d_b = \frac{0.02(60{,}000)(1.0)}{(1.0)\sqrt{4000}}(0.875) = 16.6 \text{ in.}$$

$$\geq 0.0003(60{,}000)(1.0)(0.875) = 15.8 \text{ in.}$$

$$\geq 8 \text{ in.}$$

Therefore, the compression development length, ℓ_{dc} = 16.6 in., say 17 in.

(c) Minimum footing thickness

$H_{min} = \ell_{dc}$ + concrete cover = 17 + 3 = 20 in.

(d) Compression Lap Splice

For $f_y = 60{,}000$ psi: $\ell_{sc} = 30 d_b = 30(0.875) = 26.3$ in., say 27 in.

≥ 8 in.

Therefore, use $\ell_{sc} = 27$ in.

EXAMPLE 5.9

Hooked Bar development length in a beam

(a) Determine the tension development length of the hooked 4 No. 9 top rebar for the 12" × 22" deep beam shown in Figure 5.25. Assume f_c' = 4000 psi and f_y = 60,000 psi.

(b) Determine the minimum width of column to accommodate the hooked rebars. Assume #3 ties are used.

(c) Check the straight embedment length of the rebar within the cantilever beam.

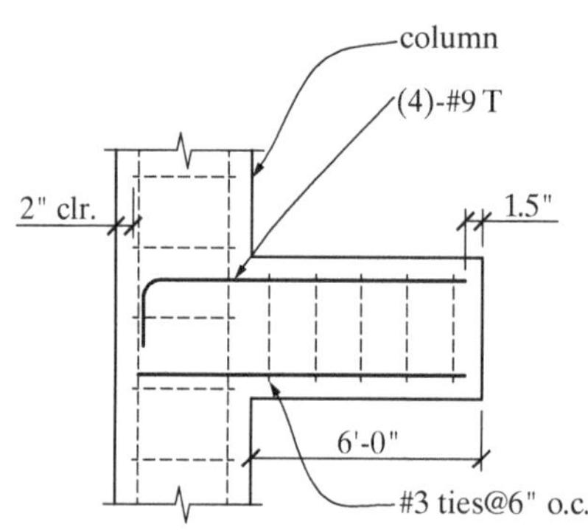

FIGURE 5.25 Detail for Example 5.9.

Solution:

a. Hooked Bar development:

For No. 9 bars, d_b = 1.128"

A_{th} = total cross-sectional area of both legs of ties or stirrups confining the hooked bars, in.^{2}, located within the hooked bar development length, ℓ_{dh}.

A_{hs} = total cross-sectional area of hooked or headed bars being developed at a critical section, in.^{2}

1. The development length ℓ_{dh} for the hooked bar shown in Figure 5.25 is calculated from

$$\ell_{dh} = \left[\frac{f_y \psi_e \psi_r \psi_o \psi_c}{55\lambda\sqrt{f_c'}}\right](d_b)^{1.5}$$
$$\geq 8d_b$$
$$\geq 6 \text{ in.}$$

The bars are uncoated, and the concrete is normal weight. Therefore, both ψ_e and λ are 1.0:

2. Uncoated rebar: ψ_e = 1.0

3. Minimum center-to-center spacing of hooked No. 9 top bars is
s =[12 in. – 2(2 in. clear cover) – (3/8 in.) – (3/8 in.) - ½ (1.128 in.) - ½ (1.128 in.)] /(4-1)
= 2.04 in. < $6d_b$ = 6(1.128 in.) = 6.8 in.,

   This implies that the hooked bars are very closely spaced , therefore, the confining reinforcement factor, ψ_r = 1.6.

   or

   A_{hs} = total cross-sectional area of hooked bars being developed at the critical section = area of 4 No. 9 top bars = 4 in.^{2}
   0.4 A_{hs} = 0.4(4) = 1.6 in.^{2}

   A_{th} = total cross-sectional area of both legs of ties or stirrups (i.e., #3 @ 6 in. o.c.) confining the hooked bars, in.^{2}, located within the hooked bar development length, ℓ_{dh}.
   = area of both legs of No. 4 stirrups = 2(0.11)(36 in./6 in.) = 1.06 in.^{2} < 0.4 A_{hs}

   Therefore, these hooked bars are very closely spaced and the confining reinforcement is not adequate, therefore, the confining reinforcement factor, ψ_r = 1.6. Note that A_{th} depends on the hooked bar development length, so the use of the second option for checking confinement is somewhat limited, since the development length is unknown at this stage.

4. The hooked bars terminate inside a column core with a side cover normal to the top bars = 2 in. clear cover + 3/8 in. stirrup = 2.38 in. < 2.5 in. Therefore, the location factor, ψ_o = 1.25.

5. For f_c' = 4000 psi: ψ_c = (4000/15,000) + 0.6 = 0.87

6. Normal-weight concrete: $\lambda = 1.0$.

Therefore, the hooked bar development length is calculated as

$$\ell_{dh} = \left[\frac{f_y \psi_e \psi_r \psi_o \psi_c}{55\lambda\sqrt{f'_c}}\right](d_b)^{1.5} = \left[\frac{(60{,}000)(1.0)(1.6)(1.25)(0.87)}{55(1.0)\sqrt{(4000)}}\right](1.128)^{1.5} = 36 \text{ in.}$$

$$\geq 8d_b = 8(1.128) = 9 \text{ in.}$$

$$\geq 6 \text{ in.}$$

(Check this with Table A.13: (1.0)(1.6)(1.25)(0.87)(20.7 in.) = 36 in.)

Therefore, the required development length of the No. 9 hooked top bars is,

$$\ell_{dh} = 36 \text{ in.}$$

b. Minimum width of column:

The minimum width of column parallel to the longitudinal axis of the beam that is required is

$$= \ell_{dh} + \text{clear concrete cover} + \text{column tie diameter}$$

$$= 36 \text{ in.} + 2 \text{ in.} + 0.375 \text{ in.} = 38.4 \text{ in., say } 39 \text{ in.}$$

Note that the beam and column width perpendicular to the longitudinal axis of the beam could be increased to achieve a spacing between the top bars of at least $6d_b$, so that the confinement factor, ψ_r, can be 1.0 instead of 1.6. The beam top bars must fit within the column footprint and column rebar cage. Alternatively, the hooked bars could be confined with closed hoops so that $A_{th} \geq 0.4\, A_{hs}$ to enable the confinement factor, ψ_r, become 1.0 instead of 1.6.

(c) Development Length of the straight bar within the beam.

Available straight bar development length is

$$\ell_{d,\ \text{available}} = (6\,\text{ft.})(12 \text{ in./ft}) - 1.5 \text{ in. clear concrete cover}$$

$$= 72 \text{ in.} - 1.5 \text{ in.} = 70.5 \text{ in.}$$

1. $f'_c = 4000$ psi and $f_y = 60{,}000$ psi

2. Establish values for the factors ψ_t, ψ_e, ψ_s, ψ_g, and λ.

a. The bars are top bars with at least 12 inches of concrete cast below the bars: $\psi_t = 1.3$.

b. The bars are uncoated; $\psi_e = 1.0$.

c. The bars are No. 9 bars: $\psi_s = 1.0$.

d. For Grade 60 reinforcement: $\psi_g = 1.0$.

e. Normal-weight concrete is used: $\lambda = 1.0$.

f. $K_{ER} = 1.0$ (assumed)

g. Stirrup is #3 @ 6 in. o.c., therefore s = 6 in. and A_{tr} = (2 legs)(0.11 in.^{2}) = 0.22 in.^{2}. The number of hooked top bars being developed, n = 4. Therefore,

$$K_{tr} = \frac{40A_{tr}}{sn} = \frac{40\left(0.22 \text{ in.}^2\right)}{(6)(4)} = 0.37$$

3. The product $\psi_t \times \psi_e = 1.3 < 1.7$ (O.K.).

4. Determine c_b. Based on cover (center of bar to nearest concrete surface), consider the clear cover, the No. 3 stirrup diameter, and one-half the diameter of the No. 9 bar:

$$c_b = \text{cover} + \text{stirrup} + \frac{d_b}{2} = 1.5 \text{ in.} + 0.375 \text{ in.} + \frac{1.128 \text{ in.}}{2} = 2.4 \text{ in.}$$

Based on bar spacing, one-half the center-to-center spacing between the top bars is,

$$c_b = \left(\frac{1}{2}\right)\left[\frac{b_w - 2(\text{cover}) - 2(\text{stirrup diameter}) - d_b}{(N_{bars} - 1)}\right]$$

$$= \left(\frac{1}{2}\right)\left[\frac{12 - 2(1.5\text{in.}) - 2(0.375 \text{ in.}) - 1.128\text{in.}}{(4-1)}\right] = 1.2 \text{ in.}$$

Therefore, using the smaller of the two values, c_b = 1.2 in.

5. Check $(c_b + K_{tr})/d_b \le 2.5$:

$$\frac{c_b + K_{tr}}{d_b} = \frac{1.2 + 0.37}{1.128} = 1.39 < 2.5 \qquad \text{(O.K.)}$$

7. Calculate the straight rebar tension development length, ℓ_d:

$$\ell_d = \frac{3}{40}\left(\frac{f_y}{\lambda\sqrt{f_c'}}\right)\left[\frac{\psi_t\psi_e\psi_s\psi_g}{\left(\frac{c_b + K_{tr}}{d_b}\right)}\right](K_{ER})d_b$$

$$= \frac{3}{40}\left(\frac{60{,}000}{(1.0)\sqrt{4000}}\right)\left[\frac{(1.3)(1.0)(1.0)(1.0)}{1.39}\right](1.0)(1.128) = 75 \text{ in.} > 12 \text{ in.}$$

$$\ell_{d,required} = 75 in. < \ell_{d,available} = 70.5 \text{ in.}, \quad \text{Not O.K.}$$

Because 75 in. > 70.5 in., the No. 9 straight top bars cannot be developed within the beam, therefore, terminate the straight bars with a standard hook, either a 90° hook or a 180° hook, at the end of the beam.

5.11 STRUCTURAL INTEGRITY REINFORCEMENT – BEAMS AND GIRDERS

The ACI Code, Section 9.7.7.1 prescribes structural integrity reinforcement for one-way slabs, beams, and girders to improve the redundancy and ductility of the structure, and to prevent progressive collapse such as what happened in the Alfred P. Murrah Federal Building Collapse in Oklahoma City in April 1995 [6]. The function of the structural integrity reinforcement is to keep any potential damage from abnormal loads localized and thus maintain overall structural stability. Structural integrity reinforcement is also required for two-way slabs (see Chapter 6). The ACI Code prescribes the following reinforcement details for beams and girders:

Positive (+ve) Moment Regions:

a. Perimeter or spandrel beams and girders must have closed stirrups with not less than 135° hooks around a continuous longitudinal rebar where spalling can occur, or a standard hook around a longitudinal rebar where spalling is restrained (ACI 318, Section 25.7.1.6).

b. At least 25% of the bottom rebar (at least 2 bars) must be continuous at the supports.

c. Top and Bottom bars should be terminated with ACI standard hook at discontinuous supports and should be capable of developing f_y at the face of the support.

d. Interior beams must satisfy criteria (b) and (c) unless closed stirrups are provided.

Negative (–ve) Moment Regions:

At least 16.7% of top rebar at supports (at least 2 bars) shall be continuous at midspan.

The above requirements are usually covered in the typical details that are a part of the structural drawings. Where splices are needed to provide rebar continuity, splice the top reinforcement at or near the midspan of the member, and splice the bottom reinforcement at or near the support. Splices shall be Class B tension splices (ACI 318, Section 25.5.2) or mechanical or welded splices satisfying ACI 318, Section 25.5.7.1. All longitudinal structural integrity reinforcement shall pass within the area in the column bounded by the column vertical reinforcement.

Structural Integrity Reinforcement for One-Way Slabs

For one-way slabs, structural integrity reinforcement shall be provided as follows in accordance with ACI Code, Section 7.7.7:

1. At least one-quarter (i.e., 25%) of the maximum positive reinforcement shall be continuous or spliced near the slab supports with Class B tension lap splice (see ACI Code, Section 25.5.2) or spliced with mechanical or welded splices (see ACI Code, Section 25.5.7).

2. The continuous structural integrity reinforcement shall be anchored at discontinuous supports (e.g., with ACI standard hook) to develop its yield strength, f_y, at the face of the support.

REFERENCES

[1] James K. Wight and James G. MacGregor, "Reinforced Concrete—Mechanics and Design," 5th Edition, 1112 pp, Pearson, 2009.

[2] Arthur H. Nilson, David Darwin, and Charles Dolan, "Design of Concrete Structures," 13th Edition, 779 pp, McGraw Hill, 2004.

[3] Mohammad Ayub, "Investigation of the October 30, 2003, Fatal Parking Garage Collapse at Tropicana Casino Resort, Atlantic City, NJ," Report No. 2004R03, Directorate of Construction, US Department of Labor, Occupational Health and Safety, April 2004, 56 pp.

[4] Eric Lipton, "Design Changes Preceded Collapse of Casino Garage," New York Times, April 24, 2004, p. 25.

[5] Dan Eschenasy. "Balcony Issues in High-Rise Buildings," Structure Magazine, November 2017

[6] HyunJin Kim. "Progressive Collapse Behavior of Reinforced Concrete Structures with Deficient Details," Ph.D. Dissertation, University of Texas, Austin, TX, 2006.

[7] Ali Ajaam, Samir Yasso, David Darwin, Matthew O'Reilly, and Jayne Sperry. "Anchorage Strength of Closely Spaced Hooked Bars," *ACI Structural Journal*, July 2018, pp. 1143-1152.

EXERCISES

For the following exercises, unless otherwise noted, concrete is normal-weight and steel is uncoated grade 60 (f_y = 60,000 psi).

5.1 Determine the tension development length required for the No. 8 bars in the T-beam shown in Figure 3.11 of Example 3.4. Use $f_c' = 3000$ psi. Assume that the No. 3 stirrups are spaced at 8 in. throughout. Concrete is normal weight. Neglect the compression steel.

5.2 A 12-in.-thick concrete wall is supported on a continuous footing as shown. Use $f_c' = 3000$ psi. Determine if the development length is adequate if the steel is No. 6 bars at 8 in. o.c. Assume that the critical section for moment (f_y is developed) is at the face of the wall.

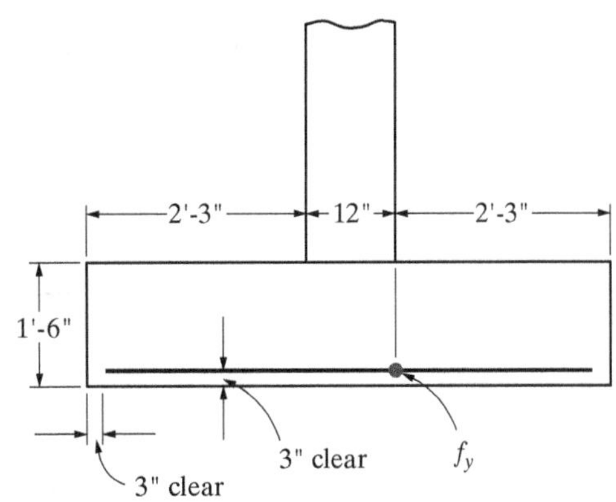

DETAIL FOR EXERCISE 5.2

5.3 The exterior balcony/canopy shown is to be constructed of lightweight concrete (f_{ct} not specified). The bars are epoxy-coated. The design-required tension steel A_s was 0.67 in.^{2}. Determine the required development length ℓ_d from the point of maximum stress in the bars. Specify the minimum required side cover. Shrinkage and temperature steel is not shown. $f'_c = 4000$ psi, $f_y = 60{,}000$ psi.

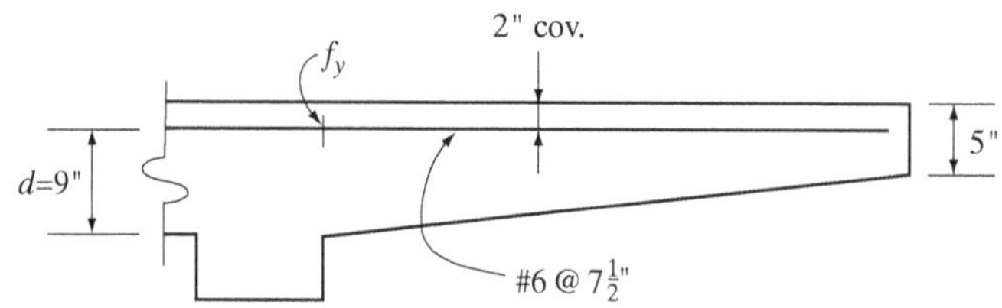

DETAIL FOR EXERCISE 5.3

5.4 Calculate the required development length (ℓ_d) into the beam for the negative moment steel shown to develop the tensile strength of the steel at the face of the column. Required A_s = 2.75 in.^{2}, and $f'_c = 4000$ psi.

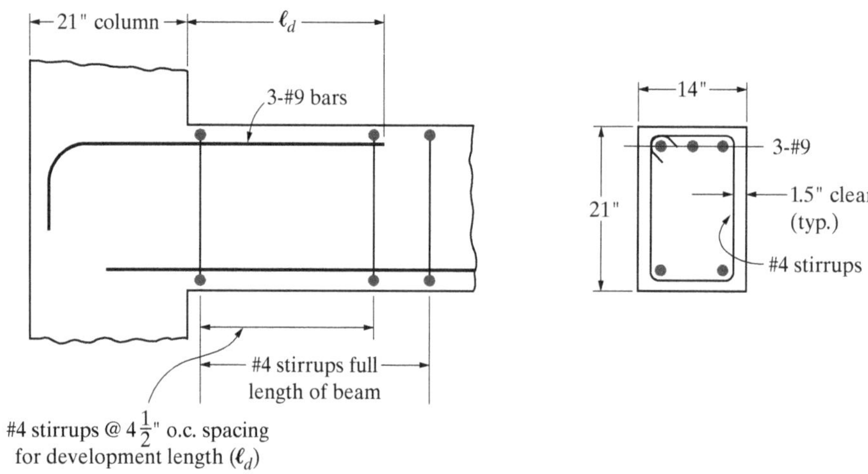

DETAIL FOR EXERCISE 5.4

5.5 Considering the anchorage of the beam bars into the column, determine the largest bar that can be used without a hook. Use $f'_c = 4000$ psi. Clear space between bars is 3 in. (minimum). The side cover is 2 in.

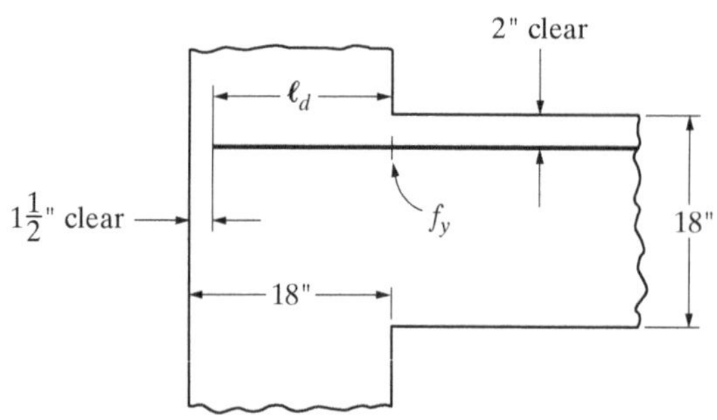

DETAIL FOR EXERCISE 5.5

5.6 Seven No. 11 vertical compression bars extend from a column into the supporting footing. Use $f'_c = 5000$ psi. The A_s required was 10.2 in.^{2} There is no lateral reinforcing enclosing the vertical bars in the footing. Determine the required compression development length.

5.7 Determine the development length required in the column for the bars shown. If the available development length is not sufficient to develop the tensile strength of the steel (f_y), design an anchorage using a 180° hook and check its adequacy. Use $f'_c = 4000$ psi. The clear space between the No. 10 bars is 3 in. with a side cover of $2\frac{1}{2}$ in.

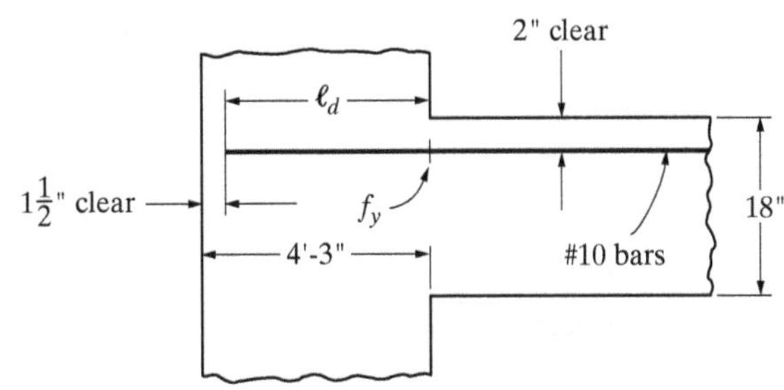

DETAIL FOR EXERCISE 5.7

5.8 The tension bars in the stem of the reinforced concrete retaining wall are No. 9 at 7 in. o.c. and are to be lap-spliced to similar dowels extending up from the footing. Required A_s = 1.50 in.^{2}/ft. Use $f'_c = 3000$ psi. Cover on the bars from the rear face of the wall is 2 in., and cover from the end of the wall to the edge bar, measured in the plane of the bars, is 3 in.

a. Calculate the required length of splice.

b. Find the hook development length ℓ_{dh} required for a 180° hook in the footing.

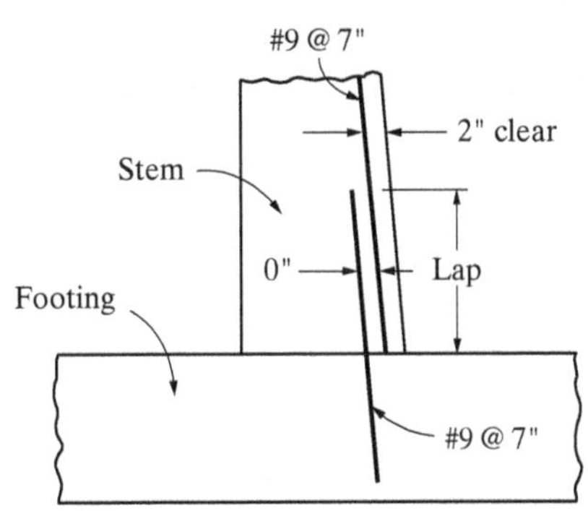

DETAIL FOR EXERCISE 5.8

5.9 Design the tension splices at points A and B in the beam shown, assuming, for purposes of this problem, that the splices must be located as shown. Positive M_u at A is 120 ft.-kip; negative M_u at B is 340 ft.-kip. Assume that 50% of the steel is spliced at the designated locations. Use $f'_c = 4000$ psi. Assume $f = 0.90$ for moment.

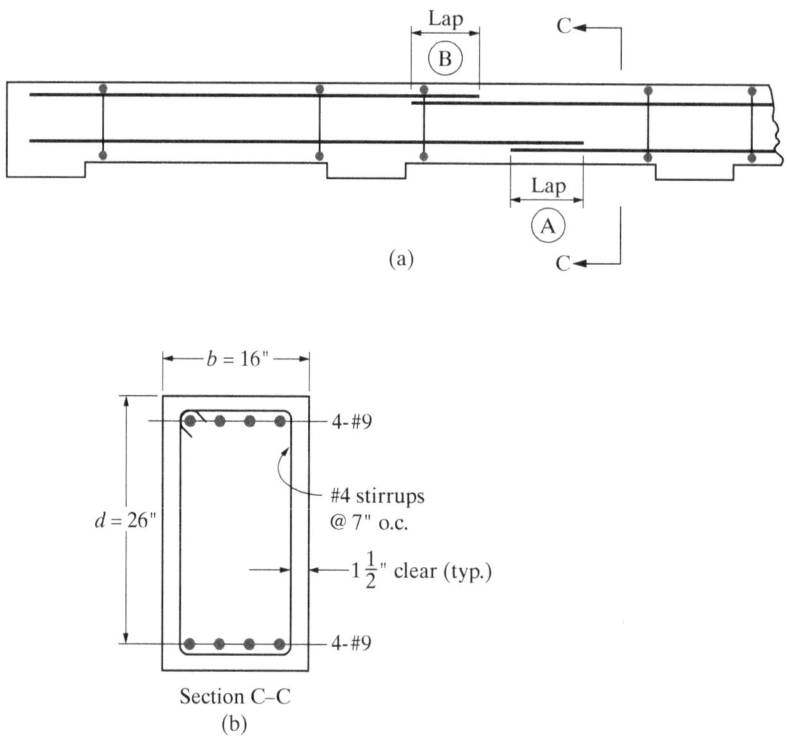

DETAIL FOR EXERCISE 5.9

5.10 A wall is reinforced with No. 7 bars at 12 in. o.c. in each face as shown. The bars are in compression and are to be spliced to dowels of the same size and spacing in the footing. Use $f'_c = 3000$ psi. Determine the required length of splice.

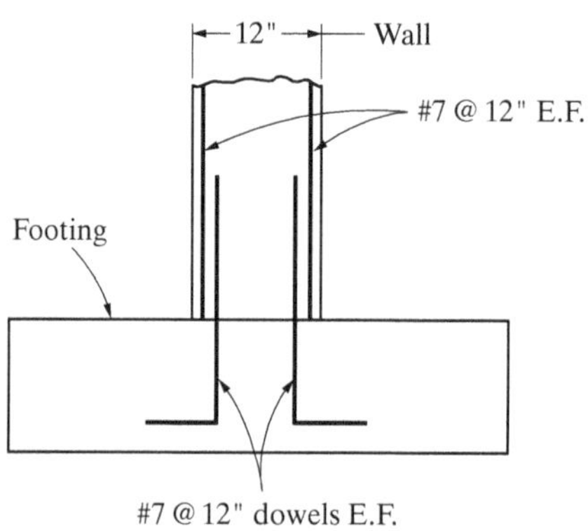

DETAIL FOR EXERCISE 5.10

5.11 Number 9 compression bars in a column are to be lap-spliced. There is no excess steel, and the bars are enclosed by a spiral of $\frac{3}{8}$-in.-diameter wire having a pitch of 3 in. Use $f'_c = 4000$ psi. Determine the required length of splice.

5.12 For the overhanging beam shown, find the theoretical cutoff point for the center No. 9 bar. Assume the section of maximum moment to be at the face of support. Use $f'_c = 5000$ psi.

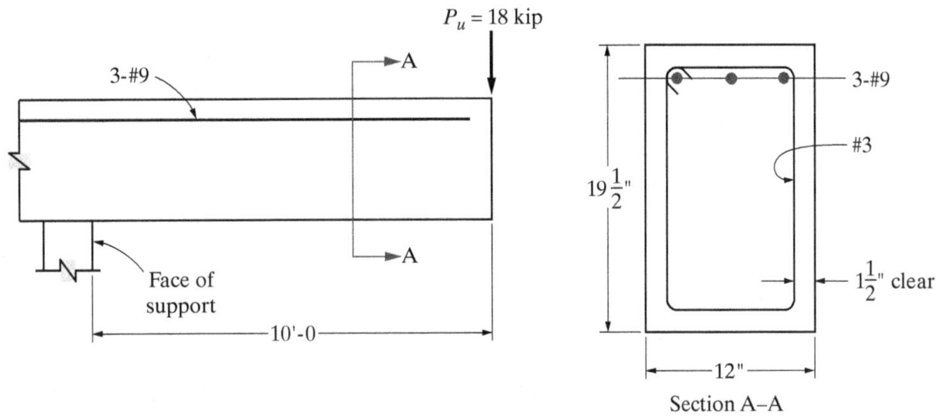

DETAIL FOR EXERCISE 5.12

5.13 A simply supported, uniformly loaded beam carries a total factored load of 4.8 kip/ft (this includes the beam weight) on a clear span of 34 ft. The cross-section has been designed and is shown. The required tension steel area for the design was 5.48 in.^{2} The supports are 12-in. wide. Use $f'_c = 3000$ psi.

a. Determine the cutoff location for the two center bars (indicated in the sketch).

b. Design the shear reinforcement and check whether code requirements are met for bars terminated in a tension zone. Redesign the stirrups if necessary.

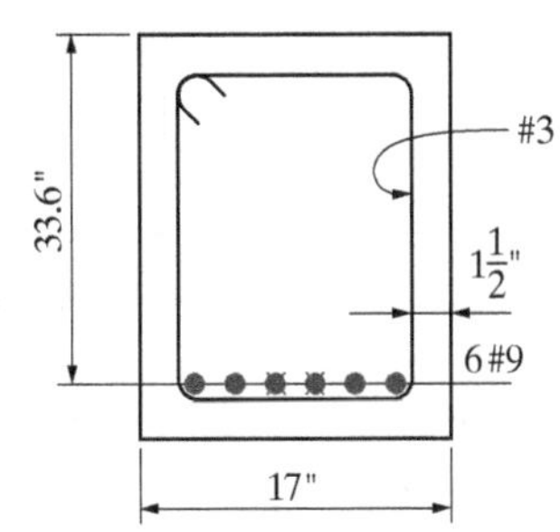

DETAIL FOR EXERCISE 5.13

5.14 The beam shown is to carry a total factored uniform load of 10 kip/ft (includes beam weight) and factored concentrated loads of 20 kip. Use $f'_c = 4000$ psi. The supports are 12-in. wide.

a. Design a rectangular beam (tension steel only).

b. Determine bar cutoffs (use the graphical approach).

c. Design the shear reinforcement and check bar cutoffs in tension zones if necessary.

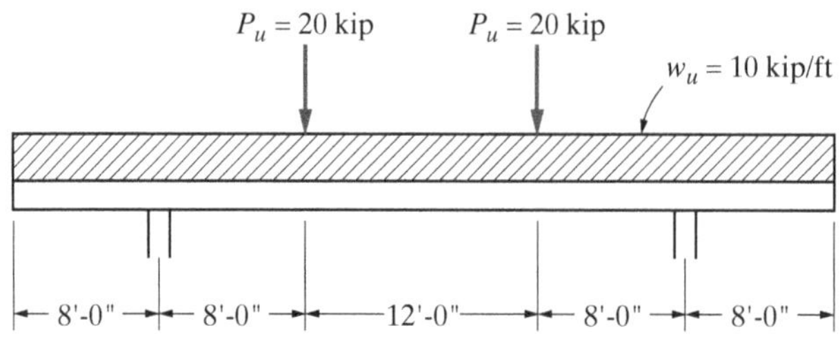

DETAIL FOR EXERCISE 5.14

5.15 For the footing detail shown below with an uplift load, assume #7 bars and use $(c+K_{tr})\,/\,d_b = 2.5$

a. Determine the required development length for both the hooked and straight bars

b. Explain why the answers are different for (a)

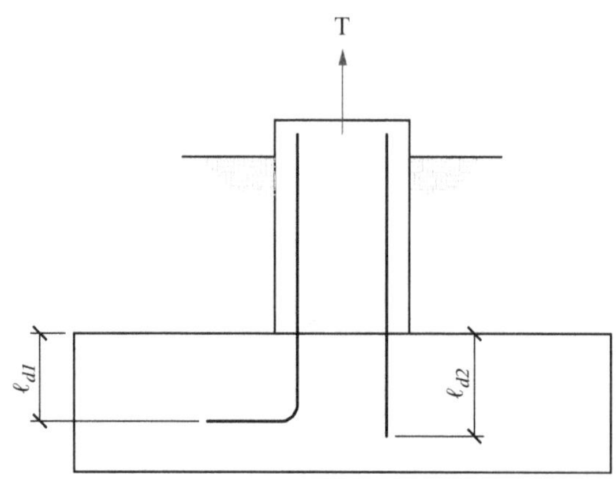

DETAIL FOR EXERCISE 5.15

5.16 For the wall and footing detail shown below, use f'_c = 3,500 psi, f_y = 60,000 psi. Also, for the #6 wall reinforcing, the design requires 0.5 in.^{2}/ft.

a. Determine the lap splice length (Note: consider the loading on the wall and determine if it is in tension or compression).

b. Determine the development length in compression and tension for the hooked bars; is the footing depth adequate?

c. Determine if the footing width "W" is adequate for the development length of the footing bars in tension.

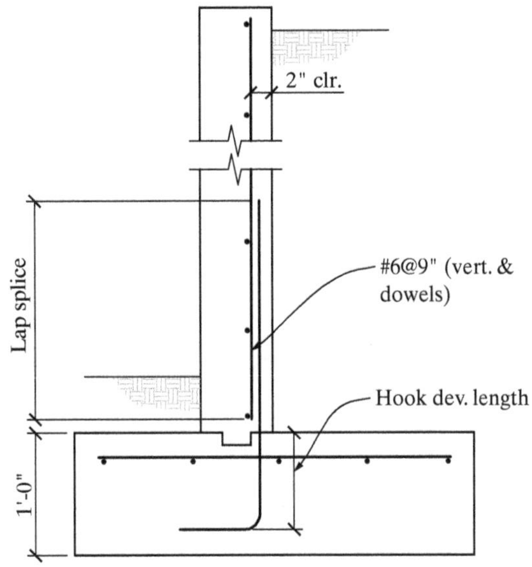

DETAIL FOR EXERCISE 5.16

5.17 For the Wall detail shown below:

a. Determine the lap splice length.

b. Determine the development length in compression and tension for the hooked bars; is the footing depth adequate?

c. Determine if the footing width "W" is adequate for the development of the footing bars in tension?

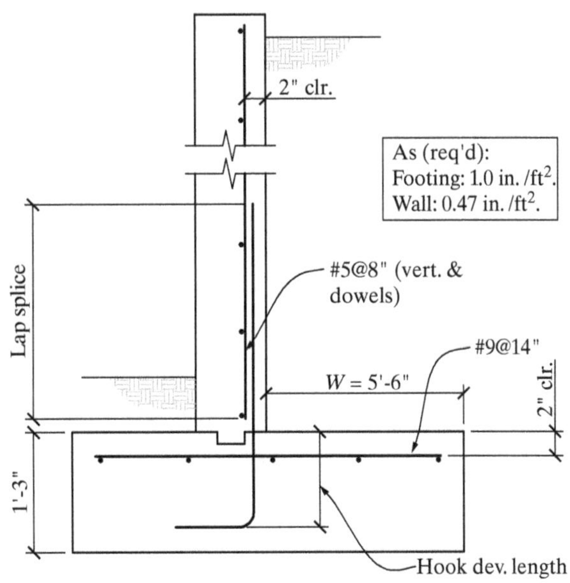

DETAIL FOR EXERCISE 5.17

CHAPTER 6

Continuous One-Way and Two-Way Floor Systems

6.1 INTRODUCTION

A common form of concrete cast-in-place building construction consists of a continuous one-way slab cast monolithically with supporting continuous beams and girders. The concrete beams and girders are not cast monolithically with the columns, but the reinforcement in the beams and girders are extended into the columns, thus making the beam/girder-to-column connections behave monolithically and, therefore, capable of resisting moments at the beam-to-column and girder-to-column joints. This contrasts with steel or wood construction where the beam-to-girder and beam/girder-to-column connections are usually simply supported connections that are incapable of resisting any appreciable moment, resulting in a statically determinate structure, unless these connections are deliberately designed and detailed as moment connections. In concrete framing systems, all members contribute in carrying the floor load to the supporting columns (see Figures 3.1 and 6.19). The slab steel runs through the beams, the beam steel runs through the girders, and the steel from both the beams and girders runs through the columns. The result is that the whole floor system is tied together, forming a highly indeterminate and complex type of rigid structure. The behavior of the members is affected by their rigid connections. Not only will loads applied directly *on* a member produce moment, shear, and a definite deflected shape, but loads applied to *adjacent* members will produce similar effects because of the rigidity of the connections. The shears and moments transmitted through a joint will depend on the relative stiffnesses of all the members framing into that joint. With this type of condition, a precise evaluation of moments and shears resulting from floor loading is excessively time-consuming and is outside the scope of this text. Several classical structural analysis methods (e.g., moment distribution and slope deflection methods), as well as several commercial structural analysis computer software (based on displacement or force methods of structural analysis), are available to facilitate these analysis computations. In the gravity load analysis of continuous one-way slabs or continuous beams, pattern live loading or skip-live loading must be considered. The aim is to vary the position of the live load (the position of the dead load is not varied) to obtain the most critical load case that produces the maximum factored shear and moment at each midspan and support location on the continuous member (see Figure 6.1*a*).

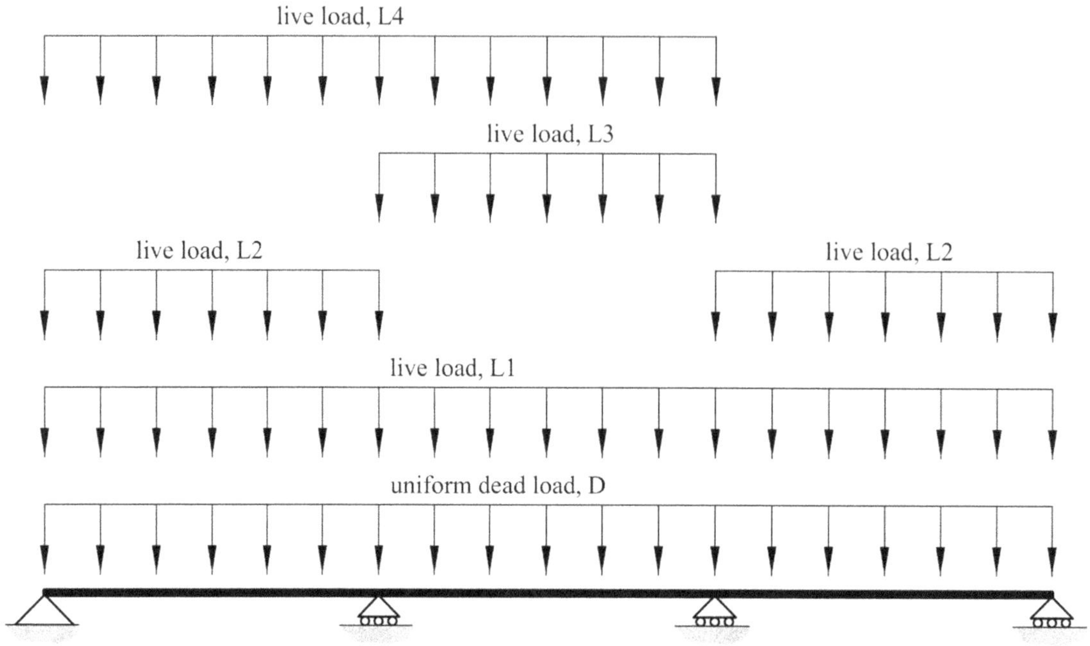

FIGURE 6.1a Pattern live loading for continuous beams and one-way slabs.

To simplify the calculation of moments and shears in continuous concrete members, approximate methods can be used for lateral load analysis (e.g., the portal **frame** method) and gravity load analysis during the preliminary design phase. These methods reduce a statically indeterminate concrete structure to a statically determinate structure by locating hinges at the inflection points on the deflected shape of the structure. The inflection points are the points on each member where a change in curvature occurs. In the portal method for lateral load analysis, the inflection points are assumed to occur at the mid-lengths of the beams and columns, and for gravity load analyses, the points of inflection or hinges can be assumed to occur at 10-to-20-percent of the beam length from each end of the beams.

To further simplify the structural analysis beyond the use of the approximate method described above and thus expedite the design phase, the ACI Code, Section 6.5, permits the use of standard moment and shear equations whenever the span and loading conditions satisfy stipulated requirements (see Figure 6.1*b*). This approach applies to continuous non-prestressed one-way slabs and beams. It is an approximate method and may be used for buildings of the usual type of construction, spans, and story heights. The ACI moment equations result from the product of a coefficient and $w_u\ell_n^2$. Similarly, the ACI shear equations result from the product of a coefficient and $w_u\ell_n$. In these equations, w_u is the factored design uniform load and ℓ_n is the *clear span* for positive moment (and shear) and the average of two adjacent clear spans for negative moment. The negative moment is obtained at the face of the support. The application of the equations is limited to the following conditions:

1. The equations can be used for two or more approximately equal spans (with the larger of two adjacent spans not exceeding the shorter by more than 20%).
2. Loads must be uniformly distributed (therefore, girders are excluded).

3. The maximum allowable ratio of live load to dead load is 3:1 (based on service loads).
4. Members must be prismatic.

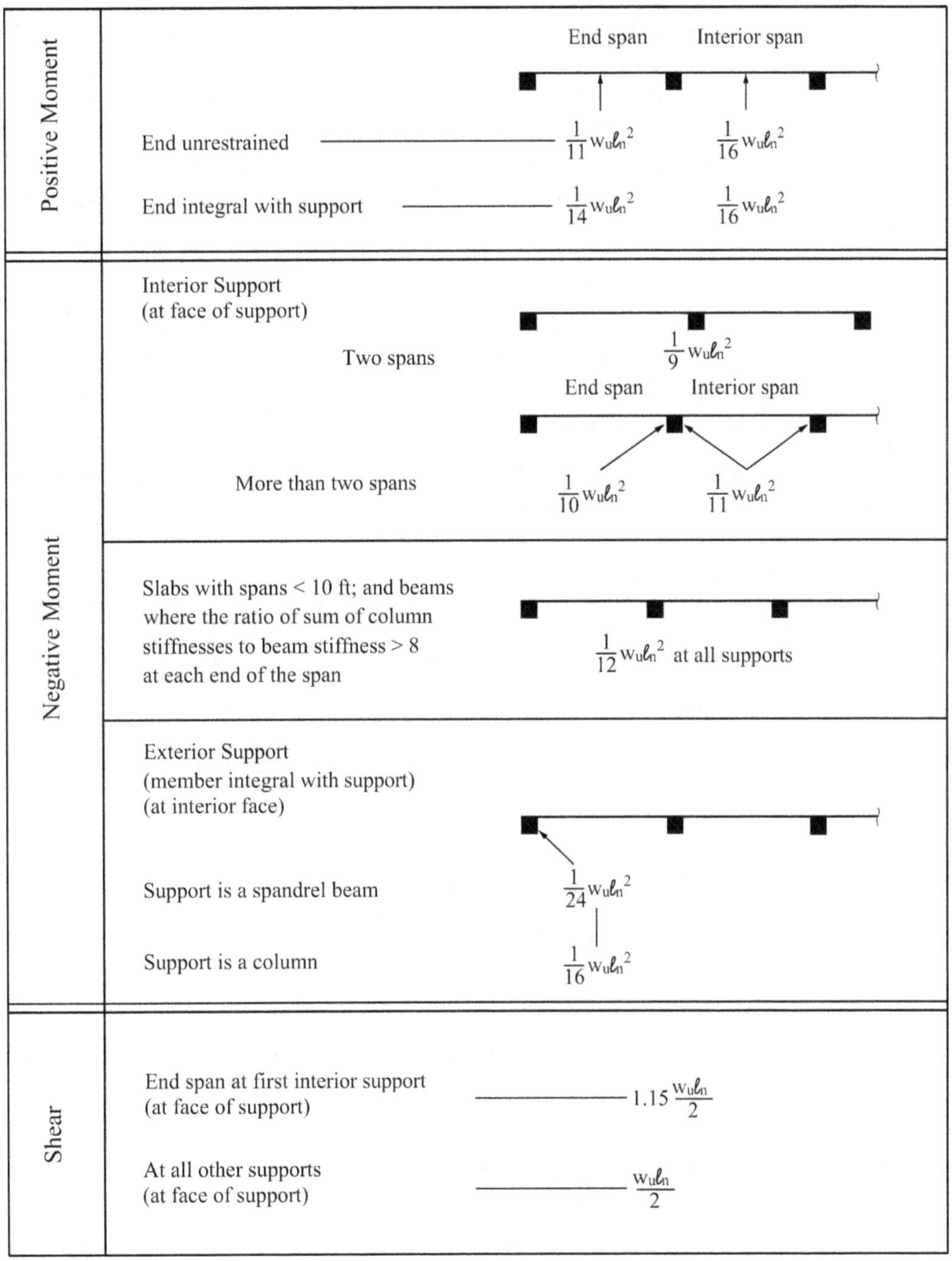

FIGURE 6.1b ACI Code approximate moments and shears for non-prestressed continuous beams and one-way slabs. (Ref: ACI 318-19 Tables 6.5.2 and 6.5.4)

The ACI approximate moment and shear equations from ACI 318 Table 6.5.2 are depicted in Figure 6.1*b*. Their use will be demonstrated later in this chapter. The ACI moment and shear coefficients give the envelopes of the maximum moments and shears, respectively, at critical locations on the flexural member. These maximum moments and shears at the different critical locations along the flexural member do not necessarily occur under the same loading condition. These shear and moment equations generally give reasonably conservative values for the stated conditions. If more precision is required, or desired, for economy, or because the stipulated conditions are not satisfied, a more theoretical and precise analysis must be carried out using a structural analysis software program or using the standard formulas for moments and shears for pattern live loading given in structural engineering handbooks (see Figure 6.1*a*).

For approximate moments and shears for continuous girders, see Section 14.2.

For reinforced concrete beams and slabs of normal weight concrete and f_y of 60,000 psi not supporting deflection sensitive elements (i.e., elements likely to be damaged by large deflections), the minimum slab thickness and minimum beam depths from ACI Code Tables 7.3.1.1. and 9.3.1.1. are given in Table 6.1.

Another common type of cast-in-place concrete construction is column-supported two-way slabs which will be discussed in Section 6.4.

TABLE 6.1 Minimum Thickness of One-Way Slabs and Minimum Depth of Beams[1,2,3,4].

Support Conditions	**Minimum h**	
	One-way slabs	**Beams**
Simply Supported	$\frac{\ell}{20}$	$\frac{\ell}{16}$
One end continuous	$\frac{\ell}{24}$	$\frac{\ell}{18.5}$
Both ends continuous	$\frac{\ell}{28}$	$\frac{\ell}{21}$
Cantilever	$\frac{\ell}{10}$	$\frac{\ell}{8}$

[1]Adapted from ACI 318-19 Tables 7.3.1.1. and 9.3.1.1

[2]For f_y other than 60,000 psi, multiply the minimum depth from the table by $(0.4 + f_y/100{,}000)$

[3]For lightweight concrete with density, w_c, between 90 and 115 psf, multiply the minimum depth from the table by the greater of (a) $1.65 - 0.005w_c$ and (b) 1.09.

[4]Note that for the minimum depths and thicknesses given in Table 6.1, the slab or beams/girders do not support partitions or deflection-sensitive non-structural elements, and the effects of creep and shrinkage, and transient construction loads are not included.

6.2 CONTINUOUS-SPAN BAR CUTOFFS

Using a design approach similar to that for simple spans, the area of main reinforcing steel required at any given point is a function of the design moment. As the moment varies along the span, the steel may be modified or reduced in accordance with the theoretical requirements of the member's strength and the requirements of the ACI Code.

Bars can theoretically be stopped or bent in flexural members whenever they are no longer needed to resist moment. A general representation of the bar cutoff requirements for continuous spans (both positive and negative moments) is shown in Figure 6.2.

In continuous members, at simple supports, the ACI Code, Sections 7.7.3.8.1 and 9.3.3.8.1, requires at least one-third of the maximum positive moment reinforcement to be extended along the slab bottom into the support. At other supports, ACI Code, Sections 7.7.3.8.2 and 9.7.3.8.2, requires that a minimum of one-fourth of the positive moment steel be extended into the support a distance of at least 6 in. The ACI Code, Sections 7.7.3.8.4 and 9.7.3.8.4, also requires that at least one-third of the negative moment steel be extended beyond the extreme position of the point of inflection a distance not less than one-sixteenth of the clear span, the effective depth of the member d, or 12 bar diameters, whichever is greater. If negative moment bars C (Figure 6.2a) are to be cut off, they must extend at least a full development length ℓ_d beyond the face of the support. In addition, they must extend a distance equal to the effective depth of the member or 12 bar diameters, whichever is larger, beyond the theoretical cutoff point defined by the moment diagram. The remaining negative moment bars D (minimum of one-third of total negative steel) must extend at least ℓ_d beyond the theoretical point of cutoff of bars C and, in addition, must extend a distance equal to the effective depth of the member, 12 bar diameters, or one-sixteenth of the clear span, whichever is the greater, past the point of inflection. Where negative moment bars are cut off before reaching the point of inflection, the situation is analogous to the simple beam cutoffs where the reinforcing bars are being terminated in a tension zone. The reader is referred to Example 5.5.

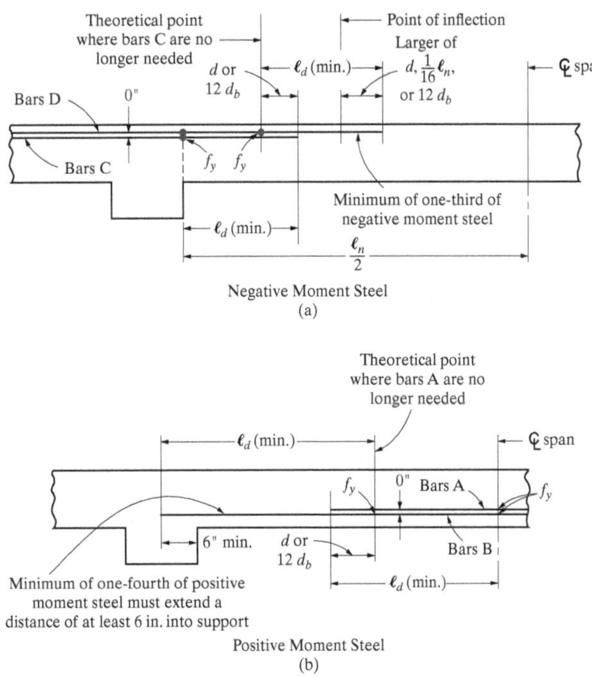

FIGURE 6.2 Bar cutoff requirements for continuous slabs (ACI Code).

In Figure 6.2b, if positive moment bars A are to be cut off, they must project ℓ_d past the point of maximum positive moment or a distance equal to the effective depth of the member or 12 bar diameters, whichever is larger, beyond their theoretical cutoff point. Recall that the location of the theoretical cutoff point depends on the amount of steel to be cut and the shape of the applied moment diagram. The remaining positive moment bars B must extend ℓ_d past the theoretical cutoff point of bars A and extend at least 6 in. into the support. Comments on terminating bars in a tension zone again apply. Additionally, the size of the positive moment bars (i.e., the bar diameter, d_b) at the point of inflection must meet the requirements of the ACI Code, Sections 7.7.3.8.3 and 9.7.3.8.3.

Because the determination of cutoff and bend points constitutes a relatively time-consuming chore, it has become customary to use defined cutoff points that experience has indicated are safe. These defined points may be used where the ACI moment coefficients have application but must be applied with judgment where parameters vary. The recommended bar details and cutoffs for continuous spans are shown in Figure 6.3.

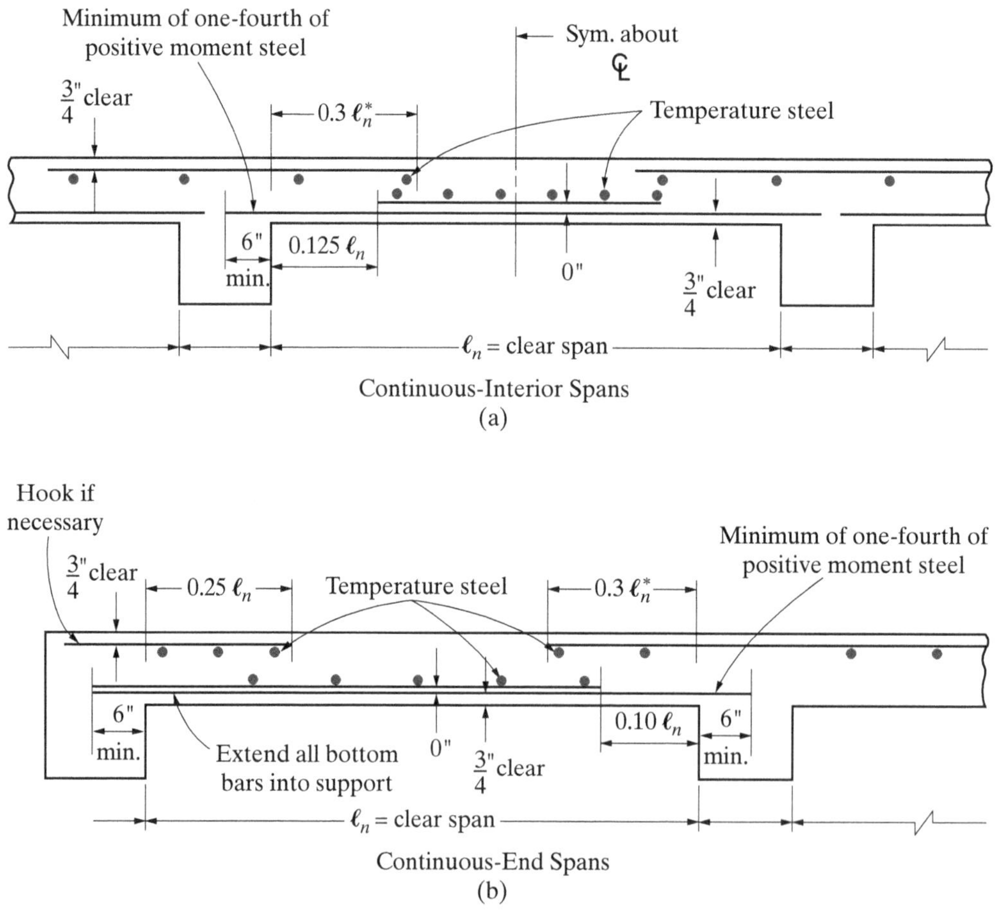

*If adjacent spans have different span lengths, use the larger of the two.

FIGURE 6.3 Recommended bar details and cutoffs, one-way slabs; tensile reinforced beams similar.

6.3 DESIGN OF CONTINUOUS FLOOR SYSTEMS

One common type of floor system consists of a continuous, cast-in-place, one-way reinforced concrete slab supported by monolithic, continuous reinforced concrete beams. If the floor system parameters and loading conditions satisfy the criteria for application of the ACI Code coefficients, the design of the system may be based on these coefficients. Example 6.1 furnishes a complete design of a typical continuous one-way slab and beam floor system. Note that the ACI moment and shear coefficients cannot be used for continuous girders since girders are subjected primarily to concentrated loads from the beam reactions.

EXAMPLE 6.1

The floor system shown in Figure 6.4 consists of a continuous one-way slab supported by continuous beams. The service loads on the floor are 25 psf dead load (does not include weight of slab) and 250 psf live load. Use $f_c' = 3000$ psi (normal-weight concrete) and $f_y = 60{,}000$ psi. The bars are uncoated.

a. Design the continuous one-way floor slab.
b. Design the continuous supporting beam.

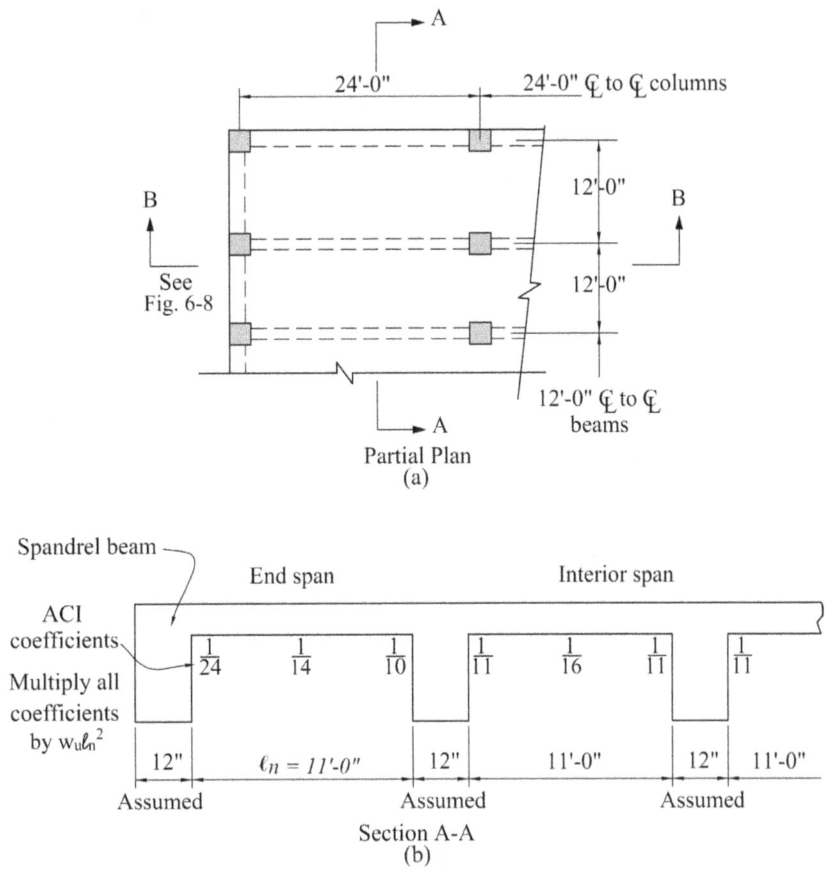

FIGURE 6.4 Sketches for Example 6.1.

Solution:

The primary difference in this design from previous flexural designs is that, because of continuity, the ACI coefficients and equations will be used to determine design shears and moments. The reader should verify that the three conditions (see Section 6.1) required for the ACI coefficients to be used are indeed satisfied for this problem.

a. Continuous one-way floor slab (see Section A-A in Figure 6.4):

1. Determine the slab thickness:
The slab will be designed to satisfy the ACI minimum thickness requirements from Table 7.3.1.1 of the Code and this thickness will be used to estimate slab weight.

With both ends continuous,

$$\text{minimum } h = \frac{1}{28}\ell_n = \frac{1}{28}(11)(12) = 4.71 \text{ in.}$$

With one end continuous,

$$\text{minimum } h = \frac{1}{24}\ell_n = \frac{1}{24}(11)(12) = 5.5 \text{ in.}$$

Try a $5\frac{1}{2}$-in.-thick slab. Design a 12-in.-wide segment (b = 12 in.).

2. Determine the load:

$$\text{slab dead load} = \frac{5.5}{12}(150) = 68.8 \text{ psf}$$

$$\text{total dead load} = 25.0 + 68.8 = 93.8 \text{ psf}$$

$$\begin{aligned} w_u &= 1.2w_{DL} + 1.6w_{LL} \\ &= 1.2(93.8) + 1.6(250) \\ &= 112.6 + 400 \\ &= 512.6 \text{ psf} \quad (\text{design load}) \end{aligned}$$

Because we are designing a slab segment that is 12-in. wide, the foregoing loading is the same as 512.6 lb/ft or 0.513 kip/ft.

3. Determine the moments and shears:
Moments are determined using the ACI moment equations. Refer to Figures 6.1 and 6.4. Thus

$$+M_u = \frac{1}{14}w_u\ell_n^2 = \frac{1}{14}(0.513)(11)^2 = 4.43 \text{ ft.-kip}$$

$$+M_u = \frac{1}{16}w_u\ell_n^2 = \frac{1}{16}(0.513)(11)^2 = 3.88 \text{ ft.-kip}$$

$$-M_u = \frac{1}{10}w_u\ell_n^2 = \frac{1}{10}(0.513)(11)^2 = 6.21 \text{ ft.-kip}$$

$$-M_u = \frac{1}{11}w_u\ell_n^2 = \frac{1}{11}(0.513)(11)^2 = 5.64 \text{ ft.-kip}$$

$$-M_u = \frac{1}{24}w_u\ell_n^2 = \frac{1}{24}(0.513)(11)^2 = 2.59 \text{ ft.-kip}$$

Similarly, the shears are determined using the ACI shear equations. In the end span at the face of the first interior support,

$$V_u = 1.15\frac{w_u \ell_n}{2} = 1.15(0.513)\left(\frac{11}{2}\right) = 3.24 \text{ kip}$$

whereas at all other supports,

$$V_u = \frac{w_u \ell_n}{2} = 0.513\left(\frac{11}{2}\right) = 2.82 \text{ kip}$$

4. Design the slab: Using the assumed slab thickness of $5\frac{1}{2}$ in., find the approximate d. Assume No. 5 bars for main steel and $\frac{3}{4}$-in. cover for the bars in the slab. Thus

$$d = 5.5 - 0.75 - 0.31 = 4.44 \text{ in.}$$

5. Design the steel reinforcing: Assume a tension-controlled section ($\varepsilon_t \geq 0.005$) and $\phi = 0.90$. Select the point of maximum moment. This is a negative moment and occurs in the end span at the first interior support, and

$$M_u = \frac{w_u \ell_n^2}{10} = 6.21 \text{ ft.-kip}$$

$$\phi M_n = \phi b d^2 \bar{k}$$

Because for design purposes $M_u = \phi M_n$ as a limit, then

$$\text{required } \bar{k} = \frac{M_u}{\phi b d^2} = \frac{6.21(12)}{0.90(12)(4.44)^2}$$
$$= 0.3500 \text{ ksi}$$

From Table A.8,

$$\rho = 0.0063 < \rho_{max} = 0.01355 \qquad \text{(O.K.)}$$
$$\text{required } A_s = \rho b d = 0.0063(12)(4.44)$$
$$A_s = 0.34 \text{ in.}^2$$

As the steel area required at all other points will be less, the preceding process will be repeated for the other points. The expression

$$\text{required } \bar{k} = \frac{M_u}{\phi b d^2}$$

can be simplified because all values are constant except M_u:

$$\text{required } \bar{k} = \frac{M_u(12)}{0.9(12)(4.44)^2} = \frac{M_u}{17.74}$$

where M_u must be in ft-kip. In the usual manner, the required steel ratio ρ and the required steel area A_s may then be determined. The results of these calculations are listed in Table 6.2.

TABLE 6.2 Slab Steel Area Requirements

Location	Moment equation	$\bar{k}$ (ksi)	Required ρ	A_s (in.^{2}/ft)
End span:				
At spandrel	$-\frac{1}{24}w_u\ell_n^2$	0.1460	0.0025	0.13
Midspan	$+\frac{1}{14}w_u\ell_n^2$	0.2497	0.0044	0.24
Interior spans:				
Interior support	$-\frac{1}{11}w_u\ell_n^2$	0.3179	0.0057	0.30
Midspan	$+\frac{1}{16}w_u\ell_n^2$	0.2187	0.0038	0.20

Minimum reinforcement for bending in slabs of constant thickness is the same as that required for shrinkage and temperature reinforcement. However, note that the reinforcement required for flexure in a slab must be placed nearest the tension face of the slab, whereas the temperature and shrinkage reinforcement are placed in the orthogonal direction to the main rebar, on top of the main bottom reinforcement and below the main top reinforcement:

$$\text{minimum required } A_s = 0.0018bh$$
$$= 0.0018(12)(5.5) = 0.12 \text{ in.}^2$$

Also, maximum $\rho = 0.01355$, corresponding to a net tensile strain ε_t of 0.005 (see Table A.8). Therefore, the slab steel requirements for flexure, as shown in Table 6.2, are within acceptable limits.

The shrinkage and temperature steel may be selected based on the preceding calculation, therefore,

$$\text{use No. 3 bars at 11 in. o.c.} \left(A_s = 0.12 \text{ in.}^2\right)$$

Recall that the maximum spacing allowed for the main reinforcement in the slab (see ACI Code, Section 7.7.2.3) is 16.5 in. (i.e., the smaller of $3h$ or 18 in.) and for temperature and shrinkage reinforcement (see ACI Code, Section 7.7.6.2.1) is the smaller of $5h$ or 18 in. Because $5h = 5(5.5) = 27.5$ in., the 18 in. would control for temperature and shrinkage steel and the spacing is acceptable.

6. Check the shear strength: From step 3, maximum $V_u = 3.24$ kip at the face of the support. A check of shear at the face of the support, rather than at the critical section that is at a distance equal to the effective depth of the member from the face of the support, is conservative. One-way slabs are not normally reinforced for shear which means the slab thickness is selected so that the shear strength is greater than or equal to the design shear; therefore,

$$\phi V_n = \phi V_c = \phi 2\sqrt{f_c'}\, b_w d$$
$$= \frac{0.75\left(2\sqrt{3000}\right)(12)(4.44)}{1000} = 4.38 \text{ kip}$$
$$\phi V_n > V_u$$

Therefore, the thickness is O.K.

7. Select the main steel: The maximum spacing for the main steel in the one-way slab is 16.5 in. (i.e., the smaller of $3h$ or 18 in.). Using Table A.4, establish a pattern in which the number of bar sizes and the number of different spacings are kept to a minimum. A work sketch (see Figure 6.5) is recommended to establish steel patterns and cutoff points. Regarding the steel selection, note that in the positive moment areas of both the end span and interior span, No. 4 bars at 9 in. could be used. If alternate bars were terminated, however, the spacing of the bars remaining would exceed the maximum spacing of 16.5 in. The use of No. 4 bars at 8 in. avoids this problem. The steel selected is conservative.

8. Check anchorage of the reinforcement into the spandrel beam: The steel is No. 4 bars at 15 in. o.c. Refer to the procedure for development length calculation in Section 5.2.

   a. From Table 5.1, $K_D = 82.2$.

   b. Establish values for the factors ψ_t, ψ_e, ψ_s, ψ_g and λ.

      1. $\psi_t = 1.3$ (the bars are top bars).
      2. The bars are uncoated; $\psi_e = 1.0$.
      3. The bars are No. 4; $\psi_s = 0.8$.
      4. Grade 60 steel; $\psi_g = 1.0$
      5. Normal-weight concrete is used; $\lambda = 1.0$.

   c. The product $\psi_t \times \psi_e = 1.3 < 1.7$. (O.K.)

   d. Determine c_b. Based on cover (center of bar to nearest concrete surface),

$$c_b = \frac{3}{4} + \frac{0.5}{2} = 1 \text{ in.}$$

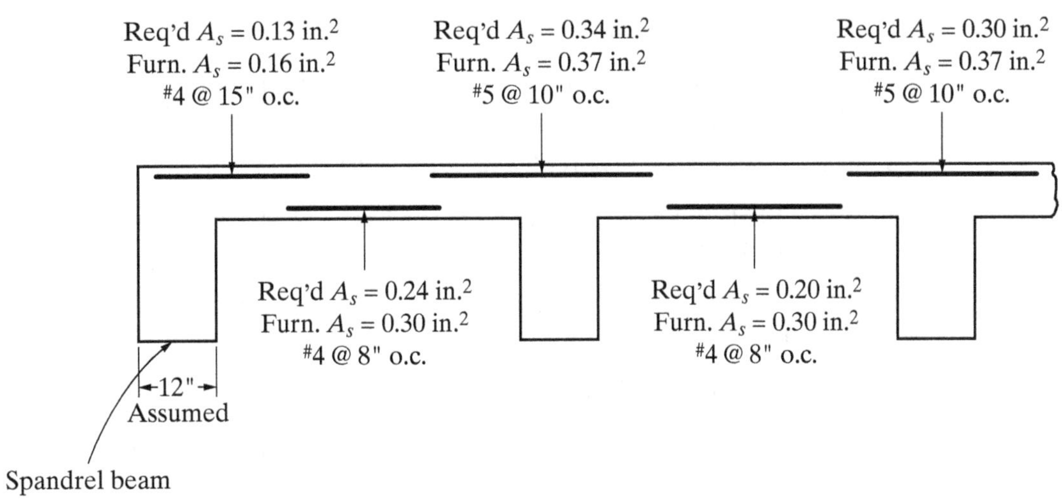

Note: Bar cutoffs and temperature steel may be observed in the design sketch, Figure 6.7.

FIGURE 6.5 Work sketch for Example 6.1.

Based on bar spacing (one-half the center-to-center distance),

$$c_b = \tfrac{1}{2}(15) = 7.5 \text{ in.}$$

Therefore, use c_b = 1.0 in.

e. K_{tr} is taken as zero. There is no transverse steel that crosses the potential plane of splitting.

f. Check $(c_b + K_{tr})/d_b \leq 2.5$:

$$\frac{c_b + K_{tr}}{d_b} = \frac{1.0 + 0}{0.5} = 2.0 < 2.5 \quad \text{(O.K.)}$$

g. Calculate the excess reinforcement factor:

$$K_{ER} = \frac{A_s \text{ required}}{A_s \text{ provided}} = \frac{0.130}{0.160} = 0.813$$

h. Calculate ℓ_d:

$$\ell_d = \frac{K_D}{\lambda}\left[\frac{\psi_t \psi_e \psi_s \psi_g}{\left(\dfrac{c_b + K_{tr}}{d_b}\right)}\right] K_{ER} d_b$$

$$= \frac{82.2}{1.0}\left[\frac{1.3(1.0)(0.8)(1.0)}{2.0}\right](0.813)(0.5)$$

$$= 17.4 \text{ in.} > 12 \text{ in.} \qquad \text{(O.K.)}$$

Use $\ell_d = 18$ in. (minimum).

9. Because the 18-in. length cannot be furnished within the width of the spandrel beam, a hook will be provided. Determine if a 180° standard hook will be adequate.

a. Calculate ℓ_{dh} (use Section 5.4 or Table A.13).

From Table A.13, the base value of the hooked bar development length for No. 4 rebar in 3000 psi concrete = 7.0 in.

b. The bars are uncoated and the concrete is normal weight. Therefore, ψ_e and λ are both 1.0.

c. Other modification factors are as follows:

1. For No. 11 or smaller hooked bars with rebar spacing, s, greater than or equal to $6d_b$, ψ_r = 1.0

2. For No. 11 or smaller hooked bars with side cover normal to the plane of the hook $\geq 6d_b$ (i.e., $6 \times 1/2 = 3$ in.). Since 3 in. > 2 in. clear cover, $\psi_o = 1.25$ (see ACI 318-19 Table 25.4.3.2).

3. For $f_c' = 3000$ psi, $\psi_c = (3000/15000) + 0.6 = 0.8$.

d. Therefore, the required development length is

$$\ell_{dh} = (7\text{ in.})(1.0)(1.0)(1.25)(0.8)$$
$$= 7\text{ in.}$$

Minimum ℓ_{dh} is 6 in. or $8d_b$, whichever is greater:

$$8d_b = 8\left(\tfrac{1}{2}\right) = 4\text{ in.}$$

Therefore, the minimum is 6 in.:

$$7\text{ in.} > 6\text{ in.} \qquad \text{(O.K.)}$$

e. Check the total width of the spandrel beam required at the discontinuous end with 2-in concrete clear cover to the hook (see Figure 6.6):

$$7 + 2 = 9\text{ in.} < 12\text{ in.} \qquad \text{(O.K.)}$$

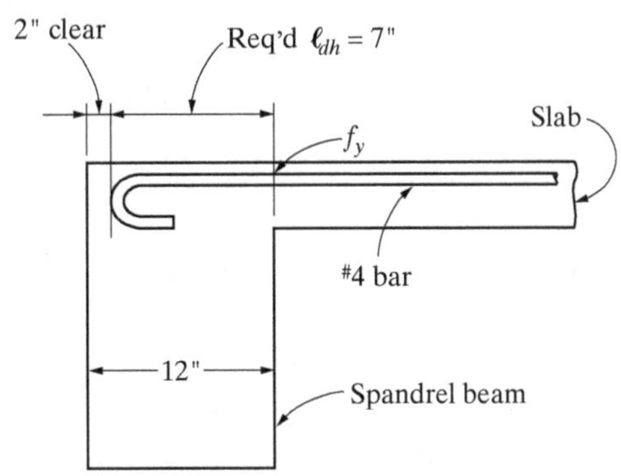

FIGURE 6.6 Hook detail for Example 6.1.

10. Determine the bar cutoff points: For the normal type of construction for which the typical bar cutoff points shown in Figure 6.3 are used, the cutoff points are located so that all bars terminate in compression zones. Thus, the requirements of the ACI Code, Section 7.7.3.5, need not be checked, and the recommended bar cutoff points, as shown in Figure 6.3, are used.

11. Prepare the design sketches:
The final design sketch for the slab is shown in Figure 6.7. For clarity, the interior and end spans are shown separately.

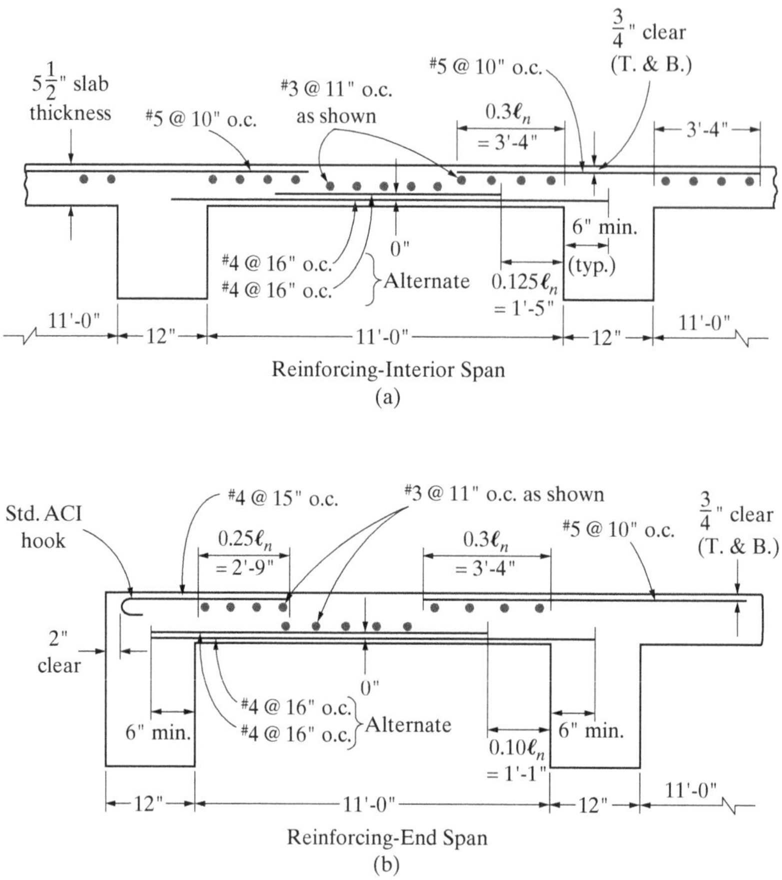

FIGURE 6.7 Design sketches for Example 6.1.

b. Continuous beam supporting the one-way slab (see Section B-B in Figures 6.4a and 6.8):

The second part of Example 6.1 involves the design of the continuous supporting beam. From Figure 6.4a, it is seen that these beams span between columns. The ACI coefficients to be used for moment determination are shown in Figure 6.8.

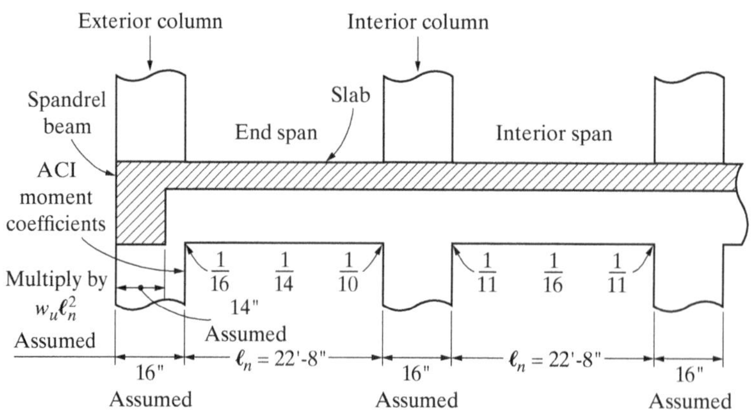

FIGURE 6.8 Section B-B from Figure 6.4a.

1. Determine the loading on the beam with a tributary width of 12-ft.:

$$\text{service live load} = 250\ \text{psf} \times 12\,\text{ft.} = 3000\ \text{lb/ft}$$
$$\text{service dead load} = 25\ \text{psf} \times 12\,\text{ft.} = 300\ \text{lb/ft}$$
$$\text{weight of slab} = \left(\frac{5.5}{12}\right)(150\,\text{pcf})(12\,\text{ft.}) = 825\ \text{lb/ft}$$

Assuming a beam width of 12 in. and an overall depth of 30 in. for purposes of member weight estimate (see Figure 6.9),

$$\text{weight of beam} = \frac{12(30 - 5.5)}{144}(150\,\text{pcf})$$
$$= 306.3\ \text{lb/ft}$$
$$\text{total service live load} = 3000\ \text{lb/ft}$$
$$\text{total service dead load} = 1431.3\ \text{lb/ft} \quad \text{say, } 1431\ \text{lb/ft}$$

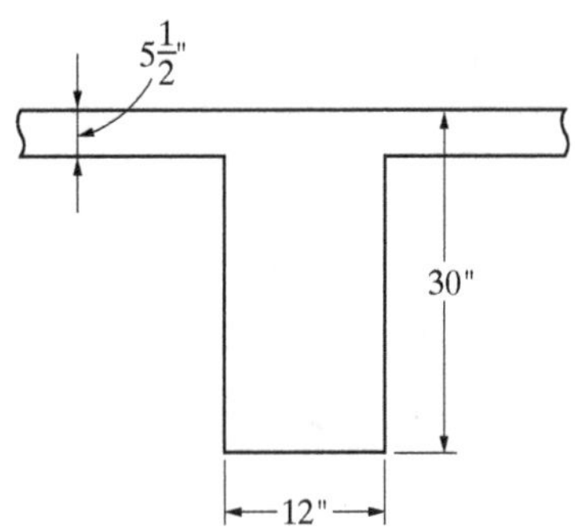

FIGURE 6.9 Sketch for Example 6.1.

2. Calculate the design load:

$$w_u = 1.2w_{DL} + 1.6w_{LL}$$
$$= 1.2(1431) + 1.6(3000)$$
$$= 1717 + 4800$$
$$= 6517\ \text{lb/ft} \quad \text{say, } 6.5\ \text{kip/ft}$$

The loaded beam is depicted in Figure 6.10.

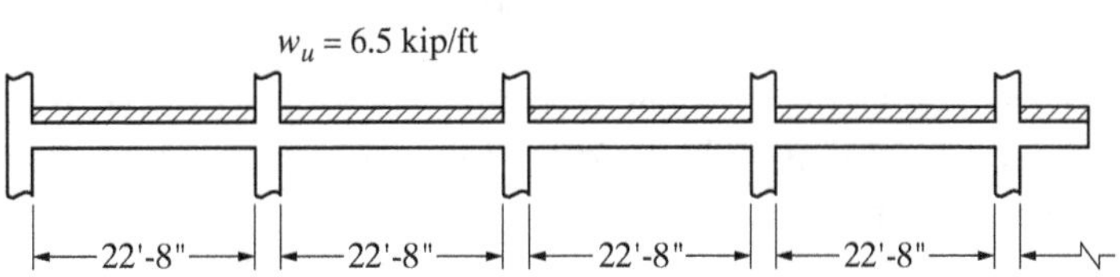

FIGURE 6.10 Beam design load and spans.

3. Calculate the design moments and shears. The design moments and shears are calculated by using the ACI equations:

$$+M_u = \frac{1}{14} w_u \ell_n^2 = \frac{1}{14}(6.5)(22.67)^2 = 238.6 \text{ ft.-kip}$$

$$+M_u = \frac{1}{16} w_u \ell_n^2 = \frac{1}{16}(6.5)(22.67)^2 = 208.8 \text{ ft.-kip}$$

$$-M_u = \frac{1}{10} w_u \ell_n^2 = \frac{1}{10}(6.5)(22.67)^2 = 334.1 \text{ ft.-kip}$$

$$-M_u = \frac{1}{11} w_u \ell_n^2 = \frac{1}{11}(6.5)(22.67)^2 = 303.7 \text{ ft.-kip}$$

$$V_u = \frac{w_u \ell_n}{2} = \frac{6.5(22.67)}{2} = 73.7 \text{ kip}$$

$$V_u = 1.15 \frac{w_u \ell_n}{2} = 1.15(6.5)\left(\frac{22.67}{2}\right) = 84.7 \text{ kip}$$

4. Design the continuous beam: Establish the concrete beam dimensions based on the maximum bending moment. This occurs in the end span at the first interior support where the negative moment $M_u = w_u \ell_n^2/10$. Since the top of the beam is in tension, the design will be that of a rectangular beam.

a. Maximum moment (negative) = 334.1 ft.-kip.

b. From Table A.5, assume that $\rho = 0.0090$ (which is less than ρ_{max} of 0.01355 from Table A.8). A check for the minimum steel required will be made shortly.

c. From Table A.8, $\bar{k} = 0.4828$ ksi.

d. Assume that b = 12 in.:

$$\text{required } d = \sqrt{\frac{M_u}{\phi b \bar{k}}} = \sqrt{\frac{334.1(12)}{0.9(12)(0.4828)}}$$

$$= 28.0 \text{ in.}$$

$$d/b \text{ ratio} = \frac{28.0}{12} = 2.34, \quad \text{which is within the acceptable range.}$$

e. Check the estimated beam weight assuming one layer of No. 11 bars and No. 3 stirrups:

$$\text{required } h = 28.0 + \frac{1.41}{2} + 0.38 + 1.5$$

$$= 30.6 \text{ in.}$$

Use h = 31 in. with an assumed d of 28 in. Also, check the minimum beam depth, h, from Table 6.1 (ACI Code, Table 9.3.1.1) in the beam end span (i.e., one end continuous):

$$\text{minimum } h = \frac{1}{18.5}(22.67 \text{ ft.})(12)$$
$$= 14.7 \text{ in.} < 31 \text{ in.} \qquad \text{(O.K.)}$$

Note that the estimated beam weight based on b = 12 in. and h = 30 in. is slightly on the low side but may be considered acceptable.

f. Design the steel reinforcing for points of negative moment as follows: At the negative moment locations, the compression zone is at the bottom of the beam, and therefore, the width of the compression zone in these regions is equal to the width of the beam web, b_w = 12 in.

1. At the first interior support, based on an assumed ρ of 0.0090 (see the preceding step 4),

$$\text{required } A_s = \rho bd$$
$$= 0.0090(12)(28) = 3.0 \text{ in.}^2$$

From Table A.5,

$$A_{s,\min} = 0.0033(12)(28) = 1.11 \text{ in.}^2$$
$$3.0 \text{ in.}^2 > 1.11 \text{ in.}^2 \qquad \text{(O.K.)}$$

2. At the other interior supports, $-M_u$ = 303.7 ft.-kip:

$$\text{required } \bar{k} = \frac{M_u}{\phi bd^2} = \frac{303.7(12)}{0.9(12)(28)^2}$$
$$= 0.4304 \text{ ksi}$$

From Table A.8, ρ = 0.0079. Check ρ with ρ_{max} (recall that ρ_{max} corresponds to a net tensile strain ε_t of 0.005 in the extreme tensile reinforcement for Grade 60 rebar with f_y = 60,000 psi):

$$0.0079 < 0.01355 \qquad \text{(O.K.)}$$

Therefore,

$$\text{required } A_s = \rho bd$$
$$= 0.0079(12)(28) = 2.65 \text{ in.}^2$$

Check the required A_s with $A_{s,\min}$. From Table A.5,

$$A_{s,\min} = 0.0033(12)(28) = 1.11 \text{ in.}^2$$
$$2.65 \text{ in.}^2 > 1.11 \text{ in.}^2 \qquad \text{(O.K.)}$$

3. At the exterior support (exterior column), $-M_u$ = 208.8 ft.-kip:

$$\text{required } \bar{k} = \frac{M_u}{\phi bd^2} = \frac{208.8(12)}{0.9(12)(28)^2}$$
$$= 0.2959 \text{ ksi}$$

From Table A.8, $\rho = 0.0053$. Check ρ with ρ_{max}:

$$0.0053 < 0.01355 \qquad \text{(O.K.)}$$

Therefore,

$$\text{required } A_s = \rho bd$$
$$= 0.0053(12)(28) = 1.78 \text{ in.}^2$$

Check the required A_s with $A_{s,min}$. From Table A.5,

$$A_{s,min} = 0.0033(12)(28) = 1.11 \text{ in.}^2$$
$$1.78 \text{ in.}^2 > 1.11 \text{ in.}^2 \qquad \text{(O.K.)}$$

g. Design steel reinforcing for points of positive moment as follows:
At points of positive moment, the top of the beam is in compression; therefore, the design will be that of a T-beam, and the width of the compression zone is equal to the effective width of the T-beam.

1. End-span positive moment:

a. Design moment = 238.6 ft.-kip = M_u.

b. Effective depth d = 28 in. (see negative moment design).

c. Effective flange width:

$$b_w + \tfrac{1}{4}\text{span length} = 12 + 0.25(22.67)(12) = 80 \text{ in.}$$
$$b_w + 16h_f = 12 + 16(5.5) = 100 \text{ in.}$$
$$\text{beam spacing} = 144 \text{ in.}$$

Use an effective flange width b = 80 in. (see Figure 6.11).

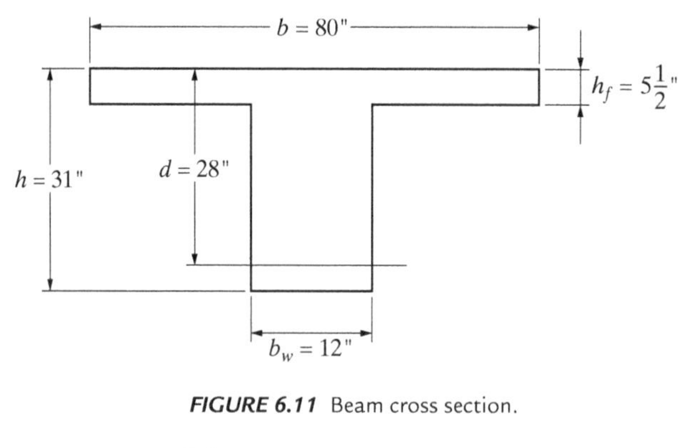

FIGURE 6.11 Beam cross section.

d. Assuming total flange in compression, $\varepsilon_t = 0.005$, and $\phi = 0.90$:

$$\phi M_{nf} = \phi(0.85f_c')bh_f\left(d - \frac{h_f}{2}\right)$$
$$= 0.9(0.85)(3)(80)(5.5)\left(\frac{28 - 5.5/2}{12}\right)$$
$$= 2125 \text{ ft.-kip}$$

e. Because 2125 > 238.6, the member behaves as a wide rectangular T-beam with b = 80 in. and d = 28 in.

f. $$\text{required } \bar{k} = \frac{M_u}{\phi b d^2}$$
$$= \frac{238.6(12)}{0.9(80)(28)^2} = 0.0507 \text{ ksi}$$

g. From Table A.8,

$$\text{required } \rho = 0.0010$$

h. The required steel area is

$$\text{required } A_s = \rho b d$$
$$= 0.0010(80)(28) = 2.24 \text{ in.}^2$$

i. Use three No. 8 bars (A_s = 2.37 in.^{2}):

The required minimum beam width to accommodate 3#8 bars (see Table A.3) = 9.0 in.

< 12 in. beam width (O.K.)

j. Check d. With a No. 3 stirrup and $1\frac{1}{2}$-in. cover,

$$d = 31 - 1.5 - 0.38 - 1.00/2$$
$$= 28.6 \text{ in.} > 28 \text{ in.} \qquad \text{(O.K.)}$$

k. Check the required A_s with $A_{s,\min}$. From Table A.5,

$$A_{s,\min} = 0.0033(12)(28) = 1.11 \text{ in.}^2$$
$$2.37 \text{ in.}^2 > 1.11 \text{ in.}^2 \qquad \text{(O.K.)}$$

i. Check ε_t and ϕ:
From Table A.8, with ρ = 0.001, $\varepsilon_t \geq 0.005$. Therefore, the assumed ϕ of 0.90 is O.K.

2. Interior span positive moment:

a. Design moment = 208.8 ft.-kip = M_u.

b. through (e) see the end-span positive moment computations. At the positive moment location, the width of the compression zone is equal to the effective width of the T-beam. Therefore, use an effective flange width b = 80 in. and an effective depth d = 28 in. Also, for total flange in compression, ϕM_n = 2125 ft.-kip > M_u. Therefore, this member also behaves as a rectangular T-beam.

c. required $\bar{k} = \dfrac{M_u}{\phi b d^2}$

$$= \frac{208.8(12)}{0.9(80)(28)^2} = 0.0444 \text{ ksi}$$

d. From Table A.8,

$$\text{required } \rho = 0.0010$$

e. The required steel area is

$$\text{required } A_s = \rho b d$$
$$= 0.0010(80)(28) = 2.24 \text{ in.}^2$$

f. Use three No. 8 bars (A_s = 2.37 in.^{2}):

The required minimum beam width to accommodate 3#8 bars (see Table A.3) = 9.0 in. < 12-in. (O.K.)

g. through i. is identical to those for the end-span positive moment.

5. Check the distribution of negative moment steel:
The ACI Code, Section 24.3.4, requires that where flanges are in tension, a part of the main tension reinforcement be distributed over the effective flange width or width equal to one-tenth of the span, whichever is smaller. The use of smaller bars spread out into part of the flange will also be advantageous where a beam is supported by a spandrel girder or exterior column and the embedment length for the negative moment steel is limited. Thus

$$\frac{\text{span}}{10} = \frac{22.67(12)}{10} = 27 \text{ in.}$$
$$\text{effective flange width} = b = 80 \text{ in.}$$

Therefore, distribute the negative moment bars over a width of 27 in. Figure 6.12 shows suitable bars and patterns to use to satisfy the foregoing code requirement and furnish the cross-sectional area of steel required for flexure.

The ACI Code also stipulates that if the effective flange width exceeds one-tenth of the span, some longitudinal reinforcement shall be provided in the outer portions of the flange. In this design, no additional steel will be furnished. In the authors' opinion, this requirement is satisfied by the slab temperature and shrinkage steel (see Figures 6.7 and 6.18).

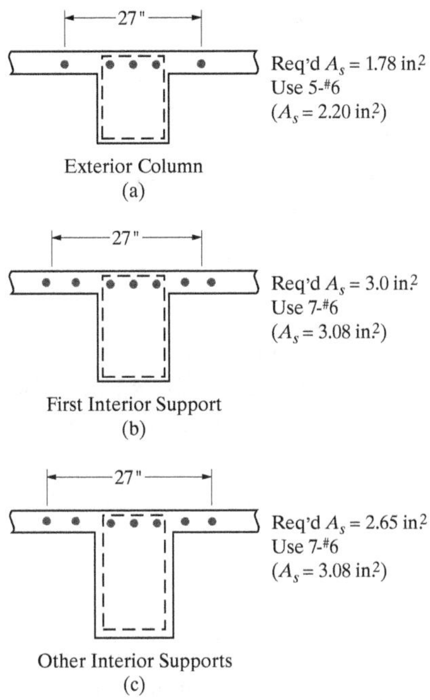

FIGURE 6.12 Negative moment steel for beam of Example 6.1.

6. Prepare the work sketch. A work sketch is developed in Figure 6.13, which includes the bars previously chosen.

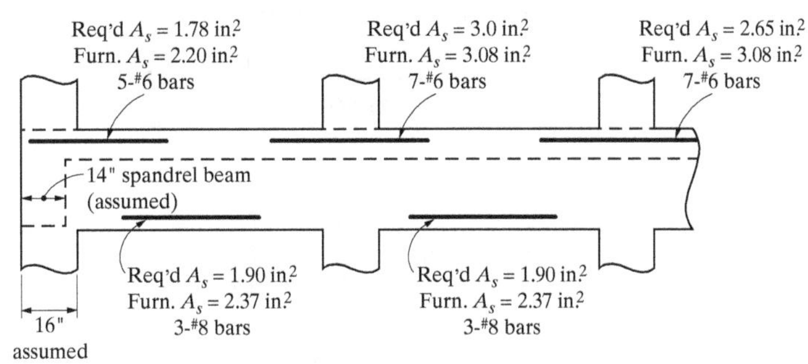

FIGURE 6.13 Work sketch for beam design of Example 6.1.

7. Check the anchorage or development of the beam top reinforcement into the exterior column:

   a. From Table 5.1, K_D = 82.2

   b. Establish values for the factors ψ_t, ψ_e, ψ_s, ψ_g and λ.

      1. ψ_t = 1.3 (the bars are top bars).

      2. The bars are uncoated; ψ_e = 1.0

      3. The bars are No. 6; ψ_s = 0.8

4. Grade 60 rebar; $\psi_g = 1.0$
5. Normal-weight concrete is used; $\lambda = 1.0$

c. The product $\psi_t \times \psi_e = 1.3 < 1.7$ (O.K.)

d. Determine c_b. Based on cover (center of bar to nearest concrete surface), assume $1\frac{1}{2}$-in. clear cover and a No. 3 stirrup:

$$c_b = 1.5 + 0.375 + \frac{0.75}{2} = 2.25 \text{ in.}$$

Based on bar spacing (one-half the center-to-center distance), refer to Figure 6.12. Consider the three No. 6 top bars in the beam that are located within the No. 3 closed loop stirrup in Figure 6.12a. Here

$$c_b = \left[\frac{12 - 2(1.5) - 2(0.375) - 0.75}{2}\right]\left(\frac{1}{2}\right)$$
$$= 1.875 \text{ in.}$$

Therefore, use $c_b = 1.875$ in.

e. K_{tr} may be conservatively taken as zero.

f. Check $(c_b + K_{tr})/d_b \leq 2.5$:

$$\frac{c_b + K_{tr}}{d_b} = \frac{1.875 + 0}{0.75} = 2.5(\leq 2.5) \qquad \text{(O.K.)}$$

g. Determine the excess reinforcement factor:

$$K_{ER} = \frac{A_s \text{ required}}{A_s \text{ provided}} = \frac{1.78}{2.20} = 0.809$$

h. Calculate ℓ_d:

$$\ell_d = \frac{K_D}{\lambda}\left[\frac{\psi_t \psi_e \psi_s \psi_g}{\left(\dfrac{c_b + K_{tr}}{d_b}\right)}\right] d_b K_{ER}$$
$$= \frac{82.2}{1.0}\left[\frac{1.3(1.0)(0.8)1.0}{2.5}\right](0.75)(0.809)$$
$$= 20.7 \text{ in.} > 12 \text{ in.} \qquad \text{(O.K.)}$$

Use $\ell_d = 21$ in. (minimum). This is the required straight bar development length.

With 2.0 in. clear concrete cover at the end of the bar, the embedment length available in the 16-in. wide column is 16.0 – 2.0 = 14.0 in.

8. Because 21 in. > 14.0 in., a hook is required. Determine if a 90° standard hook will be adequate.

a. From Table A.13 (for No. 6 bars), $\ell_{dh} = 12.9$ in.

b. Modification factors (MF) to be used are

1. Uncoated rebar; $\psi_e = 1.0$.
2. With 5 No. 6 bars within a column width of 16 in., the spacing of the hook bars, $s < 6d_b = 6(0.75) = 4.5$ in.
   Without confining reinforcement like that shown in Figure 5.9 (see also ACI 318-19 Figures 25.4.3.3a and 25.4.3.3b), $\psi_r = 1.6$. But assuming confining reinforcement will be provided, $\psi_r = 1.0$.
3. Assume concrete side cover normal to the plane of the hook within the column core $\geq 2\frac{1}{2}$ in.; $\psi_o = 1.0$.
4. For $f_c' = 3000$ psi, $\psi_c = (3000/15000) + 0.6 = 0.8$

c. The required development length for the hook is

$$\ell_{dh} = (12.9)(1.0)(1.0)(1.0)(0.8) = 10.3 \text{ in.}$$

Minimum ℓ_{dh} is 6 in. or $8d_b$, whichever is greater:

$$8d_b = 8\left(\tfrac{3}{4}\right) = 6 \text{ in.}$$

$$10.3 \text{ in.} > 6 \text{ in.} \qquad \text{(O.K.)}$$

Note that since the concrete side cover and top cover to the hooked bars within the column core is at least 2.5 in., ACI 318-19, Section 25.4.3.4 need not be satisfied.

d. Check the total width of the column required with 2-in. clear concrete cover to the hook (see Figure 6.14):

$$10.3 + 2 = 12.3 \text{ in.} < 16 \text{ in.} \qquad \text{(O.K.)}$$

Note: If confining reinforcement were not provided, the reader should verify that the hooked bar development length will be 16.5 in. and can be developed within a 20 in. column core.

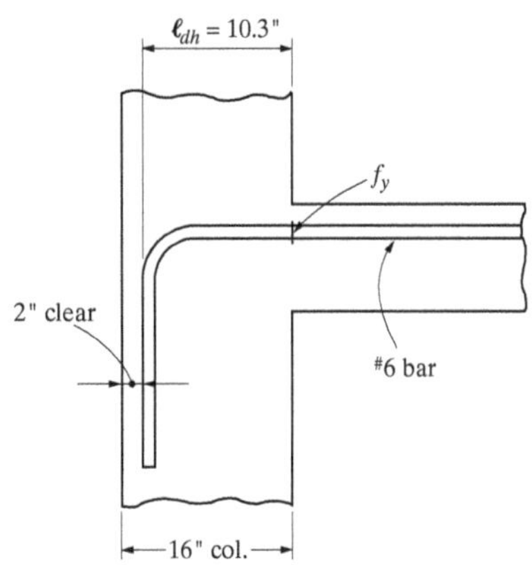

FIGURE 6.14 Anchorage of the beam reinforcement at column.

For other points along the continuous beam, use bar cutoff points recommended in Figure 6.3 and as shown in Figure 6.18.

9. Prepare the stirrup design for the beam:
   Established values are $b_w = 12$ in., effective depth $d = 28$ in., $f_c' = 3000$ psi, and $f_y = 60{,}000$ psi.

a. The shear force V_u diagram may be observed in Figure 6.15. Note that the shear diagram is unsymmetrical with respect to the centerline of the span, and the point of zero shear does not occur at the midspan of the beam. Furthermore, the point of zero shear for the right-hand section of the end span of the beam occurs at 13.03 ft. (that is, 84.7 kip/w_u = 84.7/6.5) from the face of the first interior support. It should be noted that the point of zero shear for the left-hand section of the beam occurs at 11.34 ft. (that is, 73.7 kip/w_u = 73.7 kip/6.5) from the interior face of the end support. The points of zero shears for both sections of the beam occur at different locations because of the approximate nature of the ACI shear coefficients, as pointed out earlier. The stirrup design will be based on shear in the interior portion of the end span where the maximum values occur. The resulting stirrup pattern will be used throughout the continuous beam. Only the applicable portion of the V_u diagram is shown.

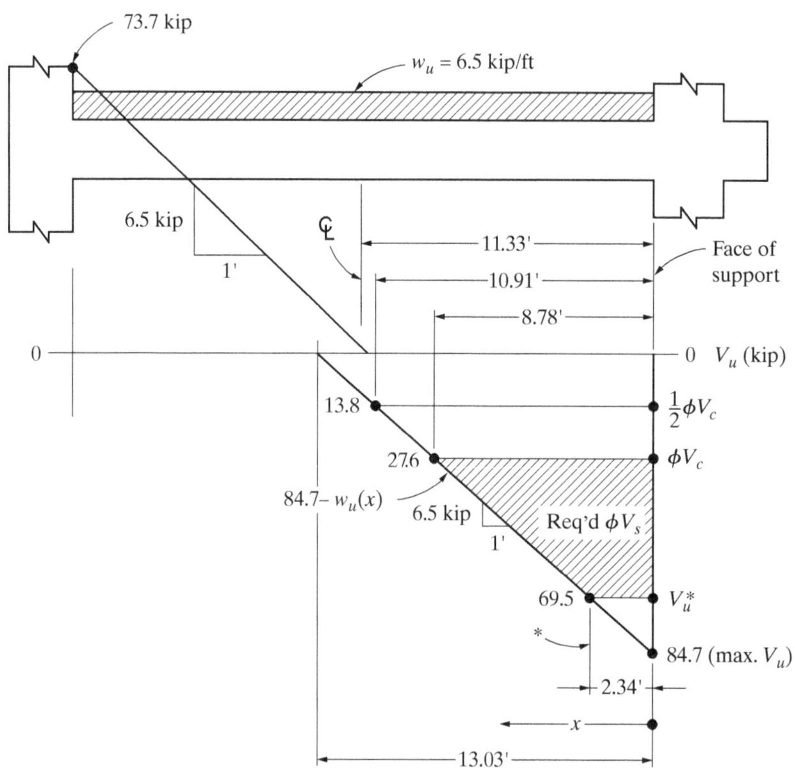

FIGURE 6.15 V_u diagram for stirrup design (Example 6.1).

b. Determine if stirrups are required:

$$\phi V_c = \phi 2\sqrt{f'_c}\, b_w d = \frac{0.75\left(2\sqrt{3000}\right)(12)(28)}{1000}$$

$$= 27.6 \text{ kip}$$

$$\tfrac{1}{2}\phi V_c = \tfrac{1}{2}(27.6) = 13.8 \text{ kip}$$

At the critical section d distance (28 in.) from the face of the support,

$$V_u^* = 84.7 - \frac{28}{12}(6.5) = 69.5 \text{ kip}$$

(Quantities at the critical section are designated with an asterisk.) Stirrups are required because

$$V_u^* > \tfrac{1}{2}\phi V_c \ \left(i.e., 69.5 \text{ kip} > 13.8 \text{ kip}\right)$$

c. Find the length of span over which stirrups are required. Stirrups are required to the point where

$$V_u = \tfrac{1}{2}\phi V_c = 13.8 \text{ kip}$$

From Figure 6.15 and referencing from the face of the support, V_u = 13.8 kip at

$$\frac{84.7 - 13.8}{6.5 \text{ kip/ft}} = 10.91 \text{ ft}$$

The distance from the face of the support to where $V_u = \phi V_c = 27.6$ kip is

$$\frac{84.7 - 27.6}{6.5 \text{ kip/ft}} = 8.78 \text{ ft}$$

d. On the V_u diagram, designate the area between the ϕV_c line, the V_u^* line, and the sloping V_u line as "Req'd ϕV_s." At locations between 2.34 ft and 8.78 ft from the face of the support, the required ϕV_s varies. Designating the slope of the V_u diagram as m (kip/ft) and taking x (ft) from the face of the support ($2.34 \le x \le 8.78$) yields

$$\begin{aligned} \text{required } \phi V_s &= \text{maximum } V_u - \phi V_c - mx \\ &= 84.7 - 27.6 - 6.5x \\ &= 57.1 - 6.5x \end{aligned}$$

e. Assume a No. 3 vertical stirrup (A_v = 0.22 in.^{2}):

$$\begin{aligned} \text{required } s^* &= \frac{A_v f_{yt} d}{V_s} \\ &= \frac{\phi A_v f_{yt} d}{\text{required } \phi V_s^*} \\ &= \frac{\phi A_v f_{yt} d}{V_u^* - \phi V_c} \\ &= \frac{0.75(0.22)(60)(28)}{69.5 - 27.6} = 6.62 \text{ in.} \end{aligned}$$

Use $6\frac{1}{2}$-in. spacing between the critical section and the face of the support.

f. Establish ACI Code maximum spacing requirements:

$$4\sqrt{f'_c}\, b_w d = \frac{4\sqrt{3000}(12)(28)}{1000} = 73.6 \text{ kip}$$

Calculating V_s^* at the critical section yields

$$\phi V_s^* = V_u^* - \phi V_c$$
$$= 69.5 - 27.6 = 41.9 \text{ kip}$$
$$V_s^* = \frac{\phi V_s^*}{\phi} = \frac{41.9}{0.75} = 55.9 \text{ kip}$$

Because 55.9 kip < 73.6 kip, the maximum spacing should be the smaller of $d/2$ or 24 in.:

$$\frac{d}{2} = \frac{28}{2} = 14 \text{ in.}$$

Also check:

$$s_{max} = \frac{A_v f_{yt}}{0.75\sqrt{f'_c}\, b_w} \leq \frac{A_v f_{yt}}{50 b_w}$$

$$\frac{A_v f_{yt}}{0.75\sqrt{f'_c}\, b_w} = \frac{0.22(60{,}000)}{0.75\sqrt{3000}(12)} = 26.7 \text{ in.}$$

and

$$\frac{A_v f_{yt}}{50 b_w} = \frac{0.22(60{,}000)}{50(12)} = 22.0 \text{ in.}$$

Therefore, use a maximum spacing of 14 in.

g. Determine the spacing requirements based on shear strength to be furnished: The denominator of the following formula for required spacing uses the expression for required ϕV_s from step d:

$$\text{required } s = \frac{\phi A_v f_{yt} d}{\text{required } \phi V_s}$$
$$= \frac{0.75(0.22)(60)(28)}{57.1 - 6.5x} = \frac{277.2}{57.1 - 6.5x}$$

The results for several arbitrary values of x are shown tabulated and plotted in Figure 6.16.

h. Using Figure 6.16, the stirrup pattern shown in Figure 6.17 is developed. Despite the lack of symmetry in the shear diagram, the stirrup pattern is symmetrical with respect to the centerline of the span. This is conservative and will be used for all spans.

The design sketches are shown in Figure 6.18. As with the slab design, the typical bar cutoff points of Figure 6.3 are used for this beam. All bars, therefore, terminate in compression zones, and the requirements of the ACI Code, Section 9.7.3.5, need not be checked. Although the beam reinforcement obtained from the preceding design has been presented in the form of sketches (see Figure 6.18), in practice, the beam reinforcement is also presented in a tabular format as a beam schedule. An example of a beam schedule is shown in Figures 13.2 and 13.3.

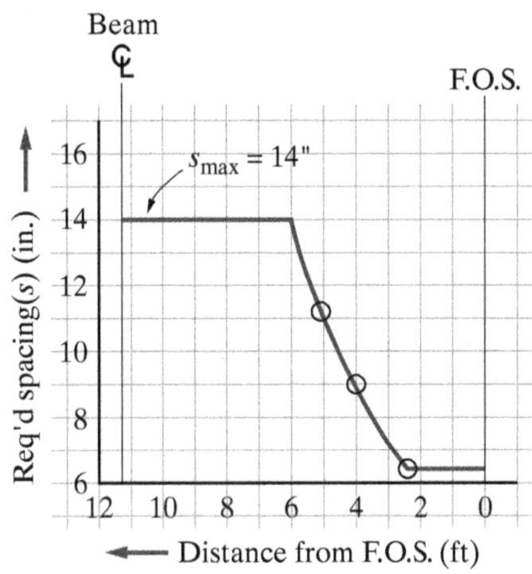

x (ft)	Req'd s (in.)
2.34	6.62
4	8.91
5	11.27
6	15.3

FIGURE 6.16 Stirrup spacing requirements for Example 6.1.

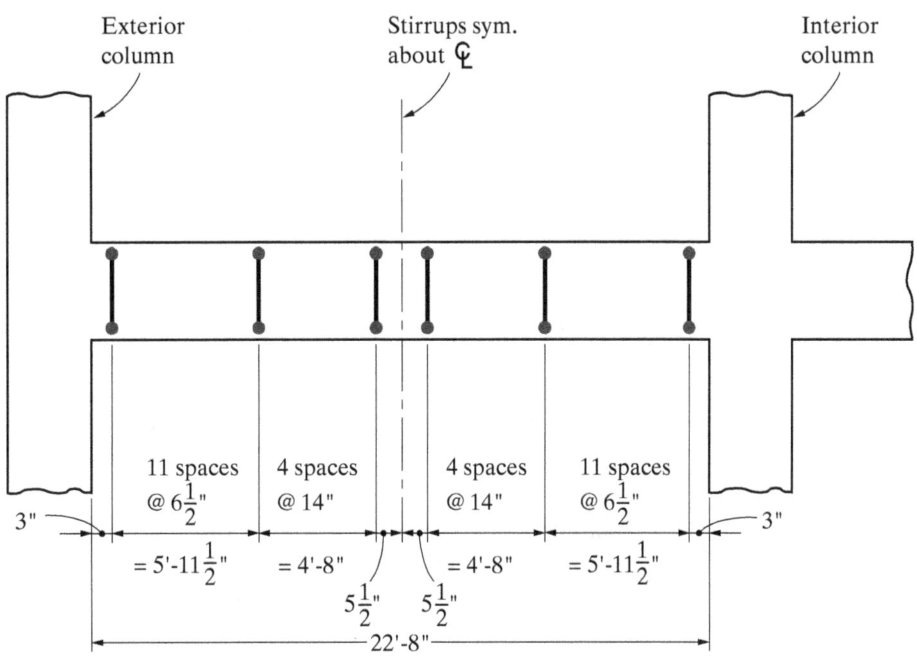

FIGURE 6.17 Stirrup spacing for Example 6.1: End span (interior spans similar).

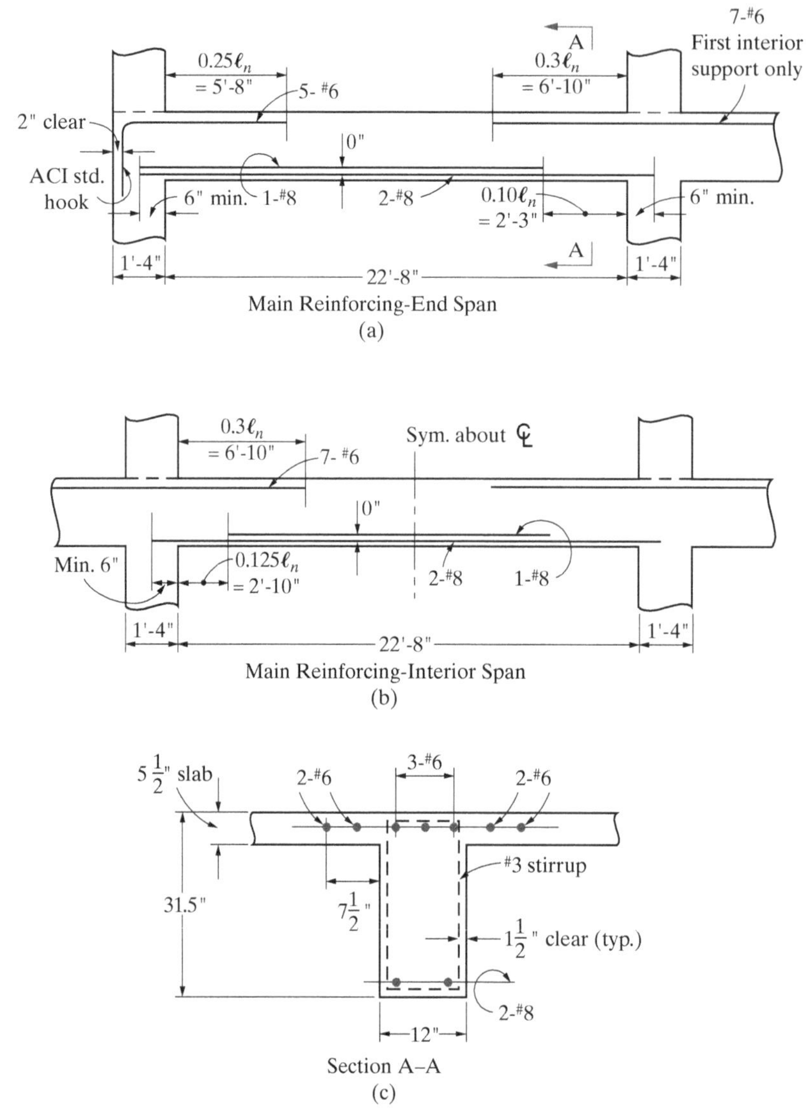

FIGURE 6.18 Design sketches for Example 6.1 (stirrup spacings not shown).

6.4 ANALYSIS AND DESIGN OF CONTINUOUS TWO-WAY SLABS

As discussed in Chapter 2, two-way slabs are horizontal or planer structural elements that support gravity loads by bending in two orthogonal directions, and the ratio of the long side to the short side (ℓ_2/ℓ_1) of the rectangular panel is less than 2. These slabs can be supported on all four sides of a panel by walls, beams, or directly by columns at the corners of the panel.

In edge-supported two-way slabs (see Figure 1.12), the uniform load on the slab is shared between the two orthogonal directions of the slab panel, and the shorter direction of the panel supports a higher proportion of the load (see Section 14.10). In contrast, for column-supported two-way slabs, 100 percent of the uniform load on the slab is supported in each orthogonal direction. The more common type of two-way slabs are those supported by columns. The thickness of column-supported two-way slabs is dependent on punching shear and deflection considerations.

6.4.1 Types of Column-supported Two-Way Slabs

There are a variety of column-supported two-way slab systems, and the basic types that are commonly used are illustrated in Figures 6.19a-f. A flat plate is the simplest and consists of a slab that is supported directly on columns with or without column capitals. A flat slab is a flat plate with drop panels with or without column capitals, and it is stronger than a flat plate because the drop panel or slab thickening at the columns increases the shear capacity of the slab. Drop panel thickness varies from 2¼-in. to 8-in. based on the size of plyform and sawn lumber that is used for forming the slab and drop panel. Figures 6.19b and 6.19c show photos of a flat plate and flat slab in a concrete building under construction. A flat slab with beams is a two-way slab with beams between columns in two orthogonal directions. A waffle slab is another variation of a two-way slab with a waffle-type soffit; they are relatively stiff and recommended for floors subject to floor vibrations, but they have higher installation costs due to the complicated forming required. Flat plates and flat slabs are the more widely used two-way slabs in high-rise building construction. Square bay sizes are more economical than rectangular bay sizes. A comparison of the practical range of sizes of flat plates and flat slabs is shown in Table 6.3 [3].

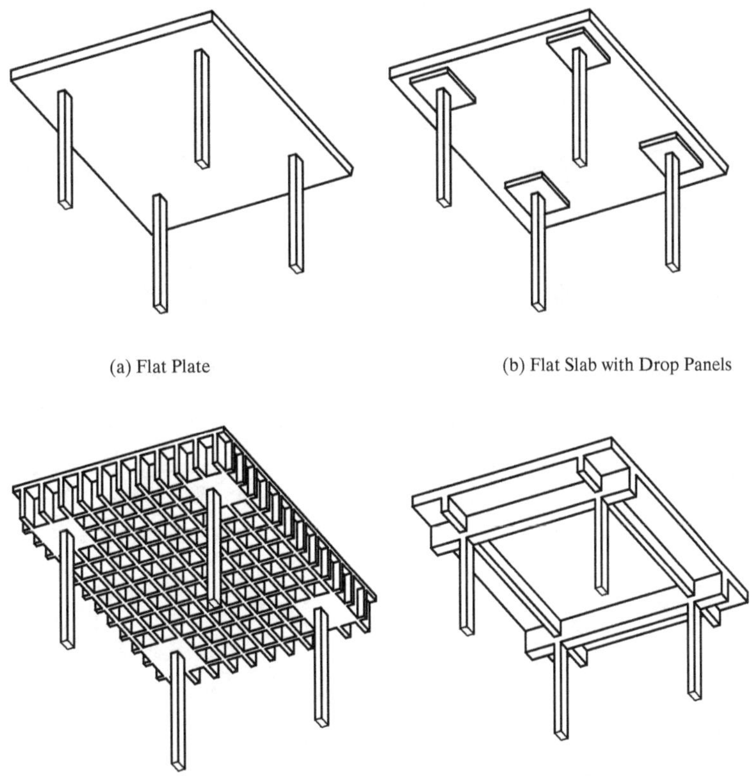

(a) Flat Plate (b) Flat Slab with Drop Panels

(c) Waffle Slab (d) Flat Slab with Beams

FIGURE 6.19 Types of column-supported two-way slabs.

FIGURE 6.19e Flat plate structure with a circular column

FIGURE 6.19f Slabs with drop panels (i.e., Flat slabs).

TABLE 6.3 Practical Range of Sizes for Flat Plates and Flat Slabs[1] [3].

Flat Plate		Flat Slab	
SDL = 20 psf L = 50 psf	SDL = 20 psf L = 100 psf	SDL = 20 psf L = 50 psf	SDL = 20 psf L = 100 psf
15 ft x 15 ft bay size 6 in. thick slab - 14 in. square column	15 ft x 15 ft bay size 6½ in. thick slab - 14 in. square column	20 ft x 20 ft bay size 7 in. thick slab 2¼ in. deep drop panel 20 in. square column	20 ft x 20 ft bay size 7 in. thick slab 2¼ in. deep drop panel 22 in. square column
30 ft x 30 ft bay size 12 in. thick slab - 32 in. square column	30 ft x 30 ft bay size 14 in. thick slab - 34 in. square column	35 ft x 35 ft bay size 12 in. thick slab 4¼ in. deep drop panel 38 in. square column	35 ft x 35 ft bay size 12 in. thick slab 6¼ in. deep drop panel 40 in. square column

[1]Assumes f_c' of 4000 psi and f_y of 60,000 psi

There are several advantages of a two-way slab system and one of them is the consistently flat soffit that allows for a cleaner installation of utility lines and other items hung from the soffit of the slab. They are also widely used in high-rise building construction because of the reduction in headroom achievable with flat plates and flat slabs, which leads to a reduction in the overall height of the building and, hence, a reduction in the overall cost of the project. Two-way slabs are often used in hospitals where a fair amount of medical equipment is hung from the underside of the structure. One of the more significant advantages of a two-way slab system is that a higher span-to-depth ratio can be achieved compared to one-way slabs since the slab bending or curvature occurs in two orthogonal directions. Per Table 6.1 (ACI Code Table 7.3.1.1) and assuming deflections are not calculated, the minimum thickness for a simple span one-way slab is $\frac{\ell}{20}$ and if the one-way slab is continuous over multiple supports, the thickness can be decreased to $\frac{\ell}{28}$ where ℓ is the clear span between supports. By comparison, Table 6.4 (ACI Code Table 8.3.1.1) requires a minimum thickness for the exterior panels of two-way slabs (with reinforcement yield strength of 60,000 psi) without drop panels of $\frac{\ell}{30}$, and $\frac{\ell}{33}$ for a slab with drop panels. Thus, two-way slabs can have spans of up to 20 percent more than the equivalent one-way slab systems with the same thickness, which results in more economical structures. Note that it is recommended to use the minimum thicknesses given in Table 6.1 (ACI Code, Tables 7.3.1.1 and 8.3.1.1) as an absolute minimum thickness of the slab, and deflections should always be calculated and checked for slabs with deflection sensitive elements or elements likely to be damaged by large deflections.

A flat slab will have relatively high shear stresses at the columns since the concentrated reaction at the column results in punching (or two-way) shear stress which results in a tendency for the column to poke through the slab under high loading (see Figures 6.20*a* and 6.20*b*). The use of a drop panel, column capital, or a shear cap helps to mitigate the punching shear issue. Figure 6.20*a* shows the classical punching shear failure where a cone of concrete is left at the top of the column after failure. The typical mechanism for the classical punching shear failure is well described in the failure investigation report of the Pipers Row Car Park partial collapse in Wolverhampton, U.K. on 20th March 1997: *"Punching shear failure at one column led to* ***progressive collapse*** *as similar failures followed at eight adjacent columns.....The sagging slab, acting as a* ***tension membrane****, pulled the edges inwards and then split along Line 2..."* [5].

TABLE 6.4 Minimum thickness of reinforced concrete two-way slabs *without* interior beams[1]

Yield strength of slab reinforcement[2], f_y (psi)	Slabs without drop panels			Slabs with drop panels[3]		
	Exterior slab panel		Interior slab panel	Exterior slab panels		Interior slab panels
	Without edge beams	With edge beams[4] $(\alpha_f \geq 0.80)$		Without edge beams	With edge beams[4] $(\alpha_f \geq 0.80)$	
40,000	$\frac{\ell_n}{33}$	$\frac{\ell_n}{36}$	$\frac{\ell_n}{36}$	$\frac{\ell_n}{36}$	$\frac{\ell_n}{40}$	$\frac{\ell_n}{40}$
60,000	$\frac{\ell_n}{30}$	$\frac{\ell_n}{33}$	$\frac{\ell_n}{33}$	$\frac{\ell_n}{33}$	$\frac{\ell_n}{36}$	$\frac{\ell_n}{36}$
80,000	$\frac{\ell_n}{27}$	$\frac{\ell_n}{30}$	$\frac{\ell_n}{30}$	$\frac{\ell_n}{30}$	$\frac{\ell_n}{33}$	$\frac{\ell_n}{33}$

Adapted from ACI 318-19, Table 8.3.1.1.

[1] ℓ_n is the clear span in the long direction of the slab panel measured face to face of the supports (in.).

[2] For f_y values between the tabulated values, linear interpolation is allowed.

[3] Drop panels must satisfy the thickness and plan length requirements in ACI Code, Section 8.2.4.

[4] If the value of the relative stiffness (α_f) of the edge beam to the corresponding slab, calculated in accordance with ACI Code, Section 8.10.2.7, is less than 0.8, the edge beam shall be neglected and the slab treated as a panel without an edge beam.

[5] For slabs ***without edge beams or with edge beams having .. less than 0.8***, increase the minimum thickness obtained from Table 6.4 by 10% (ACI Code, Section 8.3.1.2.1).

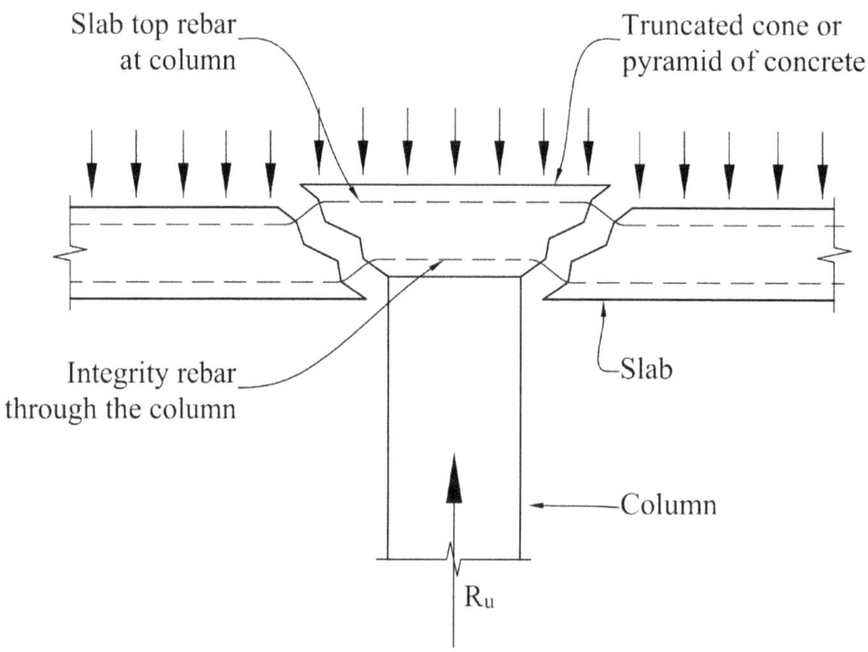

FIGURE 6.20a Punching shear failure of a column-supported two-way slab.

FIGURE 6.20b Flexure-driven punching failure of the Champlain Towers pool deck slab at the columns, with the slab pulled away from the south basement wall.*(Robert Lisman)

It has been found recently through tests that lightly reinforced two-way slabs (i.e., $\rho = A_s/bd < 1\%$) are susceptible to **flexure-induced or flexure-driven punching failure** which occurs at a lower load than the classical punching shear failure shown in Figure 6.20*a* [6, 7, 8]. In flexure-driven punching failure, the slab top flexural rebar over the columns yields and wide flexure cracks extends deep into the slab, joining with the shear cracks to induce punching failure and *"the final circumferential crack tends to be very close to the column."* Only the rebar centered on the column within a slab width of **$c + 3h$** is effective in resisting the punching load, where c is the width of column and h is the slab thickness). Corrosion of the slab top rebar over the columns will further reduce the flexure-driven punching capacity as it further reduces the top steel rebar ratio. In the flexure-driven punching failure, there is no cone of concrete remaining at the top of the column. An example of flexure-driven punching failure was the failure of the pool deck slab in the Champlain Towers South collapse that occurred in Surfside, Florida on June 24, 2021 (see Figures 6.20*b* and Figure 6.20*c*). Note the absence of the cone of concrete at the top of the columns.

The elastic behavior of two-way slabs under gravity loads can be described by a fourth-order differential equation [1]. The solution of the differential equation that defines plate behavior under gravity loads is complex and not trivial. This is further complicated when the two-way

***FIGURE 6.20c** Flexure-induced punching failure in the Champlain Towers South pool deck slab at column L-15
*(Robert Lisman)

slabs are supported on columns rather than on stiff beams or walls on all four sides of the slab panel. In contrast to beams or one-way slabs where closed-form solutions can be found that describe the distribution of moments and shears in the beam or one-way slab, no such solutions are available for column-supported two-way slabs where 100 percent of the slab load is carried in each orthogonal direction of the slab [1]. In lieu of any available "exact" methods of analysis, several methods of analysis for two-way slabs subject to gravity loads have been developed over the years, and these include elastic analysis methods, the yield line method, the Hillerborg strip method, the finite element method, the Direct Design method (DDM), and the Equivalent Frame method [1]. The ACI Code, Section 8.2.1, permits the use of "any procedure satisfying equilibrium and geometric compatibility, provided the design strength at every section is at least equal to the required strength, and all serviceability requirements are satisfied." Both the Direct Design method and the Equivalent Frame method are permitted in ACI Code, Section 8.2.1, but they are limited to two-way slabs under gravity loads only. Previous editions of the ACI Code prior to ACI 318-19 provided detailed requirements for both the Direct Design and Equivalent Frame methods, but they have been removed from ACI 318-19 because many structural design offices now predominantly use some form of finite element analysis for two-way slab analysis and design. To aid the reader in understanding the behavior of two-way slabs, we will focus in this text on the Direct Design method which can be used as a back-check for the more sophisticated finite element analysis. A step-by-step procedure for the equivalent frame method of analysis will also be provided.

To use the Direct Design Method, several conditions stated in previous ACI Codes that must be satisfied are summarized below:

- The slab has at least three continuous spans in each direction.
- The slab panels are generally rectangular with a maximum aspect ratio (long side to short side) of 2.0
- Column offsets are limited to 10 percent of the span in the offset direction from either axis between successive column centerlines. That is, the offset of a column from the regular rectangular grid must not exceed 10% of the regular column-to-column dimension in the direction of the offset within the slab panels where the column offset occurs (see Figure 6.40).
- The length of successive spans is such that the shorter span is at least two-thirds of the longer span.
- Only uniformly distributed gravity loads are applied.
- The ratio of unfactored live load to the unfactored dead load is no greater than 2.0. This obviates the need to check for the effects of pattern live loading. Where pattern live loading analysis needs to be considered (i.e., $L/D > 2$), the equivalent frame method or a finite element analysis method should be used.
- When the slab is supported on beams on all four sides of the panel, the relative stiffness of the beams to the stiffness of the floor slab should be between 0.2 and 5.0

When analyzing two-way slabs, the slab should be divided into design strips in both orthogonal directions, with the design strips centered on the column grid lines; the design strip is the slab strip bounded by one-half the distance to the columns to the right and left of the column line under consideration. The design strip consists of a column strip and two half-middle strips. The column strip is centered on the column line and has a width of the smaller of $0.25\ell_1$ or $0.25\ell_2$, where ℓ_2 is the width of the design strip and ℓ_1 is the span of the slab in the longitudinal direction of the slab (see Figure 6.21). The moments in the middle strip are generally smaller than those in the column strip because the column strip includes the columns and are, therefore, stiffer than the middle strip and thus attracts more moments.

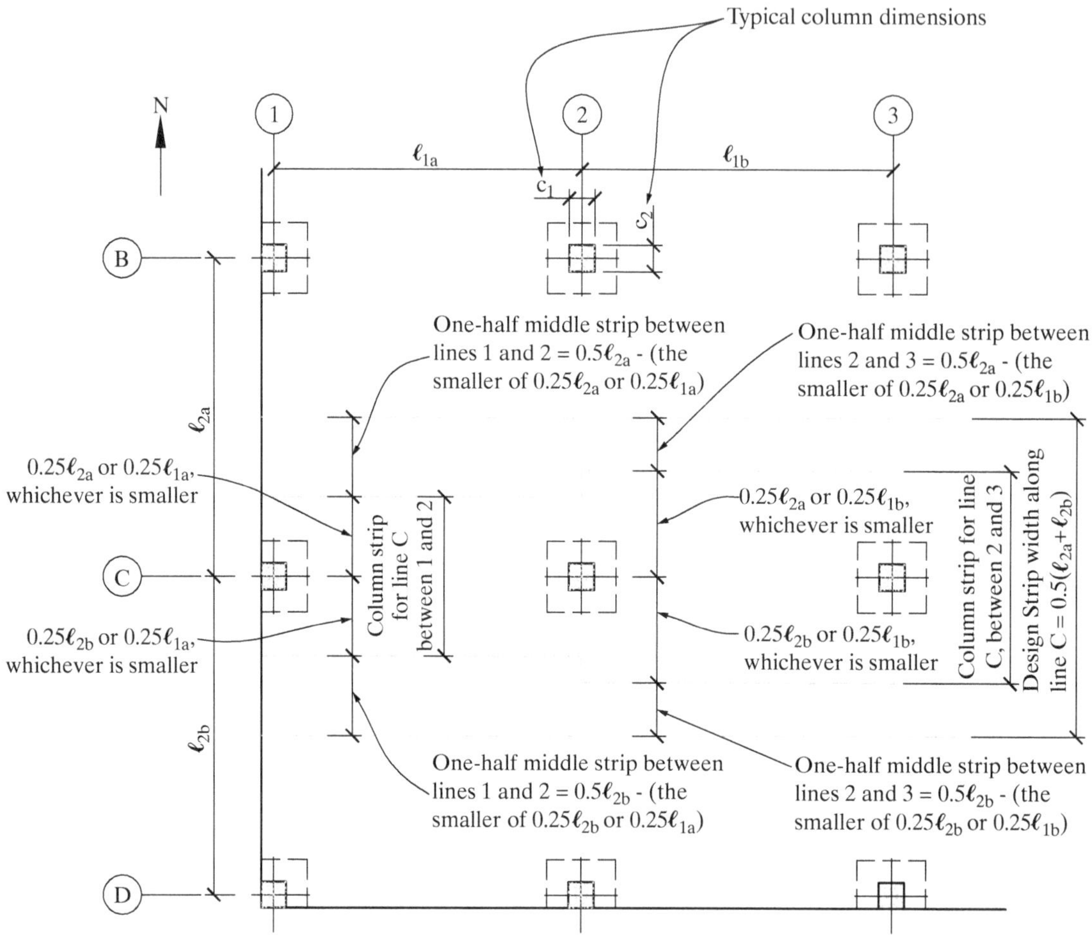

FIGURE 6.21 Column and middle strip widths.

The direct design method allows for the quick calculation of approximate moments and shears at critical locations in the column and middle strips of a two-way slab subject to gravity loads without the need for complex computer models or the calculations that are needed for the moment distribution analysis in the equivalent frame method. In general, the critical locations are the points of maximum stress for regular loading and framing conditions using the previously stated limitations. Those critical locations are on the faces of supports and at the midspan.

The steps involved in the direct design method (DDM) from previous editions of ACI 318 are as follows:

1. Determine the minimum slab thickness, h, to avoid deflection calculations (see *Minimum Thickness of Two-Way Slabs*). Where deflection-sensitive elements are supported by the slab, the deflections should be calculated and checked against the Code specified limits.

2. Determine the total factored uniform load on the slab in psf.

3. Check one-way shear and two-way (punching) shear and ensure that the slab has adequate thickness to resist direct shear and make allowance in the slab thickness to resist additional shear stresses due to the transfer of unbalanced moments from the slab to the column. The unbalanced moment will be calculated later.

   In addition to the classical punching shear failure, flexure-induced punching failure should be mitigated by providing the minimum top steel reinforcement at the column locations specified in ACI 318, Section 8.6.1.2.

4. Divide the slab into design strips in both orthogonal directions. The design strips are centered on the column grid lines. For each column grid line, determine the design strip width (i.e., the width, ℓ_2, in the transverse direction). This is the distance bounded by one-half the distance to the adjacent column lines to the right and to the left of the column grid line under consideration.

5. Calculate the total factored uniform load in (kip/ft) for each design strip. Since pattern live load analysis is not critical—and therefore not considered—in the direct design method, the factored uniformly distributed load is assumed to be applied to all the spans.

6. For each span in the longitudinal direction (i.e., in the ℓ_1 direction), calculate the total factored static moment, M_o.

7. Distribute the total factored moment in the longitudinal direction to the positive and negative moment regions of the slab.

8. At each negative and positive moment section in the slab in the longitudinal direction, distribute the longitudinal moments transversely to the column strip (which is centered on the column grid line), and the half middle strips on each side of the column strip, and to any longitudinal beams. This process is carried out for each design strip in both orthogonal directions.

9. Using the moments from step (8) and the width of each strip, determine the required area of tension steel in the column strip and the middle strip using the methods discussed in Chapter 2.

10. Show the slab reinforcement provided in both orthogonal directions on a plan of the two-way slab like that shown in Figure 6.39. Also, provide the structural integrity reinforcement required per ACI Code, Section 8.7.4.2.

6.4.2 Column and Middle Strips (ACI Code, Sections 8.4.1.5 and 8.4.1.6)

The widths of the column strips are first determined as shown in Figure 6.21. The column strip width on each side of the column line is the smaller of $0.25\ell_1$ or $0.25\ell_2$. The determination of the column strip width for the E-W design strips is illustrated in Figure 6.21. It can be seen that the width of the column strip may sometimes vary from span to span if the span of the slab varies, but in practice and to ease the difficulty of placing the slab reinforcement on site, some designers prefer to use a uniform column strip width throughout the entire length of the building.

The longitudinal moment coefficients for the direct design method as well as the transverse distribution coefficients for these longitudinal moments between the column and middle strips provided in previous editions of the ACI Code will be used in this text. The procedure for calculating the moments is as follows:

1. For each column line in both orthogonal directions, using the factored uniform load calculated previously, determine the total factored static moment, M_o (see *Static Longitudinal Moment, M_o*).

2. Using the transverse distribution coefficients from previous editions of the ACI Code, calculate the factored longitudinal design moments at the midspan and at the supports for each design strip in both orthogonal directions (see *Longitudinal Distribution of the Total Factored Static Moment, M_o*).

3. Distribute the factored longitudinal moments transversely to the column and middle strips and to any beams parallel to the span of the slab using the ACI Code coefficients (see *Longitudinal Distribution of the Total Factored Static Moment, M_o*, and Tables 6.6 through 6.9).

4. Design the reinforcement required to resist the assigned design moments from step (3) using the methods discussed in Chapter 2.

6.4.3 Minimum Thickness of Two-Way Slabs

The ACI Code, Section 8.3.1, gives the minimum slab thicknesses that obviate the need for slab deflection calculations, provided the slab does not support deflection-sensitive elements.

For two-way slabs with an aspect ratio of less than or equal to 2.0, *without interior beams* spanning between supports, the minimum slab thickness shall meet the requirements of Table 6.4 and the requirements in (a) or (b) below unless the calculated deflection satisfies the deflection limits in ACI Code, Section 8.3.2:

(a) Slabs with drop panels that do *not meet* the length and depth requirements of the ACI Code, Section 8.2.4 (see *Column Capitals, Drop Panels, and Shear Caps*): $h \geq 5$ in.

(b) Slabs with drop panels that *meet* the length and depth requirements of the ACI Code: $h \geq 4$ in.

The minimum thickness obtained from Tables 6.4 and 6.5 should be rounded to the nearest ½ inch. Note that the above minimum slab thicknesses do not apply to slabs with heavy sustained live loads such as storage loads, or mechanical room loads. For such slabs, the two-way slab deflections should be calculated in accordance with ACI Code, Section 8.3.2. For two-way slabs with beams between supports, the minimum slab thickness shall conform to Table 6.5.

TABLE 6.5 Minimum thickness of reinforced concrete two-way slabs *with* beams spanning between supports on all four sides of the panel[1-5].

α_{fm}	Minimum thickness, h	Expressions
$\alpha_{fm} \leq 0.2$	Use Table 6.4, but ≥ 5 in. for slab without drop panels ≥ 4 in. for slab with drop panels	(a)
$0.2 \leq \alpha_{fm} \leq 2.0$	$\dfrac{\ell_n\left(0.8+\dfrac{fy}{200{,}000}\right)}{36+5\beta\left(\alpha_{fm}-0.2\right)}$	(b)[1,3]
	≥ 5 in.	(c)
$\alpha_{fm} > 2.0$	$\dfrac{\ell_n\left(0.8+\dfrac{fy}{200{,}000}\right)}{36+9\beta}$	(d)
	≥ 3.5 in.	(e)

Adapted from ACI 318-19, Table 8.3.1.2.

[1] ℓ_n is the clear span in the *long direction* of the slab panel measured face to face of the supports (in.).

[2] β is the ratio of the clear spans in the long direction to the short direction of the slab panel.

[3] α_{fm} is the average value of the relative stiffness of beam to the slab (i.e., α_f) for all beams on the four edges of a slab panel.

[4] For slab panels with a long-to-short span ratio greater than 2.0, the minimum thickness in Expressions (b) and (d) in Table 6.5 may give unreasonably large values because of the use of the long span in the equations. For such cases, the minimum thickness expressions for one-way slabs will be more appropriate to use (ACI Code, Section R8.3.1.2). See Table 6.1.

[5] For slab panels with beams between supports on all four sides of the panel, provide edge beams with a relative stiffness, α_f, greater than or equal to 0.80; otherwise, the minimum thickness from Expressions (b) and (d) in Table 6.5 shall be increased by 10% in the exterior slab panel (ACI Code, Section 8.3.1.2.1).

6.4.4 Static Longitudinal Moment, M_o

In general, the total factored static moment is calculated as follows:

In the ℓ_1 direction, the factored static moment, $M_{u,o} = \dfrac{w_u \ell_2 \left(\ell_{n1}\right)^2}{8}$

In the ℓ_2 direction, the factored static moment, $M_{u,o} = \dfrac{w_u \ell_1 \left(\ell_{n2}\right)^2}{8}$

where ℓ_{n1} is the clear span (face-to-face of the columns) in the ℓ_1 direction

ℓ_{n2} is the clear span (face-to-face of the columns) in the ℓ_2 direction

For the two-way slab in Figure 6.21, the calculation of the total static longitudinal moment, $M_{u,o}$, for the East-West design strips and for the North-South design strips will now be illustrated. For example, the static moment along grid line C in the E-W direction is given as follows (see Figure 6.21):

Between grid lines 1 and 2:

$$M_{u,o} = w_u \frac{1}{2}(\ell_{2a} + \ell_{2b}) \frac{\left(\ell_{1a} - \frac{c_1}{2} - \frac{c_1}{2}\right)^2}{8}$$

where $\left(\ell_{1a} - \frac{c_1}{2} - \frac{c_1}{2}\right) \geq 0.65\ell_{1a}$

Between grid lines 2 and 3:

$$M_{u,o} = w_u \frac{1}{2}(\ell_{2a} + \ell_{2b}) \frac{\left(\ell_{1b} - \frac{c_1}{2} - \frac{c_1}{2}\right)^2}{8}$$

where $\left(\ell_{1b} - \frac{c_1}{2} - \frac{c_1}{2}\right) \geq 0.65\ell_{1b}$

In the orthogonal (N-S) direction, the static moments along grid line 2 are given as follows (see Figure 6.21):

Between grid lines B and C:

$$M_{u,o} = w_u \frac{1}{2}(\ell_{1a} + \ell_{1b}) \frac{\left(\ell_{2a} - \frac{c_2}{2} - \frac{c_2}{2}\right)^2}{8}$$

where $\left(\ell_{2a} - \frac{c_2}{2} - \frac{c_2}{2}\right) \geq 0.65\ell_{2a}$

Between grid lines C and D:

$$M_{u,o} = w_u \frac{1}{2}(\ell_{1a} + \ell_{1b}) \frac{\left(\ell_{2b} - \frac{c_2}{2} - \frac{c_2}{2}\right)^2}{8}$$

where $\left(\ell_{2b} - \frac{c_2}{2} - \frac{c_2}{2}\right) \geq 0.65\ell_{2b}$

Similar static moments can be calculated for all the other E-W and N-S design strips.

For each span, the static equilibrium relationship between the static longitudinal moment, $M_{u,o}$, and the longitudinal moments at the critical sections (i.e., at the supports and midspan) of the slab in the ℓ_1 direction is as follows:

$$M_{u,o} = \frac{w_u \ell_2 \ell_{n1}^2}{8} = \frac{1}{2}\left[M_{u,-ve(\textit{left end support})} + M_{u,-ve(\textit{right end support})}\right] + M_{u,+ve\,(\textit{midspan})}$$

$$\leq \phi M_n$$

Similarly, in the ℓ_2 direction, the static equilibrium relationship is as follows:

$$M_{u,o} = \frac{w_u \ell_1 \ell_{n2}^2}{8} = \frac{1}{2}\left[M_{u,-ve(\text{left end support})} + M_{u,-ve(\text{right end support})}\right] + M_{u,+ve\,(\text{midspan})}$$

$$\leq \phi M_n$$

In the static equilibrium equations above, use only the magnitude of the moments.

6.4.5 Column Capitals, Drop Panels, and Shear Caps

Column Capital: These are conical-shaped concrete cast with the slab on top of the column with the sole aim of increasing the slab punching shear capacity (ACI Code, Section 8.2.5). The column capital increases the shear perimeter of the critical section for punching shear at the slab-column joint (see Figure 6.22).

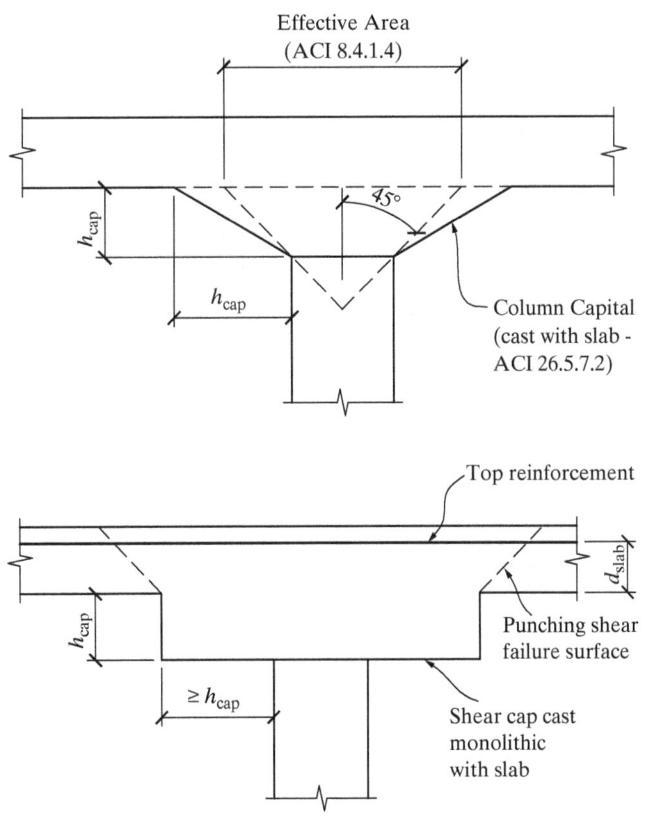

FIGURE 6.22 Column capital and shear cap.

Drop Panels: This is an additional slab thickening at the column locations between the soffit of the slab and the top of the column in the negative moment regions of the slab. The ACI Code specifies the minimum depth of the drop panel and minimum plan dimensions in both orthogonal directions. Drop panels are used for the following purposes:

- To increase the perimeter of the critical section for punching shear at the column and thus reduce the shear stress or increase the shear capacity.
- To increase the effective depth of the slab top reinforcement at the columns, and thus lead to a reduction in the required area of negative reinforcement over the columns.

To qualify as a drop panel, the following conditions must be satisfied (see Figures 6.23(a) and 6.23(b)):

- The thickness of the drop panel below the slab must be at least one-fourth the slab's thickness.
- The plan dimension of the drop panel in each orthogonal direction shall be at least one-sixth the span length of the slab in each direction. Where the drop panels do not meet the minimum dimension limits imposed by ACI Code, Section 8.2.4, the drop panel cannot be used to increase the effective depth, but a projection of the drop panel can be considered as a shear cap which is used to increase the shear strength (ACI Code, Sections 8.2.4 and 8.2.5).

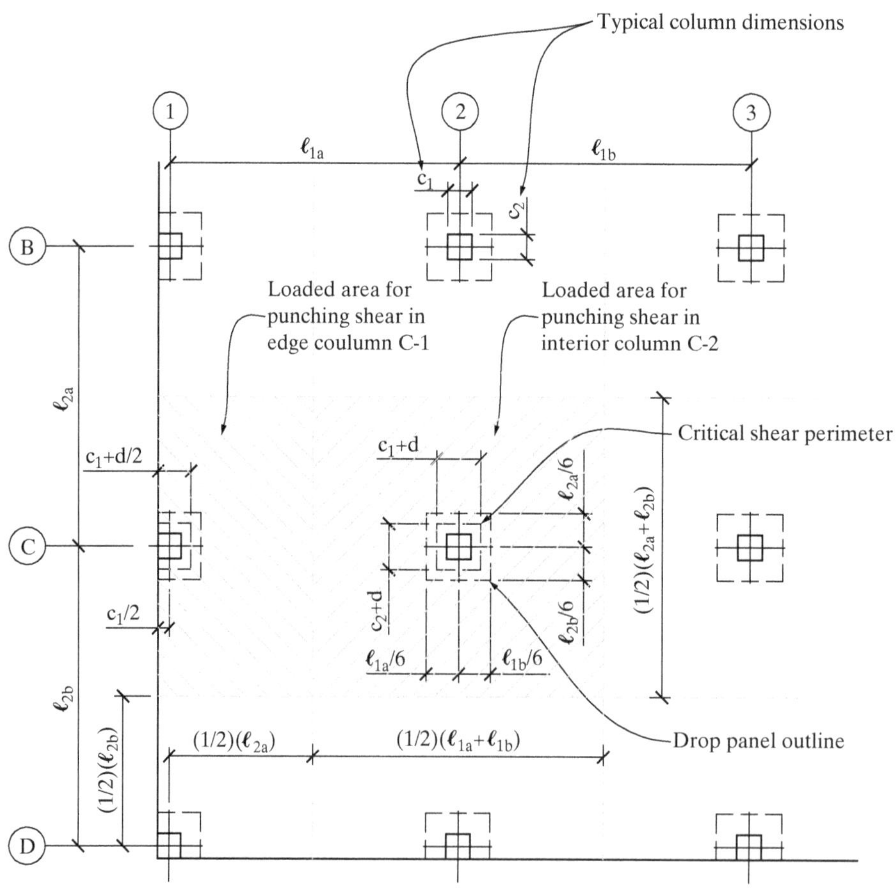

FIGURE 6.23a Loaded area for punching shear and drop panel dimensions.

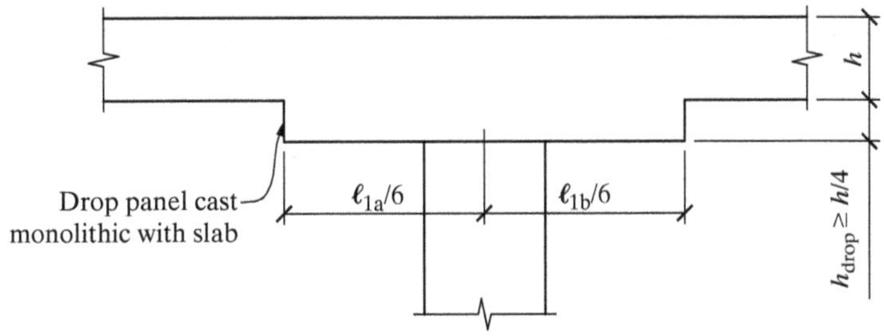

FIGURE 6.23b Drop panel thickness.

Shear Cap: These are slab thickenings that do not satisfy the drop panel minimum plan dimensions in both orthogonal directions specified in ACI Code, Section 8.2.4, required in the slab at the column locations. Shear caps must extend horizontally beyond the face of the column a distance equal to at least the depth of the shear cap below the soffit of the slab.

Shear caps can only be used to increase the c_1 and c_2 dimensions in the critical punching shear perimeter calculations, and thus will increase the critical perimeter for two-way shear at the slab-column connection, therefore leading to a reduction in the punching shear stress. Shear caps cannot be used to increase the effective depth of the slab for calculating the amount of negative reinforcement required in the slab at the column locations, and they also cannot be used in deflection calculations.

6.4.6 Relative Stiffness of Beam and Slab

To account for the effect of beam stiffness on the deflection of two-way slabs supported by beams, the stiffnesses of the beams relative to the slab can be calculated as follows:

The relative stiffness of the longitudinal beam with respect to the slab in $\ell_1$direction (i.e., the longitudinal direction in which the moment is being considered) is given as

$$\alpha_{f1} = \frac{E_{cb} I_{b1}}{E_{cs} I_{s1}}$$

The relative stiffness of the transverse beam with respect to the slab in ℓ_2 direction (i.e., the transverse direction in which the moment is being considered) is given as

$$\alpha_{f2} = \frac{E_{cb} I_{b2}}{E_{cs} I_{s2}}$$

where

E_{cb} = modulus of elasticity of the beam concrete

E_{cs} = modulus of elasticity of the slab concrete

I_{b1} = moment of inertia about the centroidal axis of the gross section of the equivalent beam that spans in the ℓ_1 direction.

I_{b2} = moment of inertia about the centroidal axis of the gross section of the equivalent beam that spans in the ℓ_2 direction.

I_{s1} = moment of inertia about the centroidal axis of the slab in the ℓ_1 direction, with a slab width equal to the dimension measured perpendicular to the longitudinal beam, from the center line of the slab panel to the right and to the left of the beam, excluding the beam width. $= \dfrac{\ell_2 h^3}{12}$

The moment of inertia expression above is for an interior design strip with a width ℓ_2. For exterior design strips, the slab width will be $0.5\ell_2 + 0.5c_2$ + distance from exterior face of the column to the slab edge.

I_{s2} = moment of inertia about the centroidal axis of the slab in the ℓ_2 direction, with a slab width equal to the dimension measured perpendicular to the longitudinal beam, from the center line of the slab panel to the right and to the left of the beam, excluding the beam width. $= \dfrac{\ell_1 h^3}{12}$

The moment of inertia expression above is for an interior design strip with a width ℓ_1. For exterior design strips, the slab width will be $0.5\ell_1 + 0.5c_1$ + distance from exterior face of the column to the slab edge.

The definitions of the equivalent beam and slab from previous editions of the ACI Code are shown in Figure 6.24 (see ACI Code, Section 8.4.1.8).

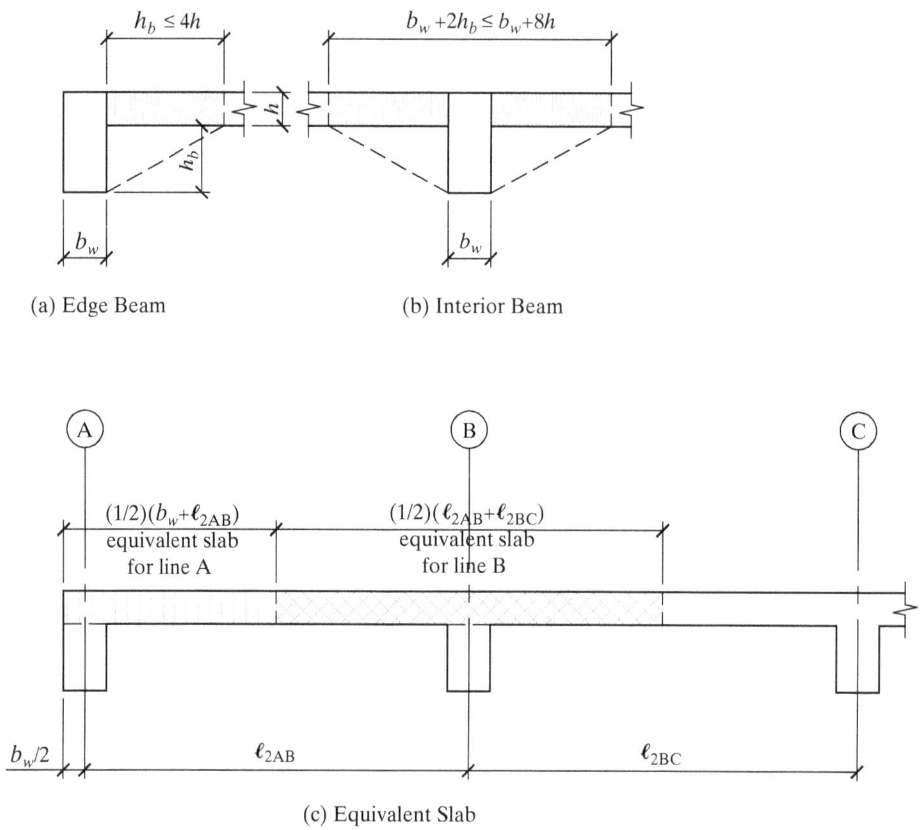

FIGURE 6.24 Dimensions of equivalent beam and slab (ACI Code, Section 8.4.1.8).

In practice, since the slab and the beam in reinforced concrete structures are cast monolithically, $E_{cb} = E_{cs}$, therefore $\frac{E_{cb}}{E_{cs}} = 1.0$ and hence the previous equations can be simplified as follows:

$$\alpha_{f1} = \frac{I_{b1}}{I_{s1}}$$

$$\alpha_{f2} = \frac{I_{b2}}{I_{s2}}$$

For flat plates, where there are no longitudinal beams, $\alpha_{f1} = 0$; and where there are no transverse or edge beams, $\alpha_{f2} = 0$. For a slab panel with beams between supports on all the four edges of the panel, the beams in the two orthogonal directions shall satisfy the following equation to ensure that the moment distribution assumed in the direct design method does not deviate significantly from the elastic distribution of moments:

$$0.2 \leq \frac{\alpha_{f1}\ell_2^2}{\alpha_{f2}\ell_1^2} \leq 5.0$$

where

ℓ_1 = length of the span in the longitudinal direction for which moments are being determined, measured center-to-center of supports (inches).

ℓ_2 = length of the span in the direction perpendicular to ℓ_1, measured center-to-center of supports (inches) (i.e., the transverse width of the design strip).

EXAMPLE 6.2

Calculate the relative stiffness of the edge beam and the interior beam for the slab-beam cross-section in Figure 6.25. The slab span being designed is ℓ_1 and is perpendicular to the ℓ_2 direction.

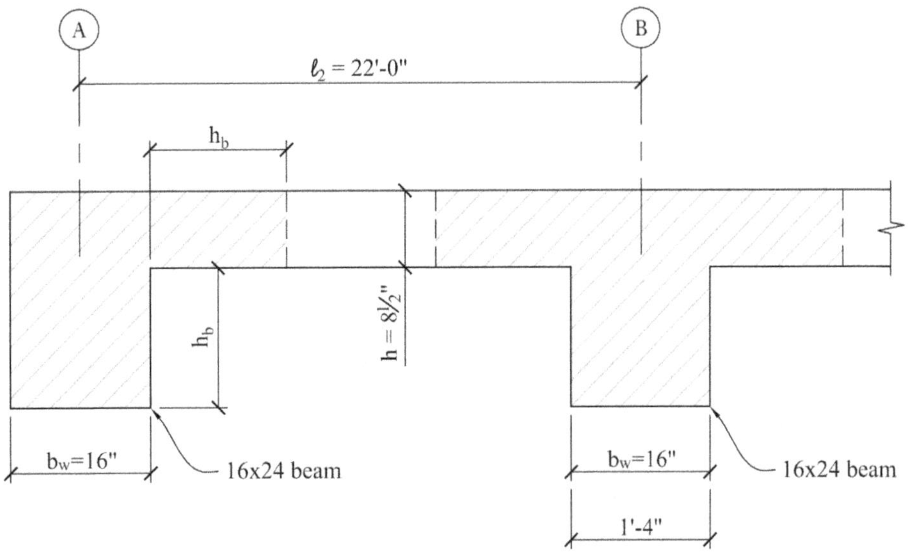

FIGURE 6.25 Relative stiffness of beams and slabs.

Solution:

Longitudinal *edge* beam along grid line A:

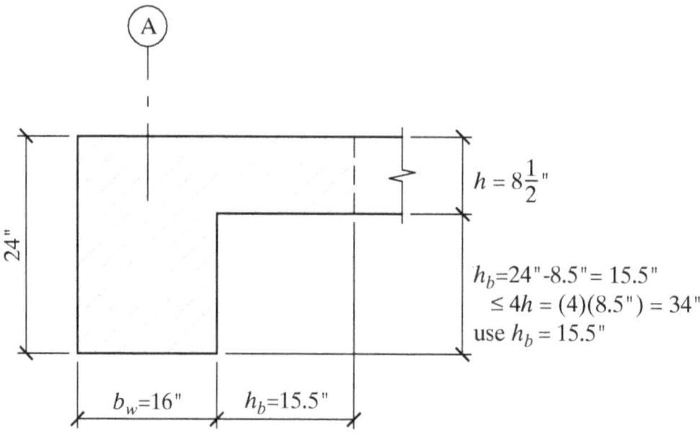

FIGURE 6.26 Equivalent longitudinal edge beam along grid line A.

The distance of the centroid of the edge beam section shown in Figure 6.26 from the top of the beam is calculated as

$$\frac{(16)(24)\left(\dfrac{24}{2}\right)+(15.5)(8.5)\left(\dfrac{8.5}{2}\right)}{(16)(24)+(15.5)(8.5)} = 10.02 \text{ in. from the top of the beam}$$

The gross moment of inertia about the centroidal axis of the gross section of the equivalent beam that spans in the ℓ_1 direction is:

$$I_{b1} = (16)\frac{(24)^3}{12}+(16)(24)\left(\frac{24}{2}-10.02\right)^2+(15.5)\frac{(8.5)^3}{12}+(15.5)(8.5)\left(10.02-\frac{8.5}{2}\right)^2$$
$$= 25{,}087 \text{ in.}^4$$

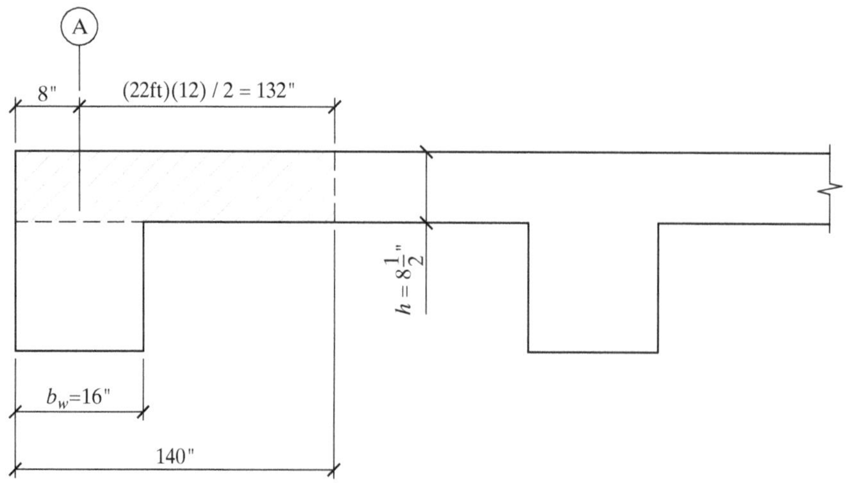

FIGURE 6.27 Equivalent longitudinal edge beam along grid line A.

The corresponding slab section for the edge beam is shown in Figure 6.27 and the gross moment of inertia of the slab section is calculated as

$$I_{s1} = (140)\frac{(8.5)^3}{12}$$
$$= 7{,}165 \text{ in.}^4$$

The relative stiffness of the edge beam along grid line A is

$$\alpha_{f1} = \frac{I_{b1}}{I_{s1}} = \frac{25{,}087}{7{,}165}$$
$$= 3.5$$

Longitudinal *interior* beam along grid line B:

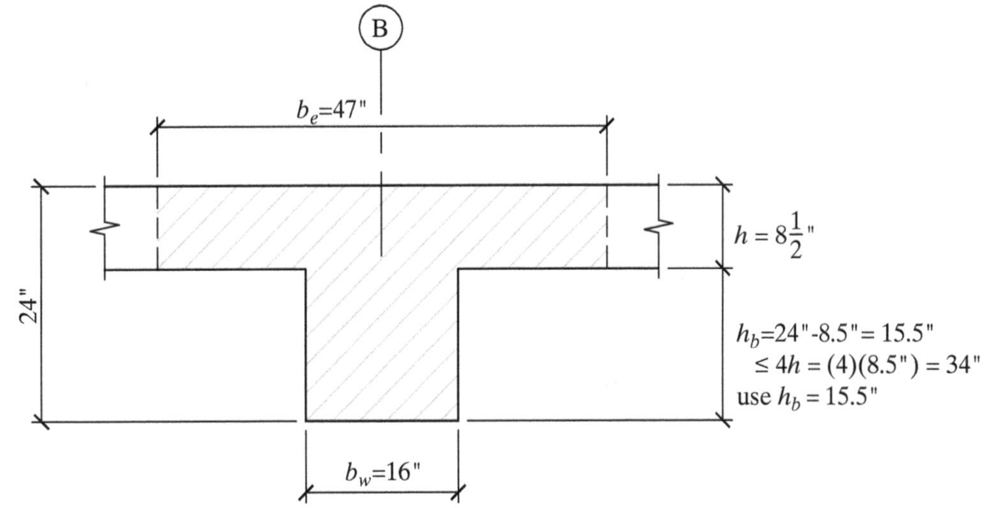

FIGURE 6.28 Equivalent longitudinal interior beam along grid line B.

The effective width of the equivalent interior beam, b_e, is calculated as (see Figures 6.24*b* and 6.28)

$$b_e = b_w + 2h_b = 16 + 2(15.5) = 47 \text{ in.}$$

$$\leq b_w + 8h = 16 + 8(8.5) = 84 \text{ in.}$$

Therefore, use b_e = 47 in.

The distance of the centroid of the interior beam section shown in Figure 6.28 is calculated as

$$\frac{(47)(8.5)\left(\frac{8.5}{2}\right)+(16)(15.5)\left(\frac{15.5}{2}+8.5\right)}{(47)(8.5)+(16)(15.5)} = 8.85 \text{ in. from the top of the beam}$$

The gross moment of inertia about the centroidal axis of the equivalent beam that spans in the ℓ_1 direction is:

$$I_{b1} = (47)\frac{(8.5)^3}{12} + (47)(8.5)\left(8.85 - \frac{8.5}{2}\right)^2 + (16)\frac{(15.5)^3}{12} + (16)(15.5)\left(\frac{15.5}{2} + 8.5 - 8.85\right)^2$$
$$= 29{,}404 \text{ in.}^4$$

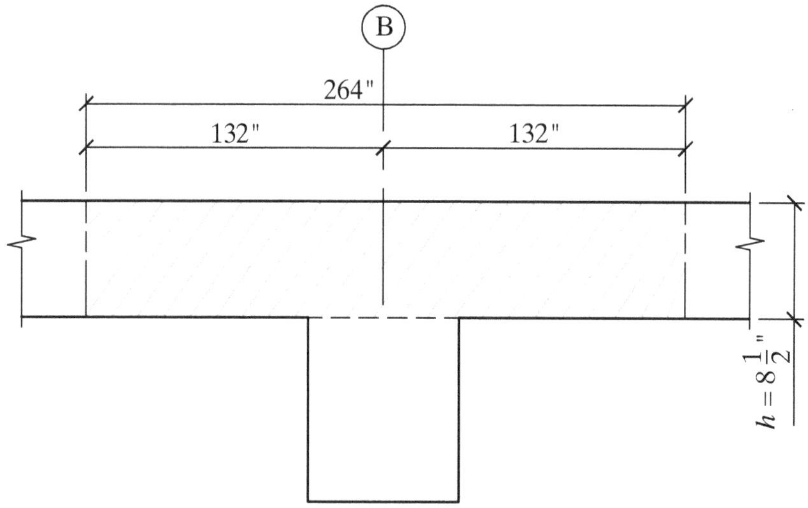

FIGURE 6.29 Equivalent slab section for interior beam along grid line B.

The corresponding slab section for the interior beam along grid line B is shown in Figure 6.29, and the gross moment of inertia of the slab section is calculated as

$$I_{s1} = (264)\frac{(8.5)^3}{12}$$
$$= 13{,}511 \text{ in.}^4$$

The relative stiffness of the interior beam is

$$\alpha_{f1} = \frac{I_{b1}}{I_{s1}} = \frac{29{,}404}{13{,}511}$$
$$= 2.18$$

6.4.7 Longitudinal Distribution of the Total Factored Static Moment, M_o

The total factored static moment, M_o, within each span is distributed in the direction of the span of the slab (i.e., in the longitudinal direction) between the positive and negative moment regions of the slab, as shown in Figures 6.30*a*, 6.30*b*, and 6.30*c*. For interior spans, the critical negative moment used in the design is the higher of the factored negative moment on each side of the slab support. The edge beams and edge of slabs shall be designed to resist their share of the exterior negative moment. Note that the moment coefficients are such that the absolute sum of the positive moment and the average of the negative moments equals the total static moment, M_o.

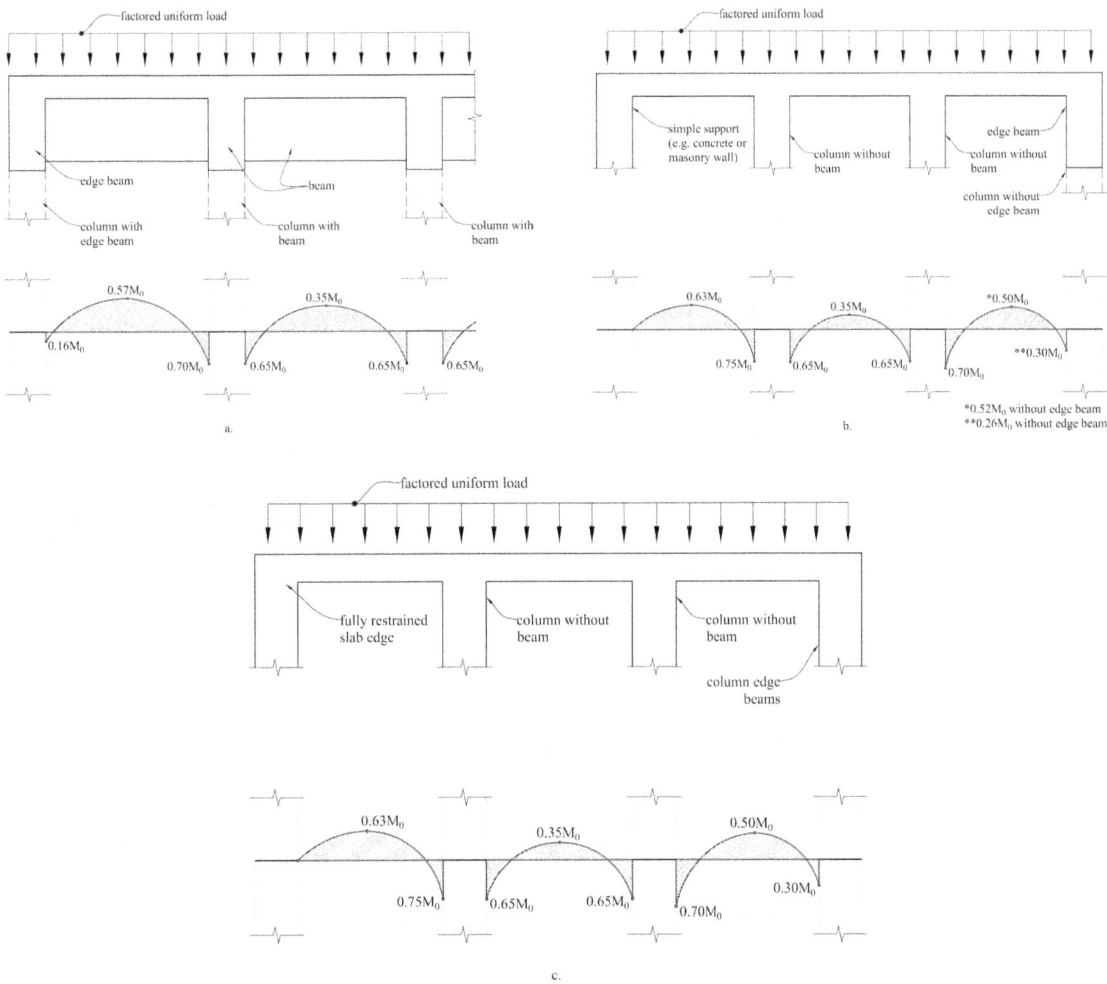

FIGURES 6.30a, 6.30b, and 6.30c Longitudinal distribution of total factored static moment, M_o.

Transverse Distribution of the Longitudinal Moments Between Column Strip, Middle Strip, and Longitudinal Beams

The longitudinally distributed moments in Figure 6.30*a* through 6.30*c* are further distributed transversely between the column strip, the middle strip, and longitudinal beams, if any. The coefficients for the transverse distributions of the interior negative moment, the exterior negative moment, and the positive moment to the column strip are given in Tables 6.6 through 6.9 (obtained from previous editions of ACI 318). For two-way slabs with longitudinal beams between supports, the portion of the column strip moment not resisted by the beams must be resisted by the slab within the column strip.

The remaining moments that are not distributed to the column strip (at the interior support, the exterior support, and the positive moment regions) are assigned to the half-middle strips on both sides of the column strip.

TABLE 6.6 The Portion of the Interior Negative Moment distributed transversely to the Column Strip.

$\alpha_{f1}\frac{\ell_2}{\ell_1}$	$\frac{\ell_2}{\ell_1}$		
	0.5	**1.0**	**2.0**
0	0.75	0.75	0.75
≥ 1.0	0.90	0.75	0.45

Adapted from ACI 318-14 Table 8.10.5.1

[1]α_{f1} is the relative stiffness of longitudinal interior beam, if any, compared to the slab

[2]Linear interpolation is permitted between values in this table

[3]$\alpha_{f1} = \frac{I_{b1}}{I_{s1}}$ (see Section 6.4.6)

TABLE 6.7 The Portion of the Exterior Negative Moment distributed transversely to the Column Strip.

$\alpha_{f1}\frac{\ell_2}{\ell_1}$	β_t	$\frac{\ell_2}{\ell_1}$		
		0.5	**1.0**	**2.0**
	0	1.0	1.0	1.0
0	≥2.5	0.75	0.75	0.75
	0	1.0	1.0	1.0
≥1.0	≥2.5	0.90	0.75	0.45

Adapted from ACI 318-14 Table 8.10.5.2

[1]Linear interpolation is permitted between the values shown

[2]$\beta_t = \frac{E_{cb}C}{2E_{cs}I_s}$

$$C = \sum\left(1 - 0.63\frac{x}{y}\right)\frac{x^3y}{3}$$

x = smaller dimension of a rectangular area; y = larger dimension of a rectangular area

I_s the moment of inertia of the slab with a width equal to the span length of the edge beam = $\frac{\ell_2 h^3}{12}$

TABLE 6.8 The Portion of the Positive Moment distributed transversely to the Column Strip.

$\alpha_{f1}\frac{\ell_2}{\ell_1}$	$\frac{\ell_2}{\ell_1}$		
	0.5	**1.0**	**2.0**
0	0.60	0.60	0.60
≥ 1.0	0.90	0.75	0.45

Adapted from ACI 318-14 Table 8.10.5.5

[1]Linear interpolation is permitted between the values shown.

TABLE 6.9 The Portion of the Column Strip Moment Distributed to the Longitudinal Beam.

$\alpha_{f1}\frac{\ell_2}{\ell_1}$	**Portion of Column Strip Moment Distributed to Longitudinal Beam**
0	0
≥ 1.0	0.85

Adapted from ACI 318-14 Table 8.10.5.7.1

[1]Linear interpolation is permitted between the values shown

[2]Beam to resist assigned moment plus moment due to the directly applied load on the beam including the weight of the beam stem.

6.4.8 One-Way Shear in Two-Way Slabs

Like one-way slabs, two-way slabs also need to be designed for one-way shear due to gravity loads, similar to what was done in Chapter 4. The critical section for one-way shear in column supported two-way reinforced concrete slabs can be taken at a distance d from the face of the column, provided the following conditions are satisfied (ACI Code, Sections 8.4.3.1 and 8.4.3.2):

- The applied loads are at the top surface of the slab.
- There are no concentrated loads on the slab within a distance d from the face of the slab support.
- The support reaction and the applied load result in compression stresses at the ends of the slab.

For each design strip, the critical loaded area for one-way shear at a column is the area bounded by the mid-span of the slab and a distance d from the face of the column. Therefore, the factored one-way shear force is given as

$$V_u = w_u \ell_2 \left(\frac{\ell_1}{2} - \frac{c_1}{2} - d \right)$$

The limit state design equation requires that the demand or factored shear be less than or equal the one-way shear capacity of the slab. Therefore, neglecting any axial loads in the slab,

$$V_u \leq \phi V_c = 2\phi\lambda\sqrt{f'_c}\, b_w d$$

where

The strength reduction factor, $\phi = 0.75$ (ACI Code, Table 21.2.1)

b_w = width of the design strip = ℓ_2

One-way shear hardly ever governs the design of two-way slabs. In any case, the thickness of the slab should be selected so that the one-way shear capacity, ϕV_c, is greater than the one-way shear demand, V_u, since it is not usual nor practical to reinforce a slab for one-way shear.

6.4.9 Punching Shear in Two-Way Slabs

In addition to one-way shear, column-supported slabs are also subject to punching or two-way shear, which is a failure mode where the column and a truncated cone or pyramid of concrete punch through the slab because of the concentrated reaction and load transfer at the column (see Figure 6.20). Shear stresses occur in the slab near the columns due to direct shear from the gravity loads and due to the unbalanced moment at the column that is transferred from the slab to the column through shear. The critical section for punching shear occurs at d/2 from the face of the column on all sides of the column (ACI Code, Section 22.6.4.1). For a flat plate or flat slab without beams, the factored direct two-way shear, V_u, is equal to the factored uniform load on the slab multiplied by the difference between the column tributary area and the area of the critical section (see the shaded areas in Figure 6.23*a*).

Critical Section for Punching Shear:

The critical section for punching shear in flat plates is at d/2 from the face of the column all around the column or from the edge of the column capital or shear cap (see Figures 6.31 and 6.32). The critical perimeter for an interior column in direct shear is,

$$\begin{aligned} b_o &= 2(c_1 + d) + 2(c_2 + d) \\ &= 2c_1 + 2c_2 + 4d \end{aligned}$$

Similar equations for the critical perimeter in direct shear can be derived for edge and corner columns as shown in Figures 6.31*b* and 6.31*c*. To allow for reserve shear capacity due to additional shear stresses from unbalanced moments (that will be calculated later), a common rule of thumb is to increase the slab depth required for direct shear by 10% at interior columns, 40% at edge columns, and 70% at corner columns [2].

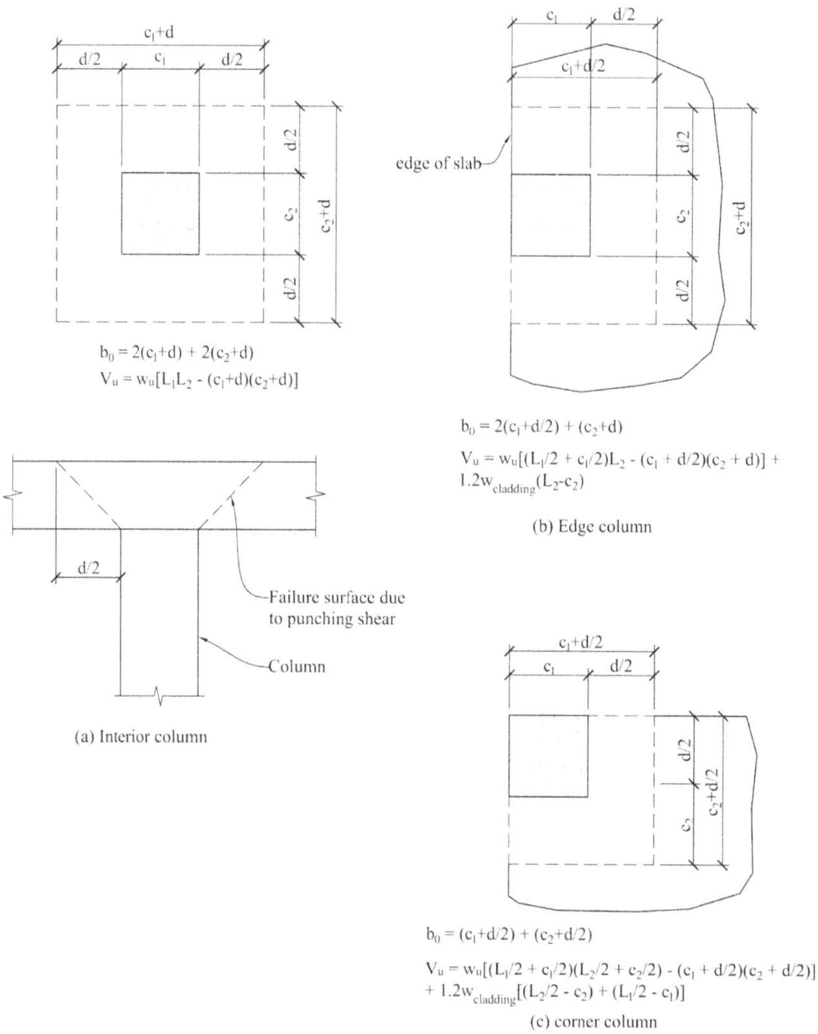

FIGURE 6.31 Critical sections for punching shear.

Critical Section at Drop Panels (ACI Code, Section 22.6.4.1)

When drop panels are used, two critical sections need to be checked: The first critical section is around the column, and the second critical section is around the drop panel, as shown in Figure 6.32.

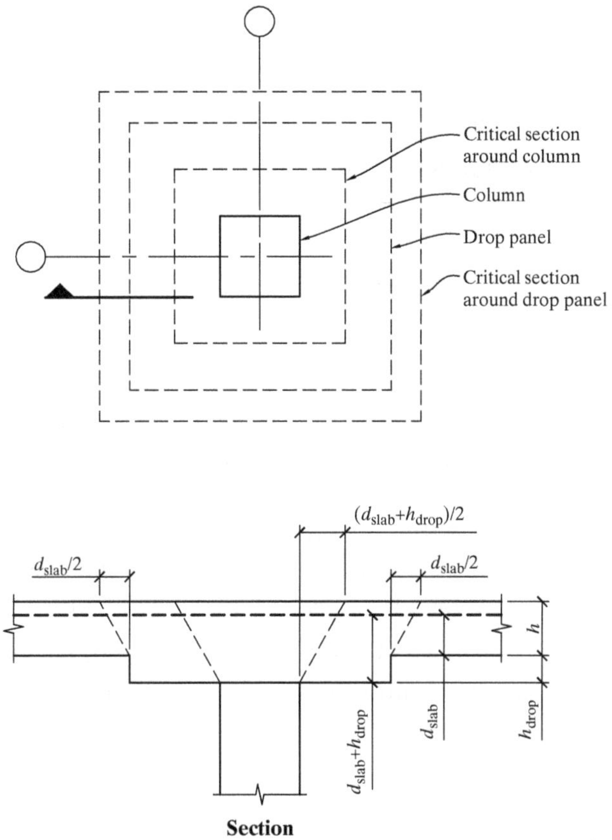

FIGURE 6.32 Critical sections around a column and drop panel.

The nominal concrete shear strength in two-way shear due to gravity loads (without unbalanced moments) is the smallest of the following equations derived from ACI Code, Table 22.6.5.2:

$$V_c = \left[4\lambda_s\lambda\sqrt{f'_c}\,\right]b_o d \qquad [ACI\,Eq.\ 22.6.5.2a]$$

$$V_c = \left[\left(2+\frac{4}{\beta}\right)\lambda_s\lambda\sqrt{f'_c}\,\right]b_o d \qquad [ACI\,Eq.\ 22.6.5.2b]$$

$$V_c = \left[\left(2+\frac{\alpha_s d}{b_o}\right)\lambda_s\lambda\sqrt{f'_c}\,\right]b_o d \qquad [ACI\,Eq.\ 22.6.5.2c]$$

where

β = ratio of the long cross-sectional dimension to the short cross-sectional dimension of the column, concentrated load, or reaction

b_o = perimeter of the critical section for punching or two-way shear

α_s = 40 for interior columns

30 for edge columns

20 for corner columns

λ = modification factor for lightweight concrete (ACI 318-19, Section 25.4.3.2)

= 1.0 for normal-weight concrete

λ_s = size-effect factor (ACI 318-19 Section 22.5.5.1.3);

$$\lambda_s = \sqrt{\frac{2}{1+\frac{d}{10}}} \leq 1$$

The size effect factor indicates a reduction in the concrete shear strength for reinforced concrete two-way slabs without minimum shear reinforcement and for effective depth, d greater than 10-inches (ACI 318, Section R22.6.5.2). Note that per ACI 318, Section 13.2.6.2, the size effect factor can be neglected in the design of "one-way shallow foundations, two-way isolated footings, or two-way combined footings, and mat foundations."

The two-way shear capacity is ϕV_c and the shear demand or factored shear force is V_u. The limit state design equation requires that the demand be less than or equal to the design capacity. Therefore,

$$V_u \leq \phi V_c$$

where the strength reduction factor for shear, ϕ = 0.75 (ACI Code, Table 21.2.1)

As can be observed from the punching shear strength equations previously presented in this section, the punching shear strength can be increased by doing one or more of the following:

- Increasing the concrete strength, f'_c.
- Increasing the effective depth of the slab at the columns. This can be achieved by using a thicker slab for flat plates or adding a drop panel for flat slabs.
- Increasing the perimeter of the critical section, b_o. Since $b_o = 2c_1 + 2c_2 + 4d$, the perimeter can be increased by increasing the size of the column (i.e., c_1 *and* c_2), and/or by increasing the effective depth of the slab at the columns.

Note that the shear demand V_u above only considers direct shear, but additional shear force (and therefore, shear stress) will be present where unbalanced moments are transferred from the slab to the column. The preceding limit states design equation for shear is used to determine the minimum thickness of the two-way slab that will satisfy the punching shear requirements without the need for shear reinforcement. For two-way slabs that do not satisfy the punching shear strength requirements above, shear reinforcement in the form of stirrups or headed studs will have to be provided. In this text, we assume that the slab thickness will be selected to avoid the need for shear reinforcement. The design of shear reinforcement is outside the scope of this text. The reader should refer to ACI Code, Sections 22.6.6 through 22.6.9, for the design of the shear reinforcement in two-way slabs.

6.4.10 Combined Shear and Unbalanced Moment Transfer in Slab-Column Connections

The transfer of unbalanced moments from the slab to the column at the slab-column connections causes non-uniform shear stress distribution, with shear stresses that are larger on one side of the column and smaller on the opposite side. The unbalanced moments are usually more critical at the corner and edge columns than at an interior column, while the direct shear is typically highest at the interior columns. In the analyses of slab-column connections for combined shear and unbalanced moments, a linear distribution of shear stress around the critical section is assumed in the ACI Code [1]. A fraction, γ_f, of the factored unbalanced moment, M_{sc}, is transferred by flexure from the slab to the column, and the effective width of the slab for transferring the moment, $\gamma_f M_{sc}$, is equal to the width of the column plus $3h$ (*i.e.*, $c_2 + 3h$), where h is the slab thickness (see Figure 6.33). The flexural negative or top reinforcement required in the slab to transfer the unbalanced moment of $\gamma_f M_{sc}$ should be concentrated and symmetrically and proportionally placed within this effective slab width, ensuring that some of these top bars fall within the column rebar cage; closer spacing of the reinforcement or additional reinforcement can be used to meet this requirement. This negative reinforcement will also aid in preventing flexure-driven punching failure. The remaining unbalanced moment, $\gamma_v M_{sc}$, is transferred by eccentricity of shear. The portion of the unbalanced moment transferred by flexure and eccentricity of shear, respectively, are defined as

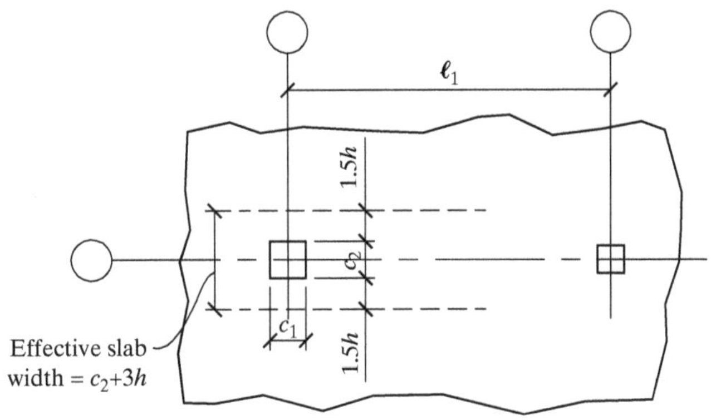

FIGURE 6.33 Effective width of slab for the transfer of the flexure component of the unbalanced moment (i.e., $\gamma_f M_{sc}$).

$$\gamma_f = \frac{1}{1 + \left(\frac{2}{3}\right)\sqrt{\frac{b_1}{b_2}}} \quad \text{(ACI Code, Eq. 8.4.2.3.2)}$$

$$\gamma_v = 1 - \gamma_f \quad \text{(ACI Code, Eq. 8.4.4.2.2)}$$

where b_1 and b_2 are the dimensions of the punching shear critical perimeter parallel and perpendicular, respectively, to the span of the slab (see Figures 6.34 and 6.35).

Interior columns: $b_1 = c_1 + d$; $b_2 = c_2 + d$

Corner columns: $b_1 = c_1 + 0.5d$; $b_2 = c_2 + 0.5d$

Edge columns: $b_1 = c_1 + 0.5d$; $b_2 = c_2 + d$

where c_1 is the column cross-sectional dimension parallel to the span of the slab, and c_2 is the column cross-sectional dimension in the transverse direction. When shear caps are used, the values of c_1 and c_2 will reflect the increased critical shear perimeter, and the revised c_1 for an interior column will be $c_{1column} + 2h_{shear\text{-}cap}$. Similarly, the revised c_2 will be $c_{2column} + 2h_{shear\text{-}cap}$.

For columns where $b_1 = b_2$, the fraction of the unbalanced moment transferred by flexure, γ_f, will be 0.6, and therefore, the fraction transferred by eccentricity of shear, γ_v, will be 0.4. This means that, 60% of the unbalanced moment is transferred by flexure from the slab to the column, and the remaining 40% is transferred by eccentricity of the shear about the centroid of the critical section. A linear distribution of the shear stress about the centroid of the critical section is assumed for the factored shear stress caused by unbalanced moment, $\gamma_v M_{sc}$ (ACI Code, Section 8.4.4.2.3) as shown in Figure 6.34 for interior and edge columns, where M_{sc} is the factored unbalanced moment transferred from the slab to the column.

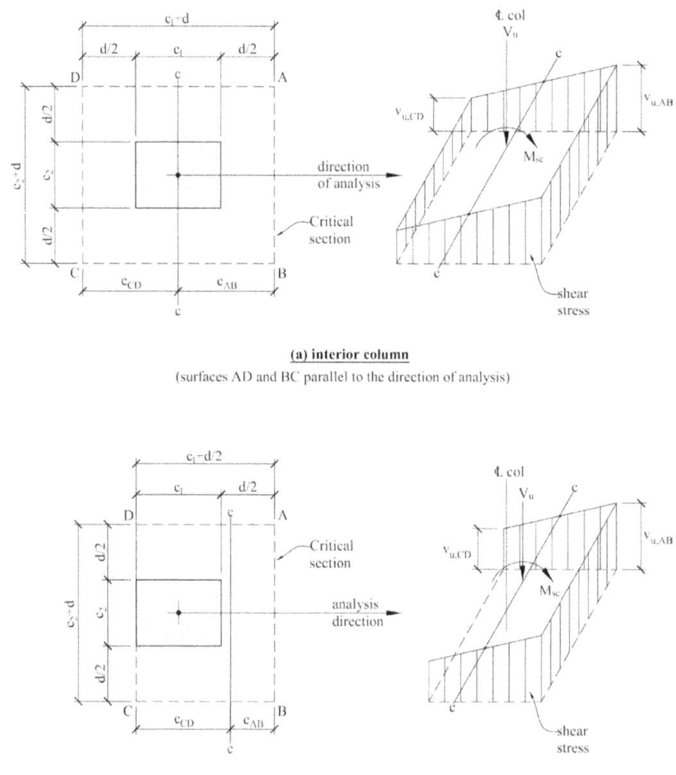

(a) interior column
(surfaces AD and BC parallel to the direction of analysis)

(b) edge column
(direction of analysis perpendicular to slab edge and surfaces AD and BC parallel to analysis direction)

FIGURES 6.34 Assumed distribution of shear stress for combined shear and moment (ACI 318-14 Figure R8.4.4.2.3).

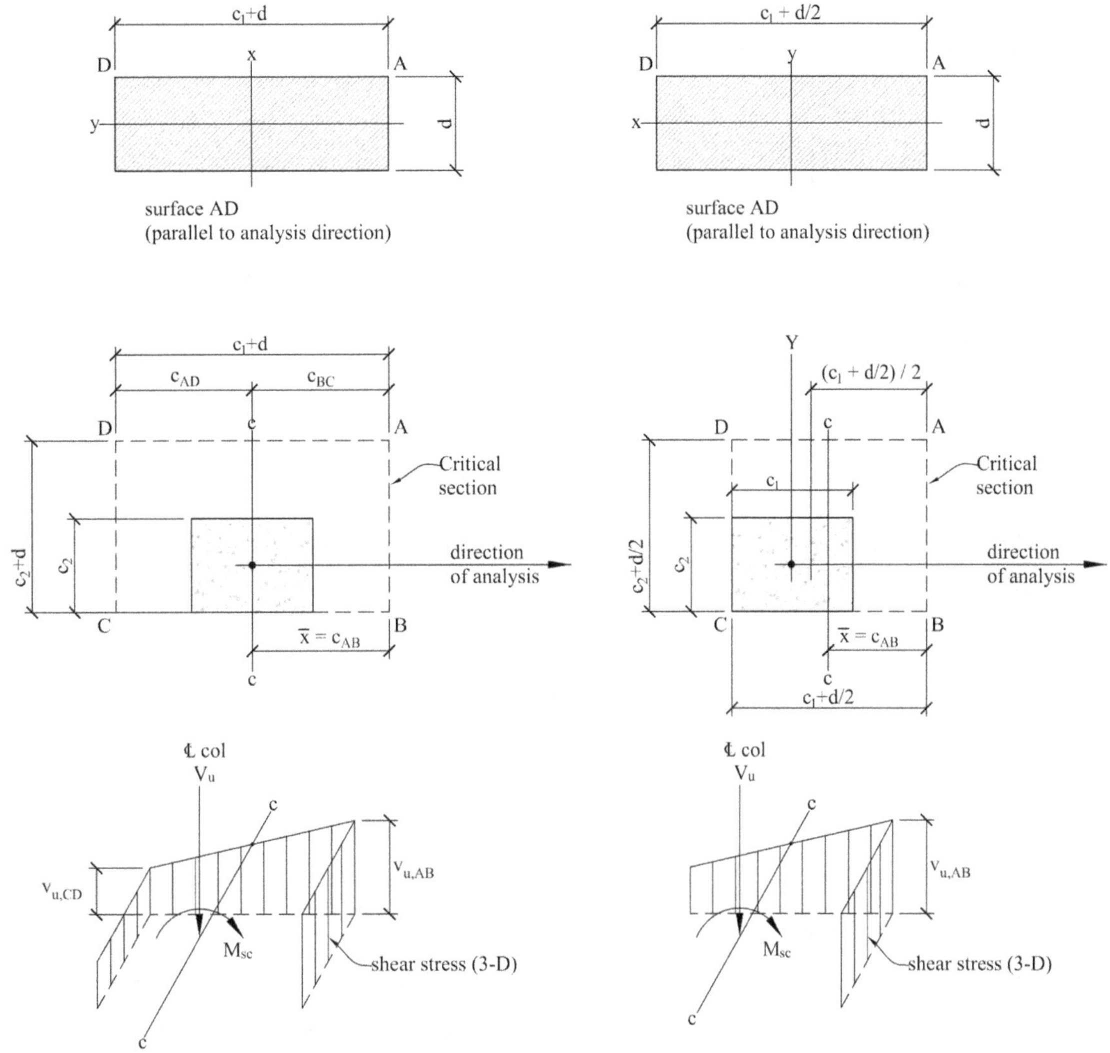

(c) edge column with direction of analysis parallel to the slab edge

(d) corner column

FIGURES 6.34 contd. Assumed distribution of shear stress for combined shear and moment (ACI 318-14 Figure R8.4.4.2.3).

For edge columns, the unbalanced gravity load moments to be transferred from the slab to the column must be at least $0.3M_o$. According to previous editions of ACI 318 and based on analyses of typical two-way slab configurations, this is a realistic upper limit between the moment values presented in Figures 6.30a through 6.30c for unrestrained and fully restrained edge conditions. Therefore, $M_{sc} \geq 0.30M_o$

Modifications to γ_f

ACI Code, Section 8.4.2.2.4 allows some flexibility in distributing the unbalanced moment, M_{sc}, transferred by shear and flexure at corner, edge, and interior columns; it allows modifications of the γ_f values as shown in Table 6.10a (see ACI Code, Table 8.4.2.2.4) provided certain conditions pertaining to the amount of direct shear stress and the net tensile strain in the extreme tension reinforcement are satisfied. Figure 6.35a shows corner, edge, and interior columns in a concrete flat plate building under construction.

TABLE 6.10a Maximum modified values of γ_f for reinforced concrete slabs (ACI Code Table 8.4.2.2.4).

Type of column	Direction of analysis	Direct shear stress, $v_{uv} = \frac{V_u}{b_o d}$	ε_t (within b_{slab})	Maximum modified γ_f
Corner column	Either direction	$\leq 0.5\phi v_c$	$\geq \varepsilon_{ty} + 0.003$	1.0
Edge column	Perpendicular to the edge	$\leq 0.75\phi v_c$	$\geq \varepsilon_{ty} + 0.003$	1.0
	Parallel to the edge	$\leq 0.4\phi v_c$	$\geq \varepsilon_{ty} + 0.008$	$\frac{1.25}{1+\frac{2}{3}\sqrt{\frac{b_1}{b_2}}} \leq 1.0$
Interior column	Either direction	$\leq 0.4\phi v_c$	$\geq \varepsilon_{ty} + 0.008$	$\frac{1.25}{1+\frac{2}{3}\sqrt{\frac{b_1}{b_2}}} \leq 1.0$

The maximum factored shear stresses under combined shear and unbalanced moment are given as:

$$v_{u,AB} = \frac{V_u}{A_c} + \frac{\gamma_v M_{sc} c_{AB}}{J_c} \leq v_{allowable}$$

$$v_{u,CD} = \frac{V_u}{A_c} - \frac{\gamma_v M_{sc} c_{CD}}{J_c} \leq v_{allowable}$$

where

V_u = factored shear acting at the centerline or centroidal axis of the column

M_u = factored unbalanced moment acting at the centerline or centroidal axis of the column

M_{sc} = factored unbalanced moment acting at the centroidal axis c-c of the critical section

For **interior columns**, V_u and M_u both act at the centroidal axis of the column (see Figure 6.34*a*), therefore: $M_{sc} = M_u$

For **edge columns** with the direction of analysis *perpendicular* to the slab edge (see Figure 6.34*b*): $M_{sc} = M_u - V_u\left(c_{CD} - \frac{c_1}{2}\right)$

For **edge columns** with the direction of analysis *parallel* to the slab edge, V_u and M_u both act at the centroidal axis of the column (see Figure 6.34*c*), therefore: $M_{sc} = M_u$

For **corner columns** (see Figure 6.34*d*):

$$M_{sc} = M_u - V_u\left[\left(c_1 + \frac{d}{2}\right) - c_{AB} - \frac{c_1}{2}\right] = M_u - V_u\left[\left(\frac{c_1 + d}{2}\right) - c_{AB}\right]$$

γ_v is as defined previously at the beginning of the section.

The allowable two-way concrete shear stress, v_c, is the smallest of the following equations from ACI Code, Table 22.6.5.2 or 22.6.5.2:

$$v_c = \left[4\lambda_s\lambda\sqrt{f'_c}\right]$$

$$v_c = \left[\left(2+\frac{4}{\beta}\right)\lambda_s\lambda\sqrt{f'_c}\right]$$

$$v_c = \left[\left(2+\frac{\alpha_s d}{b_o}\right)\lambda_s\lambda\sqrt{f'_c}\right]$$

$\phi = 0.75$ for shear

The parameter, J_c, in the factored applied shear stress equations is analogous to the polar moment of inertia of the critical section about the centroidal axis and can be calculated as follows [1, 4]:

$$J_c = \begin{pmatrix} \sum I_x \text{ about the centrodal axis for faces parallel to the analysis direction} + \\ \sum I_y \text{ about the centrodal axis for faces parallel to the analysis direction} + \\ \sum A\bar{x}^2 \text{ for all faces} \end{pmatrix}$$

For **Interior Columns** – ACI Code, Section R8.4.4.2.3 (see Figure 6.34(a) and Figure 6.35):

FIGURES 6.35a Corner, edge, and interior columns in a concrete flat plate structure.

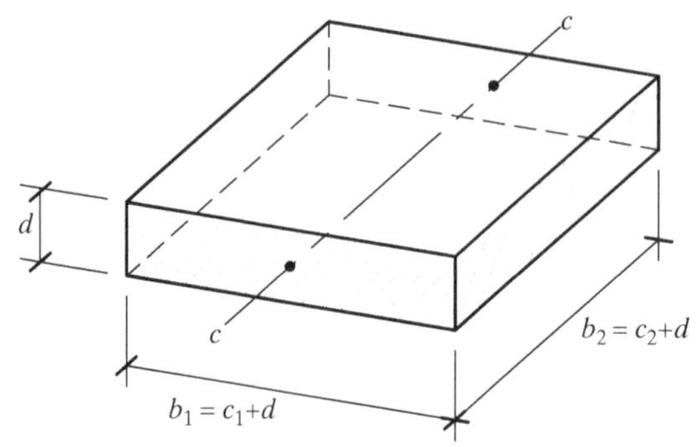

FIGURES 6.35b Critical section for punching shear at interior column.

A_c = area of the critical surface for two-way shear

$$= (b_1 + b_2 + b_1 + b_2)(d) = 2(b_1 + b_2)d$$

$$= 2(c_1 + d + c_2 + d)$$

$$= 2d(c_1 + c_2 + 2d)$$

$$c_{AB} = c_{CD} = (c_1 + d)/2$$

$$J_c = \frac{d(c_1 + d)^3}{6} + \frac{(c_1 + d)d^3}{6} + \frac{d(c_2 + d)(c_1 + d)^2}{2}$$

For **Edge Columns** with the direction of analysis **perpendicular** to the slab edge; that is, the axis of the applied moment is parallel to the slab edge (see Figure 6.34(b)):

A_c = area of the critical surface for two-way shear

$$A_c = \left(c_1 + \frac{d}{2}\right)d + \left(c_1 + \frac{d}{2}\right)d + (c_2 + d)d$$

$$= (2c_1 + c_2 + 2d)d$$

$$c_{AB} = \frac{(c_2 + d)(d)(0) + \left(c_1 + \frac{d}{2}\right)(d)\left(\frac{c_1 + \frac{d}{2}}{2}\right)(2\,sides)}{(c_2 + d)(d) + \left(c_1 + \frac{d}{2}\right)(d)(2\,sides)}$$

$$= \frac{\left(c_1 + \frac{d}{2}\right)^2 (d)}{d(2c_1 + c_2 + 2d)}$$

$$c_{CD} = c_1 + \frac{d}{2} - c_{AB}$$

$$J_c = \frac{d\left(c_1 + \frac{d}{2}\right)^3 (2\,sides)}{12} + \left(c_1 + \frac{d}{2}\right)(d)\left(\frac{c_1 + \frac{d}{2}}{2} - c_{AB}\right)^2 (2\ sides) + \frac{\left(c_1 + \frac{d}{2}\right)(d)^3 (2\,sides)}{12}$$

$$+ (c_2 + d)(d)(c_{AB})^2$$

$$= \frac{d\left(c_1 + \frac{d}{2}\right)^3}{6} + 2d\left(c_1 + \frac{d}{2}\right)\left(\frac{c_1 + \frac{d}{2}}{2} - c_{AB}\right)^2 + \frac{\left(c_1 + \frac{d}{2}\right)(d)^3}{6} + (c_2 + d)(d)(c_{AB})^2$$

Edge Columns with the direction of analysis **parallel** to the slab edge; that is, the axis of the applied moment is perpendicular to the slab edge (see Figure 6.34c) [1, 4]:

$$A_c = \text{Area of the critical vertical surfaces for two-way shear}$$

$$= (c_1 + d)d + 2\left(c_2 + \frac{d}{2}\right)d$$

Summing moment of area of the vertical surfaces about AB:

$$\bar{x} = \frac{\sum Ax}{\sum A} = \frac{(c_1 + d)d\frac{(c_1 + d)}{2} + \left(c_2 + \frac{d}{2}\right)d(c_1 + d) + \left(c_2 + \frac{d}{2}\right)d(0)}{(c_1 + d)d + 2\left(c_2 + \frac{d}{2}\right)d}$$

$$= c_{AB} = \frac{\frac{(c_1 + d)}{2}\left[d(c_1 + d) + 2\left(c_2 + \frac{d}{2}\right)d\right]}{(c_1 + d)d + 2\left(c_2 + \frac{d}{2}\right)d} = c_1 + \frac{d}{2} \quad \text{(as expected)}$$

$$= c_{CD}$$

$$J_c = \begin{pmatrix} \sum I_x \text{ about the centroidal axis for vertical surfaces parallel to the analysis direction} + \\ \sum I_y \text{ about the centroidal axis for vertical surfaces parallel to the analysis direction} + \\ \sum A\bar{x}^2 \text{ for all vertical surfaces} \end{pmatrix}$$

$$= \begin{pmatrix} \sum I_x \text{ about the centrodal axis for surface AD} + \sum I_y \text{ about the centrodal axis for surface AD} \\ + \sum A\bar{x}^2 \text{ for all surfaces} \end{pmatrix}$$

$$= \left(\begin{array}{c} (c_1+d)\dfrac{d^3}{12}+d\dfrac{(c_1+d)^3}{12}+\left[d(c_1+d)\right]\left(\dfrac{c_1+d}{2}-c_{AB}\right)^2 \\ +\left[d\left(c_2+\dfrac{d}{2}\right)\right](c_{AB})^2+\left[d\left(c_2+\dfrac{d}{2}\right)\right](c_{CD})^2 \end{array} \right)$$

$$= (c_1+d)\frac{d^3}{12}+d\frac{(c_1+d)^3}{12}+2d\left(c_2+\frac{d}{2}\right)\left(\frac{c_1+d}{2}\right)^2$$

Corner Columns: See Figure 6.34*d* [1, 4].

$$A_c = \text{Area of the critical vertical surfaces for two-way shear}$$

$$= \left(c_1+\frac{d}{2}\right)d+\left(c_2+\frac{d}{2}\right)d$$

Summing moment of area of the vertical surfaces about AB:

$$\overline{x} = c_{AB} = \frac{\sum Ax}{\sum A} = \frac{\left(c_1+\dfrac{d}{2}\right)d\dfrac{\left(c_1+\dfrac{d}{2}\right)}{2}+\left(c_2+\dfrac{d}{2}\right)d(0)}{\left(c_1+\dfrac{d}{2}\right)d+\left(c_2+\dfrac{d}{2}\right)d}$$

$$= \frac{\left(c_1+\dfrac{d}{2}\right)d\left(c_1+\dfrac{d}{2}\right)}{2\left(c_1+\dfrac{d}{2}\right)d+\left(c_2+\dfrac{d}{2}\right)d} = \frac{\left(c_1+\dfrac{d}{2}\right)^2 d}{2A_c}$$

$$J_c = \left(\begin{array}{l} \sum I_x \text{ about the centroidal axis for vertical surfaces parallel to the analysis direction} + \\ \quad \sum I_y \text{ about the centroidal axis for vertical surfaces parallel to the analysis direction} + \\ \quad \sum A\overline{x}^2 \text{ for all vertical surfaces} \end{array} \right)$$

$$= \left(\begin{array}{l} \sum I_x \text{ about the centroidal axis for surface AD} + \sum I_y \text{ about the centrodal axis for surface AD} \\ \quad + \sum A\overline{x}^2 \text{ for all surfaces} \end{array} \right)$$

$$= \left(c_1+\frac{d}{2}\right)\frac{d^3}{12}+d\frac{\left(c_1+\dfrac{d}{2}\right)^3}{12}+\left[d\left(c_1+\frac{d}{2}\right)\right]\left(\frac{c_1+\dfrac{d}{2}}{2}-c_{AB}\right)^2+\left[d\left(c_2+\frac{d}{2}\right)\right](c_{AB})^2$$

6.4.11 Corner Reinforcement in Beam or Wall-Supported Two-Way Slabs (ACI 318-19, Section 8.7.3)

When a two-way slab is supported on perimeter walls or beams with α_f greater than 1.0, additional reinforcement shall be provided at the top and bottom of the slab at the exterior corners of the slab to resist a factored moment equal to the maximum factored positive moment per unit width in the slab panel. The reinforcement shall be extended in both orthogonal directions from the edge of the slab for a distance of 20% of the longer end span of the slab panel, as shown in Figure 6.36 (ACI Code, Section R8.7.3.1).

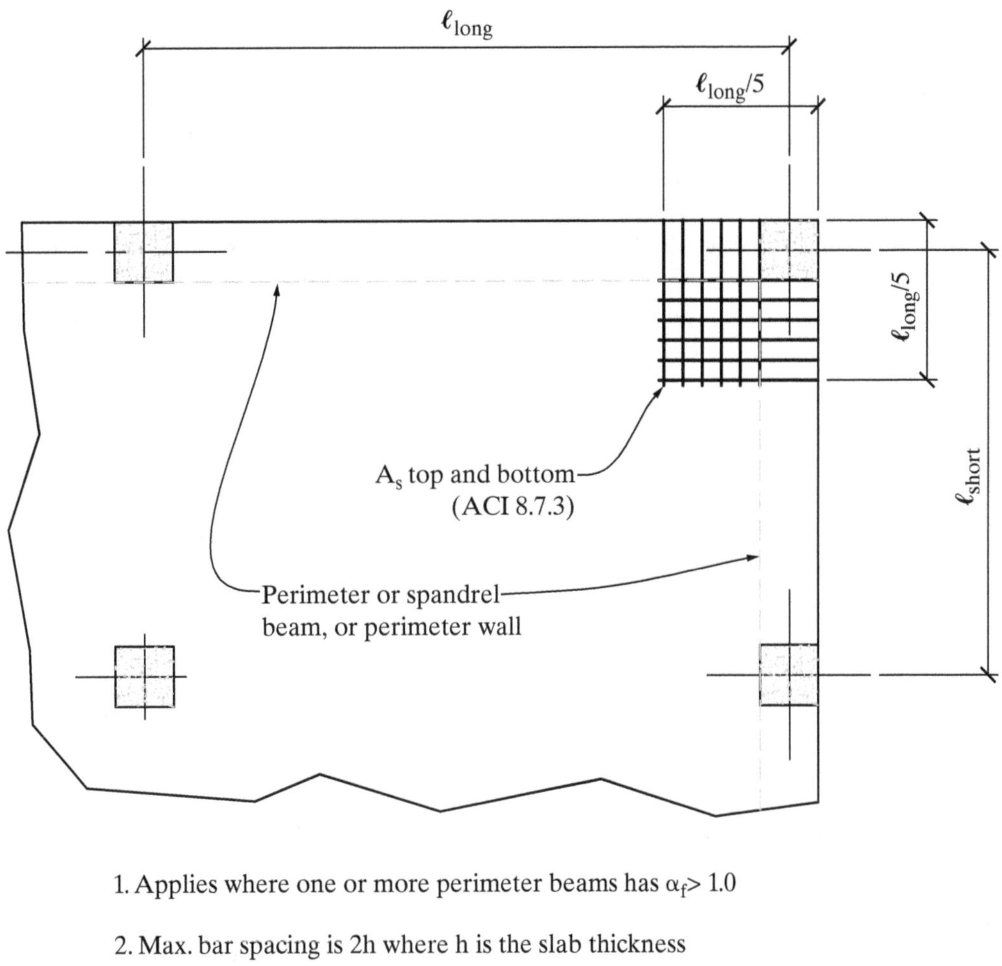

FIGURE 6.36 Plan view of slab corner reinforcing.

6.4.12 Effective Depth of Two-Way Slabs

Since the top and bottom reinforcement in two-way slabs are placed in two orthogonal directions, the effective depth in the two directions will be different (see Figure 6.37).

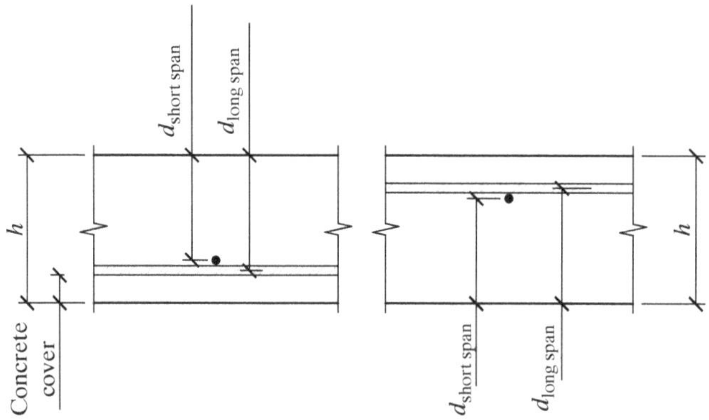

FIGURE 6.37 Effective depth in two-way slabs.

Typically, the reinforcement spanning the long direction of the slab panel is placed nearest to the bottom and top surfaces of the slab (i.e., nearest the tension faces of the slab), and therefore will have a larger effective depth to resist the larger moments in the longer span direction. Since the effective depth is a variable that is needed to design two-way slabs, an effective depth needs to be assumed for both directions. Assuming interior clear cover to reinforcement of ¾-inch and assuming #6 rebar in both directions, the approximate effective depths in the short and long directions of the slab panel are (see Figure 6.37):

$$d_{long} = h - \frac{3''}{4} - \frac{0.75''}{2} = h - 1.13''$$

$$d_{short} = h - \frac{3''}{4} - 0.75 - \frac{0.75''}{2} = h - 1.9''$$

$$d_{average} = h - 1.5''$$

For the calculation of the shear strength, ϕV_c, the average effective depth should be used (ACI Code, Section 22.6.2.1).

6.4.13 Structural Integrity Reinforcement in Two-Way Slabs (ACI 318-19, Section 8.7.4.2)

To prevent the progressive collapse of column-supported two-way slabs, all the bottom reinforcement within the column strip in each direction shall be continuous. Continuity can be achieved by using class B tension lap splice or spliced with a fully mechanical or fully welded tension splice located in the regions shown in Figure 6.38 (ACI Code, Sections 8.7.4.2.1 and 8.7.4.1.3). A new requirement for thicker two-way slabs ($\ell_n/h < 15$) such as podium slabs and mat foundations has been introduced in the ACI Code (see ACI Code, Section R8.7.4.1.3), and it calls for the extension of the negative bending reinforcement over the supports to extend at least 5d beyond the face of the supports to ensure that potential punching shear cracks which can occur at angles as small as 20-degrees can be intercepted by the tension reinforcement.

At least two bottom bars in the column strip must pass within the area bounded by the column vertical reinforcement, and this structural integrity reinforcement, first introduced in the 1989 ACI Code, must be adequately anchored at the exterior supports, and standard hooks are typically used to provide the anchorage. The structural integrity reinforcement provides the two-way slab with some residual strength or capacity that prevents progressive collapse if punching shear failure occurs at one column. Note that the minimum rebar lengths and extensions prescribed by the ACI Code in Figure 6.38 are for slabs subjected to gravity loads only. These rebar lengths and extensions may not be sufficient to resist the moments in flat plates and flat slabs subjected to combined gravity and lateral loads. The typical arrangement of two-way slab reinforcement is shown in Figure 6.39.

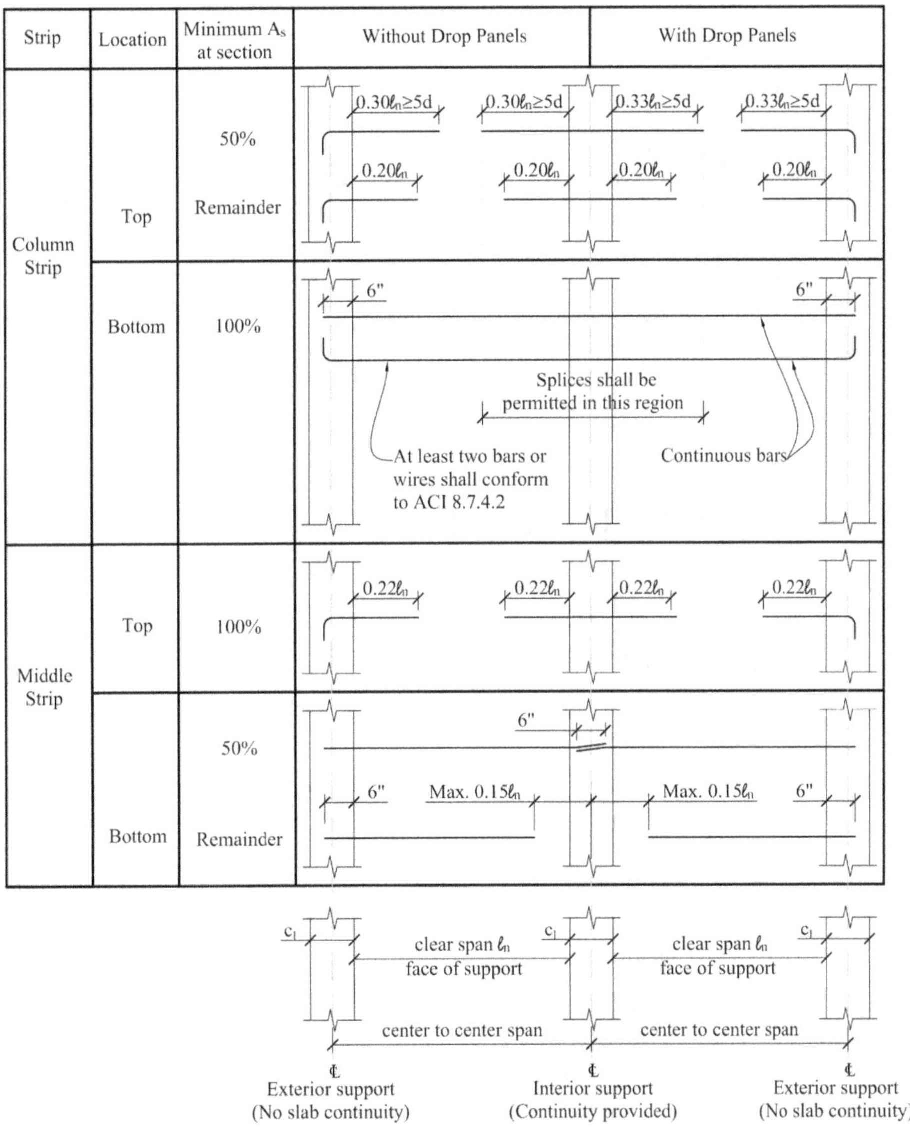

FIGURE 6.38 Details of reinforcement in two-way slabs without beams (Adapted from ACI 318-19 Figure 8.7.4.1.3).

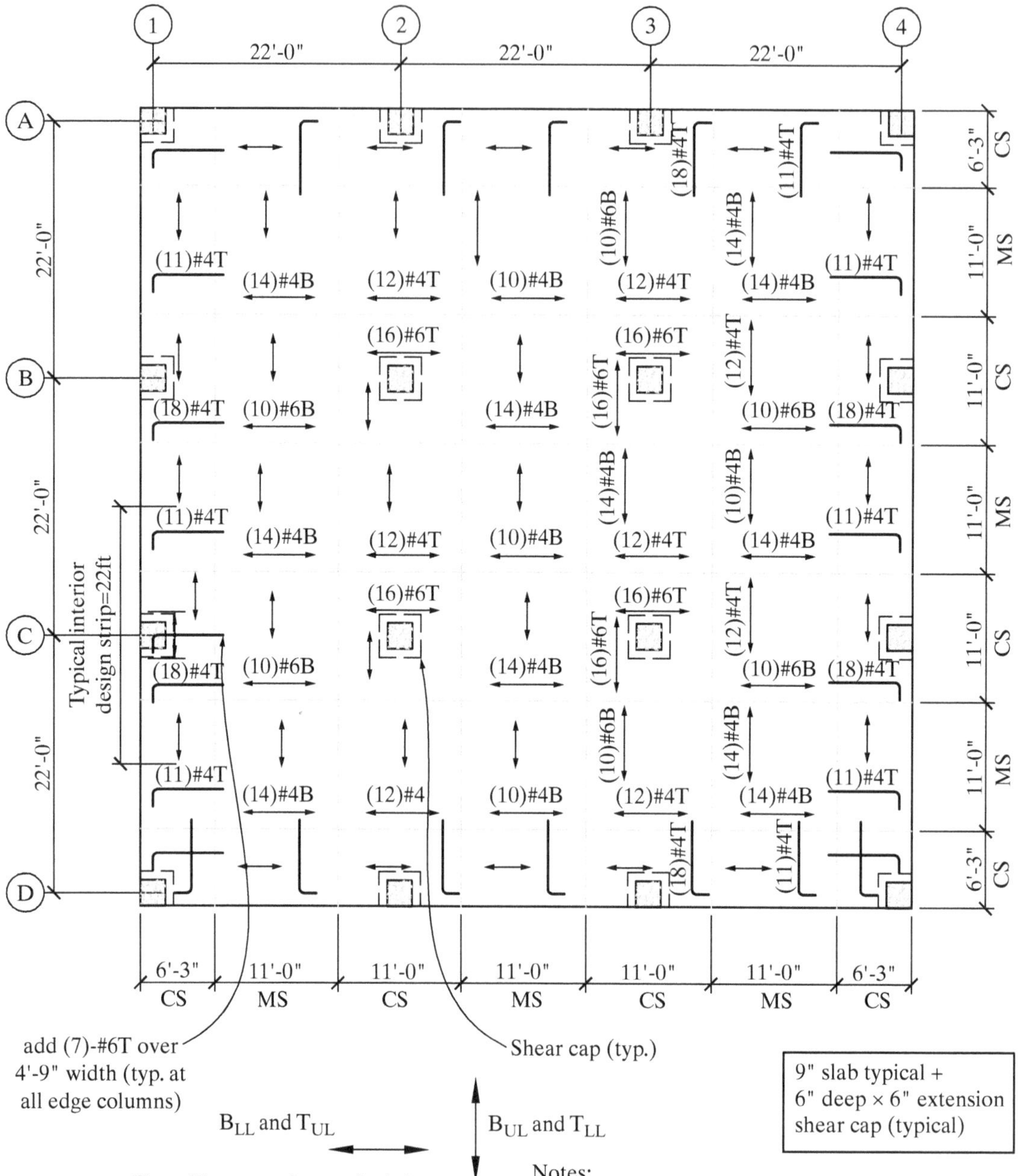

T_{UL} = Top upper layer of reinf.
T_{LL} = Top lower layer of reinf.
B_{UL} = Bottom upper layer of reinf.
B_{LL} = Bottom lower layer of reinf.
CS = Column Strip
MS = Middle Strip

Notes:

1. Areas where rebar not shown must also be filled in, in both directions. See figure 6-38 for rebar details.

2. For structural integrity reinforcement, at least two bottom bars in the column strip must pass within the column width.

FIGURE 6.39 Typical layout of reinforcement in a two-way slab.

Minimum Reinforcement in Two-Way Slabs and Flexure-Induced Punching Failure

The minimum area of tension reinforcement, which must be placed nearest to the tension face and helps to control cracking in the slab, is $0.0018A_g$ (ACI 318 Section 8.6.1.1):

In addition, if $v_{uv} > \phi 2\lambda_s \lambda \sqrt{f'_c}$, to prevent flexure-driven punching failure in lightly reinforced concrete slabs, ACI 318 Section 8.6.1.2 prescribes the minimum area of slab top reinforcement required over the slab width b_{slab} at the column-to-slab connections as

$$A_{s\min} \geq \frac{5v_{uv}b_{slab}b_o}{\phi \alpha_s f_y} = \frac{5\frac{V_{uv}}{b_o d}b_{slab}b_o}{\phi \alpha_s f_y} = \frac{5V_{uv}b_{slab}}{\phi \alpha_s f_y d}$$

$\phi = 0.75$

v_{uv} = factored direct shear stress on the slab critical section for two-way action at a column (i.e., that is, without moment transfer)

$= \frac{V_u}{b_o d}$

α_s = 40 for interior columns,

= 30 for edge column

= 20 for corner columns (ACI 318 Section 22.6.5.3)

b_o = perimeter of the critical section for punching or two-way shear

b_{slab} = effective width of slab at column location, defined in Table 6.10b (ACI 318 Table 8.4.2.2.3).

TABLE 6.10b Effective slab widths.

		Distance on each side of the column or capital
Without drop panel or shear cap	Lesser of	1.5h of slab Distance to edge of slab
With drop panel or shear cap	Lesser of	1.5h of drop panel or capital Distance to edge of the drop panel or capital plus 1.5h of slab

Note that flexure-driven punching failure, discussed earlier in Section 6.4.1, occurs at a smaller load than the classical punching shear failure discussed previously in Section 6.4.9.

For thick two-way slabs such as transfer slabs, podium slabs, and mat foundations, continuous reinforcement should be provided in both orthogonal directions at the top and bottom faces of the thick slab (ACI Code, Section R8.6.1.1). This ensures that any potential punching shear cracks are intercepted by the tension reinforcement. For mat foundations, the minimum area of flexural reinforcement of $0.0018bh$ in each principal direction must be placed near the tension faces of the mat. Since the mat is subjected to positive and negative moments in the two orthogonal directions, this means that in each face and in each orthogonal direction, a minimum area of reinforcement equal to at least 0.0018*bh* must be provided, per ACI 318 Sections 13.3.4.4 and 8.6.1.1. The maximum spacing of the flexural reinforcement is the lesser of 3h or 18 inches. As an example, for a 6 ft. deep mat foundation, the minimum reinforcement area required on each face and in each principal direction is $A_{s,min}$ = 0.0018(12")(6 ft x12) = 1.56 in^2/ft. Therefore, the minimum steel that should be provided is #9 at 7½ in. on centers, Top and Bottom Each Way (i.e., #9 @ 7½-in. o.c. BEW + #9 @ 7½-in. o.c. TEW)

Maximum Reinforcement Spacing in Two-Way Slabs (ACI 8.7.2.2)

The maximum spacing of the reinforcement in two-way slabs at *critical* sections is the smaller of 2*h* and 18 inches, where *h* is the thickness of the slab. At other sections, the maximum spacing of reinforcement is limited to the lesser of 3*h* and 18 inches. Limiting the spacing of the reinforcement helps to control cracking and provides resistance against heavy concentrated loads (e.g., forklift wheel loads) acting over small areas (ACI Code, Section R8.7.2.2) (see Figure 6.39).

The layout of the column strip and middle strip top reinforcement in two-way slabs with column offset and the maximum permitted column offset for the direct design method are shown in Figure 6.40.

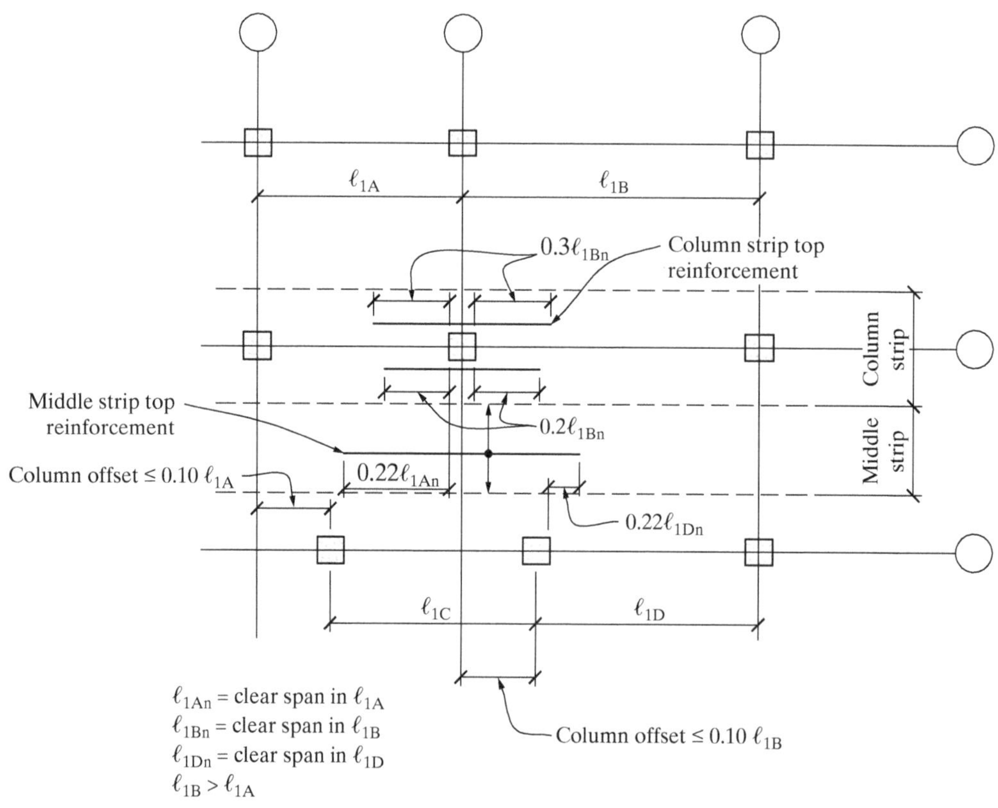

FIGURE 6.40 Reinforcement layout in two-way slabs with column offset and maximum column offset.

6.4.14 Openings in Two-Way Slabs

The reduction effects of openings and free edges on the critical perimeter for punching shear in a two-way slab are illustrated in Figure 6.41. The effect of openings in two-way slabs located within a column strip or located within $10h$ (h is the slab depth) from the column are obtained by projecting two radial lines originating from the center of the closest column to the outermost edges of the opening, and the length of the critical perimeter within the radial lines is considered ineffective and is neglected in the shear strength calculations. Slabs reinforced with shear heads and with openings located within a column strip or located within $10h$ of the concentrated load or column, must meet the requirements of ACI Code, Section 22.6.9.9 (ACI Code, Section 8.5.4.2d).

The allowable sizes of openings in two-way slabs are shown in Figure 6.42. Typical reinforcement at slab openings is shown in Figure 6.43. In practice, tower cranes may need to be

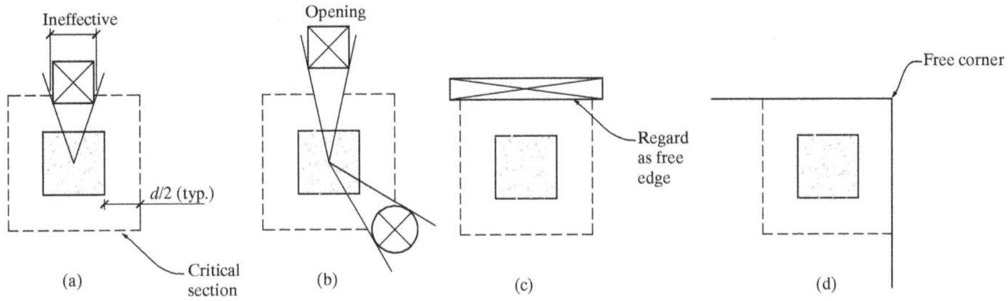

FIGURE 6.41 Effects of slab openings and free edges on the punching shear critical perimeter (Adapted from ACI 318-19 Figure R22.6.4.3).

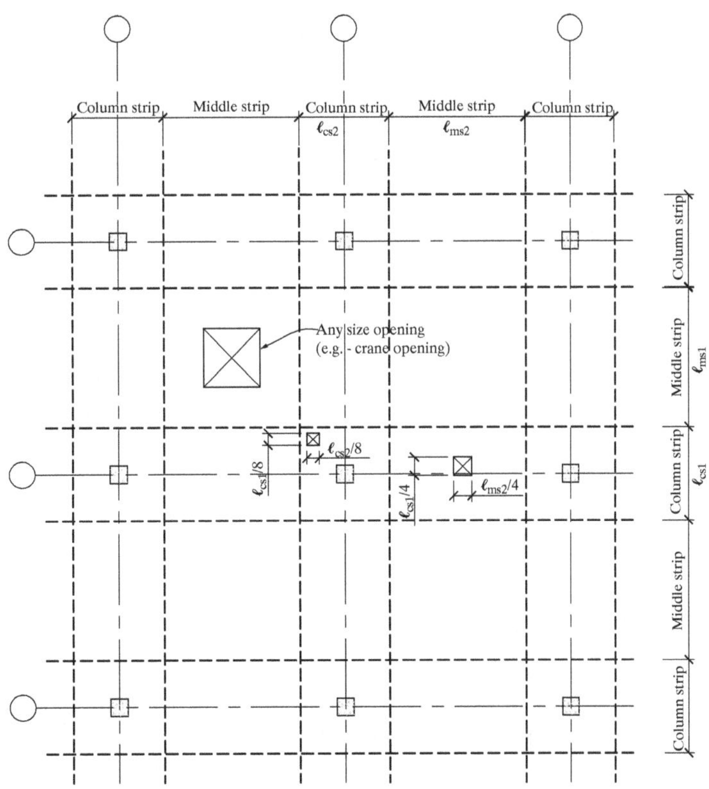

FIGURE 6.42 Allowable sizes of openings in two-way slabs.

located within the footprint of a high-rise building, and in such cases, the temporary crane opening must be located within the middle strips of the two-way slabs. It is good practice to locate the edges of the crane opening at least 6 ft. from the center of the nearest column to avoid a reduction in the critical punching shear perimeter at the column.

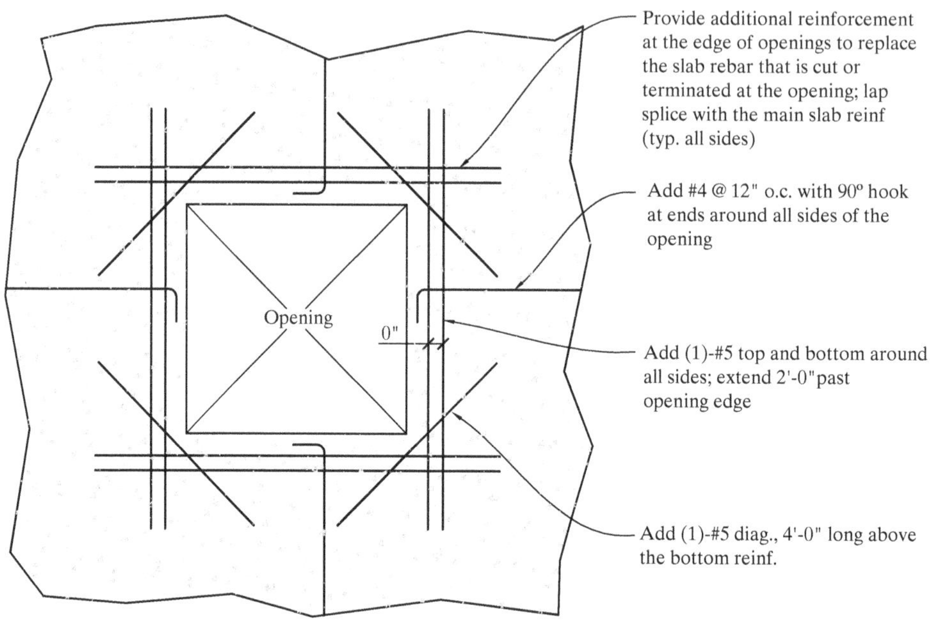

(a) Plan View Showing Reinforcement at Openings

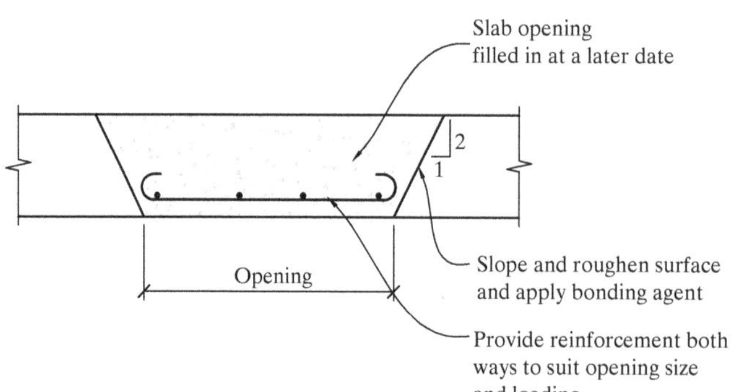

(b) Section

FIGURE 6.43 Typical reinforcement layout at slab openings.

6.4.15 Effect of Transverse Edge Beams

Where there is no transverse edge beam, all the exterior support negative moment in the ℓ_1 direction is assigned to the column strip (see Table 6.7). When there is an edge beam, the value of the torsional stiffness parameter, β_t , of the edge beam determines how much of the exterior negative moment is assigned to the middle strip. The torsional stiffness parameter, β_t , accounts for the torsional stiffness of the edge beam or wall; for the moment in the ℓ_1 direction in Figure 6.44, this would be the beam or wall along grid line 1. For a slab supported along its edge on a masonry wall or on a concrete wall that is monolithic with the slab, the values of β_t are given in Figure 6.44. If the width of the edge column or wall is at least $0.75\ell_2$, the exterior negative moment shall be distributed uniformly across the transverse length, ℓ_2. The torsional stiffness parameter for edge beams is defined as the ratio of the torsional rigidity of the edge beam that is transverse to the direction of analysis (i.e., edge beam is perpendicular to the ℓ_1 direction) to the flexural stiffness of a width of slab equal to the span length of the transverse edge beam, center-to-center of supports.

$$\beta_t = \frac{E_{cb}C}{2E_{cs}I_s}$$

where

$$C = \sum\left(1 - 0.63\frac{x}{y}\right)\frac{x^3y}{3}$$

x = smaller dimension of a rectangular area

y = larger dimension of a rectangular area

I_s = moment of inertia of the slab = $\ell_2 h^3/12$.

The other terms in the equations have been defined previously (See Section 6.4.6).

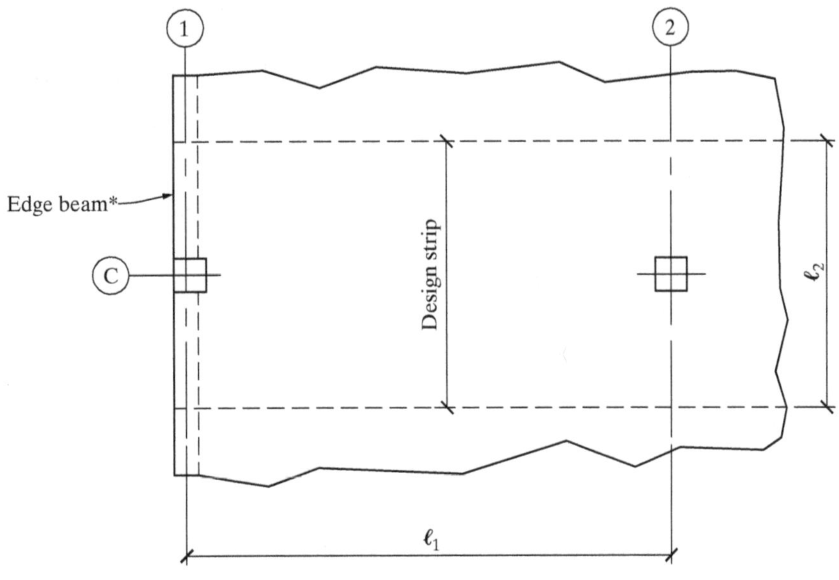

*For slab supported on wall at the edge (Line C)

β_t = 0 for masonry wall (since wall has no torsional resistance)
= 2.5 for concrete wall

ℓ_1 = length of span in the direction moment is being determined

ℓ_2 = transverse width of the design strip, perpendicular to the ℓ_1 direction

FIGURE 6.44 Effect of transverse edge beam on moment distribution in the slab.

6.4.16 Slab Systems with Beams along All Four Edges of a Panel

For slabs with supporting beams along the four edges of each rectangular slab panel, the distribution of the uniform gravity loads to the supporting beams is shown in Figure 6.45. In addition, each beam shall resist the shears caused by the factored loads directly supported by the beams (e.g., wall loads) including the weight of the beam stem.

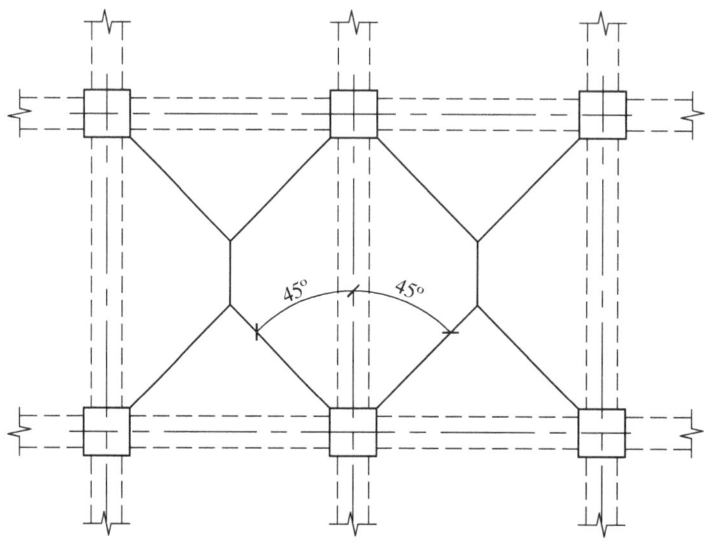

FIGURE 6.45 Tributary area for shear on interior beams.

When a beam has a stiffness ratio $\alpha_{f1}\,{}^{\ell_2}/_{\ell_1}$ greater than or equal to 1.0, the beam resists 100 percent of the shear force from the tributary, as in Figure 6.45. Therefore, the portion of shear resisted by a beam in beam-supported slabs is $\alpha_{f1}\,{}^{\ell_2}/_{\ell_1} \leq 1.0$. When the beam stiffness ratio, $\alpha_{f1}\,{}^{\ell_2}/_{\ell_1}$, is less than 1.0, only a portion of the shear from the tributary area in Figure 6.45 plus the direct load on the beam will be resisted by the beam. The uniform load on the remaining portion of the tributary area will be resisted by the slab, and this will result in shear stresses at the slab-column connection like what is obtained for flat plates and flat slabs.

EXAMPLE 6.3

A flat plate shown in Figure 6.46 is supported on 18 in. square columns with 6 in. deep shear caps on a 22 ft x 22 ft column grid, and the slab is continuous over at least three spans in both orthogonal directions. f_c' is 4000 psi and f_y is 60,000 psi. Assuming a superimposed dead load of 30 psf (includes partitions, mechanical and electrical, ceiling, and floor finishes) and an occupancy live load of 100 psf,

(a) Check if the direct design method can be used for the analysis of this two-way slab.

(b) Select the slab thickness.

(c) Calculate the total factored load on the slab in psf.

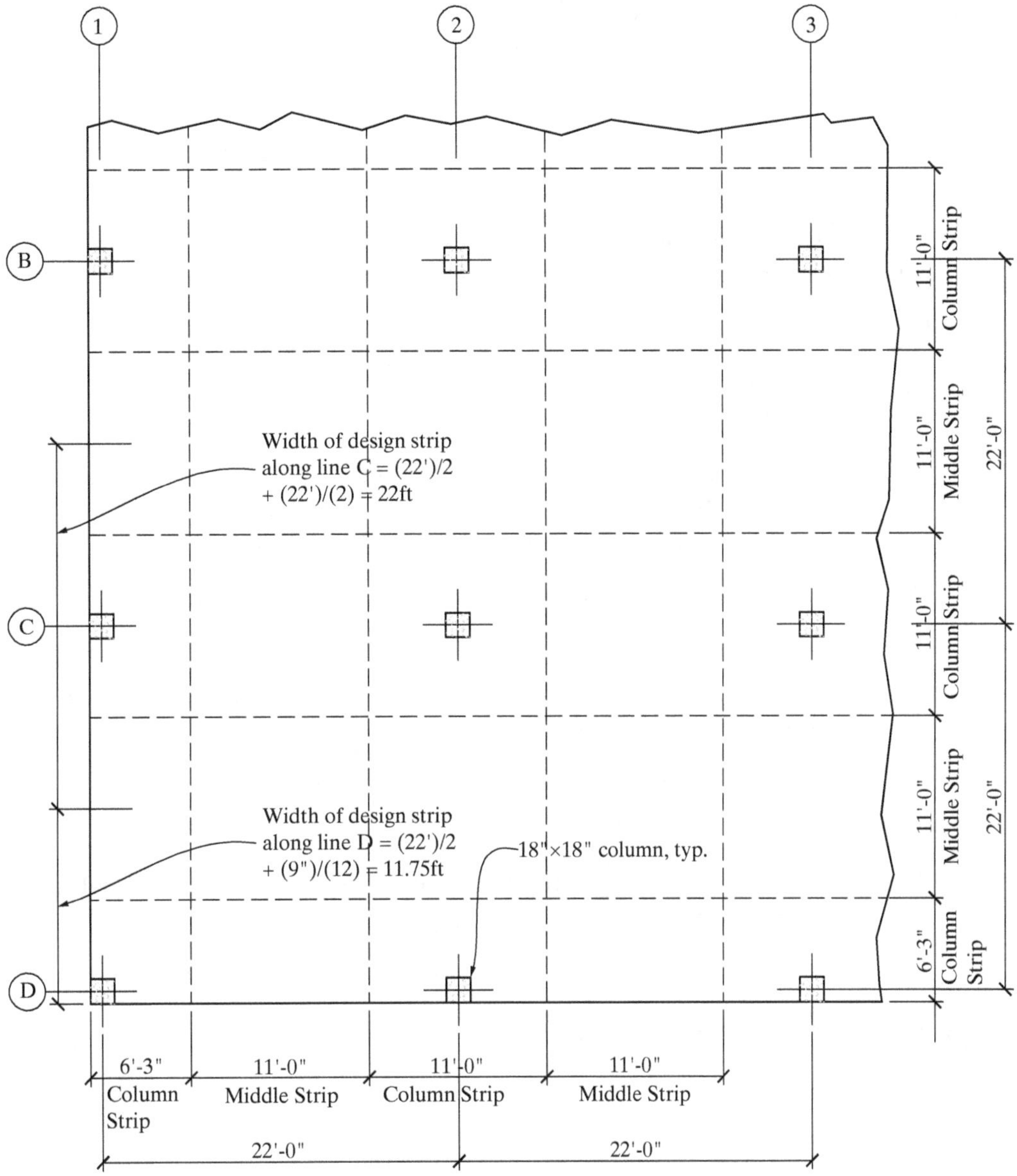

FIGURE **6.46** Plan dimensions for Example 6.3.

(d) Divide the slab into column and middle strips and determine the widths of the column and middle strips.

(e) Calculate the factored static moment for the East-West design strip along grid line C.

(f) Calculate the effective depths of the slab reinforcement and the average effective depth.

(g) Calculate the direct shear demand, V_u, at a typical interior column and check against the shear strength ϕV_c.

(h) Calculate the factored longitudinal design moments for each negative and positive regions of the slab for the East-West design strip along grid line C.

(i) Using the moments from part (h), calculate the factored design moments distributed transversely to the column and middle strips, and to the longitudinal beams, if any.

(j) Select the reinforcement for the column and middle strip for the design strip along line C.

(k) Check the shear stress due to direct shear plus unbalanced moment transfer at the typical interior column C-2.

(l) Check the shear plus unbalanced moment transfer at the typical edge column C-1. Assume an exterior cladding load of 25 psf and a floor-to-floor height of 10 ft.

Solution:

(a) Check if the direct design method can be used (see *Analysis and Design of Continuous Two-Way Slabs*)

1. The ratio of the unfactored live load to the unfactored dead load (see the load calculations in step c below) is w_L/w_D, = 100 psf/143 psf = 0.70 < 2.0. OK, therefore, pattern live load is not critical and does not need to be checked.
2. There are at least three continuous spans in each direction. OK
3. There is no column offset; therefore, the offset is less than 10%. OK
4. Ratio of long side to short side of rectangular panel measured center to center of supports = 22 ft/22 ft = 1.0 < 2.0 OK.
5. The lengths of successive spans are equal, therefore OK.
6. The loads are uniformly distributed.
7. The slab is not supported on beams on all four sides of the panels; therefore, the relative stiffness criteria does not apply.
8. Therefore, a direct design method can be used for the analysis of this two-way slab.

(b) Select slab minimum thickness:

f_y = 60,000 psi

Minimum thickness for two-way slabs without beams:

The clear span in the long direction which is the span face-to-face of columns is,

$$\ell_n = 22\text{ ft} - \left(\frac{18''/2}{12}\right) - \left(\frac{18''/2}{12}\right) = 20.5\text{ ft.}$$

Since there are no beams spanning between supports on all sides of a slab panel, use Table 6.4 (see Section 6.4.3 – *Minimum Thickness of Two-Way Slabs*)

h_{min} for slab *without* drop panels per ACI Code, Section 8.2.4 = 5 in.

For an interior panel, $h_{min} = \frac{\ell_n}{33} = \frac{20.5(12)}{33} = 7.5\text{in.}$

For an exterior panel (without edge beams), $h_{min} = \frac{\ell_n}{30} = \frac{20.5(12)}{30} = 8.2\text{ in.}$

Rounding to the nearest ½ inch, h = 8.5 inches, but use h = 9 inches to allow for the expected increase in the applied shear stress from unbalanced moment transfer at the slab-column connections.

(c) Calculate the factored uniform load on the slab in psf:

Slab self-weight = (9"/12)(150 pcf) = 113 psf

Superimposed dead load = 30 psf

Total dead load = 143 psf

Live load = 100 psf

The factored uniform load, w_u = 1.2(143 psf) +1.6(100 psf) = **332 psf**

(d) For column and middle strips, see Figure 6.45:

The width of the typical interior column strip on each side of the column line (ACI Code, Section 8.4.1.5)

= smaller of $0.25\ell_1$ and $0.25\ell_2$= 0.25(22 ft) = 5.5 ft

Therefore, the total width of a typical interior column strip = 5.5 ft + 5.5 ft = 11 ft

The width of each interior design strip = (22 ft/2) + (22 ft/2) = 22 ft

Therefore, the total width of the middle strip for each design strip = 22 ft – 11 ft = 11 ft

That is, for each 22 ft wide design strip, there is a 5.5 ft wide middle strip (i.e., 11 ft/2) on each side of the column strip.

See Figure 6.46 for the layout of the column and middle strips.

(e) Factored static design moment, M_o:

$$\ell_{n1} = 20.5 \text{ ft}$$
$$\geq 0.65\ell_1 = (0.65)(22 \text{ ft}) = 14.4 \text{ ft}$$

Therefore, use $\ell_{n1} = 20.5$ ft

From Section 6.4.4 – *Static Longitudinal Moment, M_o*, the factored static moment is calculated as

$$M_o = \frac{w_u \ell_2 \ell_n^2}{8} = \frac{(332 \text{ psf})(22 \text{ ft})(20.5 \text{ ft})^2}{8}$$
$$= 384 \text{ ft} - \text{kip}$$

This is the factored total static moment for a typical 22 ft wide interior design strip along grid line C.

(f) Effective depths of reinforcement (see Section 6.4.12 – *Effective Depths of Two-Way Slabs*):

$$d_{E-W} = h - 1.1" = 9 - 1.1 = 7.9"$$
$$d_{N-S} = h - 1.9" = 9 - 1.9 = 7.1"$$
$$d_{average} = h - 1.5" = 9 - 1.5 = 7.5"$$

(g) Direct shear at typical interior column (see Figure 6.47):

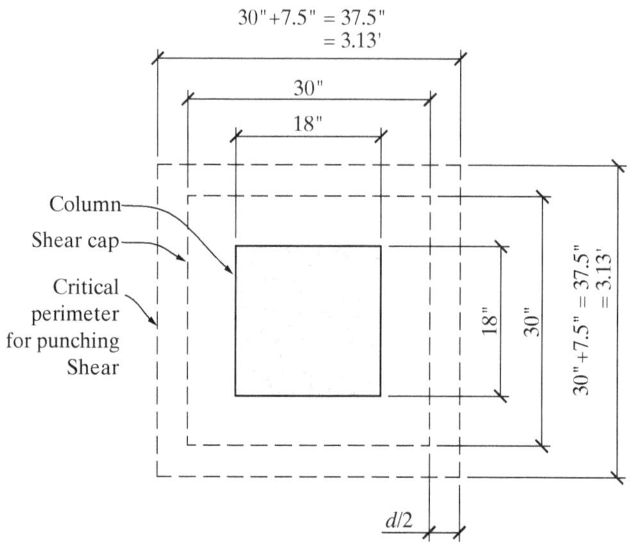

FIGURE 6.47 Critical section for two-way shear.

Without the shear cap, $c_1 = c_2 = 18"$ (i.e., the size of the column), but because of the shear cap that is 6" wide (all around the column), the revised parameters for calculating the perimeter of the critical section for punching shear are as follows;

$c_1 = 6" + 18" + 6" = 30"$

Also, $c_2 = 6" + 18" + 6" = 30"$

$d = d_{average} = 7.5"$

$b_1 = c_1 + d = 30 + 7.5 = 37.5" = 3.13$ ft

$b_2 = c_2 + d = 30 + 7.5 = 37.5" = 3.13$ ft

b_o = perimeter of the critical section for punching shear

$= 2(c_1 + d) + 2(c_2 + d) = 2(30 + 7.5) + 2(30 + 7.5) = 150$ inches

$\lambda = 1.0$ for normal weight concrete

The shear force on the tributary area of a typical interior column excluding the load on the critical perimeter is

$$V_u(2 - way\ shear) = w_u \ell_1 \ell_2 - w_u(c_1 + d)(c_2 + d)$$

$$= (332 \text{ psf})[(22 \text{ ft})(22 \text{ ft}) - (3.13 \text{ ft})(3.13 \text{ ft})] = 157{,}435 \text{ Ib} = 157 \text{ kip}$$

β = ratio of the long cross-sectional dimension to the short cross-sectional dimension of the column, concentrated load, or reaction = 18"/18" = 1.0

$\alpha_s = 40$, for interior columns

From ACI Code, Section 22.5.5.1.3, the size effect factor is calculated as

$$\lambda_s = \sqrt{\frac{2}{1+\frac{d}{10}}} \leq 1$$

$$= \sqrt{\frac{2}{1+\frac{7.5}{10}}} = 1.07 \leq 1, \text{ therefore, } \lambda_s = 1.0$$

Note that the size effect factor is always 1.0 for slab effective depths less than or equal to 10 inches. The concrete shear strength in two-way shear is the smallest of the following equations derived from ACI Code, Table 22.6.5.2:

$$\phi V_c = 4\phi\lambda_s\lambda\sqrt{f_c'}\, b_o d = (4)(0.75)(1.0)(1.0)\left(\sqrt{4000 \text{ psi}}\right)(150")(7.5") = 214 \text{ kip}$$

$$\phi V_c = \phi\left(2+\frac{4}{\beta}\right)\lambda_s\lambda\sqrt{f_c'}\, b_o d = (0.75)\left(2+\frac{4}{1}\right)(1.0)(1.0)\left(\sqrt{4000 \text{ psi}}\right)(150")(7.5") = 320 \text{ kip}$$

$$\phi V_c = \phi\left(2+\frac{\alpha_s d}{b_o}\right)\lambda_s\lambda\sqrt{f_c'}\, b_o d = (0.75)\left(2+\frac{40(7.5)}{150}\right)(1.0)(1.0)\left(\sqrt{4000 \text{ psi}}\right)(150")(7.5") = 214 \text{ kip}$$

Therefore, use $\phi V_c = 214$ kip > $V_u = 157$ kip OK.

The slab is adequate for direct shear at the interior column.

(h) Longitudinal distribution of the static Moment, M_o, parallel to the span of the slab:

$M_o = 384$ ft-kip (see part (e)).

$0.26M_o = 100$ ft-kip (see Figure 6.48)

$0.52\,M_o = 200$ ft-kip

$0.70M_o = 269$ ft-kip

$0.65M_o = 250$ ft-kip

$0.35M_o = 134$ ft-kip

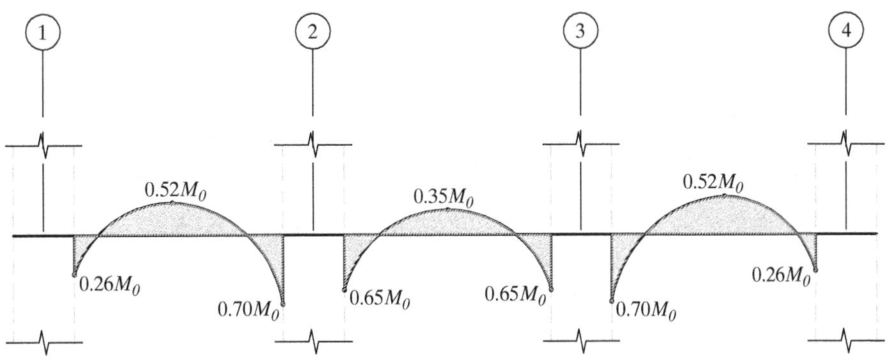

FIGURE 6.48 Longitudinal distribution of static moment, M_o.

The longitudinal moments are summarized in Table 6.11.

TABLE 6.11 Longitudinal distribution of the static Moment, M_o, parallel to the span of the slab (i.e., M_u).

	Grid line 1	Midspan	Grid line2 (left)	Grid line 2(right)	Midspan	Grid line 3(left)	Grid line 3(right)	Midspan	Grid line 4
Negative moment (ft-kip)	100		269	250		250	269		100
Positive moment (ft-kip)		200			134			200	

(i) Transverse distribution of the longitudinal moments from part (h) to the column strip and longitudinal beams, if any, and to the middle strips (see Section 6.4.7 – *Longitudinal Distribution of the Total Factored Static Moment, M_o*):

There are no interior longitudinal beams in this slab; therefore, $\alpha_{f1} = 0$

Since there are no edge beams in this slab, $\beta_t = 0$

$$\frac{\ell_2}{\ell_1} = \frac{22 \text{ ft}}{22 \text{ ft}} = 1.0$$

Portion of **interior** negative moment, M_u, in column strip = 0.75 (see Table 6.6)

Therefore, the portion of **interior** negative moment in the middle strip (i.e., the two half-middle strips) = 1 – 0.75 = 0.25

Portion of **exterior** negative moment, M_u, in column strip = 1.0 (see Table 6.7)

Therefore, the portion of **exterior** negative moment in the middle strip (i.e., the two half middle strips) = 1 – 1 = 0

Portion of **positive** moment, M_u, in column strip = 0.6 (see Table 6.8)

Therefore, the portion of **positive** moment in the middle strip (i.e., the two half-middle strips) = 1 – 0.6 = 0.4

There are no longitudinal beams; therefore, no moments are transferred to the beams. The transverse distribution of moments for the design strip along grid line C is summarized in Table 6.12.

TABLE 6.12 Transverse Distribution of Moment, M_u, Perpendicular to the Span of the Slab.

	Grid line 1	Midspan	Grid line2 (left)	Grid line 2(right)	Midspan	Grid line 3(left)	Grid line 3(right)	Midspan	Grid line 4
Transverse distribution factors for column strip	1.0	0.6	0.75	0.75	0.6	0.75	0.75	0.6	1.0
Column Strip moments in ft-kip. (width = 11 ft = 132 in)	100	120	202	202	81	202	202	120	100

	Grid line 1	Midspan	Grid line2 (left)	Grid line 2(right)	Midspan	Grid line 3(left)	Grid line 3(right)	Midspan	Grid line 4
Transverse distribution factors for middle strip	0	0.4	0.25	0.25	0.4	0.25	0.25	0.4	0
Middle Strip moment in ft-kip. (width = 11 ft = 132")	0	80	68	68	54	68	68	80	0
Area of reinforcement, A_s, required [1] (Column strip)	18#4T	10#6B	16#6T	16#6T	14#4B	16#6T	16#6T	10#6B	18#4T
Area of reinforcement, A_s, required [1] (Middle strip)	11#4T	14#4B	12#4T	12#6T	10#4B	12#4T	12#4T	14#4B	11#4T

[1] Calculate the area of reinforcement using the method presented in Chapter 2. See the reinforcement layout in Figure 6.39.

(j) For the column strip and middle strip reinforcement, see Table 6.12.

(k) Shear plus unbalanced moment transfer at the **typical interior column C-2**:

The unbalanced factored moment transferred at a typical interior column C-2 is,

$M_{sc} = M_u$ = 269 ft-kip – 250 ft-kip = 19 ft-kip (see Table 6.11 and Figure 6.49)

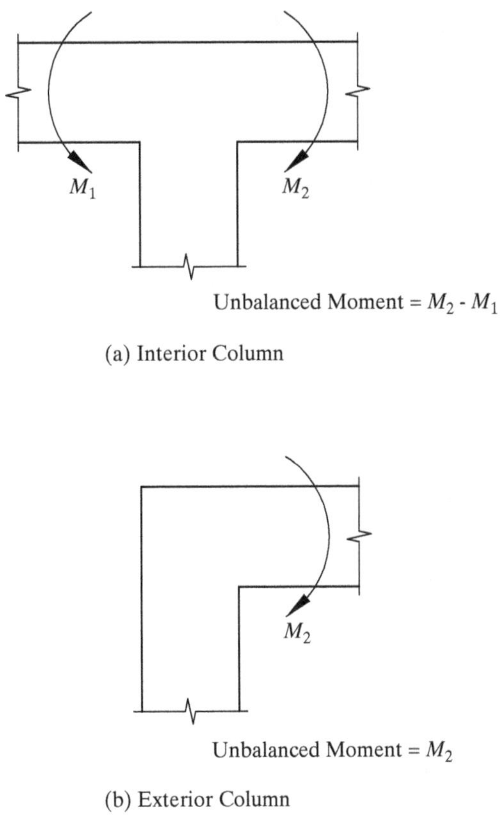

FIGURE 6.49 Unbalanced moments at columns in Example 6.3.

From part (g), the factored shear at the centroidal axis of the column, V_u = 157 kip

Without the shear cap, $c_1 = c_2$ = 18" (i.e., the size of the column), but because of the shear cap that is 6" wide (all around the column), the revised parameters are

c_1 = 6" + 18" + 6" = 30".

c_2 = 6" +18" + 6" = 30"

c_1/c_2 = 30/30 = 1.0

$d = d_{average}$ = 7.5"

$b_1 = c_1 + d$ = 30 + 7.5 = 37.5" = 3.13 ft

$b_2 = c_2 + d$ = 30 + 7.5 = 37.5" = 3.13 ft

b_1/b_2 = 1.0

b_o = perimeter of the critical section for punching shear

= $2(c_1 + d) + 2(c_2 + d)$ = 2(30 + 7.5) + 2(30 + 7.5) = 150 inches

$$\gamma_f = \frac{1}{1+\left(\frac{2}{3}\right)\sqrt{\frac{b_1}{b_2}}} = \frac{1}{1+\left(\frac{2}{3}\right)\sqrt{1.0}} = 0.60 \text{ (ACI Code, Eq. 8.4.2.2.2)}$$

$$\gamma_v = 1 - 0.60 = 0.40 \text{ (ACI Code, Eq. 8.4.4.2.2)}$$

A_c = area of the critical surface for two-way shear = $2d(c_1 + c_2 + 2d)$

= (2)(7.5)[30 + 30 + 2(7.5)] = 1125 in^2.

$c_{AB} = c_{CD} = (c_1 + d)/2$ = (30 + 7.5)/2 = 18.75 in.

$$J_c = \frac{d(c_1+d)^3}{6} + \frac{(c_1+d)d^3}{6} + \frac{d(c_2+d)(c_1+d)^2}{2}$$

$$= \frac{(7.5)(30+7.5)^3}{6} + \frac{(30+7.5)(7.5)^3}{6} + \frac{(7.5)(30+7.5)(30+7.5)^2}{2}$$

$$= 266{,}309 \text{ in.}^4$$

β = ratio of the long cross-sectional dimension to the short cross-sectional dimension of the column, concentrated load, or reaction = 18"/18" = 1.0

From ACI Code, Section 22.5.5.1.3, the size effect factor is calculated as

$$\lambda_s = \sqrt{\frac{2}{1+\frac{d}{10}}} \le 1$$

$$= \sqrt{\frac{2}{1+\frac{7.5}{10}}} = 1.07 \le 1, \textit{ therefore, } \lambda_s = 1.0$$

The allowable concrete shear stress in two-way shear is the smallest of the following equations from ACI Code, Table 22.6.5.2:

$$v_c = 4\phi\lambda_s\lambda\sqrt{f'_c} = 4(0.75)(1.0)(1.0)\sqrt{4000\,\text{psi}} = 190\,\text{psi}$$

$$v_c = \phi\left(2+\frac{4}{\beta}\right)\lambda_s\lambda\sqrt{f'_c} = (0.75)\left(2+\frac{4}{1.0}\right)(1.0)(1.0)\sqrt{4000\,\text{psi}} = 285\text{ psi}$$

$$v_c = \phi\left(2+\frac{\alpha_s d}{b_o}\right)\lambda_s\lambda\sqrt{f'_c} = (0.75)\left(2+\frac{40(7.5)}{150}\right)(1.0)(1.0)\sqrt{4000\,\text{psi}} = 190\,\text{psi}$$

$\phi = 0.75$ for shear

Therefore, $v_{allowable} = 190$ psi

The maximum factored shear stresses at interior column C-2 under combined shear and unbalanced moment are:

$$v_{u,AB} = \frac{V_u}{A_c} + \frac{\gamma_v M_{sc}\, c_{AB}}{J_c} = \frac{(157)(1000)}{1125} + \frac{(0.40)(19)(12{,}000)(18.75)}{266{,}309}$$
$$= 146\text{ psi} \le v_{allowable} = 190\text{ psi} \quad O.K.$$

$$v_{u,CD} = \frac{V_u}{A_c} - \frac{\gamma_v M_{sc}\, c_{CD}}{J_c} = \frac{(157)(1000)}{1125} - \frac{(0.40)(19)(12{,}000)(18.75)}{266{,}309}$$
$$= 133\text{ psi} \le v_{allowable} = 190\text{ psi} \quad O.K.$$

Therefore, the shear stress at the interior column C-2 due to combined direct shear and unbalanced moment is adequate.

Calculate the moment transferred by flexure at **slab-column connection C-2**:

Since $\gamma_f = 0.60$, the unbalanced moment transferred from the slab to the column through flexure is

$$M_f = \gamma_f M_{sc} = (0.6)(19\text{ ft-kip}) = 11.4\text{ ft-kip}$$

This is the moment transferred by flexure at the slab-column connection at column C-2, and the effective transfer width for transfer of this moment by flexure is

$= c_2 + 3h = 30" + (3)(9") = 57" = 4.75$ ft (see ACI Code, Section 8.4.2.3.3)

Therefore, $M_f = 11.4$ ft-kip/(4.75 ft) = 2.4 ft-kip/ft width

By inspection, this moment is small and does not require any additional reinforcement.

(l) Shear plus unbalanced moment transfer at the **typical edge column C-1**

The cladding load, $w_{cladding}$ = 25 psf(10 ft height) = 250 Ib/ft

Without the shear cap, $c_{1,column} = c_{2,column} = 18"$ (i.e., the size of the column), but because of the shear cap that is 6" wide (all around the edge column), the revised

$c_1 = 6" + 18" = 24" = 2.0$ ft

$c_2 = 6" + 18" + 6" = 30" = 2.5$ ft

$c_1/c_2 = 24/30 = 0.8$

$d = d_{average} = 7.5"$

$b_1 = c_1 + d/2 = 24" + 7.5"/2 = 27.75" = 2.31$ ft

$b_2 = c_2 + d = 30" + 7.5" = 37.5" = 3.13$ ft

$b_1/ b_2 = 27.75"/37.5" = 0.74$

b_o = perimeter of the critical section for punching shear at the edge column C-1 is

$= 2(c_1 + d/2) + (c_2 + d) = 2(24 + 7.5/2) + (30 + 7.5) = 93$ inches

$$\gamma_f = \frac{1}{1+\left(\frac{2}{3}\right)\sqrt{\frac{b_1}{b_2}}} = \frac{1}{1+\left(\frac{2}{3}\right)\sqrt{0.74}} = 0.64 \text{ (ACI Code, Eq. 8.4.2.2.2)}$$

$\gamma_v = 1 - 0.64 = 0.36$ (ACI Code, Eq. 8.4.4.2.2)

The factored shear at the **exterior edge column, C-1**, including the slab uniform load and the exterior cladding load (assuming each level of the cladding is supported by the slab) is given as

$$V_u = w_u \ell_2\left(\frac{\ell_1}{2}+\frac{c_{1,column}}{2}\right) - w_u\left(c_1+\frac{d}{2}\right)(c_2+d) + 1.2(w_{cladding})(\ell_2 - c_{2.column})$$

$$= (332\text{ psf})(22\text{ ft})\left(\frac{22\text{ ft}}{2}+\frac{1.5\text{ ft}}{2}\right) - (332\text{ psf})\left(2.5\text{ ft}+\frac{(7.5"/12)}{2}\right)(2.5\text{ ft}+(7.5"/12))$$

$$+1.2(250\text{ Ib/ft})(22\text{ ft} - 1.5\text{ft})$$

$$= 91038\text{ Ib} = 91\text{ kip}$$

When using the equivalent frame method, the factored moments, M_u, and the factored shear, V_u, are usually obtained from the frame analysis at the centerline or centroidal axis of the column in which case the unbalanced moment at the edge column at the centroid of the critical section (see Figure 6.50*a*) is

$$M_{sc} = M_u - V_u e = M_u - V_u\left(\frac{c_1+d}{2} - c_{AB}\right) \geq 0.3M_o$$

Where, M_o is the static longitudinal moment calculated in Part (e).

In contrast, for the direct design method (DDM) that is being used in this example, the approximate factored moments, M_u, and the factored shear, V_u, occur at the interior face of the edge column; therefore, the unbalanced moment in the edge column at the centroid of the critical section (see Figures 6.50*b* and 6.34*b*) is

$$M_{sc} = M_u + V_u e = M_u + V_u\left(c_{AB} - \frac{d}{2}\right) \geq 0.3M_o$$

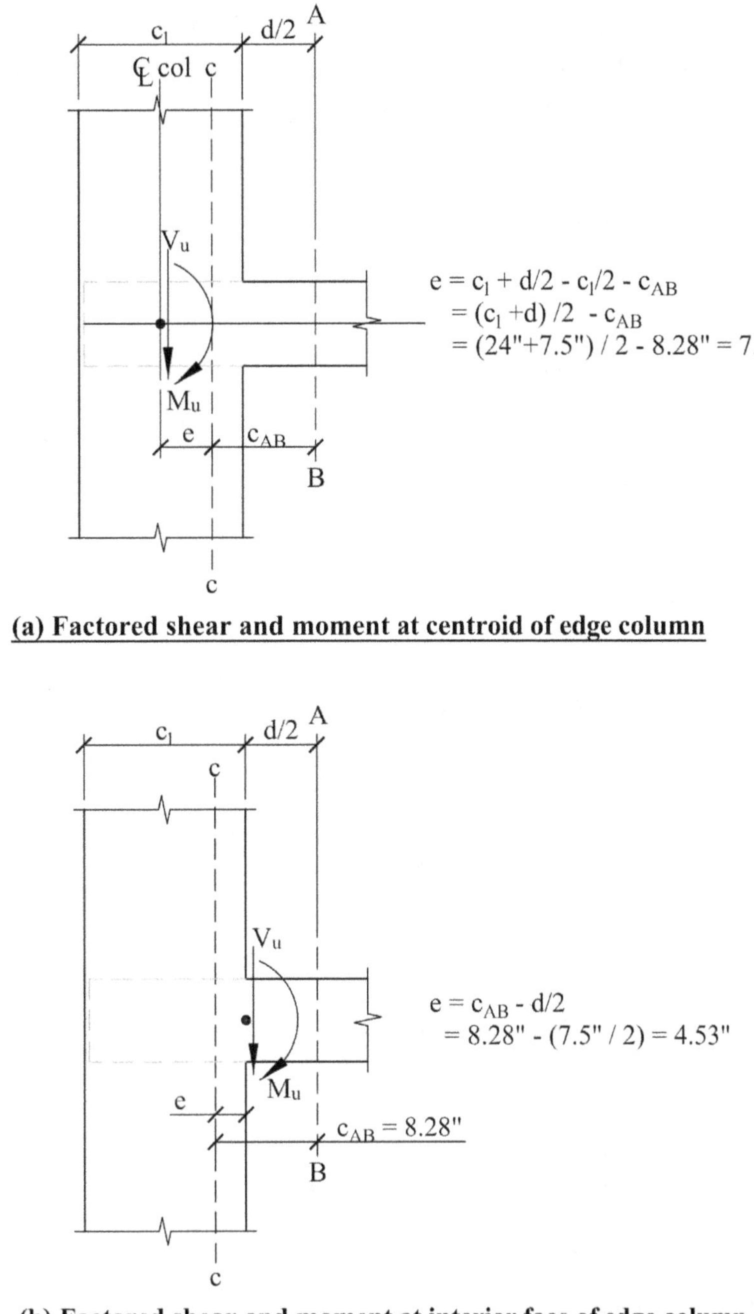

FIGURE 6.50 Elevation of slab-column connection at edge column showing moment and shear force for Example 6.3.

Unbalanced moment transferred at a typical exterior edge column C-1:

The negative moment, M_u, at the interior face of the edge column C-1 on Grid line 1 is obtained from Table 6.11 as 100 ft-kip (see Figure 6.50*b*)

The geometric properties of the critical section are calculated as follows:

For **Edge Columns with the direction of analysis perpendicular to the slab edge, that is, the moment acts about an axis parallel to the slab edge** (see Figure 6.34*b*):

A_c = area of the critical surface for two-way shear

$$A_c = \left(2c_1 + c_2 + 2d\right)d$$
$$= \left[2(24") + 30" + 2(7.5")\right](7.5") = 698 \text{ in.}^2$$

$$c_{AB} = \frac{\left(c_1 + \frac{d}{2}\right)^2 (d)}{d\left(2c_1 + c_2 + 2d\right)}$$

$$= \frac{\left(24" + \frac{7.5"}{2}\right)^2 (7.5")}{(7.5")\left(2(24") + 30" + 2(7.5")\right)} = 8.28 \text{ in.}$$

$$c_{CD} = c_1 + \frac{d}{2} - c_{AB} = 24" + \frac{7.5"}{2} - 8.28" = 19.47"$$

$$J_c = \frac{d\left(c_1 + \frac{d}{2}\right)^3}{6} + 2d\left(c_1 + \frac{d}{2}\right)\left(\frac{c_1 + \frac{d}{2}}{2} - c_{AB}\right)^2 + \frac{\left(c_1 + \frac{d}{2}\right)(d)^3}{6} + \left(c_2 + d\right)(d)\left(c_{AB}\right)^2$$

$$= \frac{(7.5")\left(24" + \frac{7.5"}{2}\right)^3}{6} + 2(7.5")\left(24" + \frac{7.5"}{2}\right)\left(\frac{24" + \frac{7.5"}{2}}{2} - 8.28"\right)^2$$
$$+ \frac{\left(24" + \frac{7.5"}{2}\right)(7.5")^3}{6} + \left(30" + 7.5"\right)(7.5")(8.28")^2$$
$$= 60{,}975 \text{ in.}^4$$

β = ratio of the long cross-sectional dimension to the short cross-sectional dimension of the column, concentrated load, or reaction = 18"/18" = 1.0

$$e = c_{AB} - \frac{d}{2} = 8.28" - \frac{7.5"}{2} = 4.53"$$

The factored static moment (previously calculated in part (e)), M_o = 384 ft-kip

The total unbalanced moment at edge column C-1 with the factored moment and shear acting at the interior face of the edge column, $M_{sc} = M_u + V_u e$ =100 ft-kip + (91 kip) (4.53"/12) = **134 ft-kip**

$\geq 0.30M_o = 0.30(384 \text{ ft-kip}) = 115 \text{ ft-kip}$

Therefore, the total unbalanced moment = 134 ft-kip

The allowable concrete shear stress in two-way shear is the smallest of the following equations from Section 6.4.10 – *Combined Shear and Unbalanced Moment Transfer in Slab-Column Connections*:

$$\nu_c = 4\phi\lambda_s\lambda\sqrt{f'_c} = 4(0.75)(1.0)(1.0)\sqrt{4000 \text{ psi}} = 190 \text{ psi}$$

$$\nu_c = \phi\left(2+\frac{4}{\beta}\right)\lambda_s\lambda\sqrt{f'} = (0.75)\left(2+\frac{4}{1.0}\right)(1.0)(1.0)\sqrt{4000 \text{ psi}} = 285 \text{ psi}$$

$$\nu_c = \phi\left(2+\frac{\alpha_s d}{b_o}\right)\lambda_s\lambda\sqrt{f'} = (0.75)\left(2+\frac{40(7.5)}{150}\right)(1.0)(1.0)\sqrt{4000 \text{ psi}} = 190 \text{ psi}$$

$\phi = 0.75$ for shear

Therefore, $\nu_{allowable} = 190$ psi

The maximum factored shear stresses under combined shear (V_u = 91 kip), and unbalanced moment (M_{sc} = 134 ft-kip) are:

$$\nu_{u,AB} = \frac{V_u}{A_c} + \frac{\gamma_v M_{sc} c_{AB}}{J_c} = \frac{(91)(1000)}{698} + \frac{(0.36)(134)(12{,}000)(8.28")}{60{,}975}$$
$$= \mathbf{209\ psi} > \nu_{allowable} = 190 \text{ psi} \quad \text{Not O.K.}$$

$$\nu_{u,CD} = \frac{V_u}{A_c} - \frac{\gamma_v M_{sc} c_{CD}}{J_c} = \frac{(91)(1000)}{698} - \frac{(0.36)(134)(12{,}000)(8.28")}{60{,}975}$$
$$= 52\,psi \leq \nu_{allowable} = 190 \text{ psi} \quad \text{O.K.}$$

There is a 10% overstress in shear at the edge column C-1 due to direct shear combined with shear from unbalanced moment. Several measures could be taken to rectify this:

1. Increase the depth and extent of the shear cap in order to increase the dimensions of the critical section (i.e., c_1 and c_2).
2. Increase the allowable shear stress by increasing the slab concrete strength, f'_c.
3. The column size could be increased to 20" square column.

Calculate the moment transferred by flexure at the slab-column connection at edge column C-1:

Since $\gamma_f = 0.64$, the unbalanced moment transferred from the slab to the column through flexure is

$M_f = \gamma_f M_{sc} = (0.64)(134 \text{ ft-kip}) = 86 \text{ ft-kip}$

This is the moment transferred by flexure at the slab-column connection at column C-1, and the effective transfer width for transfer of this moment by flexure (noting the shear cap width of 30 inches, which is now the revised value of c_2) is

$= c_2 + 3h = 30" + (3)(9") = 57" = 4.75$ ft (see Section 6.4.10 – *Combined Shear and Unbalanced Moment Transfer in Slab-Column Connections*)

Therefore, $M_f = 86$ ft-kip/(4.75 ft) = 18.1 ft-kip/ft width

Using the methods from Chapter 2 and the limit states design equation,

$M_u \leq \phi M_n = \phi_{bending} bd^2\bar{k}$

ϕ for bending = 0.9

b = 12" since the moment is calculated above in ft-kip per ft width

d = 7.5"

$M_u \leq \phi M_n = \phi\, bd^2\bar{k}$

Therefore, $\phi bd^2\bar{k} \geq M_u = (18.1 \text{ ft-kip/ft})(12)$

$$\bar{k} = \frac{18.1(12)}{(0.9)(12")(7.5")^2} = 0.358$$

From Table A10 in Appendix A, $\rho = 0.0063$

$A_{s,reqd} = \rho\, bd = 0.0063(12 \text{ in})(7.5") = 0.567 \text{ in}^2/\text{ft}$

Total area of steel required over the transfer width 4.75 ft = $(0.567 \text{ in}^2/\text{ft})(4.75 \text{ ft}) = 2.7 \text{ in}^2$

Provide additional 7#6T over the 4.75 ft width centered on the edge columns.

See Figure 6.39 for the layout of the two-way slab reinforcement.

EXAMPLE 6.4

1. If drop panels are to be used in the two-way slab in Example 6.3, determine the minimum size of the drop panel for a typical interior column that meets the requirements of the ACI Code.

2. Discuss the implications of a drop panel in an existing building not meeting the Code-specified length and depth requirements.

Solution:

1. The drop panel minimum size that satisfies the ACI Code is illustrated in Figure 6.51.

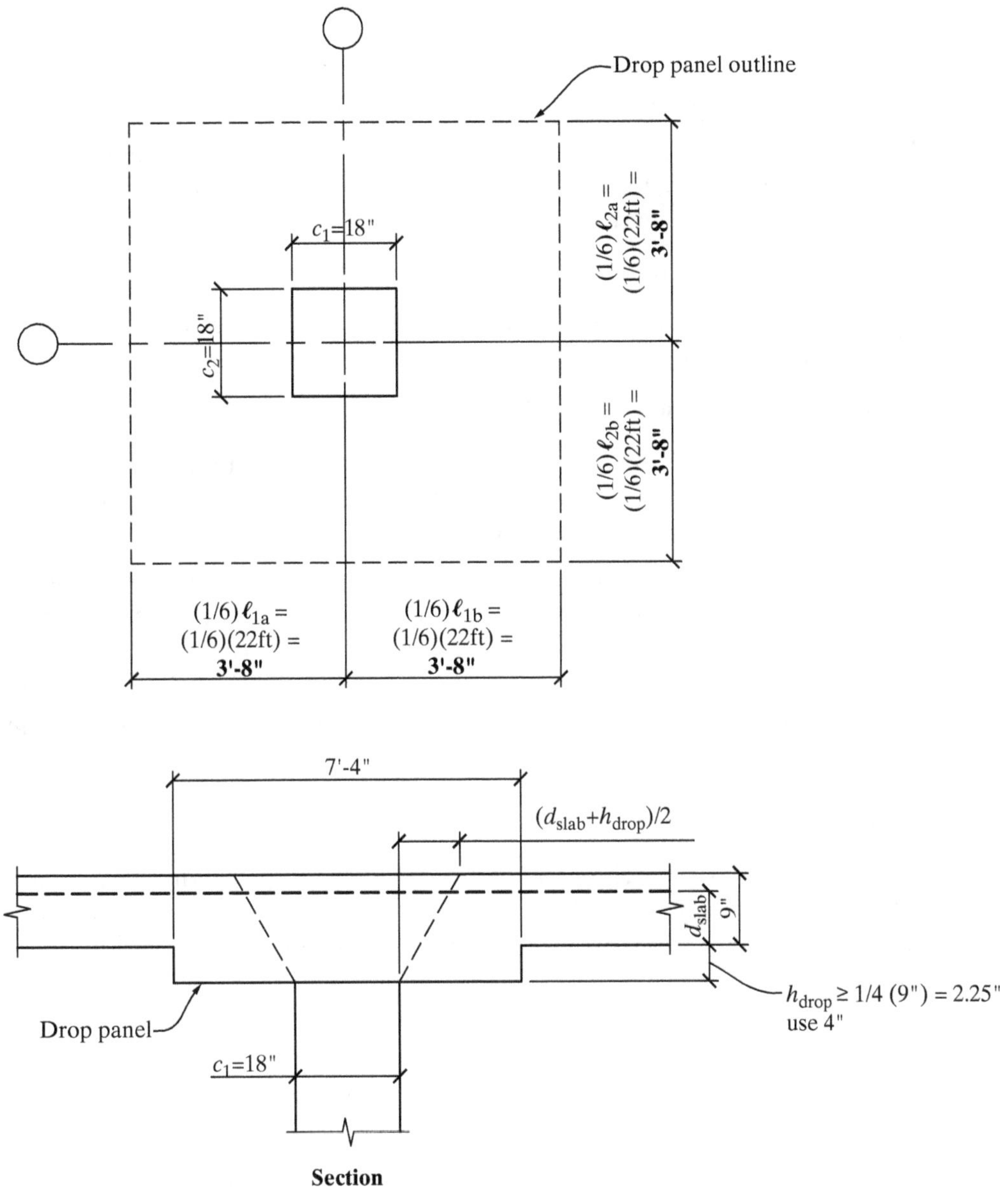

FIGURE 6.51 Drop panel dimensions for Example 6.4.

Note that in contrast to the shear cap that was used in Example 6.3 which did not meet the drop panel requirements and thus could only be used to increase the critical shear perimeter, the drop panel in this example on the other hand meets the length and depth requirements specified in the Code, therefore, the drop panel can be used to:

a. Increase the effective depth in the negative moment region and thus reduce the amount of negative reinforcement required, in addition to increasing the shear perimeter and helping to reduce the shear stress. With the 4" deep drop panel (see Figure 6.51), the average effective depth in the negative moment region, d_{-ve} = 9" + 4" drop panel – 1.5" = 11.5" (Compare to d = 7.5" in Example 6.3)

b. Due to the increased effective depth and the code specified length of the drop panel, the critical perimeter for shear around the column, b_o, is increased because of the drop panel (see Figure 6.51).

$b_o = 2\,(c_1 + c_2 + 2d) = 2(18 + 18 + 2(11.5)) = 118$ in.

c. The drop panel can be included in the slab deflection calculations, and the increased depth reduces the slab deflections.

2. If the code specified length and depth of the drop panel is not satisfied, then:

- The depth of the drop panel cannot be counted upon to increase the effective depth of the slab and thus reduce the negative reinforcement at the column. The effective depth will be that of the slab alone.
- The drop panel that does not meet the requirements of ACI Code, Section 8.2.4 will essentially act as a shear cap, provided the shear cap extension beyond the column face is at least equal to the shear cap depth. Shear caps cannot be included in the deflection calculations, nor can they be used to increase the effective depth of the slab in the negative moment regions. Shear caps will increase the shear perimeter by increasing the value of c_1 and c_2.

EXAMPLE 6.5

Assume the two-way slab in Example 6.3 has 12" x 20" deep spandrel beams along grid lines 1, 4, A, and D (see Figure 6.52). For the typical interior design strip centered on grid line C, determine the portion of the exterior longitudinal moments transferred to the column strip and the middle strip.

Solution:

The torsional constant for the 12" x 20" edge beam along line 1 is calculated as

$$C = \sum\left(1-0.63\frac{x}{y}\right)\frac{x^3y}{3} = \left(1-0.63\frac{12"}{20"}\right)\frac{(12")^3(20")}{3} + \left(1-0.63\frac{9"}{11"}\right)\frac{(9")^3(11")}{3}$$

$$= 8461 \text{ in.}^4$$

The longitudinal direction of the effective slab is perpendicular to grid line C and parallel to grid line 1. Therefore, the width of the effective slab = edge beam span length = (22 ft)(12) = 264 in.

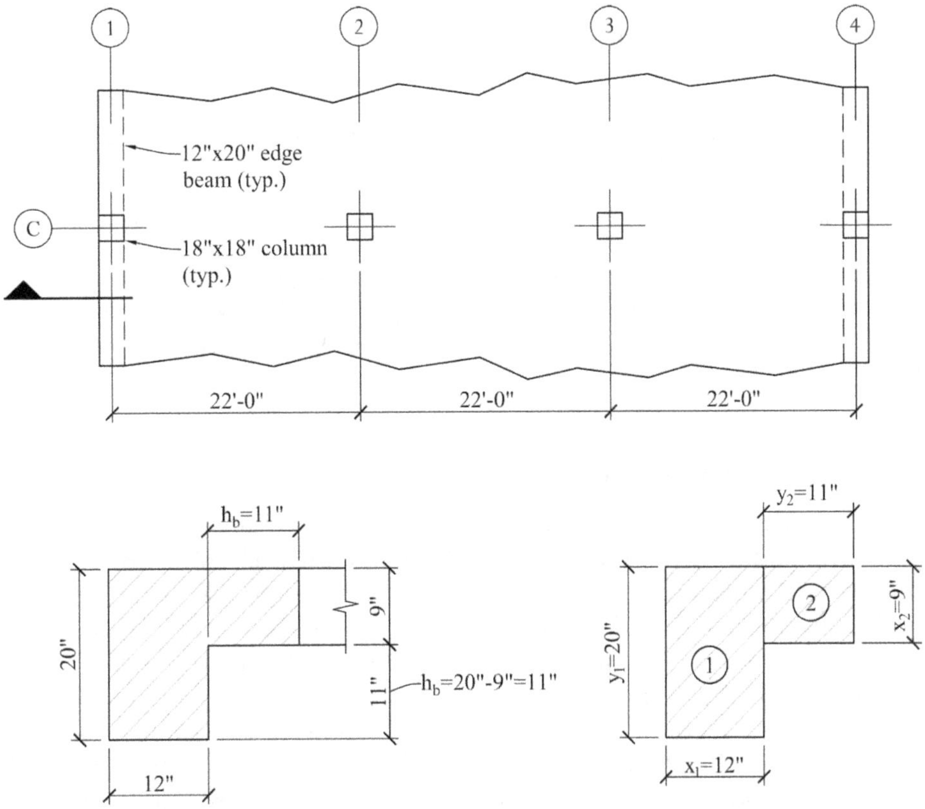

(b) edge beam (ACI Code, Section R8.4.1.8)

FIGURE 6.52 Effective edge beam section for Example 6.5.

The corresponding moment of inertia of the effective slab with a width of 264 inches (i.e., with slab width equal to the edge beam span length) is

$$I_s = (264")\frac{(9")^3}{12} = 16{,}038 \text{ in.}^4$$

The torsional stiffness parameter for the edge beam is the ratio of the torsional rigidity of the edge beam section to the flexural rigidity of the effective slab that has a width equal to the span of the edge beam,

$$\beta_t = \frac{E_{cb}C}{2E_{cs}I_s} = \frac{8461}{2(16{,}038)} = 0.26$$

Since there is no longitudinal beam along grid line C in the end span between grid lines 1 and 2,

$$\alpha_{f1} = 0$$

$$\textit{therefore, } \alpha_{f1}\frac{\ell_2}{\ell_1} = 0$$

Using Table 6.7, and entering the table with $\frac{\ell_2}{\ell_1} = 1.0$, $\alpha_{f1}\frac{\ell_2}{\ell_1} = 0$, and $\beta_t = 0.26$, and interpolating yields the portion of the exterior negative moment, M_u, for the design strip along grid line C that is assigned to the column strip along grid line C, which is -(0.1)(0.26) + 1.0 = **0.974**

Therefore only 2.6% of the exterior negative moment, M_u, is transferred to the middle strips.

NOTE *Where an edge beam has high torsional stiffness, a higher percentage of the exterior negative moment is transferred to the middle strip. According to Table 6.7, for a square slab panel, depending on the torsional stiffness of the edge beam, up to 25% of the exterior negative moment can be transferred to the middle strip.*

6.4.17 Equivalent Frame Method

The direct design method is not applicable to two-way slabs that do not meet the previously stated requirements necessary to use the DDM. The DDM is also not applicable to mat foundations which are analysed as inverted two-way slabs (ACI Code, Section 13.3.4.2). For such slabs where the DDM is not applicable, the equivalent frame method of analysis (ACI Code, Section 8.2.1) or computer software that uses the finite element analysis method can be used. In the equivalent frame method, the design strip, like the DDM, is centered on the column line with a width bounded by the centerlines of adjacent panels. The equivalent frame method differs from the direct design method in the analytical approach used to calculate the longitudinal distribution of the moment along the span of the slab. The two-way slab analysis using the equivalent frame method, as laid out in previous editions of the ACI Code, lends itself to the use of Hardy Cross's moment distribution method with its stiffness coefficients, distribution factors, carry-over factors, and fixed end moments. For gravity loads, the longitudinal moments in the slab can be determined from an analysis of a subframe along the design strip consisting of the slab and beams (if any) at the floor level under consideration, and the columns above and below the slab, with the columns assumed to be fixed at their far ends.

The equivalent frame consists of the following component parts (see Figure 6.53):

- The horizontal slab strip including any longitudinal beams
- The columns or walls extend a story height above and below the slab (for roof slabs the column or wall only extends below the slab). The columns are usually considered fixed at their ends. Concrete walls whose longitudinal direction is perpendicular to the longitudinal direction of the slab do not provide any restraint to the slab; instead, they act like knife-edge supports. Concrete walls whose longitudinal direction is parallel to the longitudinal direction of the slab provide moment restraint to the slab like columns, since the walls will be bending about their strong axis.
- Transverse elements at the columns help to transfer the moment from the slab to the column through the torsional resistance of the transverse beams and the slab.

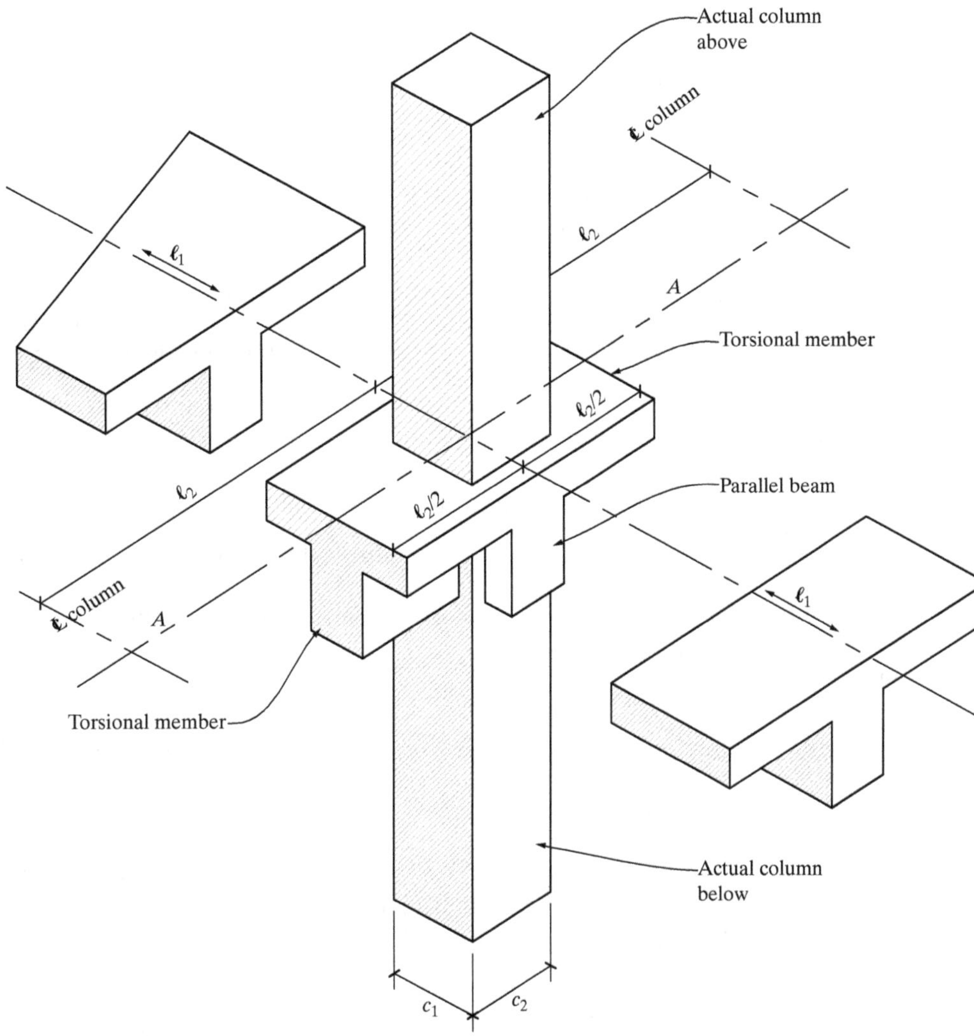

FIGURE 6.53 Components of the equivalent frame (Equivalent columns and torsional members). (Adapted from ACI 318-14 Figure R8.11.4)

The procedure for the equivalent frame method of analysis for two-way slabs is as follows:

1. Determine the minimum slab thickness, h, to avoid deflection calculations (see Section 6.4.3 – *Minimum Thickness of Two-Way Slabs*).

2. Check one-way shear and two-way (punching) shear and ensure that the slab has adequate thickness to resist direct shear and make allowance in the slab thickness to resist additional shear stresses due to the transfer of unbalanced moments from the slab to the column; the unbalanced moment will be calculated later. To allow for some reserve shear capacity to resist the additional shear stresses caused by unbalanced moments. Ref. [2] recommends increasing the required depth for perimeter shear by 10% at interior columns, 40% at edge columns and 70% at corner columns.

3. Divide the slab into equivalent frames (design strips) in both orthogonal directions, and for each design strip and at each slab-column connection, the stiffness parameters are calculated.

4. Different live load patterns are considered in addition to a uniform dead load over the entire length of the design strip, and the fixed end moments are calculated that will be used in the moment distribution analyses.

5. Calculate the slab stiffness, $K_s = \frac{4E_cI_s}{\ell_1}$, where E_c is the modulus of elasticity of the slab concrete, ℓ_1 is the span of the slab, and the slab moment of inertia, $I_s = \frac{(\ell_2)h^3}{12}$. Note that the gross sectional properties are used to calculate the moment of inertia for all members of the equivalent frame.

6. Calculate the stiffness of the column above and below the floor under consideration, $K_{column} = \frac{4E_{col}I_{column}}{\ell_{column}}$, where ℓ_{column} is the floor-to-floor height of the column, E_{col} is the modulus of elasticity of the column concrete, and the moment of inertia of the column, $I_{column} = \frac{(c_2)(c_1)^3}{12}$.

7. Calculate the torsional constant, C, for the transverse torsional member, where $C = \sum\left(1-0.63\frac{x}{y}\right)\frac{x^3y}{3}$.

   a. If there is no transverse beam, the transverse torsional member would consist of the slab with a width equal to the dimension of the column, c_1, parallel to the span of the slab, and a depth equal to the slab depth, h.

   b. Where there is a transverse beam framing into the column, the equivalent beam dimensions are determined in accordance with ACI Code, Section 8.4.1.8, like the direct design method (see Figure 6.24).

   c. For a typical interior column, there will be two torsional members, one on the near side and the second on the far side of the column.

8. Calculate the torsional stiffness, K_t, of the torsional member where

$$K_t = \sum \frac{9E_cC}{\ell_2\left(1-\frac{c_2}{\ell_2}\right)^3}\left(\frac{I_{sb}}{I_s}\right).$$

where

I_{sb} = moment of inertia of the slab and the longitudinal or parallel beam over the width of the equivalent frame (see Figure 6.53).

I_s = moment of inertia of the slab alone over the width of the equivalent frame (see Figure 6.53) $= \frac{(\ell_2)h^3}{12}$.

Where there is no longitudinal or parallel beam that frames into a column (i.e., along the direction of analysis, which means the beam is parallel to the ℓ_1 direction), then $I_{sb} = I_s$

The approximate expression above for the stiffness of the torsional member was derived based on the analyses of various slab configurations by W.G. Corley and James Jirsa [9]. Note that for typical interior columns, there will be two torsional members, one on the near side and one on the far side of the column. There are two columns at the joint, one above the level under consideration and one below.

9. Calculate the stiffness of the equivalent column, K_{ec}. The equivalent column consists of the columns above and below the level under consideration and the torsional members on the near side and far side of the column. $\frac{1}{K_{ec}} = \frac{1}{\sum K_{column}} + \frac{1}{\sum K_t}$ and thus, $K_{ec} = \frac{\sum K_{column}}{1 + \sum \frac{K_{column}}{K_t}}$

Note:

a. For a roof slab, there will be one column below and no column above; for a typical floor slab, there will be two columns, one below the floor and a second column above the floor.

b. For an exterior design strip (e.g., line D in Figure 6.46), there will be just one torsional member (on the far side), whereas for a typical interior design strip (e.g., line C in Figure 6.46), there will be two torsional members, one on the near face of the column and the second on the far face of the column.

10. Knowing the modulus of elasticity $E_c = 57{,}000\sqrt{f'_c}$, calculate the equivalent column stiffness, K_{ec}, and the slab stiffness, K_s, in units of kip-ft/unit rotation. A typical exterior joint will consist of the equivalent column, K_{ec} and a slab stiffness, K_s. A typical interior column will consist of the equivalent column stiffness and two slab stiffnesses (i.e., the slab to the left and to the right of the column). These stiffnesses are used to determine the distribution factors, DF, at each joint that is used in the moment distribution analysis.

11. These stiffness factors can then be used in the moment distribution process after all the fixed end moments are calculated. A carry-over factor of 0.5 is assumed, ignoring the non-prismatic column section, and the final-end moments and the midspan positive moments in the design strip are obtained from the moment distribution analysis.

12. The sum of the absolute value of the average negative moments at the ends of a span and the midspan positive moment should be no greater than the maximum static moment within the span, $M_o = \frac{w_u \ell_2 \ell_{n1}^2}{8}$.

13. The longitudinal moments obtained from the equivalent frame analyses can be distributed transversely to the column strips, longitudinal beams, and the middle strips using the same coefficients that are used for the direct design method (see Tables 6.6 through 6.9) provided the following condition is satisfied:

$$0.2 \le \frac{\alpha_{f1}\ell_2^2}{\alpha_{f2}\ell_1^2} \le 5.0 \ ,$$

where $\alpha_{f1} = \frac{I_{b1}}{I_{s1}}$ and $\alpha_{f2} = \frac{I_{b2}}{I_{s2}}$ (see Section 6.4.6 – *Relative Stiffness of Beam and Slab*).

The analysis of two-way slab by the equivalent frame method is more complex and time consuming than the direct design method, therefore, in design practice, two-way slab analyses using the equivalent frame method are carried out using computer analysis and design software.

REFERENCES

[1] R. Park and W.L. Gamble "Reinforced Concrete Slabs," John Wiley & Sons, New York, 1980

[2] ACI Committee 340, "Slab Design in Accordance with ACI 318-77," ACI Publication SP-17(73)(S), American Concrete Institute, Detroit, 1978

[3] David A. Fanella "Concrete Floor Systems: Guide to Estimating and Economizing," Portland Cement Association, Skokie, Illinois, 2000

[4] K. Dirk Bondy and Bryan Allred "Post-Tensioned Concrete – Principles and Practice," 3rd Edition.

[5] J.G.M. Wood "Quantitative Study of the Partial Collapse of Pipers Row Car Park, Wolverhampton on 20th March 1997," Structural Studies and Design Ltd., Surrey, U.K.

[6] Teng et al, "Punching shear strength of slabs and influence of low reinforcement ratios," ACI Structural Journal, January 2018, pp. 139-150.

[7] Ramez Gayed and Amin Ghali, "Design of concrete flat slabs to resist flexure-induced punching failure," STRUCTURE Magazine, July 2019

[8] Amin Ghali and Ramez Gayed, "Universal design of punching resistance of concrete slabs," ACI Structural Journal, January 2019.

[9] W.G. Corley and J.O. Jirsa, "Equivalent frame analysis for slab design," *ACI Journal Proceedings*, V. 67, No. 11, November 1970, pp. 875-884.

EXERCISES

For the following exercises, all concrete is normal weight, and all steel is grade 60 (f_y = 60,000 psi).

6.1 For the one-way slab shown, determine all moments and shears for service loads of 100 psf dead load (including the slab weight) and 300 psf live load.

6.2 For the continuous beam shown, determine all moments and shears for service loads of 2 kip/ft dead load (includes weight of beam and slab) and 3 kip/ft live load.

6.3 Using the ACI coefficients, design a continuous reinforced concrete beam that will span four supports. The end spans are 24-ft long; the center span is 28-ft long (spans are measured center to center of supports). The exterior support (a spandrel beam) and the interior supporting girders have widths b of 18 in. The service loads are 0.90 kip/ft dead load (not including the weight of the beam) and 1.10 kip/ft live load. Use $f_c' = 3000$ psi. only tension reinforcing. Design for the moment only. Use the recommended bar cutoffs shown in Figure 6.3.

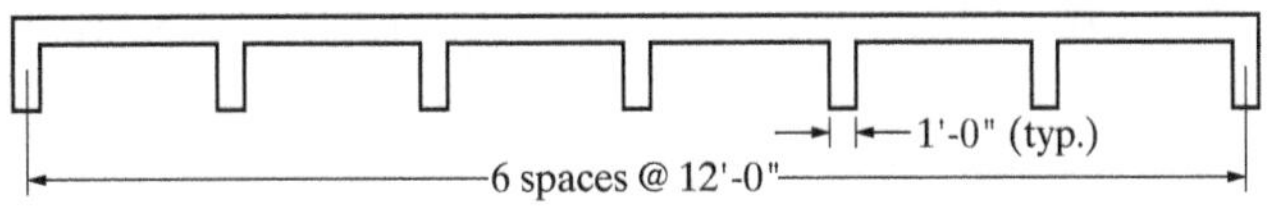

EXERCISE 6.1

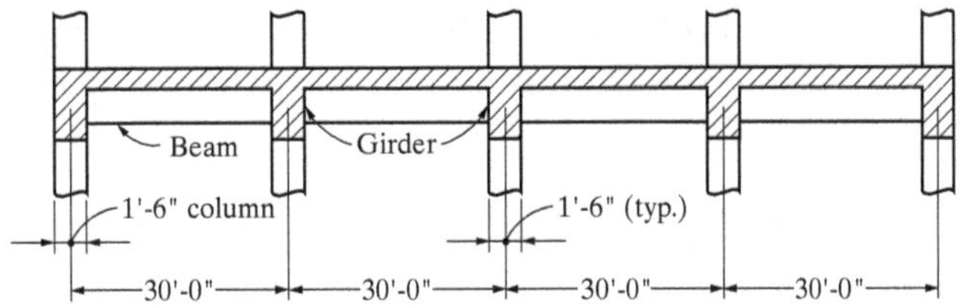

EXERCISE 6.2

6.4 A floor system consists of beams, girders, and a slab; a partial floor plan is shown. Service loads are to be 45 psf dead load (does *not* include the weight of the floor system) and 160 psf live load. Use $f_c' = 4000$ psi.

a. Design the continuous one-way slab.

b. Design the continuous beam along column line 2.

Be sure to include complete design sketches.

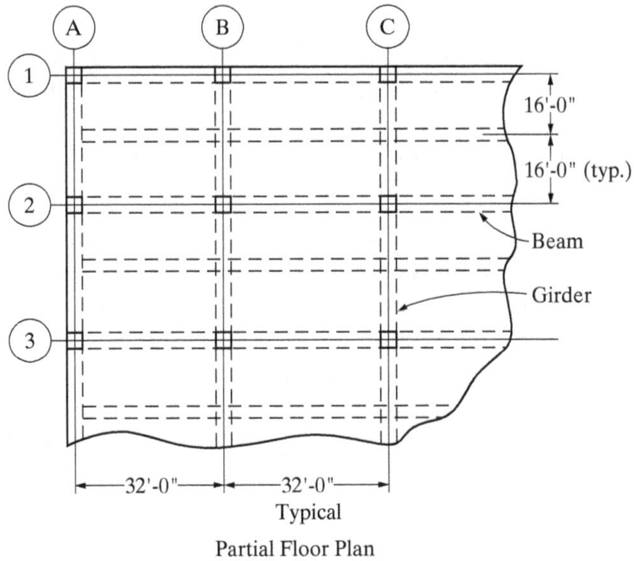

EXERCISE 6.4

6.5 Assume an 8" one-way slab continuous over four total supports. The Clear span between supports is ℓ_n = 16'-0". Use f_c' = 3000 psi and f_y = 60,000 psi. Assume d = 6" to the main reinforcing.

a. Determine the moment and shear values using the ACI coefficients. Draw each diagram assuming knife-edge support (i.e., ignore the width of the supporting beams).

b. Determine the required flexural reinforcing (bar size and spacing) in the slab at the first interior support as shown below.

6.6 Assume a 4" one-way slab continuous over three or more supports with a clear span between supports of ℓ_n = 7 ft. The total factored uniform load supported by the slab is 520 psf. Assume the main reinforcing is centered in the slab section and use f_c' = 3000 psi.

a. Determine the required flexural reinforcing in the slab at the first interior support.

b. Sketch your design showing the main reinforcing and the slab span.

6.7 The floor plan for an office building calls for a two-way flat slab supported on 18-in. square columns on a 30 ft x 30 ft column grid. Design the two-way floor slab: select the slab thickness, the dimensions of the drop panel and column capital, if any, and the slab reinforcement. Lay out the slab reinforcement in the column and middle strips in both orthogonal directions. Assume f_c' is 4000 psi and f_y is 60,000 psi.

6.8 A 10-in. thick two-way flat plate is supported on 20-in. square columns on 22 ft x 22 ft column grid, and the edge of the slab coincides with the exterior face of the columns. Assume f_c' is 4000 psi and f_y is 60,000 psi. Determine the punching shear capacity, ϕV_c, and the factored direct shear, V_u, for a typical interior column, a typical edge column, and a typical corner column.

6.9 The floor plan for an office building calls for a two-way flat plate supported on 24-in. square columns on a 24 ft x 24 ft column grid. Design the two-way floor slab: select the slab thickness and the slab reinforcement. Lay out the slab reinforcement in the column and middle strips in both orthogonal directions. Assume f_c' is 4000 psi and f_y is 60,000 psi.

6.10 An existing 12 in. thick flat slab in an industrial building is supported on 48 in (E-W). x 24 in. (N-S) columns on a 33 ft (E-W) x 27 ft (N-S) grid with 4" deep drop panels. The plan dimensions of the drop panels in the two orthogonal directions satisfy the Code-defined minimum limits. The superimposed dead load on the slab including concrete topping and mechanical and electrical equipment is 60 psf, and the live load is unknown and is to be determined. The concrete strength, f_c', is 3000 psi and the steel yield strength is 40,000 psi. The existing slab reinforcement in the exterior bay are given as follows:

E-W Direction (i.e., center to center span = 33 ft)

Width of column strip = 13.5 ft

Width of two-half middle strips = 13.5 ft

Exterior discontinuous end of slab: Column strip: 10#7; Middle strip: 10#6
Midspan of end span of slab: Column strip: 16#8; Middle strip: 14#8
Interior continuous end of slab: Column strip: 11#8; Middle strip: 14#7

N-S Direction (i.e., center to center span = 27 ft)
Width of column strip = 16.5 ft
Width of two-half middle strips = 16.5 ft
Exterior discontinuous end of slab: Column strip: 12#7; Middle strip: 12#6
Midspan of end span of slab: Column strip: 16#8; Middle strip: 14#8
Interior continuous end of slab: Column strip: 12#8; Middle strip: 14#7

1. Calculate the total unfactored dead load on the slab in psf
2. Does the depth of the drop panels satisfy the ACI Code requirements?
3. Considering only direct shear at a typical interior column, determine the live load capacity of the slab in psf.
4. For the **E-W direction**, determine the following:
   a. The moment capacity, ϕM_n, in the *column strip* at the *exterior discontinuous end* of the slab.
   b. The moment capacity, ϕM_n, in the *middle strip* at the *exterior discontinuous end* of the slab.
   c. The total moment capacity, ϕM_n, at the *exterior discontinuous end* of the slab.
   d. The moment capacity, ϕM_n, in the *column strip* at the *midspan of the end span* of the slab.
   e. The moment capacity, ϕM_n, in the *middle strip* at the *midspan of the end span* of the slab.
   f. The total moment capacity, ϕM_n, at the *midspan of the end span* of the slab.
   g. The moment capacity, ϕM_n, in the *column strip* at the *first interior continuous end* of the slab.
   h. The moment capacity, ϕM_n, in the *middle strip* at the *first interior continuous end* of the slab.
   i. The total moment capacity, ϕM_n, at the *first interior continuous end* of the slab.
   j. Using the results from above and the following relationship between the static moment, M_o, and the moments (moment capacities) at the critical sections of the slab: $M_{u,o} = \frac{w_u \ell_n^2}{8} \leq \phi M_{n,o} = \frac{1}{2}\left(\phi M_{n,-ve(\textit{left end support})} + \phi M_{n,-ve(\textit{right end support})}\right),$

$+ \phi M_{n,+ve\ (midspan)}$ determine the unfactored live load capacity of the slab in psf in the E-W direction.

5. For the **N-S direction**, determine the following:

   a. The moment capacity, ϕM_n, in the *column strip* at the *exterior discontinuous end* of the slab.

   b. The moment capacity, ϕM_n, in the *middle strip* at the *exterior discontinuous end* of the slab.

   c. The total moment capacity, ϕM_n, at the *exterior discontinuous end* of the slab.

   d. The moment capacity, ϕM_n, in the *column strip* at the *midspan of the end span* of the slab.

   e. The moment capacity, ϕM_n, in the *middle strip* at the *midspan of the end span* of the slab.

   f. The total moment capacity, ϕM_n, at the *midspan of the end span* of the slab.

   g. The moment capacity, ϕM_n, in the *column strip* at the *first interior continuous end* of the slab.

   h. The moment capacity, ϕM_n, in the *middle strip* at the *first interior continuous end* of the slab.

   i. The total moment capacity, ϕM_n, at the *first interior continuous end* of the slab.

   j. Using the results from above and the following static equilibrium relationship between the static moment, M_o, and the moments (or moment capacities) at the critical sections of the slab: $M_{u,o} = \frac{w_u \ell_n^2}{8} \leq \phi M_{n,o} = \frac{1}{2}(\phi M_{n,-ve(left\ end\ support)} + \phi M_{n,-ve(right\ end\ support)}) + \phi M_{n,+ve\ (midspan)}$, determine the unfactored live load capacity of the slab in psf in the N-S direction.

6. Based on your analyses for both the E-W and N-S directions, what is your recommended live load capacity (in psf) for this two-way slab?

6.11 Calculate the maximum two-way shear stress (in psi) in a 10 in. thick flat plate at a 40 in. x 20 in. corner column and compare to the design concrete shear strength. The 40 in. dimension of the column is parallel to the y-axis while the 20 in. dimension is parallel to the x-axis. Assume the following parameters:

- The edge distance of the slab is 2 in. all around from the exteriors edges of the column.
- $f_c' = 6000$ psi
- The factored applied shear force are $V_u = 150$ kip.

- The factored moment about the x-axis, M_{ux} = 250 ft.-kip.
- The factored moment about the y-axis, M_{uy} = 180 ft.-kip.
- Clear cover to the No.4 rebar in the slab = 3/4 in.

If the two-way shear stress is not adequate, list at least three ways to mitigate the structural inadequacy.

CHAPTER 7

SERVICEABILITY

7.1 INTRODUCTION

The ACI Code requires that bending members have adequate structural strength to support the anticipated factored design loads (ACI Code Sections 4.4.4 and 4.6) and that they have adequate performance at service load levels (ACI Code Section 4.7). Adequate performance, or *serviceability,* relates to deflections and cracking in reinforced concrete beams and slabs. It is important to realize that serviceability is to be assured at *service load levels*, not at ultimate strength level. At service loads, deflections should be held to specified limits because of many considerations, among which are aesthetics, effects on nonstructural elements such as windows, exterior cladding, and partitions, undesirable vibrations, and proper functioning of roof drainage systems. Any cracking should be limited to hairline cracks for reasons of appearance and to ensure protection of reinforcement against corrosion.

7.2 DEFLECTIONS

The deflection of concrete flexural members is affected by the following factors: the magnitude and distribution of the load on the member; the span of the member and support conditions; the percentage of reinforcement; member section properties; properties of the concrete material, and the degree and extent of flexural cracking in the member. Guidelines for the control of deflections are found in the ACI Code, Section 24.2. In addition, Table 24.2.2 of the Code indicates the maximum permissible deflections. Note that the permissible live load deflections for spandrel or edge beams supporting exterior cladding may need to be more stringent than those given in ACI Code Table 24.2.2, depending on the span of the beam and the performance requirements of the exterior cladding. A specific live load deflection limit of, say 3/8-inch for spandrel beams, regardless of the span, is recommended to ensure adequate performance of exterior cladding [4]. For the purpose of following the Code guidelines, either of two methods may be used: (1) using the minimum thickness (or minimum depth of member) criteria as established in Tables 7. 3.1.1 and 9.3.1.1 of the Code or Table 6.1 of this text, which will result in sections that are sufficiently deep and stiff enough so that short-term deflections will not be excessive (note that for the minimum depths and thicknesses given in Table 6.1, the slab or beams/girders do not support partitions or deflection sensitive nonstructural elements and the effects of creep and shrinkage and transient construction loads are not included); and (2) calculating expected deflections using standard deflection formulas in combination with the Code provisions for the moment of inertia and including the effects of cracking, creep, and shrinkage on long-term deflections. In both cases, it is assumed that the concrete flexural member will be properly supported temporarily or shored, thus allowing it to develop its specified strength before stripping the formwork.

Minimum thickness (depth) guidelines are simple and direct and should be used whenever possible. Note that the tabulated minimum thicknesses apply to non-prestressed, one-way members that do not support and are not attached to partitions or other construction likely to be damaged by large deflections. For members not within these guidelines, deflections *must* be calculated.

For the second method, in which deflections are calculated, the ACI Code stipulates that the members should have their deflections checked at *service load levels*. Therefore, the *properties* at service load levels must be used. Under service loads, concrete flexural members still generally exhibit elastic-type behavior (see Figure 1.1) but will have been subjected to cracking in tension zones at any point where the applied moment is large enough to produce tensile stress in excess of concrete tensile strength. The cross-section for moment-of-inertia determination then has the shape shown in Figure 7.1.

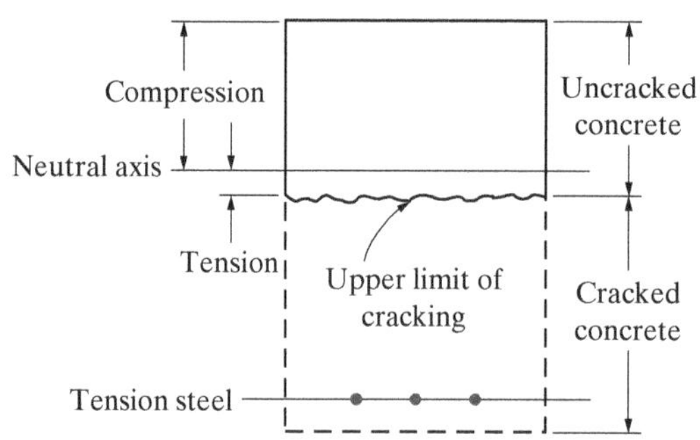

FIGURE 7.1 Typical cracked cross section.

The moment of inertia of the cracked section of Figure 7.1 is designated as I_{cr}. It is determined based on the assumption that the concrete is cracked to the neutral axis. In other words, the concrete is assumed to have no tensile strength, and the small tension zone below the neutral axis and above the upper limit of cracking is neglected.

The moment of inertia of the cracked section described represents one end of a range of values that may be used for deflection calculations. At the other end of the range, as a result of a small bending moment and a maximum flexural stress less than the modulus of rupture, the full uncracked section may be considered in determining the moment of inertia to resist deflection. This is termed the *moment of inertia of the gross cross-section* and is designated as I_g.

In reality, both I_{cr} and I_g occur in a bending member in which the maximum moment is in excess of the cracking moment. The I_{cr} occurs at or near the cracks, whereas the I_g occurs between the cracks. Research has indicated that due to the presence of the two extreme conditions, a more realistic value of a moment of inertia lies somewhere between these values. The ACI Code recommends that deflections be calculated using an *effective moment of inertia*, I_e, where

$$I_g > I_e > I_{cr}$$

Once the effective moment of inertia is determined, the member deflection may be calculated by using standard deflection expressions. The effective moment of inertia will depend on the values of the moment of inertia of the gross section and the moment of inertia of the cracked section. The gross (or uncracked) moment of inertia for a rectangular shape may be easily calculated from

$$I_g = \frac{bh^3}{12}$$

This expression neglects the presence of any reinforcing steel.

7.3 CALCULATION OF I_{CR}

The moment of inertia of the cracked cross-section can also be calculated in the normal way once the problem of the differing materials (steel and concrete) is overcome. To accomplish this, the steel area will be replaced by an equivalent area of concrete A_{eq}. This is a fictitious concrete that *can* resist tension. The determination of the magnitude of A_{eq} is based on the theory (from strength of materials) that when two differing elastic materials are subjected to equal strains, the stresses in the materials will be in proportion to their moduli of elasticity.

In Figure 7.2, ε_1 is the compressive strain in the concrete at the top of the beam, and ε_2 is a tensile strain at the level of the steel. Using the notation

f_s = tensile steel stress

$f_{c(\text{tens})}$ = theoretical tensile concrete stress at the level of the steel

E_s = steel modulus of elasticity (29,000,000 psi)

E_c = concrete modulus of elasticity

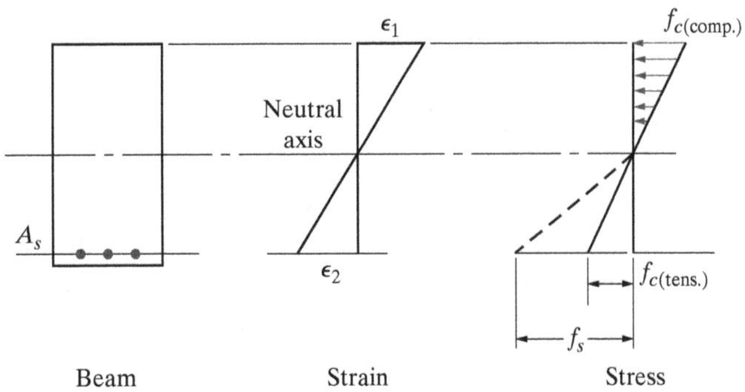

FIGURE 7.2 Beam bending: elastic theory.

The following relationships can be established:

$$\epsilon_2 = \frac{f_s}{E_s} \quad \text{and} \quad \epsilon_2 = \frac{f_{c(\text{tens})}}{E_c}$$

Equating these two expressions and solving for f_s yields

$$\frac{f_s}{E_s} = \frac{f_{c(\text{tens})}}{E_c}$$

$$f_s = \frac{E_s}{E_c}\left[f_{c(\text{tens})}\right]$$

The ratio E_s/E_c is normally called the *modular ratio* and is denoted n. Therefore,

$$f_s = nf_{c(\text{tens})}$$

Values of n may be taken as the nearest whole number (but not less than 6). Values of n for normal-weight concrete are tabulated in Table A.6.

As we are replacing the steel (theoretically) with an equivalent concrete area, the equivalent concrete area A_{eq} must provide the same tensile resistance as that provided by the steel.

Therefore,

$$A_{eq}f_{c(\text{tens})} = f_sA_s$$

Substituting, we obtain

$$A_{eq}f_{c(\text{tens})} = nf_{c(\text{tens})}A_s$$

from which

$$A_{eq} = nA_s$$

This defines the equivalent area of concrete with which we are replacing the steel. Another way of visualizing this is to consider the steel to be transformed into an equivalent concrete area of nA_s. The resulting *transformed concrete cross section* is composed of a single (although hypothetical) material and may be dealt with in the normal fashion for neutral-axis and moment-of-inertia determinations. Figure 7.3 depicts the transformed section. Because the steel is normally assumed to be concentrated at its centroid, which is a distance d from the compression face, the replacing equivalent concrete area must also be assumed to act at the same location. Therefore, the representation of this area is shown as a thin rectangle extending out past the beam sides. Recalling that the cross-section is assumed to be cracked between the tension face of the beam and the neutral axis, the resulting effective area is as shown in Figure 7.4 and the neutral axis will be located a distance $\bar{y}$ down from a reference axis at the top of the section.

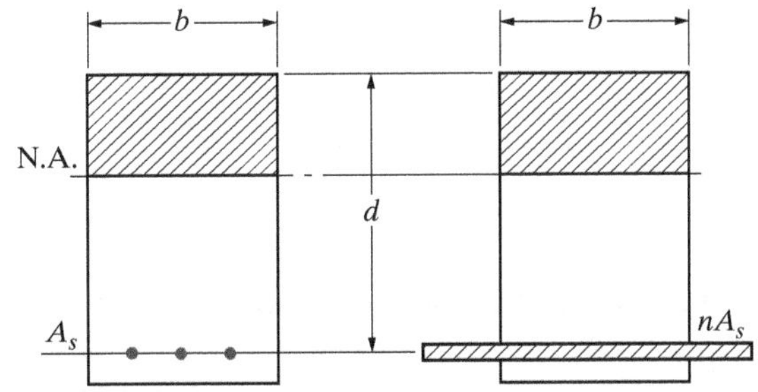

FIGURE 7.3 Method of transformed section.

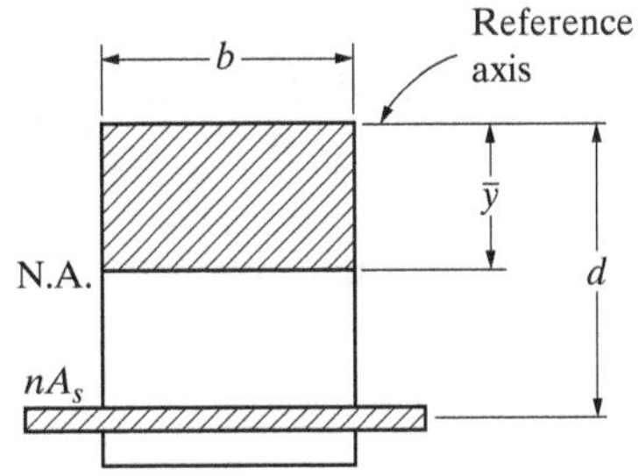

FIGURE 7.4 Effective area.

The neutral axis may be determined by taking a summation of moments of the effective areas about the reference axis:

$$\bar{y} = \frac{\Sigma(Ay)}{\Sigma A} = \frac{(b\bar{y})\frac{\bar{y}}{2} + nA_s d}{b\bar{y} + nA_s}$$

$$b(\bar{y}^2) + nA_s\bar{y} = \frac{b(\bar{y}^2)}{2} + nA_s d$$

$$\frac{b(\bar{y}^2)}{2} + nA_s\bar{y} - nA_s d = 0$$

This is a quadratic equation of the form $ax^2 + bx + c = 0$, and it may be solved either by completion of the square, as was done in Example 3.7, or by using the formula for roots of a quadratic equation, which will result in the following useful expression:

$$\bar{y} = \frac{nA_s\left[\sqrt{1 + 2\frac{bd}{nA_s}} - 1\right]}{b}$$

Once the neutral axis is located, the moment of inertia (I_{cr}) may be found using the familiar transfer formula from mechanics of materials. The equations for the gross and cracked section properties for rectangular and flanged sections are summarized in Section 7.11.

EXAMPLE 7.1

Find the cracked moment of inertia for the cross-section shown in Figure 7.5a. Use A_s = 2.00 in.^{2}, n = 9 (Table A.6), and $f_c' = 3000$ psi.

Solution:

With reference to the transformed section, shown in Figure 7.5b, the neutral axis is located as follows:

$$\bar{y} = \frac{nA_s\left[\sqrt{1+2\frac{bd}{nA_s}}-1\right]}{b}$$

$$= \frac{9(2.0)\sqrt{12\frac{(8)(17)}{9(2.0)}}1}{8}$$

$$= 6.78 \text{ in.}$$

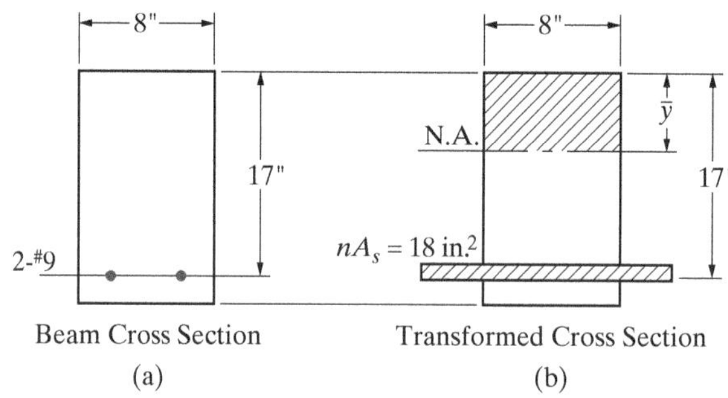

FIGURE 7.5 Sketches for Example 7.1.

The moment of inertia of the cracked section may now be found (all units are inches):

$$I_{cr} = \frac{b\bar{y}^3}{3} + nA_s\left(d-\bar{y}\right)^2$$

$$= \frac{8(6.78)^3}{3} + 18(17-6.78)^2$$

$$= 831.1 + 1880$$

$$= 2711 \text{ in.}^4$$

If the beam cross section contains compression steel, this steel may also be transformed, and the neutral-axis location and cracked moment-of-inertia calculations can be carried out as before. Because the compression steel displaces concrete that is in compression, it should theoretically be transformed using $(n-1)A_s'$ rather than nA_s'. For deflection calculations, which are only approximate, however, the use of nA_s' will not detract from the accuracy expected. The resulting transformed section will appear as shown in Figure 7.6.

The neutral axis location $\overline{y}$ may be determined from the solution of

$$\frac{b}{2}\overline{y}^2 + nA_s'\overline{y} - nA_s'd' - nA_sd + nA_s\overline{y} = 0$$

and the moment of inertia with respect to the neutral axis from

$$I_{cr} = \frac{b\overline{y}^3}{3} + nA_s\left(d-\overline{y}\right)^2 + nA_s'\left(\overline{y}-d'\right)^2$$

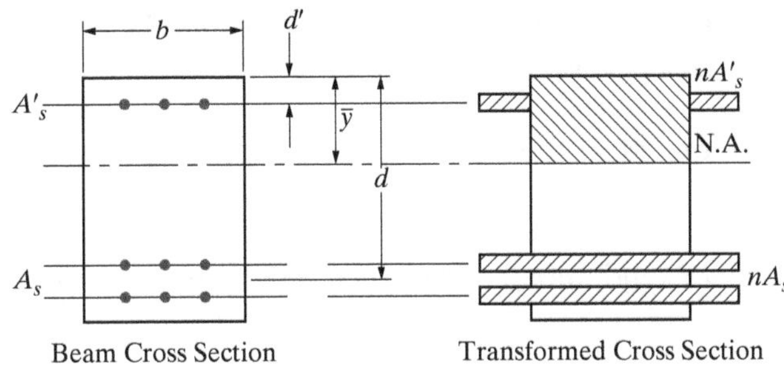

FIGURE 7.6 Doubly reinforced beam.

7.4 IMMEDIATE DEFLECTION

Immediate deflection is the deflection that occurs as soon as the load is applied to the member. For all practical purposes, the member is elastic. The ACI Code, Section 24.2.3.4, states that this deflection may be calculated using a concrete modulus of elasticity E_c as specified in Section 19.2.2 and an effective moment of inertia I_e computed using the Bischoff equations as follows:

$$\text{For } M_a \le \frac{2}{3}M_{cr},\ I_e = I_g \qquad \text{[ACI Eq. (24.2.3.5a)]}$$

$$\text{For } M_a > \frac{2}{3}M_{cr},\ I_e = \frac{I_{cr}}{1-\left(\frac{2}{3}\frac{M_{cr}}{M_a}\right)^2\left(1-\frac{I_{cr}}{I_g}\right)} \le I_g \qquad \text{[ACI Eq. (24.2.3.5b)]}$$

The revised equations above for the effective moment of inertia, which account for the effects of concrete cracking and steel reinforcement, replace the Branson equation that has been used in previous editions of the ACI 318 Code since 1965. The Branson equation was found to underestimate deflections for non-prestressed concrete members with low reinforcement ratios such as slabs. The Branson equation and the revised equations above result in approximately the same deflections for non-prestressed concrete members with at least 1% reinforcement ratio and with a service moment, M_a, greater than or equal to the cracking moment, M_{cr}. Note that the Branson equation (ACI Eqn 24.2.3.9a), presented below, is still used for calculating deflections in prestressed concrete members.

$$I_e = \left\{ \left(\frac{M_{cr}}{M_a} \right)^3 I_g + \left[1 - \left(\frac{M_{cr}}{M_a} \right)^3 \right] I_{cr} \right\} \leq I_g \qquad \text{[ACI Eq. (24.2.3.9a)]}$$

The parameters in the equations above are defined as follows:

I_e = effective moment of inertia

I_{cr} = moment of inertia of the cracked section transformed into concrete

I_g = moment of inertia of the gross (uncracked) concrete cross-section about the centroidal axis, neglecting all steel reinforcement

M_a = maximum moment in the member at the stage for which the deflection is being computed

M_{cr} = moment that would initially crack the cross section computed from

$$M_{cr} = \frac{f_r I_g}{y_t} \qquad \text{[ACI Eq. (24.2.3.5)]}$$

where

f_r = modulus of rupture for the concrete = $7.5\,\lambda\sqrt{f'_c}$ [ACI Code Eq. (19.2.3.1)]

λ = 1.0 for normal-weight concrete

= 0.85 for sand-lightweight concrete

= 0.75 for all-lightweight concrete

Values for the modulus of rupture for normal-weight concrete are tabulated in Table A.6.

y_t = distance from the neutral axis of the uncracked cross section (neglecting steel) to the extreme tension fiber

In this text, the Bischoff equations as precribed in the ACI Code for reinforced concrete members will be used. Inspection of the formula for the effective moment of inertia, which accounts for the effect of concrete cracking and steel reinforcement, will show that if the maximum moment is low with respect to the cracking moment M_{cr}, the moment of inertia of the gross section I_g will be the dominant factor. If the maximum moment is large with respect to the cracking moment, however, the moment of inertia of the cracked section I_{cr} will be dominant. In any case, I_e will lie somewhere between I_{cr} and I_g. For continuous beams, the use of the average value of the effective moments of inertia existing at sections of critical positive and negative moments is recommended. The use of midspan sectional properties for simple and continuous spans and sectional properties at the support for cantilevers will also give satisfactory results.

The actual calculation of deflections will be carried out using the standard deflection formulas for elastic members. Deflection formulas of the type found in standard handbooks may be suitable, or more rigorous deflection calculation techniques may be used when necessary.

7.5 LONG-TERM DEFLECTION

In addition to deflections that occur immediately, reinforced concrete members are subject to additional deflections that occur gradually over a long period of time. These additional deflections are due mainly to creep and shrinkage and may eventually become excessive. The additional (or long-term) deflections are computed based on two items: (1) the amount of sustained dead and live load and (2) the amount of compression reinforcement in the beam. The additional long-term deflections may be estimated as follows:

$$\Delta_{LT} = \lambda_{\Delta}\Delta_i = \left(\frac{\xi}{1+50\rho'}\right)\Delta_i$$

where

Δ_{LT} = additional long-term deflection

Δ_i = immediate or instantaneous deflection due to sustained loads

λ_{Δ} = a multiplier for additional long-term deflections (i.e., λ_{Δ} accounts for the effect of creep of concrete)

$$= \left(\frac{\xi}{1+50\rho'}\right) \qquad \text{[ACI Eq. (24.2.4.1.1)]}$$

ρ' = non-prestressed compression reinforcement ratio (A'_s/bd)

ξ = time-dependent factor for sustained loads, that is, the creep factor measured relative to time $t = 0$ (ACI 318 Table 24.2.4.1.3):

5 years or more	2.0
12 months	1.4
6 months	1.2
3 months	1.0

NOTE *At 5 years and beyond, the incremental long-term deflection of the member remains constant and is calculated using a creep or time-dependent factor, ξ, of 2.0. "Deflection sensitive elements" (DSEs) that are supported by the member which are installed during construction will undergo this long-term deflection. If, however, the installation of the DSEs were delayed till, for example, after 12 months from when the concrete member was initially cast, the incremental deflections that the DSEs will undergo would be the difference between the long-term deflection calculated with a creep factor, ξ, of 2.0 (i.e.,* *at 5 years) and the long-term deflection calculated with a creep factor, ξ, of 1.4 (i.e., at 12 months). Thus, it is advantageous to install DSEs after much of the creep deflection in a concrete member has occurred. Examples of DSEs include sensitive equipment such as medical equipment, plastered ceiling, unreinforced masonry partition walls, masonry exterior walls or cladding, glazing, expensive flooring such as terrazzo or marble, and operable or movable partition walls.*

Δ_i = immediate or instantaneous deflection due to sustained loads = $\Delta_{DL} + \alpha\Delta_{LL}$

α = percentage of live load sustained.

The total long-term deflection (*which excludes the instantaneous dead load deflection*) is the sum of the instantaneous live load deflection plus the long-term deflection due to the dead load and the sustained portion of the live load and is given as

$$\begin{aligned}\Delta_{TT} &= \Delta_{LL} + \Delta_{LT} \\ &= \Delta_{LL} + \left(\frac{\xi}{1+50\rho'}\right)\Delta_i \\ &= \Delta_{LL} + \lambda_\Delta\left(\Delta_{DL} + \alpha\Delta_{LL}\right)\end{aligned}$$

Some judgment will be required in determining just what portion of the live loads should be considered *sustained*. In a residential application, 20% sustained live load (i.e., α = 0.20) might be a logical estimate, whereas in storage facilities or mechanical rooms, 100% sustained live load (i.e., α = 1.0) would be reasonable.

The calculated deflections must not exceed the maximum permissible deflections that are found in the ACI Code, Table 24.2.2 (see Table 7.1) or the deflection limits of the non-structural elements provided by the manufacturer or supplier in their performance specifications. This table sets permissible deflections in terms of fractions of the flexural member span length, ℓ. These limitations guard against damage to the various parts of the system (both structural and nonstructural parts) as a result of excessive deflection. In the case of attached nonstructural elements, only the deflection that takes place after such attachment needs to be considered. For concrete floor slabs, beams and girders supporting medical equipment such as X-ray or MRI machines, or microelectronics manufacturing equipment, or operable partition walls, the deflection limit will be far more stringent than the limits prescribed in Table 7.1. Some operable partition walls require a deflection limit as low as span/1200.

TABLE 7.1 Maximum permissible deflections.

Member	Condition	Deflection considered	Deflection limit
Flat roofs		Immediate deflection due to maximum of L_r, S, or R	$\frac{\ell}{180}$
Floors	Not supporting or attached to non-structural elements likely to be damaged by large deflections.	Immediate deflection due to floor live load, L	$\frac{\ell}{360}$

Member	Condition		Deflection considered	Deflection limit
Roof or floors	Supporting or attached to non-structural elements	**Likely** to be **damaged** by large deflections	That part of the total deflection occurring after attachment of non-structural elements = sum of the time-dependent or long-term deflection due to all sustained loads plus the immediate deflection to any additional live load	$\frac{\ell}{480}$
		Not likely to be damaged by large deflections		$\frac{\ell}{240}$

Adapted from ACI 318-19 Table 24.2.2.

EXAMPLE 7.2

Calculate the maximum deflection of the simply supported non-prestressed reinforced concrete beam having a cross-section shown in Figure 7.7 and determine which of the ACI Code deflection criteria will be satisfied. The beam is subjected to maximum unfactored moments of M_{DL} = 20 ft.-kip and M_{LL} = 15 ft.-kip. Assume a 50% sustained live load and a sustained load time period of more than 5 years. Assume normal-weight concrete. Use $f_c' = 3000$ psi and f_y = 60,000 psi. The beam has on a simple span of 30 ft.

Solution:

1. The maximum *service* moments at midspan are

$$M_{DL} = 20 \text{ ft.-kip} \quad \text{and} \quad M_{LL} = 15 \text{ ft.-kip}$$

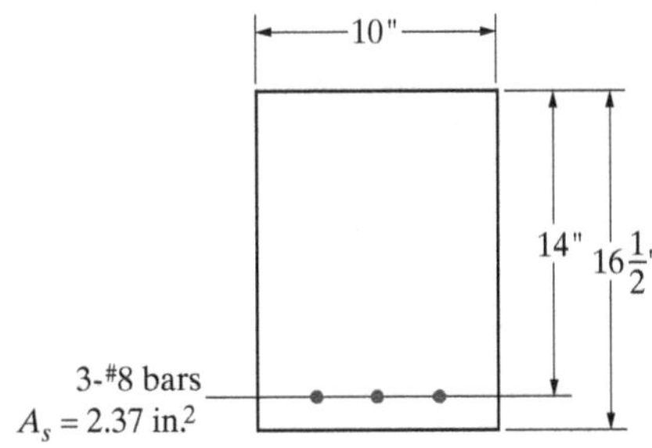

FIGURE 7.7 Sketch for Example 7.2.

2. Check the beam depth based on the ACI Code Table 9.3.1.1:

$$\text{minimum } h = \frac{\ell}{16} = \frac{30(12)}{16} = 22.5 \text{ in.}$$

Because 22.5 in. > 16.5 in., deflection must be calculated.

3. The effective moment of inertia will be calculated using

$$\text{For } M_a \leq \frac{2}{3} M_{cr}, \; I_e = I_g \qquad \text{[ACI Eq. (24.2.3.5a)]}$$

For $M_a > \frac{2}{3}M_{cr}$, $I_e = \dfrac{I_{cr}}{1-\left(\dfrac{2}{3}\dfrac{M_{cr}}{M_a}\right)^2\left(1-\dfrac{I_{cr}}{I_g}\right)} \le I_g$ [ACI Eq. (24.2.3.5b)]

Therefore, we first must compute the various terms within the expression. The moment of inertia of the cracked transformed section will be determined with reference to Figure 7.8.

The steel area of 2.37 in.^{2} and a modular ratio n of 9 (see Table A.6) will result in a transformed area of

$$nA_s = 9(2.37) = 21.3 \text{ in.}^2$$

The neutral-axis location is determined as follows:

$$\begin{aligned}\bar{y} &= \frac{nA_s\left[\sqrt{1+2\dfrac{bd}{nA_s}}-1\right]}{b}\\ &= \frac{21.3\left[\sqrt{1+2\dfrac{(10)(14)}{21.3}}-1\right]}{10}\\ &= 5.88 \text{ in.}\end{aligned}$$

The moment of inertia of the cracked transformed section is then determined:

$$\begin{aligned}I_{cr} &= \frac{10(5.88)^3}{3} + 21.3(14-5.88)^2\\ &= 2082 \text{ in.}^4\end{aligned}$$

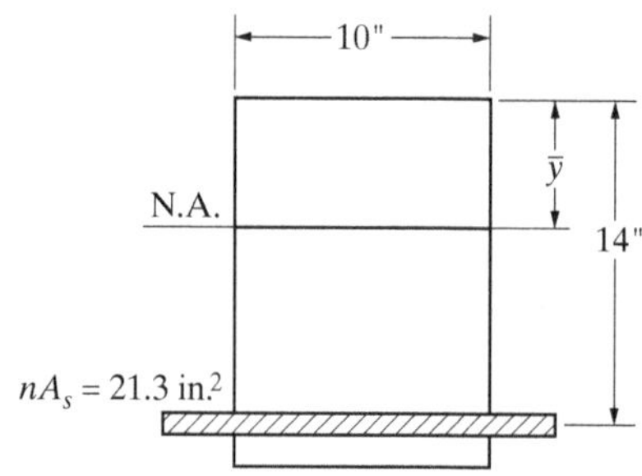

FIGURE 7.8 Transformed section for Example 7.2.

Determination of the moment of inertia of the gross section results in

$$I_g = \frac{bh^3}{12} = \frac{1}{12}(10)(16.5)^3 = 3743 \text{ in.}^4$$

The moment that would initially crack the cross section may be determined next:

$$M_{cr} = \frac{f_r I_g}{y_t}$$

where

$$y_t = \frac{16.5}{2} = 8.25 \text{ in.}$$

$$f_r = 7.5\lambda\sqrt{f_c'} = 0.411 \text{ ksi} \text{ (from Table A.6 for } \lambda = 1)$$

Therefore,

$$M_{cr} = \frac{0.411(3743)}{8.25(12)} = 15.5 \text{ ft.-kip}$$

We will assume that the maximum total unfactored moment in the beam at the midspan, $M_{am} = M_a = 20 + 15 = 35$ ft.-kip will establish the crack pattern in the beam. This is not in strict accordance with the ACI Code, which indicates that M_a should be the maximum moment occurring at the stage the deflection is computed. It is logical that, as the cracking pattern is irreversible, the use of the effective moment of inertia based on the full maximum moment is more realistic. This approach is conservative and will furnish a lower I_e, which will subsequently result in a larger computed deflection.

4. Determine the effective moment of inertia:

Since $M_a = 35$ ft-kip > (2/3) M_{cr},

$$\text{Since } M_a > \frac{2}{3}M_{cr},\; I_e = \frac{I_{cr}}{1-\left(\frac{2}{3}\frac{M_{cr}}{M_a}\right)^2\left(1-\frac{I_{cr}}{I_g}\right)} \le I_g \quad \text{[ACI Eq. (24.2.3.5b)]}$$

$$= \frac{2082}{1-\left(\frac{2}{3}\frac{15.5}{35}\right)^2\left(1-\frac{2082}{3743}\right)} \le 3743 \text{ in.}^4$$

$$= 2166 \text{ in.}^4$$

$$\text{For } M_a \le \frac{2}{3}M_{cr},\; I_e = I_g \quad \text{[ACI Eq. (24.2.3.5a)]}$$

$$\text{For } M_a > \frac{2}{3}M_{cr},\; I_e = \frac{I_{cr}}{1-\left(\frac{2}{3}\frac{M_{cr}}{M_a}\right)^2\left(1-\frac{I_{cr}}{I_g}\right)} \le I_g \quad \text{[ACI Eq. (24.2.3.5b)]}$$

Compared to I_{cr}, calculated previously as 2082 in.^{4}, the effective moment of inertia, I_e, is calculated to be 2166 in.^{4}

5. Compute the immediate or instantaneous dead load deflection (M_{DL} = 20 ft.-kip):

From structural analysis, the maximum deflection of a simply supported beam is,

$$\Delta_{DL} = \frac{5w_{DL}\ell^4}{384E_cI_e} = \frac{5M_{DL}\ell^2}{48E_cI_e}$$

where M is the moment due to a uniform load and E_c may be found in Table A.6. Thus

$$\Delta_{DL} = \frac{5(20\text{ ft.-kip})(30\text{ ft})^2(1728\text{ in.}^3/\text{ft}^3)}{48(3120\text{ kip/in.}^2)(2166\text{ in.}^4)} = 0.48\text{ in.}$$

6. Compute the immediate or instantaneous live load deflection (M_{LL} = 15 ft.-kip). By proportion,

$$\Delta_{LL} = \frac{M_{LL}}{M_{DL}}\Delta_{DL} = \frac{15}{20}(0.48) = 0.36\text{ in.}$$

7. The total immediate or instantaneous dead plus live load (DL + LL) deflection is

$$0.48 + 0.36 = 0.84\text{ in.}$$

TABLE 7.2 Permissible Versus Actual Deflection (Example 7.2).

Case	Maximum permissible deflection*	Actual computed deflection
1	$\frac{\ell}{180} = \frac{30(12)}{180} = 2$ in.	Δ_{LL} = 0.36 in. (immediate LL) O.K.
2	$\frac{\ell}{360} = \frac{30(12)}{360} = 1$ in.	Δ_{LL} = 0.36 in. (immediate LL) O.K.
3	$\frac{\ell}{480} = \frac{30(12)}{480} = 0.75$ in.	$\Delta_{LL} + \Delta_{LT}$ = 0.36 + 1.32 = 1.68 in. Not good
4	$\frac{\ell}{240} = \frac{30(12)}{240} = 1.5$ in.	$\Delta_{LL} + \Delta_{LT}$ = 0.36 + 1.32 = 1.68 in. Not good

*From ACI 318, Table 24.2.2.

8. The long-term (LT) deflection (DL + sustained LL) multiplier is

$$\lambda_\Delta = \frac{\xi}{1+50\rho'} = \frac{2.0}{1+0} = 2.0$$

Because M_{DL} = 20 ft.-kip and 50% M_{LL} = 7.5 ft.-kip, the sustained moment for long-term deflection = 27.5 ft.-kip. Then

$$\Delta_{LT} = \frac{27.5}{20}(0.48)(2.0) = 1.32\text{ in.}$$

9. A comparison of actual deflections to maximum permissible deflections may now be made. In the comparison, made in Table 7.2, the maximum permissible deflections are from the ACI Code, Table 24.2.2. Case 1 in Table 7.2 applies to flat roofs not supporting or attached to nonstructural elements likely to be damaged by large deflections, while Case 2 applies to floors not supporting or attached to nonstructural elements likely to be damaged by large deflections. Cases 3 and 4 apply to long-term deflections of roofs and floors (see Table 7.1). As the permissible deflection is not exceeded in case 1 or case 2, the beam of Example 7.2 is limited to usage as defined by Cases 1 and 2.

7.6 PROCEDURE FOR CALCULATING THE DEFLECTION OF SIMPLY SUPPORTED AND CONTINUOUS BEAMS AND SLABS

The step-by-step procedure for calculating the long-term deflections of simply supported and continuous beams and slabs subject to uniformly distributed loads is as follows [4]:

1. Calculate the service dead and live load moments at the midspan of the most critical span or the span of interest. (i.e., calculate $M_{DL,m}$ and $M_{LL,m}$). For continuous beams and slabs, also calculate the unfactored moments at the supports. For continuous members, these service load moments are usually determined using the ACI moment coefficients (see Chapter 6).
   a. Assume that a portion of the live load is permanently sustained. The percentage of live load sustained is denoted as α (see Section 7.5).
2. Calculate the cracked moment of inertia, I_{cr}, and the gross moment of inertia, I_g, at the midspan (and at the supports for continuous members).
3. Calculate the effective moment of inertia, I_e, using ACI Equations 24.2.3.5a or 24.2.3.5b.
   a. Calculate the cracking moment, M_{cr}, for the beam at the midspan using ACI Equations 24.2.3.5a or 24.2.3.5b and the effective moment of inertia, $I_{e,m\ beam}$, at the midspan using M_{cr}/M_{am} and ACI Equation 24.2.3.5a.
      i. The *total* service load moment at midspan, $M_{a,m} = M_{DL,m} + M_{LL,m}$.
      ii. If $M_{a,m} \leq \frac{2}{3}M_{cr}$, the member is uncracked and therefore $I_e = I_g$ (ACI Equation 24.2.3.5a)
   b. For continuous members, repeat Step 3a at the member supports and obtain the effective moment of inertia at the supports, $I_{e,s}$.
   c. The effective moment of inertia at the midspan and supports, respectively (i.e., $I_{e,m}$ and $I_{e,s}$), are combined as follows to obtain a weighted average moment of inertia for the continuous member, I_e (average) [4]:
      i. Beam or slab with one end continuous: I_e (average) = $0.85I_{e,m} + 0.15\,I_{e,s}$
      ii. Beam or slab with both ends continuous: I_e (average) = $0.70I_{e,m} + 0.15(I_{e,s1} + I_{e,s2})$
   d. Use the I_e (average) in the member deflection calculations. For simply supported members, there are no moments at the end supports; therefore, I_e (average) = $I_{e,m}$.

   where,

   $I_{e,m}$ = effective moment of inertia at the **midspan**, calculated using the section properties and the unfactored moment at the midspan,

   $I_{e,s}$ = effective moment of inertia at the **one continuous end**, calculated using the section properties and the unfactored moment at this support,

$I_{e,s1}$ = effective moment of inertia at the **left continuous end**, calculated using the section properties and the unfactored moments at the left end support,

$I_{e,s2}$ = effective moment of inertia at the **right continuous end**, calculated using the section properties and the moments at the right end support.

NOTE *For many practical situations, especially for the end spans of continuous beams and girders where the deflection is usually largest, it is accurate enough to use only the midspan properties and midspan moments in calculating the effective moment of inertia because the member stiffness at midspan predominates. In this case,* $I_e(\text{average}) \approx I_{e,m}$. *This simplified approach is adopted in the examples in this text.*

4. Calculate the instantaneous maximum deflection:

- For **simply supported beams** or slabs with uniformly distributed loads, the instantaneous maximum deflections, calculated using the midspan moments, are

$$\Delta_{DL} = K\left(\frac{5}{48}\right)\frac{M_{DL,m}\ell^2}{E_c I_e}$$

$$\Delta_{LL} = K\left(\frac{5}{48}\right)\frac{M_{LL,m}\ell^2}{E_c I_e} = \frac{M_{LL,m}}{M_{DL,m}}\Delta_{DL}$$

where,

K = deflection constant = 1.0 for simply supported members

$M_{DL,m}$ = maximum service dead load moment at midspan = $\dfrac{w_{DL}\ell^2}{8}$

$M_{LL,m}$ = maximum service live load moment at midspan = $\dfrac{w_{LL}\ell^2}{8}$

- For **continuous beams** and slabs with uniformly distributed loads, the instantaneous maximum deflections at the midspan deflection are calculated as

$$\Delta_{DL} = K\left(\frac{5}{48}\right)\frac{M_{DL,m}\ell^2}{E_c I_e}$$

$$\Delta_{LL} = K\left(\frac{5}{48}\right)\frac{M_{LL,m}\ell^2}{E_c I_e} = \frac{M_{LL,m}}{M_{DL,m}}\Delta_{DL}$$

$$\Delta_{DL+LL} = K\left(\frac{5}{48}\right)\frac{M_{a,m}\ell^2}{E_c I_e}$$

$$K = 1.2 - 0.2\frac{M_o}{M_{am}}$$

$$M_o = \frac{w_{DL+LL}\ell^2}{8}$$

where M_o is the statical moment.

The net midspan moment, $M_{a,m} = M_o - 0.5(M_1 + M_2)$

$M_{a,m}$ can also be obtained directly using the ACI moment coefficients.

5. The total long-term deflection (*which excludes the instantaneous dead load deflection*) is the sum of the instantaneous live load deflection plus the long-term deflection due to dead load and the sustained portion of the live load and is given as

$$\Delta_{TT} = \Delta_{LL} + \lambda_{\Delta}\left(\Delta_{DL} + \alpha\Delta_{LL}\right)$$

6. Check the permissible deflection limits from ACI 318 Table 24.2.2:
   a. For *roof or floor* flexural members *not supporting* deflection-sensitive elements, $\Delta_{TT} \leq \dfrac{\ell}{240}$
   b. For *roof or floor* flexural members *supporting* deflection-sensitive elements, $\Delta_{TT} \leq \dfrac{\ell}{480}$
   c. For *roof* flexural members *not supporting* deflection-sensitive elements, $\Delta_{LL} \leq \dfrac{\ell}{180}$
   d. For *floor* flexural members *not supporting* deflection-sensitive elements, $\Delta_{LL} \leq \dfrac{\ell}{360}$

Some examples of deflection-sensitive elements include brittle floor finishes, sensitive equipment, plastered ceilings, glazing, operable walls, external cladding, or partitions.

7.7 PROCEDURE FOR CALCULATING THE DEFLECTION OF CONTINUOUS GIRDERS

The ACI coefficients are only applicable to flexural members supporting uniformly distributed loads. For girders which typically support concentrated loads due to the reactions from the beams framing into them, the ACI moment coefficients are not applicable. Therefore, plane frame structural analysis software is typically used to analyze the girder-column sub-frame with the appropriate effective moment of inertia considered. The procedure for calculating deflections for girders is as follows:

1. Perform a computer-based structural analysis of the girder-column sub-frame under service dead and live loads (i.e., unfactored loads) using the gross moment of inertia of the girders and columns. The loads on the girder usually consist of unfactored uniform self-weight of the girder stem and the unfactored concentrated dead and live load reactions from the beams framing into the girder. The maximum deflection in a continuous girder with equal spans will typically occur in the end span. Pattern or skipped live loading must be considered.
2. From the computer analysis results, obtain the service dead load moment M_{DL} and the service live load moment M_{LL} at the midspan of the end span of the continuous girder. Similarly, obtain the moments in the girder at the girder support locations.

a. Midspan moments: $M_{a,m} = M_{DL,m} + M_{LL,m}$

b. Support Moments: $M_{a,s} = M_{DL,s} + M_{LL,s}$

3. Calculate the effective moment of inertia, I_e.

a. Calculate the cracking moment M_{cr} for the girder at the midspan using ACI Equation 24.2.3.5 and the girder effective moment of inertia $I_{e,m\ girder}$ at the midspan using M_{cr}/M_{am} and ACI Equations 24.2.3.5a or 24.2.3.5b.

i. The *maximum* service load moment at midspan, $M_{a,m} = M_{DL,m} + M_{LL,m}$

ii. If $M_{a,m} \leq \frac{2}{3} M_{cr}$, the member is uncracked and therefore, $I_e = I_g$

b. Repeat Step 3a at the girder supports and obtain the effective moment of inertia at the supports, $I_{e,s}$.

c. The effective moment of inertia at the midspan and supports, respectively (i.e., $I_{e,m}$ and $I_{e,s}$), are combined as follows to obtain a weighted average moment of inertia for the girder, I_e (average) [4]:

i. Girder with one end continuous: I_e (average) $= 0.85\, I_{e,m} + 0.15\, I_{e,s}$

ii. Girder with both ends continuous: I_e (average) $= 0.70\, I_{e,m} + 0.15\, (I_{e,s1} + I_{e,s2})$

Use the I_e(average) in the girder deflection calculations.

where,

$I_{e,m}$ = effective moment of inertia at the **midspan**, calculated using the section properties and moments at the midspan

$I_{e,s}$ = effective moment of inertia at the **one continuous end**, calculated using the section properties and the moments at this support

$I_{e,s1}$ = effective moment of inertia at the **left continuous end**, calculated using the section properties and the moments at the left end support

$I_{e,s2}$ = effective moment of inertia at the **right continuous end**, calculated using the section properties and the moments at the right end support

As discussed previously, since the girder midspan stiffness is dominant, the effective moment of inertia calculated at midspan can be used and will yield sufficiently accurate results. Thus, I_e (girder) $\approx I_{e,m}$.

4. The computer analysis in Step 1 is repeated using **the effective moment of inertia $I_{e,\ girder}$ obtained in step 3** for all the girders with the moment of inertia of the columns at service loads assumed as the gross moment of inertia of the columns (i.e., $I_{e\ columns} = I_{g\ columns}$) per ACI Table 6.6.3.2.2. From this computer analysis, the instantaneous dead load deflection, Δ_{DL}, and the instantaneous live load deflection, Δ_{LL}, at the midspan of the end-span of the girder are obtained.

5. The total long-term deflection (*which excludes the instantaneous dead load deflection*) is the sum of the instantaneous live load deflection plus the long-term deflection due to dead load and the sustained portion of the live load and is given as

$$\Delta_{TT} = \Delta_{LL} + \lambda_{\Delta}\left(\Delta_{DL} + \alpha\Delta_{LL}\right)$$

where α is the percentage of the live load sustained.

6. Check the permissible deflection limits from ACI 318, Table 24.2.2 or Table 7.1

EXAMPLE 7.3: Deflection of Continuous T-beam or L-Beams

A reinforced concrete T-beam in an office building is continuous over three equal center-to-center spans of 30 ft (assume a clear span of 29 ft). The beam is 14″ × 22″ deep with a 6″ thick slab and is reinforced with 4#8 bottom bars at the midspan of the end span, and 4#8 top bars at the first interior support. The service dead load on the beam, which includes the beam self-weight, is 1.40 kip/ft, and the service live load is 0.5 kip/ft. Calculate the maximum deflection at the midspan of the end span of the continuous beam and compare it with the ACI Code deflection limits, assuming the beam is supported by girders. As an approximation, use only the midspan section properties in calculating the beam deflections and assume the effective flange width of the beam at midspan, b = 87″. Use f_c' = 4000 psi and f_y = 60,000 psi.

Solution:

Assuming knife-edge supports for the continuous beam, and since all the three requirements for using the ACI coefficients (see Chapter 6) are satisfied, the ACI moment coefficients can be used to calculate the service load moments (i.e., unfactored moments) at the supports and at midspan. Furthermore, as stated in the problem, only the effective moment of inertia at the midspan will be used. The procedure is as follows:

1. Calculate the service dead and live load moments, M_{DL} and M_{LL} at the midspan of the end span.

   Clear span of the beam, ℓ_n = 29 ft

   Using the ACI moment coefficients from Chapter 6, the unfactored moment at midspan due to dead load is,

$$M_{DL,m} = \frac{1}{11}(1.40\,\text{kip/ft})(29\,\text{ft})^2 = 107\,\text{ft.k}$$

   Similarly, the unfactored moment at midspan due to live load is,

$$M_{LL,m} = \frac{1}{11}(0.5\,\text{kip/ft})(29\,\text{ft})^2 = 39\,\text{ft.k}$$

2. Calculate the T-section properties at midspan

   Effective flange width of the T-beam at midspan is given as, b = 87″

$$n = \frac{E_s}{E_c} = \text{modular ratio}$$

$$= \frac{(29)\left(10^6\right)}{57000\sqrt{4000}} = 8.1$$

   The area of tension rebar, A_s for 4#8 bars at midspan = 3.16 in.2

The effective depth at midspan and at the support, d = 22″ – 2.5″ = 19.5″. The pertinent beam dimensions are:

b_w = 14″

h_f = 6″

h = 22″

d = 19.5″

Using the equations for T-beams from Section 7.11, calculate the following sectional properties for the T-beam:

$$C=\frac{b_w}{nA_s}=\frac{14"}{(8.1)(3.16)}=0.55$$

$$Z=\frac{(b-b_w)h_f}{nA_s}=\frac{(87-14)(6)}{(8.1)(3.16)}=17.1$$

$$y_t=22-0.5\left(\frac{[(87-14)(6)^2+14(22)^2]}{[(87-14)(6)+14(22)]}\right)=15.7"$$

$$\overline{y}=\frac{\left[\sqrt{0.55(2[19.5]+6[17.1])+(1+17.1)^2}\right]-(1+17.1)}{0.55}=3.7"$$

$$I_g=\frac{(87-14)(6)^3}{12}+\frac{14(6)^3}{12}+6(87-14)(22-0.5[6]-15.7)^2+14(22)(15.7-0.5[22])^2$$
$$=25{,}310\text{ in.}^4$$

$$I_{cr}=\frac{(87-14)(6)^3}{12}+\frac{14(3.7)^3}{3}+(87-14)(6)(3.7-0.5[6])^2+(8.04)(3.16)(19.5-3.7)^2$$
$$=8155\text{ in.}^4$$

3. Calculate the effective moment of inertia, I_e at the midspan, using the section properties calculated at the midspan

$I_{g,m}$ = 25310 in.^{4}

$I_{cr,m}$ = 8155 in.^{4} (see Step 2)

y_t at midspan = 15.7″ (calculated previously in Step 2)

Cracking stress, $f_r=7.5\sqrt{4000\text{ psi}}=474\text{ psi}$

The cracking moment at midspan is,

$$M_{cr}=\frac{f_rI_g}{y_t}=\frac{(474)(25{,}310)}{15.7}$$
$$=764{,}136\text{ in.-Ib}=64\text{ ft.-kip}$$

$2/3M_{cr}$ = (2/3)(64) = 42.7 ft.-kip

Total service load moment in the beam at midspan, $M_{a,m} = M_{DL,m} + M_{LL,m}$

= 107 ft.-kip + 39 ft.-kip = 146 ft.-kip

> $2/3M_{cr}$

Therefore, use ACI 318, Equation 24.2.3.5b from Section 7.4 to calculate the effective moment of inertia, I_e:

Therefore, the effective moment of inertia at the midspan is,

$$I_{e,m} = \frac{I_{cr}}{1-\left(\frac{2}{3}\frac{M_{cr}}{M_a}\right)^2\left(1-\frac{I_{cr}}{I_g}\right)} \le I_g$$

$$= \frac{8155}{1-\left(\frac{2}{3}\frac{64}{146}\right)^2\left(1-\frac{8155}{25310}\right)} \le 25{,}310$$

$$= 8656 \text{ in.}^4 < 25{,}310 \text{ in.}^4 \qquad O.K.$$

This effective moment of inertia will be used to calculate the beam deflections.

4. Calculate the midspan deflections at the end span of the continuous beam

Instantaneous dead load deflection: The statical moment due to dead load is

$$M_{o,DL} = \frac{w_{DL}\ell^2}{8} = \frac{(1.40)(29)^2}{8} = 148 \text{ ft.-kip}$$

The service dead load moment at midspan due to service loads was calculated in Step 1 and is

$M_{a,DL,m} = 107$ ft.-kip

For a continuous beam with a uniformly distributed load, the deflection constant, K, is

$$K = 1.2 - 0.2\frac{M_o}{M_{a,m}} = 1.2 - 0.2\left(\frac{148}{107}\right) = 0.93$$

The instantaneous or immediate dead load deflection, Δ_{DL}, including the effect of cracking is,

$$\Delta_{DL} = K\left(\frac{5}{48}\right)\frac{M_{DL,m}\ell^2}{E_c I_e} = 0.93\left(\frac{5}{48}\right)\frac{(107)(12{,}000)(29x12)^2}{\left(57{,}000\sqrt{4000}\right)(8656)} = 0.48''$$

Instantaneous Live Load Deflection:

The statical moment due to live load is,

$$M_{o,LL} = \frac{w_{LL}\ell^2}{8} = \frac{(0.5)(29)^2}{8} = 53 \text{ ft.-kip}$$

The service live load moment at midspan due to service loads was calculated in Step 1 using the ACI moment coefficients and is

$M_{a,LL,m} = 39$ ft.-k

For a continuous beam with a uniformly distributed load, the deflection constant, *K*, is

$$K = 1.2 - 0.2\frac{M_o}{M_{a,m}} = 1.2 - 0.2\left(\frac{53}{39}\right) = 0.93$$

The instantaneous or immediate live load deflection, Δ_{LL}, including the effect of cracking is,

$$\Delta_{LL} = K\left(\frac{5}{48}\right)\frac{M_{LL,m}\ell^2}{E_c I_e} = 0.93\left(\frac{5}{48}\right)\frac{(39)(12{,}000)(29x12)^2}{\left(57{,}000\sqrt{4000}\right)(8656)} = 0.18"$$

Note that the instantaneous live load deflection could have been more quickly and easily calculated by assuming that the ratio of the dead load moment at midspan to the live load moment at midspan is equal to the ratio of the dead load deflection to the live load deflection. Therefore, the instantaneous live load deflection can be calculated from the following equation: $\Delta_{LL} = \frac{M_{a,LL,m}}{M_{a,DL,m}}\Delta_{DL} = \frac{(39)}{(107)}(0.48") = 0.18"$

The live load deflection obtained above is found to be exactly the same as the deflection that was calculated earlier by the long approach.

5. For an office building, the percentage of live load sustained can be assumed to be approximately 25%.

Therefore, $\alpha = 0.25$

For long-term deflections, use the creep factor after 5 years, therefore, $\xi = 2.0$. Since there is no compression steel, $\rho' = 0$, the long-term deflection multiplier is

$$\lambda_\Delta = \left(\frac{\xi}{1+50\rho'}\right) = \frac{2.0}{1+50(0)} = 2.0$$

The total long-term deflection (*which excludes the instantaneous dead load deflection*) is calculated as

$$\begin{aligned}\Delta_{TT} &= \Delta_{LL} + \Delta_{LT}\\ &= \Delta_{LL} + \lambda_\Delta\left(\Delta_{DL} + \alpha\Delta_{LL}\right)\\ &= 0.18" + 2.0\left(0.48" + 0.25[0.18"]\right) = 1.23"\end{aligned}$$

This long-term deflection as well as the instantaneous live load deflection will now be compared to the ACI Code deflection limits (ACI 318-14, Table 24.2.2 or Table 7.1).

a. For *roof and floor* flexural members *not supporting* deflection-sensitive elements,

$\Delta_{TT} = 1.23"$ must be $\leq \frac{\ell}{240} = \frac{29(12)}{240} = 1.45"$ *O.K.*

b. For *roof and floor* flexural members *supporting* deflection-sensitive elements,

$\Delta_{TT} = 1.23"$ must be $\leq \frac{\ell}{480} = \frac{29(12)}{480} = 0.73"$ *Not Good*

c. For *roof* flexural members *not supporting* deflection-sensitive elements,

$\Delta_{LL} = 0.18"$ must be $\leq \frac{\ell}{180} = \frac{29(12)}{180} = 1.93"$ *O.K.*

d. For *floor* flexural members *not supporting* deflection-sensitive elements, $\Delta_{LL} = 0.18"$ must be $\leq \dfrac{\ell}{360} = \dfrac{29(12)}{360} = 0.97"$ *O.K.*

The deflection at the end span of this continuous beam satisfies all the ACI Code deflection limits except the second limit with regards to roof and floor flexural members supporting deflection sensitive elements. To reduce the long-term deflections, we could add compression steel to the beam at the midspan and recalculate the long-term deflection.

Assume 4#8 bars are added at the top of the beam at midspan; therefore, the compression reinforcement ratio, $\rho' = \dfrac{A_s}{b_w d} = \dfrac{3.16\ in^2}{(14")(19.5")} = 0.0116$

The revised long-term multiplier is

$$\lambda_\Delta = \left(\frac{\xi}{1 + 50\rho'}\right) = \frac{2.0}{1 + 50(0.0116)} = 1.27$$

There is a reduction of the long-term multiplier from 2.0 (without compression steel) to 1.27 with compression steel. To prevent buckling of the compression reinforcement as previously discussed in Section 3.13, the compression steel will have to be confined with closed stirrups in accordance with ACI Code, Section 25.7.2 (see Figure 3.23).

With compression steel, the long-term deflection (excluding the instantaneous dead load deflection) is,

$$\Delta_{TT} = 0.18" + 1.27\left(0.48" + 0.25[0.18"]\right) = 0.85"$$

The revised long-term deflection still exceeds by 16% the ACI Code limit of 0.73 inch if deflection-sensitive elements are supported by the beam. Several measures that could be taken to meet the ACI Code deflection limit include one or more of the following: (1) increase the beam depth and/or (2) provide additional tension and compression reinforcement in the beam.

7.8 DEFLECTION CONTROL MEASURES IN REINFORCED CONCRETE STRUCTURES

There are several ways to control deflections in reinforced concrete flexural members; some are feasible during the design phase, and others are feasible during the construction phase. The control measures include the following:

- Add compression steel to the beam, but the compression steel must be tied with closed stirrups to prevent buckling of the compression bars, and the tie spacing must conform with ACI 318, Section 25.7.2 (see Section 3.13). Note that compression steel cannot be used to control deflections in slabs since the compression bars cannot be confined. The advantageous effect of compression steel in reducing long-term deflections is reflected by the compression rebar ratio, ρ', being in the denominator in the equation for the additional long-term deflection multiplier, λ_Δ, in Section 7.5.

- Provide additional tension reinforcement, which increases the cracked moment of inertia, I_{cr}, which in turn leads to a higher effective moment of inertia, I_e.
- Increase the slab thickness and the beam depth. Increasing the width of a beam or girder is not as effective as increasing the beam or girder depth in achieving an increase in the effective moment of inertia.
- To reduce the impact of incremental long-term deflections, the installation of the deflection-sensitive elements (DSEs) should be delayed allowing a significant proportion of the creep deflection of the member to occur prior to the installation.
- Provide a structural detail showing a deflection gap between the concrete beams/girders and the non-load-bearing vertical elements (such as exterior backup walls, partitions, or cladding) located just below the beams. This is done to avoid any unintended gravity load transfer from the floor or roof beams/girders to the non-load-bearing vertical elements.
- The above measures are design phase deflection mitigation strategies, but deflections can also be controlled through the following construction practices:
  - Do not load the concrete structure until it reaches its design strength.
  - Provide adequate shoring to the concrete member during construction until it reaches its design strength.
  - Ensure that the reinforcement in the concrete member is not displaced or misplaced during the concrete pour.

7.9 CRACK CONTROL

With the advent of higher strength reinforcing steels, where more strain is required to produce higher stresses, cracking of reinforced concrete flexural members has become more troublesome. Some of the reasons why crack control is necessary include aesthetics considerations, preventing the intrusion of deleterious materials such as salts and chemicals through the cracks into the concrete that could lead to degradation of the concrete and corrosion of the steel reinforcement, and the need to avoid the perception in the public's mind that the structure is in imminent danger of collapse due to the presence of noticeable cracks.

It seems logical that cracking would have an effect on corrosion of the reinforcing steel. However, there is no clear correlation between corrosion and surface crack widths in the usual range found in structures with reinforcement stresses at service load levels (see ACI 318-19, Section R24.3.2). Further, there is no clear experimental evidence available regarding the crack width beyond which a corrosion danger exists. Exposure tests indicate that concrete quality, adequate consolidation, and ample concrete cover may be more important in corrosion considerations than crack width.

Rather than a small number of large cracks, it is more desirable to have only hairline cracks and to accept more numerous cracks, if necessary. To achieve this, the current ACI Code (Section 24.3) directs that the flexural tension reinforcement be well distributed in the maximum tension zones of a member. ACI Code Section 24.3.2 contains a provision for maximum spacing s that is intended to control surface cracks to a width that is generally acceptable in practice.

The maximum spacing of flexural non-prestressed reinforcement (see ACI Code Table 24.3.2) is limited to

$$s = 15\left(\frac{40{,}000}{f_s}\right) - 2.5c_c \le 12\left(\frac{40{,}000}{f_s}\right)$$

where

s = center-to-center spacing of flexural tension reinforcement nearest to the tension face, in.

f_s = calculated stress, psi. This may be taken as $\frac{2}{3}$ of the specified yield strength (i.e., 40,000 psi for f_y = 60,000 psi)

c_c = clear cover from the nearest surface in tension to the surface of the flexural tension reinforcement, in.

ACI 318, Section 24.3.5, cautions that if a structure is designed to be watertight or if it is to be subjected to very aggressive exposure, the provisions of Section 24.3.2 are not sufficient, and special investigations and precautions are required. ACI 224-R-01 "Control of Cracking in Concrete Structures" specifies maximum reasonable crack widths for reinforced concrete under service loads: 0.016 inch for concrete exposed to dry air or with a protective membrane; 0.012 inch for concrete exposed to humidity, moist air, or soil; 0.007 inch for concrete exposed to deicing chemicals; 0.006 inch for concrete exposed to seawater, and 0.004 inch for water retaining structures. See ACI 350—"Code Requirements for Environmental Engineering Concrete Structures" for cover and other requirements for liquid retaining structures, and ACI 357.3—"Guide for Design and Construction of Waterfront and Coastal Concrete Marine Structures" for cover requirements for structures exposed to seawater.

EXAMPLE 7.4

Check the steel distribution for the beam shown in Figure 7.9 to establish whether reasonable control of flexural cracking is accomplished in accordance with the ACI Code, Section 24.3. Use f_y = 60,000 psi. Assume d = 30 in.

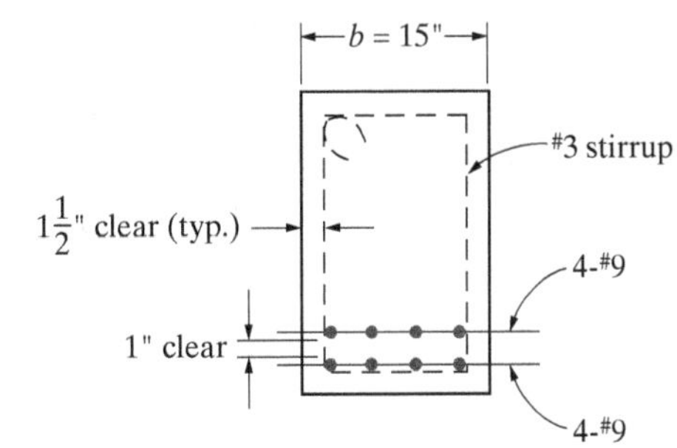

FIGURE 7.9 Sketch for Example 7.3.

Solution:

1. Calculate the center-to-center spacing between the No. 9 bars:

$$s = \frac{15 - 2(1.5) - 2(0.375) - 2\left(\frac{1.128}{2}\right)}{3} = 3.37 \text{ in.}$$

2. Assume a positive moment and calculate the concrete, clear cover from the bottom (tension) face of the beam to the surface of the nearest tension reinforcement:

$$c_c = 1.5 + 0.375 = 1.875 \text{ in.}$$

3. Calculate f_s using $\frac{2}{3}$ of f_y:

$$f_s = \frac{2}{3} f_y = \frac{2}{3}(60{,}000) = 40{,}000 \text{ psi}$$

4. Calculate the maximum spacing allowed for deformed bars using ACI Code Table 24.3.2:

$$\begin{aligned} s &= 15\left(\frac{40{,}000}{f_s}\right) - 2.5c_c \le 12\left(\frac{40{,}000}{f_s}\right) \\ &= 15\left(\frac{40{,}000}{40{,}000}\right) - 2.5(1.875) \le 12\left(\frac{40{,}000}{40{,}000}\right) \\ &= 10.31 \text{ in.} \le 12 \text{ in.} \end{aligned}$$

Therefore, the maximum spacing = 10.31 in.

And lastly, the actual spacing of the reinforcement is compared to the maximum spacing:

$$3.37 \text{ in.} < 10.31 \text{ in.} \qquad \text{(O.K.)}$$

When beams are relatively deep, there exists the possibility for surface cracking in the tension zone areas away from the main reinforcing. For beams and girders with depths h in excess of 36 in., ACI 318, Section 9.7.2.3 requires the placing of longitudinal skin reinforcing along both side faces for a distance $h/2$ from the tension face of the beam. The vertical spacing s between these surface longitudinal bars or wires (also called skin reinforcement) shall not exceed the spacing specified in ACI 318 Table 24.3.2, as presented at the beginning of this section. Bar sizes ranging from No. 3 to No. 5 (or welded wire reinforcement with a minimum area of 0.1 in.^{2} per foot of depth) are typically used as skin reinforcement.

EXAMPLE 7.5

Select skin reinforcement for the cross-section shown in Figure 7.10a. Flexural tension reinforcement is 5 No. 9 bars and f_y = 60,000 psi.

Solution:

Because $h > 36$ in., skin reinforcement is required. The skin reinforcing must cover the tension surface for a minimum of $h/2$, or 20.5 in., up from the tension face of the beam. This is shown in Figure 7.10a. Assume No. 3 bars (A_b = 0.11 in.^{2}) for the skin reinforcing and calculate the maximum spacing s as follows:

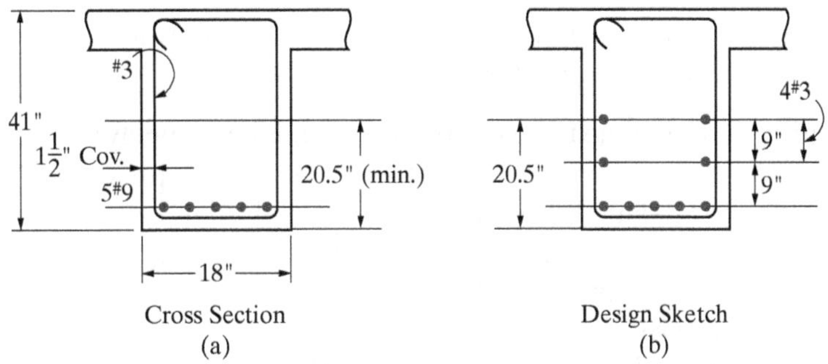

FIGURE 7.10 Sketch for Example 7.4.

$$c_c = 1.5 + 0.375 = 1.875 \text{ in.}$$

$$f_s = \frac{2}{3} f_y = \frac{2}{3}(60{,}000) = 40{,}000 \text{ psi}$$

$$\begin{aligned} s &= 15\left(\frac{40{,}000}{f_s}\right) - 2.5c_c \le 12\left(\frac{40{,}000}{f_s}\right) \\ &= 15\left(\frac{40{,}000}{40{,}000}\right) - 2.5(1.875) \le 12\left(\frac{40{,}000}{40{,}000}\right) \\ &= 10.31 \text{ in.} \le 12 \text{ in.} \end{aligned}$$

The distance from the tension face of the beam (i.e., the bottom of the beam) to the centroid of the tension reinforcement is

$$1.5 + 0.375 + \frac{1.128}{2} = 2.44 \text{ in.}$$

Therefore, the required number of spaces N of the longitudinal surface bars is

$$N = \frac{20.5 - 2.44}{10.31} = 1.75 \text{ spaces} \quad \text{(Use 2 spaces)}$$

The actual spacing provided is

$$\frac{20.5 - 2.44}{2 \text{ spaces}} \approx 9 \text{ in.} < 10.31 \text{ in.} \qquad \text{(O.K.)}$$

The design is shown in Figure 7.10b.

7.10 FLOOR VIBRATIONS

Concrete floors – because of their inherent mass - are not as susceptible to issues related to floor vibrations compared to structures framed with wood or steel, but attention needs to be given to the design of concrete floors for floor vibrations under certain conditions. The use of thinner floor slabs because of higher concrete strengths, and the use of slabs with complicated geometry or long spans will necessitate checking the floor vibrations. There are several factors that need to be considered in the assessment of floor vibrations, and some of the basic terms used in the assessment are as follows [1, 2]:

Damping: This is a measure of how much and how quickly vibrations will dissipate within a structure. Damping in a structural system is usually expressed as a percentage of critical damping, and critical damping is that which is required to bring the system to rest in one-half cycle. In general, the presence of full or partial height partitions is the critical factor in the value assumed for damping. For floors with few or no partitions, the damping ratio is 0.02. For partial height partitions, the value is 0.03, and for full height partitions, the value is 0.05. This damping ratio is directly proportional to the acceleration of the floor, so its value needs to be carefully determined.

Period: Time, in seconds, for one complete cycle of oscillation.

Frequency: Number of oscillations per second in hertz (Hz) or cycles per second, $f = 1/T$. A person walking at a pace of two steps per second is said to walk at a frequency of 2 Hz or two cycles per second.

Forcing Frequency: The frequency, in hertz, of the applied dynamic force. For a rhythmic activity such as dancing or aerobics, the forcing frequency of the activity can occur between 1.5 Hz and 3.0 Hz for the first harmonic, 3.0 Hz–5.5 Hz for the second harmonic, and 6.0Hz–8.25Hz for the third harmonic.

Harmonic: An integer multiple of a forcing frequency. Any forcing frequency can have an infinite number of harmonics, but human activities are generally limited to a maximum of three harmonics. For example, for an applied forcing frequency of 2 Hz, the first harmonic is 2 Hz, the second harmonic is 4 Hz, and the third harmonic is 6 Hz.

Natural Frequency: The frequency at which a structure vibrates when it is displaced and then suddenly released from the at-rest state. This is also called free vibration since no other external forces are applied. The natural frequency of a structure is proportional to its stiffness. The natural frequency of a structure should be greater than the forcing frequency causing the vibration excitation, to avoid resonance.

Resonance: A phenomenon where the forcing frequency (or a harmonic multiple of the forcing frequency), of the dynamic activity coincides with one of the natural frequencies, f_n, of the structure. This causes very large displacements, velocity, acceleration, and stress. For a person walking at a pace of two steps per second, the floor will have to be checked for the first three harmonics of this forcing frequency (i.e., 2 Hz, 4 Hz, and 6 Hz)

There are three basic types of vibrations that are common in buildings [1, 2]: walking, rhythmic, and sensitive equipment vibrations. Most concrete structures satisfy walking vibrations requirements [3], and so this criterion is typically not a critical concern in the design process. To satisfy walking vibrations, the peak acceleration of the floor system needs to be less than a prescribed limit that has been established through research. The acceleration limit is described in terms of a percentage of the acceleration due to gravity, g. This limit, which is defined as a_0/g, is limited to 0.005 or 0.5% for basic walking vibrations. For shopping malls and indoor footbridges, this limit is increased to 0.015, and for outdoor footbridges, the limit is 0.05. For dining and dancing, this acceleration limit is 1.5%g; for rhythmic activities, the acceleration limit is 5%g [1]. The most critical component of the vibration analysis is the stiffness of the floor system, which is captured in the natural frequency equation of the floor system, defined as:

$$f_n = 0.18\sqrt{\frac{g}{\Delta_c + \Delta_j + \Delta_g}}$$

where

f_n = natural frequency of the floor system

g = acceleration due to gravity, 386 in./s^2

Δ_c = column deflection (in.)

Δ_j = joist or beam deflection (in.)

Δ_g = girder deflection (in.)

The girder and joist (or beam) deflections can be calculated using the dynamic elastic modulus of the concrete due to the transient nature of vibrations [2, 3]. The dynamic elastic modulus of concrete can be taken as $1.2E_c$ [10].

The natural frequency of more complicated systems such as flat plates, two-way joists, and grillage systems are found in reference [2] or they can be obtained using a finite element analysis, where the slab is modeled with plate elements.

The peak acceleration as a fraction of the acceleration due to gravity is given as

$$\frac{a_p}{g} = \frac{P_o e^{-0.35f}}{\beta W} \leq \frac{a_o}{g}$$

Where

P_o = amplitude of the force causing the vibration = DLF x Weight of one person taken as 150 Ib

g = acceleration due to gravity = 32.2 ft/s^2

β = damping ratio

f = fundamental natural frequency of vibration of the floor, Hz

a_o/g = acceleration tolerance limit as a fraction of the acceleration due to gravity

W = effective weight of the floor = self-weight of the slab panel plus any superimposed dead load, Ib

The calculated acceleration is compared to the prescribed limit for the given human activity. For more information and examples on vibrations due to walking, dining and dancing, and rhythmic and aerobic activities, the reader should refer to Ref. [4].

The most common variables used to control vibrations are the span and the cross-sectional size of the members. Adjusting the modulus of elasticity, for example, would help with vibration issues, but typically it is more economical to adjust the beam and girder depths to control vibrations.

It is always advisable to locate any rhythmic activity on the ground floor of a structure where possible. For activities on elevated floors that could cause rhythmic vibrations, such as dancing, jumping exercises, aerobics, concerts, and sporting events, utilizing a two-way floor system is often more economical [3]. To avoid floor vibration problems during rhythmic activities, the natural frequency of the floor system should be greater than a specified minimum value. For any given activity, there are several harmonics of vibration. For example, the participants in a lively concert might induce a forcing frequency of 2.0 Hz for the first harmonic and 4.0 Hz for the second harmonic. The floor would have to be checked for both harmonics. For most rhythmic activities, not more than three harmonics would need to be checked. The natural frequency limits are the same as stated before for the forcing frequencies of certain activities.

H. Bachmann [5] recommends the following minimum natural frequencies for reinforced concrete structures subject to human-induced vibrations:

- Pedestrian bridge: 4.6 Hz
- Office buildings: 7.5 Hz
- Gymnasiums and sports halls: 7.5 Hz
- Dance halls and concert halls *without* fixed seating: 6.5 Hz
- Concert halls with fixed seating: 3.4 Hz

For floor structures that support sensitive equipment, the floor motion is expressed in terms of the velocity of vibration because the design criteria for equipment often correspond to a constant velocity over the frequency range considered. As was the case with rhythmic vibrations, it is often more economical to utilize a two-way floor system to control equipment vibrations. The natural frequency of the floor system is still a critical factor in the assessment of a floor for the support of sensitive equipment.

The maximum velocity of vibration experienced by the equipment is given as [4]:

$$V = \frac{P_o g e^{-0.35f}}{2\pi f \beta W}$$

Where

V = maximum velocity of vibration of the floor panel in micro-inch per second (mips) or μ-in./s

P_o = the force causing the excitation for a person walking = 45 Ib

W = self-weight of the slab panel where the equipment is located, plus any superimposed dead load, Ib

f_f = forcing frequency or the frequency of the excitation, Hz

DLF = dynamic load factor for the first harmonic of the forcing function $\approx 0.5f_f - 0.45$ (derived from Figure X.4.1-1 in Ref. [4] for $f_f \geq 1.0$ Hz)

With a walking speed of 1.5 steps per second (i.e., f_f = 1.5 Hz), DLF ≈ 0.5(1.5) - 0.45 = 0.30, therefore, P_o = (0.3)(150 Ib weight of a person) = 45 Ib.

Other terms are as previously defined. The maximum vibrational velocity is calculated and compared to the prescribed allowable vibrational velocity for the given equipment. For medical equipment, see Table X-8.2-1 in Ref. [4] for the prescribed maximum velocities by the National Institutes of Health (NIH). The manufacturers of sensitive equipment typically have the vibration limits for their equipment available to the design engineer.

Common ways to mitigate vibrations due to walking would be to locate the equipment away from corridors or to limit the length of corridors, which has the effect of slowing the walking speed of the occupants.

A floor will be considered acceptable to support sensitive equipment if the maximum velocity of the floor system is less than the prescribed value, which can be obtained from the manufacturer of the equipment. Computer floors, for example, can have a velocity limit of 8000 μ-in/s, and operating rooms may have a vibrational velocity limit of as low as 1000 μ-in/s. The most sensitive equipment can be found in a Class E microelectronics manufacturing facility, where the maximum allowable vibrational velocity is about 130 μ-in./sec [6, 7]. This would require a floor structure that has a natural frequency close to 50 Hz [6]. Floor structures that support microelectronics manufacturing are often designed with a two-way system such as a waffle slab [6, 7, 8]. For preliminary design, the minimum thicknesses and maximum spans for flat plate systems for various vibration excitations can be obtained from Tables 1, 3, and 5 of Ref. [7]. For more stringent vibrational limits, as would be required in microelectronics facilities where thicker waffle slabs or two-way joists must be used, the reader should refer to Ref. [8].

7.11 GROSS AND CRACKED SECTION PROPERTIES OF CONCRETE SECTIONS

The principle of transformed sections from mechanics of materials was used to derive the cracked section properties of concrete sections in Section 7.3. The equations for gross moment of inertia and cracked moment of inertia for rectangular and flanged (i.e., T- or L-shape) concrete sections are summarized below [9]:

Section properties for *rectangular* concrete sections with tension reinforcement only:

$$I_{cr} = \frac{b\bar{y}^3}{3} + nA_s(d-\bar{y})^2$$

$$\bar{y} = \frac{\left[\sqrt{(2dB+1)}-1\right]}{B}$$

$$B = \frac{b}{nA_s}$$

$n = \frac{E_s}{E_c}$ = modular ratio

A_s = area of tension reinforcement

$$I_g = \frac{bh^3}{12}$$

$$y_t = 0.5h$$

$\overline{y}$ is the distance from the compression face to the neutral axis.

Section properties for *T or L*-sections with tension reinforcement only:

$$I_{cr} = \frac{(b - b_w)h_f^3}{12} + \frac{b_w \overline{y}^3}{3} + (b - b_w)(h_f)(\overline{y} - 0.5h_f)^2 + nA_s(d - \overline{y})^2$$

$$\overline{y} = \frac{\left[\sqrt{C(2d + h_f Z) + (1+Z)^2}\right] - (1+Z)}{C}$$

$$C = \frac{b_w}{nA_s}$$

$$Z = \frac{(b - b_w)h_f}{nA_s}$$

$$y_t = h - 0.5\left(\frac{[(b - b_w)h_f^2 + b_w h^2]}{[(b - b_w)h_f + b_w h]}\right)$$

$$I_g = \frac{(b - b_w)h_f^3}{12} + \frac{b_w h^3}{12} + h_f(b - b_w)\left(h - 0.5h_f - y_t\right)^2 + b_w h\left(y_t - 0.5h\right)^2$$

$n = \frac{E_s}{E_c}$ = modular ratio

b = effective flange width of the compression zone of the T- or L-beam

A_s = area of tension reinforcement

$\overline{y}$ = distance from *compression* face to the neutral axis of the cracked section

y_t = distance from *tension* face to the neutral axis of the un-cracked section

REFERENCES

[1] Thomas M. Murray, David E. Allen, and Eric E. Ungar. American Institute of Steel Construction. *Steel Design Guide Series 11: Floor Vibrations Due to Human Activity*. Chicago, IL: AISC., 2003.

[2] Fanella, D.A. and Mota, M. *Design Guide for Vibrations of Reinforced Concrete Floor Systems*. Schaumburg, IL: Concrete Reinforcing Steel Institute, 2014.

[3] Fanella, D.A. and Mota, M. "Vibration of Reinforced Concrete Floor Systems." *Structure Magazine*, April 2015.

[4] B. Aalami, "Vibration Design of Concrete Floors," P-T Structures Technical Note TN290, https://ptstructures.com/uploads/pdf/V1.pdf, 29 pp., 2014. (accessed December 11, 2023)

[5] H. Bachmann. "Case Studies of Structures with Man-induced Vibrations," ASCE Journal of Structural Engineering, Vol. 118, Issue 3, March 1992.

[6] Hal Amick and Paulo J. M. Monteiro. "Construction of Nanotechnology Facilities," Concrete International, pp. 68-72, March 2004.

[7] David A. Fanella and Michael Mota. "Vibration Excitations – Part 1: How to Select a Reinforced Concrete Floor System," Structure Magazine, September 2017.

[8] David A. Fanella and Michael Mota. "Vibration Excitations – Part 2: How to Select a Reinforced Concrete Floor System," Structure Magazine, September 2017.

[9] David A. Fanella, Javeed A. Munshi, and Basile G. Rabbat. *Notes on ACI 318-99*. Portland Cement Association, Skokie, IL., 1999.

EXERCISES

For the following exercises, unless otherwise noted, assume normal-weight concrete.

7.1 Locate the neutral axis and calculate the moment of inertia for the cracked transformed cross sections shown. Use $f'_c = 4000$ psi and $f_y = 60{,}000$ psi.

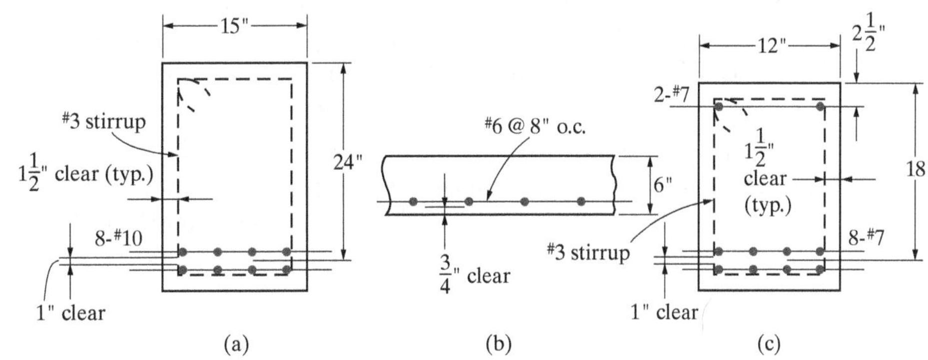

EXERCISE 7.1

7.2 Find I_g and I_{cr} for the T-beam shown. The effective flange width is 60 in., and $f'_c = 4000$ psi.

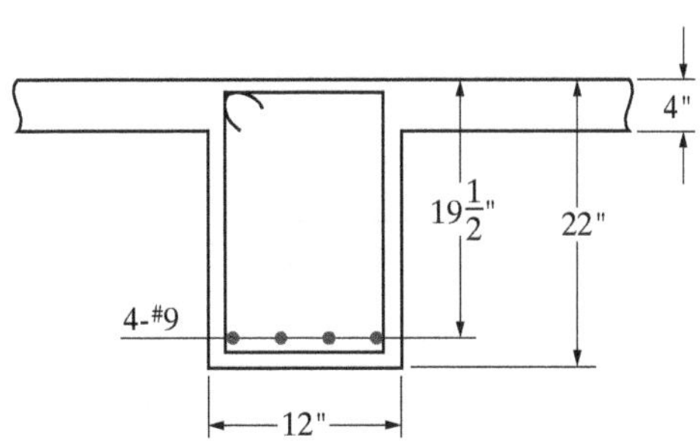

EXERCISE 7.2

7.3 The beam of cross section shown is on a simple span of 20 ft and carries service loads of 1.5 kip/ft dead load (includes beam weight) and 1.0 kip/ft live load. Use $f'_c = 3000$ psi and $f_y = 60{,}000$ psi.

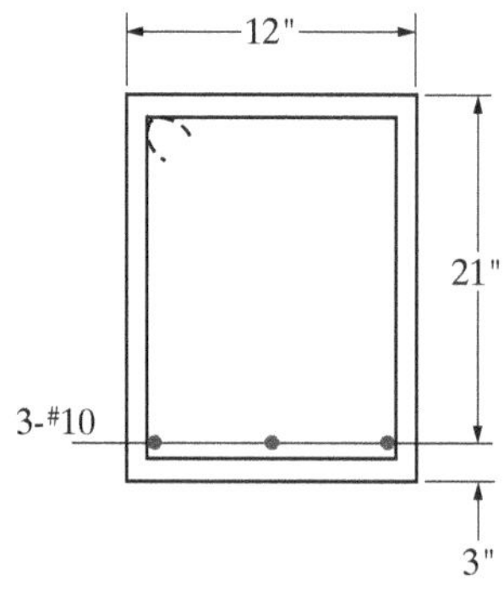

EXERCISE 7.3

a. Compute the immediate deflection due to dead load and live load.

b. Compute the long-term deflection due to the dead load. Assume the time period for sustained loads to be in excess of 5 years.

c. Develop a spreadsheet to solve this exercise and to confirm your calculations.

7.4 Rework Exercise 7.3 using a beam on a simple span of 26 ft, service load of 0.8 kip/ft dead load (includes beam weight) over the full span, and a point live load of 12 kip at midspan. Assume tensile reinforcing to be three No. 9 bars and $f'_c = 4000$ psi.

7.5 The floor beam shown is on a simple span of 16 ft. The beam supports nonstructural elements likely to be damaged by large deflections. The service loads are 0.6 kip/ft dead load (does not include the beam weight) and 1.40 kip/ft live load. Assume that the live load is 60% sustained for 6-month periods. Use $f'_c = 3000$ psi and $f_y = 60{,}000$ psi.

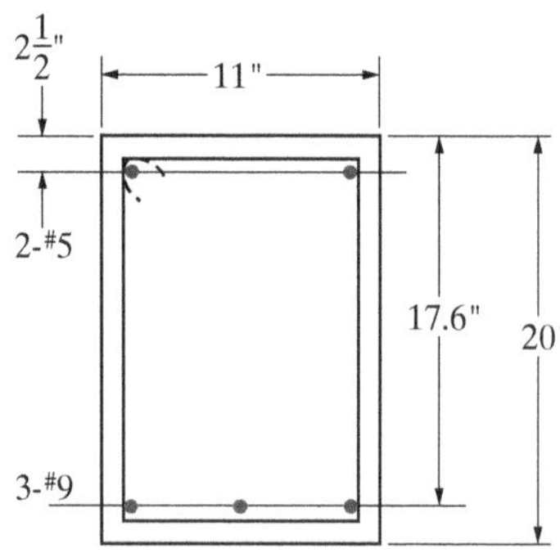

EXERCISE 7.5

a. Check the beam for deflections.

b. If the beam is unsatisfactory, redesign it so that it meets both flexural strength and deflection requirements.

7.6 Calculate the expected long-term deflection due to dead load and sustained live load for (a) the slab and (b) the beam of the floor system designed for Exercise 6.4. Assume 10% sustained live load and a time period in excess of 5 years.

7.7 Check the cross sections of Problems 7.1(a) and (b) for acceptability under the ACI Code provisions for the distribution of flexural reinforcement (crack control).

7.8 Check the distribution of flexural reinforcement for the members designed in Exercises 2.21 and 2.25. If necessary, redesign the steel.

7.9 Refer to the Beam diagram and detail shown. The loads shown are service level and the self-weight is included in the dead load. Assume 20% of the live load is permanently sustained for more than 5 years. Use f'_c = 4000 psi, f_y = 60,000 psi

a. Find the effective moment of inertia, I_e

b. Determine the deflections due to Dead, Live, and Total loads; compare with a deflection limit of $L/480$, where L is the span of the beam.

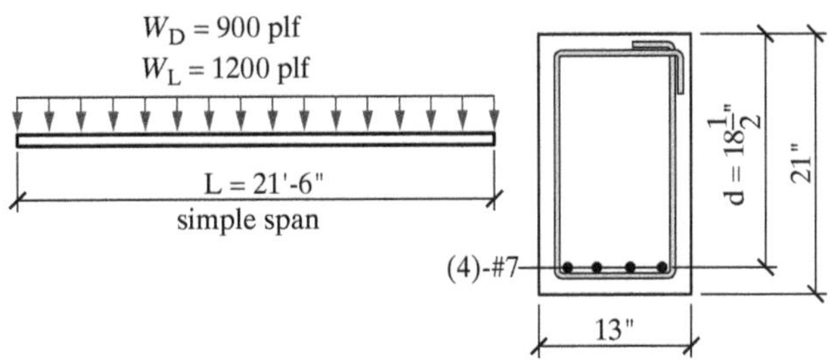

EXERCISE 7.9

7.10 Refer to the concrete beam diagram and detail shown. The loads shown are service level, and the self-weight is included in the dead load. Assume 25% of the live load is permanently sustained for more than 5 years. Use f'_c = 4000 psi, f_y = 60,000 psi

a. Find the effective moment of inertia, I_e

b. Determine the deflections due to Dead, Live, and Total loads; is the beam adequate for total load deflection? Compare with a deflection limit of $L/240$, where L is the span of the beam.

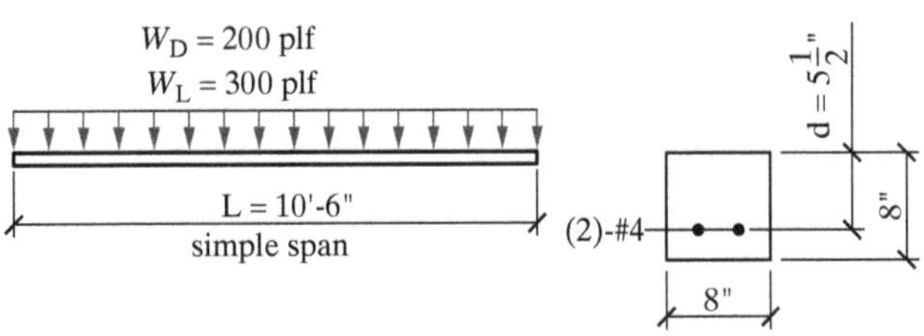

EXERCISE 7.10

CHAPTER 8

Structural Walls

8.1 INTRODUCTION

Structural walls are used to support gravity loads (i.e., axial force), provide lateral support for an earth fill, embankment, or fluids such as water or sewage when acting as retaining walls (i.e., results in out-of-plane moments and shears), and provide sliding and overturning moment resistance to lateral forces (e.g., wind and seismic) when acting as shear walls (i.e., results in in-plane moments and shears). See ACI Figure R11.4.1.3 for the in-plane and out-of-plane forces that can act on structural walls. Some common types of structural walls are shown in Figure 8.1. When acting as retaining walls, one primary purpose for these walls is to maintain a difference in the elevation of the ground surface on each side of the wall. The soil whose ground surface is at a higher elevation is commonly called the *backfill*, and the wall is said to retain this backfill. Hence, the structural walls in Figure 8.1 may be broadly termed *retaining structures* or *retaining walls*.

The structural walls shown in Figure 8.1 have applications in either building or bridge projects. They do not necessarily behave in an identical manner under load but still serve the same basic function of providing lateral support for a mass of soil or other material that is at a higher elevation behind the wall than the soil or other material in front of the wall. Some retaining walls may support vertical loads and lateral loads from the retained materials. Note that structural walls resisting out-of-plane loads like lateral earth pressures, hydrostatic pressures, or wind loads perpendicular to its surface behave like one-way or two-way slabs. In contrast, structural walls that resist inplane shear forces from lateral wind and seismic forces behave like shear walls, which will be discussed later in this chapter. In both scenarios, the minimum transverse and longitudinal reinforcement for structural walls presented in ACI Code, Sections 11.6.1 and 11.6.2 will usually govern over that for slabs presented in ACI Code, Sections 7.6.4.1 and 24.4.3 or 24.4.3.2.

The gravity wall (Figure 8.1a) depends mostly on its own weight for stability. It is usually made of plain concrete and is used for walls up to approximately 10 ft in height. The semigravity wall is a modification of the gravity wall in which small amounts of reinforcing steel are introduced. This, in effect, reduces the massiveness of the wall.

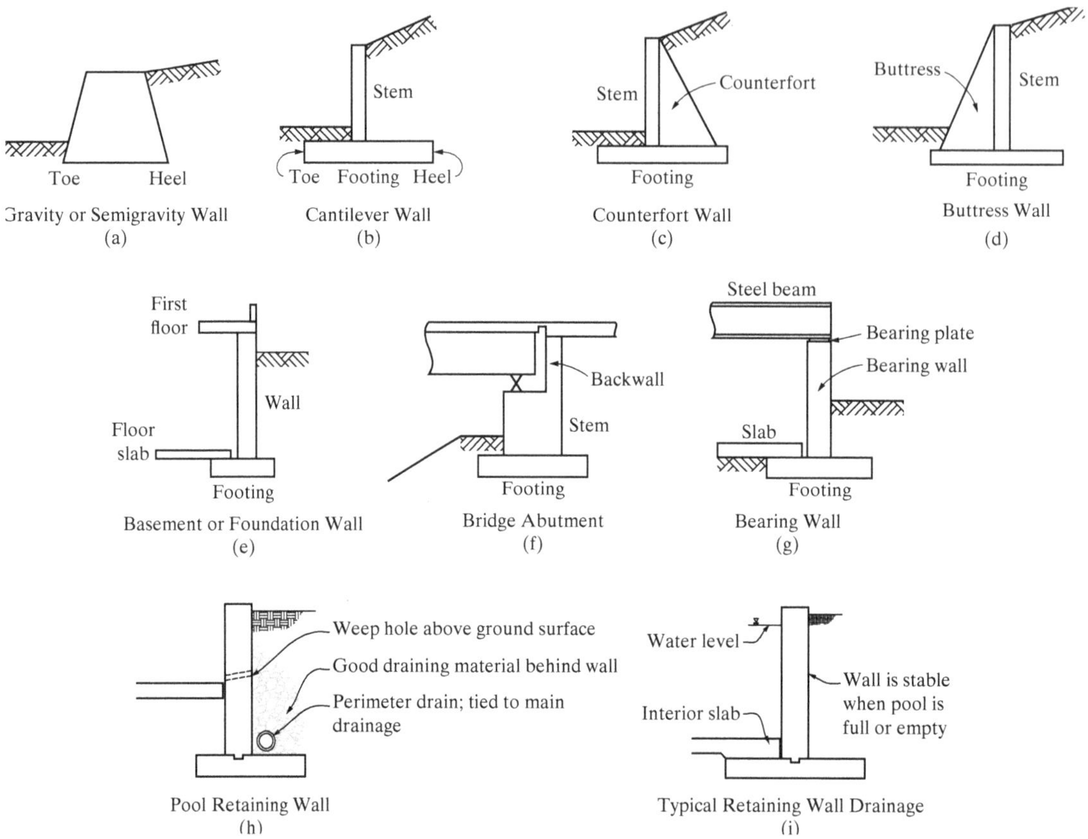

FIGURE 8.1 Common types of retaining walls.

The cantilever wall (Figures 8.1b and 8.2) is the most common type of retaining structure and is generally used for walls in the range from 10 to 25 ft in height. It is so named because its individual parts (toe, heel, and stem) behave as, and are designed as cantilever beams. Aside from its stability, the capacity of the wall is a function of the strength of its individual parts.

The counterfort wall (Figure 8.1c) may be economical when the wall height is more than 25 ft. The counterforts are spaced at intervals and act as tension members to support the stem. The stem is then designed as a continuous member spanning horizontally between the counterforts.

The buttress wall (Figure 8.1d) is similar to the counterfort wall except that the buttresses are located on the side of the stem opposite to the retained material and act as compression members to support the stem. The counterfort wall is more commonly used because it has a clean, uncluttered, exposed face and allows more efficient use of space in front of the wall.

The basement or foundation wall (Figure 8.1e) may act as a cantilever retaining wall. The first floor may provide an additional horizontal reaction similar to the basement floor slab, however, thereby making the wall act as a vertical beam. This wall would then be designed as a simply supported member spanning between the first floor and the basement floor slab.

The bridge abutment (Figure 8.1f) is similar in some respects to the basement wall. The bridge superstructure induces horizontal as well as vertical loads, thus altering the normal cantilever behavior.

The bearing wall (Figure 8.1g) may exist with or without lateral loads. A bearing wall may be defined as a wall that supports any vertical load in addition to its own weight. Depending on the magnitudes of the vertical and lateral loads, the wall may have to be designed for combined bending and axial compression. Bearing walls and basement walls are further discussed later in this chapter.

Figure 8.1h shows a swimming pool retaining wall and Figure 8.1i shows the typical drainage provided at the perimeter of retaining walls. The drainage is provided by means of weep holes that are spaced at regular intervals along the length of the wall.

8.2 LATERAL FORCES ON RETAINING WALLS

The design of a retaining wall must account for all the applied loads. The load that presents the greatest problem and is of primary concern is the lateral earth pressure induced by the retained soil. Comprehensive earth pressure theories evolving from the original Coulomb and Rankine theories can be found in almost any textbook on soil mechanics.

The magnitude and direction of the pressures as well as the pressure distribution exerted by a soil backfill upon a wall are affected by many variables. These variables include, but are not limited to, the type of backfill used, the drainage of the backfill material, the level of the water table, the slope of the backfill material, added loads applied on the backfill, the degree of soil compaction, and movement of the wall caused by the action of backfill.

FIGURE 8.2 Cantilever retaining wall, h_w = 12'-0" (George Limbrunner)

Figure 8.2 shows a retaining wall under construction. An important consideration is that water must be prevented from accumulating in the backfill material. Walls are rarely designed to retain saturated material, which means that proper drainage must be provided. It is generally agreed that the best backfill material behind a retaining structure is a well-drained, cohesionless material. Hence, it is the condition that is usually specified and designed for. Materials that contain combinations of types of soil will act like the predominant material. An example of an acceptable backfill used behind basement and retaining walls is "well graded sand and gravel or crusher stone having a maximum size of 3 in. and no more than 7% of particles passing the #200 sieve." For basement walls, the perimeter drainage system could consist of a 4-in. diameter perimeter drain wrapped in geotextile material that acts as a filter and located in a bed of gravel just above the underside of the wall footing are provided around the outside perimeter of the building near the basement wall footing. Drainage for retaining walls can be provided using a sloped 3-in. diameter pipe through the thickness of the stem of the retaining wall near the bottom of the retaining wall stem that act as weep holes to drain water from the back of the retaining wall to the front of the wall. Typical horizontal spacing of the weep holes is 8 ft on centers and located vertically within 1 ft from the top of the retaining wall footing. These drainage systems for basement and retaining walls help to prevent the buildup of hydrostatic pressures behind these walls as well as help to prevent uplift pressures on the slab-on-grade in buildings with basement walls. Where the natural water table is only a few feet below grade, the retained soil behind retaining walls and basement walls will be saturated; in this situation, the walls must be designed for the hydrostatic pressures in addition to the lateral pressures from the saturated soil.

The lateral earth pressure can exist and develop in three different categories: active state, at rest, and passive state. If a wall is rigid, earth pressure at rest will develop. If the wall deflects or moves a very small amount away from the backfill, active earth pressure will develop and, in effect, reduce the lateral earth pressure occurring in the at-rest state. Should the wall be forced to move toward the backfill for some reason, passive earth pressure will develop and increase the lateral earth pressure appreciably above that occurring in the at-rest state. As indicated, the magnitude of earth pressure at rest lies somewhere between active and passive earth pressures.

Under normal conditions, earth pressure at rest is of such a magnitude that the wall deflects slightly, thus relieving itself of the at-rest pressure, and resulting in active earth pressure. For this reason, retaining walls are generally designed for active earth pressure due to the retained soil.

Because of the involved nature of a rigorous analysis of an earth backfill and the variability of the material and conditions, assumptions and approximations are made with respect to the nature of lateral pressures on a retaining structure. It is common practice to assume linear active and passive earth pressure distributions. The pressure intensity is assumed to increase with depth as a function of the weight of the soil in a manner similar to that which would occur for a retained fluid. Hence, this horizontal pressure of the earth against the wall is frequently called an *equivalent fluid pressure*. Experience has indicated that walls designed based on these assumptions and those of the following discussion are safe and relatively economical.

Level Backfill

If we consider a level backfill (of well-drained cohensionless soil), the assumed pressure diagram is shown in Figure 8.3. The unit pressure intensity *py* in any plane a distance *y* down from the top is

$$p_y = K_a w_e y$$

Therefore, the total active earth pressure acting on a 1-ft width of wall may be calculated as the product of the average pressure on the total wall height *hw* and the area on which this pressure acts:

$$H_a = \tfrac{1}{2} K_a w_e h_w^2 \qquad \textbf{(8.1)}$$

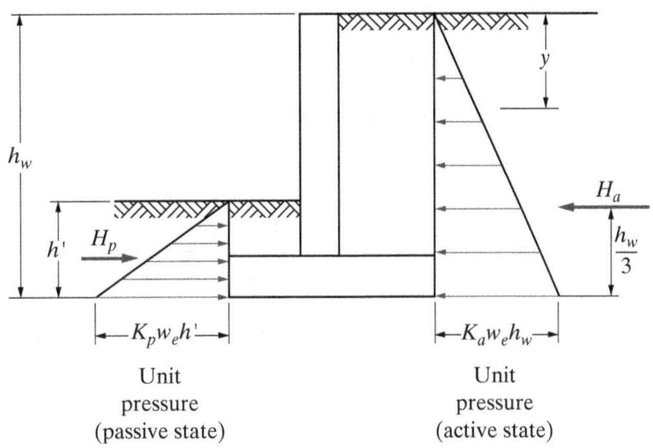

FIGURE 8.3 Analysis of forces acting on walls: level backfill.

where K_a, the *coefficient of active earth pressure*, has been established by both Rankine and Coulomb to be

$$K_a = \frac{1 - \sin\phi}{1 + \sin\phi} = \tan^2\left(45° - \frac{\phi}{2}\right) \qquad \textbf{(8.2)}$$

and

$$w_e = \text{unit weight of earth } (\text{lb/ft}^3)$$
$$\phi = \text{angle of internal friction (soil on soil)}$$

The coefficient of active earth pressure, K_a, usually varies from 0.27 to 0.40. The term $K_a w_e$ in Equation (8.1) is generally called an *equivalent fluid weight* because the resulting pressure is identical to that which would occur in a fluid of that weight (units are lb/ft^3).

In a similar manner, the total passive earth pressure force may be established as

$$H_p = \tfrac{1}{2} K_p w_e (h')^2 \qquad (8.3)$$

where h' is the height of earth or soil in front of the wall, and K_p is the *coefficient of passive earth pressure:*

$$K_p = \frac{1+\sin\phi}{1-\sin\phi} = \tan^2\left(45° + \frac{\phi}{2}\right) = \frac{1}{K_a}$$

Note that K_p usually varies from 2.5 to 4.0.

The total force in each case is assumed to act at one-third of the height of the triangular pressure distribution from the underside of the retaining wall footing, as shown in Figure 8.3.

Sloping Backfill

If we consider a sloping backfill, the assumed active earth pressure distribution is shown in Figure 8.4, where $H_s = \frac{1}{2}K_a w_e h_b^2$

h_b = the height of the backfill at the back of the footing, and

K_a = the coefficient of active earth pressure.

Thus

$$K_a = \cos\theta\left(\frac{\cos\theta - \sqrt{\cos^2\theta - \cos^2\phi}}{\cos\theta + \sqrt{\cos^2\theta - \cos^2\phi}}\right)$$

where θ is the slope angle of the backfill and ϕ is as previously defined. Note that H_s is shown acting parallel to the slope of the backfill.

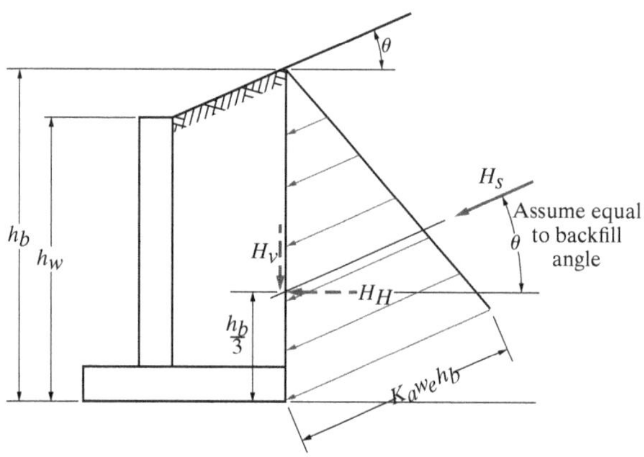

FIGURE 8.4 Analysis of forces acting on walls: sloping backfill.

For walls approximately 20 ft in height or less, it is recommended that the horizontal force component H_H simply be assumed equal to H_s and be assumed to act at $h_b/3$ above the bottom of the footing, as shown in Figure 8.4. The effect of the vertical force component H_V is neglected. This is a conservative approach.

Assuming a well-drained, cohensionless soil backfill that has a unit weight of 110 lb/ft^3 and an internal friction angle ϕ of 33°, values of equivalent fluid weight for sloping backfill may be determined as listed in Table 8.1.

Level Backfill with Surcharge

Loads are often imposed on the backfill surface behind a retaining wall. They may be either live loads or dead loads. These loads are generally termed a *surcharge* and theoretically may be transformed into an equivalent height of earth.

TABLE 8.1 $K_a w_e$ Values for Sloping Backfill

θ (deg)	$K_a w_e$ (lb/ft3)
0	32
10	33
20	38
30	54

A uniform surcharge over the adjacent area adds the same effect as an additional (equivalent) height of earth. This equivalent height of earth h_{su} may be obtained by

$$h_{su} = \frac{w_s}{w_e}$$

where

w_s = surcharge load (lb/ft^2)

w_e = unit weight of earth or soil (lb/ft^3)

In effect, this adds a rectangular pressure behind the wall with a total lateral surcharge force assumed acting at its midheight, as shown in Figure 8.5. Surcharge loads that are far enough removed from the wall cause no additional lateral pressure acting on the wall.

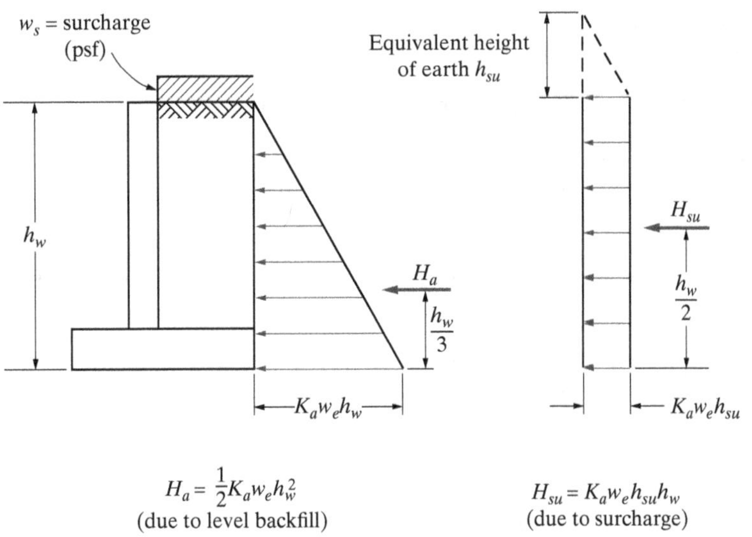

FIGURE 8.5 Forces acting on wall: level backfill and surcharge.

8.3 DESIGN OF REINFORCED CONCRETE CANTILEVER RETAINING WALLS

A retaining wall must be stable and must have sufficient strength to resist the forces acting on it. Four possible *modes of failure* will be considered:

- *Overturning about the toe*, point *O*, as shown in Figure 8.6, could occur due to lateral loads. The stabilizing moment must be sufficiently greater than the overturning moment so that an adequate factor of safety against overturning is provided. The factor of safety against overturning should never be less than 1.5 and should preferably be 2.0 or more.
- *Sliding on the base of the footing*, surface *OM* in Figure 8.6, could also occur due to lateral loads. The resisting force is based on an assumed coefficient of friction between the concrete and the soil beneath the retaining wall footing. The factor of safety against sliding should never be less than 1.5 and should preferably be 2.0 or more.
- *Excessive soil pressure* under the footing will lead to undesirable settlements and possible rotation of the wall. Actual soil pressures should not be allowed to exceed specified allowable pressures, which depend on the characteristics of the underlying soil.
- The *structural failure of component parts of the wall* such as the stem, toe, and heel, each acting as a cantilever beam, could occur. These must be designed to have sufficient strength to resist all anticipated loads.

A *general design procedure* for specific, known conditions may be summarized as follows:

1. Establish the general shape of the wall based on the desired height and function.
2. Establish the site soil conditions, loads, and other design parameters. This includes the determination of allowable soil pressure, earth-fill properties for active and passive pressure calculations from the soil or geotechnical report, and the amount of surcharge, and the desired factors of safety.
3. Establish the tentative proportions of the wall.
4. Analyze the stability of the wall. Check factors of safety against overturning and sliding and compare the actual soil pressure with the allowable soil bearing pressure from the soil report.
5. Assuming that all previous steps are satisfactory, design the component parts of the cantilever retaining wall - stem, toe, and heel - as cantilever beams.

Using a procedure similar to that used for one-way slabs, the analysis and design of cantilever retaining walls is based on a 12-in. (1-ft)-wide strip measured along the length of wall. The preliminary proportions of a cantilever retaining wall may be obtained from the following rules of thumb (see Figure 8.6):

1. Footing width L: Use $\frac{1}{2}h_w$ to $\frac{2}{3}h_w$.
2. Footing thickness h: Use $\frac{1}{10}h_w$.

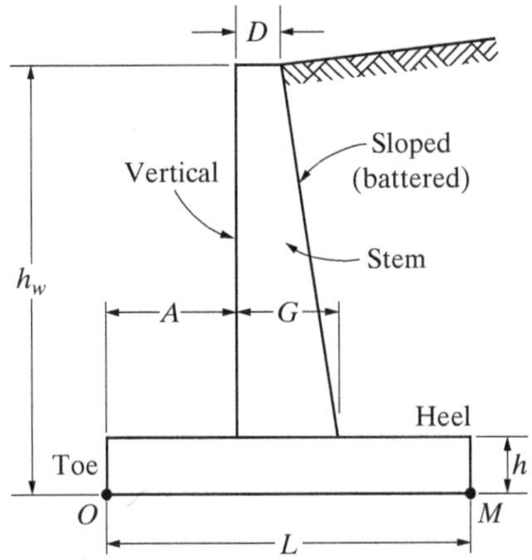

FIGURE 8.6 Cantilever retaining wall proportions.

3. Stem thickness G (at top of footing): Use $\frac{1}{12}h_w$.
4. Toe width A: Use $\frac{1}{4}L$ to $\frac{1}{3}L$.
5. Use a minimum wall batter of $\frac{1}{4}$ in./ft to improve the efficiency of the stem as a bending member and to decrease the quantity of concrete required.
6. The top of stem thickness D should not be less than 10 in.

The given rules of thumb will usually result in walls that can reasonably be designed. Depending on the specific conditions, however, dimensions may have to be adjusted somewhat to accommodate such design criteria as reinforcement limits, shear strength, anchorage, and development. One common alternative design approach is to assume a footing thickness and then immediately design the stem thickness for an assumed steel ratio. Once the stem thickness is established, the wall stability can be checked. Whichever procedure is used, adjustment of dimensions during the design is not uncommon.

EXAMPLE 8.1

Design a retaining wall for the conditions shown in Figure 8.7. Use $f_c' = 3000$ psi and f_y = 60,000 psi. Other design data are given in step 2. Assume normal-weight concrete.

Solution:

1. The general shape of the wall, as shown, is that of a cantilever wall, because the overall height of 18 ft is within the range in which this type of wall is normally economical.
2. The design data are unit weight of earth we = 100 lb/ft^3, allowable soil pressure = 4000 psf, equivalent fluid weight $K_a w_e$ = 30 lb/ft^3, and surcharge load w_s = 400 psf. The desired minimum factor of safety against overturning is 2.0 and against sliding is 1.5.
3. Establish preliminary proportions for the wall.

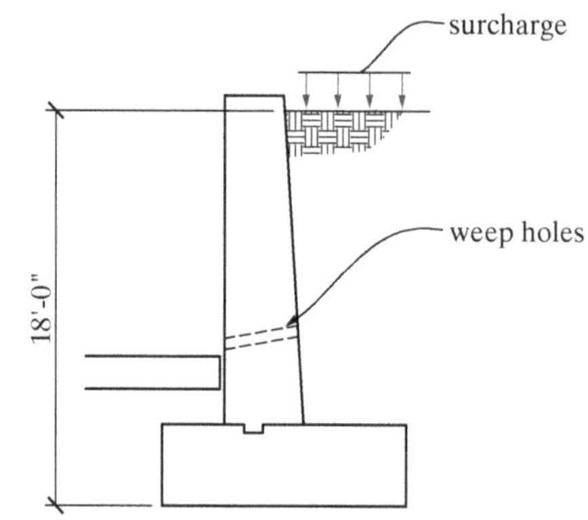

FIGURE 8.7 Sketch for Example 8.1.

a. Footing width:

$$\tfrac{1}{2} \text{ to } \tfrac{2}{3} \text{ of wall height}$$
$$\tfrac{1}{2}(18) \text{ to } \tfrac{2}{3}(18) = 9 \text{ to } 12 \text{ ft}$$

Use 11 ft-0 in.

b. Footing thickness:

$$\tfrac{1}{10}(18) = 1.8 \text{ ft}$$

Use 1 ft-9 in.

c. Stem thickness at top of footing:

$$\tfrac{1}{12}(18) = 1.5 \text{ ft}$$

Use 1 ft-6 in.

d. Toe width:

$$\tfrac{1}{4} \text{ to } \tfrac{1}{3} \text{ footing width}$$
$$\tfrac{1}{4}(11) \text{ to } \tfrac{1}{3}(11) = 2.75 \text{ to } 3.67 \text{ ft}$$

Use 3 ft-0 in.

e. Use a batter for the rear face of wall approximately $\tfrac{1}{2}$ in./ft.

f. Top of stem thickness, based on G and a batter of $\tfrac{1}{2}$ in./ft:

$$\begin{aligned} D &= G - (h_w - h)\tfrac{1}{2} \\ &= 18 \text{ in.} - (18 \text{ ft} - 1.75 \text{ ft})\tfrac{1}{2} \text{ in./ft} = 9.88 \text{ in.} \end{aligned}$$

Use 10 in. Therefore, the calculated batter is

$$\frac{\text{total batter}}{\text{stem height}} = \frac{18 \text{ in.} - 10 \text{ in.}}{18 \text{ ft} - 1.75 \text{ ft}} = 0.492 \text{ in./ft}$$

The preliminary wall proportions are shown in Figure 8.8.

4. For the stability analysis, use unfactored weights and loads in accordance with the ACI Code, Section 13.3.1.1 or 13.4.1.1.

a. *Factor of safety against overturning:* The tendency of the wall to overturn is a result of the horizontal loads acting on the wall. An assumption is made that in overturning, the wall will rotate about the toe, and the horizontal loads are said to create an *overturning moment* about the toe. Any vertical loads will tend to create rotation about the toe in the opposite direction and are, therefore, said to provide a *stabilizing moment.*

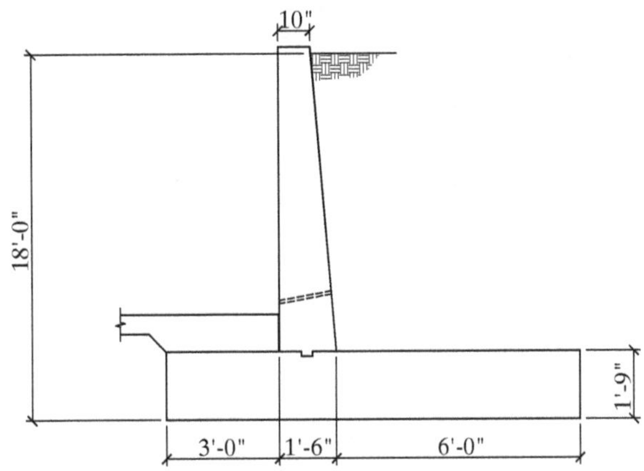

FIGURE 8.8 Preliminary wall proportions, Example 8.1.

The factor of safety (FS) against overturning is then expressed as

$$\text{FS} = \frac{\text{stabilizing moment}}{\text{overturning moment}}$$

The required minimum factor of safety against overturning is normally governed by the applicable building code. A minimum of 1.5 is generally considered good practice. The passive earth resistance of the soil in front of the wall is generally neglected in stability computations because of the possibility of its removal by erosion or excavation.

As discussed previously, the surcharge may be converted into an equivalent height of earth,

$$h_{su} = \frac{w_s}{w_e} = \frac{400}{100} = 4 \text{ ft}$$

thus, adding a rectangle of earth pressure behind the wall.

The various vertical and horizontal forces and their associated moments are shown in Tables 8.2 and 8.3 (see Figure 8.9). Note that the soil on the toe is neglected. Both stabilizing and overturning moments are calculated with respect to point *O* in Figure 8.9.

TABLE 8.2 Stabilizing Moments (Vertical Forces)

Force	Magnitude (lb)	Lever arm (ft)	Moment (ft-lb)
W_1	0.833(16.25)(150) = 2030	3.42	6940
W_2	7.17(16.25 + 4.0)(100) = 14,520	7.42	107,700
W_3	$\left(\frac{1}{2}\right)$ (16.25)(0.67)(50) = 272	4.05	1102
W_4	11.0(1.75)(150) = 2890	5.5	15,900
	ΣW = 19,710		ΣM = 131,600

TABLE 8.3 Overturning Moments (Horizontal Forces)

Force	Magnitude (lb)	Lever arm (ft)	Moment (ft-lb)
H_1	$1/2(30)(18)^2$ = 4860	6.0	29,200
H_2	4(30)(18) = 2160	9.0	19,440
	ΣH = 7020		ΣM = 48,600

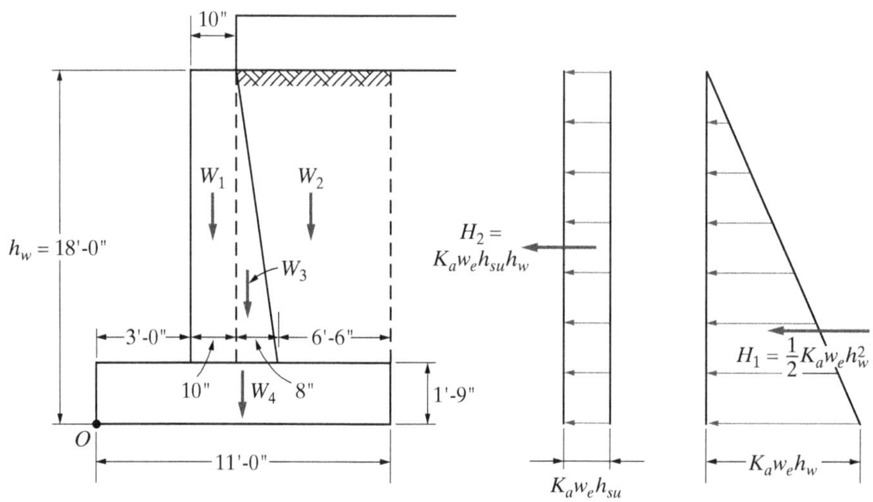

FIGURE 8.9 Stability analysis diagram.

In Table 8.2, note that W_2 is the weight of soil and surcharge on the heel from the right-hand edge of the heel to the vertical dashed line in the stem. The weight difference between reinforced concrete and soil (50 lb/ft^3) in the triangular portion that is the back of the stem is represented by W_3.

The factor of safety against overturning is

$$FS = \frac{131,600}{48,600} = 2.71 > 2.0 \qquad \text{(O.K.)}$$

b. *Factor of safety against sliding:* The tendency of the wall to slide is primarily a result of the horizontal forces from the lateral earth pressures and surcharge loads, whereas the vertical forces on the retaining wall and its footing cause the frictional resistance against sliding.

The total frictional force (or resisting force) available may be expressed as

$$F = \mu(\Sigma W)$$

where μ = coefficient of friction between the concrete and soil and ΣW = summation of vertical forces (see Figure 8.9). A typical value for the coefficient of friction is $\mu = 0.50$. Then, the resisting force F can be calculated:

$$F = \mu(\Sigma W)$$
$$0.50(19,710) = 9860 \text{ lb}$$

The factor of safety against sliding may be expressed as

$$FS = \frac{\text{resisting force } F}{\text{actual horizontal force } \Sigma H}$$
$$= \frac{9860}{7020} = 1.40$$

The required minimum factor of safety for this problem is 1.5; hence, the resistance against sliding is *inadequate.*

One solution to this problem is to reproportion the wall until the requirements are met. Rather than change the wall, we will use a base shear key to mobilize the passive resistance of the soil and in effect increase the resisting force F and subsequently increase the factor of safety. The design of the base shear key will be one of the last steps in the design of the retaining wall.

c. *Soil pressures and location of resultant force:* The vertical soil pressure under the retaining wall footing is a function of the location of the resultant force, which in turn is a function of the vertical forces and horizontal forces acting on the retaining wall. It is generally desirable and required that for footings on soil, the resultant of all the forces acting on the footing must lie within the middle third of the base of the footing. When this occurs, the resulting vertical soil pressure distribution on the base of the footing could be either triangular, rectangular, or trapezoidal in shape, with the soil in compression under the entire width of the footing. The resulting maximum foundation pressure must not exceed the safe bearing capacity of the soil.

The first step is to locate the point at which the resultant of the vertical forces ΣW and the horizontal forces ΣH intersects the bottom of the footing. This may be visualized with reference to Figure 8.10. Note that ΣW and ΣH are first combined to form the resultant force R. Because R may be moved anywhere along its line of action, it is moved to the bottom of the footing, where it is then resolved back into its components ΣW and ΣH. The moment about point O due to the resultant force acting at the bottom of the footing must be the same as the moment effect about point O of the components ΣW and ΣH. Therefore, using the components in the plane of the bottom of the footing, the location of the resultant, which is a distance x from point O may be determined. Note that the moment due to ΣH at the bottom of the footing is zero:

$$\Sigma W(x) = \Sigma Wm - \Sigma Hn$$

As may be observed, ΣWm is the stabilizing moment of the vertical forces with respect to point O, and ΣHn is the overturning moment of the horizontal forces with respect to the same point. Hence, this expression may be rewritten as

$$\Sigma W(x) = \text{stabilizing } M - \text{overturning } M$$

$$x = \frac{\text{stabilizing } M - \text{overturning } M}{\Sigma W}$$

With reference to Figure 8.9,

$$x = \frac{131{,}600 - 48{,}600}{19{,}710} = 4.21 \text{ ft from toe}$$

Therefore, the eccentricity e with respect to the centerline of the footing is

$$e = 5.5 - 4.21 = 1.29 \text{ ft}$$

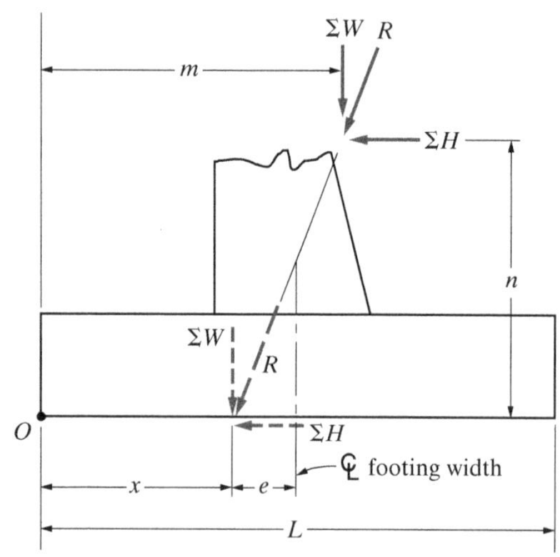

FIGURE 8.10 Resultant of forces acting on wall.

With a footing length of 11 ft, the middle third has a length of $\frac{11}{3} = 3.667$ ft. Eccentricity e is measured from the centerline of the footing, and because

$$e = 1.29 \text{ ft} \le \frac{3.667 \text{ ft}}{2} = 1.834 \text{ ft}$$

The resultant lies in the middle third of the footing, and the resulting pressure distribution is trapezoidal. If the resultant intersected the base at the edge of the middle third (i.e., $e = \frac{L}{6} = 1.833$ ft), the pressure distribution would have been triangular, and if $e = 0$, the pressure distribution would have been rectangular, indicating a uniform soil pressure distribution.

The pressures may now be calculated considering ΣW applied as an eccentric load on a rigid rectangular section 11 ft long by 1 ft wide (this is a *typical* 1-ft-wide strip of the wall footing). The pressures are obtained by using the basic equations for bending and axial compression:

$$p = \frac{P}{A} \pm \frac{Mc}{I}$$

where

p = unit soil pressure intensity under the footing

P = total vertical load (ΣW)

A = footing cross-sectional area $[(L)(1.0)]$

M = moment due to eccentric load $[(\Sigma W)(e)]$

c = distance from the centerline of footing to outside edge ($L/2$)

I = moment of inertia of footing with respect to its centerline $[(1.0)(L)^3/12]$

The expression given previously may be rewritten as

$$p = \frac{\Sigma W}{(L)(1.0)} \pm \frac{(\Sigma W)(e)(L/2)}{1.0(L)^3/12}$$

Simplifying and rearranging yields the soil vertical pressure distribution as

$$p = \frac{\Sigma W}{L}\left(1 \pm \frac{6e}{L}\right)$$

Substituting, we obtain

$$p = \frac{19{,}710}{11}\left[1 \pm \frac{6(1.29)}{11}\right]$$

$$= 1792(1 \pm 0.704)$$

$$\text{maximum } p = 1792(1.704) = 3050 \text{ psf}$$

$$\text{minimum } p = 1792(0.296) = 530 \text{ psf}$$

The resulting pressures are less than the allowable soil bearing pressure of 4000 psf, and are, therefore, satisfactory. The resulting pressure distribution beneath the footing is shown in Figure 8.11.

5. Design the component parts.

a. ***Design of heel:*** The load on the heel is primarily the soil dead load and the surcharge, if any, acting vertically downward. With the assumed straight-line pressure distribution under the footing, the downward load is reduced somewhat by the upward-acting pressure. For design purposes, the heel is assumed to be a cantilever beam 1 ft in width and with a span length equal to 6 ft-6 in. fixed at the rear face of the wall (point A in Figure 8.11).

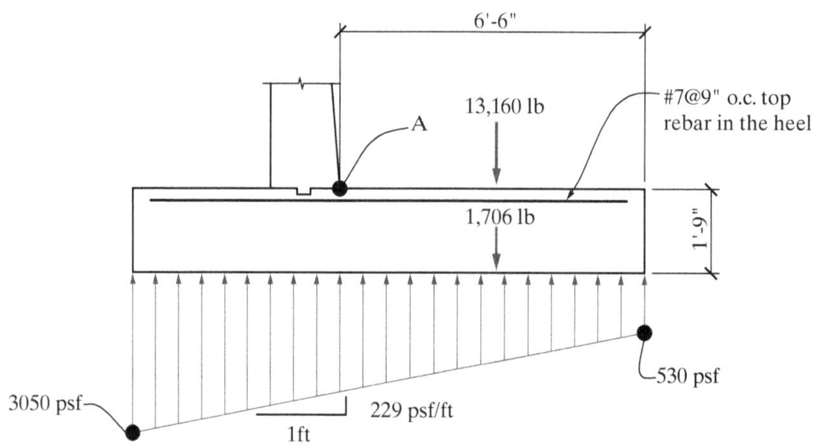

FIGURE 8.11 Pressure distribution under the footing, and top rebar in the heel

The weight of the footing is

$$1.75(150)(6.50) = 1706 \text{ lb}$$

The soil and surcharge weight is

$$6.50(20.25)(100) = 13{,}160 \text{ lb}$$

The slope of pressure distribution under the footing is

$$\frac{3050 - 530}{11} = 229 \text{ psf/linear foot}$$

Thus far, all analysis has been based on service (unfactored) loads because allowable soil pressures are determined using a certain factor of safety against reaching pressures that will cause unacceptable settlements. Reinforced concrete, however, is designed at the ultimate limit state using factored loads. The ACI Code, Section 5.3.8, specifies that where lateral earth pressure H must be included in the design and adds to the primary load effect, the strength U shall be at least equal to $1.2D + 1.6L + 1.6H$. Where the lateral earth pressure H opposes the primary load effect, a load factor of 0.9 should be used on H when the lateral earth pressure is permanent, and a load factor of zero on H should be used for all other conditions.

A conservative approach to the design of the heel is to use factored loads, as required, and to ignore the relieving effect of the upward pressure under the heel. This is the unlikely condition that would exist if there occurred a lateral force overload (and no associated increased vertical load) causing uplift of the heel.

The *maximum moment* may be obtained by taking a summation of moments about point A (utilizing service loads):

$$M = (13{,}160 + 1706)\ \frac{6.50}{2} = 48{,}300 \text{ ft-lb}$$

The bending of the heel is such that tension occurs at the top of the footing.

The *maximum shear* may be obtained by taking a summation of vertical forces on the heel side of point A (using service loads):

$$V = 13{,}160 + 1706 = 14{,}870 \text{ lb}$$

The maximum moment and shear must be modified for strength design. As the loads on the heel are predominantly dead load, a load factor of 1.2 is used. This, in effect, considers the surcharge to be a dead load. This is acceptable, however, because of the conservative nature of the design. Thus

$$M_u = 48{,}300(1.2) = 58{,}000 \text{ ft-lb}$$
$$V_u = 14{,}870(1.2) = 17{,}840 \text{ lb}$$

The *footing size and reinforcement* (for heel) will be determined next. Because it commonly occurs that shear strength will be the controlling factor with respect to footing thickness, the shear will be checked first. The heel effective depth available, assuming 2-in. cover to the top rebar in the heel (ACI Code, Section 7.7.1) and No. 8 bars, is

$$d = 21\,in. - 2\,in. - \frac{1}{2}(1\,in.) = 18.5 \text{ in.}$$

The shear strength ϕV_n of the heel, if no shear reinforcing is provided as is usually the case, is the shear strength of the concrete alone:

$$\phi V_n = \phi V_c = \phi(2\sqrt{f_c'})bd$$
$$= 0.75(2)\sqrt{3000}\,(12)(18.5)$$
$$\phi V_n = 18{,}240 \text{ lb}$$

Therefore, $\phi V_n > V_u$ (O.K.).

The *tensile reinforcement* requirement may now be determined in the normal way. Assuming $\phi = 0.90$:

$$\text{required } \bar{k} = \frac{M_u}{\phi bd^2} = \frac{58.0(12)}{0.9(12)(18.5)^2}$$
$$= 0.1883 \text{ ksi}$$

Therefore, from Table A.8, the required $\rho = 0.0033$, $\varepsilon_t > 0.005$, and $\phi = 0.90$ (O.K.). The required steel area is then calculated from

$$\text{required } A_s = \rho bd = 0.0033(12)(18.5) = 0.73 \text{ in.}^2$$

The minimum area of steel required by the ACI Code, Section 9.6.1, may be obtained using Table A.5:

$$A_{s,\min} = 0.0033(12)(18.5) = 0.73 \text{ in.}^2$$

As discussed in Section 2.8, $A_{s,\min}$ must be provided wherever reinforcement is needed, except where such reinforcement is at least one-third greater than that required by analysis (see the ACI Code, Section 9.6.1.3).

The ACI Code, Sections 7.6.1.1 and 7.7.2.3, also permits the use of a minimum reinforcement equal to that required for shrinkage and temperature steel in structural slabs of uniform thickness as furnished in the ACI Code, Section 24.4.1. This will always be somewhat less than that required by the ACI Code, Section 9.6.1, but not necessarily less than that specified by the ACI Code, Section 9.6.1.3. For footings, the shrinkage and temperature steel requirement for slabs of uniform thickness will be used as an absolute minimum in this text.

For the heel reinforcement for this retaining wall, because the required steel area is equal to $A_{s,\min}$, we will select No. 7 bars at 9 in. o.c. ($A_s = 0.80$ in.^{2}).

b. ***Design of toe:*** The load on the toe is primarily a result of the soil pressure distribution on the bottom of the footing acting in an upward direction. For design purposes, the toe is assumed to be a cantilever beam 1 ft in width and with a span length equal to 3 ft-0 in. fixed at the front face of the wall (point B in Figure 8.12). The soil on the top of the toe is conservatively neglected. Forces and pressures are calculated with reference to Figure 8.12.

As the reinforcing steel will be placed in the bottom of the footing, the effective depth available, assuming a 3-in. cover and No. 8 bars, is

$$d = 21 \text{ in.} - 3 \text{ in.} - \frac{1}{2}(1 \text{ in.}) = 17.5 \text{ in..}$$

The weight of footing for the toe design is

$$1.75(3.0)(150) = 788 \text{ lb}$$

The soil pressure directly under point B, recalling that the slope of the pressure diagram is 229 psf/ft, is

$$3050 - 3.0(229) = 2360 \text{ psf}$$

The *design moment* M_u may be obtained by a summation of moments about point B in Figure 8.12. Both the moment and the shear must be modified for strength design. According to the ACI Code, Section 5.3.8, a load factor of 1.6 should be used for horizontal earth pressure and for live load, whereas 1.2 should be used for dead load. Because the soil pressure under the toe is largely the result of horizontal earth pressure, a conservative procedure of using 1.6 is recommended. In addition, as the weight of the footing reduces the effect of the horizontal earth pressure, a factor of 0.9 is used for the footing dead load, as recommended by the ACI Code, Section 5.3.8.

The design moment M_u is then calculated as

$$\begin{aligned} M_u &= 1.6(\tfrac{1}{2})(3050)(3.0)^2(\tfrac{2}{3}) \\ &\quad + 1.6(\tfrac{1}{2})(2360)(3.0)^2(\tfrac{1}{3}) - 0.9(788)\left(\frac{3.0}{2}\right) \\ &= 19{,}240 \text{ ft-lb} \end{aligned}$$

The bending of the toe is such that tension occurs at the bottom of the footing.

The *design shear* V_u is obtained by a summation of vertical forces on the toe side of point B and applying load factors, as previously discussed:

$$\begin{aligned} V_u &= 1.6(\tfrac{1}{2})(3050)(3.0) \\ &\quad + 1.6(\tfrac{1}{2})(2360)(3.0) - 0.9(788) \\ &= 12{,}270 \text{ lb} \end{aligned}$$

Footing size and reinforcement, based on the requirements of the toe, are treated as they were for the heel. The shear strength ϕV_n of the toe is calculated from

$$\begin{aligned} \phi V_n &= \phi V_c = \phi(2\sqrt{f_c'})bd \\ &= 0.75(2)\sqrt{3000}(12)(17.5) \\ \phi V_n &= 17{,}250 \text{ lb} \\ \phi V_n &> V_u \end{aligned}$$

Therefore, the thickness of the footing is satisfactory.

Based on the determined M_u, and assuming $\phi = 0.90$:

$$\text{required } \bar{k} = \frac{M_u}{\phi bd^2} = \frac{19.24(12)}{0.9(12)(17.5)^2} = 0.0698 \text{ ksi}$$

Therefore, from Table A.8,

$$\text{required } \rho = 0.0012$$

Note that $\varepsilon_t > 0.005$. Therefore, $\phi = 0.90$. The required steel area is

$$\text{required } A_s = \rho bd = 0.0012(12)(17.5) = 0.25 \text{ in.}^2$$

The minimum area of steel required may be obtained using Table A.5:

$$A_{s,\min} = 0.0033(12)(17.5) = 0.69 \text{ in.}^2$$

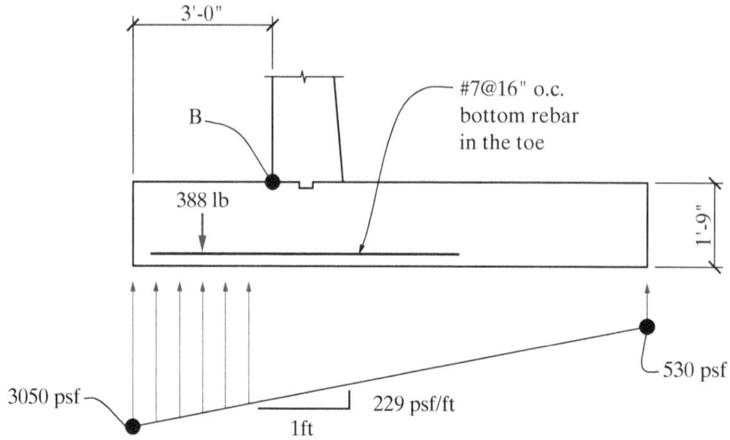

FIGURE 8.12 Pressure distribution for toe design.

Because required $A_s < A_{s,min}$, other minimum steel criteria should be checked to establish a controlling minimum value:

1. Provide one-third additional reinforcing as outlined in the ACI Code, Section 9.6.1.3:

$$A_{s,min} = 1.33(0.25) = 0.33 \text{ in.}^2$$

2. Check the required steel area based on the absolute minimum of shrinkage and temperature steel required for structural slabs of uniform thickness (ACI Code, Section 24.4.3.1):

$$\begin{aligned} \text{required } A_s &= 0.0018\text{bh} \\ &= 0.0018(12)(21) = 0.45 \text{ in.}^2 \end{aligned}$$

As we consider the shrinkage and temperature steel requirement as an absolute minimum, use No. 7 bars at 16 in. o.c. ($A_s = 0.45$ in.^{2}).

c. ***Design of stem***: The load on the stem is primarily lateral earth pressure acting (in this problem) horizontally. For design purposes, the stem is assumed to be a vertical cantilever beam 1 ft in width and with a span length equal to 16 ft-3 in. fixed at the top of the footing.

The loads acting on the stem from the top of the wall to the top of the footing are depicted in Figure 8.13. The magnitudes of the horizontal forces are

$$H_{s1} = \tfrac{1}{2}(30)(16.25)^2 = 3960 \text{ lb}$$
$$H_{s2} = 4(30)(16.25) = 1950 \text{ lb}$$

The design moment M_u in the stem may be obtained by a summation of moments about the top of the footing. Both shears and moments must be modified for strength design. Because the forces are due to lateral earth pressure, a factor of 1.6 should be used. M_u may then be calculated as

$$\begin{aligned} M_u &= 1.6(3960)(5.42) + 1.6(1950)(8.125) \\ &= 59{,}700 \text{ ft-lb} \end{aligned}$$

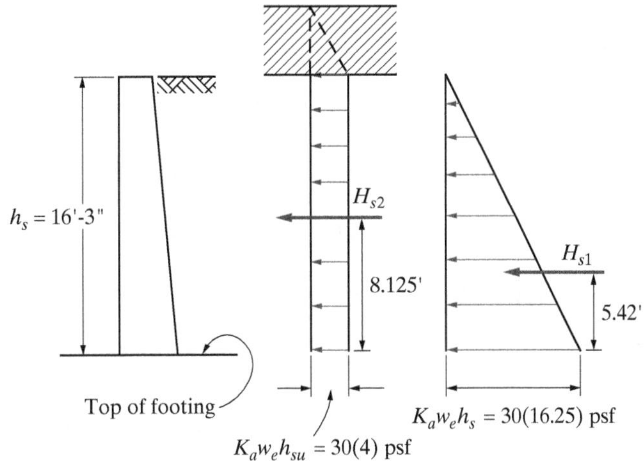

FIGURE 8.13 Forces acting on stem.

The design shear V_u is obtained by a summation of the horizontal forces acting on the stem above the top of the footing:

$$\begin{aligned} V_u &= 1.6(3960) + 1.6(1950) \\ &= 9460 \text{ lb} \end{aligned}$$

The *stem size and reinforcement* requirements will be determined next. Because the stem is assumed to be a cantilever slab, V_u as a limit may equal ϕV_n, where $V_n = V_c$. It is not practical to reinforce the stem for shear; therefore, if shear strength is inadequate, stem thickness must be increased. The effective depth of the stem at the top of the footing, assuming 2-in. cover and No. 8 bars, is

$$d = 18 \text{ in.} - 2 \text{ in.} - \frac{1}{2}(1 \text{ in.}) = 15.5 \text{ in.}$$

The shear strength ϕV_n of the stem (at the top of the footing), if no shear reinforcement is provided – as is usually the case, is the shear strength of the concrete alone:

$$\begin{aligned} \phi V_n &= \phi V_c = \phi 2\sqrt{f_c'}\, bd \\ &= 0.75(2)\sqrt{3000}\,(12)(15.5) \\ \phi V_n &= 15{,}280 \text{ lb} \\ \phi V_n &> V_u \end{aligned}$$

Thus, the stem thickness need not be increased. Based on M_u at the bottom of the stem and assuming $\phi = 0.90$,

$$\begin{aligned} \text{required } \bar{k} &= \frac{M_u}{\phi bd^2} = \frac{59.7(12)}{0.9(12)(15.5)^2} \\ &= 0.2761 \text{ ksi} \end{aligned}$$

Therefore, from Table A.8, the required ρ = 0.0049, and ε_t > 0.005, so ϕ = 0.90. The required steel area is

$$\text{required } A_s = \rho bd = 0.0049(12)(15.5) = 0.91 \text{ in.}^2$$

The minimum area of steel required may be obtained using Table A.5:

$$A_{s,\min} = 0.0033(12)(15.5) = 0.61 \text{ in.}^2$$

Because required $A_s > A_{s,\min,}$ the calculated required A_s controls. Use No. 7 bars at $7\frac{1}{2}$ in. o.c. (A_s = 0.96 in.^{2}).

The *stem reinforcement pattern* will be investigated more closely. As the moment and shear vary along the height of the wall, it is only logical that the steel requirements also vary. A pattern is usually created whereby some of the stem reinforcement is cut off where it is no longer required. The actual pattern used is the choice of the designer based on various practical constraints.

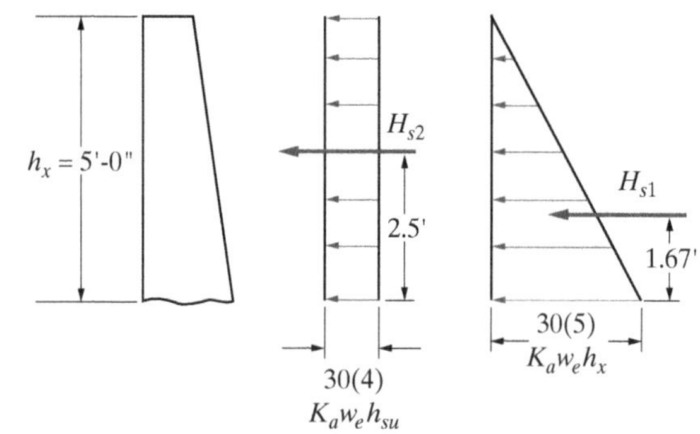

FIGURE 8.14 Stem analysis for the top 5 ft. of the retaining wall.

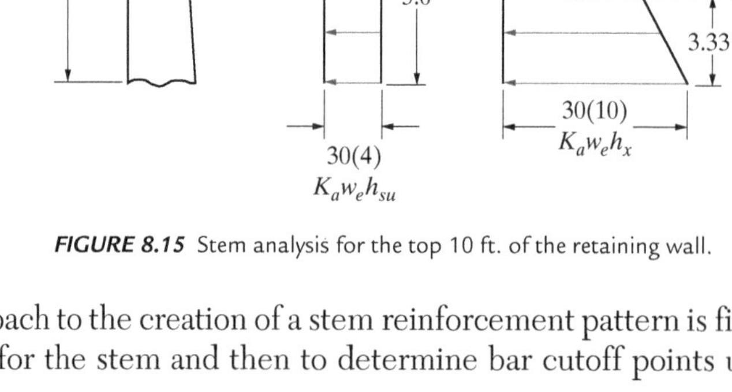

FIGURE 8.15 Stem analysis for the top 10 ft. of the retaining wall.

A typical approach to the creation of a stem reinforcement pattern is first to draw the complete M_u diagram for the stem and then to determine bar cutoff points using the procedure discussed in Chapter 5. Stem moments due to the external loads will be calculated at 5 ft from the top of the wall and at 10 ft from the top of the wall. These distances are arbitrary and should be chosen so that enough points will be available to draw the M_u diagram.

With reference to Figure 8.14, the moment at 5 ft from the top of the wall is calculated by first determining the horizontal forces:

$$H_{s1} = \tfrac{1}{2}(30)(5.0)^2 = 375 \text{ lb}$$
$$H_{s2} = 4(30)(5.0) = 600 \text{ lb}$$

Summing the moments about a plane 5 ft below the top of the wall and introducing the appropriate load factor,

$$M_u = 1.6(375)\left(\frac{5.0}{3}\right) + 1.6(600)\left(\frac{5.0}{2}\right)$$
$$= 3400 \text{ ft-lb}$$

With reference to Figure 8.15, the moment at 10 ft below the top of the wall may be calculated:

$$H_{s1} = \tfrac{1}{2}(30)(10.0)^2 = 1500 \text{ lb}$$
$$H_{s2} = 4(30)(10.0) = 1200 \text{ lb}$$

and

$$M_u = 1.6(1500)\left(\frac{10.0}{3}\right) + 1.6(1200)\left(\frac{10.0}{2}\right)$$
$$= 17{,}600 \text{ ft-lb}$$

The three M_u values may now be plotted as the solid curve in Figure 8.16. This diagram will reflect both the design moment and moment strength with respect to the distance from the top of the wall and must be plotted to scale.

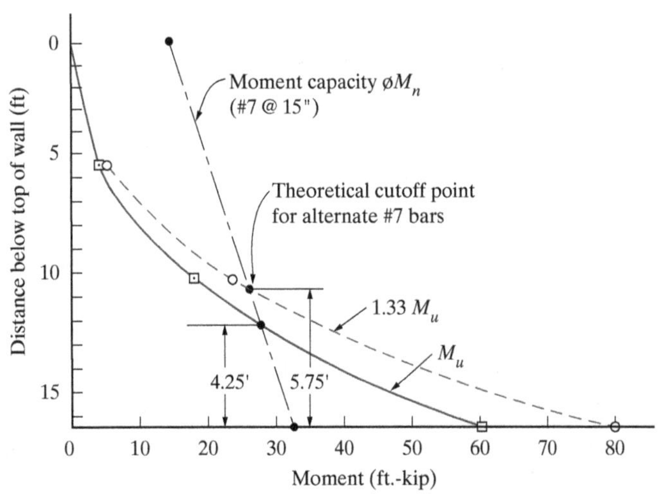

FIGURE 8.16 Stem steel design.

The moment strength ϕM_n of the stem at various locations may be computed from the expression

$$\phi M_n = \phi bd^2\overline{k}$$

Because we are primarily attempting to establish a cutoff location for some vertical stem reinforcement, the moment strength will be based on a pattern of alternate vertical bars being cut off. With alternate bars cut off, the remaining reinforcement pattern is No. 7 bars at 15 in. o.c., which furnishes an A_s = 0.48 in.^{2} Caution must be exercised that the spacing of the remaining bars does not exceed three times the wall thickness ($3h$) or 18 in., whichever is less.

The moment strength will be computed at the top of the wall and at the top of the footing, neglecting minimum steel requirements at this time. As the rear face of the wall is battered, the effective depth varies and may be computed as follows:

$$\text{top of wall: } d = 10\text{ in.} - 2\text{ in.} - \frac{1}{2}(1\text{ in.}) = 7.5\text{ in.}$$

$$\text{bottom of wall or at top of footing: } 18\text{ in.} - 2\text{ in.} - \frac{1}{2}(1\text{ in.}) = 15.5\text{ in.}$$

At the top of the wall, with half of the stem bars cut off:

$$\rho = \frac{A_s}{bd} = \frac{0.48}{12(7.5)} = 0.0053$$

Therefore, from Table A.8, $\overline{k} = 0.2982\text{ ksi}$. Then

$$\begin{aligned}\phi M_n &= \frac{0.9(12)(7.5)^2}{12}(0.2982)\\ &= 15.10\text{ ft.-kip}\end{aligned}$$

At the bottom of the wall or at the top of the footing,

$$\rho = \frac{A_s}{bd} = \frac{0.48}{12(15.5)} = 0.0026$$

Therefore, from Table A.8, $\overline{k} = 0.1512\text{ ksi}$. Then

$$\begin{aligned}\phi M_n &= \frac{0.9(12)(15.5)^2}{12}(0.1512)\\ &= 32.7\text{ ft.-kip}\end{aligned}$$

These two values (ϕM_n) may now be plotted in Figure 8.16. A straight line will be used to approximate the variation of ϕM_n for No. 7 bars at 15 in. o.c. from the top to the bottom of the wall.

The intersection of the ϕM_n line and the M_u solid curve establishes a theoretical cutoff point for alternate vertical bars based on strength requirements. This location may be determined graphically by scaling in Figure 8.16. The scaled distance is 12.0 ft below the top of the wall or 4 ft-3 in. above the top of the footing. The No. 7 bar must be extended past this theoretical cutoff point a distance equal to the effective depth of the member, d or 12 bar diameters ($12d_b$), whichever is larger.

Because the effective depth of the stem steel at the top of the footing is 15.5 in. and the wall batter is approximately 0.492 in./ft, the effective depth of the stem steel at 4.25 ft above the top of the footing may be calculated as follows:

$$d = 15.5 - 4.25(0.492) = 13.4 \text{ in.}$$
$$12d_b = 12(0.875) = 10.5 \text{ in.}$$

Therefore, the required length of bar above the top of the footing would be

$$4.25 + \frac{13.4}{12} = 5.37 \text{ ft}$$

Use 5 ft 5 in. (5.42 ft).

Check the required minimum area of steel, $A_{s,\min}$, at the theoretical cutoff point to ensure that discontinuing half of the steel will not violate this code requirement. Once alternate bars have been terminated, the remaining steel will be No. 7 bars at 15 in. o.c. (A_s = 0.48 in.^{2}). At the theoretical cutoff point 4.25 ft above the top of the footing, the effective depth d = 13.4 in. From Table A.5,

$$A_{s,\min} = 0.0033bd$$
$$= 0.0033(12)(13.4) = 0.53 \text{ in.}^2$$

This is unsatisfactory because the steel area provided is less than $As_{,\min}$. However, recall from previous discussions that the minimum reinforcing requirement need not be applied if the reinforcing provided is one-third greater than that required by analysis. Therefore, a simple solution is to move the cutoff point upward to where the reinforcing provided is one-third greater than required. This can be accomplished by drawing a 1.33 × M_u curve and reestablishing the theoretical cutoff point for alternate No. 7 bars. At this cutoff point, one-third more steel is provided than is required; therefore, the minimum steel requirement does not apply.

Multiplying the previously determined design moments M_u by 1.33, the following moments are obtained:

At 5 ft below the top of the wall:

$$M_u = 1.33(3400) = 4520 \text{ ft-lb}$$

At 10 ft below the top of the wall:

$$M_u = 1.33(17{,}600) = 23{,}400 \text{ ft-lb}$$

At the bottom of the wall or at the top of the footing:

$$M_u = 1.33(59{,}700) \text{ ft-lb} = 79{,}400 \text{ ft-lb}$$

Plotting this moment curve (see the dashed line in Figure 8.16) and scaling the distance above the top of the footing to theoretical cutoff point, we arrive at a value of 5.75 ft, or 5 ft-9 in.

The bars must extend past this theoretical cutoff point a distance equal to the greater of the effective depth of the member or 12 bar diameters.

$$d = 15.5 - 5.75(0.492) = 12.67 \text{ in.}$$
$$12d_b = 12(0.875) = 10.5 \text{ in.}$$

Adding 12.67 in. to the previously determined theoretical cutoff point above the top of the footing, we have

$$5.75 + \frac{12.67}{12} = 6.81 \text{ ft}$$

Use 6 ft-10 in. (6.83 ft).

Therefore, terminate alternate No. 7 bars at 6 ft-10 in. above the top of the footing.

The ACI Code, Sections 7.7.3.5 and 9.7.3.5, stipulate that flexural reinforcement must not be terminated in a tension zone unless one of several conditions is satisfied. One of these conditions is that the shear at the cutoff point does not exceed two-thirds of the shear permitted (i.e., $V_u \leq (2/3)\phi V_n$). Therefore, check the shear at the actual cutoff point, 6.83 ft above the top of the footing (9.42 ft below the top of the wall; see Figure 8.17).

For the shear strength calculation, the wall thickness at a height of 6.83 ft above the top of the footing is

$$18.0 - 6.83(0.492) = 14.64 \text{ in.}$$

from which d may be calculated as

$$d = 14.64 - 2 - 0.5 = 12.14 \text{ in.}$$
$$H_{s1} = \tfrac{1}{2}(30)(9.42)^2 = 1331 \text{ lb}$$
$$H_{s2} = 4(30)(9.42) = 1130 \text{ lb}$$
$$\text{total} = 2460 \text{ lb}$$
$$V_u = 2460(1.6) = 3940 \text{ lb}$$
$$\tfrac{2}{3}\phi V_n = \tfrac{2}{3}\phi V_c = \tfrac{2}{3}\phi\left(2\sqrt{f_c'}\right)bd$$
$$= \tfrac{2}{3}(0.75)(2)\sqrt{3000}\,(12)(12.14)$$
$$\tfrac{2}{3}\phi V_n = 7980 \text{ lb}$$
$$\tfrac{2}{3}\phi V_n > V_u$$

In summary, alternate vertical stem reinforcing bars may be stopped at 6 ft-10 in. above the top of the footing (see the typical wall section shown in Figure 8.21 later in this chapter).

d. ***Additional design details***: Check the anchorage requirements for the stem steel (No. 7 bars at $7\frac{1}{2}$ in. o.c.). Assume all steel is uncoated. Note that anchorage requirements into the stem and into the footing may differ for these bars. Anchorage length required in the stem will be equal to or greater than that required in the footing and will also impact the splice length. Therefore, stem anchorage will be checked.

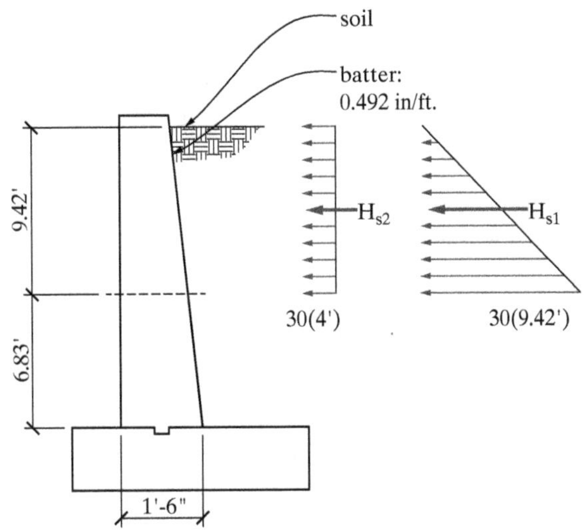

FIGURE 8.17 Shear at bar cutoff point.

1. From Table 5.1, let us assume that the value of $K_D = 82.2$.
2. Establish values for the factors ψ_t, ψ_e, ψ_s, ψ_g, and λ.
   a. $\psi_t = 1.0$ (the bars are not top bars).
   b. The bars are uncoated: $\psi_e = 1.0$.
   c. The bars are No. 7: $\psi_s = 1.0$.
   d. Grade 60 steel: $\psi_g = 1.0$.
   e. Normal-weight concrete is used: $\lambda = 1.0$.
3. The product $\psi_t \times \psi_e = 1.0 < 1.7$. (O.K.)
4. Determine c_b. Based on cover (center of bar to nearest concrete surface), consider the clear cover and one-half the diameter of the No. 7 bar:

$$c_b = 2 + \frac{0.875}{2} = 2.44 \text{ in.}$$

   Based on bar spacing of one-half the center-to-center distance between bars:

$$c_b = \left(\frac{1}{2}\right)(7.5) = 3.75 \text{ in.}$$

   Using the smaller of the two values, therefore, $c_b = 2.44$ in.
5. K_{tr} is taken as zero. There is no transverse steel crossing the potential plane of splitting.
6. Check $(c_b + K_{tr})/d_b \leq 2.5$:

$$\frac{c_b + K_{tr}}{d_b} = \frac{2.44 + 0}{0.875} = 2.79 > 2.5$$

   Use 2.5.

7. Calculate the excess reinforcement factor:

$$K_{ER} = \frac{A_s \text{ required}}{A_s \text{ provided}} = \frac{0.91}{0.96} = 0.95$$

8. Calculate the tension development or anchorage length, ℓ_d.

a. Omitting K_{ER} (this calculated value of ℓ_d will be used shortly for splice length determination):

$$\begin{aligned}\ell_d &= \frac{K_D}{\lambda}\left[\frac{\psi_t\psi_e\psi_s\psi_g}{\left(\dfrac{c_b + K_{tr}}{d_b}\right)}\right]d_b \\ &= \frac{82.2}{1.0}\left(\frac{1.0(1.0)(1.0)(1.0)}{2.5}\right)(0.875) \\ &= 28.8 \text{ in.} > 12 \text{ in.} \qquad \text{(O.K.)}\end{aligned}$$

b. Including K_{ER} from step 7, the required anchorage length is

$$\ell_d = 28.8(0.95) = 27.4 \text{ in.} > 12 \text{ in.} \quad \text{(O.K.)}$$

Use $\ell_d = 28$ in. (minimum).

With the footing thickness equal to 21 in. and a minimum of 3 in. of cover required for the steel at the bottom of the footing, the anchorage length available is 21 – 3 = 18 in. This anchorage length is not adequate. Rather than hook the bars; however, the bars will be extended into a footing base shear key that will be used to increase the sliding resistance of the wall (see the check on factor of safety against sliding).

The *length of splice* required for main stem reinforcing steel (see the typical wall section shown in Figure 8.21 later in this chapter) may be calculated recognizing that the class B splice is applicable for this condition (see Chapter 5, Section 5.7). Therefore, the required length of the splice is calculated from

$$1.3(28.8) = 37.4 \text{ in.}$$

Use 38 in. Note that the preceding calculation omits the effect of excess reinforcement.

Stem face steel in the form of horizontal and vertical reinforcement will be provided as per the ACI Code, Section 11.6. The minimum horizontal reinforcement is specified in ACI Code, Section 11.6.1. Although the minimum vertical reinforcement requirement of ACI Code, Section 11.6.1, does not strictly apply to reinforced concrete cantilever retaining walls, it is considered good practice to provide some vertical bars in the exposed face of the wall. The minimum recommended steel for deformed bars not larger than No. 5 with a minimum specified yield strength of 60,000 psi is as follows (ACI 318-19 Table 11.6.1):

$$\text{horizontal bars (per foot of height of wall):}$$
$$A_s = 0.0020bt$$
$$\text{vertical bars (per foot of wall horizontally):}$$
$$A_s = 0.0012bt$$

where t = thickness of the wall.

The code also stipulates that walls more than 10-in. thick, except basement walls, must have reinforcement for each direction on each face of the wall. The exposed face must have a minimum of one-half and a maximum of two-thirds of the total steel required for each direction. The maximum spacing for the steel must not exceed three times the wall thickness nor 18 in. (ACI Code, Sections 11.7.2.1 and 11.7.3.1) for both vertical bars and horizontal bars. For the front face of the wall (exposed face), use two-thirds of the total steel required and an average stem thickness = 14 in.

For horizontal steel:

$$0.0020(12)(14) = 0.34 \text{ in.}^2/\text{ft of height}$$
$$= 0.34(0.67) = 0.23 \text{ in.}^2$$

Use No. 4 bars at 10 in. on centers (A_s = 0.24 in.^{2}).

For vertical steel:

$$0.0012(12)(14) = 0.20 \text{ in.}^2/\text{ft of horizontal length}$$
$$= (0.20)(0.67) = 0.14 \text{ in.}^2$$

Use No. 4 bars at 17 in. on centers (A_s = 0.14 in.^{2}) in the front face. No additional vertical steel is needed in the rear face.

Rear face of wall, not exposed, for horizontal steel,

$$0.34(0.33) = 0.11 \text{ in.}^2$$

Use No. 4 bars at 18 in. on centers (A_s = 0.13 in.^{2}).

Longitudinal reinforcement in the footing should provide a steel area equal to that required for shrinkage and temperature in slabs. This is a conservative and acceptable approach because temperature and shrinkage exposure are ordinarily less severe for footings than for slabs. These bars serve chiefly as bar supports and spacers to hold the main steel in place during construction. Thus

$$\text{required } A_s = 0.0018(11 \text{ ft})(1.75 \text{ ft})(144 \text{ in.}^2/\text{ft})$$
$$= 4.99 \text{ in.}^2$$

Use 12 No. 6 bars (A_s = 5.28 in.^{2}), as shown in Figure 8.20 later in this chapter.

Transverse reinforcement in the footing need not run the full width of the footing. Proper anchorage length must be provided from the point of maximum tensile stress, however.

Check anchorage for the top transverse (heel) steel in the footing (No. 7 bars at 9 in. o.c.). This calculation for ℓ_d follows the eight-step procedure presented in Chapter 5, Section 5.2, and is summarized as follows:

1. $K_D = 82.2$
2. $\psi_t = 1.3$, $\psi_e = 1.0$, $\psi_s = 1.0$, $\psi_g = 1.0$, and $\lambda = 1.0$
3. $\psi_t \times \psi_e = 1.3$ (O.K.)
4. $c_b = 2.44$ in.
5. $K_{tr} = 0$
6. $(c_b + K_{tr})/db = 2.79$ in. Use 2.5 in.
7. $K_{ER} = 0.91$
8. $\ell_d = 34$ in.

Check the anchorage of the bottom transverse (toe) steel in the footing (No. 7 bars at 16 in. o.c.). As before, this ℓ_d calculation follows the procedure presented in Chapter 5 and is summarized as follows:

1. $K_D = 82.2$
2. $\psi_t = 1.0$, $\psi_e = 1.0$, $\psi_s = 1.0$, $\psi_g = 1.0$, and $\lambda = 1.0$
3. $\psi_t \times \psi_e = 1.0$ (O.K.)
4. $c_b = 3.44$ in.
5. $K_{tr} = 0$
6. $(c_b + K_{tr})/d_b = 3.93$ Use 2.5
7. $K_{ER} = 1.0$
8. $\ell_d = 28.8$ in. Use 29 in.

These details are included in the typical wall section shown in Figure 8.21.

The stem and the footing are elements cast at different times and a *shear key* will be used between the two. This is common practice. We will use a depressed key formed by a 2 × 6 lumber (dressed dimensions $1\frac{1}{2} \times 5\frac{1}{2}$), as shown in Figure 8.21. The need for a shear key is questionable because considerable slip is required to develop the key for purposes of lateral force transfer. It may be considered as an added mechanical factor of safety, however.

The shear-friction design method of ACI Code, Section 22.9, should be used to design the required contact surface and reinforcement for the transfer of the horizontal force between the stem of the retaining wall and the retaining wall footing. This approach eliminates the need for the traditional shear key. The shear-friction approach assumes that all the horizontal force will be transferred through friction that develops on the contact surface between the two elements. The magnitude of the force that can be transmitted will depend on the characteristics of the contact surfaces and on the existence of adequate shear-friction reinforcing A_{vf} crossing those surfaces. The angle at which the shear-friction reinforcement crosses the contact surface also plays a role. For our example, we assume the dowels will be placed perpendicular to the top of the footing.

The Code allows for two possible contact-surface conditions that could exist in a situation such as between stem and footing in our retaining wall, either not intentionally roughened, though clean and free of laitance, or intentionally roughened. The latter assumption requires the interface to be roughened to a full amplitude of approximately $\frac{1}{4}$ in. This may be accomplished by raking of the fresh concrete or by some other means.

Assuming normal-weight concrete and shear-friction reinforcement placed perpendicular to the interface, the nominal shear strength (or friction force that resists sliding) may be computed from

$$V_n = A_{vf} f_y \mu \qquad \textbf{[ACI Eq. (22.9.4.2)]}$$

where

A_{vf} = area of shear-friction reinforcement

μ = coefficient of friction in accordance with the ACI Table 22.9.4.2; it may be taken as 1.0 for normal-weight concrete placed against hardened concrete roughened to ¼-in. amplitude and a clean surface free of dirt or debris.

Note that A_{vf} is steel that is provided for the shear-friction development and that it is in addition to any other steel already provided. This additional steel will be provided in the form of *dowels*. (A *dowel* is defined as a short bar that connects two separately cast sections of concrete.) The dowels will be placed approximately perpendicular to the shear plane.

As a limit,

$$V_u = \phi V_n$$

where V_u is a shear force applied at the cracked plane. Substituting for V_n,

$$V_u = \phi A_{vf} f_y \mu$$

Solving for A_{vf},

$$\text{required } A_{vf} = \frac{V_u}{\phi f_y \mu}$$

$$= \frac{9460}{(0.75)(60{,}000)(1.0)}$$

$$= 0.21 \text{ in.}^2 \text{ per ft}$$

It is desirable to distribute the shear-friction reinforcement across the width of the contact surface. Therefore, provide 0.21/2 = 0.11 in.^{2}/ft in each face. Use No. 4 bars at 18 in. in each face. This will provide 0.13 in.^{2}/ft in each face for a total of 0.26 in.^{2}/ft. Then

$$\begin{aligned}\text{shear strength } V_n &= A_{vf} f_y \mu\\ &= (0.26)(60{,}000)(1.0)\\ &= 15{,}600 \text{ lb}\end{aligned}$$

The ACI Code, Sections 22.9.4.1 and 22.9.4.4, stipulate that the maximum V_n for concrete placed monolithically or against hardened concrete intentionally roughened as previously described shall not exceed the smallest of

$$0.2 f_c' A_c, (480 + 0.08 f_c') A_c, \quad \text{and} \quad 1600 A_c$$

where A_c is the contact area resisting the shear transfer, and for all other cases, V_n shall not exceed the smaller of

$$0.2 f_c' A_c \quad \text{and} \quad 800\ A_c$$

Thus, checking the upper limit for V_n:

$$\begin{aligned}0.2 f_c' A_c &= 0.2(3000)(12)(18) = 129{,}600 \text{ lb}\\ (480 + 0.08 f_c') A_c &= [480 + 0.08(3000)](12)(18)\\ &= 155{,}500 \text{ lb}\\ 1600 A_c &= 1600(12)(18) = 346{,}000 \text{ lb}\\ \text{Calculated } V_n &= 15{,}600 \text{ lb} << 129{,}600 \text{ lb} \quad (\text{O.K.})\end{aligned}$$

The tension bar development length, ℓ_d, for the No. 4 dowels (at 18 in. o.c.) must be furnished into both the stem and the footing of the retaining wall. These dowels are tension bars, and the calculation for ℓ_d follows the procedure presented in Chapter 5, Section 5.2, and is summarized as follows:

1. $K_D = 82.2$
2. $\psi_t = 1.0$, $\psi_e = 1.0$, $\psi_s = 0.8$, $\psi_g = 1.0$, and $\lambda = 1.0$
3. $\psi_t \times \psi_e = 1.0$ (O.K.)

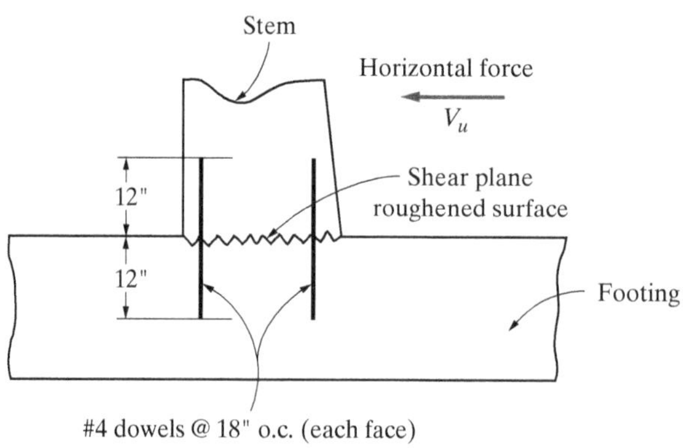

FIGURE 8.18 Shear-friction reinforcement.

4. $c_b = 2.25$ in.
5. $K_{tr} = 0$
6. $(c_b + K_{tr})/d_b = 4.50$ Use 2.5
7. $K_{ER} = 0.808$
8. $\ell_d = 10.6 \text{ in.} < 12 \text{ in. Use 12 in.}$

6. The No. 4 dowels are depicted in Figure 8.18.

Design the footing base shear key. The *footing base shear key* (sometimes called a *bearing lug*) is primarily used to prevent a sliding failure. The magnitude of the additional resistance to sliding offered by the key is questionable and is a function of the subsoil material. The key, cast in a narrow trench excavated below the bottom of the footing elevation, becomes monolithic with the footing. The excavation for a key will generally disturb the subsoil during construction and conceivably will do more harm than good. Hence, the use of the key for the purpose intended is controversial.

An acceptable design approach is to use the passive earth resistance H_p in front of the key (from the bottom of the footing to the bottom of the key) as the additional resistance to sliding (see Figure 8.19). This neglects any earth in front of the footing and reflects the case where excavation or scour has removed the earth to the level of the bottom of the footing.

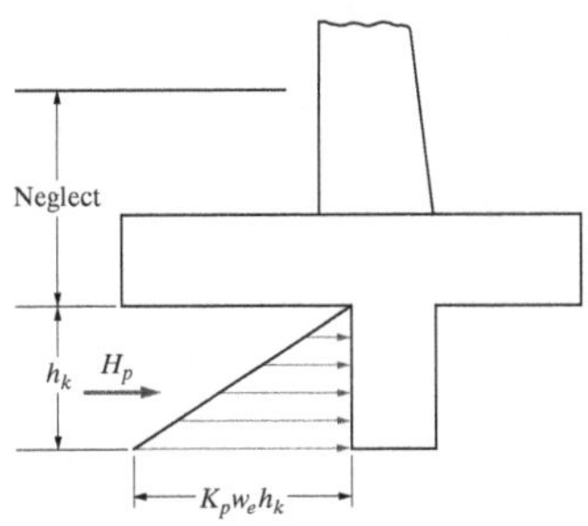

FIGURE 8.19 Base shear key force analysis.

The passive earth resistance may be expressed in terms of an equivalent fluid weight. Because $K_a w_e = 30 \text{ lb/ft}^3$ and $w_e = 100 \text{ lb/ft}^3$,

$$K_a = \frac{30}{100} = 0.3$$

The coefficient of passive earth resistance is

$$K_p = \frac{1 + \sin\phi}{1 - \sin\phi} = \frac{1}{K_a} = \frac{1}{0.3} = 3.33$$

and

$$K_p w_e = 3.33(100) = 333 \text{ lb/ft}^3$$
$$H_p = \tfrac{1}{2} K_p w_e h_k^2 = \tfrac{1}{2}(333) h_k^2$$

The unfactored horizontal force $\Sigma H = 7020$ lb. Therefore, the required resistance to sliding that will furnish a factor of safety of 1.5 is

$$1.5(7020) = 10{,}530 \text{ lb}$$

The frictional resistance furnished is 9860 lb (see the discussion of stability analysis). Thus, the resistance that must be furnished by the base shear key and the passive earth resistance is

$$10{,}530 - 9860 = 670 \text{ lb}$$

With reference to Figure 8.19, the height h_k required to furnish this resistance may be obtained by establishing horizontal equilibrium ($SH = 0$).

$$\tfrac{1}{2}(333)(h_k)^2 = 670$$
$$h_k^2 = 4.02$$
$$\text{required } h_k = 2.01 \text{ ft}$$

Use 2 ft-0 in.

The key must be designed for moment and shear. The worst case would be the situation where excavation had *not* taken place, and the full height of earth in front of the wall was available to develop passive pressure. This is shown in Figure 8.20. A 1-ft depth of soil has been assumed on top of the footing. Assuming the key as a vertical cantilever beam, taking a summation of moment about the plane of the bottom of footing, and applying a load factor of 1.6,

$$M_u = 1.6(916)(2.0)(1.0) + 1.6\tfrac{1}{2}(666)(2.0)(\tfrac{2}{3})(2.0)$$
$$= 4350 \text{ ft-lb}$$

Assume 10-in. width for the concrete key and a No. 8 bar, which provides an effective depth of d = 10 in. – 3 in. cover – ½(1 in.) = 6.5 in. Assume $\phi = 0.90$.

$$\text{required } \bar{k} = \frac{M_u}{\phi b d^2} = \frac{4350}{0.9(12)(6.5)^2} = 0.1144 \text{ ksi}$$

From Table A.8, the required $\rho = 0.0020$, $\varepsilon_t > 0.005$; therefore, $\phi = 0.90$. Now, we can calculate the required steel area:

$$\text{required } A_s = \rho b d = 0.0020(12)(6.5) = 0.16 \text{ in.}^2$$

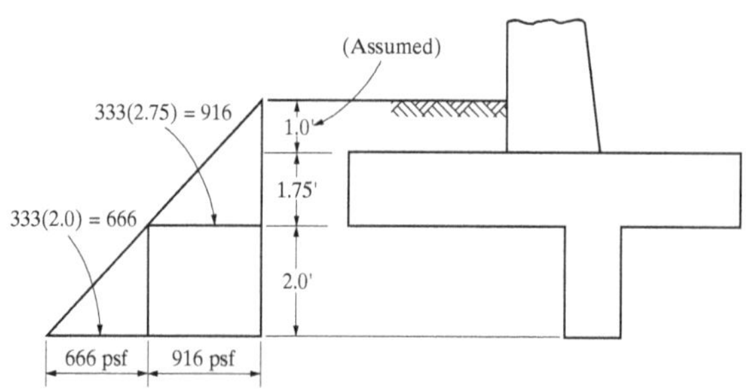

FIGURE 8.20 Base shear key forces.

The minimum area of steel required by the ACI Code, Sections 9.6.1.1 and 9.6.1.2, may be obtained using Table A.5:

$$A_{s,\min} = 0.0033(12)(6.5) = 0.26 \text{ in.}^2$$

Therefore, $A_{s,\min}$ of 0.26 in.^{2} controls. This is a very small amount of steel, and it will be more practical to extend some existing stem bars into the key. All stem bars must be extended into the key for anchorage reasons, however. Therefore, all the bars will be extended to within 3 in. of the bottom of the key. This provides 42 in. of anchorage length below the bottom of the stem for the No. 7 bars and exceeds the requirement of 28 in. (see typical wall section in Figure 8.21). The steel furnished in the key, therefore, is No. 7 bars at $7\frac{1}{2}$ in. o.c. (A_s = 0.96 in.^{2}).

If we conservatively neglect the effect of the greater cover (3.0 in.) in the key, calculations for the required anchorage length at the top of the key are identical to those for the stem, except that the excess reinforcement factor K_{ER} = 0.26/0.96 = 0.27, from which the final ℓ_d calculation yields

$$\ell_d = 28.8(0.27) = 7.8 \text{ in.} < 12 \text{ in.}$$

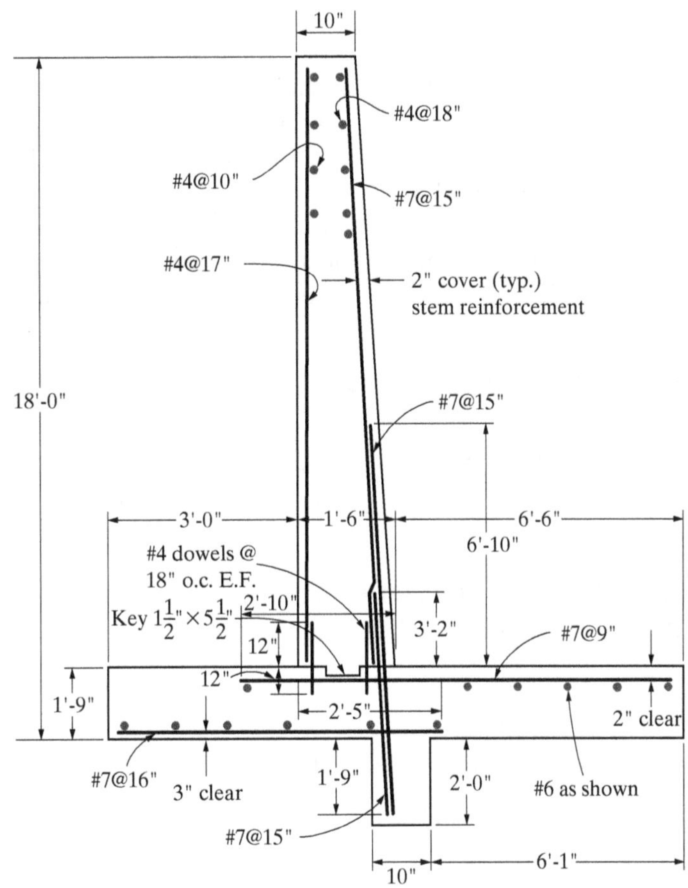

FIGURE 8.21 Typical wall section.

Use $\ell_d = 12$ in. The available anchorage length is 24 – 3 = 21 in., which exceeds the requirements of 12 in.

The shear strength of the key is

$$\begin{aligned}\phi V_n &= \phi V_c \\ &= \phi\left(2\lambda\sqrt{f'_c}\right)bd \\ &= 0.75(2)(1.0)\left(\sqrt{3000}\right)(12)(6.5) \\ &= 6410 \text{ lb}\end{aligned}$$

The factored shear is

$$V_u = (1.6)(916)(2.0) + 1.6\left(\tfrac{1}{2}\right)(666)(2) = 4000 \text{ lb}$$

$$\phi V_n > V_u \qquad \text{(O.K.)}$$

8.4 DESIGN CONSIDERATIONS FOR BEARING WALLS

Bearing walls (Figure 8.1g) were briefly described at the beginning of this chapter as those walls that carry vertical load in addition to their own weight. Recommendations for the empirical design of such walls are presented in Chapter 11 of the ACI Code (318-19) and apply primarily to relatively short walls spanning vertically and subject to vertical loads only, such as those resulting from the reactions of floor or roof systems supported on walls. Walls, other than short walls carrying "reasonably concentric" loads, should be designed as compression members for combined axial load and flexure in accordance with ACI 318-19, Chapter 22. "Reasonably concentric" implies that the resultant factored load falls within the middle third of the cross-section.

The design axial load strength or capacity of such a wall will be

$$\phi P_n = 0.55\phi f'_c A_g\left[1-\left(\frac{k\ell_c}{32h}\right)^2\right] \qquad \textbf{[ACI Eq. (11.5.3.1)]}$$

where

ϕ = strength-reduction factor corresponding to compression-controlled sections in accordance with ACI Code, Section 21.2.2; 0.75 for members with spiral reinforcement and 0.65 for other reinforced members

h = thickness of wall (in.)

ℓ_c = vertical distance between supports (in.)

A_g = gross area of section (in.^{2})

k = effective length factor

The effective length factor k shall be as follows:

1. For walls braced top and bottom against lateral translation and
   a. Restrained against rotation at one or both ends (top and/or bottom) 0.8
   b. Unrestrained against rotation at both ends 1.0
2. For walls not braced against lateral translation 2.0

Where the wall is subject to concentrated loads, the effective length of wall for each concentration must not exceed the center-to-center distance between loads nor exceed the width of the bearing plus four times the wall thickness.

The following requirements applicable to bearing walls are prescribed, among others, by the ACI Code, Chapter 11:

1. Reinforced concrete bearing walls must have a thickness of at least $\frac{1}{25}$ of the unsupported height or width, whichever is shorter, and not less than 4 in.
2. Thickness of nonbearing walls shall not be less than 4 in. nor less than $\frac{1}{30}$ times the least distance between members that provide lateral support.
3. Assuming the inplane shear in the bearing or structural wall $V_u \leq 0.5\phi\alpha_c\lambda\sqrt{f_c'}A_{CV}$, the minimum area of horizontal reinforcement in the wall (i.e., transverse reinforcement), for rebar sizes larger than No. 5 bars, must be at least 0.0025 times the area of the wall (i.e., $0.0025bh$ per foot width of the wall). The minimum area of vertical reinforcement in the wall (i.e., longitudinal reinforcement) should not be less than 0.0015 times the area of the wall (i.e., $0.0015bh$ per foot width of the wall), where b = 12 in. and h is the wall thickness. These minimum areas of reinforcement may be reduced to 0.002bh and 0.0012bh, respectively, if the wall uses deformed bars *not* larger than No. 5 with f_y not less than 60,000 psi or if the reinforcement is welded wire reinforcement larger than W31 or D31. (See ACI Code, Table 11.6.1).

   Where,

   $\phi = 0.75$ (shear)

   $\lambda = 1.0$ for normal-weight concrete

   $$\alpha_c = 3 \quad for \ \frac{h_w}{\ell_w} \leq 1.5$$

   $$\alpha_c = 2 \quad for \ \frac{h_w}{\ell_w} \geq 2.0$$

   $$\alpha_c \text{ varies linearly between 2 and 3} \quad for \ 1.5 < \frac{h_w}{\ell_w} < 2.0$$

   $A_{CV} = \ell_w h$

   ℓ_w = length of wall

   h_w = height of wall

   h = thickness of wall

4. Exterior basement walls and foundation walls must not be less than $7\frac{1}{2}$-in. thick.
5. Reinforced concrete walls must be anchored to intersecting elements such as floors and roofs or to columns, pilasters, buttresses, intersecting walls, and footings.
6. Walls more than 10-in. thick, except for basement walls, must have reinforcement in each direction for each face. The exterior surface shall have a minimum of one-half and a maximum of two-thirds of the total steel required, with the interior surface having the balance of the reinforcement.
7. Vertical reinforcement in the wall must be enclosed by lateral ties, if the area of the vertical rebar is greater than 0.01 times the gross concrete area, or when the vertical rebar serves as compression reinforcement.

Additionally, from the ACI Code, Section 22.8.3, the design-bearing strength of concrete under a bearing plate may be taken as $\phi\,(0.85\,f_c'A_1)$, where A_1 is the loaded area. An exception occurs when the supporting surface is wider on all sides than the loaded area. In that case, the foregoing expression for bearing strength may be multiplied by a confinement factor, $\sqrt{A_2/A_1} \leq 2.0$, where A_2 is a concentric and geometrically similar support area that is the lower base of a frustum (upper base of which is A_1) of a pyramid having 1:2 sloping sides and fully contained within the support.

EXAMPLE 8.2

Design a reinforced concrete bearing wall to support a series of steel wide-flange beams spaced at 8 ft-0 in. o.c. Each beam rests on a bearing plate 6 in. × 12 in. (the 12 in. dimension of the bearing plate is parallel to the wall while the 6 in. dimension is parallel to the steel beams or perpendicular to the wall). The wall is braced top and bottom against lateral translation. Assume the bottom end fixed against rotation. The wall height is 15 ft and the design (factored) load P_u from each beam is 115 kip. Use $f_c' = 3000$ psi and $f_y = 60{,}000$ psi. See Figure 8.22.

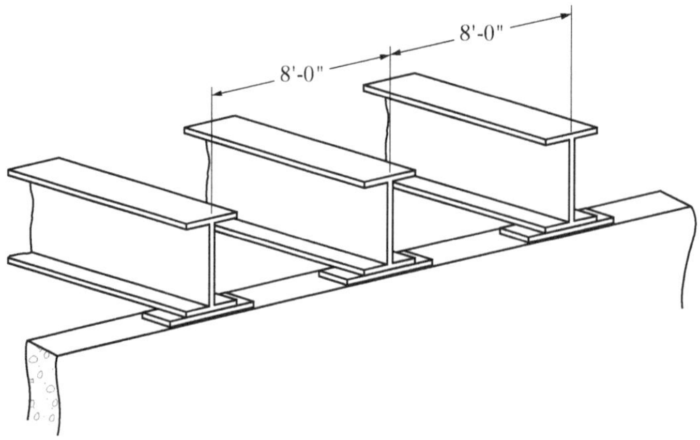

FIGURE 8.22 Bearing wall.

Solution:

1. Assume an 8-in.-thick wall with full concentric bearing.
2. From our previous discussion, the bearing strength of the concrete under the bearing plate, neglecting the concrete confinement multiplier, $\sqrt{A_2/A_1}$, is

$$\phi(0.85)f'_cA_1 = 0.65(0.85)(3000)(6)(12)$$
$$= 119{,}300 \text{ lb}$$
$$\text{factored bearing load } P_u = 115{,}000 \text{ lb}$$
$$119{,}300 > 115{,}000 \qquad \text{(O.K.)}$$

3. The effective length of the wall (ACI Code, Section 11.2.3.1) must not exceed the center-to-center distance between loads nor the width of bearing plus four times the wall thickness. Beam spacing = 96 in. Thus

$$12 + 4(8) = 44 \text{ in.}$$

Therefore, use 44 in.

4. The minimum thickness required is $\frac{1}{25}$ times the shorter of the unsupported height or width. Assume, for our case, that the width does not control. Then

$$h_{\min} = \frac{\ell_c}{25} = \frac{15(12)}{25} = 7.2 \text{ in.}$$

Also, $h_{\min}$ = 4 in. Therefore, the 8-in. wall is satisfactory.

5. The axial compression load capacity of wall is calculated from

$$\phi P_n = 0.55\phi f'_cA_g\left[1-\left(\frac{k\ell_c}{32h}\right)^2\right]$$

The strength-reduction factor ϕ for compression-controlled sections is discussed in Section 2.9 of this text. In this case, because the wall will not contain spiral lateral reinforcing, ϕ is taken as 0.65.

$$\phi P_n = 0.55(0.65)(3)(44)(8)\left\{1-\left[\frac{0.8(15)(12)}{32(8)}\right]^2\right\}$$
$$= 378(1-0.316)$$
$$= 258 \text{ kip}$$
$$\phi P_n > P_u \qquad \text{(O.K.)}$$

6. The reinforcing steel (ACI Code, Table 11.6.1), assuming No. 5 bars or smaller and assuming V_u = 0 kip, can be calculated as follows: For the vertical reinforcement per foot of wall length,

$$\text{required } A_s = 0.0012bh$$
$$= 0.0012(12)(8) = 0.12 \text{ in.}^2$$

For the horizontal reinforcement per foot of wall height,

$$\text{required } A_s = 0.002bh$$
$$= 0.002(12)(8) = 0.19 \text{ in.}^2$$

The maximum spacing of reinforcement must not exceed three times the wall thickness, $3h$, nor 18 in. (ACI Code, Sections 11.7.2.1 and 11.7.3.1). Thus

$$3h = 3(8) = 24 \text{ in.}$$

Use 18 in.

To select reinforcing,

vertical steel: Use No. 4 bars at 18 in.

$$(A_s = 0.13 \text{ in.}^2)$$

horizontal steel: Use No. 4 bars at 12 in.

$$(A_s = 0.20 \text{ in.}^2)$$

The steel may be placed in one layer, as the wall is less than 10-in. thick (see Figure 8.23).

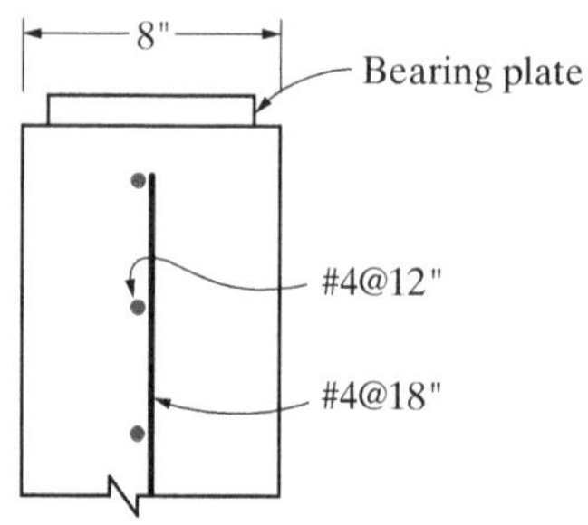

FIGURE 8.23 Section of wall for Example 8.2.

8.5 DESIGN CONSIDERATIONS FOR BASEMENT WALLS

Basement walls are used to retain soil in the basement levels of a building, and they are laterally supported by the floor slabs at the ground floor and basement levels (see Figure 8.23b). The basement walls are cast monolithically with the columns below the ground floor level as shown in Figure 8.32b.

The active soil pressure coefficient, K_a, is used for the design of retaining walls, while the at-rest lateral pressure

coefficient, K_o, is used for the design of basement walls due to the relatively smaller displacement of a

FIGURE 8.23b Basement walls

basement wall compared to a retaining wall. K_a is typically smaller than K_o. As an example, for a granular

native soil with angle of internal friction of 30 degrees at a particular site, the specified K_a was 0.33, the passive

pressure coefficient, K_p was 2.50, and the at-rest lateral soil pressure coefficient, K_o was 0.50.

A basement wall is a type of retaining wall in which lateral support is assumed to be provided at the bottom and top of the wall by the basement floor slab and first-floor construction, respectively. As previously mentioned, the wall would be designed as a simply supported member with a loading diagram and moment diagram as shown in Figure 8.24.

The designer must ensure that the basement wall is indeed laterally supported at the top of the wall. In residential wood construction where the wood floor joists may be spanning parallel to the concrete basement wall, no lateral support of the top of the basement wall is assured unless there is blocking at periodic spacings or intervals provided between adjacent floor joists and the perimeter or rim joists to transfer the lateral reaction due to the soil pressure from the top of the concrete basement wall to the wood floor diaphragm. Also, the sill plate and the anchor bolts embedded into the concrete basement wall, and the toenails connecting the blocking to the sill plates must be adequate to transfer the lateral reaction from the top of the basement wall into the floor diaphragm.

The design approach for the wall is similar to that of one-way slabs that was discussed in Chapter 2. For buildings with multi-level basements, the basement wall is analyzed similar to a continuous one-way slab subject to a linearly varying load due to the lateral soil pressure. The basement floor slabs act as the lateral supports for the wall. The main vertical reinforcement in the wall needed to resist the lateral soil pressure will be located on the inside face of the basement wall between the basement floor levels, and on the outside face of the wall at the basement floor supports.

If the wall is part of a bearing wall, the vertical load will relieve some of the tension in the vertical reinforcement. This may be neglected because its effect may be small compared with the uncertainties in the assumption made regarding the loads. If the vertical load is of a permanent nature and of significant magnitude, its effect should be considered in the design. The wall should then be designed as a member subject to combined bending and axial compression loads using the methods presented in Chapter 9.

When a part of the basement wall is above ground, the lateral bending moment may be small and may be computed as shown in Figure 8.25. This assumes that the wall is spanning in a vertical direction. Depending on the type of construction, the basement wall may also span in a horizontal direction or may behave as a slab spanning in one or two directions. If the basement wall design assumes two horizontal reactions (i.e., lateral support assumed at the top and bottom of the wall), as shown, the two lateral supports must be in place prior to backfilling behind the basement wall.

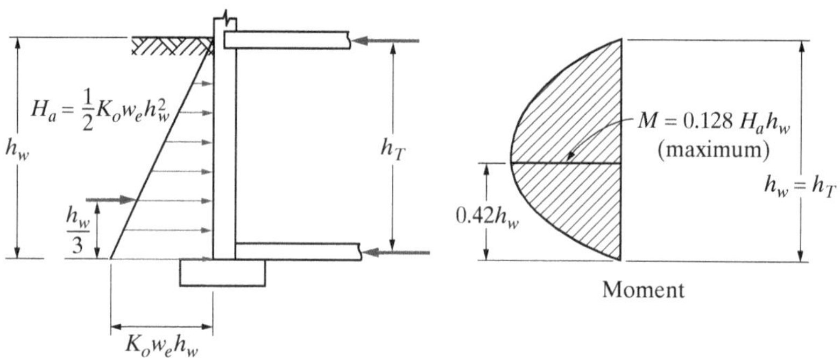

FIGURE 8.24 Basement wall with full height backfill: forces and moment diagram.

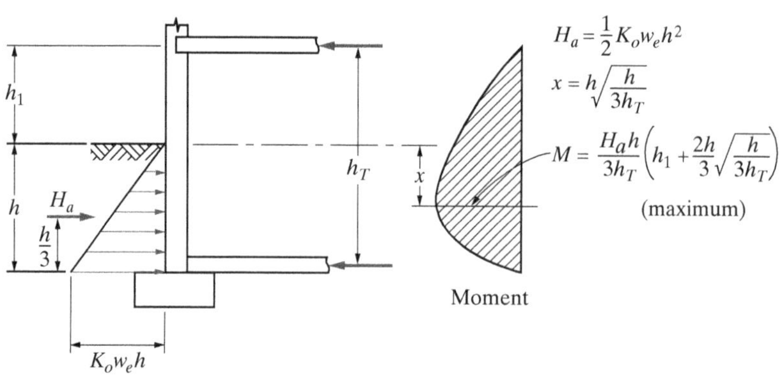

FIGURE 8.25 Basement wall with partial backfill: forces and moment diagram.

8.6 LATERAL FORCE RESISTING SYSTEMS IN CONCRETE BUILDINGS

The lateral forces that act on structures include wind loads, seismic loads, unbalanced lateral earth pressure, hydrostatic pressure (and uplift), and flood loads, but the most common lateral forces that act on building structures are wind loads, seismic loads, and lateral earth pressure. For a discussion of the methods for calculating the lateral forces acting on buildings, the reader should refer to Ref. [3, 4]. The two main types of lateral force resisting systems (LFRS) used in concrete buildings are *reinforced concrete moment frames* and *reinforced concrete shear walls*. These lateral-force-resisting-systems (LFRS) may be combined in a building, or one or the other LFRS may be the sole system used. The lateral-force-resisting-elements can be located within the interior of the building or along the building perimeter, subject to architectural constraints. Thus, concrete moment resisting frames will be acceptable along the building perimeter, whereas the use of shear walls on the exterior of a building could negatively impact the architectural features by blocking views and preventing the location of windows at the wall location.

8.7 CONCRETE MOMENT FRAMES

The connections between the beams and columns in concrete buildings are usually built integrally such that the beam-to-column joints behave like "rigid" joints and thus have the capacity to resist moments caused by lateral loads. Moment frames consists of girders or beams rigidly

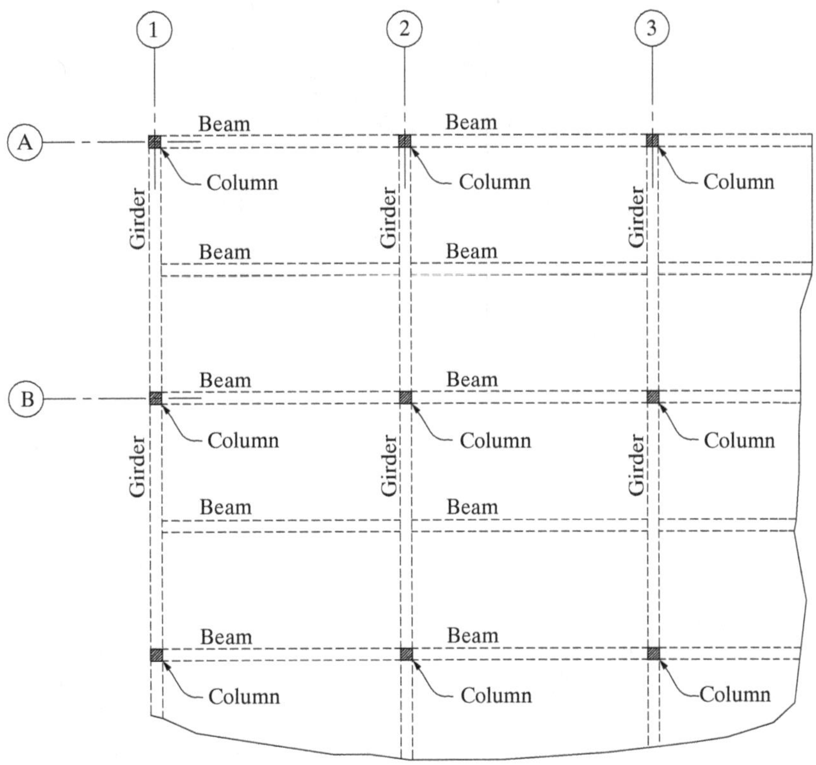

FIGURE 8.26 Floor plan showing reinforced concrete moment frames (along column lines only)

connected to the columns (see Figures 8.26 and 8.27) and they resist the lateral load moments through the bending of the concrete beams/girders and columns. The beam-to-column joints must be adequately detailed to ensure their ability to transfer the induced moments from the lateral loads. Naturally, moment frames occur only along the column lines in a building (e.g., lines 1, 2 and 3 in Figure 8.26). The distribution of the lateral load to each moment frame in each orthogonal direction will depend on the stiffness of the floor/roof diaphragms and the relative lateral stiffness of the moment frames and any shear walls that may be present. The elevation of the reinforced concrete moment frame along Grid Line 2 is shown in Figure 8.27. In the analysis of this moment frame, any plane frame analysis software can be used or an approximate method like the portal method could be used. In modelling the foundation supports, it is usual to model these as pinned supports unless the foundations are deliberately fixed against rotation by, for example, socketing and dowelling the footing into a rock foundation with rock anchors where the footing bears directly on bedrock, or by supporting the building on piles or caissons (see Chapter 10).

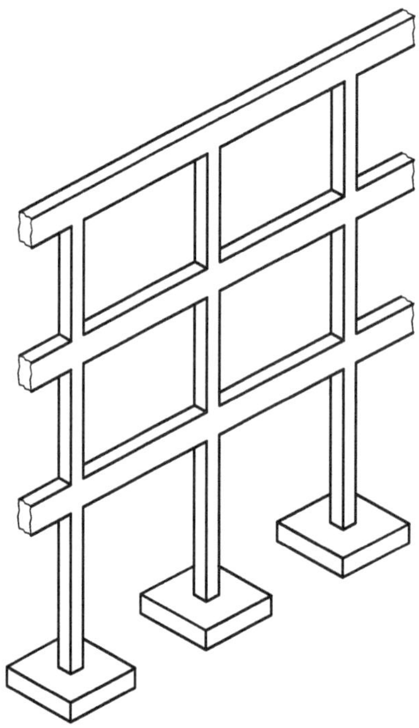

FIGURE 8.27 Elevation of reinforced concrete moment frame along Grid Line 2

8.8 SHEAR WALLS

Concrete or masonry walls fixed at their base are used to resist lateral wind and seismic loads in building structures with the lateral loads parallel to the plane of the shear wall, in addition to supporting gravity loads. These walls are subject to overturning moments, sliding shear at the base of the wall, and inplane shear in the wall. These lateral force resisting elements may consist of single walls located internally within the building or on the exterior face of the building, or they could be the stair and elevator core walls. Shear walls are very efficient lateral load resisting elements that resist lateral loads by acting as a vertical cantilever. The shear wall thickness depends on the height of the building, the lateral base shear and overturning moment, the applied gravity loads, and the concrete compressive strength; core wall thicknesses of up to 48 inches at the base of the building and high strength concrete with compression strengths of up to 12000 psi were recently used for a 55-story building in Chicago.

Sometimes, the shear walls may be perforated by door, window, or corridor openings, and when the vertical wall segments or piers on both sides of an opening are connected with horizontal wall or deep beams spanning over the door, window or corridor openings, the shear walls are called "coupled shear" walls (see Figure 8.29c). The coupling beams (or link beams) connecting the shear wall segments or piers reduce the flexural moment demand in the shear wall piers, help with energy dissipation in the coupled shear wall and help to enhance the efficiency of the shear wall.

These coupling beams are subjected to large moments and shear forces, and deep coupling beams with clear span-to-depth ratios less than 4.0 usually require, in addition to transverse and other reinforcement in the beam, two intersecting groups of confined diagonal reinforcement symmetrically placed about the midspan of the coupled beams, because they are controlled by shear (see ACI 318-19, Section 18.10.7, Figures R18.10.7a and R18.10.7b). Coupling beams with clear span-to-depth ratios greater than 4.0 are controlled by flexure and usually require, in addition to transverse and other reinforcement in the beam, at least two continuous bars at the top and bottom faces of the beams (see ACI 318-19, Sections 18.6.3 and 18.6.4).

The strength of a coupled shear wall depends on the stiffness of the coupled walls and the stiffness of the coupling beams, and its strength lies between the strength of a moment frame and that of an unperforated shear wall. The analysis of coupled shear walls, which is much more cumbersome than the analysis of segmented cantilevered shear walls, can be accomplished using structural analysis software, or the equivalent frame method if an approximate method is desired [1]. Figures 8.28 and 8.29 show some typical shear wall layout and typical shear wall elevations, respectively, in building structures. Single shear walls could be laid out between column lines as shown in Figures 8.29a and 8.29b. When the shear wall extends to and is built integrally with the columns, the column at both ends of the wall will serve as the boundary members for the shear wall and the vertical reinforcement in the columns will serve as the vertical end reinforcement in the shear wall (see Figure 8.29a).

One prominent concrete building where coupled shear walls have been used is the Burj Khalifa (formerly known as the Burj Dubai) in the United Arab Emirates, the tallest building in the world (since 2009) with 163 floors and a height of over 2700 ft. The coupling or link beams were 650 mm wide x 825 mm deep, with maximum reinforcement of 27-32 mm diameter continuous top and bottom bars, 16 mm diameter 10-leg stirrups at 75 mm on centers, plus 4-12 mm diameter skin reinforcement on each vertical face. The link beams have a clear span of 1400 mm giving a clear span-to-depth ratio of 1.70. Results from nonlinear finite element analyses showed that the link beams were actually stronger than predicted by the 2005 ACI Code [5].

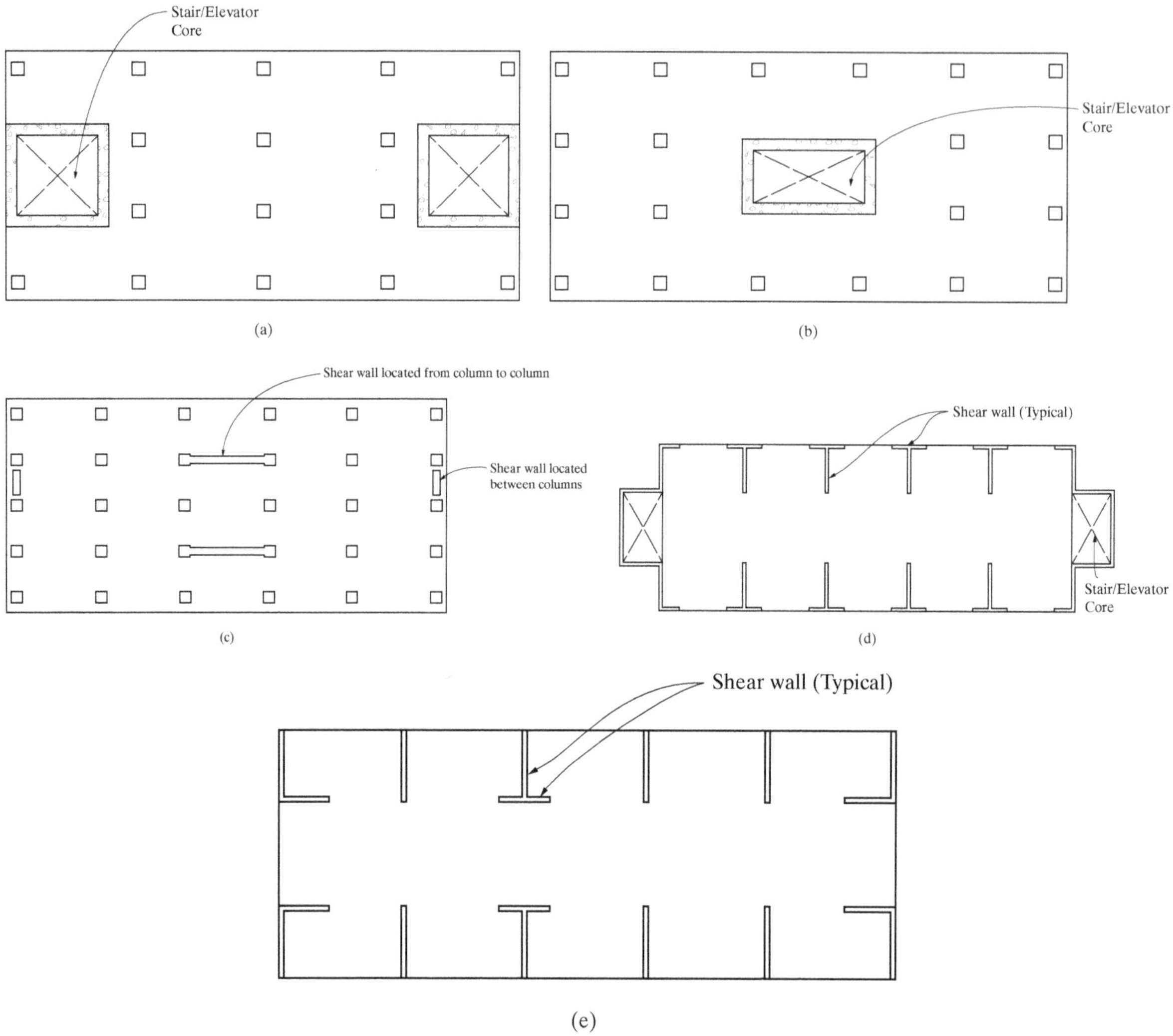

FIGURE 8.28 Typical shear wall layout in buildings.

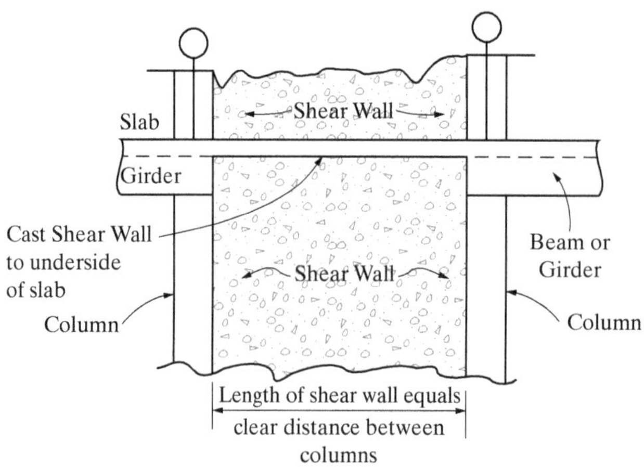

(a)

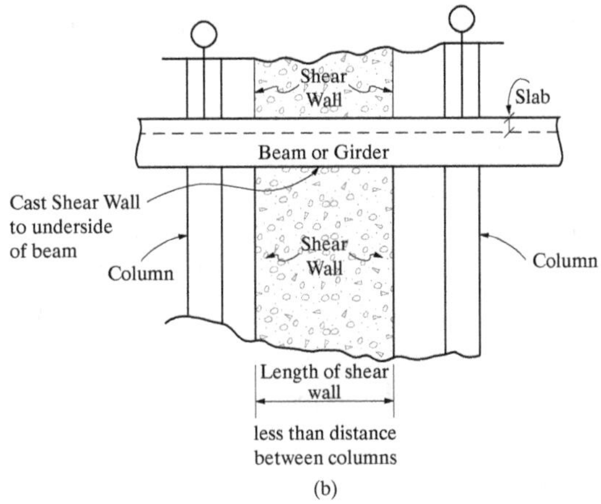

FIGURE 8.29 Typical shear wall elevations in concrete buildings.

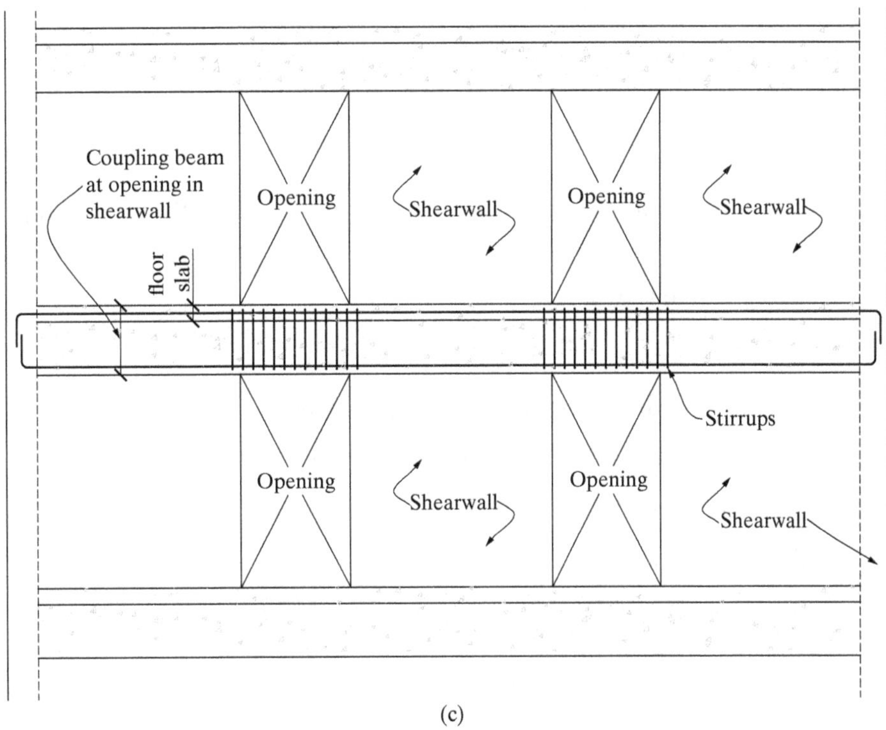

FIGURE 8.29c Elevation of Shear wall with coupling beams.

The lateral wind load on buildings act on the exterior cladding perpendicular to the wind direction, and the cladding transfers the wind loads to the roof and floor diaphragms, which in turn transfers the lateral loads - through in-plane bending of the diaphragm - to the lateral force resisting systems - LFRS (i.e., shear walls and moment frames) that are parallel to the direction of the wind load. The LFRS transfers the lateral loads to the foundations. The load path for seismic loads starts with ground acceleration which causes lateral movement in the structure. In the Code, the effect of the ground acceleration on the building is modeled as a set of equivalent lateral forces acting on the building at the floor and roof levels. The lateral forces in the diaphragms are transmitted to the foundation through the LFRS similar to the load path for wind loads.

Typical concrete buildings have floors and roof slabs that can be considered rigid in the horizontal plane, and thus the floors and roofs can be modeled as rigid diaphragms. See Chapter 14 for further discussion of rigid and flexible diaphragms. For buildings with rigid diaphragms, the lateral load acting on the building is distributed to the lateral force resisting elements parallel to the lateral load in proportion to the stiffness of the lateral force resisting element. For shear walls, the stiffness, K, consists of a flexural stiffness component and a shear stiffness component, and is calculated as

$$K = \frac{3EI}{h_w^3} + \frac{GA}{1.2h_w}$$

If the total lateral force on the building at level x is F_x, the lateral force at level x is distributed to each shear wall that is parallel to the lateral load as follows:

$$F_{\text{shear wall},x} = \frac{K_{\text{wall}}}{\Sigma K_{\text{wall}}} F_x$$

where

h_w = overall or total height of shear wall from top of footing to the top of the wall (see Figure 8.30a)

E = modulus of elasticity

I = moment of inertia of the wall about the strong axis $= \dfrac{h\ell_w^3}{12}$

ℓ_w = overall length of shear wall (see Figure 8.30a)

h = wall thickness (minimum practical wall thickness is 8 in.; see Figure 8.30b)

A = gross cross-sectional area of the wall $= h\ell_w$

G = shear modulus of elasticity $= \dfrac{E}{2(1+\nu)}$

ν = Poisson's ratio for concrete ≈ 0.20.

K_{wall} = stiffness of the shear wall being considered

ΣK_{wall} = sum of the stiffnesses of all the shear walls *parallel* to the lateral load

The following practical considerations should be considered when laying out shear walls in concrete buildings:

- Locate shear walls to minimize the effect on architectural features in the building such as doors and windows.
- Utilize stair and elevator shaft walls as shear walls. Shear walls can also be located on the outer perimeter of a building, but this may reduce the number of available windows in a building and, therefore, lead to a reduction in natural light and exterior views.

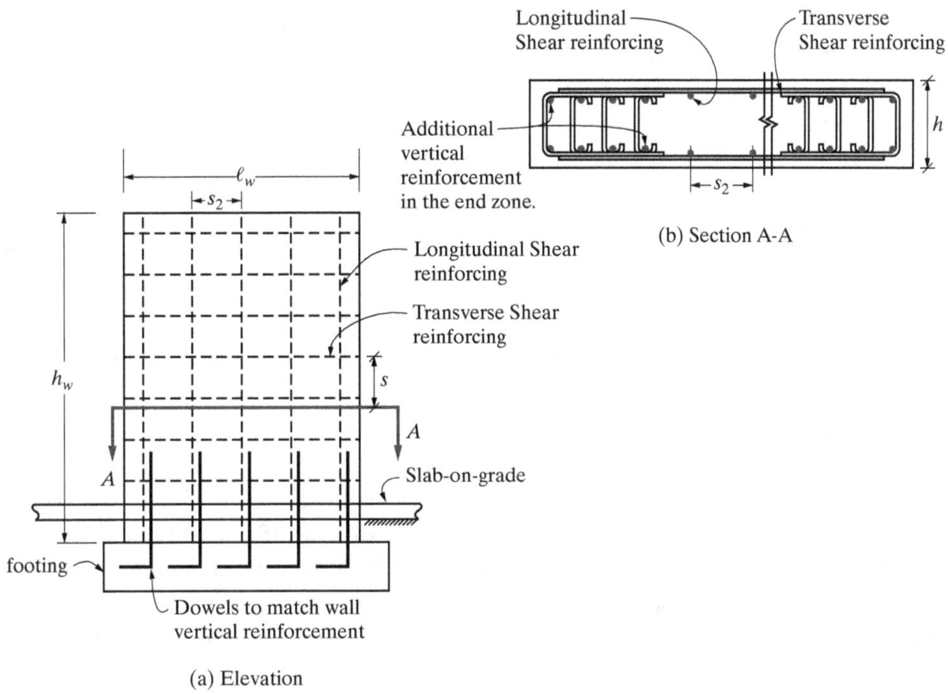

FIGURE 8.30 Reinforcement in shear wall without boundary element.

- Locate shear walls in each orthogonal direction as symmetrically as possible to minimize twisting or torsional deformations of the building from lateral loads. If a symmetrical arrangement is *not* feasible because of architectural or other constraints, the building should be analyzed for the resulting inplane *twisting forces,* and these would lead to additional lateral forces in the shear walls.
- Shear walls and/or moment frames are required in both orthogonal directions of concrete buildings (see Chapter 14 of Ref. [3]).

Shear Wall Design Considerations

The shear wall design considerations in this chapter pertain to ordinary reinforced concrete shear walls as presented in ACI Code, Section 11.5.4. The design considerations for ductile or special shear walls used in highly seismic zones, which is outside the scope of this text, are presented in ACI Code, Section 18.10. The following load effects should be considered in the design of shear walls:

- The horizontal *shear* force which is maximum at the base of the wall.
- The *bending moment* which is maximum at the base of the wall. This produces compression at the end zone at one end of the wall and tension at the end zone at the opposite end. The location of the tension and compression forces will change depending on the direction of the lateral load.
- The *gravity* or *vertical loads* (i.e., roof and floor dead and live loads) that cause compression on the wall.

For low-rise concrete buildings, the effect of the interaction of the gravity loads with the shear or bending moment capacity of the wall will be minimal, and therefore, it is practical to consider each load effect separately for low-rise buildings. However, for high-rise buildings where the gravity load on the wall could be substantial, the interactions between the axial load and bending moment and shear capacities of the wall must be considered, and in that case, the shear wall is usually designed as a "column" with combined axial load and moment, and a column-type interaction diagram is used (see Chapter 9).

Reinforcement in Shear Walls

The reinforcement in shear walls consists of distributed horizontal reinforcement used to resist shear forces in the wall, distributed vertical reinforcement used to resist gravity loads and to control shrinkage and cracking, and concentrated vertical end reinforcement used to resist the bending moment due to lateral loads. To avoid congestion of the vertical reinforcement at the end of the shear wall, mechanical splices are used for No. 10 and larger bars, and the splices are located above every other floor level [2]. The typical shear wall reinforcement is shown in the wall elevation and section in Figure 8.30, and the typical reinforcement details at the corners and at the ends of shear walls are shown in Figures 8.31(a) and 8.31(b). Figure 8.31(c) shows shear walls with boundary elements at the ends of the wall which act integrally with the shear wall stem to resist the lateral forces. Note that the boundary elements could be the same thickness as, or thicker than, the shear wall. For shear walls with flanges, the effective flange width can be determined using ACI 318-19, Section 18.10.5.2. (See Figure 8.31(c)). The effective flange width (for both tension and compression zones) beyond the face of the wall web is equal to the smaller of one-half the distance to an adjacent wall web and 25% of the overall wall height. Note that for compound structural walls such as box-shaped shear wall core or walls with flanges, only the wall segments parallel to the lateral force is assumed to resist the lateral shear, while the tension and compression force couple arising from the overturning moments are resisted by the farthest wall segments that are perpendicular to the lateral load direction,

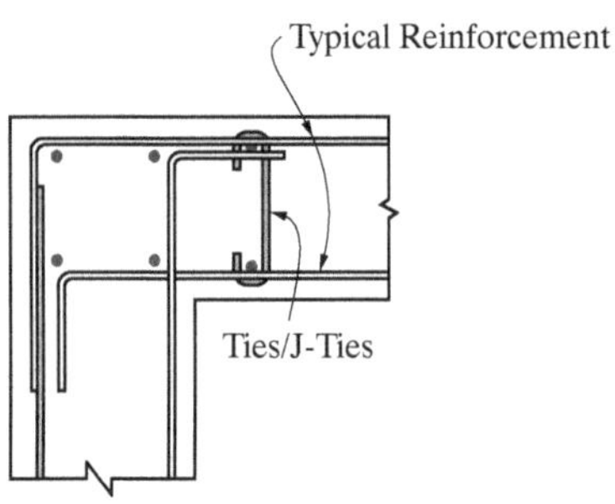

(a) Wall Corner Reinforcement

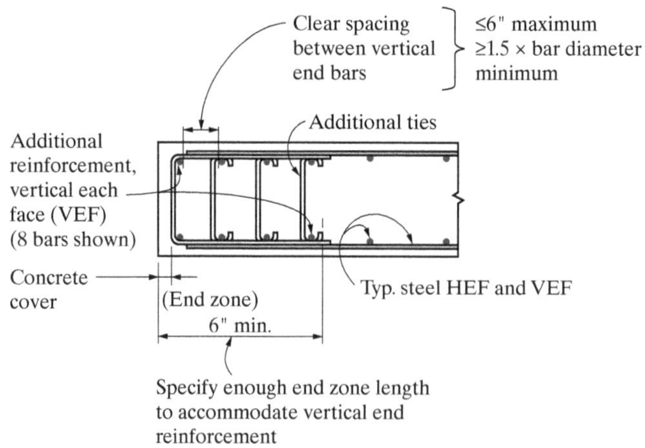

(b) Wall end zone reinforcement

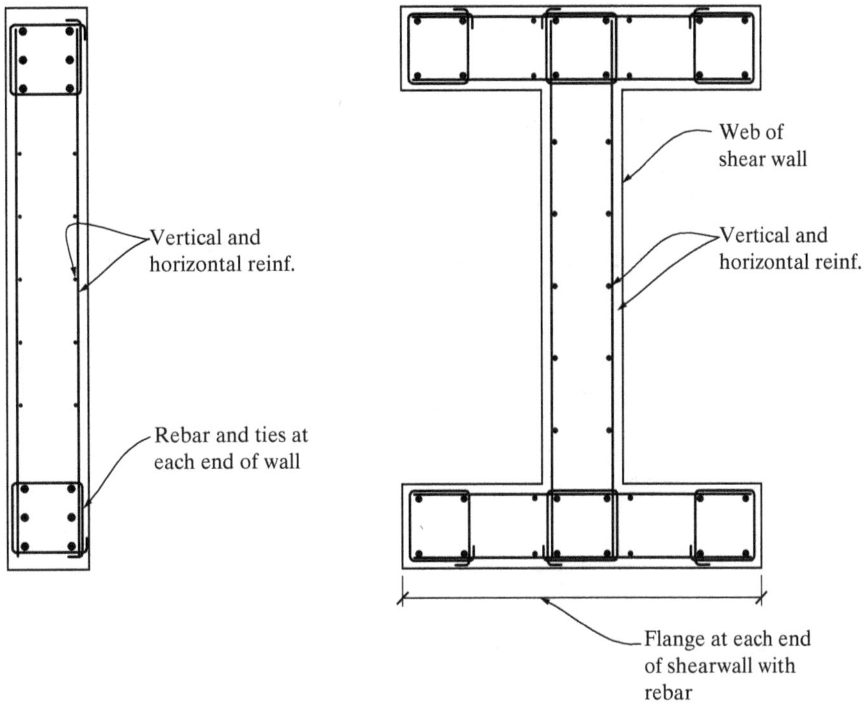

FIGURE 8.31 Typical added reinforcement at the corners and end zones of shear walls.

Minimum Reinforcement in Shear Walls

The amount of minimum traverse or horizontal reinforcement required in structural walls is a function of the applied inplane shear force, V_u. When the inplane shear force in a shear wall $V_u \leq 0.5\phi\alpha_c\lambda\sqrt{f_c'}A_{CV}$, the minimum area of horizontal reinforcement and vertical reinforcement is as previously presented for bearing walls in Section 8.4. When the inplane shear force in a structural wall $V_u > 0.5\phi\alpha_c\lambda\sqrt{f_c'}A_{CV}$, the minimum reinforcement should be determined as follows (ACI Code, Section 11.6.2):

The minimum ratio of distributed *transverse* or *horizontal* reinforcement, ρ_t in the plane of the wall to the gross cross-sectional area of the wall perpendicular to the reinforcement is given in ACI Code, Section 11.6.2:

$$\rho_t = \frac{A_v}{sh} \geq 0.0025$$

The minimum ratio of distributed *vertical* or *longitudinal* reinforcement, ρ_ℓ in the wall to the gross cross-sectional area of the wall perpendicular to the reinforcement is given in ACI 11.6.2:

$$\rho_\ell = \frac{A_v}{s_2 h} = 0.0025 + 0.5\left(2.5 - \frac{h_w}{\ell_w}\right)(\rho_t - 0.0025) \geq 0.0025 \qquad \textbf{[ACI Eq. (11.6.2)]}$$

but need not be greater than ρ_t.

where

A_v = area of distributed transverse reinforcement

s = center-to-center vertical spacing of the transverse or horizontal reinforcement (see Figure 8.30)

s_2 = center-to-center horizontal spacing of the longitudinal or vertical reinforcement (see Figure 8.30)

ϕ = 0.75 (shear)

λ = 1.0 for normal-weight concrete

$$\alpha_c = 3 \quad for \ \frac{h_w}{\ell_w} \leq 1.5$$

$$\alpha_c = 2 \quad for \ \frac{h_w}{\ell_w} \geq 2.0$$

$$\alpha_c \ \text{var} \, ies\, linearly\, between\, 2 \ and \ 3 \quad for \ 1.5 < \frac{h_w}{\ell_w} < 2.0$$

$$A_{CV} = \ell_w h$$

ℓ_w = length of wall

h_w = height of wall

h = thickness of wall

The size of the distributed horizontal and vertical reinforcement is usually No. 4 or larger bars.

The maximum spacing of the *horizontal* or transverse reinforcement allowed by the Code (s_{maximum}) is the smallest of $\ell_w/5$, $3h$, or 18 in, and the maximum spacing of the *vertical* or longitudinal reinforcement allowed by the Code ($s_{2\ \text{maximum}}$) is the smallest of $\ell_w/3$, $3h$, or 18 in.

For 8-in.- or 10-in.-thick walls, although ACI Code, Section 11.7.2.3 implies that the horizontal and vertical reinforcement can be placed in a single layer at the center of the wall, it is common in design practice to use two layers of reinforcement (i.e., reinforcement on both faces of the wall) for 8-in.- and 10-in.-thick shear walls.

Shear Strength of Shear Walls

In the limit states design for shear, the ACI Code requires the design shear strength, ϕV_n to be greater than or equal to the required shear strength or factored shear, V_u. That is,

$$\phi V_n \geq V_u$$

where

ϕ = 0.75

The nominal shear strength (ACI Code, Sections 11.5.4.2 and 11.5.4.3) is

$$V_n = V_c + V_s = \left(\alpha_c \lambda \sqrt{f'_c} + \rho_t f_{yt}\right) A_{CV}$$

$$\leq 8\sqrt{f'_c} A_{CV}$$

$$\phi V_n = \phi\left(\alpha_c \lambda \sqrt{f'_c} + \rho_t f_{yt}\right) A_{CV} \geq V_u$$

$$\rho_t = \frac{A_v}{sh} = \frac{V_u - \phi \alpha_c \lambda \sqrt{f'_c} A_{CV}}{\phi f_{yt} A_{CV}} \geq 0.0025$$

$$\rho_\ell = \frac{A_v}{s_2 h} = 0.0025 + 0.5\left(2.5 - \frac{h_w}{\ell_w}\right)(\rho_t - 0.0025) \geq 0.0025$$

λ = lightweight concrete modification factor defined in Chapter 1 and is 1.0 for normal-weight concrete

Provide adequate wall thickness, wall length, compressive strength, and transverse reinforcement in the wall so that the limit states design equation for shear, $\phi V_n \geq V_u$, is satisfied, and the minimum horizontal and vertical rebar ratios and their maximum spacing requirement are satisfied.

Flexural Strength of Shear Walls

After selecting the longitudinal and transverse reinforcement (usually No. 4 or larger bars), the end zone or boundary element vertical reinforcement at the extreme ends of the wall are determined. These end zone vertical reinforcement are required to resist the overturning moment in the shear wall caused by the lateral loads. They are determined using the rectangular beam design procedure previously presented in Chapter 2, if the gravity loads on the shear wall can be neglected. The goal is to ensure that the limit states design equation for moment, $\phi M_n \geq M_u$, is satisfied. If the vertical end reinforcement in the shear wall is required as compression reinforcement, then the vertical end reinforcement must be enclosed by lateral ties in accordance with ACI Code, Section 25.7.2, similar to columns (see Chapter 9). Otherwise, lateral ties are not required, though it is good practice to provide additional lateral ties for the vertical end reinforcement in shear walls. The vertical reinforcement in boundary elements in shear walls should be enclosed by lateral ties satisfying ACI Code, Section 25.7.2. The ties help to prevent buckling of the vertical end bars when subjected to compression. If the gravity loads on the shear wall is significant and therefore cannot be neglected, the axial load-moment interaction diagram approach for column design under combined axial load and large moment presented in Sections 9.7 through 9.10 of chapter 9 can be used to design the vertical end reinforcement in the shear wall.

For box-shaped concrete core walls similar to those shown in Figures 8.28a and 8.28b, the following method can be used to determine the vertical reinforcement in the walls, in lieu of using the Chapter 2 method: The factored overturning moment, M_{OTu}, at the base of the wall due to lateral loads is divided by the moment arm or center to center distance between opposite walls of the core to obtain the vertical force-couple, T and C. That is, T = C = M_{OTu}/(length of core wall parallel to the lateral load – wall thickness). Therefore, the area of the tension steel in the walls, $A_s = T/(\phi f_y) = T/(0.9f_y)$. The vertical reinforcement can be placed in multiple rows within the wall thickness. Similarly, to resist the compression force, C, the required thickness of the core wall can be determined by equating the compression force to the uniform compression, $0.85f_c'bh$, where b is the width of the box-shaped core wall perpendicular to the lateral load, and h is the wall thickness. Therefore, the required wall thickness to resist the compression force, C (ignoring the vertical rebar in the wall) is $h = C/(0.85f_c'b)$.

EXAMPLE 8.3

Shear Wall Design

The shear wall layout for a three-story building and the unfactored north–south lateral seismic loads acting on the building are shown in Figure 8.32. Design the north–south shear walls for the seismic lateral loads shown. Assume normal-weight concrete; f_c' is 4000 psi and f_y is 60,000 psi. Assume all shear walls are of equal thickness.

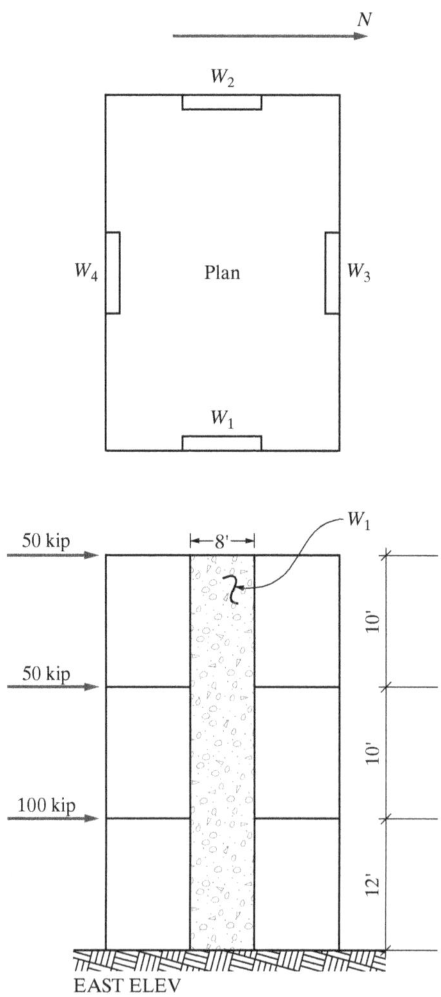

FIGURE 8.32 Shear wall layout and unfactored lateral loads for Example 8.3.

Solution:

1. The maximum load factor for seismic loads from ACI Section 5.3.1 is 1.0. Because this is a low-rise building, the effect of the vertical or gravity loads on the moment capacity of the shear walls will be small; therefore, the interaction between the gravity loads and the moment is assumed to be negligible and hence the load effects on the wall will be considered separately.

2. As the length, thickness, and total height of the north–south shear walls, W1 and W2, are equal, their stiffness, K, will also be equal. The factored seismic lateral forces for shear walls $W1$ or $W2$, including the seismic load factor, are calculated as follows:

$$\text{Roof: } F_r = 1.0 \times \frac{K}{2K}(50 \text{ kip}) = 25 \text{ kip}$$

$$\text{Third floor: } F_3 = 1.0 \times \frac{K}{2K}(50 \text{ kip}) = 25 \text{ kip}$$

$$\text{Second floor: } F_2 = 1.0 \times \frac{K}{2K}(100 \text{ kip}) = 50 \text{ kip}$$

For high-rise buildings, the wall reinforcement is typically specified or designed for two- or three-story lifts, but for this low-rise building, only the reinforcement at the base of the wall will be designed and this reinforcement will be used throughout the full height of the shear wall. The factored shear wall lateral loads and load effects are shown in Figure 8.33.

3. Select wall thickness: Assume h = 8 in. (the reinforcement will be placed on both faces of the wall).

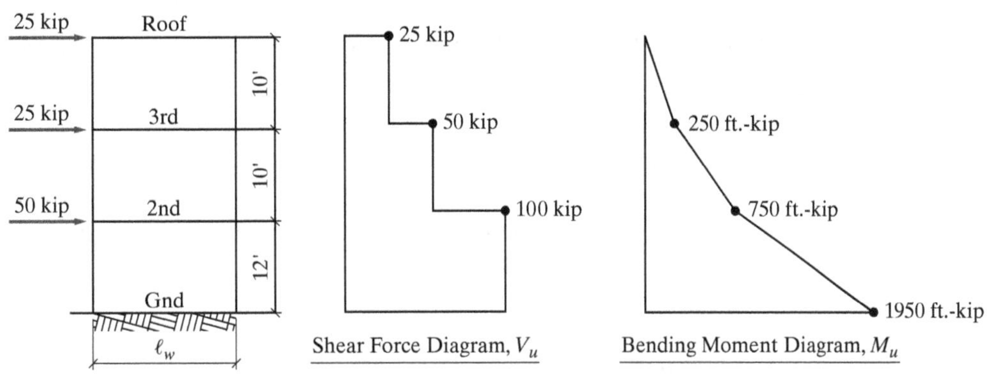

FIGURE 8.33 Shear wall lateral loads and load effects.

4. Calculate the maximum allowable shear strength.

$$\phi 8\sqrt{f_c'}A_{CV} = (0.75)(8)\left(\sqrt{4000}\right)(768) = 291.4\ kip >> V_u = 100\ kip\ \ (O.K.)$$

ϕ = 0.75 (shear)

λ = 1.0 for normal-weight concrete

ℓ_w = length of wall = 8 ft.

h_w = height of wall = 32 ft.

h = thickness of wall = 8 inches

$$\alpha_c = 3 \quad for \ \frac{h_w}{\ell_w} \le 1.5$$

$$\alpha_c = 2 \quad for \ \frac{h_w}{\ell_w} \ge 2.0$$

$$\alpha_c \text{ varies linearly between 2 and 3} \quad for \ 1.5 < \frac{h_w}{\ell_w} < 2.0$$

$$\frac{h_w}{\ell_w} = \frac{32\ ft}{8\ ft} = 4.0 > 2.0$$

$$\alpha_c = 2$$

$$A_{CV} = \ell_w h = (8\ ft)(12\ in./\ ft)(8\ in.) = 768\ in.^2$$

5. Calculate the minimum area of reinforcement in the wall:

When the inplane shear in a structural wall, $V_u \le 0.5\phi\alpha_c\lambda\sqrt{f_c'}A_{CV}$, the minimum area of horizontal reinforcement and vertical reinforcement is as previously presented for bearing walls in Section 8.4. When the inplane shear in a structural wall, $V_u > 0.5\phi\alpha_c\lambda\sqrt{f_c'}A_{CV}$, the minimum reinforcement should be as follows (ACI Code, Section 11.6.2):

$$0.5\phi\alpha_c\sqrt{f_c'}A_{CV} = (0.5)(0.75)(2.0)(1.0)\left(\sqrt{4000}\right)(768) = 36.4 \text{ kip} < V_u = 100 \text{ kip}$$

Therefore, the required minimum area of transverse or horizontal reinforcement ratio is

$$\rho_t \ge 0.0025$$

The required minimum area of longitudinal or vertical reinforcement ratio is

$$\rho_\ell = \frac{A_v}{s_2 h} = 0.0025 + 0.5\left(2.5 - \frac{h_w}{\ell_w}\right)(\rho_t - 0.0025)$$

$$= 0.0025 + 0.5\left(2.5 - \frac{32 \text{ ft}}{8 \text{ ft}}\right)(0.0028 - 0.0025)$$

$$= 0.00228$$

ρ_ℓ must not be less than 0.0025 and need not be greater than ρ_t. Therefore, use $\rho_\ell = 0.0025$.

6. Determine the required *horizontal* shear reinforcement to resist the applied shear, V_u:

$$\phi V_n = \phi\left(\alpha_c \lambda \sqrt{f'_c} + \rho_t f_{yt}\right) A_{CV} \geq V_u$$

$$\rho_t = \frac{A_v}{sh} = \frac{V_u - \phi\alpha_c\lambda\sqrt{f'_c}A_{CV}}{\phi f_{yt} A_{CV}} = \frac{100(1000) - (0.75)(2)(1)\sqrt{4000}(768)}{(0.75)(60{,}000)(768)}$$

$$= 0.00079 \geq 0.0025$$

Therefore, $\rho_t = 0.0025$

The maximum spacing of the horizontal reinforcement that is allowed by the Code (s_{maximum}) is the smallest of the following:

- $\dfrac{\ell_w}{5} = \dfrac{96 \text{ in.}}{5} = 19.2 \text{ in.}$
- $3h = (3)(8 \text{ in.}) = 24$ in.
- 18 in. (Controls)

Therefore, try s = 18 in. and assume No. 4 Horizontal Each Face (HEF)

The corresponding horizontal reinforcement ratio provided is

$$\rho_t = \frac{A_v}{sh} = \frac{(2\ legs)(0.2 \text{ in.}^2)}{(18 \text{ in.})(8 \text{ in.})} = 0.0028 \geq 0.0025 \quad \text{(O.K.)}$$

For the distributed horizontal reinforcement, provide No. 4 HEF @ 18 in. o.c. (HEF = horizontal each face of wall.)

7. Determine distributed vertical shear reinforcement.

$$\rho_\ell = 0.0025 = A_v / s_2 h$$

$$\frac{A_v}{s_2} = (0.0025)(h) = (0.0025)(8 \text{ in.}) = 0.02 \text{ in.}$$

Try No. 4 vertical bar both faces of wall: therefore, A_v = 2 faces × 0.2 in.^{2} = 0.4 in.^{2}

Therefore, the required spacing of the vertical shear reinforcement is

$$s_2 \text{ required} = \frac{0.4}{0.02} = 20 \text{ in.}$$

The maximum spacing of the vertical reinforcement allowed by the Code (s_2 maximum) is the smallest of the following:

- $\frac{\ell_w}{3} = \frac{96 \text{ in.}}{3} = 32 \text{ in.}$
- $3h = (3)(8 \text{ in.}) = 24 \text{ in.}$
- 18 in. (Controls)

Because maximum spacing = 18 in. < s_2 required = 20 in.; therefore, use s_2 = 18 in.

For the distributed vertical reinforcement, provide No. 4 VEF @ 18 in. o.c. (VEF = vertical each face of wall.)

8. Design the shear wall for flexure or bending and determine the end zone vertical reinforcement. The maximum factored bending moment at the base of the wall due to the factored seismic lateral load is

$$M_u = 1950 \text{ ft.-kip.}$$

The limit states design equation for flexure requires that $\phi M_n = \phi h d^2 \bar{k} \geq M_u$. Initially, assume ϕ = 0.9 (this will be checked later after the net tensile strain in the extreme tension steel, ε_t, is determined) and then calculate the required $\bar{k}$ as follows:

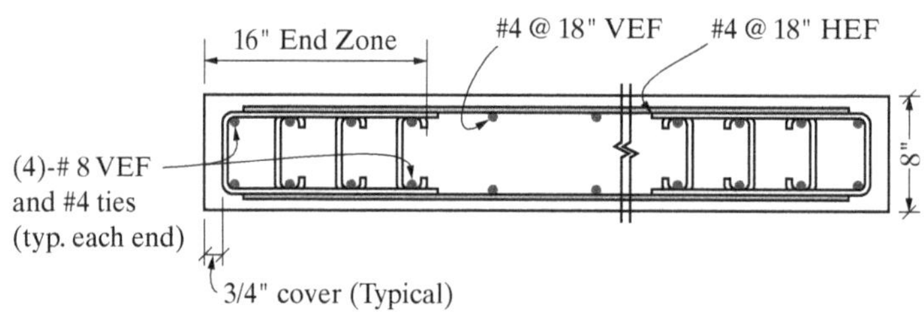

FIGURE 8.34 Shear wall reinforcement detail.

Assume the effective flexural depth, $d = 0.8\ell_w = (0.8)(96 \text{ in.}) = 76.8 \text{ in.}$

$$\bar{k} = \frac{M_u}{\phi h d^2} = \frac{1950(12)}{(0.9)(8 \text{ in.})(76.8 \text{ in.})^2} = 0.55$$

From Table A.10, we obtain $\rho = 0.0101$ and $\varepsilon_t \gg 0.005$; therefore, $\phi = 0.9$, as initially assumed. The concentrated vertical reinforcement required at each end zone of the shear wall is

$$A_s \text{ required} = \rho h d = (0.0101)(8 \text{ in.})(76.8 \text{ in.}) = 6.21 \text{ in.}^2$$

The minimum area of concentrated vertical steel required for bending at the ends of the shear wall is

$$A_{s,\min} = \frac{3\sqrt{f'_c}}{f_y}hd \geq \frac{200}{f_y}hd$$

$$= \frac{3\sqrt{4000}}{60,000}(8)(76.8) = 1.94 \text{ in.}^2$$

$$\geq \frac{200}{60,000}(8)(76.8) = 2.05 \text{ in.}^2$$

Because A_s required = 6.21 in.^{2} > $A_{s,\min}$ = 2.05 in.^{2}, A_s required = 6.21 in.^{2}

Try 8 No. 8 vertical reinforcement at each end of the wall (i.e., 4 No. 8 VEF). Total area of steel provided at each end of the wall = 6.32 in.^{2} > A_s required (O.K.).

A plan detail showing the reinforcement provided in the shear wall is shown in Figure 8.34. An end zone length of 16 in. has been assumed. This will be checked below. The concentrated vertical reinforcement must be provided at both ends of the shear wall as the lateral load can reverse direction and each end of the shear wall will be subjected to tension or compression forces depending on the direction of the lateral load. From Figure 8.34, the clear spacing between the concentrated vertical bars is

$$\frac{16\text{in.} - 0.75 \text{ in. cover} - 0.5 \text{ in. tie} - 0.5 \text{ in. tie} - (4 \text{ bars})(1 \text{ in. diameter})}{3 \text{ spaces}}$$

$$= 3.42 \text{ in.} > 1.5(1 \text{ in. diameter bar}) = 1.5 \text{ in.} \qquad \text{(O.K.)}$$

Therefore, the assumed end zone length of 16 in. is O.K. because it provides adequate spacing between the vertical bars to allow for aggregates to fit between the bars and allow for adequate vibration of the concrete between the vertical bars. Note that the vertical end reinforcement in the shear wall must be enclosed by transverse ties since the end reinforcement is required to resist compression forces and the area of steel provided, i.e., 6.32 in.^{2}, is greater than $0.01A_g$ or 0.01(8 in.)(16 in.) = 1.28 in.^{2}

Note that the building in the preceding example had no basement floor levels, so the overturning moment at the base of the shear wall was the sum of the moments of the lateral forces about the base of the shear wall. Similarly, the total base shear was the sum of all the lateral forces acting on the shear wall. However, for multi-story buildings with basement floor levels, some horizontal restraint will be provided to the shear wall at the ground floor level by the ground floor and the basement floor slabs acting as diaphragms in transferring some of the lateral force to the basement walls which act as shear walls in a direction parallel to its length. The basement shear walls are typically much longer in length compared to the core walls of the building; therefore, they are much stiffer than the shear walls and thus attract a larger portion of the lateral load at the ground floor and basement levels. The amount of reaction provided at the ground floor level and hence the corresponding moment and lateral shear at the base of the building shear wall will depend on the in-plane stiffness of the ground floor slab diaphragm and the relative in-plane stiffness of the basement walls compared to the in-plane stiffness of the building shear walls. (See Figure 8.35). The use of the basement walls in resisting a portion of the lateral loads reduces the otherwise large moments at the base of the core or shear walls in the building.

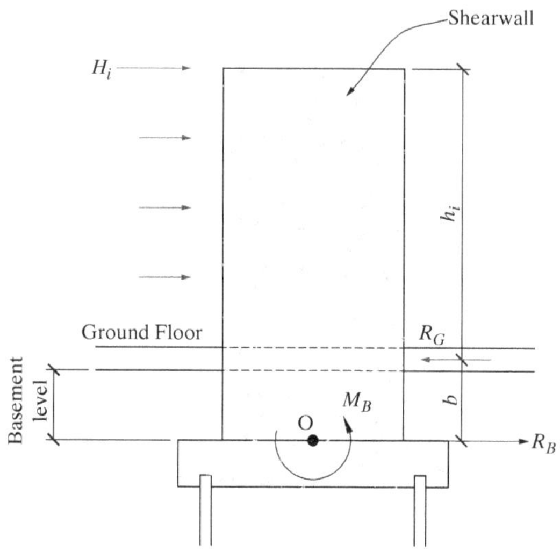

FIGURE 8.35 Shear wall forces due to lateral restraint at ground floor level.

Using equilibrium equations $\sum M_O = 0$ *and* $\sum F_x = 0$ in Figure 8.35 gives the ground floor lateral restraint and base shear as

$$R_G = \frac{\sum H_i(h_i + b) - M_B}{b} \quad and \quad R_B = \frac{\sum H_i(h_i) - M_B}{b}$$

Note that for simplicity we have not shown the vertical forces on the wall and the vertical reaction at the foundation level. These do not impact the above analysis. Theoretically, R_G can vary from a small value up to a maximum of $\sum H_i$ depending on the ratio of the in-plane stiffness of the basement wall to the in-plane stiffness of the building core or shear walls. The drag strut reinforcement in the ground floor diaphragm required to transfer the reaction, R_G, from the shear walls into the ground floor slab or diaphragm is critical and must be properly developed or extended into and through the shear walls, so that the lateral force transfer can indeed occur. Further discussions of drag struts and diaphragms are covered in Chapter 14.

The base shear in the basement walls will be resisted by a combination of friction between the basement wall footing and the soil (or by the lateral resistance of piles or caissons where the basement wall is supported on deep foundations) and by the passive soil pressure against the basement walls from the surrounding soil (see Chapter 10).

Spread or strip footings may only be adequate for shear walls in low rise buildings because of the detrimental effect of the wall rotation caused by the overturning moments, given the fact that the base of a shear wall is usually assumed to be fixed in the structural analysis. For mid- and high-rise buildings, the shear walls are usually supported on mat, pile, or caisson foundations, rather than on shallow spread footings [2].

The foundations for shear walls in midrise and highrise buildings are typically supported on mat or pile or drilled shaft (or caisson) foundations to enable resistance to the base shear and overturning moments caused by the lateral loads on the shear walls [2]. Depending on the height of the building and the nature of the soil, the core walls may be supported by a concrete mat that in turn is supported on piles or drilled shafts or caissons. The design of foundations is discussed in Chapter 10.

REFERENCES

[1] Angelo Mattacchione, "Equivalent Frame Method Applied to Concrete Shearwalls," *Concrete International*, November 1991, pp. 65-72.

[2] David G. Kittridge, "High Rise Shearwall Analysis, Design and Constructibility Guidelines," ASCE Structures Congress, 2002.

[3] Abi Aghayere and Jason Vigil, Structural Steel Design: A Practice-oriented Approach, 2nd Edition, Pearson, Hoboken, NJ, 499 pp.

[4] David A. Fanella, Structural Loads – 2012 IBC and ASCE/SEI 7, International Code Council.

[5] Ho Jung Lee, Daniel A. Kuchma, William Baker, and Lawrence C. Novak, "Design and Analysis of Heavily Loaded Reinforced Concrete Link Beams for Burj Dubai," ACI Structural Journal, July-August 2008.

EXERCISES

8.1 Compute the active earth pressure horizontal force on the wall shown for the following conditions. Use $w_e = 100$ lb/ft^3.

Case	ϕ	θ	w_s (psf)	h_w(ft)
(a)	25	0	400	15
(b)	28	10	0	18
(c)	30	0	200	20
(d)	33	20	0	25

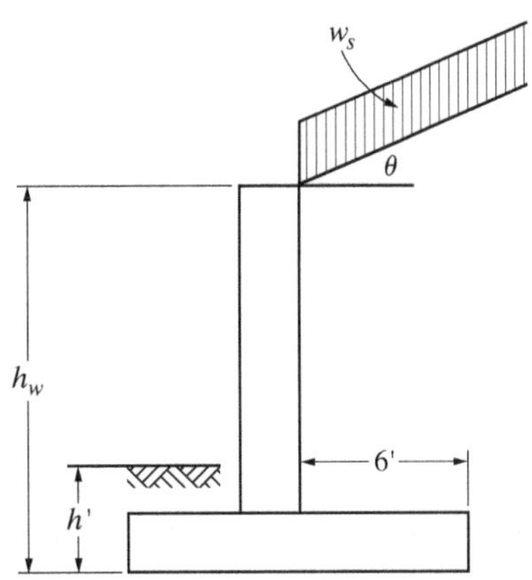

FIGURE FOR EXERCISE 8.1

8.2 Find the passive earth pressure force in front of the wall for Exercise 8.1(a), if $h' = 4$ ft.

8.3 For the wall shown, determine the factors of safety against overturning and sliding and determine the soil pressures under the footing. Use $K_a = 0.3$ and $w_e = 100$ lb/ft^3. The coefficient of friction $f = 0.50$.

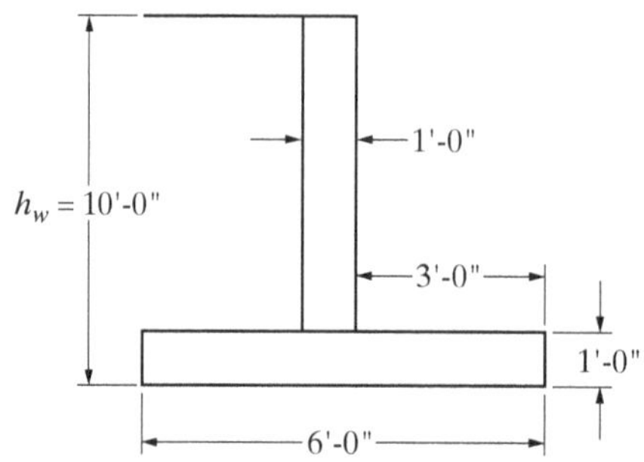

FIGURE FOR EXERCISE 8.3

8.4 Same as Exercise 8.3, but the toe is 2 ft-0 in.

8.5 In Problem 8.1(c), assume a footing depth of 2 ft. Design the stem steel for M_u at the top of the footing. Check the anchorage into the footing. Disregard other stem steel details. Use $f_c' = 4000$ psi and $f_y = 60{,}000$ psi. Use thickness at top of stem = 12 in. and thickness at bottom of stem = 1 ft-9 in.

8.6 Completely design the cantilever retaining wall shown. The height of wall h_w is 22 ft, the backfill is level, the surcharge is 600 psf, $w_e = 100$ lb/ft^3, $h' = 4$ ft, $K_a = 0.30$, and the allowable soil pressure is 4 ksf. Use a coefficient of friction of concrete on soil of 0.5, $f_c' = 3000$ psi, and $f_y = 60{,}000$ psi. The required factor of safety against overturning is 2.0 and against sliding is 1.5. The wall batter should be between $\frac{1}{4}$ and $\frac{1}{2}$ in./ft. The design is to be in accordance with the ACI 318-19.

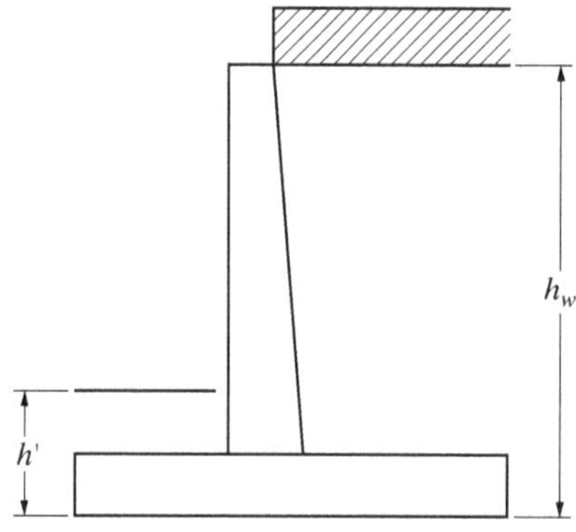

FIGURE FOR EXERCISE 8.6

8.7 Develop a spreadsheet application that will determine the location of the theoretical cutoff point for alternate bars in the stem of Example 8.1 (refer to Figure 8.16). Set up the spreadsheet so that a user could input any typical design data for a reinforced concrete cantilever retaining wall.

8.8 Design a reinforced concrete bearing wall to support a series of precast single tees spaced 7 ft-6 in. on centers. The stem of each tee section is 8-in. wide and bears on the full thickness of wall. The wall is braced top and bottom against lateral translation. Assume the bottom end to be fixed against rotation and the top to be unrestrained against rotation. The wall height is 14 ft, and P_u from each tee is 65 kip. Use $f'_c = 4000$ psi and $f_y = 60{,}000$ psi.

8.9 Design the first-story shear wall for a three-story building with unfactored north–south lateral wind loads of 20 kip, 40 kip, 40 kip at the roof, third floor, and second floor, respectively. The floor-to-floor height is 12 ft, and all shear walls have equal thickness and an equal length of 15 ft. Assume normal-weight concrete; f'_c is 3000 psi and f_y is 60,000 psi. Draw the detail of the shear wall reinforcement showing the required vertical end zone reinforcement and the distributed vertical and horizontal face reinforcement.

8.10 Refer to the given details for a shear wall. Use $f'_c = 5000$ psi, $f_y = 60{,}000$ psi

a. Determine the required reinforcing for bending. Sketch a plan detail of your design.

b. Determine the required horizontal and vertical reinforcing (bar size and spacing) for the wall. Sketch a wall elevation with the reinforcing.

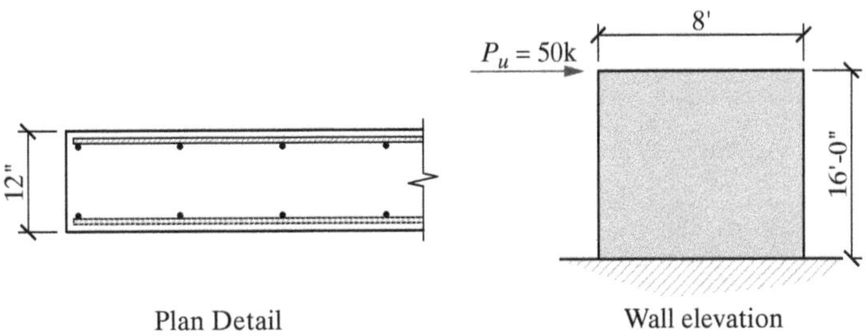

FIGURE FOR EXERCISE 8.10

8.11 Refer to the given plan detail and loading diagram. The Loads shown are factored.

a. Determine if the shear wall is adequate for flexural strength

b. Determine if the shear wall is adequate for shear strength

c. Determine if the wall has adequate horizontal and vertical reinforcing at each face of the wall.

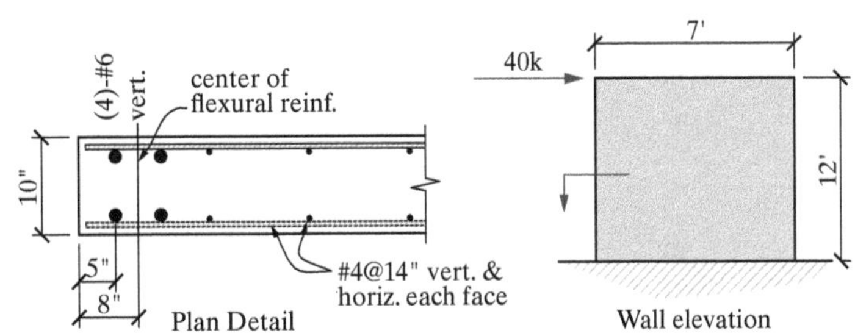

FIGURE FOR EXERCISE 8.11

CHAPTER 9

COLUMNS

9.1 INTRODUCTION

The main vertical load-carrying members in buildings are called *columns*. The ACI Code defines a column as a member used primarily to support axial compressive loads and with a height at least three times its least lateral dimension. The code further defines a *pedestal* as an upright compression member having a ratio of unsupported height to least lateral dimension of 3 or less. The code definition for columns will be extended to include members subjected to combined axial compression load and bending moment (in other words, eccentrically applied compressive loads), because, for all practical purposes, no column is truly only axially loaded.

The three basic types of reinforced concrete columns are shown in Figure 9.1. *Tied columns* (Figures 9.1a and 9.2) are reinforced with longitudinal bars enclosed by horizontal or lateral ties placed at specified spacings. *Spiral columns* (Figure 9.1b) are reinforced with longitudinal bars enclosed by a continuous, rather closely spaced, steel spiral. The spiral is made up of either wire or bar and is formed in the shape of a helix. A third type of reinforced concrete column, a *composite column*, is shown in Figure 9.1c. This type of column encompasses compression members reinforced longitudinally with structural steel shapes, pipes, or tubes with or without longitudinal bars. The requirements for composite compression members are no longer covered in the ACI Code, but are now covered in the AISC 360 Code (ACI Code, Section R10.1.1). Figures 9.1d and 9.1e shows rebar cages for tied circular columns and tied rectangular columns. Our discussion in this text is limited to tied and spiral columns. Tied columns are generally square, rectangular, or circular, whereas spiral columns are normally circular. This is not a hard-and-fast rule, however, as square, spirally reinforced columns, and circular tied columns do exist, as do other shapes, such as octagonal and L-shaped columns.

We initially discuss the analysis and design of columns that are *short*. A column is said to be short when its length is such that lateral or flexural buckling need not be considered. The ACI Code does, however, require that the length of columns be a design consideration. It is recognized that as length increases, the usable strength of a given cross-section is decreased because of the buckling problem. By their very nature, concrete columns are more massive and, therefore, stiffer than their structural steel counterparts. For this reason, slenderness is less of a problem

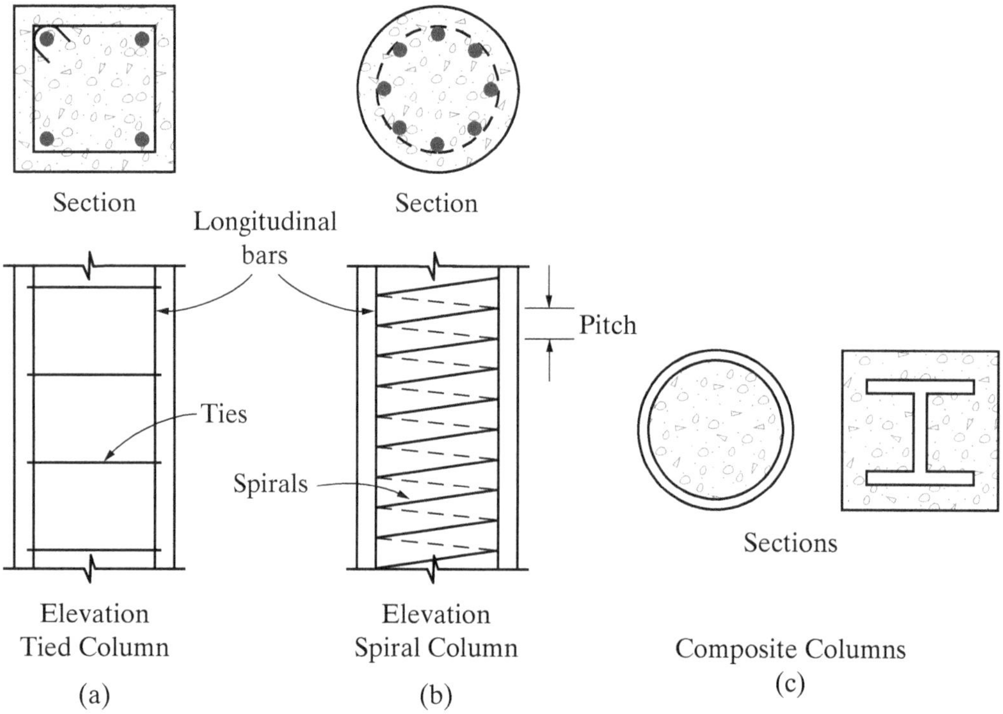

FIGURE 9.1 Column types.

(d)

FIGURE 9.1d

(e)

FIGURE 9.1d, e Rebar cages for tied circular and rectangular columns (Abi Aghayere)

FIGURE 9.2 Foundation columns, turbine generator building. Seabrook Station, New Hampshire. (George Limbrunner).

in reinforced concrete columns. It has been estimated that more than 90% of typical reinforced concrete columns existing in concrete braced frame buildings (i.e., concrete buildings braced by shear walls) may be classified as short columns, and slenderness effects may be neglected.

9.2 STRENGTH OF REINFORCED CONCRETE COLUMNS: SMALL ECCENTRICITY

If a compressive load P is applied coincident with the longitudinal axis of a symmetrical column (i.e., the plastic centroid of a reinforced concrete column), it theoretically induces a uniform compressive stress over the cross-sectional area. For symmetrically reinforced concrete columns, the plastic centroid coincides with the longitudinal axis or geometric centroid of the column. If the compressive load is applied a small distance e away from the longitudinal axis, however, there is a tendency for the column to bend due to the moment $M = Pe$. This distance e is called eccentricity. Unlike the zero-eccentricity condition, the compressive stress is not uniformly distributed over the cross-section but is greater on one side than the other. This is analogous to our previously discussed eccentricity of applied loads with respect to retaining wall footings, where the eccentrically applied load resulted in a nonuniform soil pressure under the footing.

We consider an *axial load* to be a load that acts parallel to the longitudinal axis of a member but *need not* be applied at any particular point on the cross-section, such as a centroid or a geometric center. The column that is loaded with a compressive axial load at zero eccentricity is probably nonexistent, and even the axial load/small eccentricity (axial load/small moment) combination is relatively rare. Nevertheless, we first consider the case of columns that are loaded with compressive axial loads at *small eccentricities*, further defining this situation as that in which the induced moments, although they are present, are so small that they are of little significance. Earlier codes have defined small eccentricity as follows:

- For *spirally reinforced columns*: $e/h \leq 0.05$
- For *tied columns*: $e/h \leq 0.10$

where h is the column dimension perpendicular to the axis of bending.

The fundamental assumptions for the calculation of column axial load strength (small eccentricities) are that at nominal strength, the concrete is stressed to a uniform compressive stress of $0.85f_c'$, and the steel is stressed in compression to its yield strength, f_y. For the cross-sections shown in Figures 9.1a and 9.1b, the nominal axial load strength at small eccentricity is a straightforward sum of the forces existing in the concrete and longitudinal steel when each of the materials is stressed to its maximum. The following ACI notation will be used:

A_g = gross area of the column section (in.^{2})

A_{st} = total area of *longitudinal* reinforcement (in.^{2})

P_0 = nominal, or theoretical, axial load strength at zero eccentricity

P_n = nominal, or theoretical, axial load strength at given eccentricity

P_u = factored applied axial load at given eccentricity

For convenience, we will use the following longitudinal steel reinforcement ratio:

$$\rho_g = \text{ratio of total longitudinal reinforcement area to cross-sectional area of column } (A_{st}/A_g)$$

The nominal, or theoretical, axial load strength of a column for the special case of zero eccentricity may be written as

$$P_0 = 0.85 f_c'(A_g - A_{st}) + f_y A_{st}$$

This theoretical strength must be further reduced to a maximum usable axial load strength using two different strength reduction factors.

Extensive testing has shown that spiral columns are tougher than tied columns, as depicted in Figure 9.3. Both types behave similarly up to the column yield point, at which time the outer shell spalls off. At this point, the tied column fails through crushing and shearing of the concrete and through outward buckling of the bars between the ties. The spiral column, however, has a core area within the spiral that is effectively laterally supported and continues to withstand load. Failure occurs only when the spiral steel yields following large deformation of the column. Naturally, the size and spacing of the spiral steel will affect the final load at failure. Although both columns have exceeded their usable strengths once the outer shell spalls off, the ACI Code recognizes the greater tenacity of the spiral column in Section 21.2.1, where it prescribes a strength-reduction factor as follows:

- ϕ = 0.75 for spiral columns
- ϕ = 0.65 for tied columns

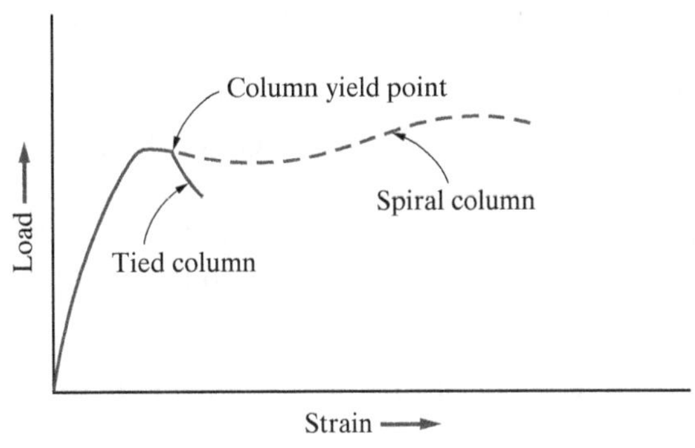

FIGURE 9.3 Load-strain relationship for columns.

The code directs that the basic load-strength relationship at the ultimate limit state be

$$\phi P_n \geq P_u$$

where P_n is the nominal axial load strength at a given eccentricity and ϕP_n is designated as the *design axial load strength*. Logically, for the case of zero eccentricity, if it could exist, P_n would equal P_0. The ACI Code recognizes that no practical column can be loaded with zero eccentricity, however. Therefore, in addition to imposing the strength reduction factor ϕ, the code directs that the nominal strengths be further reduced by factors of 0.80 and 0.85 for tied and spiral columns, respectively, to account for moments in the column caused by accidental or small eccentricities. This results in the following expressions for usable axial load strengths (see ACI Code Tables 22.4.2.1).

For spiral columns,

$$\phi P_{n(\max)} = 0.85\phi[0.85 f_c'(A_g - A_{st}) + f_y A_{st}]$$ **[ACI Eq. (22.4.2.2)]**

For tied columns,

$$\phi P_{n(\max)} = 0.80\phi[0.85 f_c'(A_g - A_{st}) + f_y A_{st}]$$ **[ACI Eq. (22.4.2.2)]**

Recall that the strength-reduction factor, ϕ, for spiral and tied columns are 0.75 and 0.65, respectively. These expressions provide the magnitude of the maximum design axial load strength that may be realized from any column cross-section. This will be the design axial load strength at small eccentricity. Should the eccentricity (and the associated moment) become larger, ϕP_n will have to be reduced, as shown in Section 9.9. It may be recognized that the code equations for $\phi P_{n(\max)}$ provide for an extra margin of axial load strength. This will, in effect, provide some reserve strength to carry small moments.

9.3 CODE REQUIREMENTS CONCERNING COLUMN DETAILS

Main (longitudinal) reinforcing should have a cross-sectional area so that ρ_g will be between 0.01 and 0.08 (ACI Code, Section 10.6.1.1). The 1% minimum reinforcement required in concrete columns is necessary to reduce the effects of creep and shrinkage under sustained compression loads, and to resist bending which may exist in the column (ACI Code, Section R10.6.1.1). The maximum vertical reinforcement ratio of 8% applies at all sections along the column, including the column splice locations. For columns with lap splices, the area of reinforcement at the lap splice locations may effectively be twice that of the area of the vertical rebar outside of the lap splice regions. The reasons for the 8% maximum reinforcement required in concrete columns are as follows (ACI Code, Section R10.6.1.1):

- To ensure adequate spaces between adjacent vertical rebar in the column and satisfy Code requirements for placement of rebar, and thus ensure proper consolidation of concrete around the rebars.
- To ensure adequate bond between the concrete and the column vertical rebar. The effectiveness of the load transfer mechanism at the lap splice is reduced when the concrete does not form proper bond with the rebar due to improper consolidation of the concrete.
- To ensure that concrete columns designed according to the Code are similar in behavior to the test specimens by which the ACI Code equations for concrete columns were calibrated. Note that there are limited tests available for axially loaded columns with vertical rebar ratios greater than 6%.

The minimum number of longitudinal bars is four within rectangular or circular ties, three within triangular ties, and six for bars enclosed by spirals. The foregoing requirements are stated in the ACI Code, Sections 10.6.1.1 and 10.7.3.1. Although not mentioned in the present code, the 1963 code recommended a minimum bar size of No. 5.

The clear distance, s, between longitudinal bars must not be less than 1.5 times the nominal bar diameter nor $1\frac{1}{2}$ in. nor $\frac{4}{3}$ times the nominal maximum coarse aggregate size (ACI Code, Sections 25.2.3). This requirement also holds true where bars are spliced.

That is

$$\begin{aligned} s &\geq 1.5d_b \\ &\geq 1.5 \text{ in.} \\ &\geq \left(\frac{4}{3}\right)d_{aggregate} \end{aligned}$$

Table A.14 may be used to determine the maximum number of bars allowed in one row around the periphery of circular or square columns.

Cover shall be $1\frac{1}{2}$ in. minimum over primary reinforcement, ties, or spirals (ACI Code, Section 20.6.1.3.1).

Tie requirements are discussed in detail in the ACI Code, Section 25.7.2 and are presented as follows:

- The minimum size of column ties is No. 3 for longitudinal bars No. 10 and smaller; otherwise, minimum tie size is No. 4 (see Table A.14 for a suggested tie size). Usually, No. 5 is the maximum tie size.
- The center-to-center spacing of ties should not exceed
  - the smaller of 16 longitudinal bar diameters,
  - 48 tie-bar diameters,
  - or the least column dimension.
- Furthermore, rectilinear ties shall be arranged so that every corner and alternate longitudinal bar will have lateral support provided by the corner of a tie having an included angle of not more than 135°, and no bar shall be farther than 6 in. clear on each side from such a laterally supported bar. Note that when the clear spacing between vertical bars, $s \leq 6$ inches, no extra ties are required, but when $s > 6$ inches, extra ties are required. See Figure 9.4 for typical rectilinear tie arrangements.

Individual circular ties are permitted per ACI Code 25.7.2.4 where longitudinal column reinforcements are located around the perimeter of a circle. The ends of the circular ties should be terminated with standard hooks around a longitudinal column bar and should overlap by at least 6 in. The overlaps around the ends of adjacent circular ties should be staggered.

Spiral requirements are discussed in the ACI Code, Section 25.7.3. The minimum spiral size is $\frac{3}{8}$ in. in diameter for cast-in-place construction ($\frac{5}{8}$ in. diameter is usually the maximum).

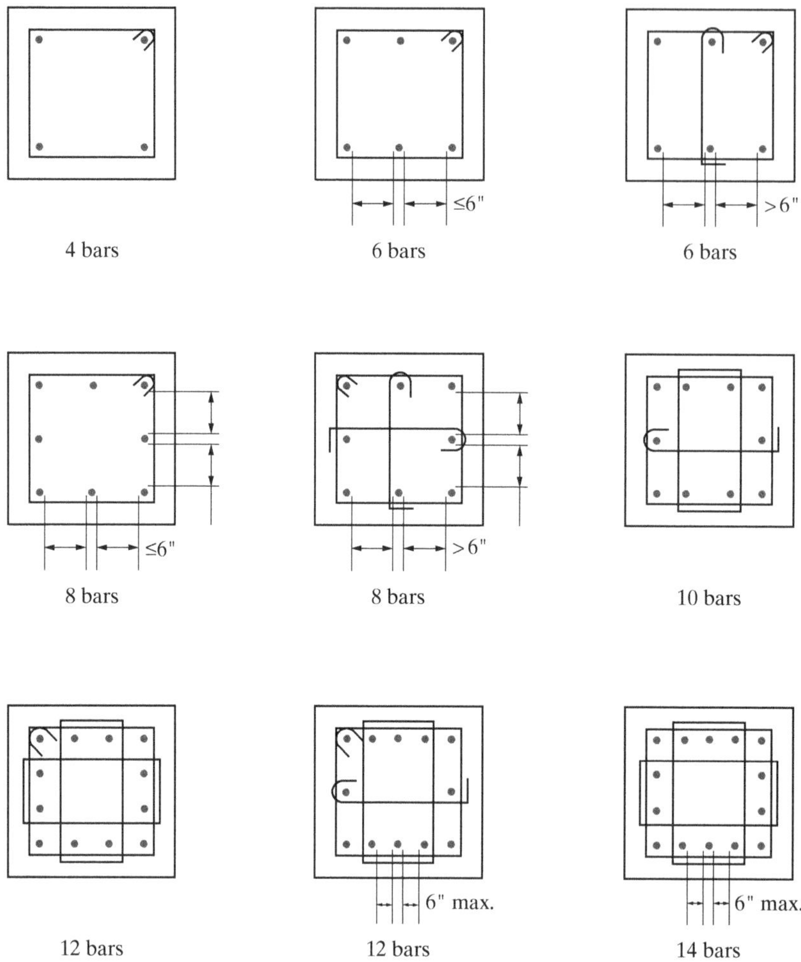

FIGURE 9.4 Typical tie arrangements.

Clear spacing between spirals must not exceed 3 in. but must not be less than 1 in. or (4/3) $d_{aggregate}$; to ensure toughness and ductility of a column, the volumetric *spiral steel ratio* ρ_s must not be less than the value given by

$$\rho_{s(\min)} = 0.45\left(\frac{A_g}{A_{ch}} - 1\right)\frac{f_c'}{f_{yt}} \qquad \textbf{[ACI Eq. (25.7.3.3)]}$$

where

$$\rho_s = \frac{\text{volume of spiral steel in one turn}}{\text{volume of column core in height } (s)}$$

s = center-to-center spacing of spiral (in.) (sometimes called the *pitch*)

A_g = gross cross-sectional area of the column (in.^{2})

A_{ch} = cross-sectional area of the core (in.^{2}) (out-to-out of spiral)

f_{yt} = *spiral* steel yield point (psi) ≤ 60,000 psi

f_c' = compressive strength of concrete (psi)

This particular spiral steel ratio will result in a spiral that will make up the strength lost due to the spalling of the outer shell (see Figure 9.3).

An approximate formula for the calculated spiral steel ratio in terms of the physical properties of the column cross-section may be derived from the preceding definition of ρ_s. In Figure 9.5, we denote the overall core diameter (out-to-out of spiral) as D_{ch} and the spiral diameter (center to center) as D_s. The cross-sectional area of the spiral bar or wire is denoted A_{sp}. From the definition of ρ_s:

$$\text{calculated } \rho_s = \frac{A_{sp}\pi D_s}{\left(\pi D_{ch}^2/4\right)(s)}$$

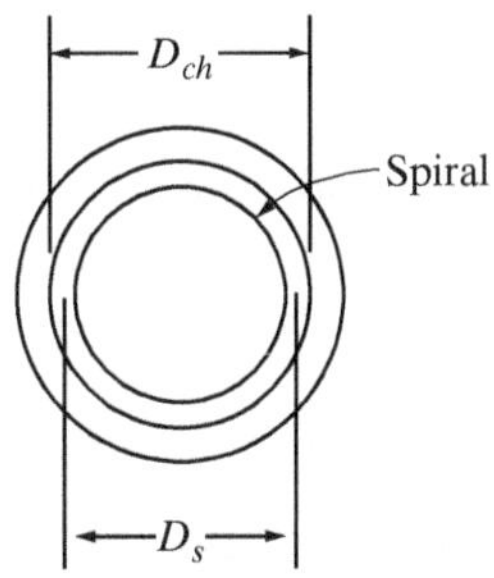

FIGURE 9.5 Definition of D_{ch} and D_s.

If the small difference between D_{ch} and D_s is neglected, then in terms of D_{ch},

$$\text{calculated } \rho_s = \frac{4A_{sp}}{D_{ch}s}$$

9.4 ANALYSIS OF SHORT COLUMNS: SMALL ECCENTRICITY

The analysis of short columns supporting axial loads with small eccentricities involves checking the maximum design axial load strength and the various details of the reinforcing. The procedure is summarized in Section 9.6.

EXAMPLE 9.1

Find the maximum design axial load strength for the tied column of cross-section shown in Figure 9.6. Check the ties. Assume a short column. Use $f_c' = 4000$ psi and f_y = 60,000 psi for both longitudinal steel and ties.

Solution:

1. Check the steel ratio for the longitudinal steel:

$$\rho_g = \frac{A_{st}}{A_g} = \frac{8.00}{(16)^2} = 0.0313$$

$$0.01 < 0.0313 < 0.08 \qquad \text{(O.K.)}$$

2. From Table A.14, using a 13-in. core (column size less cover on each side), the maximum number of No. 9 bars is eight. Therefore, the number of longitudinal bars is satisfactory.

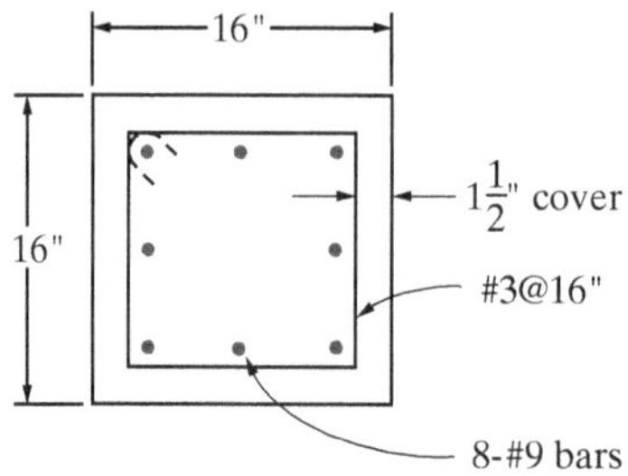

FIGURE 9.6 Sketch for Example 9.1.

3. The maximum design axial load strength may now be calculated:

$$\begin{aligned}\phi P_{n(\max)} &= 0.80\phi[0.85 f_c'(A_g - A_{st}) + f_y A_{st}] \\ &= 0.80(0.65)\left[0.85(4)(256-8)+60(8)\right] \\ &= 688 \text{ kip}\end{aligned}$$

4. Check the ties. Tie size of No. 3 is acceptable for longitudinal bar size up to No. 10. The spacing of the ties must not exceed the smaller of

$$\begin{aligned}\text{48 tie-bar diameters} &= 48\left(\tfrac{3}{8}\right) = 18 \text{ in.} \\ \text{16 longitudinal-bar diameters} &= 16(1.128) = 18 \text{ in.} \\ \text{least column dimension} &= 16 \text{ in.}\end{aligned}$$

Therefore, the tie spacing of 16 inches is O.K. The tie arrangement for this column may be checked by ensuring that the clear distance between longitudinal bars does not exceed 6 in. Clear spacing between adjacent vertical bars in excess of 6 in. would require additional ties in accordance with the ACI Code, Section 25.7.2.3. Thus

$$\text{clear distance} = \frac{16\text{ in.} - 2(1\tfrac{1}{2}\text{ in. cover}) - 2\left(\tfrac{3}{8}\text{ in. stirrups}\right) - (3\text{ bars})(1.128)}{(3\text{ bars} - 1)}$$

$$= 4.4 \text{ in.} < 6 \text{ in.}$$

Therefore, no extra ties are needed.

EXAMPLE 9.2

Determine whether the spiral column of the cross-section shown in Figure 9.7 is adequate to support a factored axial load (Pu) of 540 kip. Assume small eccentricity. Check the spiral. Use $f_c' = 4000$ psi and $f_y = 60{,}000$ psi.

Solution:

1. From Table A.2, $A_{st} = 5.53$ in.^{2}, and from Table A.14, a diameter of 15 in. results in a circular area $A_g = 176.7$ in.^{2}. Therefore,

$$\rho_g = \frac{5.53}{176.7} = 0.0313$$

$$0.01 < 0.0313 < 0.08 \qquad \text{(O.K.)}$$

2. From Table A.14, using a 12-in. core (i.e., 15 in. – 1.5 in. cover – 1.5 in. cover), seven No. 8 bars are satisfactory.

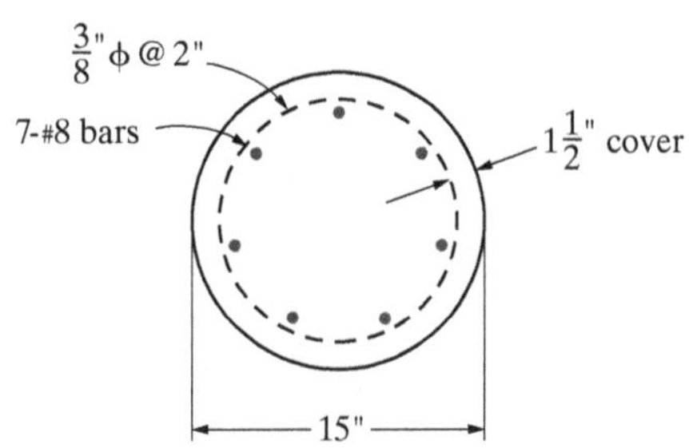

FIGURE 9.7 Sketch for Example 9.2.

3. Find the maximum design axial load strength:

$$\begin{aligned}\phi P_{n(\max)} &= 0.85\phi\left[0.85 f_c'(A_g - A_{st}) + f_y A_{st}\right]\\ &= 0.85(0.75)\left[0.85(4)(176.7 - 5.53) + 60(5.53)\right]\\ &= 583 \text{ kip}\end{aligned}$$

The strength is O.K. because 583 kip > 540 kip.

4. Check the spiral steel. The $\frac{3}{8}$-in. diameter is O.K. (ACI Code, Section 25.7.3.2; and Table A.14). The minimum ρ_s is calculated using Table A.14 for the value of Ach:

$$\rho_{s(\min)} = 0.45\left(\frac{A_g}{A_{ch}} - 1\right)\frac{f_c'}{f_{yt}}$$

$$= 0.45\left(\frac{176.7}{113.1} - 1\right)\frac{4}{60} = 0.0169$$

$$\text{actual } \rho_s = \frac{4A_{sp}}{D_{ch}s} = \frac{4(0.11)}{12(2)} = 0.0183$$

$$0.0183 > 0.0169 \qquad \text{(O.K.)}$$

The clear distance between spirals must not be more than 3 in. nor less than 1 in.:

$$\text{clear distance} = 2.0 - \tfrac{3}{8} = 1\tfrac{5}{8}\text{ in.} \qquad \text{(O.K.)}$$

The column is satisfactory for the specified conditions.

9.5 DESIGN OF SHORT COLUMNS: SMALL ECCENTRICITY

The design of reinforced concrete columns involves the proportioning of the steel and concrete areas and the selection of properly sized and spaced ties or spirals. Because the ratio of steel to concrete area must fall within a given range ($0.01 \le \rho_g \le 0.08$), the strength equation given in Section 9.2 is modified to include this term. For a tied column,

$$\phi P_{n(\max)} = 0.80\phi\left[0.85 f_c'(A_g - A_{st}) + f_y(A_{st})\right]$$

$$\rho_g = \frac{A_{st}}{A_g}$$

from which

$$A_{st} = \rho_g A_g$$

Therefore,

$$\begin{aligned}\phi P_{n(\max)} &= 0.80\phi[0.85 f_c'(A_g - \rho_g A_g) + f_y \rho_g A_g] \\ &= 0.80\phi A_g\left[0.85 f_c'(1-\rho_g) + f_y\rho_g\right]\end{aligned}$$

From the limit state design equation,

$$P_u \le \phi P_{n(\max)}$$

an expression can be written for the required A_g in terms of the material strengths, P_u and ρ_g.

For tied columns,

$$\begin{aligned}\text{required } A_g &= \frac{P_u}{0.80\phi\left[0.85 f_c'(1-\rho_g) + f_y\rho_g\right]} \\ &= \frac{P_u}{\left[0.442 f_c'(1-\rho_g) + 0.52 f_y\rho_g\right]}\end{aligned}$$

Similarly for spiral columns,

$$\begin{aligned}\text{required } A_g &= \frac{P_u}{0.85\phi\left[0.85 f_c'(1-\rho_g) + f_y\rho_g\right]} \\ &= \frac{P_u}{\left[0.542 f_c'(1-\rho_g) + 0.638 f_y\rho_g\right]}\end{aligned}$$

It should be recognized that there can be many valid choices for the size of column that will provide the necessary strength to carry any load P_u. A low ρ_g will result in a larger required A_g and vice versa. Other considerations will normally affect the practical choice of column size. Among them are architectural requirements and the desirability of maintaining the same column size from floor to floor for several stories so that forms may be reused, since the cost of formwork is a significant portion of the cost of reinforced concrete construction.

The procedure for the design of short columns for loads at small eccentricities is summarized in Section 9.6.

EXAMPLE 9.3

Design a square tied column to carry axial service loads of 320 kip dead load and 190 kip live load. There is no identified applied moment. Assume that the column is short. Use ρ_g about 0.03, $f'_c = 4000$ psi, and $f_y = 60{,}000$ psi.

Solution:

1. Material strengths and approximate ρ_g are given.
2. The factored axial load is

$$P_u = 1.6(190) + 1.2(320) = 688 \text{ kip}$$

3. The required gross column area is

$$\begin{aligned}\text{required } A_g &= \frac{P_u}{0.80\phi\left[0.85f'_c(1-\rho_g)+f_y\rho_g\right]} \\ &= \frac{688}{0.80(0.65)\left[0.85(4)(1-0.03)+60(0.03)\right]} \\ &= 260 \text{ in.}^2\end{aligned}$$

4. The required size of the square column would be

$$\sqrt{260} = 16.1 \text{ in.}$$

Use a 16-in.-square column. This choice will require that the actual ρ_g be slightly in excess of 0.03:

$$\text{actual } A_g = (16 \text{ in.})^2 = 256 \text{ in.}^2$$

5. The load on the concrete area (this is approximate since ρ_g will increase slightly) is

$$\begin{aligned}\text{load on concrete} &= 0.80\phi(0.85f'_c)A_g(1-\rho_g) \\ &= 0.80(0.65)(0.85)(4)(256)(1-0.03) \\ &= 439 \text{ kip}\end{aligned}$$

Therefore, the load to be carried by the steel is

$$688 - 439 = 249 \text{ kip}$$

Because the maximum design axial load strength of the steel is $(0.80\phi A_{st} f_y)$, the required steel area may be calculated as

$$\text{required } A_{st} = \frac{249}{0.80(0.65)(60)} = 7.98 \text{ in.}^2$$

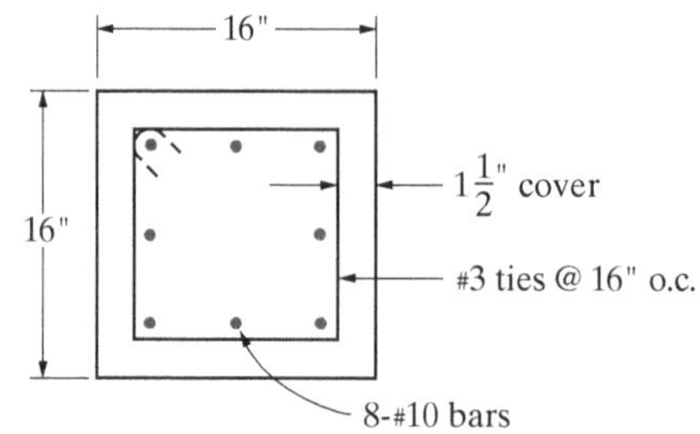

FIGURE 9.8 Design sketch for Example 9.3.

We will distribute bars of the same size evenly around the perimeter of the column and must, therefore, select bars in multiples of four. Use eight No. 9 bars ($A_{st} = 8.0$ in.^{2}). Table A.14 indicates a maximum of eight No. 9 bars for a 13-in. core (O.K.).

6. Design the ties. From Table A.14, select a No. 3 tie. The spacing must not be greater than

$$48 \text{ tie-bar diameters} = 48\left(\tfrac{3}{8}\right) = 18 \text{ in.}$$

$$16 \text{ longitudinal-bar diameters} = 16(1.128) = 18.0 \text{ in.}$$

$$\text{least column dimension} = 16 \text{ in.}$$

Use No. 3 ties spaced 16 in. o.c. Check the arrangement with reference to Figure 9.8. The clear spacing between adjacent bars in the same face is

$$\frac{16 - 1.5 \text{in. cover} - 1.5 \text{in. cover} - \frac{3}{8}\text{in.} - \frac{3}{8}\text{in.} - (3 \text{bars})(1.13)}{(3 \text{bars} - 1)} = \frac{16 - 3 - 0.75 - 3(1.13)}{2} = 4.43 \text{in.} < 6.0 \text{ in.}$$

Therefore, no additional ties are required by the ACI Code, Section 25.7.2.3.

7. The design sketch is shown in Figure 9.8.

EXAMPLE 9.4

Redesign the column of Example 9.3 as a circular, spirally reinforced column.

Solution:

1. Use $f'_c = 4000$ psi, $f_y = 60{,}000$ psi, and ρ_g approximately 0.03.

2. $P_u = 688$ kip, as in Example 9.3.

3. Required gross cross-sectional area of the column

$$\text{required } A_g = \frac{P_u}{0.80\phi\left[0.85f'_c(1-\rho_g)+f_y\rho_g\right]}$$
$$= \frac{688}{0.80(0.75)\left[0.85(4)(1-0.03)+60(0.03)\right]}$$
$$= 212 \text{ in.}^2$$

4. From Table A.14, use an 18-in.-diameter column. Thus

$$A_g = 254.5 \text{ in.}^2$$

5. Load on the concrete

$$\text{Load on concrete} = 0.85\phi(0.85)f'_cA_g(1-\rho_g)$$
$$= 0.85(0.75)(0.85)(4)(254.5)(1-0.03)$$
$$= 535 \text{ kip}$$

$$\text{load on steel} = 688 - 535 = 153 \text{ kip}$$
$$\text{required } A_{st} = \frac{153}{0.85\phi f_y}$$
$$= \frac{153}{0.85(0.75)60} = 4.00 \text{ in.}^2$$

Use seven No. 7 bars (A_{st} = 4.20 in.^{2}). Table A.14 indicates a maximum of 13 No. 7 bars for a circular core of 15 in. in diameter (O.K.).

6. Design the spiral. From Table A.14, select a $\frac{3}{8}$-in.-diameter spiral. The spacing will be based on the required spiral steel ratio. Here A_{ch} is from Table A.14. Thus

$$\rho_{s(\min)} = 0.45\left(\frac{A_g}{A_{ch}} - 1\right)\frac{f'_c}{f_{yt}}$$
$$= 0.45\left(\frac{254.5}{176.7} - 1\right)\frac{4}{60} = 0.0132$$

The maximum spiral spacing may be found by setting the calculated ρ_s equal to $\rho_{s(\min)}$:

$$\text{calculated } \rho_s = \frac{4A_{sp}}{D_{ch}s}$$

from which

$$s_{max} = \frac{4A_{sp}}{D_{ch}\rho_{s(min)}} = \frac{4(0.11)}{15(0.0132)} = 2.22 \text{ in.}$$

Use a spiral spacing of 2 in. The clear space between spirals must not be less than 1 in. nor more than 3 in.:

$$\text{clear spacing} = 2 - \tfrac{3}{8} = 1\tfrac{5}{8} \text{ in.} \qquad \text{(O.K.)}$$

7. The design sketch is shown in Figure 9.9.

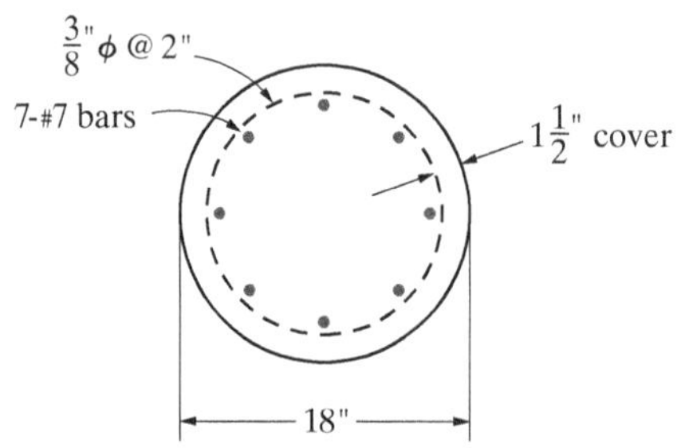

FIGURE 9.9 Design sketch for Example 9.4.

9.6 SUMMARY OF PROCEDURE FOR ANALYSIS AND DESIGN OF SHORT COLUMNS WITH SMALL ECCENTRICITIES

Analysis

1. Check ρ_g within acceptable limits:

$$0.01 \leq \rho_g \leq 0.08$$

The above maximum column reinforcing ratio of $\rho_{g(maximum)}$ of 0.08 applies at all sections of the column, including the splice regions. When lap splices are used and if all the rebars are spliced at the same location, the area of rebar in the lap splice regions will effectively be twice that of the rebar area in the column outside of the lap splice regions. Therefore, satisfying the 8% maximum rebar ratio in the lap splice regions would imply that the maximum vertical reinforcement ratio in the column outside of the lap splice regions is one-half of 8%, or 4%. When mechanical splices are used, the maximum vertical reinforcement ratio both within and outside of the splice regions will be 8%.

2. Check the number of bars within acceptable limits for the clear spacing between adjacent bars (see Table A.14). The minimum number is four bars for columns with rectangular or circular ties, and six bars for columns with vertical reinforcement enclosed by spirals.
3. Calculate the maximum design axial load strength $\phi P_{n(max)}$. See Section 9.2.
4. Check the lateral reinforcing. For ties, check size, spacing, and arrangement. For spirals, check size, ρ_s, and clear distance.

Design

1. Establish the material strengths. Establish the desired ρ_g (if any).
2. Calculate the factored axial load, P_u.
3. Determine the required gross column area, A_g.
4. Select the column dimensions. Use full-inch increments.
5. Find the load carried by the concrete and the load required to be carried by the longitudinal steel. Determine the required longitudinal steel area. Select the longitudinal steel.
6. Design the lateral reinforcing (ties or spiral).
7. Sketch the design or include the design information in a column schedule.

9.7 THE LOAD-MOMENT RELATIONSHIP FOR CONCRETE COLUMNS

The equivalency between an eccentrically applied load and an axial load-moment combination is shown in Figure 9.10. Assume that a force P_u is applied to a cross-section at a distance e (eccentricity) from the centroid, as shown in Figures 9.10a and 9.10b. Add equal and opposite forces P_u at the centroid of the cross-section (Figure 9.10c). The original, eccentric force P_u may now be combined with the upward force P_u to form a couple, P_ue, that is a pure moment. This will leave the remaining one force, P_u, acting downward at the centroid of the cross-section. It can, therefore, be seen that if a force P_u is applied with an eccentricity e, the situation that results is identical to the case where an axial load of P_u at the centroid and a moment of P_ue are simultaneously applied (Figure 9.10d). If we define M_u as the factored moment to be applied on a *compression* member along with a factored axial load of P_u at the centroid, the relationship between the two is

$$e = \frac{M_u}{P_u}$$

We have previously presented information on the nominal axial load strength of columns at zero eccentricity, P_0. Temporarily neglecting the strength reduction factors that must be applied to P_0, let us assume that we may apply a load P_u with zero eccentricity on a column of nominal axial load strength P_0 (where $P_u = P_0$). If we now move the load P_u away from the zero-eccentricity position at a distance of e, the column must resist the load P_u and, in addition, a moment $P_u e$. Because $P_u = P_0$ (in our hypothetical case), it is clear that there is no additional strength to carry the moment and that the column is overloaded. For this particular column to not be overloaded when subjected to load at eccentricity e, we must reduce P_u to the point where the column can carry both P_u and $P_u e$. The amount of the required decrease in P_u will depend on the magnitude of the eccentricity.

The preceding discussion must be modified because the ACI Code imposes $\phi P_{n(\max)}$ as the upper limit of axial load strength for any column. Nevertheless, the strength of any column cross-section is such that it will support a broad spectrum of load and moment (or load and eccentricity) combinations. In other words, we may think of a column cross-section as having many different axial load strengths, each with its own related moment strength.

9.8 COLUMNS SUBJECTED TO AXIAL LOAD AT LARGE ECCENTRICITY

At one time, the ACI Code stipulated that compression members be designed for an eccentricity e of not less than $0.05h$ for spirally reinforced columns or $0.10h$ for tied columns, but at least 1 in. in any case. Here h is defined as the overall dimension of the column. These specified minimum eccentricities were originally intended to serve as a means of reducing the axial load design strength of a section in pure compression. The effect of the minimum eccentricity requirement was to limit the *maximum* axial load strength of a compression member.

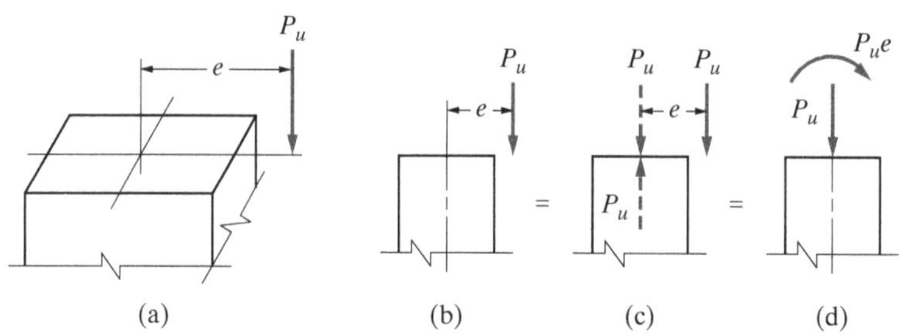

FIGURE 9.10 Load-moment-eccentricity relationship.

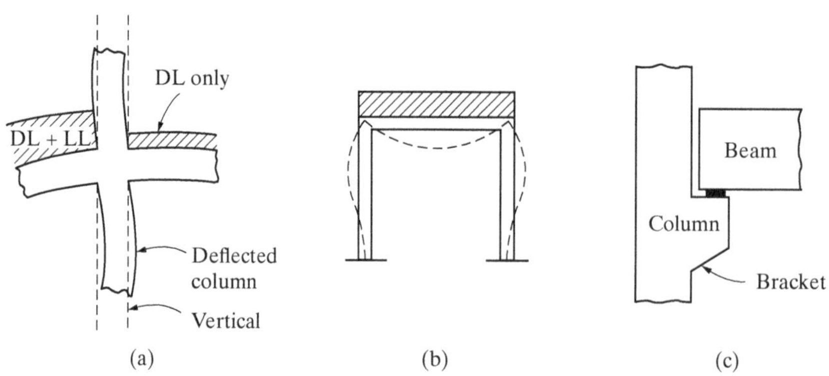

FIGURE 9.11 Eccentrically loaded columns.

In the ACI Code, as we have discussed, the maximum design axial load strength $\phi P_{n(\max)}$ is obtained from ACI Equation (22.4.2.2). These two equations apply when eccentricities are not in excess of *approximately* the $0.10h$ and $0.05h$ minimum eccentricity limits previously discussed. Therefore, *small eccentricities* may be considered as those eccentricities up to about $0.10h$ and $0.05h$ for tied and spiral columns, respectively. We will consider cases of *large eccentricities* as those in which ACI Equation (22.4.2.2) no longer applies and where ϕP_n must be reduced below $\phi P_{n(\max)}$.

The occurrence of columns subjected to eccentricities sufficiently large so that moment must be a design consideration is common. Even interior columns supporting beams of equal spans will receive unequal loads from the beams due to applied live load patterns. These unequal loads could mean that the column must carry both load and moment, as shown in Figure 9.11a, and the resulting eccentricity of the loads could be appreciably in excess of our definition of small eccentricity. Another example of a column carrying both load and moment is shown in Figure 9.11b. In both cases, the rigidity of the joint will require the column to rotate along with the end of the beam that it is supporting. The rotation will induce a moment in the column. A third and very practical example can be found in precast work (Figure 9.11c), where the beam reaction can clearly be seen to be eccentrically applied on the column through the column bracket. Figure 9.11d show multi-story columns in a high rise building that will be subjected to axial loads and large moments.

FIGURE 9.11d Columns in a high rise building that are subjected to axial loads and large moments

9.9 ϕ FACTOR CONSIDERATIONS

Columns discussed so far have had strength-reduction factors applied in a straightforward manner. That is, $\phi = 0.75$ for spiral columns, and $\phi = 0.65$ for tied columns. These ϕ factors correspond to the compression-controlled strain limit or a net tensile strain in the extreme tension reinforcement, $\varepsilon_t \leq 0.002$. Eccentrically loaded columns, however, carry both axial load and moment. For values of ε_t larger than 0.002, the ϕ equations from ACI Code, Table 21.2.2, discussed in Chapter 2, will give higher values than indicated above. The ϕ equations from Chapter 2 are repeated here as follows:

Tied columns:

$$\phi = 0.65 + (\varepsilon_t - 0.002)\left(\frac{250}{3}\right)$$

$$0.65 \leq \phi \leq 0.90$$

Spiral columns:

$$\phi = 0.75 + (\varepsilon_t - 0.002)\left(\frac{200}{3}\right)$$

$$0.75 \leq \phi \leq 0.90$$

9.10 ANALYSIS OF SHORT COLUMNS: LARGE ECCENTRICITY

The first step in our investigation of short columns carrying loads at large eccentricity is to determine the strength of a given column cross-section that carries loads at various eccentricities. This may be thought of as an analysis process that will result in a column load-moment interaction diagram. For this development, we will find the design axial load strength ϕP_n, where P_n is defined as the nominal axial load strength at a given eccentricity.

EXAMPLE 9.5

Find the design axial load strength ϕP_n for the tied column for the following conditions: (a) small eccentricity ($e = 0$ to $0.10h$); (b) $e = 5$ in.; (c) the balanced strain condition or compression-controlled strain limit, $\varepsilon_t = 0.002$; (d) $\varepsilon_t = 0.004$; (e) the tension-controlled strain limit, $\varepsilon_t = 0.005$; and (f) pure moment. The column cross-section is shown in Figure 9.12. Assume a short column. Bending is about the Y-Y axis. Use $f'_c = 4000$ psi and $f_y = 60{,}000$ psi.

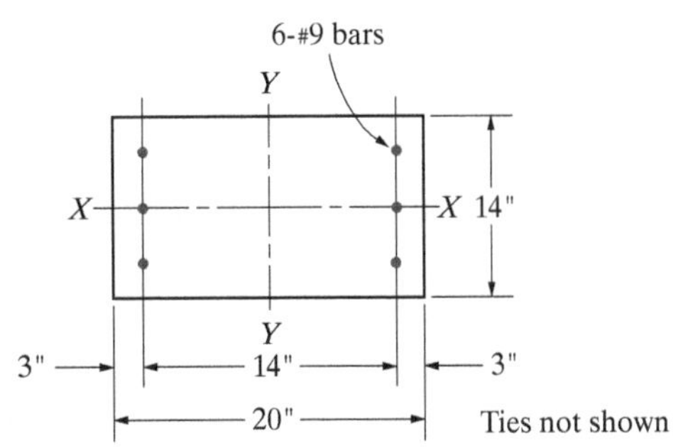

FIGURE 9.12 Column cross-section for Example 9.5.

Solution:

a. The analysis of the small eccentricity condition is similar to the analyses of Examples 9.1 and 9.2. We can calculate the design axial load strength from

$$\begin{aligned}\phi P_n &= \phi P_{n(\max)}\\ &= 0.80\phi\left[0.85 f_c'(A_g - A_{st}) + f_y A_{st}\right]\\ &= 0.80(0.65)\left[0.85(4)(280-6)+60(6)\right]\\ &= 672 \text{ kip}\end{aligned}$$

The corresponding maximum moment:

$$\phi M_n = \phi P_n e = 672(0.10)\left(\frac{20}{12}\right) = 112 \text{ ft.-kip}$$

b. The situation of $e = 5$ in. is shown in Figure 9.13. In part (a) of this example, all steel was in compression. As eccentricity increases, the steel on the side of the column away from the load is subjected to less compression. Therefore, there is some value of eccentricity at which this steel will change from compression to tension. Because this value of eccentricity is not known, the strain situation shown in Figure 9.14 is *assumed* and will be verified (or disproved) by calculation.

The assumptions at nominal strength are

1. Maximum concrete strain = 0.003.
2. $\varepsilon_s' > \varepsilon_y$ Therefore, $f_s' = f_y$.
3. ε_s is tensile.
4. $\varepsilon_s < \varepsilon_y$. Therefore, $f_s' < f_y$.

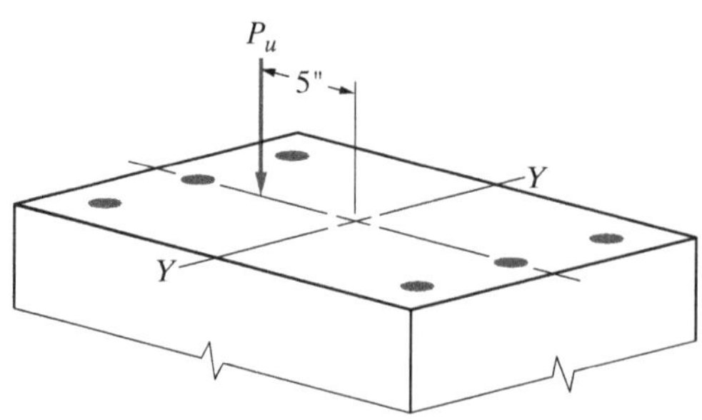

FIGURE 9.13 Example 9.5b, e = 5 in.

The unknown quantities are P_n and c.

Using basic units of kip and inches, the tensile and compressive forces are evaluated. Force C_2 is the force in the compressive steel accounting for the concrete displaced by the steel. With reference to Figure 9.14c,

$$C_1 = 0.85f'_c ab = 0.85(4)(0.85c)(14) = 40.46c$$
$$C_2 = f_y A'_s - 0.85f'_c A'_s = A'_s\left(f_y - 0.85f'_c\right)$$
$$= 3\left[60 - 0.85(4)\right] = 169.8$$
$$T = f_s A_s = \varepsilon_s E_s A_s = 87\left(\frac{d-c}{c}\right)A_s$$
$$= 87\left(\frac{17-c}{c}\right)3 = 261\left(\frac{17-c}{c}\right)$$

From Σ forces = 0 in Figure 9.14,

$$P_n = C_1 + C_2 - T$$
$$= 40.46c + 169.8 - 261\left(\frac{17-c}{c}\right)$$

From Σ moments = 0, taking moments about T in Figure 9.14,

$$P_n(12) = C_1\left(d - \frac{a}{2}\right) + C_2(14)$$
$$P_n = \frac{1}{12}\left[40.46(c)\left(17 - \frac{0.85c}{2}\right) + 169.8(14)\right]$$

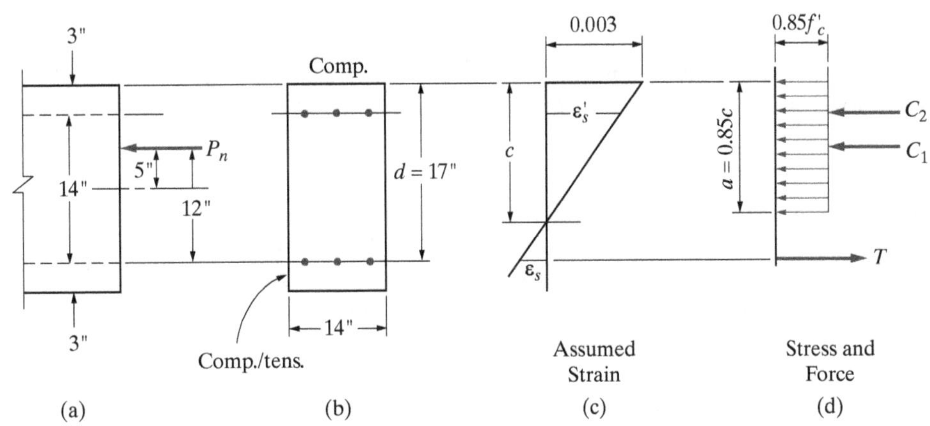

FIGURE 9.14 Stress and strain distribution for Example 9.5b, e = 5 in.

The preceding two equations for P_n may be equated, and the resulting cubic equation solved for c by trial or by some other iterative method. The solution will yield c = 14.86 in., which will result in a value for P_n (from either equation) of 733 kip. The net tensile strain in the extreme tension reinforcement may be calculated from

$$\varepsilon_t = 0.003\left(\frac{d-c}{c}\right) = 0.003\left(\frac{17-14.86}{14.86}\right)$$
$$= 0.00043 < 0.002$$

For $\varepsilon_t \leq 0.002$, the corresponding tied column strength-reduction factor ϕ is 0.65. Therefore,

$$\phi P_n = 0.65(733)$$
$$= 476 \text{ kip}$$

Check the assumptions that were made:

$$\varepsilon_s' = \left(\frac{14.86-3}{14.86}\right)(0.003) = 0.0024$$
$$\varepsilon_y = 0.00207$$

Because $\varepsilon_s' > \varepsilon_y$,

$$f_s' = f_y \qquad \text{(O.K.)}$$

Based on the location of the neutral axis, the steel away from the load *is* in tension and

$$f_s = 87\left(\frac{17-14.86}{14.86}\right) = 12.53 \text{ ksi} < 60 \text{ ksi} \quad \text{(O.K.)}$$

All assumptions are verified.

We may also determine the design moment strength for an eccentricity of 5 in. as follows:

$$\phi P_n e = \frac{476(5)}{12}$$
$$= 198 \text{ ft.-kip}$$

Therefore, the given column has a design load-moment combination strength of 476 kip axial load and 198 ft.-kip moment. This assumes that the moment is applied about the Y-Y axis.

c. The compression-controlled strain limit (balanced condition) exists when the concrete reaches a strain of 0.003 at the same time the extreme tension steel reaches a strain of 0.002, as shown in Figure 9.15c. Here P_b is defined as nominal axial load strength at the balanced condition, e_b is the associated eccentricity, and c_b is the distance from the compression face to the balanced neutral axis.

Using the strain diagram in Figure 9.15, we may calculate the value of c_b:

$$\frac{0.003}{c_b} = \frac{0.002}{17-c_b}$$

from which $c_b = 0.6(17) = 10.2$ in.

For $\varepsilon_t = 0.002$, the tied column strength-reduction factor ϕ is 0.65.

We then may determine ε_s':

$$\varepsilon_s' = \frac{7.2}{10.2}(0.003) = 0.0021$$

Because 0.0021 > 0.00207, the compression steel has yielded and $f_s' = f_y = 60$ ksi.

Summarize the forces in Figure 9.15d and let C_2 account for the force in the concrete displaced by the three No. 9 bars:

$$C_1 = 0.85(4)(0.85)(10.20)(14) = 413 \text{ kip}$$
$$C_2 = 60(3) - 0.85(4)(3) = 170 \text{ kip}$$
$$T = 60(3) = 180 \text{ kip}$$
$$P_b = C_1 + C_2 - T = 413 + 170 - 180$$
$$= 403 \text{ kip}$$

The value of e_b may be established by summing moments about T:

$$P_b(e_b + 7) = C_1\left(d - \frac{0.85c_b}{2}\right) + C_2(14)$$
$$403(e_b + 7) = 413\left[17 - \frac{0.85(10.20)}{2}\right] + 170(14)$$

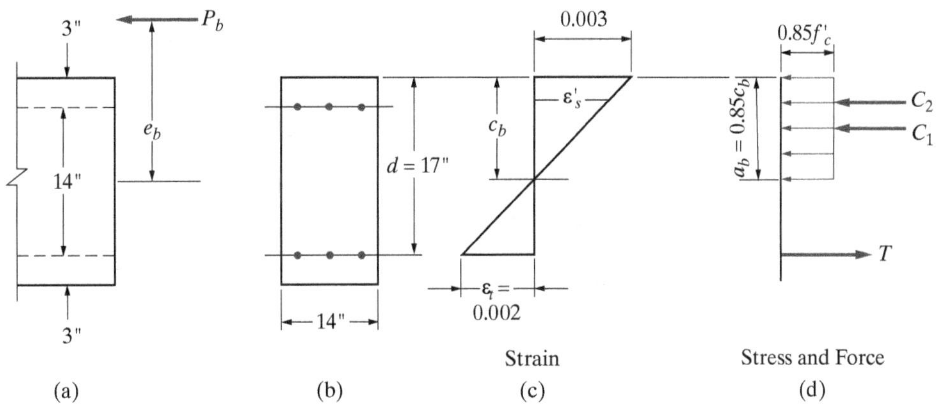

FIGURE 9.15 Example 9.5c, balanced condition.

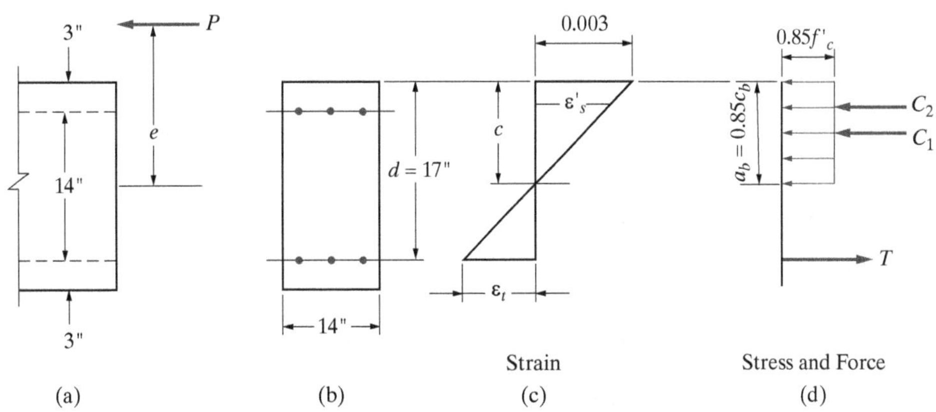

FIGURE 9.16 Examples 9.5d and 9.5e.

from which e_b = 11.88 in. Therefore, at the balanced condition,

$$\phi P_b = 0.65(403) = 262 \text{ kip}$$

$$\phi P_b e_b = \frac{262}{12}(11.88) = 259 \text{ ft.-kip}$$

d. Net tensile strain in the extreme tension steel, $\varepsilon_t = 0.004$ and a corresponding compression strain of 0.003 in the compression face.

Using the strain diagram in Figure 9.16, the neutral axis depth, c is calculated as follows:

$$\frac{0.003}{c} = \frac{0.004}{17-c}$$

from which c = 0.429(17) = 7.29 in.

We may then determine ε_s':

$$\varepsilon_s' = 0.003\frac{(7.29-3)}{7.29} = 0.00177 < \varepsilon_y$$

$$f_s' = E\,\varepsilon_s' = 29{,}000\,(0.00177) = 51.2 \text{ ksi}$$

Using similar equations from case (b), the forces in the concrete, compression steel, and tension steel are as follows:

$$C_1 = 0.85(4)(0.85)(7.29)(14) = 295 \text{ kip}$$

$$C_2 = 51.2(3) - (0.85)(4)(3) = 143 \text{ kip}$$

$$T = 60(3) = 180 \text{ kip}$$

From Σ forces = 0 in Figure 9.16,

$$P_n = C_1 + C_2 - T$$
$$= 295 + 143 - 180 = 258 \text{ kip}$$

From Σ moments = 0, taking moments about T in Figure 9.16,

$$P_n(e+7) = C_1\left(d - \frac{0.85c}{2}\right) + C_2(14)$$

$$258(e+7) = 295\left[d - \frac{0.85(7.29)}{2}\right] + 143(14)$$

from which e = 16.7 in.

For $\varepsilon_t = 0.004$,

$$\phi = 0.65 + (0.004 - 0.002)\left(\frac{250}{3}\right) = 0.82$$

$$0.65 < 0.82 < 0.9 \qquad \text{(O.K.)}$$

Therefore, at $\varepsilon_t = 0.004$,

$$\phi P_n = 0.82(258) = 212 \text{ kip}$$
$$\phi M_n = \phi P_n e = 212\left(\frac{16.7}{12}\right) = 295 \text{ ft.-kip}$$

e. Net tensile strain in the extreme tension steel ε_t is 0.005, and a corresponding compression strain of 0.003 exists in the compression face.

Using the strain diagram in Figure 9.16, the neutral axis depth, c, is calculated as follows:

$$\frac{0.003}{c} = \frac{0.005}{17-c}$$

from which $c = 0.375(17) = 6.38$ in.

We may then determine ε_s':

$$\varepsilon_s' = 0.003\left(\frac{6.38-3}{6.38}\right) = 0.00159 < \varepsilon_y$$
$$f_s' = E\,\varepsilon_s' = 29{,}000\,(0.00159) = 46.1 \text{ ksi}$$

Using similar equations from case (b), the forces in the concrete, compression steel, and tension steel are as follows:

$$C_1 = 0.85(4)(0.85)(6.38)(14) = 258 \text{ kip}$$
$$C_2 = 46.1(3) - (0.85)(4)(3) = 128 \text{ kip}$$
$$T = 60(3) = 180 \text{ kip}$$

From Σ forces = 0 in Figure 9.16,

$$P_n = C_1 + C_2 - T$$
$$= 258 + 128 - 180 = 206 \text{ kip}$$

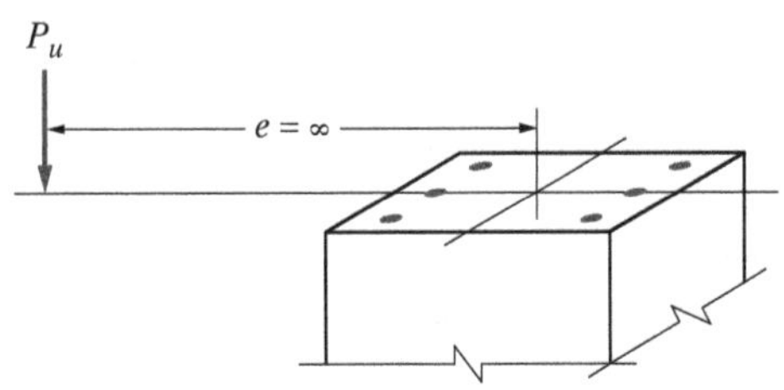

FIGURE 9.17 Column of Example 9.5 loaded with pure moment.

From Σ moments = 0, taking moments about T in Figure 9.16,

$$P_n(e+7)=C_1\left(d-\frac{0.85c}{2}\right)+C_2(14)$$

$$206(e+7)=258\left[d-\frac{0.85(6.38)}{2}\right]+128(14)$$

from which $e = 19.6$ in.

For $\varepsilon_t \geq 0.005$, $\phi = 0.90$. Therefore, at $\varepsilon_t = 0.005$,

$$\phi P_n = 0.90(206) = 185 \text{ kip}$$

$$\phi M_n = \phi P_n e = 185\left(\frac{19.6}{12}\right) = 302 \text{ ft.-kip}$$

f. The analysis of the pure moment condition is similar to the analysis of the case where eccentricity e is infinite, shown in Figure 9.17. We will find the design moment strength ϕM_n, because P_u and ϕP_n will both be zero.

With reference to Figure 9.18d, notice that for pure moment the bars on the load side of the column are in compression, whereas the bars on the side away from the load are in tension. The total tensile and compressive forces must be equal to each other. Since $A_s = A'_s$, A'_s *must* be at a stress less than yield. Assume that A_s is at yield stress. Then

$$C_1 = \text{concrete compressive force}$$
$$C_2 = \text{steel compressive force}$$
$$T = \text{steel tensile force}$$

Referring to the compressive strain diagram portion of Figure 9.18c and noting that the basic units are kip and inches,

$$\varepsilon'_s = 0.003\frac{(c-3)}{c}$$

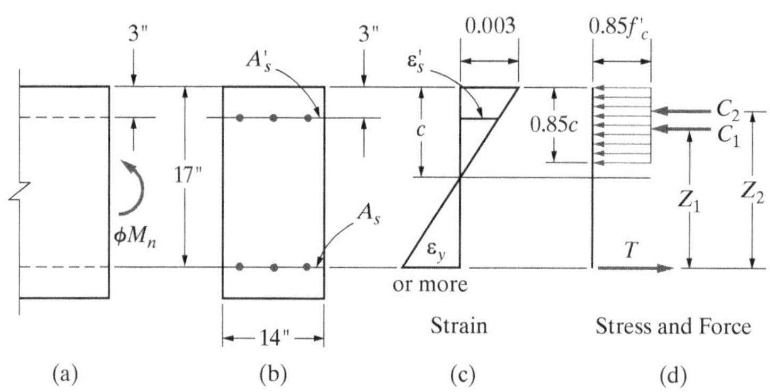

FIGURE 9.18 Example 9.5f, $e = \infty$.

Because

$$f'_s = E_s\varepsilon'_s$$

substituting yields

$$f_s' = 29{,}000(0.003)\frac{(c-3)}{c}$$
$$= 87\frac{(c-3)}{c}$$

For equilibrium in Figure 9.18d,

$$C_1 + C_2 = T$$

Substituting into the foregoing and accounting for the concrete displaced by the compression steel, we obtain

$$(0.85f_c')(0.85c)(b) + f_s'A_s' - 0.85f_c'A_s' = f_yA_s$$
$$(0.85)(4)(0.85c)(14) + 87\left(\frac{c-3}{c}\right)(3) - 0.85(4)(3) = 3(60)$$

Solving the preceding equation for the one unknown quantity c yields

$$c = 3.62 \text{ in.}$$

The net tensile strain in the extreme tension reinforcement may be calculated from

$$\varepsilon_t = 0.003\left(\frac{d-c}{c}\right) = 0.003\left(\frac{17-3.62}{3.62}\right) = 0.011$$

For $\varepsilon_t \geq 0.005$, the corresponding strength reduction factor ϕ is 0.90.

Therefore,

$$f_s' = 87\left(\frac{3.62-3}{3.62}\right) = 14.90 \text{ ksi} \qquad \text{(compression)}$$

Summarizing the forces,

$$C_1 = 0.85f_c'(0.85)cb = 0.85(4)(0.85)(3.62)(14) = 146.5 \text{ kip}$$
$$\text{displaced concrete} = 0.85f_c'A_s' = 0.85(4)(3) = -10.2 \text{ kip}$$
$$C_2 = f_s'A_s' = 14.90(3) = \underline{\ 44.7 \text{ kip}}$$
$$181.0 \text{ kip}$$
$$T = f_yA_s = 60(3.0) = 180 \text{ kip}$$

The slight error between T and $(C_1 + C_2)$ will be neglected.

Summarizing the internal couples,

$$\begin{aligned} M_{n1} &= C_1 Z_1 \\ &= \frac{146.5}{12}\left[17 - \frac{0.85(3.62)}{2}\right] \\ &= 188.8 \text{ ft.-kip} \end{aligned}$$

$$\begin{aligned} M_{n2} &= C_2 Z_2 \\ &= (44.7 - 10.2)\left(\frac{14}{12}\right) \\ &= 40.3 \text{ ft.-kip} \end{aligned}$$

$$\begin{aligned} M_n &= M_{n1} + M_{n2} \\ &= 188.8 + 40.3 \\ &= 229 \text{ ft.-kip} \end{aligned}$$

The design moment strength becomes

$$\begin{aligned} \phi M_n &= 0.90(229) \\ &= 206 \text{ ft.-kip} \end{aligned}$$

The results of the six parts of Example 9.5 are tabulated (see Table 9.1) and plotted in Figure 9.19. All design axial load strengths are denoted ϕP_n, and all design moment strengths are denoted $\phi P_n e$. This plot is commonly called a *column interaction diagram*. It applies *only* to the column analyzed, but it is a representation of *all* combinations of axial load and moment strengths for that column cross-section.

In Figure 9.19, any point *on* the solid line represents an allowable combination of load and moment. Any point *within* the solid line represents a load-moment combination that is also allowable but for which this column is *overdesigned*. Any point *outside* the solid line represents an unacceptable load–moment combination or a load-moment combination for which this column is *underdesigned*. The value of $\phi P_{n(\max)}$, which we calculated in part (a), is superimposed on the plot as the horizontal line.

Radial lines from the origin represent various eccentricities. (Actually, the slopes of the radial lines are equal to $\phi P_n/\phi P_n e$ or $1/e$.) The intersection of the $e = e_b$ line with the solid line represents the balanced condition. Any eccentricity *less* than e_b will result in compression controlling the column. For eccentricities between e_b (11.88 in.) and 19.6 in., the column will be in the transition zone ($0.002 \le \varepsilon_t \le 0.005$). For eccentricities greater than 19.6 in., the column will be tension controlled.

TABLE 9.1 Column Axial Load-Moment Interaction for Example 9.5

Eccentricity, *e*	**Net tensile strain in extreme tension steel, ε_t**	**Strength-reduction factor ϕ**	**Axial load strength (ϕP_n, kip)**	**Moment strength ($\phi P_n e$, ft.-kip)**
Small		0.65	672	112
(i.e., 0 to 0.10h) or 5 in.	0.00043	0.65	476	198
11.88 in. (balanced)	0.002	0.65	262	259
16.7 in.	0.004	0.82	212	295
19.6 in.	0.005	0.90	185	302
Infinite (pure moment)	>>0.005	0.90	0	206

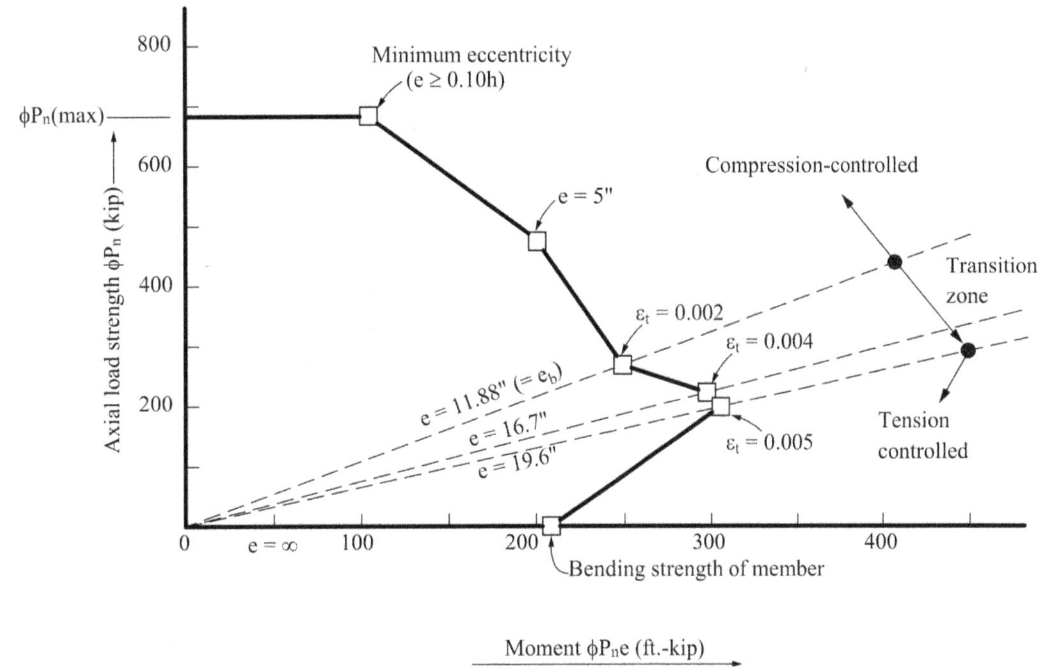

FIGURE 9.19 Column interaction diagram for Example 9.5.

The calculations involved with column loads at large eccentricities are tedious. The previous examples were analysis examples. Design of a cross-section using the calculation approach would be a trial-and-error method and would become exceedingly tedious. Therefore, design and analysis aids have been developed that shorten the process to a great extent. These aids may be found in the form of tables and charts. A chart approach is developed in ACI Publication SP-17DA(14), *ACI Design Handbook*—Design Aids [1]. The design aids are based on the assumptions of ACI 318-19 and on the principles of static equilibrium; they are developed in a fashion similar to what was done in Example 9.5. No ϕ factors are incorporated into the diagrams. Eight interaction diagrams are included in Appendix A (Diagrams A.15 through A.22.)

The diagrams take on the general form of Figure 9.19 but are generalized to be applicable to more situations. Referring to Diagram A.15, which corresponds to our Figure 9.19, the following definitions will be useful:

$$\rho_g = \frac{A_{st}}{A_g}$$

h = column dimension perpendicular to the bending axis (see the sketch included on Diagram A.15)

γ = ratio of the distance between the centroids of the outer rows of bars to the column dimension perpendicular to the bending axis

Note that the vertical axis and the horizontal axis are in general terms of K_n and R_n, where

$$K_n = \frac{P_n}{f'_c A_g}$$

$$R_n = \frac{P_n e}{f'_c A_g h}$$

Note also that P_n and $P_n e$ are nominal axial load strength and nominal moment strength. Recall that at the ultimate limit state, $\phi P_n = P_u$ and $\phi M_n = M_u$, thus $P_n = P_u/\phi$ and $M_n = M_u/\phi$, in the limit. The slope of a radial line from the origin can be represented as follows:

$$\text{slope} = \frac{\text{rise}}{\text{run}} = \frac{\left(\dfrac{P_n}{f'_c A_g}\right)}{\left(\dfrac{P_n e}{f'_c A_g h}\right)} = \frac{h}{e}$$

Curves are shown for the range of allowable ρ_g values from 0.01 to 0.08. A line near the horizontal axis labeled $\varepsilon_t = 0.0050$ indicates the limit for tension-controlled sections. Columns with load-moment-strength combinations below this line are tension-controlled ($\phi = 0.90$). The line labeled $f_s/f_y = 1.0$ indicates the balanced condition. Columns with load-moment-strength combinations above this line are compression-controlled ($\phi = 0.65$ for tied columns; 0.75 for spiral columns). Columns with load-moment-strength combinations between these two lines are in the transition zone. The line labeled K_{max} indicates the maximum allowable nominal load strength [$\phi P_{n(max)}$] for columns loaded with small eccentricities. A horizontal line drawn through the intersection of the K_{max} line and a ρ_g curve corresponds to the horizontal line near the top of the interaction diagram in Figure 9.19.

The interaction diagrams can be used for both design and analysis of reinforced concrete columns. For column design, plot the values of K_n and R_n on the vertical and horizontal axes, respectively, of the applicable interaction diagram; the intersection of the horizontal line through the K_n value and the vertical line through the R_n value gives the required column reinforcement ratio. Use linear interpolation where necessary.

For analysis, there are two possible cases: Cases 1 and 2.

Case 1: Determine the moment capacity of a concrete column for a given or known factored axial load. Calculate K_n. Using the applicable column interaction diagram, plot the value of K_n on the vertical axis and obtain the value of R_n at the point where a horizontal line through the K_n value intersects the interaction curve for the given or known column reinforcement ratio. Then, use the equation for $R_n = M_n/f_c'A_g h$ to calculate M_n from which the design moment capacity, ϕM_n, can be obtained. Use linear interpolation where necessary.

Case 2: Determine the axial load capacity of a concrete column for a given or known factored moment. Calculate R_n. Using the applicable column interaction diagram, plot the value of R_n on the horizontal axis and obtain the value of K_n at the point where a vertical line through the R_n value intersects the interaction curve for the given or known column reinforcement ratio. Then, use the equation for $K_n = P_n/f_c'A_g$ to calculate P_n from which the design axial load capacity, ϕP_n, can be obtained. Use linear interpolation where necessary.

The following three examples illustrate the use of the ACI interaction diagrams for analysis and design of short reinforced concrete columns.

EXAMPLE 9.6

Using the interaction diagrams of Appendix A, find the axial load strength ϕP_n and the moment strength ϕM_n for the column cross-section with six No. 9 bars, as shown in Figure 9.20. Eccentricity e = 5 in., and use $f_c' = 4000$ psi and f_y = 60,000 psi. Compare the results with Example 9.5b.

Solution:

First, determine which interaction diagram to use, based on the type of cross-section, the material strengths, and the factor γ.

$$\gamma h = 14 \text{ in.}$$

$$\gamma = \frac{14}{20} = 0.7$$

Therefore, use Interaction Diagram A.15.

$$\rho_g = \frac{6.00}{14(20)} = 0.0214$$

$$0.01 \le 0.0214 \le 0.08 \qquad \text{(O.K.)}$$

Next, calculate the slope of the radial line from the origin, which relates h and e:

$$\text{slope} = \frac{h}{e} = \frac{20}{5} = 4$$

Recall that $K_n = P_n/f_c'A_g$ and $R_n = M_n/f_c'A_g h = P_n e/f_c'A_g h$. Therefore, since $h/e = 4$ and $e/h = 1/4$, we can rewrite $R_n = P_n/4f_c'A_g = (1/4)K_n$.

A straight edge and some convenient values (e.g., $K_n = 1.0$ and $R_n = 0.25$) may be used to intersect this radial line with an estimated $\rho_g = 0.0214$ curve. At this intersection, we read $K_n \approx 0.64$ and $R_n \approx 0.16$. Because this combination of load and moment is above the $f_s/f_y =$ 1.0 line on the column interaction diagram A.15, this is a compression-controlled section and $\phi = 0.65$.

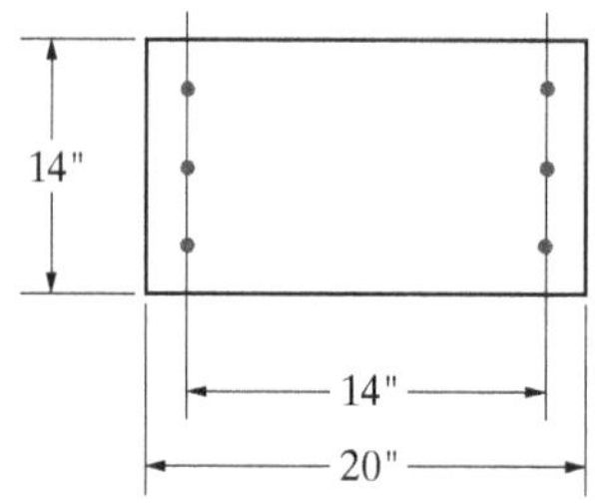

FIGURE 9.20 Sketch for Example 9.6.

$$\begin{aligned}\phi P_n &= \phi K_n f_c' A_g \\ &= 0.65\,(0.64)(4)(20)(14) = 466 \text{ kip} \\ \phi M_n &= \phi R_n f_c' A_g h \\ &= \frac{0.65(0.160)(4)(20)(14)(20)}{12 \text{ in./ft}} = 194 \text{ ft.-kip}\end{aligned}$$

or

$$\phi M_n = \phi P_n e = \frac{466 \text{ ft.-kip}\,(5 \text{ in.})}{12 \text{ in./ft}} = 194 \text{ ft.-kip}$$

This compares reasonably well with the results of Example 9.5b: $\phi P_n = 476$ kip and $\phi M_n = 198$ ft.-kip.

EXAMPLE 9.7

Design a circular spirally reinforced concrete column to support a design load $P_u = 1{,}100$ kip and a design moment $M_u = 285$ ft.-kip. Use $f_c' = 4000$ psi and $f_y = 60{,}000$ psi.

Solution:

Estimate the column size required based on $\rho_g = 1\%$ and axial load only.

$$\begin{aligned}\text{required } A_g &= \frac{P_u}{0.85\phi\left[0.85 f_c'(1-\rho_g) + f_y\rho_g\right]} \\ &= \frac{1100}{0.85\,(0.75)\left[0.85(4)(0.99) + 60(0.01)\right]} \\ &= 435 \text{ in.}^2\end{aligned}$$

Try a 24-in.-diameter column ($A_g = 452$ in.^{2}).

If No. 9 bars are eventually chosen (refer to Figure 9.21),

$$\gamma h = 24 - 2(1\tfrac{1}{2}) - 2\left(\frac{3}{8}\right) - 1.13 = 19.12 \text{ in.}$$

$$\gamma = \frac{19.12}{h} = \frac{19.12}{24} = 0.797$$

Therefore, use Diagram A.21 from Appendix A (ACI Interaction Diagram C4-60.8). Next, determine the required ρ_g. Assume that this column will be compression-controlled ($\phi = 0.75$) and subject to later check.

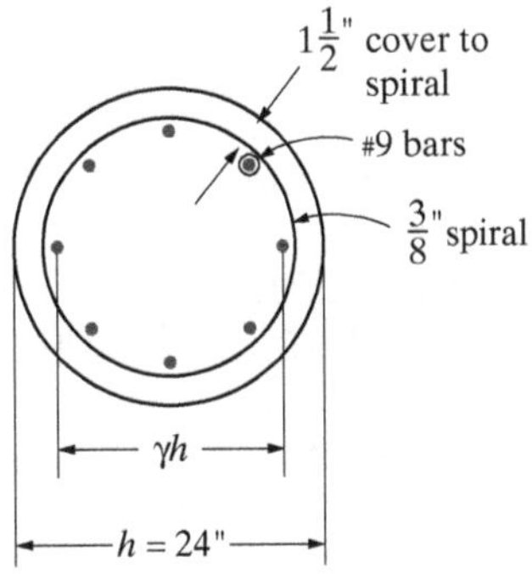

FIGURE 9.21 Sketch for Example 9.7.

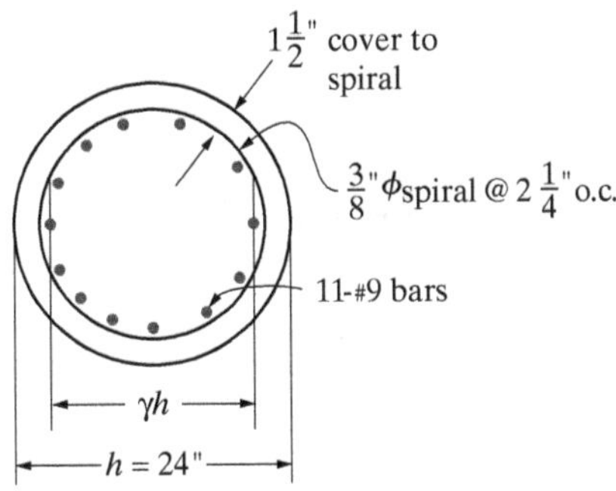

FIGURE 9.22 Design sketch for Example 9.7.

Recognizing that in the limit state design method, the required $P_n = P_u/\phi$ and the required $M_n = P_n e = M_u/\phi$, we can calculate the values of required K_n and R_n:

$$\text{required } K_n = \frac{P_u}{\phi f'_c A_g} = \frac{1100}{0.75(4)(452)} = 0.811$$

$$\text{required } R_n = \frac{M_u}{\phi f'_c A_g h} = \frac{285(12)}{0.75(4)(452)(24)} = 0.105$$

From the Column Interaction Diagram A.21, we find, $\rho_g = 0.024$. Note that this is well above the $f_s/f_y = 1.0$ line; therefore, the column is compression-controlled and the assumption that $\phi = 0.75$ is O.K.

$$\text{required } A_s = \rho_g A_g = 0.024(452) = 10.85 \text{ in.}^2$$

Select 11 No. 9 bars (As = 11.00 in.^{2}). Check the maximum number of No. 9 bars from Table A.14: 15 (O.K.).

Design the spiral. Use a $\frac{3}{8}$-in.-diameter spiral.

The concrete core diameter is $D_{ch} = 24 \text{ in.} - 2\left(1\frac{1}{2} \text{ in. cover}\right) = 21 \text{ in.}$

$$A_{ch} = \frac{\pi(21 \text{ in.})^2}{4} = 346 \text{ in.}^2$$

$$\text{required } \rho_s = 0.45\left(\frac{A_g}{A_{ch}} - 1\right)\frac{f'_c}{f_{yt}} = 0.45\left(\frac{452}{346} - 1\right)\frac{4}{60}$$

$$= 0.0092$$

$$\text{required } s = \frac{4A_{sp}}{D_{ch}\rho_s} = \frac{4(0.11)}{21(0.0092)} = 2.27 \text{ in.}$$

Use $2\frac{1}{4}$-in. spacing for the spirals. The design is shown in Figure 9.22.

EXAMPLE 9.8

Design a square-tied reinforced concrete column to support a design load P_u = 1,300 kip and a design moment M_u = 550 ft.-kip. Use f'_c = 4000 psi and f_y = 60,000 psi.

Solution:

Estimate the column size required based on ρ_g = 1% and axial load only.

$$\text{required } A_g = \frac{P_u}{0.80\,\phi\left[0.85\, f'_c\left(1-\rho_g\right) + f_y\rho_g\right]}$$

$$= \frac{1300}{0.80\,(0.65)\left[0.85(4)(0.99) + 60(0.01)\right]}$$

$$= 631 \text{ in.}^2$$

Try a 26-in.-square column (A_g = 676 in.^{2}).

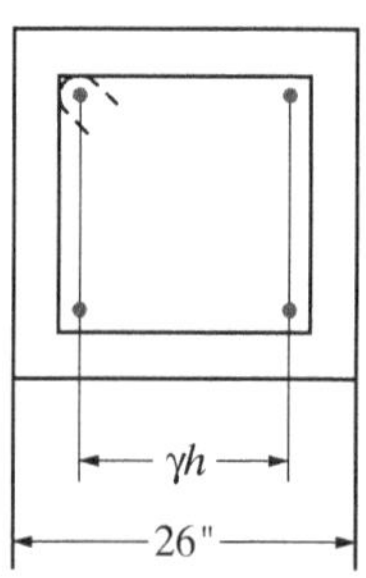

FIGURE 9.23 Sketch for Example 9.8.

If No. 9 bars are eventually chosen (refer to Figure 9.23):

$$\gamma h = 26 - 2(1\tfrac{1}{2}) - 2\left(\frac{3}{8}\right) - 1.13 = 21.12 \text{ in.}$$

$$\gamma = \frac{21.12}{h} = \frac{21.12}{26} = 0.812$$

Therefore, use Diagram A.18 from Appendix A (ACI Interaction Diagram R4-60.8).

Next, determine the required ρ_g. Assume that this column will be compression-controlled ($\phi = 0.65$) subject to later check.

Recognizing that in the limit state design method, the required $P_n = P_u/\phi$ and the required $M_n = P_n e = M_u/\phi$, we can calculate the values of required K_n and R_n:

$$\text{required } K_n = \frac{P_u}{\phi f'_c A_g} = \frac{1300}{0.65(4)(676)} = 0.740$$

$$\text{required } R_n = \frac{M_u}{\phi f'_c A_g h} = \frac{550\,(12)}{0.65(4)(676)(26)} = 0.144$$

From Column Interaction Diagram A.18, $\rho_g \approx 0.023$. Note that this is well above the $f_s/f_y = 1.0$ line; therefore, the column is compression-controlled and the assumption that ϕ = 0.65 is O.K.

$$\text{required } A_s = \rho_g A_g = 0.023(676) = 15.55 \text{ in.}^2$$

Select 16 No. 9 bars (A_s = 16.00 in.^{2}). Check the maximum number of No. 9 bars from Table A.14: 20 (O.K.).

Design the ties: Use a $\frac{3}{8}$-in.-diameter tie because the vertical bar size (No. 9 bar) is not greater than a No. 10.

The maximum tie spacing is the smallest of the following:

$$16\,(\text{bar diameter}) = 16 \times 1.13 = 18 \text{ in.}$$

$$48\,(\text{tie diameter}) = 48 \times \frac{3}{8} = 18 \text{ in.}$$

$$\text{least column dimension} = 26 \text{ in.}$$

Therefore, use No. 3 ties at 18-in. spacing. The design is shown in Figure 9.24.

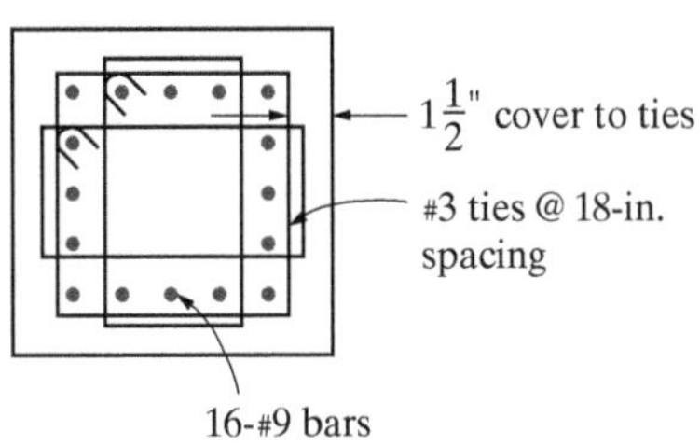

FIGURE 9.24 Design sketch for Example 9.8.

9.11 BIAXIAL BENDING IN CONCRETE COLUMNS

In previous sections of this chapter, we discussed the analysis and design of columns subjected to axial compression loads plus bending moment about one axis only. In buildings, columns are often subjected to axial loads plus simultaneous bending about the two orthogonal axes. Examples of such columns include corner columns and edge columns. Even interior columns may be subjected to unbalanced moments about both orthogonal axes due to unbalanced loads on the beams and girders framing into the column. The unbalanced loads and the resulting unbalanced moments may occur due to skipped or pattern live loading and/or because of differences in adjacent span lengths of the beams or girders framing into the column. There are several methods available that can be used for the design of biaxially loaded columns. For the analysis of reinforced concrete columns under biaxial bending, the Bresler reciprocal load method will be used in this text. For design, a simplified approach using the equivalent eccentricity method will be used [3]. In this method, the two orthogonal moments are converted into an equivalent uni-axial moment about the weaker axis. The column is then designed for the axial load and the uniaxial moment about the weaker axis. The design procedure for biaxial bending in columns is as follows:

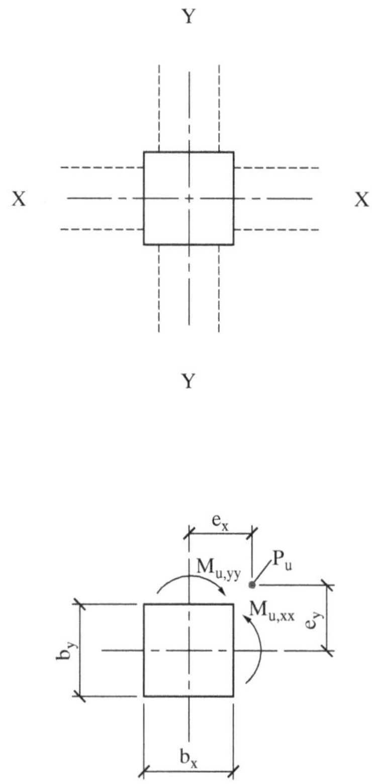

FIGURE 9.25 Biaxial moments in columns.

- Convert the moments $M_{u,xx}$ and $M_{u,yy}$ to an equivalent uniaxial moment about the Y-Y axis, $M_{uo,yy}$.
- $M_{u,xx}$ is the factored moment about the X-X or strong axis, and $M_{u,yy}$ is the factored moment about the Y-Y or weak axis
- The eccentricity $e_x = \dfrac{M_{u,yy}}{P_u}$ and $e_y = \dfrac{M_{u,xx}}{P_u}$ (see Figure 9.25)
    - b_x = column dimension parallel to the *X-X* axis
    - b_y = column dimension parallel to the *Y-Y* axis
    - e_x = eccentricity of the axial load for bending about the *Y-Y* axis
    - e_y = eccentricity of the axial load for bending about the *X-X* axis
- The method assumes reinforcement on all four faces of the column.
- The method is valid for b_x/b_y between 0.5 and 2.0.
- *For* $\dfrac{P_u}{f'_c A_g} \leq 0.4, \quad \alpha = \left(0.5 + \dfrac{P_u}{f'_c A_g}\right)\left(\dfrac{f_y + 40{,}000}{100{,}000}\right) \geq 0.60$
- *For* $\dfrac{P_u}{f'_c A_g} > 0.4, \quad \alpha = \left(1.3 - \dfrac{P_u}{f'_c A_g}\right)\left(\dfrac{f_y + 40{,}000}{100{,}000}\right) \geq 0.50$

Where,

f_y = rebar tensile yield strength in psi

f'_c = concrete compressive strength at 28 days

- The equivalent eccentricity for bending about the *Y-Y* axis, $e_{ox} = e_x + \alpha\left(\dfrac{b_x}{b_y}\right)$
- Design the column for the axial load, P_u, and an equivalent uni-axial moment about the weaker axis, $M_{uo,yy} = P_u\, e_{ox}$

EXAMPLE 9.9

Column Design with Biaxial Moments

Design the reinforcement for a 16 in. × 16 in. square reinforced concrete column to resist the following loads: P_u = 450 kip, M_{ux} = 70 ft.-kip, M_{uy} = 45 ft.-kip. The rebar tensile strength is 60,000 psi, and the concrete compressive strength at 28 days is 4,000 psi. Assume a short column and use the equivalent eccentricity method to account for biaxial bending in the column.

Solution:

The equivalent eccentricity method assumes reinforcement on all four faces of the column.

Calculating the eccentricities about both orthogonal axes, we obtain the eccentricity of the axial load for bending about the *Y-Y* axis $e_x = \dfrac{M_{u,yy}}{P_u} = \dfrac{45\text{ ft.-kip}}{450\text{ kip}} = 0.10\text{ ft}$

The eccentricity of the axial load for bending about the *X-X* axis

$$e_y = \frac{M_{u,xx}}{P_u} = \frac{70 \text{ ft.-kip}}{450 \text{ kip}} = 0.156 \text{ ft}$$

b_x = column dimension parallel to the *X-X* axis = 16 in.

b_y = column dimension parallel to the *Y-Y* axis = 16 in.

The method is valid for b_x/b_y between 0.5 and 2.0: b_x/b_y = 16/16 = 1, which lies between 0.5 and 2.0, therefore O.K.

$$\frac{P_u}{f_c' A_g} = \frac{(450 \text{ kip})(1000)}{(4000 \text{ psi})(16 \text{ in})(16 \text{ in.})} = 0.44 > 0.4$$

$$\alpha = \left(1.3 - \frac{P_u}{f_c' A_g}\right)\left(\frac{f_y + 40{,}000}{100{,}000}\right) \geq 0.50$$

$$= (1.3 - 0.44)\left(\frac{60{,}000 + 40{,}000}{100{,}000}\right) = 0.86 \geq 0.50,$$

Therefore, $\alpha = 0.86$

The equivalent eccentricity for bending about the *Y-Y* axis is

$$e_{ox} = e_x + \alpha\left(\frac{b_x}{b_y}\right) = 0.10 \text{ ft.} + (0.86)(0.156 \text{ ft.})\left(\frac{16 \text{ in.}}{16 \text{ in}}\right) = 0.234 \text{ ft.}$$

Design the column for a factored axial load, P_u = 450 kip and a uni-axial factored moment about the *Y-Y* axis, $M_{uo,yy} = P_u e_{ox} = (450 \text{ kip})(0.234 \text{ ft.}) = 105$ ft.-kip

Assume strength reduction factor $\phi = 0.65$ (i.e., assuming compression-controlled section)

γh = 16" – 1.5" × 2 sides – (2)(3/8 in. ties) – (1 in bar diameter) = 11.25 inches,

Therefore, γ = (11.25/16) = 0.70.

Calculate the interaction parameters as follows:

$$K_n = \frac{P_u}{\phi f_c' A_g} = \frac{450 \text{ kip}}{(0.65)(4 \text{ ksi})(16 \text{ in.})(16 \text{ in.})} = 0.676$$

$$R_n = \frac{M_u}{\phi f_c' A_g h} = \frac{(105 \text{ ft.-kip})(12 \text{ in/ft})}{(0.65)(4 \text{ ksi})(16 \text{ in.})(16 \text{ in.})(16 \text{ in.})} = 0.118$$

From the Column Interaction Diagram A.17, the required gross reinforcement ratio,

$$\rho_{g,\ required} \approx 0.016$$

Therefore, A_g required = 0.016 (16 in.)(16 in.) = 4.10 in.^{2}

This falls within the compression-controlled region of the interaction diagram, therefore, $\phi = 0.65$ as assumed.

Select **8 # 7** vertical bars for the column (A_g = 4.80 in.^{2}). The reinforcement ratio provided is

$$\rho_{g,\ provided} = \frac{4.80\,\text{in.}^2}{(16\,\text{in.})(16\,\text{in.})} = 0.019$$

$$> 0.01 \text{ (minimum column vertical rebar ratio)} \qquad \text{O.K.}$$

$$< 0.08 \text{ (maximum column vertical rebar ratio)} \qquad \text{O.K.}$$

At the lap-spliced regions (assuming the same amount of column reinforcement above the splice), the rebar ratio will be effectively doubled, that is,

$$\rho_{g,\ provided} = (2)(0.019) = 0.038 \text{ (at the lap splice location)}$$

$$> 0.01 \text{ (minimum column vertical rebar ratio)} \qquad \text{O.K.}$$

$$< 0.08 \text{ (maximum column vertical rebar ratio)} \qquad \text{O.K}$$

Check the clear spacing between the adjacent column vertical reinforcement:

$$\text{clear spacing} = \frac{16\text{ in.} - (2)(1.5\text{ in. side cover}) - (2)(3/8\text{ in. stirrup}) - (7/8\text{ in. rebar diamter})}{2\text{ spaces}} - (7/8\text{ in rebar diamter})$$

$$= 4.81\text{ in.}$$

$$> 1.5\left(\frac{7}{8}\text{ in.}\right) = 1.3\text{ in.} \qquad \text{O.K.}$$

$$> 1.5\text{ in.} \qquad \text{O.K.}$$

$$< 6\text{ in.} \qquad \text{O.K., no extra intermediate tie required.}$$

Since the clear spacing between adjacent bars is less than 6 inches, no extra ties are required.

Design the column ties:

Select #3 ties since column vertical rebar size is #7 bar < #10 O.K.

Tie spacing ≤ 16 bar diameter = (16)(7/8 in.) = **14 in.**

≤ 48 tie diameter = (48)(3/8 in.) = 18 in.

≤ least column dimension = 16 in.

Therefore, use #3 ties at 14 in. o.c.

See the column rebar detail in Figure 9.26.

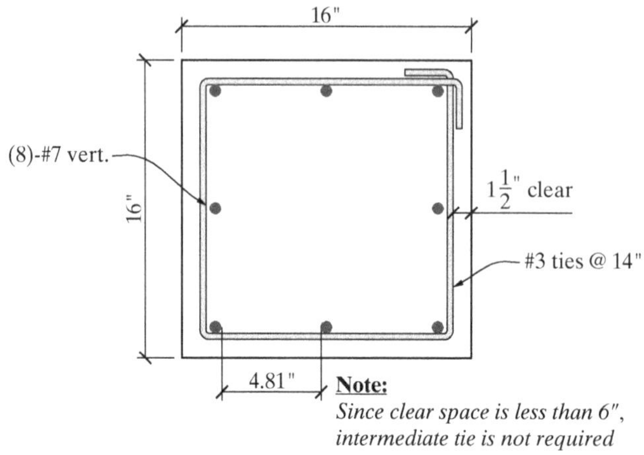

FIGURE 9.26 Column rebar detail.

Biaxial Bending in Reinforced Concrete Columns - The Bresler Reciprocal Load Method:

The equivalent eccentricity method presented previously in this section is more suited for the design of biaxially loaded columns for which the factored axial load and factored moments about both orthogonal axes are known. For the analysis of biaxially loaded columns, the Bresler reciprocal load method [4] is more suitable. The reciprocal load method models the interaction between the nominal axial load capacities as shown in the following equation:

$$\frac{1}{P_{ni}} = \frac{1}{P_{nx}} + \frac{1}{P_{ny}} - \frac{1}{P_{n0}}$$

Where,

The nominal concentric axial load capacity of the column with zero moments (from ACI Code Equation 22.4.2.2), $P_{n0} = 0.85f_c'\,(A_g - A_{st}) + f_yA_{st}$

P_{nx} = nominal axial load capacity of the column when subjected to the factored moment M_{ux} (with $M_{uy} = 0$)

P_{ny} = nominal axial load capacity of the column when subjected to the factored moment M_{uy} (with $M_{ux} = 0$)

P_{ni} = nominal axial load capacity of the column when subjected to the factored biaxial moments, M_{ux} and M_{uy}

We will now analyze the column in Example 9.9 using the Bresler Reciprocal Load method:

Analyzing the 16 in. x 16 in. column in Example 9.9 for the given factored load and biaxial moments, we calculate the following nominal capacities:

$$\begin{aligned} P_{n0} &= 0.85f_c'\left(A_g - A_{st}\right) + f_yA_{st} \\ &= 0.85(4\text{ ksi})\left[(16)(16) - 4.80\right] + (60\text{ ksi})(4.80) \\ &= 1142\text{ kip} \end{aligned}$$

Calculate P_{nx}:

$$R_{nx} = \frac{P_{nx}e}{f'_c A_g h} = \frac{M_{ux}}{\phi f'_c A_g h}$$

$$= \frac{(70\,\text{ft.-kip})(12)}{0.65(4\,\text{ksi})(16)(16)(16)} = 0.079$$

Entering the column interaction diagram (Table A.17) with the R_{nx} value above and the column vertical rebar ratio, ρ_g of 0.019 obtained in Example 9.9, we obtain

$$K_{nx} = 0.88$$

$$= \frac{P_{nx}}{f'_c A_g} = \frac{P_{nx}}{(4\,\text{ksi})(16)(16)}$$

Therefore, $P_{nx} = 901$ kip

Calculate P_{ny}:

$$R_{ny} = \frac{P_{ny}e}{f'_c A_g h} = \frac{M_{uy}}{\phi f'_c A_g h}.$$

$$= \frac{(45\,\text{ft.-kip})(12)}{0.65(4\,\text{ksi})(16)(16)(16)} = 0.051$$

Entering the column interaction diagram (Table A.17) with the R_{ny} value above and the column vertical rebar ratio, ρ_g of 0.019 obtained in Example 9.9, we obtain

$$K_{ny} = 0.96$$

$$= \frac{P_{ny}}{f'_c A_g} = \frac{P_{ny}}{(4\,\text{ksi})(16)(16)}$$

Therefore, $P_{ny} = 983$ kip

Using the Bresler reciprocal load equation yields

$$\frac{1}{P_{ni}} = \frac{1}{P_{nx}} + \frac{1}{P_{ny}} - \frac{1}{P_{n0}}$$

$$= \frac{1}{901} + \frac{1}{983} - \frac{1}{1142}$$

Therefore, $P_{ni} = 799$ kip

For a tied column, the strength reduction factor, $\phi = 0.65$ (i.e., assuming compression-controlled zone)

The axial load capacity for this column is

$\phi P_{ni} = (0.65)(799) = 519 \text{ kip} > P_u = 450 \text{ kip}$. O.K.

This indicates that the column as designed in Example 9.9 is indeed adequate to resist the applied factored axial compression load and biaxial moments.

9.12 THE SLENDER COLUMN

Columns in building structures are typically braced at every floor level in both orthogonal directions. The lateral bracing of the column is provided by the beams and girders framing into the column. Thus far, our design and analysis have been limited to short columns that require no consideration of necessary strength reduction due to the possibility of buckling. All compression members will experience the buckling phenomenon as they become longer and more flexible. These are termed *slender columns* and they occur when, for aesthetic or other architectural reasons, the column is continuous over two or more floor levels (e.g., tall columns spanning multiple stories in the atrium area of a building). A column may be categorized as *slender* if its cross-sectional dimensions are small in comparison to its unsupported length. The degree of slenderness may be expressed in terms of the *slenderness ratio*

$$\frac{k\ell_u}{r}$$

where

k = effective length factor for compression members

ℓ_u = the unsupported length of a compression member, which shall be taken as the clear distance between floor slabs, beams, or other members capable of providing lateral support in the direction being considered (ACI Code, Section 2.2)

r = radius of gyration of the cross-section of the compression members, which may be taken as $0.30h$, where h is the overall dimension of a rectangular column in the direction of the moment, or $0.25D$, where D is the diameter of a circular column (ACI Code, Section 6.2.5.1)

The numerator $k\ell_u$ is termed the *effective length*. It is not only a function of the unsupported length and end conditions of the column but also a function of whether or not *sidesway* exists. Sidesway may be described as a kind of deformation whereby one end of a member moves laterally with respect to the other. Sidesway is also termed *lateral drift*.

The ACI Code, Section 6.6.4.4.3, states that for compression members braced against sidesway, the effective length factor, k may be taken as 1.0. This is conservative. The ACI Code, Section 6.6.4.4.3, states that for compression members *not* braced against sidesway, the effective length must be greater than 1.0. Therefore, as a rule, compression members free to buckle in a sidesway mode are appreciably weaker than when braced against sidesway. In concrete buildings, braced frame systems typically imply that lateral resistance to wind and seismic or other lateral loads is provided by shear walls. Unbraced (or sway) frame buildings imply that lateral resistance to wind and seismic or other lateral loads is provided by concrete moment frames.

A simple example is a column fixed at one end and entirely free at the other (cantilever column or flagpole). Such a column will buckle, as shown in Figure 9.27. The upper end would move laterally with respect to the lower end. This lateral movement is the sidesway (or lateral drift). In reinforced concrete structures, it is common to deal with indeterminate rigid frames, such as illustrated by the simple portal frame in Figure 9.28. The upper end of the frame can move sideways as it is unbraced. This type of frame is sometimes termed a *sway frame*, and it depends on the rigidity of the joints for stability. The lower ends of the columns may be theoretically pin-connected (i.e., pin-support), fully restrained against rotation (i.e., fixed support), or lie somewhere in between.

As an example of how the effective length of a column is influenced by sidesway, consider the simple case of a single member, as shown in Figure 9.29. The member braced against sidesway (Figure 9.29a) has an effective length half that of the member without sidesway bracing (Figure 9.29b) and has four times the axial-load capacity based on the Euler critical column buckling load theory.

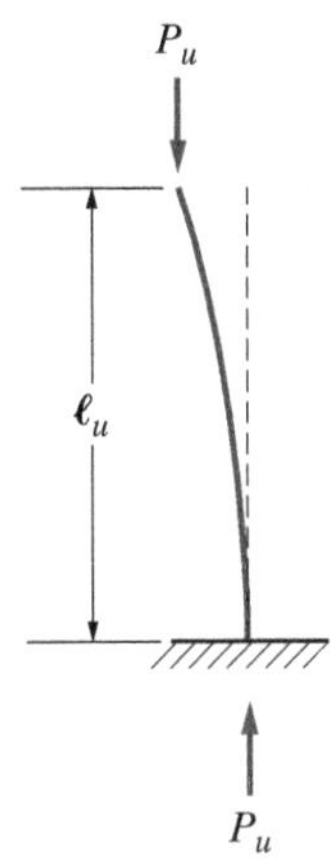

FIGURE 9.27 Fixed-free column.

If we consider the column shown in Figure 9.29b to be part of a frame and give the sidesway the notation Δ as shown in Figure 9.30, it is seen that the axial load now acts eccentrically and creates end moments of $P_u\Delta$. This is referred to as the *P-delta effect*. These moments are also referred to as "second-order end moments" because they are in addition to any primary (first order) end moments that result from applied loads on the frame with no consideration of geometry change (sidesway) or geometric non-linearities. Note that material non-linearities due to cracking also must be considered in the analysis and design of slender concrete columns; this will be discussed later.

Actual structures are rarely completely braced (nonsway) or completely unbraced (sway). Sidesway may be minimized in various ways. The common approach is to use walls sufficiently strong and rigid in their own planes to minimize the lateral or horizontal displacement. Another method is to use a rigid central concrete core that is capable of resisting lateral loads and lateral displacements due to unsymmetrical loading conditions. ACI 318-19 Code,

Section 6.2.5 states that columns in a story may be considered braced if the total stiffness of the lateral force-resisting elements within the story (e.g., total stiffness of all the shear walls in a particular direction) is at least 12 times the total lateral stiffness of all the columns in that story in the direction considered. For those cases when it is not readily apparent whether a structure is braced or unbraced, the ACI Code (Section 6.2.5) provides analytical methods to aid in the decision.

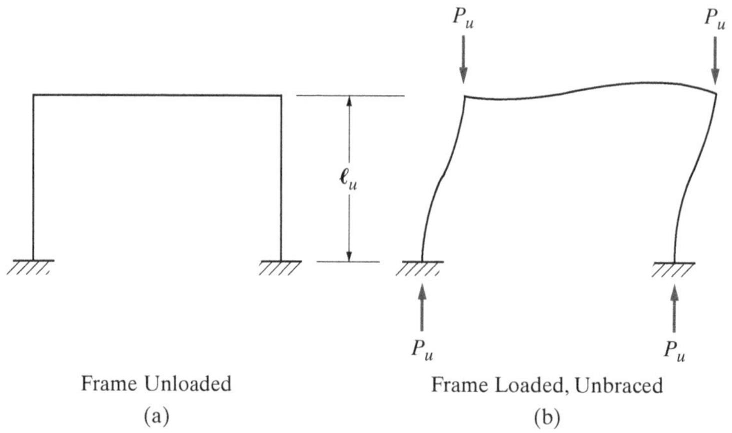

FIGURE 9.28 Sidesway on portal frame.

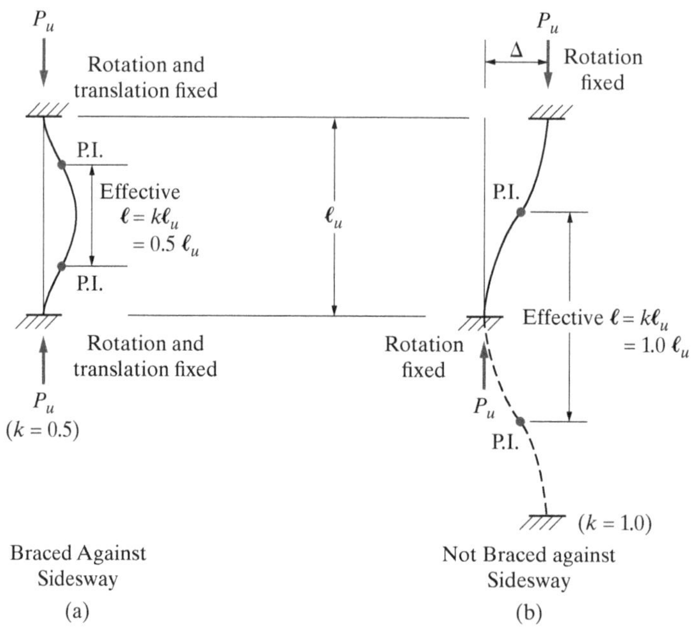

FIGURE 9.29 Sidesway and effective length.

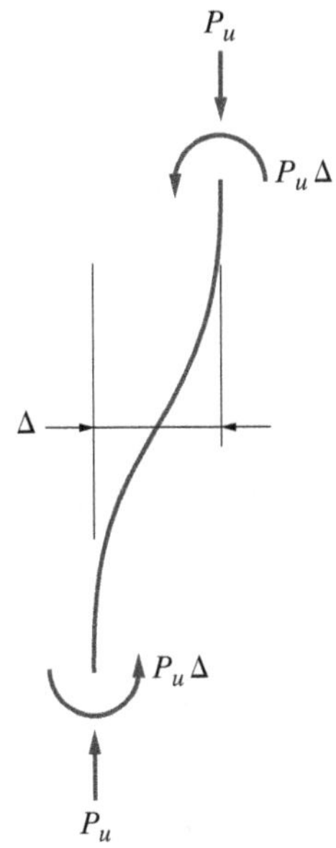

FIGURE 9.30 Column subjected to sidesway.

For columns in sway or moment frames (i.e., unbraced frames or moment frames), slenderness effects may be neglected when

$$\frac{k\ell_u}{r} \leq 22 \qquad \textbf{[ACI Eq. (6.2.5.1(a))]}$$

For braced frame columns, slenderness effects may be neglected when

$$\frac{k\ell_u}{r} \leq 34 + 12\left(\frac{M_1}{M_2}\right) \qquad \textbf{[ACI Eq. (6.2.5.1(b))]}$$

where M_1 is the smaller end moment and M_2 is the larger end moment, both obtained by an elastic frame analysis. The ratio M_1/M_2 is negative if the column is bent in *single curvature,* and is positive if bent in *double curvature* (see Figure 9.31), and the term $[34 + 12M_1/M_2]$ shall *not* be taken greater than 40. At this maximum slenderness ratio and assuming an effective length factor, k, of 1.0, the maximum column unbraced height ℓ_u would be approximately twelve times the plan dimension of the column perpendicular to the axis of bending (i.e., $\ell_u \leq 12h$)

An example of a column in single curvature bending include ground floor columns in braced frames that are supported on spread footings with minimal restraint to moment at the base of the column, thus the moment in the column at the spread footing, M_1, is zero, and therefore, the M_1/M_2 ratio is zero. Double curvature bending occurs in a ground floor column if the column is supported on a foundation that can provide restraint to bending moment (e.g., mat or pile foundation). Double curvature bending also occurs in the columns above the 2nd floor level, and Table 9.2 presents some examples of the $k\ell_u/r$ limits for various M_1/M_2 ratios in a braced-frame column:

TABLE 9.2 Maximum slenderness ratio for columns in braced frames.

M_1/M_2 ratio	Type of curvature	Maximum kL_u/r
-1	Single curvature	22
-0.5	Single curvature	28
0	Single curvature	34
+1	Double curvature	40
+0.5	Double curvature	40

As previously indicated, for columns in sway or moment frames (i.e., in buildings where the lateral load is resisted by the bending of the beams or girders and columns and where there are no shear walls), slenderness effects may be neglected when $k\ell_u/r$ is less than 22 (ACI 318-19, Section 6.2.5.1(a)). Assuming a rectangular column and an effective length factor, k, of approximately 1.2 for columns in moment frames, the preceding requirement for columns in sway frames would indicate that the effects of slenderness can be neglected for sway or moment frames if the ℓ_u/h ratio does not exceed 5.3 (i.e., 22(0.288)/1.2).

Fortunately, for ordinary beam and column sizes and typical story heights of concrete framing systems, effects of slenderness may be neglected in more than 90% of columns in braced (nonsway) frames and in about 40% of columns in unbraced (sway) frames [2].

In cases where column slenderness must be considered, the ACI Code prescribes the methods that can be used for the design of these columns. The design approaches range from an approximate method (ACI 318-19, Section 6.6.4) in which moments are magnified to account for slenderness to a nonlinear second-order analysis method. The approximate method uses a moment magnification factor that amplifies the factored moment computed from a conventional elastic analysis. In turn, the combination of the magnified factored moment and the factored axial load is used in the design of the compression member. Although the approximate analyses have been the traditional design methods of the past, more exact analyses have become possible and practical with the increased availability of sophisticated computer methods.

Moment Magnification for Slender Columns in Braced Frames

The effect of buckling in slender columns results in a magnification of the first order moments, the moments that are applied to the column before buckling. These first order moments may be as a result of the unbalanced end moments from the beams and girders framing into the column or from the moments acting on the column due to lateral loads.

The second order or magnified moment for a column in a braced frame is

$$M_c = \delta_{ns} M_2 \qquad \text{(ACI Eq. 6.6.4.5.1)}$$

The first order moment, M_2, is the unbalanced first-order factored moment from the beams and girders framing into the column, but must be greater than or equal to $P_u(0.6 + 0.03h)$ about each orthogonal axis of the column (ACI 318-19, Section 6.6.4.5.4); this is the minimum moments for which slender columns should be designed, and it accounts for geometric imperfections in the column. The minimum moment should *not* be applied to both orthogonal axes of the column simultaneously (ACI Code, Section R6.6.4.5.4). Also, if minimum moments govern, the ratio M_1/M_2 in ACI Code, Equation 6.2.5.1(b) is determined using

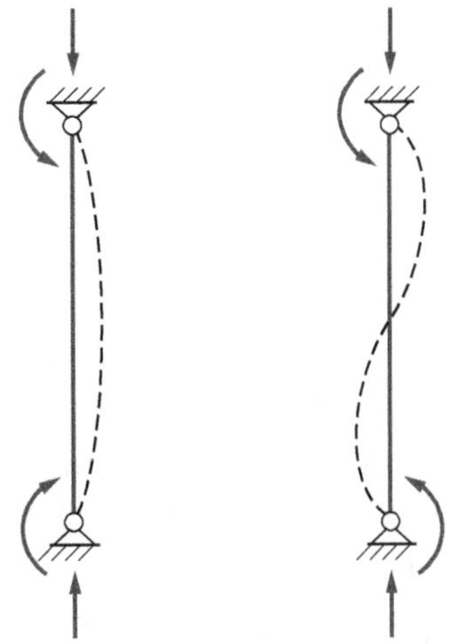

FIGURE 9.31 Column curvature.

the actual factored column end moments from the structural analysis (ACI Code, Section R6.6.4.5.4).

The non-sway moment magnification factor is given as

$$\delta_{ns} = \frac{C_m}{1 - \frac{P_u}{0.75P_c}} \geq 1.0 \qquad [ACI\,Eq. 6.6.4.5.2]$$

Where the critical buckling load, P_c, is given as

$$P_c = \frac{\pi^2 (EI)_{eff}}{(k\ell_u)^2} \qquad [ACI\,Eq. 6.6.4.4.2]$$

The effective length factor, k, for columns in braced frame buildings (i.e., buildings that are braced with shear walls) is usually taken as 1.0 since the column is lateral supported at both ends (ACI 318 Code, Section 6.6.4.4.3).

The effective stiffness of the column, $(EI)_{eff}$, which is meant to account for material non-linearities such as cracking and creep in concrete, is calculated using any one of the equations from ACI 318-19, Section 6.6.4.4.4 as follows:

$$(EI)_{eff} = \frac{0.4E_c I_g}{1+\beta_{dns}} \qquad \text{(ACI Eq. 6.6.4.4.4a)}$$

$$(EI)_{eff} = \frac{0.2E_c I_g + E_s I_{se}}{1+\beta_{dns}} \qquad \text{(ACI Eq. 6.6.4.4.4b)}$$

$$(EI)_{eff} = \frac{E_c I}{1+\beta_{dns}} \qquad \text{(ACI Eq. 6.6.4.4.4c)}$$

β_{dns} = ratio of maximum factored sustained axial load to the maximum factored axial load for the same load combination

The moment of inertia values, I, for columns and walls to be used in ACI Equation 6.6.4.4.4c are given in ACI Table 6.6.3.1.1(b).

A simplified value of the effective stiffness can be obtained by assuming β_{dns} = 0.60 in ACI Eq. 6.6.4.4.4a (ACI 318-19, Section R6.6.4.4.4). This gives $(EI)_{eff} = 0.25E_cI_g$. This simplified approach will be followed in this text.

The C_m correction factor accounts for the fact that the actual moment diagram in the column is different than an equivalent uniform moment diagram that was used to derive the moment magnification equations (ACI 318-19, Section 6.6.4.5.3).

For columns with transverse loads between supports, $C_m = 1.0$

For columns where $M_{2min} > M_2$, $C_m = 1.0$ (ACI Eq. 6.6.4.5.4)

For all other columns, $C_m = 0.6 - 0.4(M_1/M_2)$ (ACI Eq. 6.6.4.5.3a)

$M_{2min} = P_u\,(0.6 + 0.03h)$ (ACI Eq. 6.6.4.5.4)

M_1 and M_2 are the factored end moments in the column, where M_2 is the larger end moment and is always positive. Where a column is subjected to transverse loads between supports, M_2 is the largest moment occurring within the column length.

M_1/M_2 = negative (–ve) for single curvature columns

M_1/M_2 = positive (+ve) for double curvature columns

For columns with $M_1 = M_2 = 0$, use $M_1/M_2 = -1$ which results in a value of $C_m = 1.0$.

EXAMPLE 9.10

A 16 inch x 16 inch ground floor interior column in a multi-story building braced with shear walls is supported on a spread footing and has an unbraced height of 15 ft. The column is subjected to a factored axial load of 600 kip and a moment about the *X-X* axis of 65 ft-kip. Determine if the column is slender and calculate the magnified or second-order moment, M_c. Determine the required area of the column vertical reinforcement. Assume f_c' = 4,000 psi and f_y = 60,000 psi.

Solution:

The column unbraced length, ℓ_u = 15 ft. = 180 inches

Since the building is braced by shear walls, the slenderness factor for the column, $k = 1.0$

Since the column is at the ground floor story, the recommended M_1/M_2 ratio = 0, which yields a $k\ell_u/r$ limit calculated as follows from **ACI Eq. (6.2.5.1(b))**:

$k\ell_u/r \le 34 + 12\,(0) = 34$

Therefore $(1.0)(\ell_u)/0.288h \le 34$

Which yields $\ell_u/h \le 9.8$

With $\ell_u = 180$ inches, this gives a minimum column size of 18.4 inches for slenderness effects to be neglected.

Since the column size is 16 inches which is less than 18.4 inches, *slenderness effects must be considered* in the design of this column. We will now calculate the moment magnification factor, δ_{ns}.

Check Minimum Moment in the Column:

$M_{min} = P_u\,(0.6{+}0.03\text{h}) = 600$ kip [0.6 + 0.03(16 inch)]= 648 in-kip = 54 ft-kip

The applied factored moment, M_2 = 65 ft-kip > M_{min}, therefore, use M_2 = 65 ft-kip.

(Note: If M_{min} were greater than M_2, then we would have used $M_2 = M_{min}$)

Since $M_1/M_2 = 0$, using ACI Code Eq. 6.6.4.5.3a we calculate $C_m = 0.6 - 0.4(0) = 0.6$

The effective stiffness of the column, $(\text{EI})_{\text{eff}}$, is approximately $0.25\text{E}_\text{c}\text{I}_\text{g}$

$$(EI)_{eff} = 0.25E_cI_g = 0.25\left(57{,}000\sqrt{4000}\right)\frac{(16)(16)^3}{12}$$
$$= 4.9x10^9\ lb.in.^2 = 4.9x10^6\ kip.in.^2$$

The critical buckling load is calculated from ACI Eq. 6.6.4.4.2 as

$$P_c = \frac{\pi^2(EI)_{eff}}{(k\ell_u)^2} = \frac{\pi^2\left(4.9x10^6\right)}{([1.0][180])^2} = 1493\ kip$$

The moment magnification factor is calculated from ACI Eq. 6.6.4.5.2 as

$$\delta_{\text{ns}} = \frac{C_{\text{m}}}{1-\dfrac{P_{\text{u}}}{0.75P_{\text{c}}}} \ge 1.0$$
$$= \frac{0.6}{1-\dfrac{600\,\text{kip}}{0.75(1493\,\text{kip})}} = 1.3 \ge 1.0$$

$$\text{Use } \delta_{\text{ns}} = 1.3$$

Therefore, the magnified moment is calculated from ACI Eq. 6.6.4.5.1 as

$M_c = \delta_{ns}M_2$ = (1.3)(65 ft-kip) = 84.5 ft kip.

The 16-inch x 16-inch column is then designed for a factored axial load of 600 kip and a factored magnified moment of 84.5 ft-kip.

Assuming #3 ties and #8 bars,

γh = 16 – 2(1.5" cover) – 2(3/8" ties) – (½") - (½") = 11.25"

Therefore, γ = 11.25/16 = 0.7

Therefore, use the Column Interaction Diagram A.17 and calculate the following parameters:

required $K_n = P_u/\phi f_c'A_g$ = 600/[0.65(4)(16)(16)] = 0.9

required $R_n = M_u/\phi f_c'A_g h$ = (84.5)(12)/[0.65(4)(16)(16)(16)] = 0.095

Therefore, the required column reinforcement ratio, ρ_g = 0.024 > 0.01

Therefore, A_g required = 0.024(16)(16) = 6.14 in^2

Use 8 #8 rebar (A_g provided = 6.28 in^2 > A_s required)

The final column vertical reinforcement ratios are as follows:

- Non-lap splice regions: column reinforcement ratio, ρ_g = 6.28/256 = 0.025 < 0.08 O.K.
- Lap splice regions: column reinforcement ratio, ρ_g = 2(0.025) = 0.05 < 0.08 O.K.

Moment Magnification for Slender Columns in Unbraced or Sway or Moment Frames

For reinforced concrete columns in moment resisting frames where the lateral load is resisted by the bending of the beams/girders and columns, the moments at the end of each individual column will be a combination of the non-sway moments caused by the factored gravity loads (i.e., dead, snow, floor live loads) and the sway moments caused by the factored lateral loads (i.e., wind or seismic loads), considering the ultimate strength design load combinations . Thus

$$M_1 = M_{1ns} + \delta_s M_{1s} \quad \text{(ACI Eq. 6.6.4.6.1a)}$$

$$M_2 = M_{2ns} + \delta_s M_{2s} \quad \text{(ACI Eq. 6.6.4.6.1b)}$$

The sway moment magnification factor is given as

$$\delta_s = \frac{1}{1-\dfrac{\sum P_u}{0.75\sum P_c}} \geq 1.0 \quad [\textit{ACI Eq. } 6.6.4.5.2b]$$

Where the critical buckling load for each column, P_c, is given as

$$P_c = \frac{\pi^2 (EI)_{eff}}{(k\ell_u)^2} \quad [\textit{ACI Eq. } 6.6.4.4.2]$$

Where,

M_{1ns} = smaller factored end moment in the column due to gravity loads

M_{2ns} = larger factored end moment in the column due to gravity loads

M_{1s} = smaller factored end moment in the column due to lateral loads

M_{2s} = larger factored end moment in the column due to lateral loads

These moments are obtained from a first order elastic frame analysis.

ΣP_u = summation of all the factored vertical loads in a story

ΣP_c = summation of all the critical buckling load of all the columns in a story

k = the effective length factor (ACI 318, Section 6.6.4.4.3). For sway or moment frames, it is calculated using the Jackson-Moreland Alignment Charts given in ACI 318, Figure R6.2.5.1b. To calculate the ratio of the stiffnesses of the columns to the stiffnesses of the beams/girders at each joint, use a concrete modulus of elasticity, E_c , from ACI 318, Section 19.2.2, and moment of inertia of the columns and beams/girders, I, from ACI 318, Table 6.6.3.1.1a.

β_{dns} = ratio of maximum factored sustained axial load to the maximum factored axial load for the same load combination = 0

For moment or sway frames subjected to lateral wind or seismic loads, the sustained load factor, β_{dns} will be zero because of the non-sustained nature of the lateral loads. In this case, substituting β_{dns} = 0 in ACI Eq. 6.6.4.4.4a (ACI 318-19, Section R6.6.4.4.4) yields $(EI)_{eff} = 0.40E_cI_g$. However, for lateral loads caused by lateral earth pressure, β_{dns} will not be zero. An example is a building subjected to unbalanced lateral earth pressures due to a sloped grade.

The approximate method for obtaining the M_{1s} and M_{2s} moments (i.e., the sway moments) using a first order elastic analysis is a two-step procedure. The method accounts for material non-linearities by using the effective moment of inertias and cross-sectional areas of columns, walls, beams, flat slabs and flat plates from ACI Code Table 6.6.3.1.1a. **Step 1:** Carry out a structural analysis of the moment frame under all the applied loads and considering the applicable load combinations that includes gravity and lateral loads, and lateral translation is prevented at each floor and roof levels by assuming vertical rollers at the roof and floor levels. This first analysis gives the non-sway moments (i.e., M_{1ns} and M_{2ns}) and axial loads, and generates fictitious horizontal reactions at the vertical rollers. Note that all the loads on the columns that are not part of the moment frames (i.e., leaning columns) must also be included since the moment frames provide lateral stability to the leaning columns. **Step 2:** The fictitious horizontal restraints introduced in the first analysis are removed from the moment frame, and lateral forces equal and opposite to the horizontal reactions obtained at the fictitious vertical rollers in the first analysis are now applied to the structure in a second analysis. Note that these fictitious horizontal reactions are the only load applied in this second step. The moments obtained from this second analysis are the sway moments (i.e., M_{1s} and M_{2s}). The effective moment of inertias and cross-sectional areas from ACI Code Table 6.6.3.1.1a are also used in this second analysis.

Most reinforced concrete buildings are laterally braced by shear walls and thus, braced frame concrete structures are more common in practice. Therefore, the design of slender reinforced concrete columns in sway frames (i.e., buildings braced with moment frames) is not as common as the design of columns in buildings braced by shear walls (i.e., non-sway or braced frames). For more information on the design of slender columns in sway frames, the reader should refer to reference [3].

9.13 CONCRETE COLUMN SCHEDULE

The column reinforcement and sizes, and the factored axial loads, are usually presented in a tabular format in a concrete column schedule as shown in Figure 9.32. For each level of the building, the concrete column schedule shows the concrete strength, the factored axial load on the column, the column cross-sectional dimensions, the column vertical reinforcement, and the transverse reinforcement (i.e., ties or spirals).

Other information on the column schedule include the bottom elevation of the column as well as the size of the pier that transfers the column load to the foundation. Other pertinent information presented in the schedule are the column pier information and any vertical dowels. It is typical to see a note on the column schedule indicating that the vertical column dowels are to match the vertical rebar in the column. That implies, as indicated previously, that the total area of column reinforcement in the lap splice regions is twice that of the area of rebar in the non-splice regions. A typical multistory column detail showing the column lap splices and dowels is shown in Figure 9.33.

As previously discussed in chapter 5, there are three types of lap splices: Class B tension lap splice, Class A tension lap splice, and compression lap splice. The type of lap splice specified in a column will depend on where the combination of applied factored axial load and factored moment on the column falls on the column interaction diagram (see Figure 9.34). If all the column vertical bars are in compression, then a compression lap splice should be specified. If the tensile stress in the vertical bars on the tension side of the column exceed $50\% f_y$, then a class B tension lap splice should be specified for the column, otherwise, a class A tension lap splice should be specified.

COLUMN SCHEDULE

MARK	f'_c	A-1, A-6, E-1, E-6	A-2, A-3, A-4, A-5 E-2, E-3, E-4, E-5	B-1, C-1, D-1, B-6, C-6, D-6	B-2, B-3, B-4, B-5 C-2, C-3, C-4, C-5 D-2, D-3, D-4, D-5
NO. OF COLUMNS		4	8	6	12
ROOF ELEV. = 138'-6" FACT. AXIAL LOAD PLAN SIZE VERT. REINF TIES	4000psi	100K 18"×18" (8)-#9 #3@18	180K 22"×22" (12)-#9 #3@18	190K 22"×22" (12)-#9 #3@18	350K 24"×24" (12)-#11 #4@22
3RD FLOOR = 125'-6" FACT. AXIAL LOAD PLAN SIZE VERT. REINF TIES	4000psi	215K 18"×18" (8)-#9 #3@18	380K 22"×22" (12)-#9 #3@18	400K 22"×22" (12)-#9 #3@18	350K 24"×24" (12)-#11 #4@22
2ND FLOOR = 112'-10" FACT. AXIAL LOAD PLAN SIZE VERT. REINF TIES	4000psi	330K 18"×18" (8)-#9 #3@18	580K 22"×22" (12)-#9 #3@18	610K 22"×22" (12)-#9 #3@18	750K 24"×24" (12)-#11 #4@22
1ST FLOOR = 100'-0" TOP OF PIER OR FTG = 99'-4" PLAN SIZE VERT. REINF TIES TOP OF FTG = 96'-0"	4000psi	18"×18" (8)-#9 #3@18	22"×22" (12)-#9 #3@18	22"×22" (12)-#9 #3@18	NO PIER

Note: Dowels from the pier, footing, or pile cap to match column vertical reinforcement (Typical).

FIGURE 9.32 Concrete column schedule.

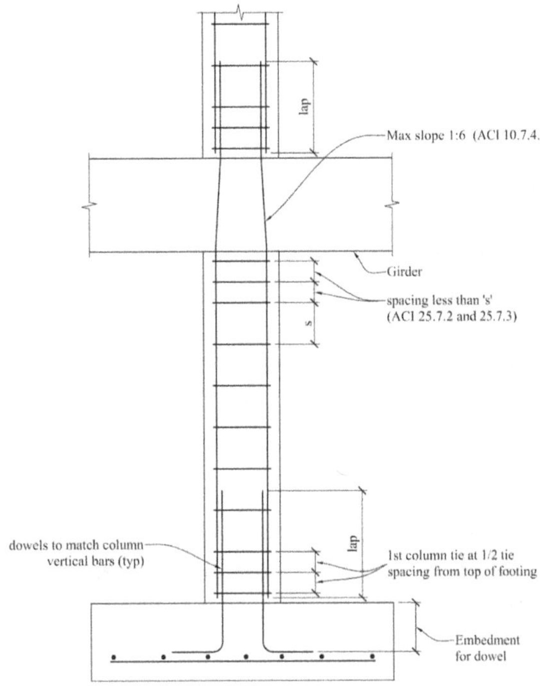

FIGURE 9.33 Multistory column detail showing column lap splices and dowels.

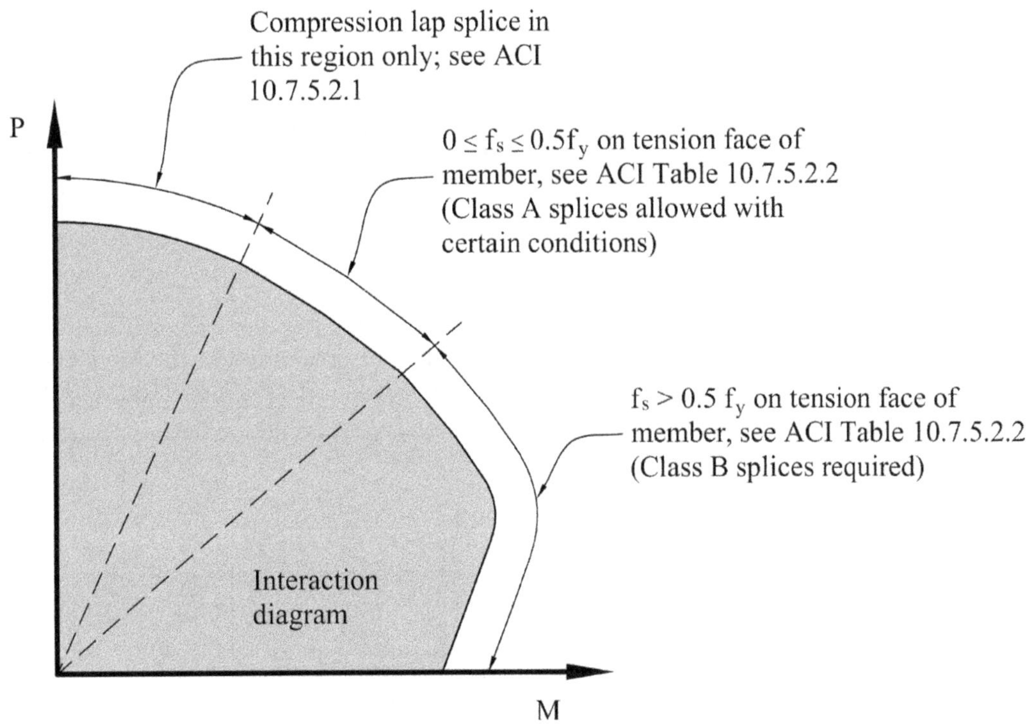

FIGURE 9.34 Lap splice requirement for columns (Adapted from ACI 318-19, Figure R10.7.5.2).

REFERENCES

[1] *ACI Design Handbook*, "Design of Structural Reinforced Concrete Elements in Accordance with Strength Design Method of ACI 318-19," Publication SP-17(14). American Concrete Institute, P.O. Box 9094, Farmington Hills, MI 48333.

[2] "Notes on ACI 318-02, Building Code Requirements for Structural Concrete, with Design Applications." Portland Cement Association, 5420 Old Orchard Road, Skokie, IL 60077-1083, 2002.

[3] "James K. Wight and James G. MacGregor, *Reinforced Concrete: Mechanics and Design*, 7th ed., New York: Pearson, 2015

[4] *ACI Committee 318*, "Building Code Requirements for Structural Concrete (ACI 318-19)," American Concrete Institute, P.O. Box 9094, Farmington Hills, MI 48333.

EXERCISES

9.1 Compute the maximum design axial load strength of the tied columns shown. Assume that the columns are short. Check the tie size and spacing. Use $f_c' = 4000$ psi and $f_y = 60{,}000$ psi.

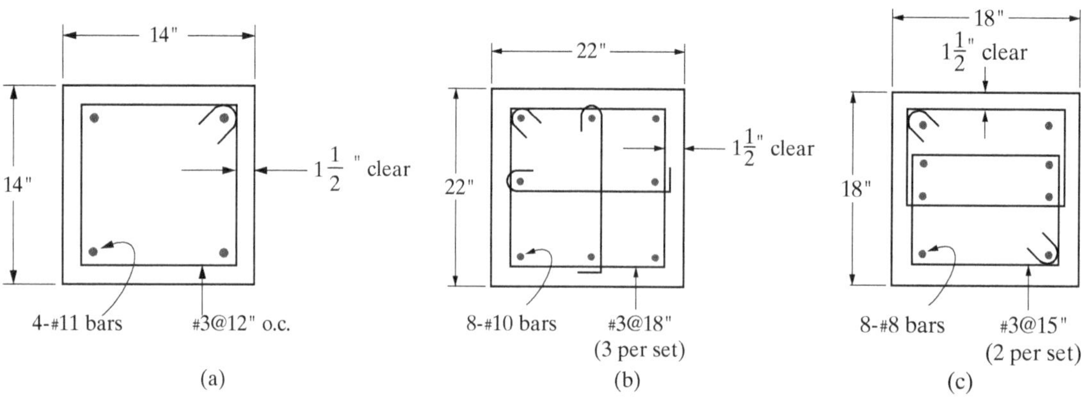

EXERCISE 9.1

9.2 Compute the maximum design axial load strength for the tied column shown. The column is short. Check the tie size and spacing. Use $f_c' = 4000$ psi and $f_y = 60{,}000$ psi.

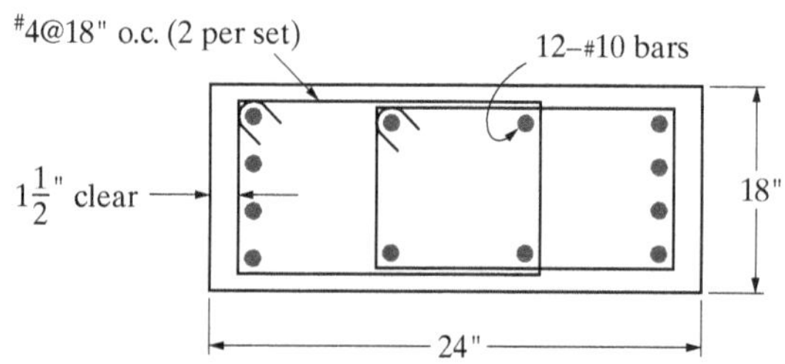

EXERCISE 9.2

9.3 Find the maximum axial compressive service loads that the column of cross-section shown can carry. The column is short. Assume that the service dead load and live load are equal. Check the ties. Use $f'_c = 4000$ psi and $f_y = 60{,}000$ psi.

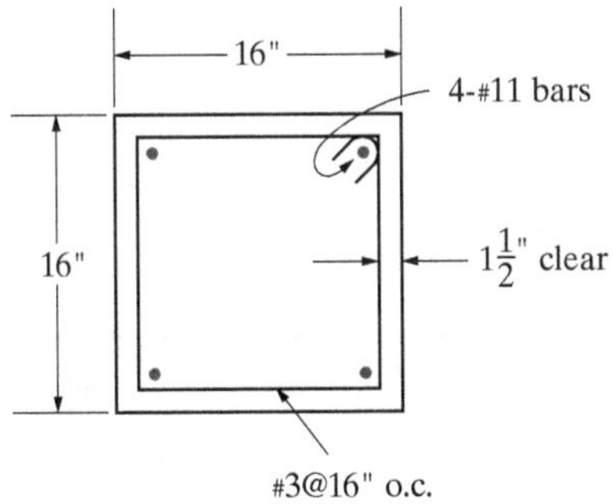

EXERCISE 9.3

9.4 A short, circular spiral column having a diameter of 18 in. is reinforced with eight No. 9 bars. The cover is $1\frac{1}{2}$ in., and the spiral is $\frac{3}{8}$ in. in diameter spaced 2 in. o.c. Find the maximum design axial load strength and check the spiral. Use $f'_c = 3000$ psi and $f_y = 40{,}000$ psi.

9.5 Same as Problem 9.4, but .. and $f_y = 60{,}000$ psi.

9.6 Compute the maximum axial compressive service live load that may be placed on the column shown. The column is short and is subjected to an axial service deadload of 200 kip. Check the ties. Use $f'_c = 3000$ psi and $f_y = 40{,}000$ psi.

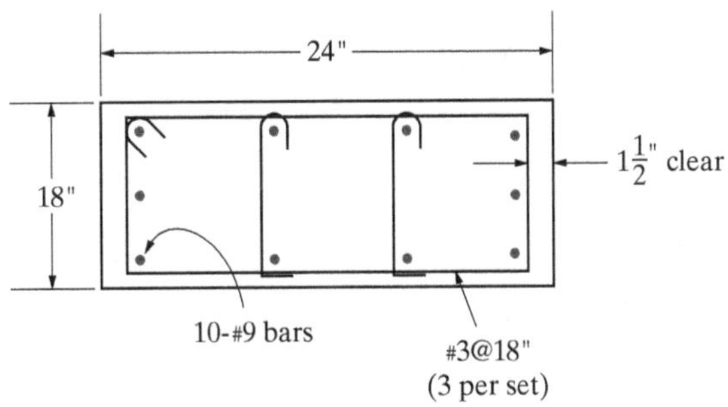

EXERCISE 9.6

9.7 Design a short, square tied column to carry a total factored design load P_u of 905 kip. Space and practical limitations require a column size of 18 in. × 18 in. Use $f'_c = 4000$ psi and $f_y = 60{,}000$ psi.

9.8 Design a short, square tied column for service loads of 205 kip dead load and 165 kip live load. Use ρ_g of about 0.04, $f'_c = 3000$ psi, and $f_y = 60{,}000$ psi. Assume that eccentricity is small.

9.9 Same as Exercise 9.8 but use $f'_c = 4000$ psi, $f_y = 60{,}000$ psi, and a ρ_g of about 0.03.

9.10 Design a short, circular spiral column for service loads of 175 kip dead load and 325 kip live load. Assume that the eccentricity is small. Use $f'_c = 4000$ psi and $f_y = 60{,}000$ psi. Make ρ_g about 3%.

9.11 For the short column of cross-section shown, find ϕP_n and eb at the balanced condition using basic principles. Use $f'_c = 4000$ psi and $f_y = 60{,}000$ psi. Bending is about the strong axis.

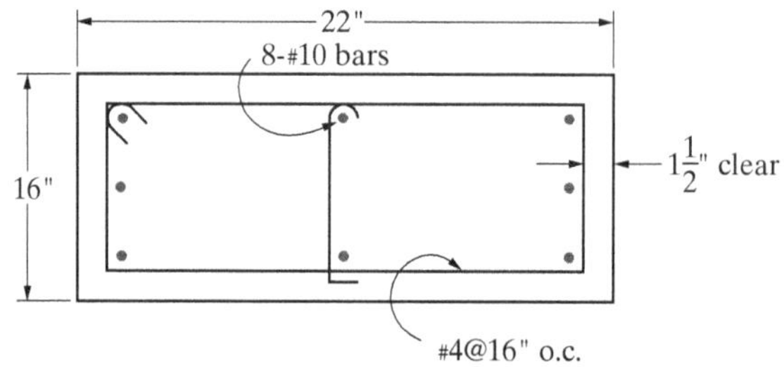

EXERCISE 9.11

9.12 Design a short square tied column to carry a factored axial design load P_u of 890 kip and a factored design moment M_u of 390 ft.-kip. Place the longitudinal reinforcing uniformly in the four faces. Use $f'_c = 4000$ psi and $f_y = 60{,}000$ psi.

9.13 Same as Exercise 9.12, but design a circular spiral column.

9.14 For the short tied column of cross-section shown, find the design axial load strength ϕP_n for an eccentricity of 14 in. Use $f'_c = 4000$ psi and $f_y = 60{,}000$ psi. Assume that the ties and bracket design are adequate.

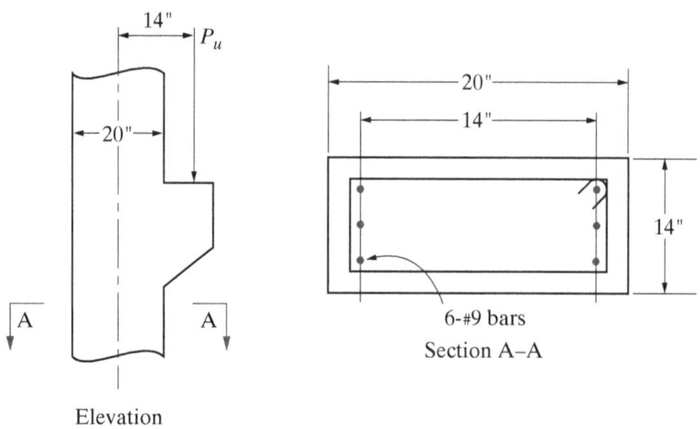

EXERCISE 9.14

9.15 Refer to the Column plan detail shown. Use $f'_c = 4{,}000$ psi and $f_y = 60{,}000$ psi

a. Does the tie arrangement and vertical reinforcing ratio meet the ACI code?

b. Without using the interaction diagram, determine ϕPn (kip) assuming minimum moments (i.e., $\phi M_n = 0.1\phi Pn$ h)

c. Confirm the answer from (b) using the interaction diagram; show calculations and draw appropriate lines on a copy of the appropriate interaction diagram to support your answer

d. Using the interaction diagram, determine the design bending strength, ϕM_n (k-ft) when the applied axial load is P_u = 325 k; show calculations and draw appropriate lines on the diagram to support your answer

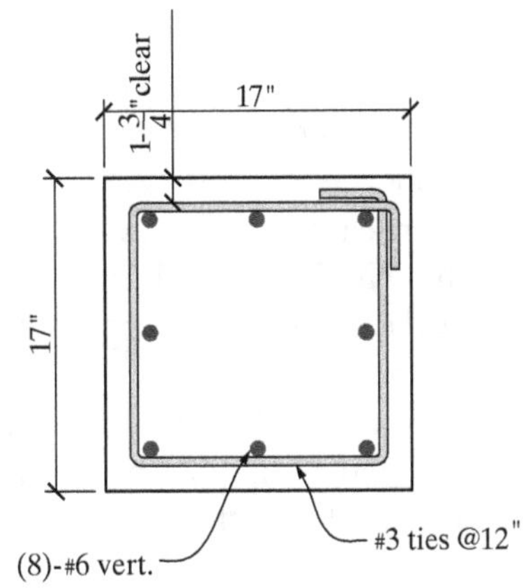

EXERCISE 9.15

9.16 Refer to the Column plan detail shown below. Use f'_c = 4,000 psi and f_y = 60,000 psi

a. Select a tie arrangement and spacing to meet ACI code.

b. Without using any interaction diagram, determine ϕP_n (kip) assuming minimum moments (i.e., $\phi M_n = 0.1\phi P_n$ h)

c. Confirm the answer from (b) using a copy of the appropriate interaction diagram; show calculations and draw appropriate lines on the diagram to support your answer

d. Using the same interaction diagram, determine the design axial strength, ϕP_n (k) when the applied factored moment is M_u = 200 k-ft; show calculations and draw appropriate lines on the diagram to support your answer

e. Identify the following regions on the diagram: ϕ = 0.65 and ϕ = 0.9

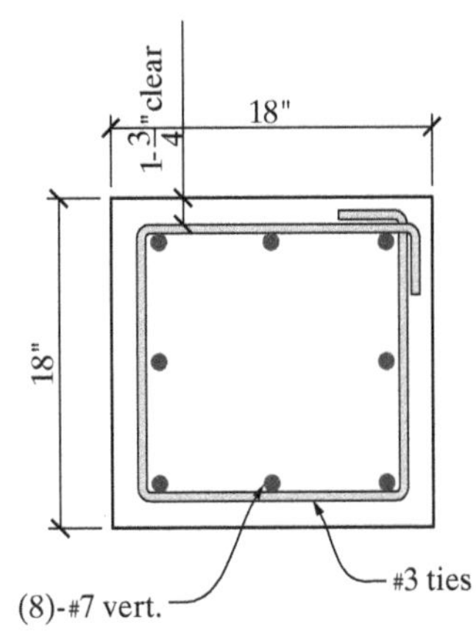

EXERCISE 9.16

9.17 A 20 inch x 20 inch ground floor interior column in a multi-story building braced with shear walls is supported on a spread footing and has an unbraced height of 20 ft. The column is subjected to a factored axial load of 650 kip and a moment about the X-X axis and the Y-Y axis of 60 ft-kip and 35 ft-kip, respectively. Determine if the column is slender about the X-X and Y-Y axis, and calculate the magnified or second order moment, M_u about the X-X and the Y-Y axis. Determine the required area of the column vertical reinforcement. Assume $f_c' = 4{,}000$ psi and $f_y = 60{,}000$ psi.

9.18 A spiral reinforced concrete column was originally designed as a 32 in. diameter column with 16 No. 7 vertical bars. However, the as-built column size turned out to be 24-in. diameter with 8 No. 7 vertical bars. Calculate the percentage reduction in the axial compression load capacity assuming small eccentricity. Discuss the implications of this construction error.

9.19 Determine the minimum practical size for a square reinforced concrete column to resist a factored axial load of 3,500 kip. Assume small eccentricity, f_c' of 15,000 psi and 2% column vertical rebar ratio.

CHAPTER 10

FOUNDATIONS

10.1 INTRODUCTION

The purpose of the structural portion of every building is to transmit applied loads safely from one part of the structure to another. The loads pass from their point of application into the *superstructure*, then to the *foundation*, and then into the underlying bearing strata. We have discussed the superstructure and foundation walls to some extent. The foundation is generally considered the entire lowermost supporting part of the structure. Normally, a *footing* or foundation is the last structural element through which the loads pass on their path to the underlying soil or bearing strata. A footing has as its function the requirement of spreading out the superimposed load so as not to exceed the safe capacity of the underlying material, usually soil, to which it delivers the load. Additionally, the design of footings must take into account certain practical and, at times, legal considerations with respect to property lines and rights-of-way. Our discussions will focus on spread footings, mat foundations, and deep foundations, as well as pile caps and deep beams.

The more common types of foundations may be categorized as follows:

1. *Individual column footings* (Figure 10.1a) are often termed *isolated spread footings* and are generally square. If space limitations exist, however, the footing may be rectangular in shape. Individual footings mainly support direct gravity load, but they are also often used to resist uplift forces, with the weight of the footing and possibly the soil on top of the footing acting as the resistance to uplift. Additional uplift resistance of spread footings can be achieved by placing the footing deeper into the soil (i.e., lowering the bottom elevation of the footing) or by increasing the footing plan size to engage more dead load from the soil on top of the footing.

   The resistance to sliding of spread footings from the base shear resulting from the lateral wind and seismic loads is provided by the friction between the underside of the footings and the underlying soil combined with the passive lateral resistance of the soil in front of each footing. Where the spread footing bears directly on the rock strata, more lateral

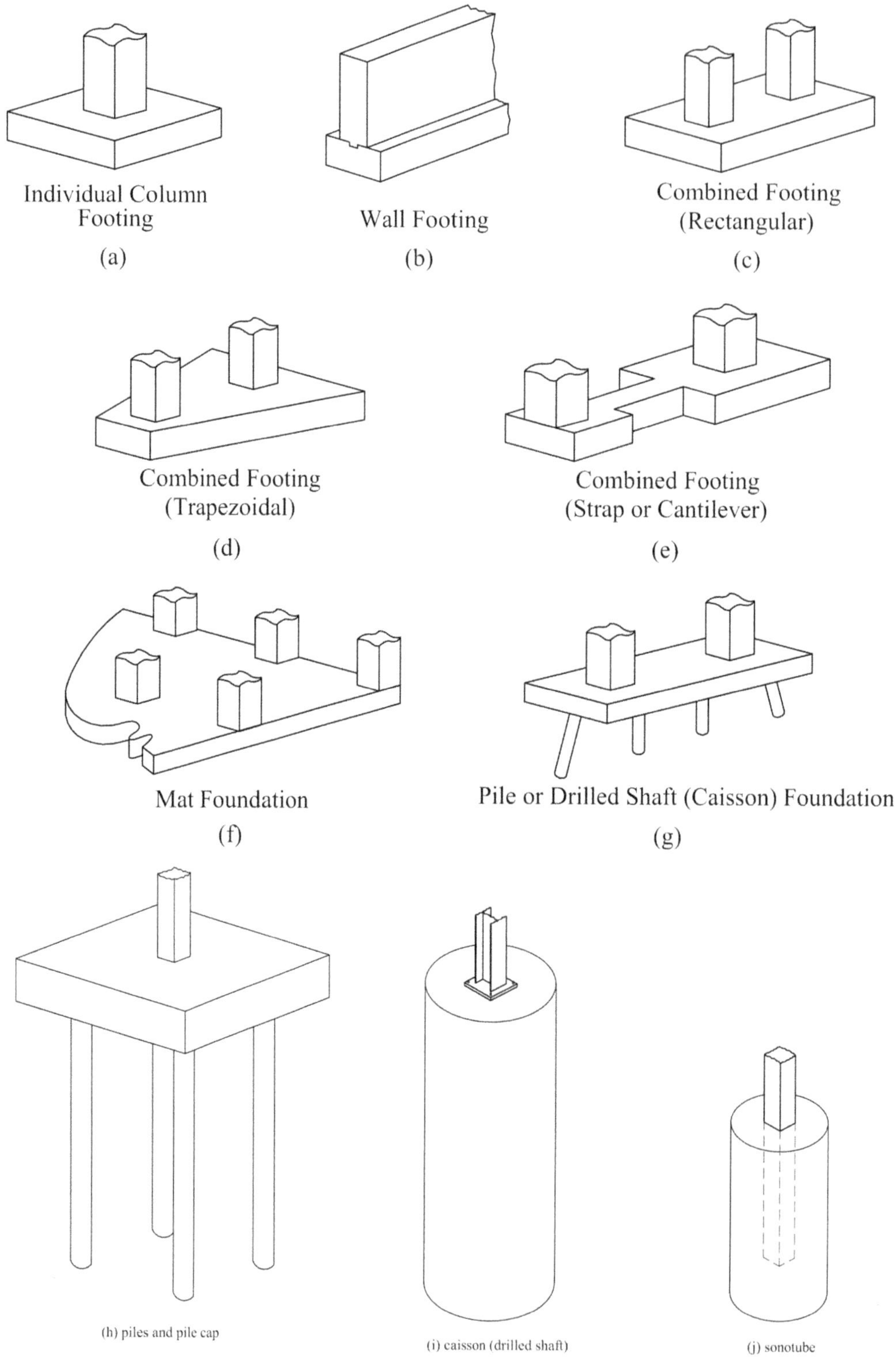

FIGURE 10.1 Foundation types.

resistance and uplift resistance of the footing can be achieved by anchoring the footings to the bedrock using rock anchors.

Overturning moment resistance at a shear wall can be provided by spread footings at each end of the shear wall, where the spread footings will be subjected to a force-couple made up of upward and downward acting loads at the ends of the shear wall due to the overturning moments.

2. *Wall footings* support walls that may be either load bearing on non-load bearing walls (Figure 10.1b).
3. *Combined footings* support two or more columns and may be either rectangular or trapezoidal in shape (Figure 10.1c and d).
4. If two isolated footings are joined by a *strap beam*, the footing is called a *cantilever or strap footing* (Figure 10.1e). Strap footings are used to prevent eccentric loading of individual column foundations by "strapping" an edge foundation - that will otherwise be eccentrically loaded - to a concentrically loaded interior foundation (see Figure 10.1e).
5. *Mat foundations* (also sometimes called raft foundations) are large, continuous footings that support all columns and walls of a structure. They are commonly used where undesirable soil conditions or low allowable soil pressures prevail (Figure 10.1f) for great depths and column loads are high, resulting in overlapping of individual column spread footing areas. The mat foundation can be supported directly on the soil if the soil bearing capacity is strong enough and will not result in excessive settlement; where the soil is very weak or soft, which could result in excessive direct and differential settlements, the mat foundation can be supported on piles or caissons (i.e., bored reinforced concrete piles) that are in turn supported on the hard bedrock strata that could be several hundred feet beneath the structure or supported through friction between the piles and the surrounding soil. In this piled mat foundation, the mat acts like a two-way pile cap to support the column loads and in turn transmit the loads to the piles or caissons below. In piled mat foundations, load sharing between the mat foundation and the piles or caisson foundations is often considered if the soil below the mat foundation has adequate bearing capacity [5]. Mat foundations are also used for moderate-height buildings with several basement levels where the foundation level is close to bedrock. For supertall buildings, a mat foundation by itself is usually not sufficient to develop the lateral and overturning moment resistance necessary to support the high lateral loads [5]. If a "compensated" or "floating" mat foundation is desired, the required excavation can be rather deep, and the excavated area serves as the multi-level basement area. In the "compensated" mat foundation, an amount of soil approximately equal to the weight of the building is excavated, resulting in a net soil pressure (after the building is constructed) at the underside of the mat foundation that will be theoretically equal to the applied soil pressure which originally existed in the native soil at the elevation of the underside of the mat before the soil was excavated. Thus, the weight of the excavated soil is approximately equal to the total dead weight of the building, and the uniform and differential settlement are reduced compared to a similar foundation located at the ground surface. The "floating" mat foundation is recommended for soils susceptible to excessive settlement and to heaving during excavation (such as soft clay and loose sand), and where the bedrock is very deep below the ground surface[5].

For mat foundations in areas with a high-water table, the mat foundation will be subjected to hydrostatic pressures (i.e., upward forces from the water pressure). Where the mat foundation sits directly on bed rock, rock anchors placed at regular intervals in both orthogonal directions will be required to tie the mat down to the bed rock and thus resist the upward hydrostatic forces. Where the mat is supported on piles or caissons, the connection between the mat foundation and the piles or caissons will be subjected to upward hydrostatic forces. The dowels between the caissons and the mat foundation must be detailed accordingly to resist these upward forces.

6. *Deep foundations* consist of piles, caissons or drilled shafts, and pile caps. The piles could be made of steel shapes (H-piles or steel tubes filled with concrete) or formed concrete piles (caissons).

   Pile caps *serve to transmit column loads to a group of piles or caissons, which will in turn transmit the loads to* the supporting soil through skin friction with the surrounding soil (i.e., friction piles) or through direct bearing on the underlying bedrock (Figures 10.1g and h).

   In expansive soils or where frost protection is required along the perimeter of a structure, void forms are typically required underneath the entire framed slab, the grade beams and the pile or caisson caps to prevent the transfer of upward forces from the soil into the structure. Slabs-on-grade should be kept free and isolated from the grade beams. Only framed structural slabs should be monolithic with, or connected to, grade beams.

   The minimum frost protection depth which is the minimum depth from the exterior grade elevation to the underside of the exterior pile cap or caisson cap or grade beam (or footing in the case of shallow foundations) is typically a minimum of about 3 ft in the eastern United States to higher than 10 ft in northern Alberta [23].

   Where the piles or caissons extend down to the bedrock strata, they are usually socketed several feet into the bedrock. For building sites with very deep bedrock-bearing strata, friction piles that rely on the skin friction between the piles and the surrounding soil may be used. Friction piles should be used with great care since they can be susceptible to excessive uniform and differential settlement, as appears to have been the case with San Francisco's 58-story, 650-foot-tall Millennium Tower. All high-rise buildings developed in San Francisco since 2009, after the problems with the Millenium Tower surfaced, have used piles bearing directly on the bedrock strata [22].

   Solid caissons could also be used without the use of a pile cap where the building column bears directly on the foundation (Figure 10.1i). The sizes of caissons can vary from 30 inches up to more than 120 inches in diameter. The foundations for shear walls in midrise and high-rise buildings usually consist of mat or pile or caisson foundations; spread footings are not used for shear wall foundations in such buildings since they are incapable of resisting the overturning moments on the shear wall without excessive rotation of the footing [1]. Uplift forces can be resisted by socketing the caisson several feet into the bedrock, the depth of which is typically provided by the geotechnical engineer in the soil report. The axial load-bearing capacity of caissons is also enhanced by socketing the caisson or drilled shaft into the bedrock.

   Lateral forces on piles can be resisted by "battering" or placing the piles at an angle. The lateral capacity for this condition would also be contained in the geotechnical engineering report. Deep foundations (i.e., caissons or piles) can also resist the base shear from lateral forces by acting as vertical cantilevers. Typically, the geotechnical consultant will provide in their soil report the level below grade at which the fixity of a caisson or pile group may be assumed. For instances where this information is not provided, the IBC 2021, Section 1810.2.1 (Lateral Support) provides some guidance as follows: "Where deep foundation elements stand unbraced in air, water, or fluid soils, it shall be permitted to consider them laterally supported at a point 5 ft. into stiff soil or 10 ft. into soft soil unless otherwise approved by the building official on the basis of a geotechnical investigation by a registered design professional." Rotational fixity at the top of the piles or caissons can be achieved by extending them deep into the pile cap or caisson cap well beyond the minimum embedment of 4 in. to 6 in. Where lateral support to the piles (or caissons) from the soil cannot be assured (e.g., in soils subject to liquifaction), the piles and the pile caps/grade beams will resist the base shear as moment resisting frames with rotational fixity at the base of the piles (since the piles are typically driven several feet into the bedrock) and moment or rigid connections at the pile cap-to-pile interface at the top of the piles.

7. Lightly loaded foundations (Figure 10.1j) are used mainly in residential construction where wood posts bear directly on them, or they are embedded into the concrete. In the case where the post is embedded into the concrete, this is often done to allow for lateral stability. These are sometimes referred to as "sonotube" type foundations as they are formed with a fiber, plastic, or similar composite material that is in the shape of a hollow tube. These foundations are commonly used for decks, pole-type buildings, and small signs and flagpoles.

10.2 THE GEOTECHNICAL REPORT

For most structures, a site exploration is carried out by a licensed geotechnical engineer who then prepares a Geotechnical report that summarizes their findings. A typical geotechnical report for a project includes the following [2]:

- The type of soil and bedrock on the site on which the structure will be built.
- The soil borings and bedrock elevation, and the elevation of the water table.
- The recommended foundation types and their net allowable bearing pressure in psf, and the pros and cons for each foundation type including estimates of the magnitude of uniform and differential settlements.
- Allowable axial (compression and tension or uplift) and lateral load capacities for steel piles and pile groups, and/or drilled concrete piles (caissons) if deep foundations are to be used, and the minimum spacing between piles within a pile group.
- The net allowable bearing pressure of the soil and the bedrock (if the foundations are to be founded on the bedrock), and the minimum depth at which interior and perimeter footings should be found. For footings bearing on rock, the allowable bond stress for any rock anchors.
- The site classification and design properties for seismic design.

- If corrosive soils or soils with sulfates are present on the site, this information should be included in the report, as this would determine the type of cement that needs to be used for the foundation elements of the structure. In this situation, a recommendation for the use of epoxy-coated rebar to prevent corrosion of the rebar will also be included in the report.
- Lateral earth pressure coefficients (active, at-rest, and passive pressure coefficients), the density of the soil, the coefficient of sliding friction between footing and soil, and the elevation of the groundwater table for the design of basement and retaining walls.
- The modulus of subgrade reaction for the design of the slab-on-grade. The geotechnical report will also include recommendations as to whether a vapor barrier is needed or not, and if needed, what type of vapor barrier should be used.
- The compaction and density requirements (i.e., whether modified or standard proctor density) for the native soil below the slab-on-grade and for all fill and backfill behind the basement and retaining walls will also be included in the geotechnical report. The soil properties needed to achieve the recommended design values should also be provided. These would include the soil type (e.g., gravel), gradation, and optimum moisture content.

Note that the allowable soil bearing pressure provided by the geotechnical engineer usually implies a factor of safety of up to 2.5 to 3.0 [3, 4]; therefore, the required area of the footing or foundation is calculated using unfactored column loads, whereas the strength design of the foundation is carried out using the factored column loads. Typical values for allowable soil bearing pressures and other soil properties used for structural design are presented in Table 10.1. When a geotechnical report is not available, it is common to assume the lowest code-specified value for bearing capacity of 1,500 psf. The soil conditions and bearing capacity would have to be verified in any case during construction.

TABLE 10.1 Typical Soil Properties for Structural Design (Adapted from Table 1806.2 of the IBC)

Type of soil	Allowable soil bearing pressure (psf)	Lateral bearing pressure (psf/ft below natural grade)	Lateral Sliding Resistance	
			Coefficient of friction[1]	Cohesion (psf)[2]
Crystalline bedrock	12,000	1,200	0.70	-
Sedimentary and foliated rock	4000	400	0.35	-
Gravel and/or Sandy gravel	3000	200	0.35	-
Clay, Sandy clay, Silty clay, Clayey silt, Silt, Sandy silt	1500	100	-	130

Data from International Building Code (IBC) Table 1806.2

[1] Friction coefficient to be multiplied by dead load

[2] Multiply cohesion value by the contact area per the limitations in IBC Section 1806.3.2.

FIGURE 10.2 Column and wall footings. The Health, Physical Education and Recreation Complex, Hudson Valley Community College, Troy, New York. (George Limbrunner)

10.3 WALL FOOTINGS

Wall footings (see Figure 10.2) are commonly required to support direct concentric loads. An exception to this is the footing for a retaining wall. A wall footing may be of either plain or reinforced concrete. Because it has bending in only one direction, it is generally designed in much the same manner as a one-way slab, by considering a typical 12-in.-wide strip along the length of wall. Footings supporting relatively light loads on well-drained cohesionless soil are often made of plain concrete—that is, concrete without reinforcing. This material is referred to as *structural plain concrete*.

A wall footing under concentric load behaves similarly to a cantilever beam, where the cantilever extends out from the wall and is loaded in an upward direction by the soil pressure. The flexural tensile stresses induced in the bottom of the footing are acceptable for a plain concrete footing.

From the ACI Code, Section 14.5.2, the nominal flexural design strength of a plain concrete cross section is calculated from

$$M_n = 5\lambda\sqrt{f'_c}S_m$$

if tension controls, and

$$M_n = 0.85 f'_c S_m$$

if compression controls. The quantity S_m is the elastic section modulus of the section. Lamda (λ) is the modification factor reflecting the lower tensile strength of lightweight concrete relative to normal-weight concrete and is described in Chapter 1. For normal-weight concrete, $\lambda = 1.0$. In this chapter, we will consider only normal-weight concrete as lightweight concrete would not likely ever be used for a foundation element; therefore, λ will be omitted in the examples. These formulas are based on the flexure formula,

$$f_b = \frac{Mc}{I}$$

rewritten

$$M = f_b S$$

where $5\lambda\sqrt{f'_c}$ and $0.85f'_c$ are limiting stress f_b values in tension and compression, respectively. Similarly, nominal shear strength for beam action for a plain concrete member is calculated from

$$V_n = \frac{4}{3}\lambda\sqrt{f'_c}\text{bh}$$

This is based on the general shear formula for a rectangular section:

$$f_v = \frac{3V}{2\text{bh}}$$

which is rewritten

$$V = \frac{2}{3} f_v \text{bh}$$

where the limiting shear stress value is $2\lambda\sqrt{f'_c}$, a familiar value from our previous discussions of shear in flexural members and $\lambda = 1.0$ for normal-weight concrete.

In each case, the basis for the design must be

$$\phi M_n \geq M_u \quad \text{and} \; \phi V_n \geq V_u$$

as applicable. The strength-reduction factor for plain concrete is $\phi = 0.60$ (ACI Code Section 21.2.1).

In a *reinforced* concrete wall footing, the behavior is identical to that just described. Reinforcing steel is placed in the bottom of the footing in a direction perpendicular to the longitudinal axis of the wall, thereby resisting the induced flexural tension, similar to a reinforced concrete beam or slab.

In either case, the cantilever action is based on the maximum bending moment occurring at the face of the wall if the footing supports a concrete wall or at a point halfway between the middle of the wall and the face of the wall if the footing supports a masonry wall. This difference is primarily because a masonry wall is somewhat less rigid than a concrete wall.

For each type of wall, the critical section for shear in the footing may be taken at a distance from the face of the wall equal to the effective depth of the footing.

EXAMPLE 10.1

Design a plain normal-weight concrete wall footing to carry a 12-in. concrete block masonry wall, as shown in Figure 10.3. The service loading may be taken as 10 kip/ft dead load (which includes the weight of the wall) and 20 kip/ft live load. Use $f_c' = 3{,}000$ psi. The gross allowable soil pressure is 5000 psf (5.0 ksf), and the weight of earth or soil, $w_e = 100$ lb/ft³.

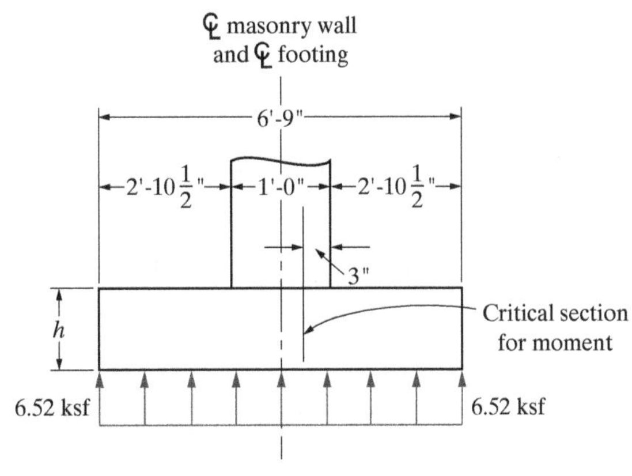

FIGURE 10.3

Solution:

1. Compute the factored load:

$$\begin{aligned} w_u &= 1.2w_{\text{DL}} + 1.6w_{\text{LL}} \\ &= 1.2(10) + 1.6(20) \\ &= 44 \text{ kip/ft} \end{aligned}$$

2. Assume a footing thickness of 3 ft-0 in. This will be checked later in the design. The footing weight is

$$0.150\ (3) = 0.450 \text{ ksf}$$

Assuming the bottom of the footing to be 4 ft-0 in. below the finished ground line, the weight of the 1-ft. height of soil on top of the footing is

$$(1)(100) = (100 \text{ psf}) = 0.100 \text{ ksf}$$

Therefore, the net allowable soil pressure for superimposed service loads is

$$5.00 - 0.45 - 0.100 = 4.45 \text{ ksf}$$

Sometimes, the geotechnical engineering report for a project may specify the net allowable soil pressure directly, thus obviating the need for the calculations in this step.

3. The maximum allowable soil pressure for strength design must now be found. It must be modified in a manner consistent with the modification of the service loads. This may be accomplished by multiplying the ratio of the total design load (44 kip) to the total service load (30 kip). We then obtain a maximum allowable soil pressure solely for use in the strength design of the footing. This soil pressure should not be construed as an actual allowable soil pressure. Thus, the factored soil pressure is obtained using the ratio of factored to service loads.

$$4.45\left(\frac{44.0}{10+20}\right) = 6.53 \text{ ksf}$$

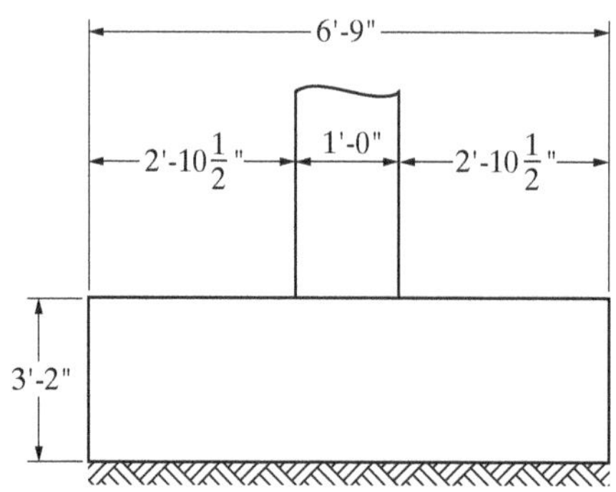

FIGURE 10.4. Detail of plain concrete wall footing for Example 10.1.

4. We may now determine the required footing width:

$$\frac{44.0}{6.53} = 6.74 \text{ ft}$$

Use 6 ft-9 in.

5. Determine the factored soil pressure to be used for the footing design if the footing width is 6 ft-9 in.:

$$\frac{44.0}{6.75} = 6.52 \text{ ksf} < 6.53 \text{ ksf}, \qquad \text{O.K.}$$

6. With the factored soil pressure known, the bending moment in the footing may be calculated. For concrete block masonry walls, the critical section for moment should be taken at the quarter-point of wall thickness (ACI Code, Section 13.2.7.1).

   With reference to Figure 10.3, the factored moment is determined as follows:

$$M_u = \frac{6.52(3.125)^2}{2} = 31.8 \text{ ft-kip}$$

7. Find the required footing thickness based on the required moment strength. Assuming that tension controls:

$$\phi M_n = \phi 5\sqrt{f'_c} S_m = \phi 5\sqrt{f'_c}\frac{bh^2}{6}$$

   Setting $\phi M_n = M_u$ and considering a typical 12-in.-wide strip:

$$\phi 5\sqrt{f'_c}\frac{(12)(h^2)}{6} = 31.8 \text{ ft-kip}$$

   from which

$$required\, h = \sqrt{\frac{(31.8 \text{ ft-kip})(12 \text{ in/ft})(1000 \text{ lb/kip})(6)}{(0.60)(5)\sqrt{3000}(12 \text{ in})}} = 34.1 \text{ in.}$$

8. It is common practice to assume that the bottom 1 or 2 in. of concrete placed against the ground may be of poor quality and, therefore, may be neglected for strength purposes. The total required footing thickness may then be determined:

$$34.1 + 2.0 = 36.1 \text{ in}$$

We will use $h = 38$ in. This checks closely with the assumed thickness of 36 in. No revision of the calculations is warranted.

9. Shear is generally of little significance in plain concrete footings because of the large concrete footing thickness. With reference to Figure 10.4, if the critical section for shear is considered a distance equal to the depth of the member h from the face of the wall and if the depth is taken as 34.1 in. (the required h), it may be observed that the critical section is outside the edge of the footing. Therefore, the shear check may be neglected.

10. It is common practice to use some *longitudinal* steel in continuous wall footings whether or not *transverse* steel is present. This will somewhat enhance the structural integrity by limiting differential movement between parts of the footing should transverse cracking occur. It will also lend some flexural strength in the longitudinal direction. The rationale for this stems from the many uncertainties that exist in both the supporting soil and the applied loads.

As a guide, this longitudinal steel may be computed in a manner similar to that for temperature and shrinkage steel in a one-way slab, where

$$\textit{required}\ A_s = 0.0018\,bh$$

The total longitudinal steel required for the 6-ft-9-in. footing width is

$$A_s = 0.0018\ (6.75)\ (12\ \text{in./ft})(\ 38\ \text{in.}) = 5.54\ \text{in.}^2$$

Use 13 No. 6 bars ($A_s = 5.72$ in.^{2}). This would satisfy any longitudinal steel requirement. It is obvious, however, that this large steel requirement does not lend itself to an economical footing design, but rather leads us to conclude that we should question the use of plain concrete wall footings when superimposed loads are heavy.

A redesign using a reinforced concrete wall footing is accomplished in the following example.

EXAMPLE 10.2

Design a normal-weight reinforced concrete wall footing to carry a 12-in. concrete block masonry wall, as shown in Figure 10.5. The service loading is 10 kip/ft dead load (which includes the weight of the wall) and 20 kip/ft live load. Use $f'_c = 3000$ psi, $f_y = 60{,}000$ psi, weight of earth = 100 lb/ft^3, and gross allowable soil pressure = 5000 psf (5.0 ksf). The bottom of the footing is to be 4 ft-0 in. below the finished ground line.

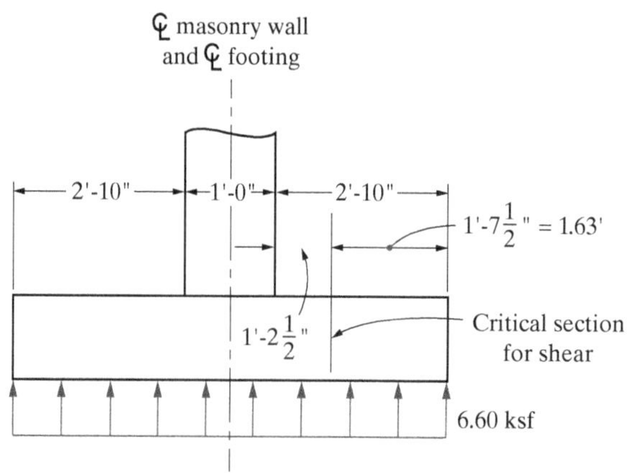

FIGURE 10.5 Reinforced concrete wall footing for Example 10.2 – factored soil pressure and critical section for shear.

Solution:

1. Compute the factored load:

$$w_u = 1.2w_{DL} + 1.6w_{LL}$$

$$= 1.2(10) + 1.6\,(20)$$

$$= 44 \text{ kip/ft}$$

2. Assume a total footing thickness = 18 in. Therefore, the footing weight is

$$0.150(1.5\text{ ft}) = 0.225\text{ ksf}$$

Because the bottom of the footing is to be 4 ft below the finished ground line, there will be 30 in. of earth or soil on top of the footing. This depth of earth has a weight of

$$\frac{30(100)}{12} = 250\text{ psf} = 0.250\text{ ksf}$$

The net allowable soil pressure for superimposed service loads is

$$5.00 - 0.225 - 0.250 = 4.53\text{ ksf}$$

3. As in Example 10.1, we use the ratio of factored load to service load to determine the soil pressure for strength design:

$$\frac{44(4.53)}{30} = 6.64\text{ ksf}$$

4. The required footing width is

$$\frac{44.0}{6.64} = 6.63\text{ ft}$$

Use 6 ft-8 in.

5. The factored soil pressure to be used for the footing design is

$$\frac{44.0}{6.67} = 6.60\text{ ksf}$$

6. The assumed effective depth d for the footing is determined by subtracting the concrete cover [see the ACI Code, Table 20.5.13.2] and one-half of the bar diameter (No. 8 assumed) from the total thickness:

$$d = 18 - 3 - 0.5 = 14.5\text{ in}$$

7. Because required thicknesses of reinforced concrete footings are generally controlled by shear requirements, the shear should be checked first.

With reference to Figure 10.5, as the wall footing carries shear in a manner similar to that of a one-way slab or beam, the critical section for shear will be taken at a distance equal to the effective depth of the footing (14.5 in.) from the face of the wall (ACI Code, Sections 7.4.3.2 or 9.4.3.2):

$$V_u = 1.63 \text{ ft.}(1 \text{ ft.})(6.60 \text{ ksf}) = 10.75 \text{ kip per foot of wall}$$

The total nominal shear strength V_n is the sum of the shear strength of the concrete V_c and the shear strength of any shear reinforcing V_s.

$$V_n = V_c + V_s$$
$$\phi V_n = \phi V_c + \phi V_s$$

Assuming no shear reinforcing,

$$\phi V_n = \phi V_c$$

In footings, shear reinforcing is not required if

$$\phi V_c > V_u$$

Computing ϕV_c,

$$\phi V_c = \phi 2\sqrt{f'_c}\, bd$$
$$= (0.75)(2)\sqrt{3000}(12)(14.5)$$
$$= 14.30 \text{ kip/ft of wall}$$
$$14.30 \text{ kip/ft} > 10.75 \text{ kip/ft}$$

Therefore,

$$\phi V_c > V_u$$

Thus, the assumed thickness of footing is satisfactory for shear, and no revisions are necessary with respect to footing weight.

8. With reference to Figure 10.6, the critical section for the moment is taken at the quarter-point of wall thickness (ACI Code, Section 13.2.7.1). The maximum factored moment, assuming the footing to be a cantilever beam, is

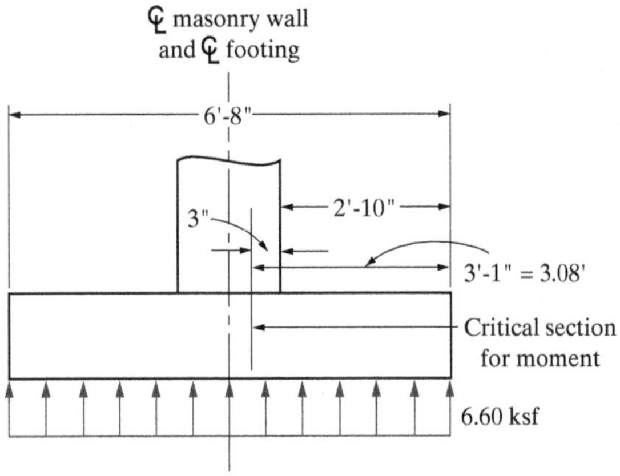

FIGURE 10.6 Reinforced concrete wall footing for Example 10.2 – factored soil pressure and critical section for moment.

$$M_u = \frac{6.60(3.08)^2}{2} = 31.3 \text{ ft-kip}$$

9. The required area of tension steel is then determined in the normal way (see Chapter 2), using $d = 14.5$ in. and $b = 12$ in., and assuming $\phi = 0.90$:

$$\textit{required } \bar{k} = \frac{Mu}{\phi bd^2} = \frac{(31.3)(12)}{(0.9)(12)(14.5)^2}$$

$$= 0.1654 \text{ ksi}$$

From Table A.8, the required $\rho = 0.0029$, and $\varepsilon_t > 0.005$; therefore, $\phi = 0.90$ (O.K.).

$$\text{required } A_s = \rho bd$$

$$= 0.0029(12)(14.5)$$

$$= 0.50 \text{ in.}^2\text{/ft of wall}$$

We will use the ACI Code minimum reinforcement requirement for beams as being applicable for footings. From Table A.5,

$$A_{s,\,\min} = 0.0033(12)(14.5)$$

$$= 0.57 \text{ in.}^2\text{/ft of wall}$$

Finally, as discussed in Chapter 8, the provisions of the ACI Code, Section 7.6.1.1, for minimum reinforcement in structural slabs of uniform thickness may be considered applicable for footings such as this that transmit vertical loads to the underlying soil. In this text, this will be used only as an absolute minimum. Checking the provision of the code for the minimum reinforcement required for grade 60 steel gives us

$$\text{required } A_s = 0.0018 \text{ bh} = 0.0018(12)(18)$$

$$= 0.39 \text{ in.}^2 \text{ f/ft of wall}$$

Therefore, considering the minimum reinforcement ratio for beams as applicable, use required $A_s = 0.57$ in.^{2} per ft of wall and use No. 6 bars at 9 in. o.c. ($A_s = 0.59$ in.^{2}).

10. The development length should be checked for the bars selected. Assume uncoated bars. This calculation for ℓ_d follows the eight-step procedure presented in Chapter 5, Section 5.2, and is summarized as follows:

1. $K_D = 82.2$

2. $\psi_g = 1.0,\ \psi_e = 1.0,\ \psi_s = 1.0,\ \psi_t = 1.0,\ \lambda = 1.0$

3. $\psi_t \times \psi_e = 1.0 < 1.7$ (O.K.)

4. $c_b = 3.38$ in.

5. $K_{tr} = 0$

6. $(c_b + K_{tr})/d_b = 4.51$ Use 2.5

7. $K_{ER} = (0.57 \text{ in.}^2)/(0.59 \text{ in.}^2) = 0.97$

8. $\ell_d = 19.1 \text{ in.} > 12 \text{ in.}$ (O.K.)

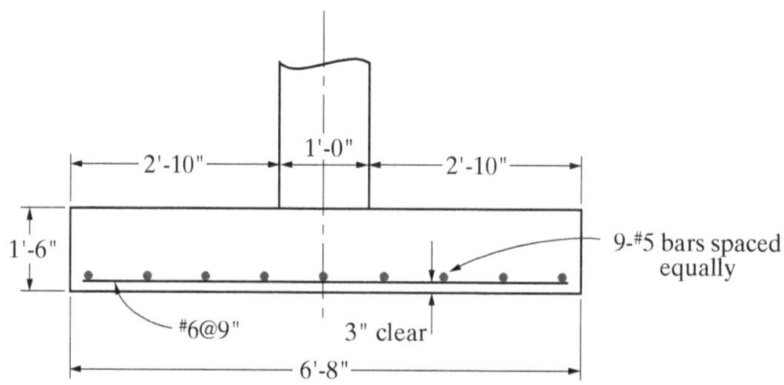

FIGURE 10.7 Design sketch for Example 10.2.

The development length provided, measured from the critical section for moment and allowing for 3-in. end cover, is 34 in. (i.e., 37 in. - 3 in.). Because 34 in. > 19.1 in., the development length provided is adequate.

11. Although not specifically required in footings by the ACI Code, longitudinal steel will be provided on the same basis as for one-way slabs (Section 24.4.1). Thus,

$$\text{required } A_s = 0.0018\,bh = 0.0018(66.7\text{ft.})(12)(18)$$
$$= 2.59 \text{ in.}^2/\text{ft of wall}$$

Use nine No. 5 bars ($A_s = 2.79$ in.^{2}) spaced equally. The footing design is shown in Figure 10.7.

10.4 WALL FOOTINGS UNDER LIGHT LOADS

A relatively common situation is one in which a lightly loaded wall is supported on average soil. As previously indicated in this chapter, a design would result in a very small footing thickness and width. In this case, it would not be necessary to provide reinforcing as the footing would be designed using the provisions for plain concrete. However, it is good practice to provide longitudinal reinforcement for temperature and shrinkage (0.0018bh).

In such a situation, experience has shown that for footings carrying plain concrete or block masonry walls, the minimum recommended thickness is 6 to 8 inches per the IBC. In order to minimize the bending stress and thus the need for transverse reinforcement in the bottom of the footing, the footing thickness, T, should be at least the thickness of the wall, and the footing width should be at least 2T. A common footing for commercial and residential wall footings is shown in Figure 10.8.

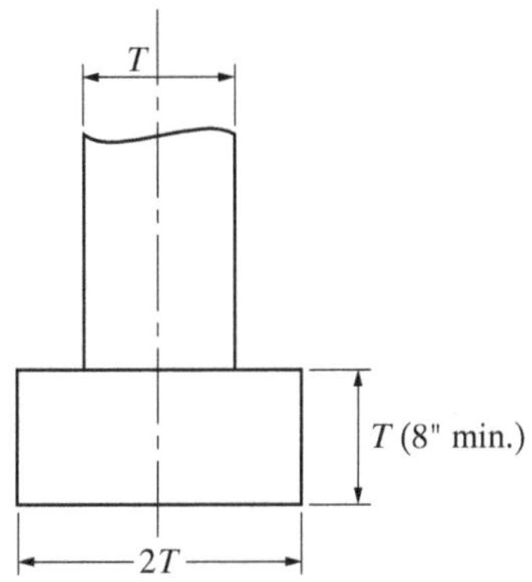

FIGURE 10.8 Recommended minimum footing dimensions for walls carrying light loads.

10.5 INDIVIDUAL REINFORCED CONCRETE FOOTINGS FOR COLUMNS

An individual reinforced concrete footing for a column, also termed an *isolated spread footing*, is probably the most common, simplest, and most economical of the various types of footings used for structures (see Figure 10.2). Individual column footings are generally square in plan. Rectangular shapes are sometimes used where dimensional limitations exist, however. The footing is a slab that directly supports a column. At times, a pedestal is placed between a column and a footing so that the base of the column need not be set below grade.

The footing behavior under concentric load is that of a two-way cantilever action extending out from the column or pedestal. The footing is loaded in an upward direction by the soil pressure. Tensile stresses are induced in each direction in the bottom of the footing. Therefore, the footing is reinforced by two bottom layers of steel perpendicular to each other and parallel to the edges. The required footing-soil contact area is a function of, and determined by, the allowable soil bearing pressure and the column loads being applied to the footing.

Shear

Because the footing is subject to two-way action, two different types of shear strength must be considered: two-way shear and one-way shear. The footing thickness (depth) is generally established by the shear requirements. The two-way shear is commonly termed *punching shear* because the column or pedestal tends to punch through the footing, inducing stresses around the perimeter of the column or pedestal. Tests have verified that, if failure occurs, the fracture takes the form of a truncated pyramid with sides sloping away from the face of the column or pedestal. The critical section for this two-way shear is taken perpendicular to the plane of the footing and located so that its perimeter, b_0, is a minimum but does not come closer to the edge of the column or pedestal than one-half the effective depth of the footing (ACI Code, Section 22.6.4.1).

The design of the footing for two-way action is based on a shear strength V_n, which is not to be taken greater than V_c unless shear reinforcement is provided. V_c may be determined from the ACI Code, Table 22.6.5.2, and shall be the smallest of

a. $V_c = \left[4\lambda_s\lambda\sqrt{f'_c}\right]b_0 d$

b. $V_c = \left[\left(2+\dfrac{4}{\beta_c}\right)\lambda_s\lambda\sqrt{f'_c}\right]b_0 d$

c. $V_c = \left[\left(2+\dfrac{\alpha_s d}{b_0}\right)\lambda_s\lambda\sqrt{f'_c}\right]b_0 d$

where

β_c = ratio of the long side to the short side of the concentrated load or reaction area (loaded area)

b_o = perimeter of critical section for two-way shear action in the footing

α_s = 40 for interior columns, 30 for edge columns, and 20 for corner columns

λ_s = size-effect factor (ACI 318-19 Section 22.5.5.1.3)

$$\lambda_s = \sqrt{\frac{2}{1+\dfrac{d}{10}}} \leq 1$$

The terms V_c, f'_c, λ, and d are as previously defined. Also, for $d \leq 10$ inches, $\lambda_s = 1.0$.

Note that according to ACI 318-19 Section 13.2.6.2, the size effect factor can be neglected in the calculation of the one-way or two-way shear strength in the design of:

- One-way shallow foundations
- Two-way isolated footings

- Two-way combined footings
- Mat Foundations

Note that the terms *interior*, *edge*, and *corner* columns in ACI 22.6.5.3 refer to the location of the column relative to the edges of the spread footing. Therefore, *interior*, *edge*, and *corner columns* will have four-, three-, and two-sided critical sections, respectively.

The introduction of shear reinforcement in footings is impractical and undesirable purely on an economic basis. It is general practice to select spread and strip footing thicknesses based solely on the shear strength of the concrete, and thus avoid the need for any shear reinforcement.

The one-way (or beam) shear may be compared with the shear in a beam or one-way slab. The critical section for this one-way shear is taken on a vertical plane extending across the entire width of the footing and located at a distance equal to the effective depth of the footing from the face of the concentrated load or reaction area (ACI Code, Section 7.4.3.2 or 9.4.3.2). As in a beam or one-way slab, the shear strength provided by the footing concrete may be taken as

$$V_c = 2\lambda\sqrt{f'_c}\,b_w d \qquad \text{[ACI Eq. (22.5.5.1)]}$$

This equation for V_c for one-way shear is for normal-weight concrete with no axial load in the plane of the footing. For both one- and two-way action, if we assume no shear reinforcement in the footing such as stirrups or studrails, the basis for the shear design will be $\phi V_n \geq V_u$, where $V_n = V_c$ and ϕ for shear = 0.75.

Moment and Development of Bars

The size and spacing of the footing reinforcing steel are primarily a function of the bending moment induced by the net upward soil pressure. As the footing is loaded by the upward soil pressure, it behaves as a cantilever beam in two directions The fixed end, or critical section for the bending moment, is located as follows (ACI Code, Section 13.2.7.1):

1. At the face of the column or pedestal, for a footing supporting a concrete column or pedestal (see Figure 10.9a).
2. Halfway between the face of the column and the edge of a steel base plate, for a footing supporting a column with a steel base plate (see Figure 10.9b).

The ACI Code, Section 13.2.8.3, stipulates that the critical section for the development length of footing reinforcement shall be assumed to be at the same location as the critical section for bending moment given in ACI Code, Table 13.2.7.1.

Transfer of Load from Column into Footing

All loads applied to a column must be transferred to the top of the footing (through a pedestal or pier, if there is one) by compression in the concrete and by the vertical dowel reinforcement.

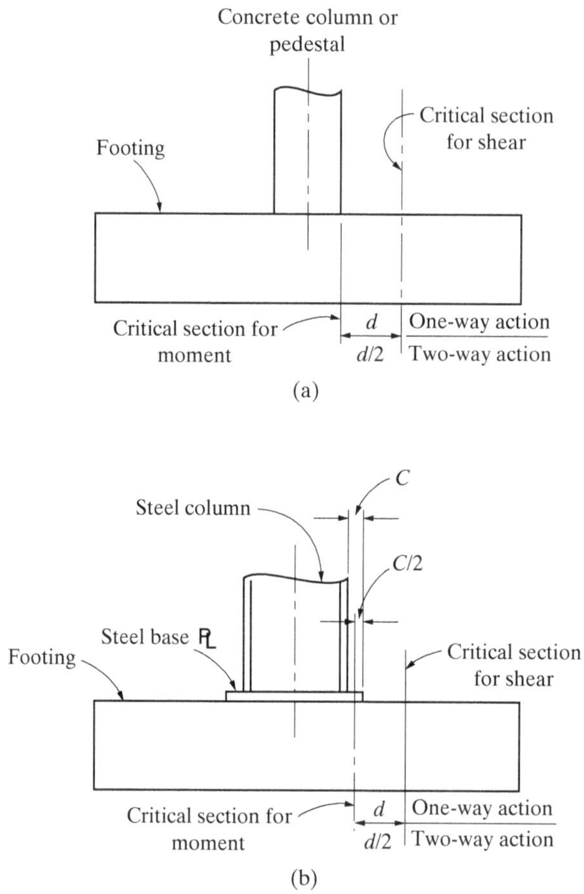

FIGURE 10.9 Critical sections for design of reinforced concrete footings supporting columns or pedestals.

The bearing strength of the concrete contact area between the supporting and supported members cannot exceed $\phi(0.85 f_c' A_1)$ as prescribed in the ACI Code, Section 22.8.3.2. When the supporting surface is wider on all sides than the loaded area, this causes confinement of the bearing area which leads to an increase in the bearing strength; therefore, the design-bearing strength on the loaded area may be multiplied by $\sqrt{\frac{A_2}{A_1}} \leq 2.0$, as previously discussed in Chapter 8, Section 8.4. A_2 is a concentric and geometrically similar support area that is the larger base of the frustrum of a pyramid or cone having a slope of 1 vertical to 2 horizontal and contained wholly within the supporting concrete member (e.g., footing, pier, wall, or pile cap), and A_1 is the smaller base or loaded area or reaction area (ACI Code, Section 22.8.3.2). Therefore, in no case can the design-bearing strength for the loaded area be in excess of

$$\phi(0.85 f_c' A_1)(2)$$

where $\phi = 0.65$ for bearing on concrete and f_c' is as previously defined.

It is common for the footing concrete to be of a lower strength (f_c') than the supported column concrete. This suggests that both supporting and supported members should be considered in determining load transfer.

Where a reinforced concrete column cannot transfer the load entirely by bearing, the excess load must be transferred by dowel reinforcement where the required A_s = (excess load)/ϕf_y. This may be accomplished by furnishing dowels, one per column bar if necessary, but not larger than No. 11 (ACI Code, Sections 16.3.5.4 and 25.5.5.4).

To provide a positive connection between a reinforced concrete column and the footing (whether dowels are required or not), the ACI Code, Section 16.3.4.1, requires a minimum area of reinforcement crossing the bearing surface of 0.005 of the column cross-sectional area. It is generally recommended that a minimum of four bars be used. These four bars should preferably be dowels for the four corner bars of a square column.

The development length of the dowels must be sufficient on both sides of the bearing surface to provide the necessary development length for bars in compression (see Chapter 5).

When the dowel carries excess load into the footing, it must be lap spliced with to the column bar using the necessary compression splice. The same procedure applies where a column rests on a pedestal or pier and where a pedestal or pier rests on a footing.

Where structural steel columns and column base plates are used, the total load is usually transferred entirely by bearing on the concrete contact area. The design-bearing strength, as stipulated previously, also applies in this case. Where a column base detail is inadequate to transfer the total load, adjustments may be made as follows:

1. Increase column base plate dimensions.
2. Use higher-strength concrete (f_c') for the pedestal or pier or footing.
3. Increase the supporting area with respect to the base plate area until the ratio reaches the maximum allowed by the ACI Code.

In building design, it is common practice to use a concrete pier between the footing and the column. The pedestal or pier as it's commonly called, in effect, distributes the column load over a larger area of the footing, thereby contributing to a more economical footing design. piers may be either plain or reinforced. If the ratio of height to least lateral dimension is in excess of 3, the member is, by definition, a column and must be designed and reinforced as a column (see Chapter 9). If the ratio is less than 3, it is categorized as a pier and theoretically may not require any reinforcement.

The cross-sectional area of a pier is usually established by the concrete-bearing strength as stipulated in ACI Code, Section 22.8.3.2, by the size of a steel column base plate, or by the desire to distribute the column load over a larger footing area. It is common practice to design a pier in a manner similar to a column, using a minimum of four corner bars (for a square or rectangular cross-section) anchored into the footing and extending up through the pier. Ties should be provided in pedestals or piers according to the same requirements as in columns, and per ACI 10.7.6.1.5, there should be at least three No. 3 ties or two No. 4 ties within the first 5 inches from the top of the pier (see Figure 10.9c).

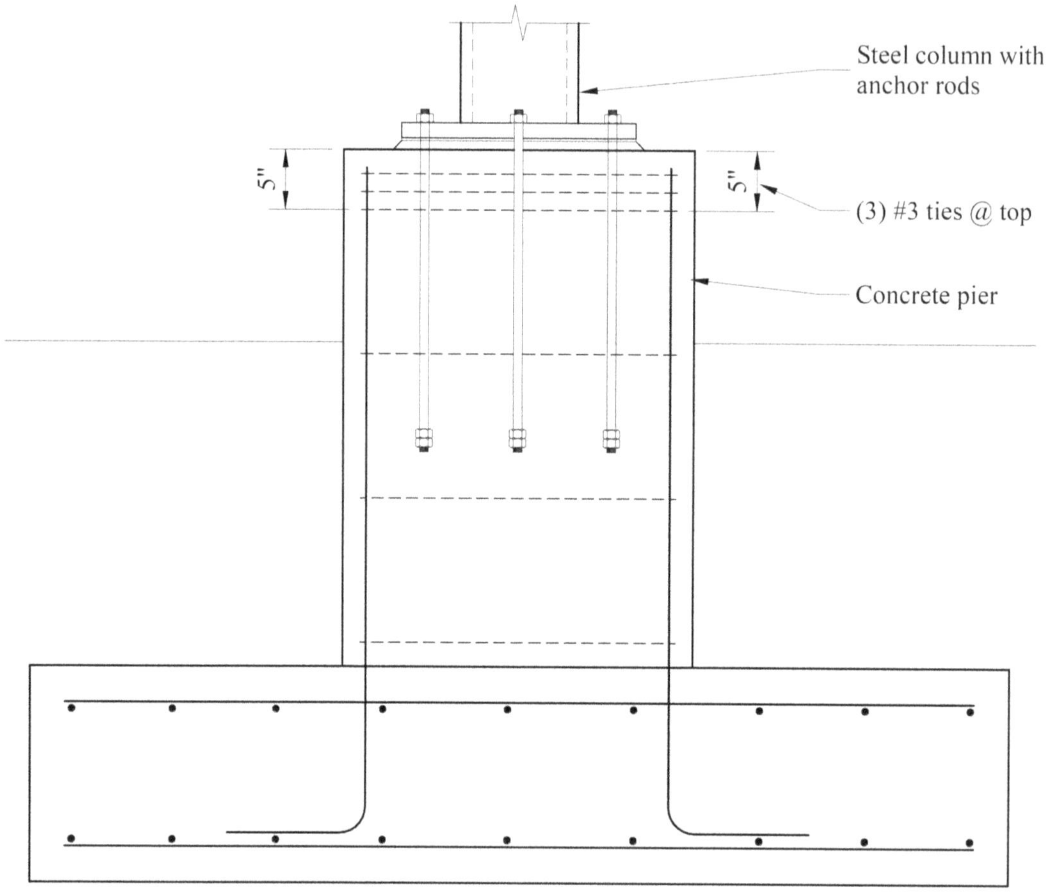

FIGURE 10.9c Added ties at concrete piers supporting steel columns with anchor rods

10.6 SQUARE REINFORCED CONCRETE FOOTINGS

In isolated square footings, the reinforcement should be uniformly distributed over the width of the footing in each direction. Because the bending moment is the same in each direction, the reinforcing bar size and spacing should be the same in each direction. In reality, the effective depth is not the same in both directions. It is common practice to use the same average effective depth for design computations for both directions, however.

It is also common practice to assume that the minimum tensile reinforcement for beams is applicable to two-way spread footings for each of the two directions unless the reinforcement provided is one-third greater than required. As discussed in Section 2.8 (and in the ACI Code, Section 9.6.1.2), the minimum tensile reinforcement is determined from

$$A_{s,\min} = \frac{3\sqrt{f'_c}}{f_y} b_w d \geq \frac{200}{f_y} b_w d$$

The use of this minimum reinforcement is conservative for footings. The ACI Code, Section 7.6.1.1, permits the use of a minimum reinforcement equal to that required for shrinkage and temperature steel in structural slabs of uniform thickness. This will always be somewhat less than that required by $A_{s,min}$, but not necessarily less than that specified by the ACI Code, Section 9.6.1.3. In this book, the criteria of the ACI Code, Section 7.6.1.1, will be used in isolated footing cases only as an absolute minimum of steel area to be provided.

EXAMPLE 10.3

Design a square reinforced concrete footing to support an 18-in.-square tied concrete column, as shown in Figure 10.10. The column is a typical interior column in a building. f_c' for the column = 4000 psi and for the footing = 3000 psi, f_y for all steel = 60,000 psi, and longitudinal column steel consists of No. 8 bars. The weight of earth = 100 lb/ft^3. Assume normal-weight concrete.

Solution:

1. The design data are as follows:

   Service dead load = 225 kip,

   Service live load = 175 kip,

   Allowable soil pressure = 5000 psf (5.00 ksf).

2. Assume a total footing thickness of 24 in. The footing weight may then be calculated as

$$0.150(2.0) = 0.300 \text{ ksf}$$

   Because the bottom of the footing is to be 4 ft below the finished ground line, there will be 24 in. of earth or soil on top of the footing. This depth of earth has a weight of

$$\left(\frac{24}{12}\right)(0.100) = 0.200 \text{ ksf}$$

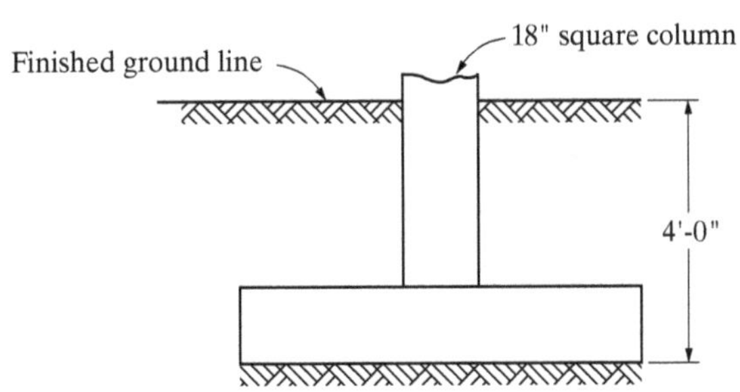

FIGURE 10.10 Sketch for Example 10.3.

Therefore, the *net allowable* soil pressure for the superimposed loads is the gross soil bearing pressure less the weight of the footing and the weight of the soil above the footing, which is

$$5.00 - 0.300 - 0.200 = 4.50 \text{ ksf}$$

The required area of the footing is determined using service loads and the allowable soil pressure or by modifying both the service loads and the allowable soil pressure with the ACI load factors. Using the service loads,

$$\text{Required A} = \frac{225 + 175}{4.50} = 88.9\,\text{ft}^2$$

Use a 9 ft-6 in. square footing. This furnishes an actual area A of 90.3 ft^2.

3. The factored uniform soil pressure from superimposed loads to be used for the footing design may now be calculated:

$$p_u = \frac{P_u}{A} = \frac{(1.2)(225) + (1.6)(175)}{90.3} = 6.09\,\text{ksf}$$

4. The footing thickness is usually determined by shear strength requirements. Therefore, using the assumed footing thickness, check the shear strength. The thickness h was assumed to be 24 in. Therefore, the average effective depth, based on a 3-in. cover for bottom steel and assuming No. 8 bars in each direction, is

$$d = 24 - 3 - 1 = 20\text{ in}$$

This constitutes an average effective depth that will be used for design calculations for both directions.

5. The shear strength of individual column footings is governed by the more severe of two conditions: two-way action (punching shear) or one-way action (beam shear). The location of the critical section for each type of behavior is depicted in Figure 10.11. *For two-way action* (Figure 10.11a),

$$\text{B} = \text{column width} + \left(\frac{d}{2}\right)2$$

$$= 18\text{ in} + \left(\frac{20\text{ in}}{2}\right)2 = 38\text{ in} = 3.17\text{ ft.}$$

The critical perimeter for punching shear, b_o = 4B

The total factored shear acting on the critical section is

$$V_u = p_u(W^2 - B^2)$$
$$= 6.09(9.5^2 - 3.17^2)$$
$$= 488\text{ kip}$$

The shear strength of the concrete (with $\lambda = 1$ for normal-weight concrete, and $\lambda_s = 1$) is taken as the *smallest* of

a. $V_c = \left[4\lambda_s\lambda\sqrt{f_c'}\right]b_0 d = (4)(1.0)(1.0)\sqrt{3000}(38)(4)(20) = 666{,}000\text{ lb}$

b.

$$V_c = \left[\left(2+\frac{4}{\beta_c}\right)\lambda_s\lambda\sqrt{f_c'}\right]b_0 d$$

$$= \left(2+\frac{4}{1}\right)(1.0)(1.0)\sqrt{3000}(38)(4)(20) = 999{,}000\,\text{lb}$$

c. With $\alpha_s = 40$ for an interior column,

$$V_c = \left[\left(2+\frac{\alpha_s d}{b_0}\right)\lambda_s\lambda\sqrt{f_c'}\right]b_0 d$$

$$V_c = \left(2+\frac{(40)(20)}{(38)(4)}\right)(1.0)(1.0)\sqrt{(3000)}(38)(4)(20) = 1{,}209{,}000\,\text{lb}$$

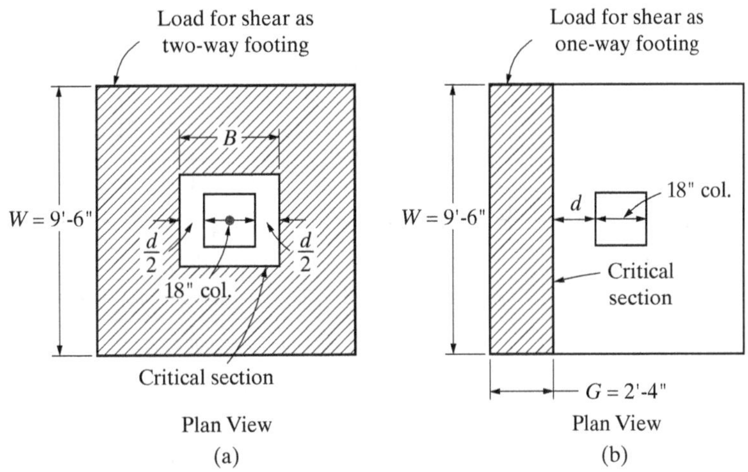

FIGURE 10.11 Footing shear analysis.

Thus, the lowest value governs and, $V_c = 666{,}000$ lb $= 666$ kip

$$\phi V_n = \phi V_c = 0.75(666) = 500\ \text{kip}$$

Therefore,

$$\phi V_n > V_u \qquad (O.K.)$$

When V_u is relatively close to the shear strength of the concrete (ϕV_c), it indicates that the assumed footing thickness is approximately equal to that required for shear. If these two values were significantly different, the assumed footing thickness should be modified.

For one-way action, the total factored shear acting on the critical section is distance d from the face of the column:

$$\begin{aligned} V_u &= p_u WG \\ &= (6.09)(9.5)(2.33) \\ &= 134.8 \text{ kip} \end{aligned}$$

The shear strength of the concrete is

$$\begin{aligned} V_c &= 2\sqrt{f'_c}b_w d = 2\sqrt{3000}(9.5)(12)(20) = 250{,}000 \text{ lb} \\ &= 250 \text{ kip} \\ \phi V_n &= \phi V_c = (0.75)(250) = 187.5 \text{ kip} \end{aligned}$$

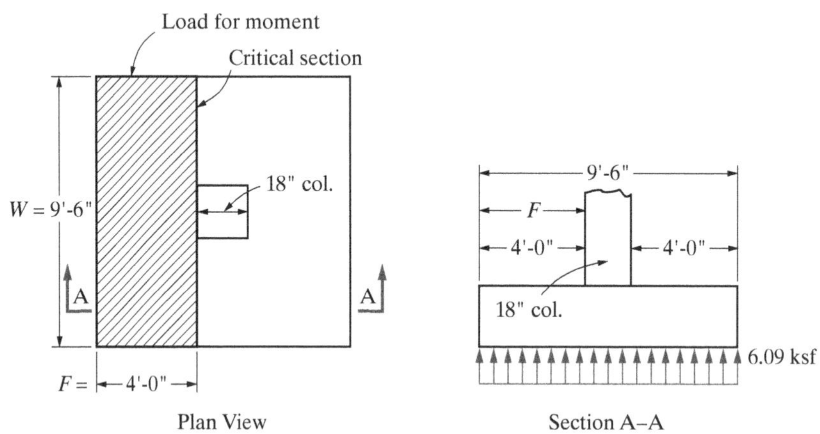

FIGURE 10.12 Footing moment analysis.

Therefore,

$$\phi V_n > V_u = 134.8 \text{ kip(O.K.)}$$

The 24-in.-deep footing is satisfactory with respect to shear. Our assumption of step 2 regarding weight of the footing and the soil on the footing is satisfactory.

6. The critical section for bending moment may be taken at the face of the column, as depicted in Figure 10.12. Using the factored soil pressure and assuming the footing to act as a wide cantilever beam in both directions, the design moment may be computed:

$$\begin{aligned} M_u &= p_u F\left(\frac{F}{2}\right)W \\ &= (6.09)(4)\left(\frac{4}{2}\right)(9.5) \\ &= 463 \text{ ft-kip} \end{aligned}$$

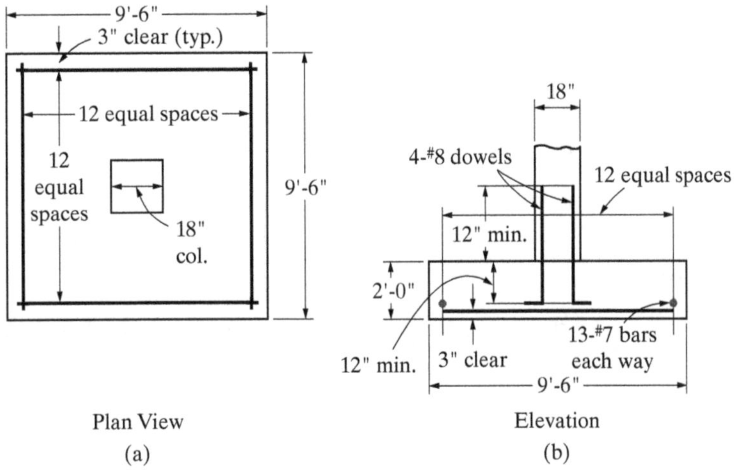

FIGURE 10.13 Design sketch for Example 10.3.

7. Assume $\phi = 0.90$ and design the tension steel at the bottom of the footing as follows:

$$required\,\bar{k} = \frac{M_u}{\phi bd^2} = \frac{(463)(12)}{(0.9)(9.5)(12)(20)^2}$$

$$= 0.1354 \text{ ksi}$$

From Table A.8, the required $\rho = 0.0024$, $\varepsilon_t > 0.005$, and $\phi = 0.90$. Therefore,

$$\text{required } A_s = \rho bd$$

$$= 0.0024(9.5)(12)(20)$$

$$= 5.47 \text{ in.}^2$$

Alternate check (see Chapter 2):

$$\left(\frac{f_y}{1.7 f'_c bd}\right) A_s^{\,2} - A_s + \frac{M_u}{\phi f_y d} = 0$$

$$\left(\frac{60}{1.7(3)(12)(20)}\right) A_s^{\,2} - A_s + \frac{(463)(12)}{(0.9)(60)(20)} = 0$$

$$0.00515 A_s^{\,2} - A_s + 5.144 = 0$$

where:

$a_1 = 0.00515$

$b_1 = -1$

$c_1 = 5.144$

$$A_s = \frac{-b_1 - \sqrt{b_1^2 - 4a_1c_1}}{2a_1}$$

$$A_s = \frac{-(-1) - \sqrt{(-1)^2 - 4(0.00515)(5.144)}}{2(0.00515)} = 5.29 \text{ in}^2$$

The solution of 5.47 in^2 is approximate as it is done by interpolation and is thus also conservative.

Check the ACI Code minimum reinforcement requirement. From Table A.5,

$$A_{s,\min} = 0.0033(9.5)(12)(20)$$
$$= 7.52 \text{ in.}^2$$

Of the two steel areas, the larger (7.52 in.^{2}) controls.

Because the footing is square and an average effective depth was used, the steel requirements in the other direction may be assumed to be identical. Therefore, use 13 No. 7 bars, bottom each way (A_s = 7.80 in.^{2} in each direction) and distribute the bars uniformly across the footing in each direction, as shown in Figure 10.13a.

Check the development length for the No. 7 bars:

This calculation for ℓ_d follows the eight-step procedure presented in Chapter 5, Section 5.2, and is summarized as follows:

1. $K_D = 82.2$
2. $\psi_g = 1.0$, $\psi_e = 1.0$, $\psi_s = 1.0$, $\psi_t = 1.0$, $\lambda = 1.0$
3. $\psi_t \times \psi_e = 1.0 < 1.7$ (O.K.)
4. $c_b = 3.44$ in.
5. $K_{tr} = 0$
6. $(c_b + K_{tr})/d_b = 3.93$ Use 2.5
7. $K_{ER} = 7.52 \text{ in.}^2/7.80 \text{ in.}^2 = 0.96$
8. $\ell_d = 27.6 \text{ in.} > 12 \text{ in.}$ (O.K.)

Use $\ell_d = 27.6$ in. (minimum). The development length provided is 45 in., which is in excess of that required (O.K.).

8. The concrete bearing strength at the base of the column cannot exceed

$$\phi(0.85 f_c' A_1)$$

except where the supporting surface is wider on all sides than the loaded area, for which case the concrete bearing strength of the supporting surface cannot exceed

$$\phi(0.85 f_c' A_1)\sqrt{\frac{A_2}{A_1}}$$

As described in Section 8.4, A_2 is the lower base of the frustum of a pyramid having 1:2 sloping sides and fully contained within the support, as shown in Figure 10.14. In this case, A_2 is the same as the area of the footing and

$$\sqrt{\frac{A_2}{A_1}} = \sqrt{\frac{90.3}{2.25}} = 6.3 > 2.0$$

Therefore, use 2.0. Then

$$\begin{aligned}\text{Footing bearing strength} &= \phi(0.85 f_c' A_1)(2.0)\\ &= (0.65)(0.85)(3.0)(18)^2(2.0)\\ &= 1074 \text{ kip}\end{aligned}$$

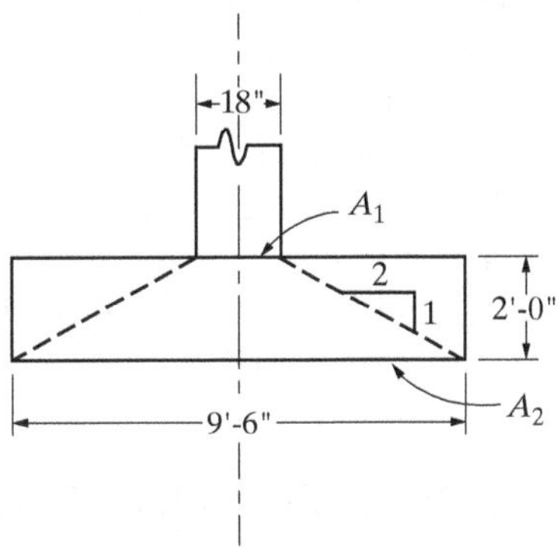

FIGURE 10.14 Determination of A_2.

The *column*-bearing strength is computed as follows:

$$\phi(0.85) f_c' A_1 = 0.65(0.85)(4.0)(18)^2 = 716 \text{ kip}$$

The *calculated* design-bearing load is

$$\begin{aligned}P_u &= 1.2(225) + 1.6(175)\\ &= 550 \text{ kip}\end{aligned}$$

Because 550 < 716 < 1074, the entire column load can be transferred by concrete alone at the column-footing interface. The ACI Code, however, requires a minimum vertical dowel area of

$$\begin{aligned}\text{Required } A_s &= 0.005 A_g\\ &= 0.005(18)^2 = 1.62 \text{ in}^2\end{aligned}$$

Use a minimum of four bars. Four No. 6 bars, $A_s = 1.76$ in.^{2}, is satisfactory. It is general practice in a situation such as this, however, to use dowels of the same diameter as the column steel. Therefore, use four No. 8 dowels, and place one in each corner ($A_s = 3.16$ in.^{2}).

The development length for dowels into the column and footing must be adequate even though full load transfer can be made without dowels. The bars are in compression. The compression development length for the No. 8 dowels into the footings is ℓ_{dc} from Table A.12 and may be reduced by any applicable modification factors. We will use the modification factor for the case where the steel provided is in excess of the steel required:

$$\frac{\text{Required } A_s}{\text{Provided } A_s} = \frac{1.62}{3.16} = 0.51$$

Therefore,

$$\text{Required } \ell_{dc} = 21.9(0.51) = 11.2 \text{ in}$$

For bars in compression, ℓ_{dc} must not be less than 8 in. Therefore, use ℓ_{dc} of 12 in. into both the column and the footing because f'_c for the column concrete is higher and the required ℓ_{dc} for the bars into the column would be less than that into the footing. The actual anchorage used may be observed in Figure 10.13b. The dowels should be placed adjacent to the corner longitudinal bars. Generally, these dowels are furnished with a 90° hook at their lower ends and are placed on top of the main footing reinforcement. This will tie the dowel in place and will reduce the possibility of the dowel being dislodged during construction. The hook cannot be considered effective as part of the required development length (ACI Code, Section 25.4.1.2).

10.7 RECTANGULAR REINFORCED CONCRETE FOOTINGS

Rectangular footings are generally used where space limitations require it. The design of these footings is very similar to that of the square column footing, with the one major exception that each direction must be investigated independently. Shear is checked for two-way action in the normal way, but for one-way action, it is checked across the shorter side only. The bending moment must be considered separately for each direction. Each direction will generally require a different area of steel. The reinforcing steel running in the long direction should be placed below the short-direction steel so that it may have the larger effective depth to carry the larger bending moments in that direction.

In rectangular footings, the *distribution* of the reinforcement is different than for square footings (ACI Code, Section 13.3.3.3). The reinforcement in the long direction should be uniformly distributed over the shorter footing width. A part of the required reinforcement in the short direction is placed in a band equal to the length of the short side of the footing. The portion of the total required steel that should go into this band is

$$\frac{2}{\beta + 1}$$

where β is the ratio of the long side to the short side of the footing. The remainder of the reinforcement in the short direction is uniformly distributed in the outer portions of the footing. This distribution is depicted in Figure 10.15. Other features of the design are similar to those for the square column footing.

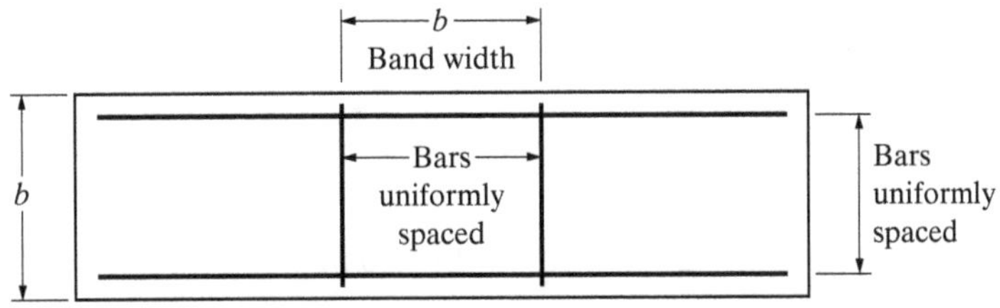

FIGURE 10.15 Rectangular footing plan.

EXAMPLE 10.4

Design a reinforced concrete footing to support an 18-in.-square tied interior concrete column, as shown in Figure 10.16. One dimension of the footing is limited to a maximum of 7 ft. f'_c for both footing and column = 3000 psi, f_y for all steel = 60,000 psi, and longitudinal column steel consists of No. 8 bars. The weight of earth = 100 lb/ft^3. Assume normal-weight concrete.

Solution:

1. The design data are as follows:

   Service dead load = 175 kip,

   Service live load = 175 kip,

   Allowable soil pressure = 5000 psf (5.00 ksf).

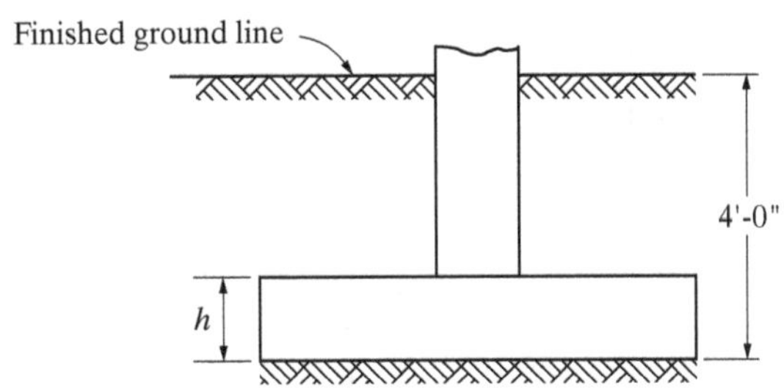

FIGURE 10.16 Sketch for Example 10.4.

2. Assume a total footing thickness h of 24 in. subject to later check. The footing weight may then be calculated as

$$0.150(2.0) = 0.300\,\text{ksf}$$

Because the bottom of the footing is to be 4 ft below the finished ground line, there will be 24 in. of earth or soil on top of the footing. This depth of earth has a weight of

$$\frac{24}{12}(0.100) = 0.200\text{ ksf}$$

Therefore, the *net allowable* soil pressure for the superimposed loads becomes

$$5.00 - 0.300 - 0.200 = 4.50\,\text{ksf}$$

Based on service loads, the required area of footing may be calculated:

$$\text{Required}\,A = \frac{175 + 175}{4.50} = 77\text{ft}^2$$

Use a rectangular footing 7 ft-0 in. by 11 ft-6 in. This furnishes an actual area A of 80.5 ft^2.

3. The factored soil pressure from superimposed loads to be used for the footing design may now be calculated:

$$p_u = \frac{P_u}{A} = \frac{1.2(175) + 1.6(175)}{80.5} = 6.09\,\text{ksf}$$

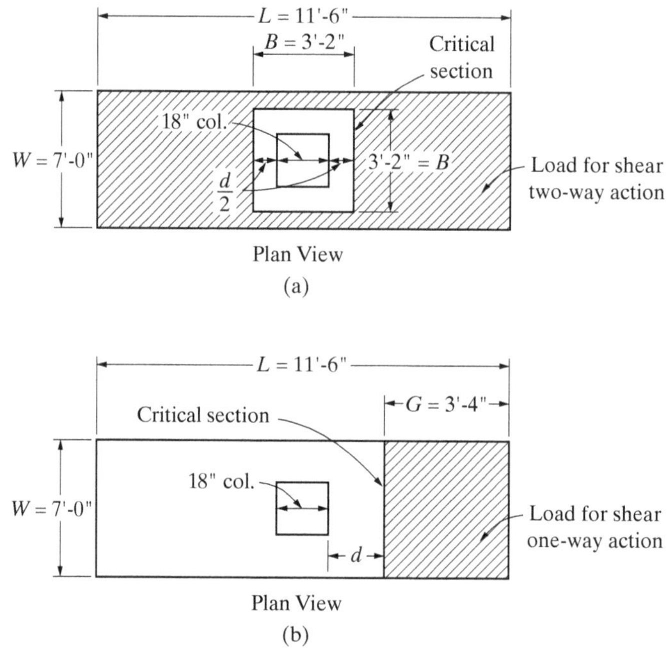

FIGURE 10.17 Footing shear analysis.

4. The footing thickness h was assumed to be 24 in. Therefore, the average effective depth, based on a 3-in. cover for bottom steel and 1-in.-diameter bars in each direction, will be

$$d = 24 - 3 - 1 = 20\,\text{in}$$

This constitutes an *average* effective depth, which will be used for design calculations for both directions.

5. Checking the shear strength *for two-way action* (with reference to Figure 10.17a),

$$B = \text{column width} + \left(\frac{d}{2}\right)2$$
$$= 18 + 20 = 38 \text{ in} = 3.17 \text{ ft}$$

The critical perimeter for punching shear, b_o = 4B (see Figure 10.17), and recall that the footing area, A = LW = 80.5 ft.^{2}

The total factored shear acting on the critical section is

$$\begin{aligned} V_u &= p_u\left(A - B^2\right) \\ &= 6.09(80.5 - 3.17^2) \\ &= 429 \text{ kip} \end{aligned}$$

The shear strength of the concrete is taken as the smallest of

a. $V_c = \left[4\lambda_s\lambda\sqrt{f_c'}\right]b_0 d = \left[4(1.0)(1.0)\sqrt{3000}\right](38)(4)(20) = 666{,}000 \text{ lb}$

b. $V_c = \left[\left(2 + \dfrac{4}{\beta_c}\right)\lambda_s\lambda\sqrt{f_c'}\right]$

$$= \left[\left(2 + \frac{4}{1}\right)(1.0)(1.0)\sqrt{3000}\right](38)(4)(20) = 999{,}000 \text{ lb}$$

c. With $a_s = 40$ for an interior column,

$$V_c = \left[\left(2 + \frac{\alpha_s d}{b_0}\right)\lambda_s\lambda\sqrt{f_c'}\right]b_0 d$$

$$V_c = \left(2 + \frac{(40)(20)}{(38)(4)}\right)(1.0)(1.0)\sqrt{(3000)}(38)(4)(20) = 1{,}209{,}000 \text{ lb}$$

The lowest value governs, which means V_c = 666,000 lb = 666 kip. Thus

$$\phi V_n = \phi V_c = 0.75(666) = 500 \text{ kip}$$

Therefore,

$\phi V_n > V_u = 429$ kip (O.K.)

$$\phi V_n > V_u \quad \text{(O.K.)}$$

For one-way action, consider shearing across the short side only. The critical section is at a distance equal to the effective depth of the member from the face of the column (see Figure 10.17b).

The total factored shear acting on the critical section is

$$V_u = p_u WG$$
$$= (6.09)(7.0)(3.33)$$
$$= 142.0 \text{ kip}$$

The nominal shear strength of the concrete is

$$V_c = 2\sqrt{f_c'}b_w d = 2\sqrt{3000}(7.0)(12)(20) = 184{,}000 \text{ lb}$$
$$= 184 \text{ kip}$$

from which

$$\phi V_n = \phi V_c = (0.75)(184) = 138 \text{ kip}$$

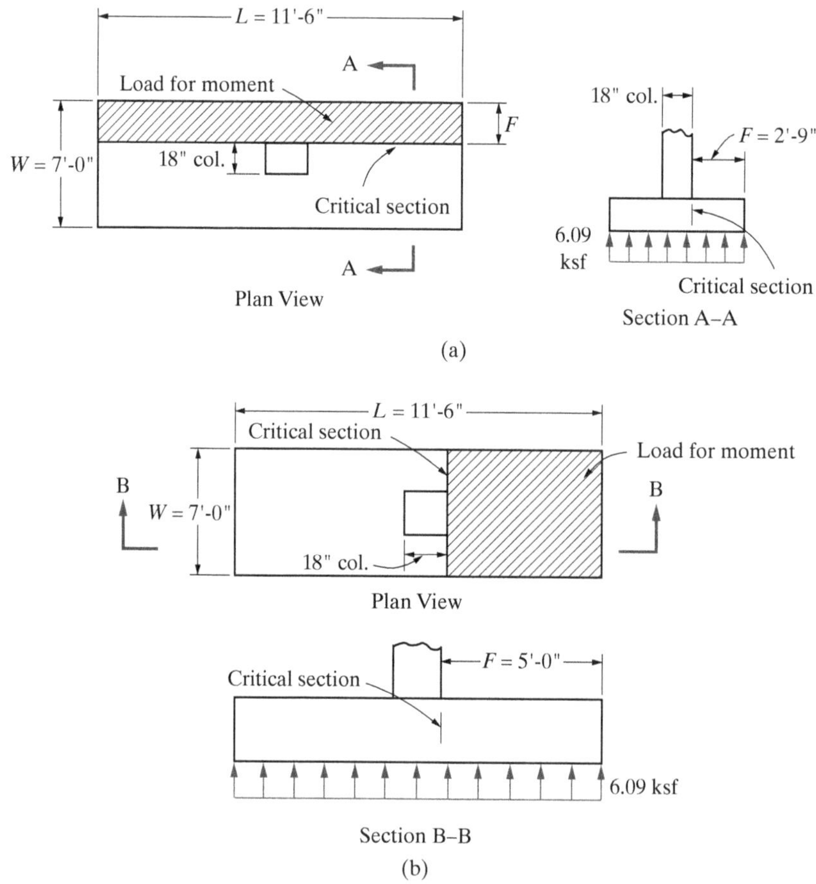

FIGURE 10.18 Footing moment analysis.

This is unsatisfactory. We will determine the required d based on one-way shear using $\phi V_c = V_u$.

$$\phi V_n = \phi 2\sqrt{f_c'}b_w d = V_u = 142.0 \text{ kip}$$

from which

$$\text{Required } d = \frac{V_u}{\phi 2\sqrt{f'_c} b_w} = \frac{142.0(1000 \text{ lb/kip})}{0.75(2)\sqrt{3000}(7.0)(12)}$$

$$= 20.6 \text{ in}$$

Therefore, the required $h = 20.6 + 3 + 1 = 24.6$ in. Use $h = 25$ in. The effect of this change on the previous calculations is small, and no other revisions are considered warranted.

$$\text{New } d = 25 - 1 - 3 = 21.0 \text{ in}$$

6. For the bending moment, each direction must be considered independently, with the critical section taken at the face of the column. Using the actual factored soil pressure and assuming the footing to act as a wide cantilever beam in each direction, the design moment may be calculated. With reference to Figure 10.18b, for the moment in the long direction, the design moment may be computed as follows:

$$M_u = p_u F\left(\frac{F}{2}\right)W$$

$$= (6.09)(5)\left(\frac{5}{2}\right)(7.0)$$

$$= 533 \text{ ft-kip}$$

For the moment in the short direction (Figure 10.18a),

$$M_u = p_u F\left(\frac{F}{2}\right)L$$

$$= (6.09)(2.75)\left(\frac{2.75}{2}\right)(11.5)$$

$$= 265 \text{ ft-kip}$$

7. Assume $\phi = 0.90$ and design the tension steel as follows. *For the long direction*, where $M_u = 533$ ft-kip, and $b = 7$ ft.,

$$\textit{required } \bar{k} = \frac{M_u}{\phi b d^2} = \frac{(533)(12)}{(0.9)(7)(12)(21)^2}$$

$$= 0.1918 \text{ ksi}$$

From Table A.8, the required $\rho = 0.0034$, $\varepsilon_t > 0.005$, and $\phi = 0.90$. Therefore,

$$\text{required } A_s = \rho b d$$

$$= 0.0034(7)(12)(21)$$

$$= 6.00 \text{ in.}^2$$

Check the ACI Code minimum reinforcement requirement. From Table A.5,

$$A_{s,\min} = 0.0033(7)(12)(21)$$
$$= 5.82 \text{ in.}^2 \text{ ft of wall}$$

Of the two steel areas, the larger (6.00 in.^{2}) controls; therefore, use 10 No. 7 bars (A_s = 6.00 in.^{2}). These bars will run in the long direction and will be distributed uniformly across the 7 ft-0 in. width. They will be placed in the bottom layer, where they will have the advantage of slightly greater effective depth. The development length must be checked for these bars. Assume uncoated bars. This calculation for ℓ_d follows the eight-step procedure presented in Chapter 5, Section 5.2, and is summarized as follows:

1. $K_D = 82.2$
2. $\psi_g = 1.0$, $\psi_e = 1.0$, $\psi_s = 1.0$, $\psi_t = 1.0$, $\lambda = 1.0$
3. $\psi_t \times \psi_e = 1.0 < 1.7$ (O.K.)
4. $c_b = 3.44$ in.
5. $K_{tr} = 0$
6. $(c_b + K_{tr})/d_b = 3.93$ Use 2.5
7. $K_{ER} = 1.00$
8. $\ell_d = 28.8 \text{ in.} > 12 \text{ in.} (\text{O.K.})$

The development length furnished is 60 – 3 = 57 in., which is in excess of that required (O.K.).

For the short direction, where M_u = 265 ft-kip, and b = 11.5 ft.,

$$\text{required } \bar{k} = \frac{M_u}{\phi bd^2} = \frac{(265)(12)}{(0.9)(11.5)(12)(21)^2}$$
$$= 0.0581 \text{ksi}$$

From Table A.8, the required $\rho = 0.0010$, $\varepsilon_t > 0.005$, and $\phi = 0.90$. Therefore,

$$\text{required } A_s = \rho bd$$
$$= 0.0010(11.5)(12)(21)$$
$$= 2.90 \text{ in.}^2$$

Check the ACI Code minimum reinforcement requirement. From Table A.5,

$$A_{s,\min} = 0.0033(11.5)(12)(1421)$$
$$= 9.56 \text{ in.}^2$$

The $A_{s,min}$ requirement need not be applied if the area of steel provided is at least one-third greater than that required (ACI Code, Section 9.6.1.3), however. Therefore, the required steel area may be calculated from

$$\text{required } A_s = \left(\frac{4}{3}\right)(2.90) = 3.86 \text{ in}^2$$

Checking further, using the ACI Code, Section 7.6.1.1, as an absolute minimum, the required steel area is

$$A_{s,min} = 0.0018bh = 0.0018(11.5)(12)(25) = 6.21\text{in}^2$$

Therefore, 6.21 in.^{2} controls. Use 15 No. 6 bars (A_s = 6.60 in.^{2}). These bars will run in the short direction but will *not* be distributed uniformly across the 11 ft-6 in. side.

In rectangular footings, a portion of the total reinforcement required in the short direction is placed in a band centered on the column and having a width equal to the length of the short side. The portion of the total required steel that goes into this band is

$$\frac{2}{\beta + 1}$$

Where

$$\beta = \frac{\text{long side dimension}}{\text{short side dimension}} = \frac{11.5}{7.0} = 1.64$$

from which

$$\frac{2}{\beta + 1} = \frac{2}{1.64 + 1} = 0.757 = 75.7\%$$

Therefore, 75.7% of 15 No. 6 bars must be placed in a band width = 7 ft-0 in. The balance of the required bars will be distributed equally in the outer portions of the footing. Thus

$(0.757)(15) = 11.4$ bars

Use 12 bars in the 7 ft-0 in. band width. Because reinforcing should be symmetrical with respect to the centerline of the footing, use two bars on each side of the 7 ft-0 in. band width: Therefore, the total steel used in the short direction will be 16 No. 6 bars (A_s = 7.04 in.^{2}). The bar arrangement is depicted in Figure 10.19.

The development length will be checked for the No. 6 bars in the center band. Assume uncoated bars. This calculation for ℓ_d follows the eight-step procedure presented in Chapter 5, Section 5.2, and is summarized as follows:

1. $K_D = 82.2$
2. $\psi_g = 1.0$, $\psi_e = 1.0$, $\psi_s = 0.8$ (for No. 6 bars), $\psi_t = 1.0$, $\lambda = 1.0$

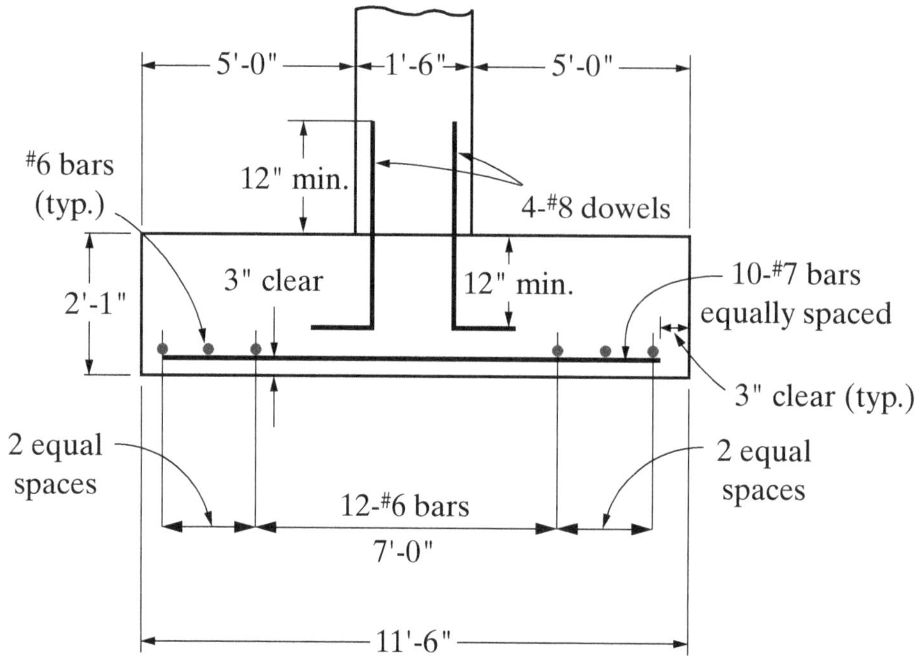

FIGURE 10.19 Design sketch for Example 10.4.

3. $\psi_t \times \psi_e = 1.0 < 1.7$ (O.K.)
4. $c_b = 3.38$ in.
5. $K_{tr} = 0$
6. $(c_b + K_{tr})/d_b = 4.51$ Use 2.5
7. $K_{ER} = 6.21/7.04 = 0.882$
8. $\ell_d = 17.4\text{in.} > 12$ in. (O.K.)

   Use $\ell_d = 17.4\text{in.}$ (minimum). The development length furnished = 33 in. – 3 in. cover = 30 in., (see Section A-A in Figure 10.18) which is in excess of that required. (O.K.)

8. Because the supporting surface is wider on all sides, the bearing strength for the footing may be computed as follows:

$$\text{Footing bearing strength} = \phi(0.85 f_c' A_1)\sqrt{\frac{A_2}{A_1}}$$

where A_2 is calculated with reference to Figure 10.20 and is seen to be a square, 7 ft-0 in. on each side.

$$\sqrt{\frac{A_2}{A_1}} = \sqrt{\frac{49.0}{22.5}} = 4.67 > 2.0$$

Therefore, use 2.0. Then,

$$\begin{aligned}\text{Footing bearing strength} &= \phi(0.85f_c'A_1)(2.0)\\ &= 0.65(0.85)(3.0)(18)^2(2.0)\\ &= 1074\ \text{kip}\end{aligned}$$

The column bearing strength is computed as follows:

$$\phi(0.85f_c'A_1) = 0.65(0.85)(3.0)(18)^2 = 537\ \text{kip}$$

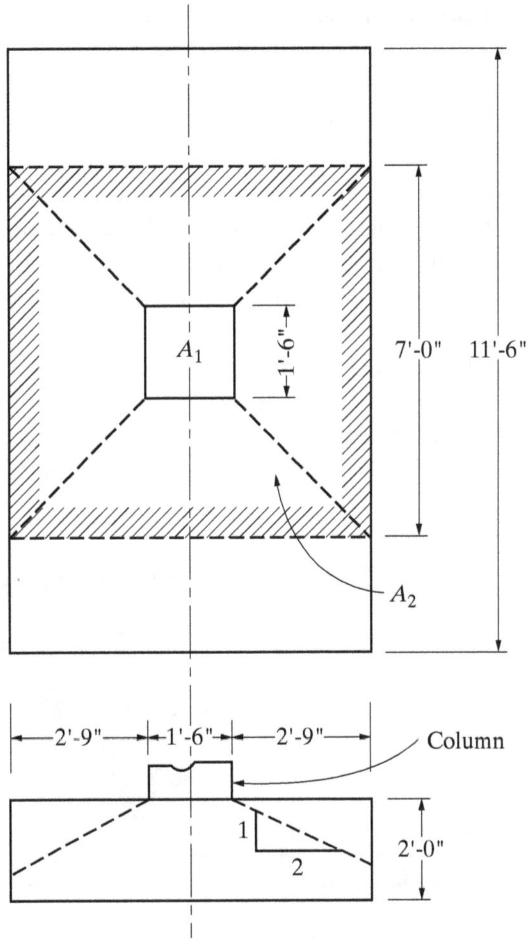

FIGURE 10.20 Determination of A_2.

The factored bearing load is

$$\begin{aligned}P_u &= 1.2(175)+1.6(175)\\ &= 490\ \text{kip}\end{aligned}$$

Because 490 kip < 537 kip < 1074 kip, the entire column load can be transferred by concrete alone.

For proper connection between column and footing, use four No. 8 dowels (to match the column steel), one in each corner. The development length requirement for the No. 8 dowels is identical to that of Example 10.3. In practice, many designers simply provide a note on the column schedule that states: "Vertical dowels to match the vertical reinforcement in the columns."

10.8 ECCENTRICALLY LOADED FOOTINGS

Isolated spread footings may sometimes be subjected to uni-axial or bi-axial moments due to eccentric gravity loads and/or lateral loads that produce moments in the footing. Where footings are subject to eccentric vertical loads or to moments transmitted by the supported column, the design of the footing varies somewhat from that of the preceding sections. The soil pressure is no longer uniform across the footing width but may be assumed to vary linearly. The resultant force should be within the middle one-third of the footing base to ensure a positive contact surface between the footing and the soil. With the soil pressure distribution known, the footing must be designed to resist all moments and shears, similar to the concentrically loaded footings. The effects of load eccentricity on isolated footings may result in an undesirable large rotation of the footing, but this can be mitigated by connecting or strapping the eccentrically loaded footing to an adjacent concentrically loaded footing, as discussed in Section 10.10. A few examples of isolated spread footings subjected to moments and gravity loads are shown in Figure 10.21, where the moments in the footing are caused by the horizontal and vertical components of the diagonal members. The soil pressure distributions in spread footings subjected to an axial load P and moment M will be as shown in Figure 10.22.

FIGURE 10.21 Eccentrically loaded footings (Abi Aghayere)

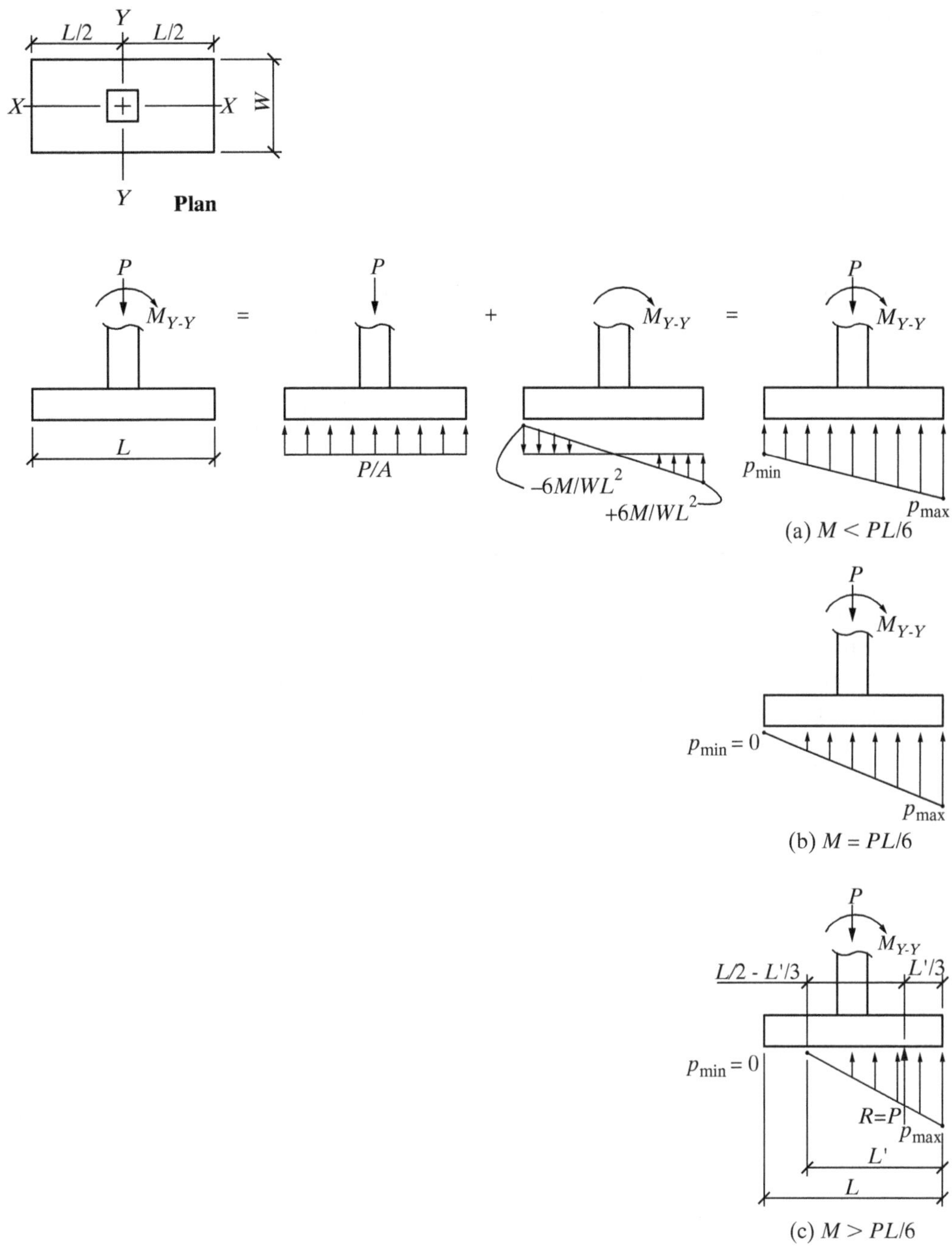

FIGURE 10.22 Pressure distributions in eccentrically loaded spread footings

Assuming a rigid footing and an elastic homogeneous soil, the pressure distribution under the footing will vary linearly from a minimum pressure of p_{min} to a maximum value of p_{max}. If the size of the footing is selected so that the footing is never under a negative pressure underneath the footing, the minimum and maximum soil pressures can be calculated, assuming a rigid footing, using the following stress equations:

$$p_{min} = \frac{P}{A} - \frac{M\left(\frac{L}{2}\right)}{\frac{WL^3}{12}} = \frac{P}{WL} - \frac{6M}{WL^2} \geq 0$$

$$\leq \text{Net allowable soil pressure}$$

$$p_{max} = \frac{P}{A} + \frac{M\left(\frac{L}{2}\right)}{\frac{WL^3}{12}} = \frac{P}{WL} + \frac{6M}{WL^2}$$

$$\leq \text{Net allowable soil pressure}$$

If the size of the footing results in a negative pressure, the area of the footing in the negative pressure zone is neglected since tension is not possible in the soil. The maximum moment on the footing to avoid negative soil pressure can be calculated by equating p_{min} to zero. Therefore, if $M \leq PL/6$ (i.e., if the load eccentricity, $e \leq L/6$), there will be no negative pressure beneath the footing. At a moment, $M = PL/6$ (i.e., the load eccentricity, $e = L/6$), the soil pressures are

$$p_{min} = 0$$

$$p_{max} = \frac{P}{WL} + \frac{6\left(\frac{PL}{6}\right)}{WL^2} = \frac{2P}{WL}$$

$$\leq \text{Net allowable soil pressure}$$

For footings with large moments where $PL/6 < M < PL/2$ (i.e., $L/6 < e < L/2$), the modified pressure distribution will be as shown in Figure 10.22(c). Since the soil cannot resist tension, the zone of the footing with negative soil pressure is neglected. Only the footing area with positive soil pressure is considered. In this case, the maximum soil pressure will be

$$p_{max} = \frac{2P}{WL'}$$

$$\leq \text{Net allowable soil pressure}$$

and

$$p_{min} = 0$$

where L' is the length of the footing in contact with the soil (i.e., length of the positive soil pressure zone). Using equilibrium of forces and noting that the resultant of the reaction due to the soil pressure is equal to the applied load, P, and summing moments of this resultant reaction about the location of the gravity load, P, yields

$$P\left(\frac{L}{2}-\frac{L'}{3}\right)=M$$

Noting that the eccentricity of the vertical load, $e = M/P$, we can solve for L' in the previous equation as follows:

$$\left(\frac{L}{2}-\frac{L'}{3}\right)=\frac{M}{P}=e$$

Therefore, $L'=3\left(\frac{L}{2}-e\right)$

The maximum soil pressure for this case of large moments can be rewritten as

$$p_{\max}=\frac{2P}{WL'}=\frac{2P}{3W\left(\frac{L}{2}-e\right)}$$

$$\leq \text{Net allowable soil pressure}$$

The pressure distributions for all the cases considered above are summarized in Table 10.2. If the eccentricity of the vertical load, e, lies within the middle third of the footing, there will be no negative soil pressures beneath the footing. This middle third zone is referred to as the kern. After the pressures are determined, the design of the footing follows a similar approach to that for concentrically loaded footings with uniform soil pressures, the only difference being that the calculation of the shear forces and the moments in the eccentrically loaded footing take into account the linearly varying soil pressures. Note that for calculating the applied soil pressures and checking that the net allowable soil pressure is not exceeded, the unfactored vertical load and unfactored moment are used. However, for the ultimate strength design of the footing for shear and flexure, the factored vertical load and factored moment are used. An isolated spread footing may also be subjected to bi-axial moments simultaneously (i.e., moments about the two orthogonal axes in plan); the moments about one axis will result in soil pressures that are additive to the soil pressures caused by the moment about the other orthogonal axis, or the soil pressures from the orthogonal moments will negate each other, depending on the direction of the applied moments. As long as the eccentricity of the vertical load lies within the middle third of the plan dimension of the footing (i.e., the kern) in both directions (i.e., $e_x=\frac{M_{Y-Y}}{P}\leq\frac{L}{6}$ and $e_y=\frac{M_{X-X}}{P}\leq\frac{W}{6}$), the minimum soil pressures below the footing will be greater than or equal to zero (see Figure 10.23). In all cases, the designer should ensure that the size of the footing selected is such that the maximum soil pressure, $p_{\max}$, does not exceed the net allowable soil pressure. The nonuniform pressure distribution underneath eccentrically loaded footings leads to differential settlement of the soil, which could cause excessive rotation of the footing and the supported structure. Therefore, isolated spread footings that are eccentrically loaded are not common in practice, unless there are no adjacent spread footings, since the eccentric loading can easily be mitigated by strapping the eccentrically loaded footing to an adjacent interior footing. Strap footings will be discussed in Section 10.10.

TABLE 10.2 Maximum and Minimum Soil Pressures in Eccentrically Loaded Spread Footings

Soil pressure distribution and applied moment, *M*	**Minimum soil pressure p_{min}**	**Maximum soil pressure p_{max}**
Positive soil pressure beneath the spread footing over entire length, *L*. ($M \le PL/6$ or $e \le L/6$)	$\frac{P}{WL} - \frac{6M}{WL^2} \ge 0$ ≤ Net allowable soil pressure	$\frac{P}{WL} + \frac{6M}{WL^2}$ ≤ Net allowable soil pressure
Positive soil pressure beneath the spread footing over entire length, *L* ($M = PL/6$ or $e = L/6$)	0	$\frac{2P}{WL}$ ≤ Net allowable soil pressure
Negative soil pressure beneath a portion of the spread footing and positive soil pressure exists over a length $L' < L$ where the footing is in contact with the soil. ($PL/6 < M < PL/2$ or $L/6 < e < L/2$)	0	$\frac{2P}{3W\left(\frac{L}{2} - e\right)}$ ≤ Net allowable soil pressure

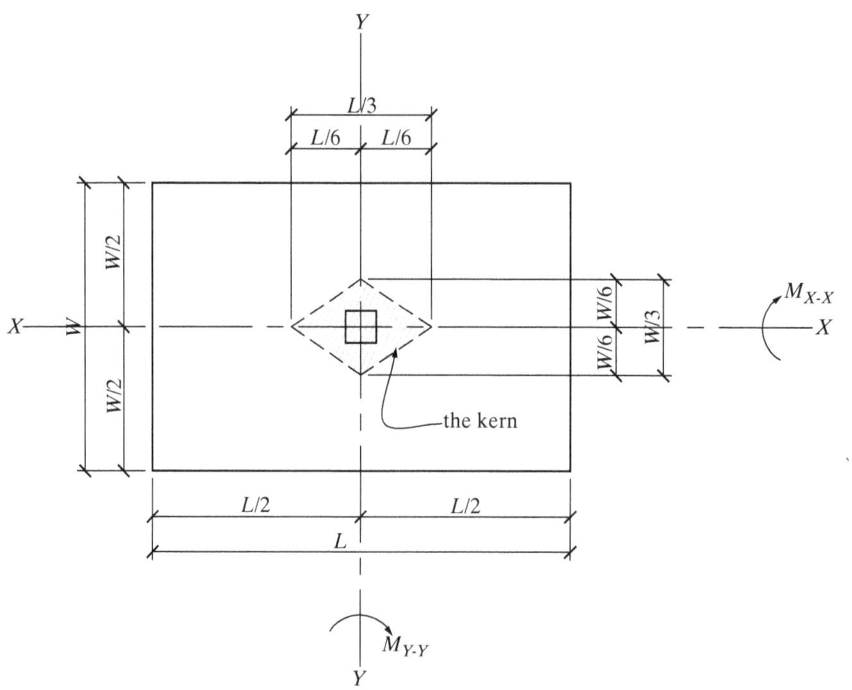

FIGURE 10.23 The kern in eccentrically loaded footings with bi-axial moments

EXAMPLE 10.5 Eccentrically Loaded Spread Footing

A 10 ft x 10 ft x 24" thick footing supports an 18" x 18" column and is subjected to the following service loads and moments: $D = 112.5$ kip, $L = 87.5$ kip, $M_{D,\,Y\text{-}Y} = 187.5$ ft-kip, $M_{L,\,Y\text{-}Y} = 145.8$ ft-kip. If the footing is reinforced with 14 #7 bars bottom each way (BEW), check the adequacy of the footing under the eccentric loading. The Geotechnical report specifies a NET allowable soil pressure of 4500 psf. Assume f_c' of 3000 psi and f_y of 60,000 psi. Neglect the weight of the footing.

Solution:

1. Calculate the unfactored vertical load and moment and check the applied soil pressure.

$$P_{D+L} = 1.0(112.5) + 1.0(87.5) = 200 \text{ kip}$$

$$M_{D+L,\,Y\text{-}Y} = 1.0(187.5) + 1.0(145.8) = 333.3 \text{ ft-kip}$$

$$e_{x-x} = \frac{M_{Y-Y}}{P} = \frac{333.3}{200} = 1.67 \text{ ft}$$

$L/6 = 10 \text{ ft}/6 = 1.67$ ft

Since $e_{X-X} = L/6$, we obtain the following from Table 10.2:

$$p_{\min} = 0$$

$$p_{\max} = \frac{2P}{WL} = \frac{2(200 \text{ kip})}{(10 \text{ ft})(10 \text{ ft})} = 4 \text{ ksf}$$

$\leq$ Net allowable soil pressure = 4.5 ksf, O.K.

For purposes of illustration, we will recalculate $p_{\min}$ and $p_{\max}$ using the general equation for the case when $e_{X-X} \leq L/6$.

$$p_{\min} = \frac{P}{WL} - \frac{6M}{WL^2} = \frac{(200 \text{ kip})}{(10 \text{ ft})(10 \text{ ft})} - \frac{6(333.3 \text{ ft-kip})}{(10 \text{ ft})(10 \text{ ft})^2}$$

$$= 2 - 2 = 0 \text{ ksf}$$

$$\geq 0$$

$\leq$ Net allowable soil pressure = 4.5 ksf (O.K.)

$$p_{\max} = \frac{P}{WL} + \frac{6M}{WL^2} = \frac{(200 \text{ kip})}{(10 \text{ ft})(10 \text{ ft})} + \frac{6(333.3 \text{ ft-kip})}{(10 \text{ ft})(10 \text{ ft})^2}$$

$$= 2 + 2 = 4 \text{ ksf}$$

$\leq$ Net allowable soil pressure = 4.5 ksf (O.K.)

2. Calculate the factored load and moment and the factored soil pressure that will be used to check the footing for shear and flexure. The factored load and moment are

$P_u = 1.2(112.5) + 1.6(87.5) = 275$ kip

$M_{u,\,Y\text{-}Y} = 1.2(187.5) + 1.6(145.8) = 458.3$ ft-kip

$$e_{u,x-x} = \frac{M_{u,Y-Y}}{P} = \frac{458.3 \text{ ft-kip}}{275 \text{ kip}} = 1.67 \text{ ft}$$

$L/6 = 10/6 = 1.67$ ft

Since $e_{u,X-X} = L/6$,

$p_{u,\min} = 0$

$$p_{u,\max} = \frac{2P_u}{WL} = \frac{2(275 \text{ kip})}{(10 \text{ ft})(10 \text{ ft})} = 5.5 \text{ ksf}$$

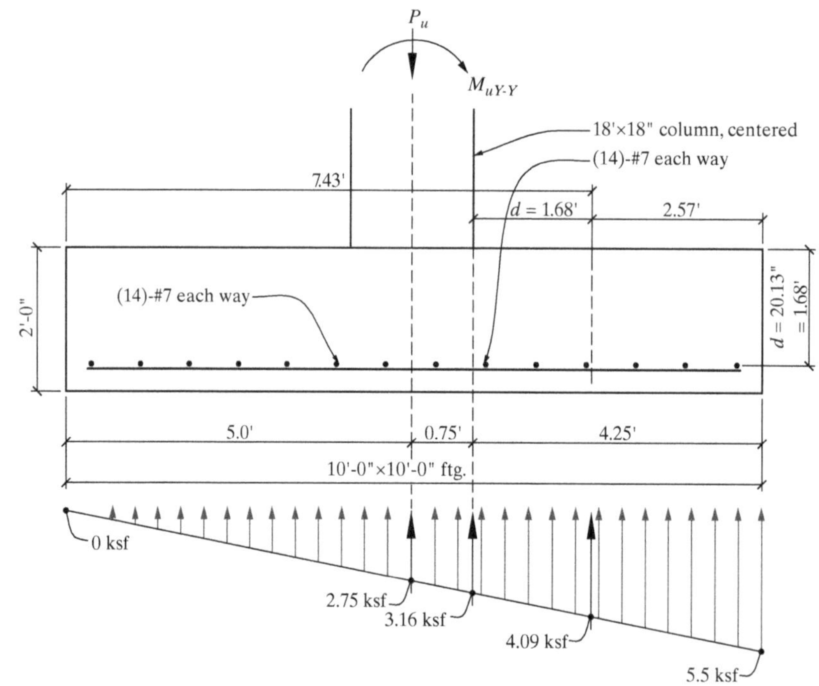

FIGURE 10.24 Factored soil pressures for Example 10.5

3. Check one-way shear at the critical section at a distance d from the face of the column using the factored pressures (see Figure 10.24).

Since #7 (i.e., 7/8" diameter) bottom bars each way is specified, the average effective depth, $d = 24" - 3"$ cover $-(7/8") = 20.13"$ will be used.

The factored shear, $V_u = (10 \text{ ft})[(4.09 \text{ ksf})(2.57 \text{ ft}) + ½\,(5.5 \text{ ksf} - 4.09 \text{ ksf})(2.57 \text{ ft})] = 123.2$ kip

The one-way shear strength of the concrete is

$$\phi V_c = \phi 2\sqrt{f'_c}b_w d$$
$$= (0.75)(2)\sqrt{3000}(10)(12)(20.13)$$
$$= 264{,}620 \text{ lb}$$
$$= 265 \text{ kip}$$
$$\phi V_n = \phi V_c = 265 \text{ kip} > V_u, \quad \text{(O.K.)}$$

The footing is satisfactory with respect to one-way shear.

4. Check the two-way or punching shear at the critical perimeter of *d*/2 from the face of the column using the average factored pressure at the centroid of the column.

The critical perimeter for two-way or punching shear is

b_o = 4(c + d) = 4(18" + 20.13") = 152.5"

c + d = 18" + 20.13" = 38.13" = 3.18 ft

At the critical perimeter located at a distance of *d*/2 from the face of the column, the average factored soil pressure (see Figure 10.24) = (1.88 + 3.63 ksf)/2 = 2.75 ksf

The factored punching shear at the critical perimeter is

$V_u = P_u - p_{avg}((c+d)(c+d)$ = 275 kip – 2.75 ksf(3.18 ft)(3.18 ft) = 247 kip

Using a similar approach to that in Problem 10.3 for calculating punching shear capacity, the governing punching shear capacity equation (with both λ and λ_s = 1.0) is

$$\phi V_n = \phi V_c = 4\phi\sqrt{f'_c}b_0 d$$
$$= 4(0.75)\sqrt{3000}(152.5")(20.13") = 672{,}565 \text{ lb}$$
$$= 673 \text{ kip} > V_u \qquad \text{(O.K.)}$$

The footing is satisfactory with respect to two-way or punching shear.

5. Using the factored soil pressures, calculate the factored moment at the critical section which occurs at the face of the column.

M_u = (10 ft) [½ (5.5 ksf – 3.16 ksf)(4.25 ft)(2/3 x 4.25 ft) + 3.16 ksf (4.25 ft)(½)(4.25 ft)]
= 567 ft-kip

Assume ϕ = 0.90 and with b = 10 ft., design the tension steel as follows:

$$required\ \bar{k} = \frac{M_u}{\phi b d^2} = \frac{(567)(12)}{(0.9)(10)(12)(20.13)^2}$$
$$= 0.1555 \text{ ksi}$$

From Table A.8, the required ρ = 0.0027, ε_t = 0.005, and ϕ = 0.90. Therefore,

$$\text{required } A_s = \rho b d$$
$$= 0.0027(10)(12)(20.13)$$
$$= 6.52 \text{ in.}^2$$

Check the ACI Code minimum reinforcement requirement. From Table A.5,

$$A_{s,\ min} = 0.0033(10\ \text{ft})(12\ \text{in/ft})(20.13\ \text{in})$$

$$= 7.97\ \text{in.}^2$$

Of the two steel areas, the larger (7.97 in.^{2}) controls.

Because the footing is square and an average effective depth was used, the steel requirements in the other direction may be assumed to be identical. Therefore, the specified 14 No. 7 bars each way, which yields an area ($A_s = 8.4$ in.^{2} in each direction), is greater than the steel area required. Therefore, the footing is adequate to resist the gravity loads and moments. All other design checks such as for bond and, anchorage, and bearing, can be carried out similarly to what was done in Example 10.3

10.9 COMBINED FOOTINGS

Combined footings are footings that support more than one column or wall. The two-column type of combined footing, which is relatively common, generally results from necessity. Two conditions that may lead to its use are (1) an exterior column that is immediately adjacent to a property line where it is impossible to use an individual footing centered on a column that would potentially cross a property line or other obstruction, such as another footing or utilities and (2) two columns that are closely spaced, causing their individual footings areas to overlap. In these situations, a rectangular or trapezoidal combined footing would usually be used. The choice of which shape to use is based on the difference in column loads as well as on the physical (dimensional) limitations. If the footing cannot be rectangular, a trapezoidal shape would then be selected. Combined footings are also used at braced frame locations to resist uplift forces; with combined footings, more column dead loads needed to resist the uplift forces can be engaged.

The physical dimensions (except thickness) of the combined footing are generally established by the allowable soil pressure. In addition, the centroid of the footing area should coincide with the line of action of the resultant of the two column loads. These dimensions are usually determined using service loads in combination with the net allowable soil pressure.

EXAMPLE 10.6

Determine the shape and proportions of a combined footing subject to two column loads, as shown in Figure 10.25.

Solution:

1. The design data are as follows: Service load on the footing from column A is 300 kip and from column B is 500 kip, and the allowable soil pressure is 6.00 ksf. Therefore, the unfactored resultant, R = 300 kip + 500 kip = 800 kip.

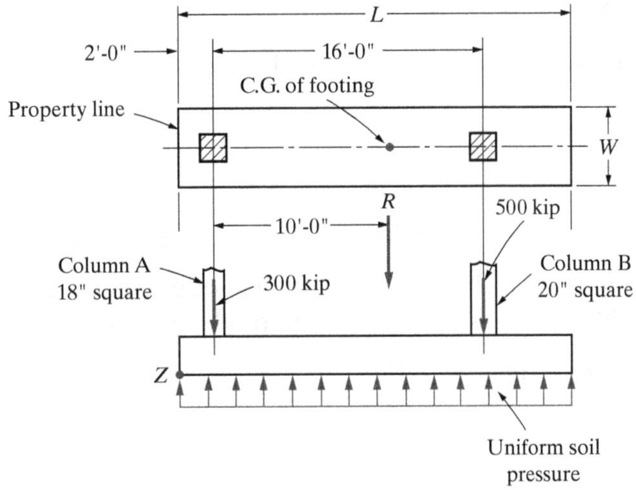

FIGURE 10.25 Sketch for Example 10.6.

2. Locate the resultant column load by a summation of moments about point Z in Figure 10.25:

$$\Sigma M_z = 300 \text{ kip}(2 \text{ ft}) + 500 \text{ kip}(18 \text{ ft}) = 800 \text{ kip}(x)$$

from which

$$x = 12 \text{ ft-0 in.}(\text{measured from Z})$$

3. Assuming a rectangular shape, establish the length of footing L so that the centroid of the footing area coincides with the line of action of the resultant force R:

$$\text{Required } L = 12 \text{ ft}(2) = 24 \text{ ft-0 in.}$$

4. Assume a footing thickness of 3 ft-0 in. Therefore, its weight = 0.150(3) = 0.450 ksf, and the net allowable soil pressure for superimposed loads = 6.00 – 0.450 = 5.55 ksf. This neglects any soil *on* the footing.

5. The footing area required is

$$\frac{R}{5.55} = \frac{800 \text{ kip}}{5.55 \text{ ksf}} = 144.1 \text{ ft}^2$$

6. With a length = 24 ft-0 in., the footing width W required is

$$\frac{144.1 \text{ ft}^2}{24 \text{ ft}} = 6 \text{ ft-0 in.}$$

7. The actual uniform soil pressure is

$$\frac{800 \text{ kip}}{(6 \text{ ft})(24 \text{ ft})} + 0.450 = 6.00 \text{ ksf} \qquad \text{(O.K.)}$$

EXAMPLE 10.7

Using the design data from Example 10.6, determine the proportions of a combined footing if the footing length is limited to 22 ft.

Solution:

1. The resultant column load is located as in Example 10.6 at a point 12 ft-0 in. from point Z.
2. Assume a footing thickness of 3 ft-0 in. Therefore, its weight = 3(0.150) = 0.450 ksf, and the allowable soil pressure for superimposed loads is 6.00 – 0.450 = 5.55 ksf.
3. The footing area required is then

$$\frac{800\,\text{kip}}{5.55\,\text{ksf}} = 144.1\ \text{ft}^2$$

4. Assume a trapezoidal shape. The area A of a trapezoid is (see Figure 10.26)

$$A = \frac{(b + b_1)L}{2}$$

from which

$$b + b_1 = \frac{2A}{L} = \frac{2(144.1\,\text{ft}^2)}{22\,\text{ft}} = 13.1\ \text{ft}$$

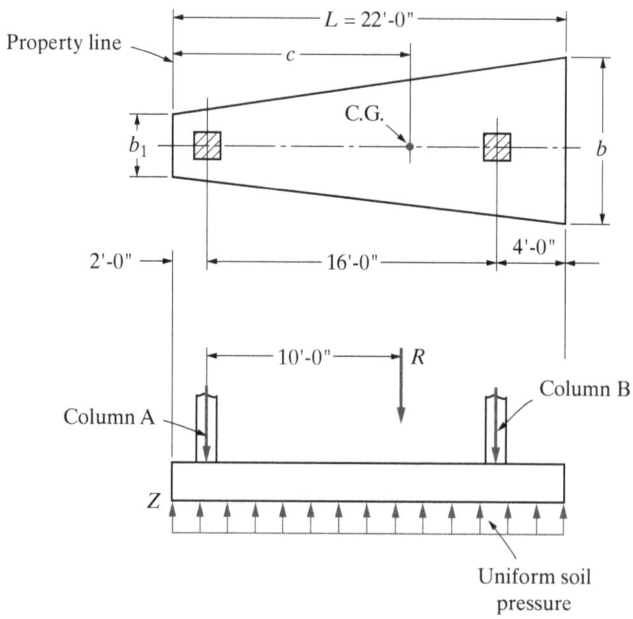

FIGURE 10.26 Sketch for Example 10.7.

5. The center of gravity of the trapezoid and the resultant force of the column loads are to coincide. The location of the center of gravity, a distance c from point Z, may be written as:

$$c = \frac{L(2b+b_1)}{3(b+b_1)} = \frac{L(b+b+b_1)}{3(b+b_1)} = 12 \text{ ft}$$

6. We now have two equations that contain b and b_1. The equation of step 5 may easily be solved for b by substitution of $L = 22$ ft and $(b + b_1) = 13.1$ ft:

$$\frac{22(b+13.1)}{3(13.1)} = 12$$

from which

$$b = \frac{12(3)(13.1)}{22} - 13.1 = 8.34 \text{ ft}$$

and

$$b_1 = 13.1 - 8.34 = 4.76 \text{ ft}$$

7. Use $b = 8$ ft-4 in. and $b_1 = 4$ ft-9 in. Thus

$$\text{Actual } A = \frac{22(4.75+8.33)}{2} = 143.9 \text{ ft}^2$$

The footing is very slightly undersized. No revision is warranted.

The structural design of rectangular and trapezoidal combined footings is generally based on uniform soil pressure, even though loading combinations will almost always introduce some eccentricity with respect to the centroid of the footing. The determination of the footing thickness and reinforcement must be based on factored loads and soil pressure to be consistent with the ACI strength method approach. The assumed footing behavior, which is briefly described, is a generally used approach to simplify the design of the footing.

In Figure 10.26, the columns are positioned relatively close to the ends of the footing. Assuming the columns as the supports and the footing subjected to an upward, uniformly distributed load caused by the uniform soil pressure, moments that create tension at the top of the footing between the columns will predominate in the longitudinal direction. Therefore, the principal longitudinal reinforcement will be placed at the top of the concrete footing between the two columns, equally distributed across the footing width. Between the columns and the ends of the combined footing in the longitudinal direction, the upward soil pressures will create tension at the bottom of the footing; therefore, longitudinal reinforcement at the bottom of the footing is also required from the edge of the footing to a point past the moment inflection point between the columns. In the transverse direction, somewhat smaller moments will cause tension at the bottom of the footing; therefore, transverse steel will be placed under each column at the bottom of the footing to distribute the column load in the transverse direction using the provisions for individual column footings. In effect, the combined footing behaves as a wide rectangular beam in the longitudinal direction, which may then be designed using the ACI Code provisions for flexure.

Pertinent design considerations may be summarized as follows:

1. Main reinforcement (uniformly distributed) is placed in a longitudinal direction in the *top* of the footing, assuming the footing to be a longitudinal beam. Longitudinal reinforcement is also required in the *bottom* of the footing to resist the cantilever moments at the columns due to the upward soil pressures.
2. Shear should be checked considering both one-way shear at a distance d from the face of the column and two-way (punching) shear on a perimeter $d/2$ from the face of the column.
3. Stirrups or bent bars are frequently required to maintain an economical footing thickness. This assumes that the shear effect is uniform across the width of the footing.
4. Transverse reinforcement is generally uniformly placed in the *bottom* of the footing within a band having a width not greater than the column width plus twice the effective depth of the footing. The design treatment in the transverse direction is similar to the design treatment of the individual column footing, assuming dimensions equal to the band width as previously described and the transverse footing width.
5. Additionally, longitudinal steel is also placed in the bottom of the footing to tie together and position the stirrups (if any) and transverse steel.

10.10 CANTILEVER OR STRAP FOOTINGS

A third type of combined footing is generally termed a *cantilever* or *strap footing* and is often used to resist the effect of eccentric gravity loading on an isolated footing or pile foundation; the eccentrically loaded footing or foundation is "strapped" to an interior concentrically loaded footing or foundation (see Figures 10.27 and 10.28). This is an economical type of footing when the proximity of a property line precludes the use of other types. For instance, an isolated column footing may be too large for the area available, and the nearest column is too distant to allow a rectangular or trapezoidal combined footing to be economical. The strap footing may be regarded as two individual column footings connected by a strap beam.

In Figure 10.27, the exterior footing (or foundation) is placed eccentrically under the exterior column so that it does not violate the property-line constraints. This would produce a nonuniform pressure distribution under the footing, which could lead to footing rotation. To balance this rotational or overturning effect, the exterior footing is connected by a stiff beam, or strap, to the nearest interior footing, and uniform soil pressures under the footings are assumed. The strap, which may be categorized as a flexural member, is subjected to both bending moment and shear, resulting from the forces P_e and R_e acting on the exterior footing. As shown in Figure 10.27, the applied moment is counterclockwise, and the shear will be positive because $R_e > P_e$. At the interior column, there is no eccentricity between the column load P_i and the resultant soil pressure force R_i. Therefore, we will assume that no moment is induced in the strap at the interior column.

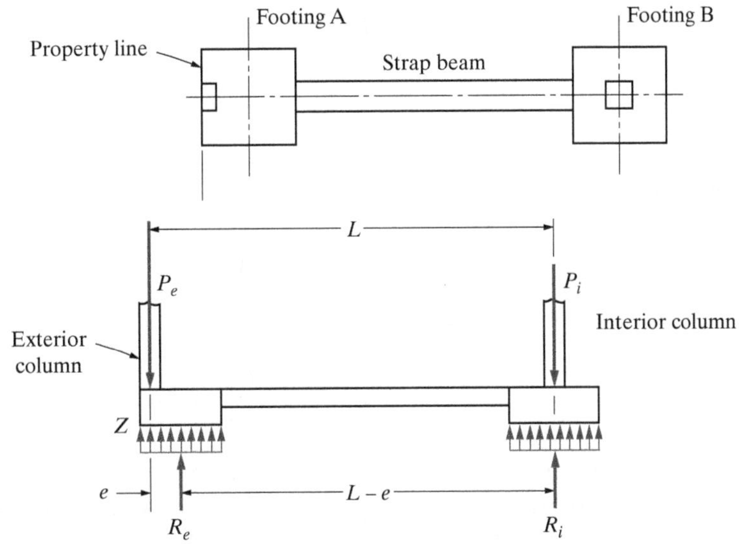

FIGURE 10.27 Cantilever footing.

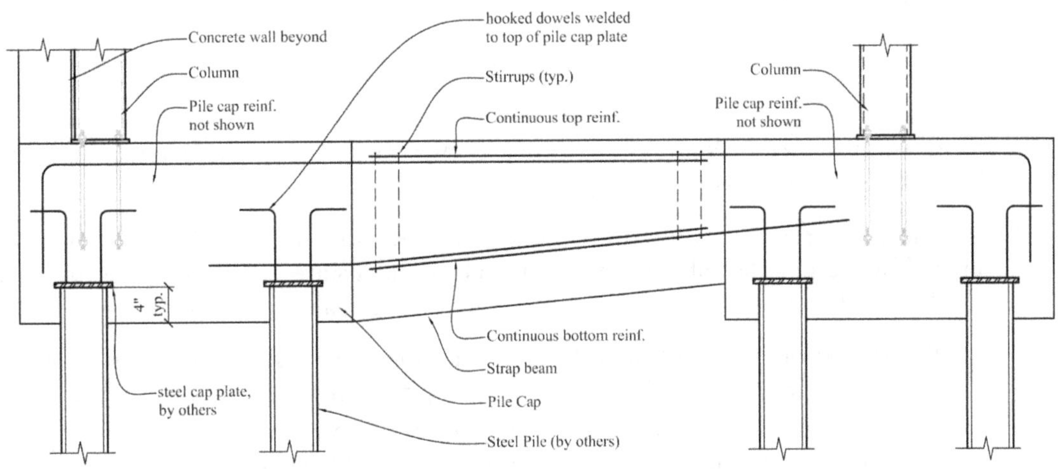

FIGURE 10.28 Strap beam at a pile or caisson foundation.

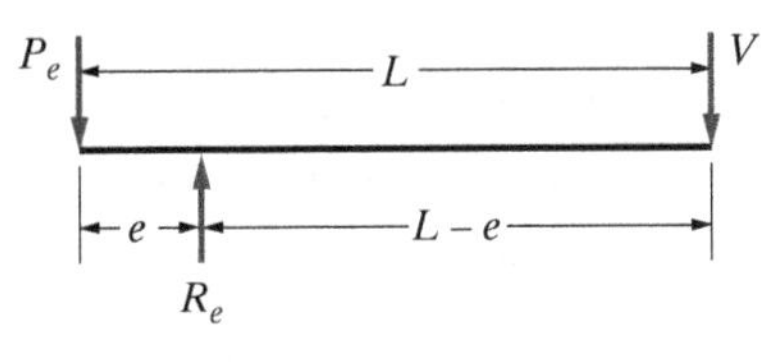

FIGURE 10.29 Free body diagram of the strap beam for the cantilever footing in Figure 10.27.

A free-body diagram of the cantilever or strap footing in Figure 10.27 is shown in Figure 10.29. We define V as the vertical shear force necessary to keep the strap in equilibrium, as shown in the free body diagram in Figure 10.29. Then, with P_e known, V and R_e may be calculated using the principles of statics. A moment summation about R_e yields

$$P_e e = V(L-e)$$

$$V = \frac{P_e e}{L-e}$$

and from a summation of vertical forces,

$$R_e = P_e + V$$

Then, by substitution,

$$R_e = P_e + \frac{P_e e}{L-e}$$

Note that V acting downward on the strap beam also means that V is an uplift force on the interior footing. Therefore, using the equilibrium of forces acting on the interior column in both Figures 10.27 and 10.29 gives

$$R_i = P_i - V$$

and, by substitution,

$$R_i = P_i - \frac{P_e e}{L-e}$$

In summary, R_e becomes greater than P_e by a magnitude of V, whereas R_i becomes less than P_i by a magnitude equal to V.

The footing areas required are merely the reactions R_e and R_i based on service loads divided by the net allowable soil pressure. These values are based on an assumed trial e and may have to be recomputed until the trial e and the actual e are the same.

The structural design of the interior column footing is simply the design of an isolated column footing subject to a load R_i. The exterior footing is generally considered as under one-way *transverse* bending similar to a wall footing, with longitudinal steel furnished by extending the strap beam steel into the footing. The selection of footing thickness and reinforcement should be based on factored loads to be consistent with the ACI Code strength design approach.

The strap beam is assumed to be a flexural member that is not bearing on the soil underneath. Many designers make a further simplifying assumption that the beam weight is carried by the underlying soil; hence, the strap is designed as a rectangular beam subject to a constant shearing force and a linearly varying negative bending moment based on factored loads. The strap beam is reinforced with longitudinal reinforcement to resist the bending moments, and stirrups to resist the shear forces.

EXAMPLE 10.8

Determine the size of the exterior and interior footings of a strap footing for the conditions and design data furnished in Figure 10.30.

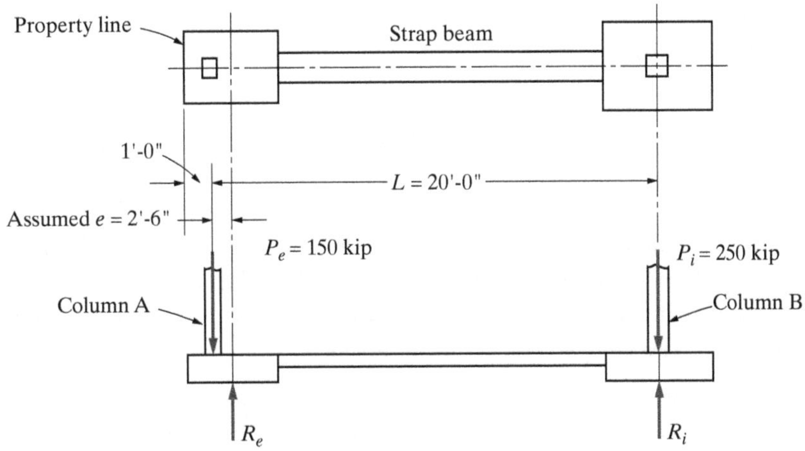

FIGURE 10.30 Sketch for Example 10.8.

Solution:

1. The design data are as follows: Service load on the footing from column A is 150 kip and from column B is 250 kip, and the allowable soil pressure is 4.00 ksf.
2. Assume that $e = 2$ ft-6 in. and that the footing thickness is 2 ft-0 in. Therefore, the footing weight = 2(0.150) = 0.300 ksf.
3. The net allowable soil pressure for superimposed loads = 4.0 – 0.300 = 3.70 ksf.
4. Determine the strap beam shear V:

$$V = \frac{P_e e}{L-e} = \frac{150(2.5)}{17.5} = 21.4 \text{ kip}$$

5. The footing reactions are

$$R_e = P_e + V = 150 + 21.4 = 171.4 \text{ kip}$$
$$R_i = P_i - V = 250 - 21.4 = 228.6 \text{ kip}$$

6. For the *exterior* footing, the required area is

$$\frac{171.4 \text{ kip}}{3.7 \text{ ksf}} = 46.3 \text{ ft}^2$$

Use a footing 7 ft-0 in. by 7 ft-0 in. ($A = 49$ ft^2). Note in Figure 10.30 that the actual e is then the same as the assumed e, and no revision of calculations is necessary.

For the *interior* footing, the required area is

$$\frac{228.6 \text{ kip}}{3.7 \text{ ksf}} = 61.8 \text{ ft}^2$$

Use a footing 8 ft-0 in. by 8 ft-0 in. ($A = 64$ ft^2).

Next, calculate the bending moments and shear forces in the strap beam, and then design the strap beam using the methods presented in previous chapters.

10.11 ANALYSIS AND DESIGN OF MAT FOUNDATIONS

As discussed previously, mat foundations are used where the structure supports heavy loads and the soil bearing capacity is low to great depths such that the column footing areas overlap in both orthogonal directions. The mat foundation slab acts like a continuous inverted two-way slab supporting all the columns and walls in the structure. There are several methods that can be used for the analysis of mat foundation slabs.

The "conventional rigid method" [6] for analyzing mat foundations assumes that the mat slab behaves as an infinitely rigid body subjected to the combined total axial load, P, and moments acting about the orthogonal plan axes of the mat slab.

A linear elastic stress distribution in the soil is also assumed. The elastic stress equation can be used to determine the soil pressure at any point underneath the mat slab. For a mat foundation with plan dimensions L and W along the X- and Y-axes, respectively, the soil pressure, q_i, at any point below the mat slab is given as [6]:

$$q_i = \frac{P}{WL} \pm \frac{M_{X-X}(y_i)}{\left(\frac{LW^3}{12}\right)} \pm \frac{M_{Y-Y}(x_i)}{\left(\frac{WL^3}{12}\right)}$$

$\leq$ net allowable soil pressure

where

P = sum of all the axial dead and live loads on the mat slab from all the columns and walls

$M_{X\text{-}X}$ is the moment about the X- axis = Pe_y

$M_{Y\text{-}Y}$ is the moment about the Y- axis = Pe_x

x_i, y_i = the coordinates of a point on the mat along the X-axis and Y- axis, respectively.

Using the elastic stress equation above, the soil pressures below the mat slab can be determined for unfactored dead plus live loads. These pressures are compared to the net allowable soil pressure to ensure that the soil bearing capacity is not exceeded. The factored soil pressures are calculated using the factored load and moments. These factored soil pressures can then be used in the equivalent frame analysis of the mat slab where the mat is treated as an inverted slab. The mat slab is divided into design strips centered on the column lines in both orthogonal directions, and for simplicity the average factored soil pressure in the strip can be used, in lieu of the linearly varying soil pressures, for the analysis and design of the strip [6]. The analyses yield the positive and negative moments in the mat slab in both orthogonal directions from which the steel reinforcement at the top and bottom of the slab, respectively, in both orthogonal directions can be calculated using the methods presented in Chapter 2. The mat is usually reinforced with continuous top and bottom reinforcement in both orthogonal directions. Given the large depth or thickness of some mat foundations (sometimes up to 6 ft to 10 ft thick), the stability of the mat foundation reinforcement cage should be carefully addressed by the contractor since they are responsible for the means and methods of construction. A note regarding the mat foundation reinforcement cage and the contractor's responsibility to ensure its stability should be included in the cast-in-place concrete specification (see Section 14.11).

An alternate approach to the analysis of mat foundations is to use one of several available finite-element-based computer analysis software [7]. In this method, the mat is modeled as a flexible slab on an elastic foundation, where the slab is supported on springs in both orthogonal directions. The stiffness of the springs, k, also known as the modulus of subgrade reaction, is supplied by the geotechnical consultant. The coefficient of subgrade reaction is not constant and may vary significantly within the footprint of the mat foundation, and this variability in the modulus of subgrade reaction would influence the moments in the mat slab [8]. The conditions under which the conventional rigid method can be used are given in ref. [9]. Note that for both methods, the mat slab thickness is selected so that the punching shear capacity is not exceeded.

For mat foundations, the minimum area of flexural reinforcement required in each principal direction and placed near each tension face of the mat is 0.0018bh. Since the mat slab is subjected to positive and negative moments in the two orthogonal directions (i.e., moments that cause tension at the bottom of the mat slab at the column locations, and at the top of the mat slab between the columns), this means that in each face and in each orthogonal direction, a minimum area of reinforcement equal to at least 0.0018bh must be provided, per ACI 318 Sections 13.3.4.4 and 8.6.1.1. The maximum spacing of the flexural reinforcement is the lesser of $3h$ or 18 inches. As an example, for a 6 ft. deep mat foundation, the minimum reinforcement area required on each face and in each principal direction is $A_s = 0.0018(12\text{ in.})(6\text{ ft.})(12\text{ in./ft}) = 1.56\text{ in}^2/\text{ft}$. Therefore, the minimum steel that should be provided is #9 at 7½ in. on centers Top and Bottom Each Way (i.e., #9 @ 7½-in. o.c. BEW + #9 @ 7½-in. o.c. TEW).

10.12 DEEP FOUNDATIONS—PILES, DRILLED SHAFT (CAISSONS), AND PILE CAPS

Piles/Caissons

Deep foundations, as discussed earlier, are typically used where the soil bearing capacity is low for great depths, and they are used to support structures that are subjected to high gravity and lateral loads, such as tall buildings and bridges. Common examples of deep foundations are piles and drilled shafts – also known as caissons or drilled pier or bored piles. Piles are installed by driving or forcing them into the ground with a hammer, and this usually causes vibrations and noise. They resist loads through end bearing and/or through skin friction between the soil and the surface of the pile. Piles can be made of steel (e.g., HP steel sections or steel pipes filled with concrete) or precast or cast-in-place concrete. The cast-in-place concrete piles can be cased or uncased [10]. For piles with permanent casing, a steel shell is driven into the ground, the soil within the shell is removed, and the shell is subsequently filled with concrete. For the uncased pile, the steel shell is removed gradually as the concrete is cast, starting from the bottom of the pile. Since the efficiency of a pile within a pile group is impacted by the spacing between adjacent piles, the geotechnical consultant typically specifies the minimum spacing between adjacent piles. Typical spacings commonly used in practice are 3 to 4 pile diameters between adjacent piles, but no less than 3 ft center-to-center spacing or 2 ft. clear spacing between adjacent piles, and an edge distance of 2-pile diameters minimum measured from the centroid of the edge pile to the edge of the pile cap. To prevent vertical splitting at the edge of the pile caps, the Concrete Reinforcing Steel Institute (CRSI) Handbook [24] prescribes the following minimum edge distances from the edge of the pile cap to the center of the edge or corner pile, as a function of the unfactored axial load capacity of the pile, as follows:

Unfactored Axial Load Capacity (D+L) of the Pile, P	Edge Distance
0 – 60 tons*	15 inches
$60 < P \leq 120$ tons	21 inches
$120 < P \leq 200$ tons	27 inches
$P > 200$ tons	30 inches

*1 ton = 2000 Ib or 2 kip

The minimum embedment depths of the piles or caissons into the pile caps are typically 6 inches for steel piles, and 4 inches for concrete piles. The clear concrete cover between the top of the steel piles and the bottom reinforcement in the pile cap is 3 inches.

Drilled shaft or caissons can support loads through skin friction or through end bearing with the bottom of the caisson enlarged or "belled" to provide a larger surface area and hence a higher axial load capacity. To construct caissons—usually without permanent casing, the hole is drilled with a steel casing, and the soil within the steel shell is removed. The hole is then filled with concrete as the steel casing is gradually withdrawn. Permanent steel casings for drilled shafts or caissons may be used in areas where there is unstable soil or a debris field in the drilled shaft that may become mixed with the concrete, or there is an existing concrete debris field with large voids within the shaft which can suck away some of the fluid concrete, thus leading to the loss of a significant portion of the concrete pour. Not using permanent casings in the above situations could lead to deficient caissons which could be very costly to remediate [17]. Caissons are quieter to install compared to driven piles, and there is less vibration during the caisson installation. Caissons can be reinforced or unreinforced. Often times, a tied or spiral type reinforcement cage might be provided for the top several feet of the caisson. Because of their larger diameter compared with piles, it is possible for a human being to be lowered down to the bottom of the caisson to inspect the bearing strata to ensure that it is as sound as assumed by the geotechnical consultant. Caisson diameters typically vary from 2 ft 6 in. to 10 ft, though caissons as large as 30 ft in diameter have been used in supertall buildings [16]. Since caissons are drilled rather than driven, and because they are oftentimes end bearing, there is no impact on adjacent caissons unlike driven piles where the efficiency of a pile group is impacted by the spacing between adjacent piles.

The axial load capacity of a concrete caisson is the lesser of the short column capacity (since the caisson is assumed to behave as a concrete column braced all along its length by the surrounding soil) and the strength based on the bearing capacity of the bearing strata [10]. The required caisson area (from which the diameter is calculated) can be determined by dividing the unfactored gravity load on the caisson by the allowable bearing capacity of the bearing strata. The shear strength of a caisson is determined from Chapter 4 using $b = D$, and $d = 0.8D$, where D is the diameter of the caisson. The maximum allowable compressive strength for deep foundation members is now also specified in ACI Code Table 13.4.2.1:

Type of Deep Foundation Member	Maximum allowable compressive strength
Uncased cast-in-place concrete drilled or augered pile (e.g., drilled caissons)	$P_a = 0.3f_c'A_g + 0.4f_yA_s$
Cast-in-place concrete pile in rock or within a pipe, tube, or other permanent metal casing that does not satisfy ACI Code Section 13.4.2.3	$P_a = 0.33f_c'A_g + 0.4f_yA_s$
Metal cased concrete pile confined in accordance with ACI Code Section 13.4.2.3 (i.e., satisfying conditions (a) through (f)). (Nominal diameter of the pile ≤ 16 inches)	$P_a = 0.4f_c'A_g$
Precast nonprestressed concrete pile (A_s is minimum of 4 bars and ≥ $0.008A_g$ per ACI Code Section 13.4.5.3)	$P_a = 0.33f_c'A_g + 0.4f_yA_s$
Precast prestressed concrete pile	$P_a = (0.33f_c' - 0.27f_{pc})\,A_g$

Adapted from ACI 318-19 Table 13.4.2.1

NOTE

- *A_s is the vertical reinforcement in the concrete pile and does not include the steel casing, pipe, or tube.*
- *A_g is the gross cross-sectional area of the concrete pile, and does not include the steel casing, pipe, or tube.*
- *f_{pc} = compressive stress in the concrete after allowance for all prestress losses, assuming total prestress losses of 30,000 psi (ACI Code Section 13.4.5.5). The minimum compressive stresses in precast prestressed piles are 400 psi for pile length no greater than 30 ft., 550 psi for pile lengths between 30 ft. and 50 ft., and 700 psi for pile lengths greater than 50 ft. (ACI Code Table 13.4.5.4). The advantage of prestressing is that it limits cracking in the concrete piles.*

To use the preceding table, the deep foundation member is assumed to be laterally supported for its entire height and subjected to a maximum moment caused by an accidental eccentricity of 5 percent of the member width or diameter (ACI Code Section 13.4.2.1). If these two foregoing conditions are not satisfied, the deep foundation member must be designed using the strength design procedures for columns that was discussed previously in Chapter 9, and using the strength reduction factors in the following table (ACI Code Section 13.4.3.2) for deep foundation members subjected to axial load *without* moments:

Type of Deep Foundation Member	Compressive strength reduction factors for Strength Design, ϕ
Uncased cast-in-place concrete drilled or augered pile (e.g., drilled caissons)	0.55
Cast-in-place concrete pile in rock or within a pipe, tube - *with wall thickness of steel pipe < 0.25* in., or other permanent metal casing that does not satisfy ACI Code Section 13.4.2.3	0.60
Cast-place concrete-filled steel pipe pile, *with wall thickness of steel pipe ≥ 0.25* in.	0.70
Metal cased concrete pile confined in accordance with ACI Code Section 13.4.2.3 (i.e., satisfying conditions (a) through (f)). (Nominal diameter of the pile ≤ 16 inches)	0.65
Precast nonprestressed concrete pile	0.65
Precast prestressed concrete pile	0.65

Adapted from ACI 318-19 Table 13.4.3.2

For deep foundations subject to tension, shear, and combined axial force and moment, the strength reduction factors presented in previous chapters (i.e., 0.65 to 0.90 for axially load concrete members and members in flexure, and 0.75 for concrete members subject to shear) should be used (ACI 318, Table 21.2.1).

The tie or spiral reinforcement requirements in Chapter 9 do not apply to precast concrete piles (ACI Code Section 13.4.3.2), but instead, the following tie arrangement should be used (to resist driving and handling stresses ACI Code Table 13.4.5.6(b)):

Location of the transverse reinforcement along the length of the pile (Use a minimum of No. 3 bar)	Maximum center to center spacing of the transverse reinforcement
First five ties or spiral at each end of the pile	1 inch
24 inches from each end of the pile	4 inches
Remainder of the pile	6 inches

Adapted from ACI 318-19 Table 13.4.5.6(b)

The top of individual caissons usually has a concrete cap that supports the column. Where there are grade beams, they are framed over the top of the caissons and serve also as the concrete cap at the caisson locations. The grade beam-caisson moment resisting frame can also serve as the lateral force resisting system to transfer the base shear and lateral forces from the superstructure lateral force resisting system into the soil. For added lateral and uplift capacity, the caissons are usually socketed into the bedrock, and socket depths of up to 15 ft. into the solid bedrock have been used in tall buildings. As previously discussed in Section 10.1, caissons or piles may be designed to function as vertical cantilevers to resist the superstructure lateral loads/base shear. In such cases, the geotechnical consultant will provide in the soil report the assumed point of fixity of the caisson or pile measured from grade, or the IBC guidance previously presented in Section 10.1 may be used.

As previously indicated, grade beams and caisson caps in heaving or expansive soils require void forms to be placed underneath them to prevent undesirable uplift forces from the expansive soil being transferred to the foundation and other structural elements. Exterior grade beams also require void forms to provide the necessary frost protection and prevent upward forces from the frozen soil from being transferred to the grade beam.

Piles are laid out in groups of 2, 3, 4, 5, 6 or more piles, depending on the column or wall loads, and they are connected together at their top ends with a concrete pile cap through which the loads from the column or wall are transmitted to the group of piles (see Figure 10.31).

The pile cap must be capable of supporting the column and wall loads while spanning between the piles. Depending on the shear span-to-depth ratio (see Figure 10.39), a pile cap may act like a deep beam. Pile caps that do not fall into the deep beam category can be designed using the methods covered in previous chapters. The design of pile caps that act as deep beams is covered later in this chapter. Figure 10.32 shows a typical pile foundation plan with different pile cap layouts. More details of some typical pile cap layouts are shown in Figure 10.34, and Figure 10.33 shows a typical section at a two-pile pile cap. Note that for steel piles with a steel cap plate, the vertical dowels in the pile cap should be welded to the top of the steel cap plate – to resist shear and uplift forces. An introduction to the design of pile caps, a topic not often covered in reinforced concrete textbooks [19], is discussed later in this chapter.

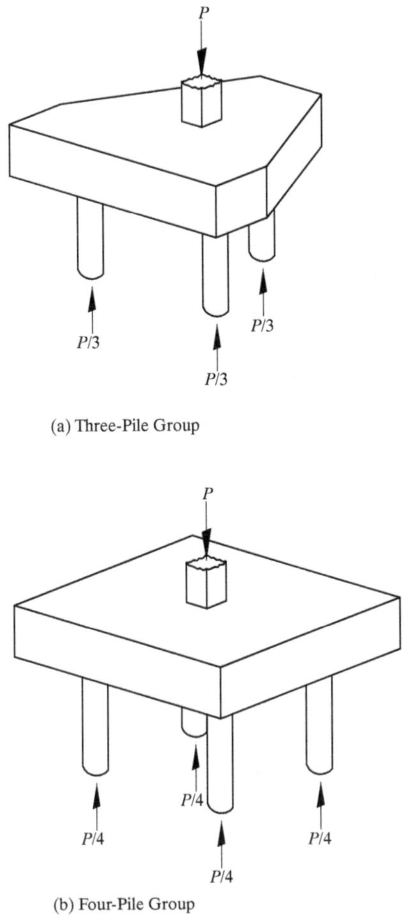

FIGURE 10.31 Piles and pile caps.

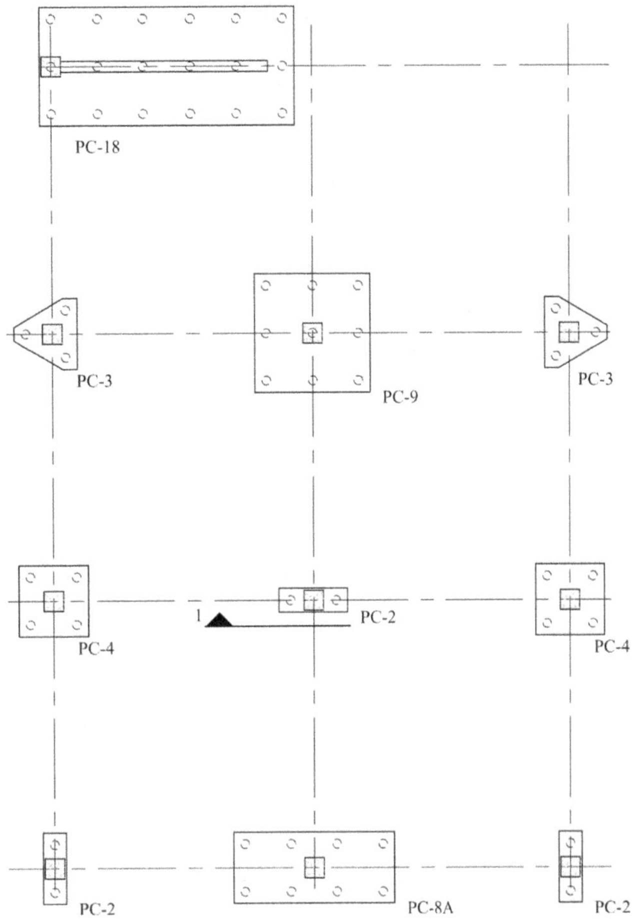

FIGURE 10.32 Typical pile foundation part plan with different pile cap layouts.

The pile reactions are assumed to act at the centroid of each pile (ACI Code, Section 13.4.2.2), and the total number of piles in a pile cap layout is obtained by dividing the total unfactored load in the column or wall by the unfactored allowable load for each pile (ACI Code, Section 13.4.1.1) - a value that is supplied by the Geotechnical consultant.

When the bedrock is closer to the surface and the column or wall loads are high, concrete caissons or drilled concrete shafts may make the most sense for the foundations. For super tall buildings where the bedrock may be hundreds of feet below grade, piles, or caissons (drilled shafts) are typically used. For caissons or piles, the gravity load is transferred from the columns or walls through a concrete pile cap, which transmits the loads to the caissons, and the caisson then transfers the load to the bearing strata through end bearing. Net uplift loads are resisted by tension reinforcement in the caissons, and by socketing the caissons or piles into the bedrock, or by the use of rock anchors.

Depending on the axial and lateral loads on a caisson, they may be unreinforced, or reinforced throughout its entire length, or only in the upper portion of the caisson. Figure 10.35 shows a typical caisson (drilled shaft) foundation plan, and Figure 10.36 shows a typical section at a caisson foundation. In Figure 10.35, the notations CA3, CA4, and CA5, etc., denote caissons with different diameters, and the properties (diameter and reinforcement) of these caissons are typically given in the caisson schedule. Similarly, the notations P1 and P2 denote the different pier or pedestal sizes, and their properties (plan size and reinforcement) are usually indicated in the pier schedule. The notation GB1 refers to grade beams which are sometimes required to tie the caissons together.

Note that in Seismic Design Categories C, D, E and F, deep foundations such as drilled piers or caissons and individual pile caps must be interconnected by ties (typically grade beams like those shown in Figures 10.33 and 10.36) in both orthogonal directions capable of resisting a tension and compression force of $0.10S_{DS}$ times the larger pile cap or column factored dead plus factored live load, where S_{DS} is the design spectral response acceleration parameter at short periods (see ASCE 7-22 Sections 12.13.7.2 and 12.13.8.2) [26]. In addition, spread footings founded on liquifiable soils in Seismic Design Categories E and F must also be interconnected by ties in both orthogonal directions (ASCE 7-22 Sections 12.13.8.2 and 12.13.9.2.1.1) [26].

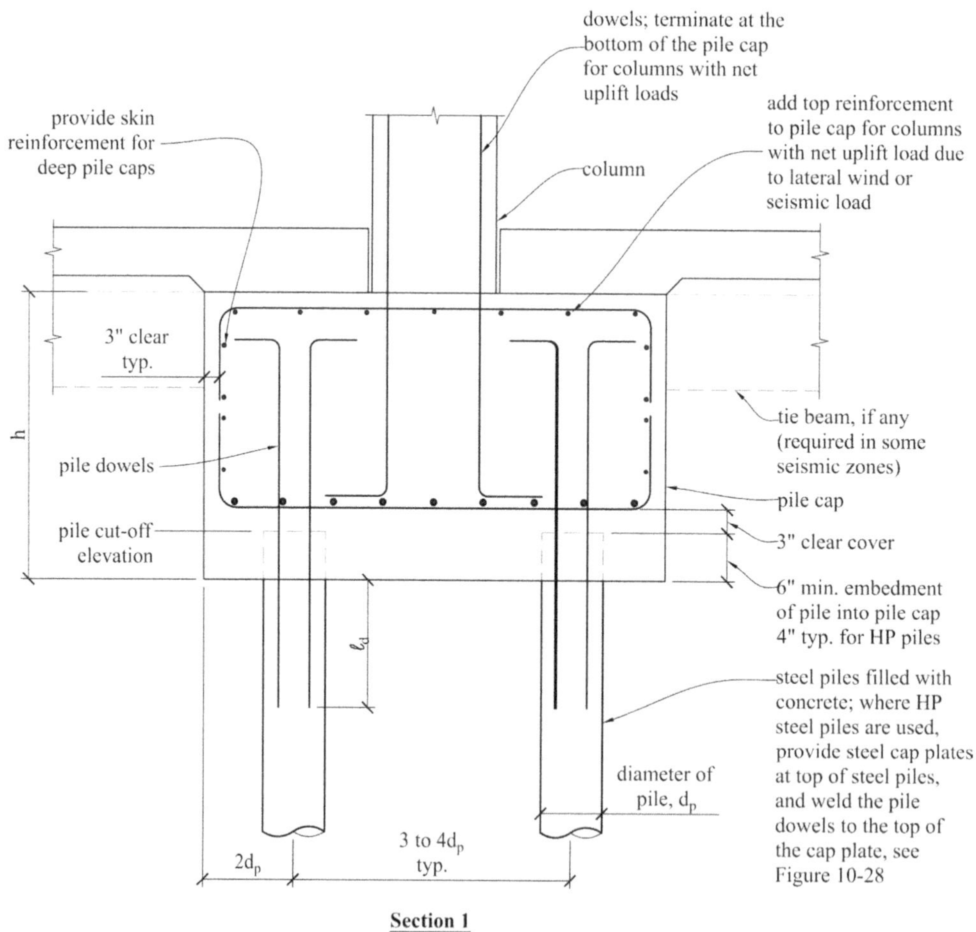

FIGURE 10.33 Typical section at two-pile cap locations.

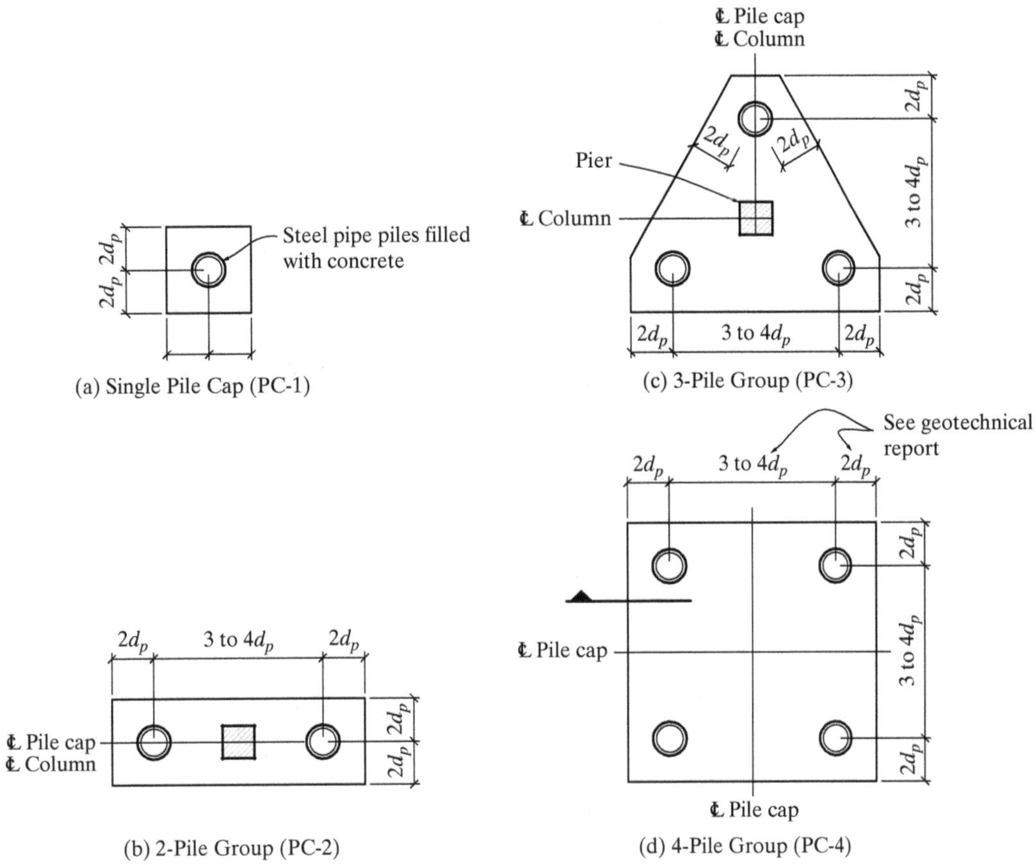

FIGURE 10.34 Typical pile cap layout.

Load Distribution in Pile Caps Subject to Axial Load and Moments

Some situations occur in practice where a pile cap is subjected to moments about both orthogonal axes in addition to the gravity loads due to the eccentricity of the column or wall load with respect to the centroid of the pile group, or due to lateral loads. The maximum axial load on a pile within a pile group that is subjected to combined gravity loads and moments about both orthogonal axes of the pile cap (see Figure 10.37) can be obtained using the following elastic equation [10].

The load on pile i,

$$P_i = \frac{P_{total}}{n} \pm \frac{M_{X-X}(y_i)}{\sum_{i=1}^{n}(y_i)^2} \pm \frac{M_{Y-Y}(x_i)}{\sum_{i=1}^{n}(x_i)^2}$$

where

n = total number of piles in the pile group

$M_{X\text{-}X}$ = moment about the X-X axis

$M_{Y\text{-}Y}$ = moment about the Y-Y axis

x_i and y_i are the coordinate locations of the centroid of the piles.

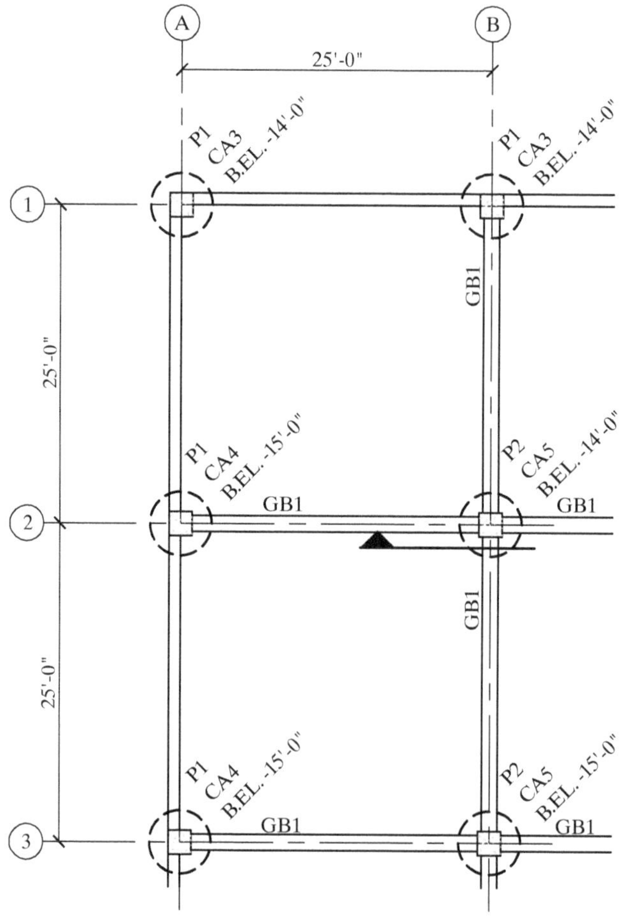

FIGURE 10.35 Typical Caisson foundation part plan.

NOTES

1. *B.EL. = Caisson bearing elevation–to be confirmed and approved by the geotechnical consultant.*
2. *Caisson design bearing pressure is 12,500 psf.*
3. *Center caisson on column or wall unless otherwise noted.*
4. *Dowels from column/wall into caisson is the same as column/wall vertical rebar (min 4-#5 dowels).*
5. *Slab-on-grade: 6" thick concrete with 6×6-W2.9×W2.9 WWF on 8"granular fill over 8mil vapor barrier. Finished floor elevation = 0'-0".*

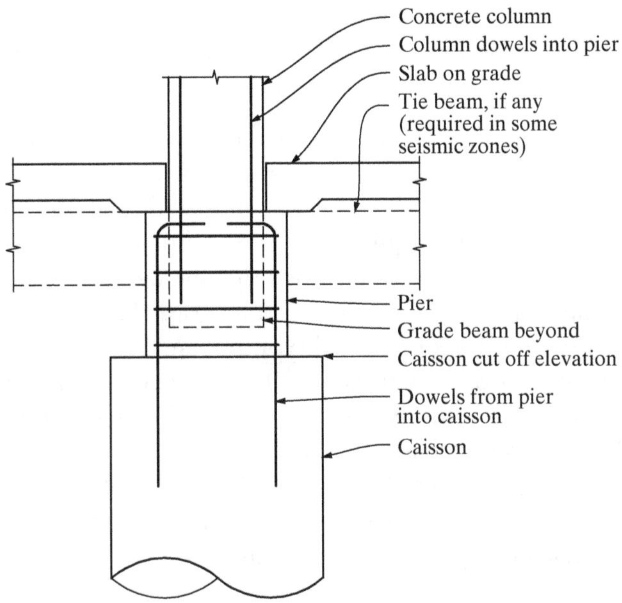

FIGURE 10.36 Typical section at Caisson foundation.

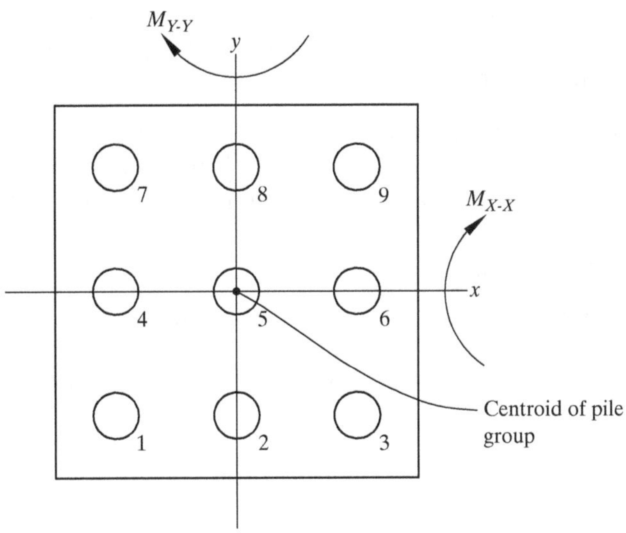

FIGURE 10.37 Pile group subject to axial load and moments.

Transfer of Lateral Base Shears in Pile and Caisson Foundations

The transfer of the lateral base shears from a building structure to the ground involves interaction between the structure and the soil, and can be accomplished through a combination of the following mechanisms [11, 12, 13].

- Passive soil pressure resistance against the vertical face of the basement walls, pile caps, and grade beams.

- Passive soil resistance against the front sides of the piles or caissons. The settlement of the soil beneath the structure should be accounted for in calculating the effective depth that will be available for passive soil resistance.
- Bending of the caissons or pile groups as a vertical cantilever or through some frame action due to the moment restraint provided by the pile caps or grade beams. To achieve rotational fixity or a rigid moment connection at the pile-to-pile cap/grade beam interface at the top of a pile, the required pile embedment depth into the pile cap/grade beam must be "equal to or greater than twice the pile depth or diameter" [28]. In addition, the pile cap/grade beam must be reinforced to resist the applied moment, shear, and torsion (if any), and the reinforcement will typically include top and bottom reinforcement, and closed stirrups to resist any torsion [29]. As previously discussed under "Deep Foundations" in Section 10.1, the pile is also assumed to have full rotational fixity somewhere along its length, or at its base if lateral support is not provided by the soil. The reduction effect of the group action of the group piles should also be accounted for. If the piles are in a river, the competent soil will be below the riverbed, which would mean that the structure will be placed on piles that act as cantilevered stilts above the riverbed.
- The use of batter piles which is quite limited in application due to the fact that many of these piles would be needed, and other batter piles will be ineffective in a direction perpendicular to the plane of inclination of the batter pile. Also, the driving of batter piles could cause major interference with the vertical piles, and some of the batter piles may extend beyond the property lines.
- For pile or caisson-supported structures, the friction resistance between the base of the structure and the soil below the structure is neglected because over time, the soil settlement below the base of the building eliminates this shearing resistance.

Critical Sections for Shear in Pile Caps

The critical sections for shear in pile caps are as follows [14]:

a. One-way deep beam shear located **at the face of the column** (see Figure 10.38b and c). The one-way shear capacity is given as

$$\phi V_c = \phi 2\sqrt{f_c'}bd \geq V_u$$

where

$\phi = 0.75$

b = width of pile cap

d = effective depth of pile cap bottom reinforcement

V_u = the factored total reaction on the piles causing the one-way shear (e.g., in Figure 10.38, this would be the factored reaction from two piles)

b. Two-way or punching shear **around the supported column**, with the critical perimeter located at a distance of $d/2$ from the face of the column, as shown in Figure 10.38a, where

d is the effective depth of the pile cap. The punching shear capacity is given as the smallest of the following (ACI Code Table 22.6.5.2):

a. $\phi V_c = \phi\left[4\lambda_s\lambda\sqrt{f_c'}\right]b_0 d \geq V_u$

b. $\phi V_c = \phi\left[\left(2+\frac{4}{\beta_c}\right)\lambda_s\lambda\sqrt{f_c'}\right]b_0 d \geq V_u$

c. $\phi V_c = \phi\left[\left(2+\frac{\alpha_s d}{b_0}\right)\lambda_s\lambda\sqrt{f_c'}\right]b_0 d \geq V_u$

where

$\phi = 0.75$

b_o = perimeter of the punching shear critical section = $2(c_1 + d + c_2 + d)$

d = effective depth of pile cap bottom reinforcement

c_1 and c_2 = dimensions of the column

V_u = the factored load on the column

Other terms are as previously defined in Section 10.5. Equation (a) typically governs for most cases.

In spite of the critical sections indicated in items (a) and (b) above, if however, the piles are located within the critical sections (see Figure 10.38a), the ACI Code sets an upper limit on the one-way shear capacity, with the critical section for one-way shear taken at the face of the column. Similarly, for two-way shear, if the piles are located within the critical section, the ACI Code sets an upper limit on the two-way shear capacity, with the critical section taken at the face of the column (ACI Code, Section 13.4.6.5). That would indicate that for punching shear in item (b) above (i.e., punching shear at the supported column), the revised perimeter of the critical section for punching shear would be $b_o = 2(c_1 + c_2)$.

c. Punching shear **around an individual edge or corner pile** with the critical perimeter located at $d/2$ from the face of the pile as shown in the 4-pile pile cap in Figure 10.38b, and assuming the pile edge distance is $2d_p$ measured from the center of the pile.

$\phi V_c = \phi 4\sqrt{f'_c}b_0 d \geq V_u$

Where, for an edge pile in a 4-pile pile cap shown in Figure 10.38b, the critical perimeter for punching shear, b_o, is calculated as

$$b_o = 4d_p + \left(\frac{1}{4}\right)2\pi\left(\frac{d_p}{2}+\frac{d}{2}\right) = 4d_p + \frac{\pi}{4}\left(d_p + d\right)$$

d = effective depth of pile cap bottom reinforcement

d_p = diameter of pile

V_u = the factored reaction in an individual pile

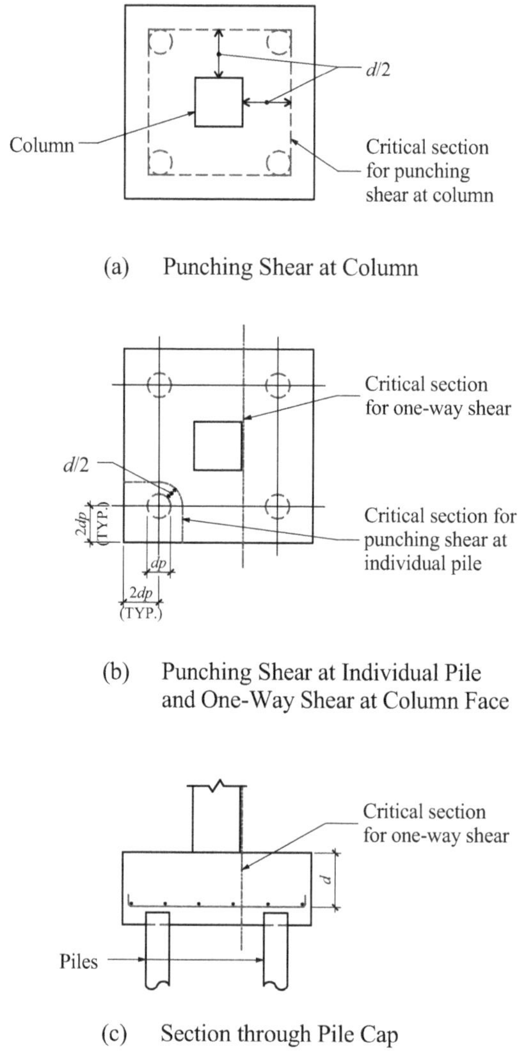

FIGURE 10.38 Critical sections in pile caps (e.g., 4-pile pile cap)

Design Requirements for Pile Caps

Pile caps usually fall into the category of deep beams and are therefore typically designed using the strut-and-tie method that will be discussed in the next section (see ACI Code, Section 13.2.6.5). The following design parameters are required for the design of pile caps:

1. The cumulative total axial load, P, and moments at the base of the supported column or wall.
2. Based on the axial load capacity of the piles or caissons that are typically specified in the Geotechnical Report, select the appropriate pile cap layout that has a sufficient number of piles to resist the column or wall loading. For the chosen pile group, check to ensure that the maximum load on each individual pile due to total gravity load and any moments about the two orthogonal axes do not exceed the axial load capacity of the pile (see Figure 10.37).

3. Check the bearing stresses at the column-pile cap interface and at the reaction points at the pile-to-pile cap interface (see Section 10.5 for discussion of concrete bearing capacity).
4. The Geotechnical Report will specify minimum spacing and edge distance to ensure maximum efficiency for the pile group and to avoid the area of influence of one pile overlapping with other adjacent piles. In practice, a minimum spacing between piles of 3 to 4 times the pile diameter is commonly recommended. To ensure confinement in the nodal zones and provide adequate space to develop the tension tie reinforcement beyond the nodal zone, an edge distance of twice the pile diameter is commonly used and recommended [14].
5. The minimum depth of pile caps should be such that the effective depth of the bottom reinforcement in the pile cap is at least 12 in. (ACI Code, Section 13.4.6.1).
6. For pile caps designed using the strut-and-tie method, the effective compressive strength of the strut, f_{ce}, should be calculated using $\beta_s = 0.60\lambda$ (ACI Code, Section 13.4.6.4) since confined reinforcement satisfying ACI Code Section 23.5.1 is not usually provided in pile caps. Confined strut reinforcement can be provided in deep beams and corbels.

10.13 STRUT-AND-TIE MODELS FOR PILE CAPS AND DEEP BEAMS

When beams or pile caps have relatively large depths with shear spans, a, such that a/h is less than 2.0 (see Figure 10.39 and ACI Code, Section 9.9.1), the load transfer is by direct strut action rather than by flexure or bending. For deep beams and pile caps, the assumptions in Chapter 2 that plane sections remain plane is no longer valid; hence, the strut-and-tie method must be used for such members [18, 19, 20]. Deep beams occur as transfer beams or girders in reinforced concrete high-rise buildings where they support the accumulated loads from discontinued multi-story columns or walls. Pile caps are “transitional elements” used to transfer column and wall loads from the superstructure—with its smaller construction tolerances—to a group or cluster of piles that are installed with larger construction tolerances.

The ACI Code permits the strut-and-tie method in ACI Code Chapter 23, which is typically used for the design of deep beam members, to be used for the design of pile caps (see ACI Code Section 13.2.6.5). The method consists of modeling the deep beam or pile cap as an idealized truss that consists of several components:

- the **interior struts**, which are in compression.
- the **boundary struts**, which are in compression and lie along the boundary of the member.
- the **tension tie**.
- the **nodal zones**.

See Figure 10.40 for 2D strut-and-tie models and Figure 10.41 for a 3D strut-and-tie model. In modeling the truss, the angle between the longitudinal axis of the strut and the horizontal must be at least 25° (ACI Code, Section 23.2.7).

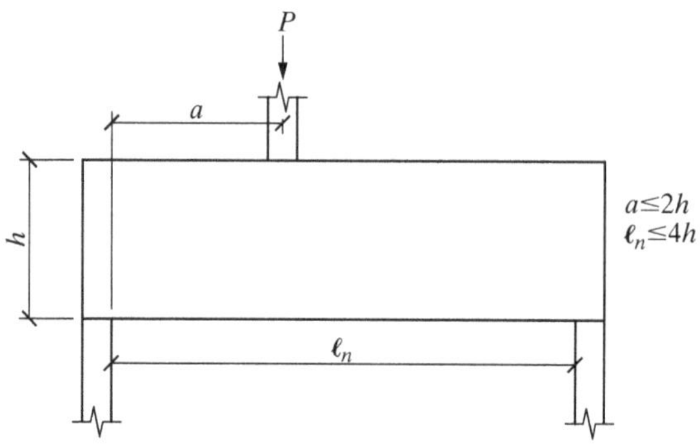

(a) Deep Beam with Single Concentrated Load

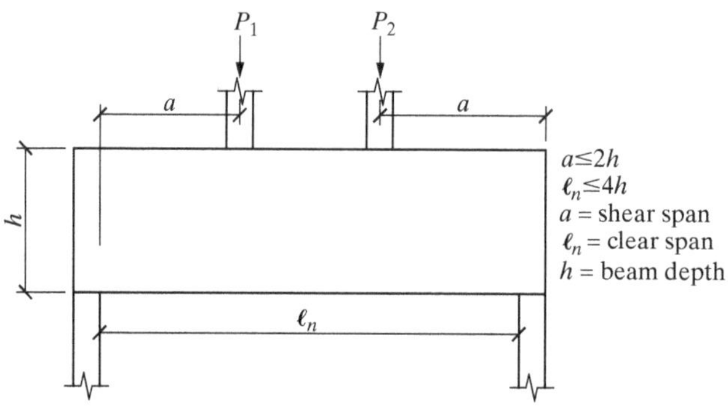

(b) Deep Beam with Single Concentrated Loads

FIGURE 10.39. Shear spans in deep beams and pile caps.

The interior and boundary struts can be idealized as uniform prismatic members. The nodes in strut-and-tie models can be classified as follows, depending on the type of members or forces meeting at the node (see Figure 10.40):

- C-C-C nodes: indicates that all the forces in all the members meeting at the node and the reaction or load induces compression on the node.
- C-C-T node: indicates that the nodal zone resists compression forces (C) on two faces, and a tension force (T) on one face.
- C-T-T node: Though not shown in Figure 10.40, this indicates that the nodal zone resists compression force (C) on one face, and tension forces (T) on the two other faces of the nodal zone.

The development of a strut-and-tie model is an iterative and graphical design process where the strut-and-tie geometry is assumed (with an angle of inclination of the strut that does not violate the Code limitations), and the forces in each strut, tension tie, and nodal zones are calculated and checked to ensure that their ultimate strengths are not exceeded. There are several possible solutions, but the best strut-and-tie model for a structure is the one that mimics the elastic stress trajectories of the structure. The strut-and-tie method - which is based on the lower bound theorem of plasticity—results in a lower bound solution, provided

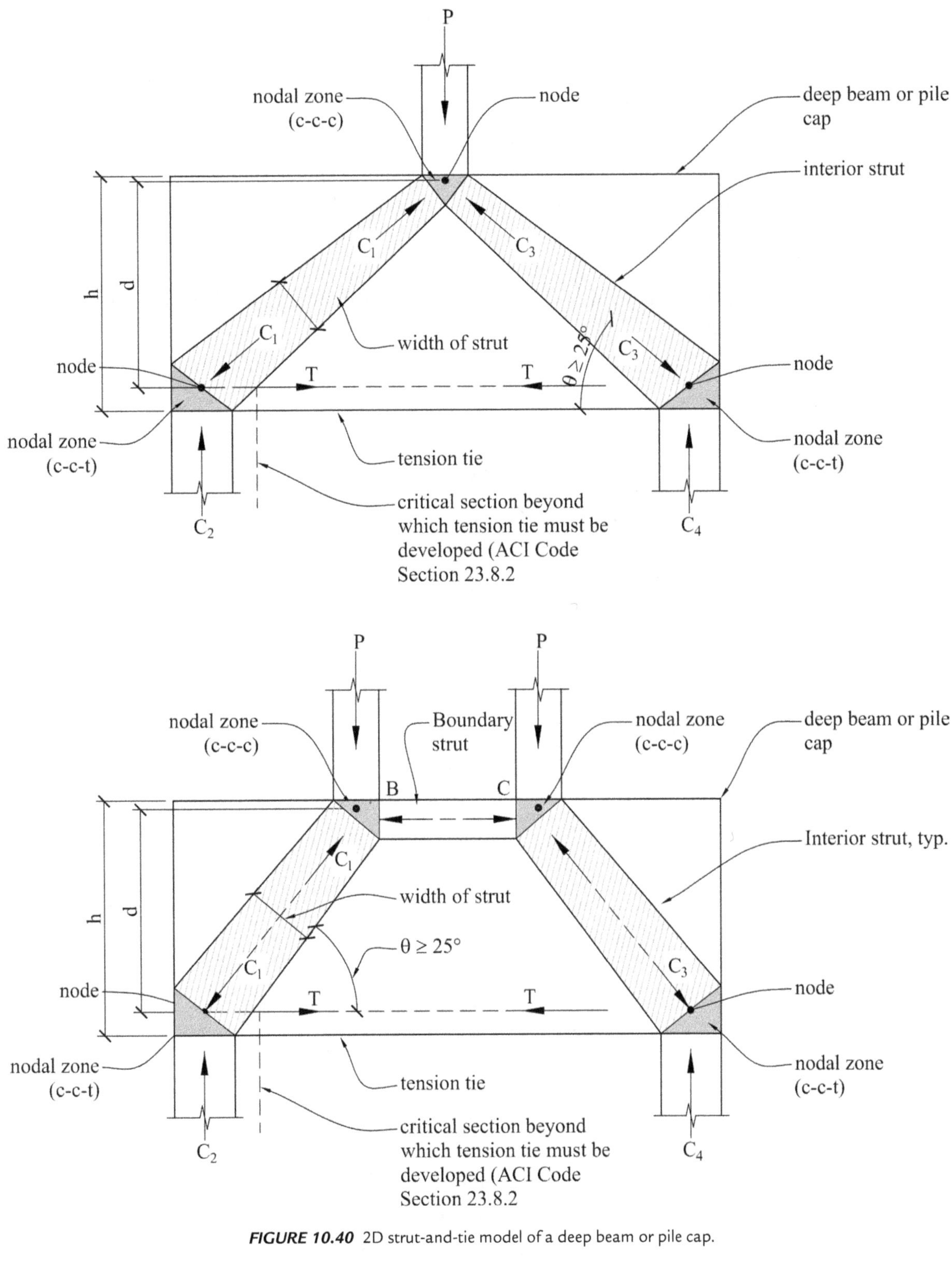

FIGURE 10.40 2D strut-and-tie model of a deep beam or pile cap.

the assumed stress field is in equilibrium with the external loads, and the failure or yield criterion of the struts, tension ties, and nodes in the model are not violated, and the member has adequate ductility to ensure any required redistribution of forces [18, 19]. The possible failure modes in the strut-and-tie model are:

- Compression failure of the strut,
- Tension failure of the tension ties,
- Failure of the nodal zones, and
- Bearing failure at the load and reaction points (See Section 10.5 for the calculation of bearing strength).

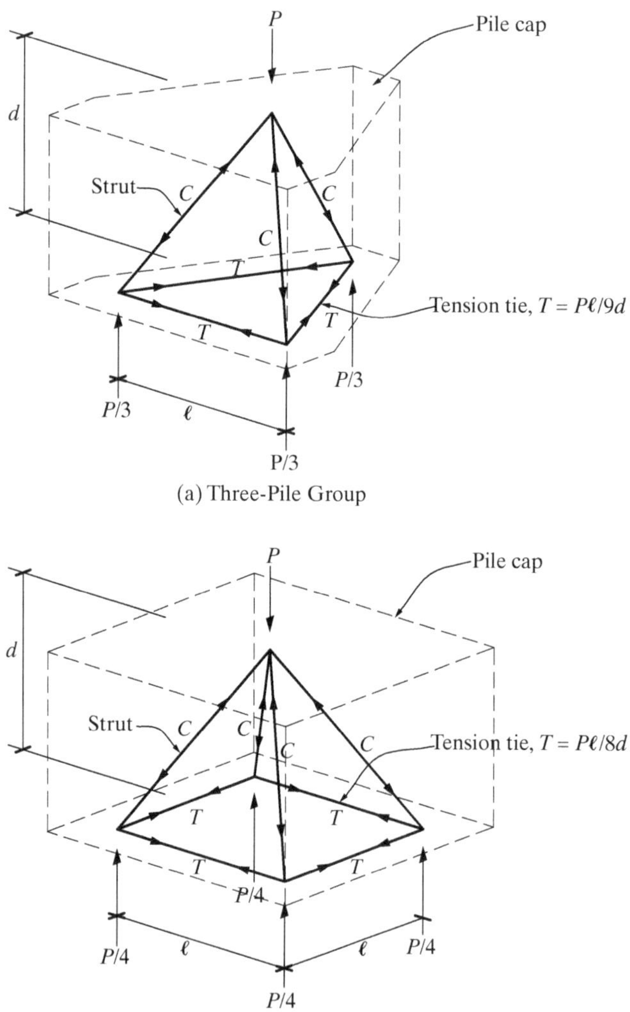

(a) Three-Pile Group

(b) Four-Pile Group

FIGURE 10.41 3D strut-and-tie model of pile caps.

TABLE 10.3 Strut Coefficient, β_s

Strut location	Strut type	Criteria	β_s	
Tension members or tension zones of members	Any	All cases	0.4	(a)
All other cases	Boundary struts	All cases	1.0	(b)
	Interior struts	Reinforcement satisfying (a) or (b) of Table 10.5 (ACI Table 23.5.1)	0.75	(c)
		Located in regions satisfying ACI Code Section 23.4.4: that is, if the deep beam or pile cap member dimensions are such that^: $V_u \le \phi 5\lambda\lambda_s \tan(\theta)\sqrt{f'_c}\, b_w d$	0.75	(d)
		Beam column joints	0.75	(e)
		All other cases (i.e., members without transverse reinforcement), but does not apply to pile caps (ACI Code Section 13.4.6.4)	0.4	(f)

Adapted from ACI 318-19 Table 23.4.3(a)

^ This requirement, if met, precludes diagonal tension failure (ACI Code, Section R23.4.4), and the steeper the angle of inclination of the interior strut, θ, the higher the shear strength of the deep beam or pile cap.

TABLE 10.4 Strut and tie node confinement modification factor, βc

Location	βc		
• End of a strut connected to a node that has a bearing surface • Node that includes a bearing surface	Lesser of	$\sqrt{A_2/A_1}$, where A_1 is defined by the bearing surface	(a)
		2.0	(b)
Other cases	1.0		(c)

Adapted from ACI 318-19 Table 23.4.3(b)

Strength of Struts

The struts are subjected to compression, and the minimum angle of inclination between the strut and a tension tie is 25° (ACI Code, Section 23.2.7) because unreasonably small angles will lead to cracking and cause "incompatibilities" between the strut and the tension tie due to the axial shortening of the strut and the elongation of the tension tie in approximately the same direction for small angle struts. The idealized struts are shown in Figure 10.40.

There are two types of struts which are subjected to compression forces:

- **Interior struts** which have a constant cross-section, and usually inclined at an angle and converge to a node. They are intersected by the tension tie and are therefore subjected to transverse tension, hence the lower β_s value (see Table 10.3).
- **Boundary struts** which have a constant cross-section and are located along the boundary of the member. They are not subject to transverse tension, and therefore, have a higher effective compressive strength, f_{ce}, than interior struts, as reflected in the higher β_s value (see Table 10.3).

As is evident from the strut-and-tie models of Figure 10.40, transverse tension does occur in interior struts as a result, and transverse reinforcement in the form of closed ties can be provided in deep beams to resist this transverse tension force. The strength of an unreinforced strut is reduced due to the presence of this transverse tension as is reflected in the β_s value of 0.40 (line f in Table 10.3) for unreinforced struts compared to β_s of 0.75 for reinforced struts that satisfies the requirements of ACI Code, Section 23.5 (see Table 10.3). For pile caps, it is not feasible or practical to provide confining reinforcement to resist the transverse tension, therefore, β_s will be 0.60λ (see ACI Code, Section 13.4.6.4). The compressive strength of an unreinforced strut (i.e., strut without longitudinal reinforcement) can be obtained from ACI Code Equation 23.4.1a as

$$\phi F_{ns} = \phi A_{cs} f_{ce} = \phi A_{cs}\left(0.85\beta_s\beta_c f_c'\right) \geq F_{us}$$

For a strut with longitudinal reinforcement provided parallel to the longitudinal direction of the strut, the compressive strength can be obtained from ACI Code Equation 23.4.1b as

$$\begin{aligned}\phi F_{ns} &= \phi\left(A_{cs} f_{ce} + A_s' f_s'\right)\\ &= \phi[A_{cs}\left(0.85\beta_s\beta_c f_c'\right) + A_s' f_s'] \geq F_{us}\end{aligned}$$

where

F_{us} = the factored compression force in the strut

$f_{ce} = 0.85\beta_s\,\beta_c f_c'$ is the effective concrete compression strength of a strut

β_s is the strut coefficient calculated from Table 10.3 (see ACI Table 23.4.3). It accounts for the effects of cracking as well as the effects of providing crack control reinforcement across the strut.

β_c = the strut and node confinement modification factor obtained from Table 10.4

$\phi = 0.75$

A_{cs} = cross-sectional area at the end of the strut

A_s' = area of longitudinal reinforcement in the strut

Since the strut dimensions may vary from node to node, the strength of the strut should be evaluated at each end of a strut, and the lower value is taken as the capacity of the strut.

The longitudinal compression reinforcement in the strut must be tied similar to the closed ties used in columns, and the reinforcement and closed ties shall conform to ACI Code, Section 23.6. The minimum distributed reinforcement shall be in accordance with Table 10.5 (ACI Code Table 23.5.1). The spacing of the distributed reinforcement shall not exceed 12 inches, and the angle shall not be less than 40 degrees (ACI Code Section 23.5.2):

α_1 = angle between the distributed vertical reinforcement and the centroidal axis of the strut

b_s = width of the strut

TABLE 10.5 Minimum Distributed Reinforcement

Lateral restraint of the strut (perpendicular to the plane of the strut)	Reinforcement configuration	Total Minimum distributed reinforcing ratio (i.e., the total area on both faces of the member)	
Not restrained	Orthogonal grid	0.0025 in each direction	(a)
	Reinforcement in one direction crossing the strut at an angle α_1	$\frac{0.0025}{\sin^2 \alpha_1}$	(b)
Restrained	Distributed reinforcement not required		(c)

Adapted from ACI 318-19 Table 23.5.1

TABLE 10.6 Nodal Zone Coefficient, β_n

Type of nodal zone	β_n
Nodal zone bounded by struts, or bearing areas or both (e.g., C-C-C nodal zone)	1.0
Nodal zone anchoring one tension tie (e.g., C-C-T nodal zone)	0.80
Nodal zone anchoring two or more tension ties (e.g., C-T-T nodal zone)	0.60

(Adapted from ACI 318-19 Table 23.9.2)

Strength of Nodal Zones

A node is the intersection of the centroidal axes of the strut, the bearing or loading area and/or the tension ties. The area surrounding a node is known as the nodal zone. There are two types of nodal zones:

- **Hydrostatic nodal zones** where the stresses on all the loaded faces of the nodes are equal, and the faces of the nodes are perpendicular to the longitudinal axes of the struts and ties intersecting at the node.
- **Non-hydrostatic nodal zones** where the face of the node is not perpendicular to the axes of the strut, and the size of the node is controlled by the face of the node with the highest stress (ACI Code, Section R23.2.6).

The following nodal zone parameters are needed to establish the geometry of the strut-and-tie model: the dimensions of the loaded area (i.e., the supported columns or piers), the dimensions of the reaction points (i.e., the supporting columns or piles), and the width (i.e., vertical dimension) of the tension tie zone. [18, 19]. The compressive strength of a nodal zone is obtained from ACI Code Equation 23.9.1 and rewritten as

$$\phi F_{nn} = \phi A_{nz} f_{ce} = \phi A_{nz}\left(0.85\beta_c\beta_n f_c'\right) \geq F_{un}$$

where

$f_{ce} = 0.85\beta_c\beta_n f_c'$ is the effective concrete compression strength at the face of the nodal zone,

β_n is calculated from Table 10.6 (see ACI Table 23.9.2)

β_c which accounts for the effect of concrete confinement on the effective compressive strength of a node that contains a bearing surface, and is determined from Table 10.4

$\phi = 0.75$

A_{nz} = area of the face of the nodal zone or a section through the nodal zone perpendicular to the direction of the force, F_{us} (i.e., the factored compression force in the strut)

F_{un} = factored or ultimate force on the face of a node

Strength of Tension Ties

Tension ties are allowed to be non-prestressed or pre-stressed reinforcement and must have adequate anchorage for non-prestressed reinforcement. Anchorages of tension tie reinforcement can be provided by end plates or by hooks at the ends of the deep beam or pile cap beyond the nodal zone. The tension capacity of a tension tie is given as (see ACI Code Section 23.7.2):

$$\phi F_{nt} = \phi\left[A_{ts} f_y + A_{tp}(\Delta f_p)\right] \geq F_{ut}$$

where,

F_{ut} = factored tension force in the tension tie

A_{ts} = area of non-prestressed tension tie reinforcement

A_{tp} = area of tension tie prestressing steel

= 0 for non-prestressed members

f_y = yield strength of the non-prestressed reinforcement

Δf_p = 60,000 psi for *bonded* prestressed tension tie reinforcement

= 10,000 psi for *unbonded* prestressed tension tie reinforcement

Δf_p must be less than or equal to $f_{py} - f_{se}$ if determined from analysis.

Note that the centroid of the tension tie assumed in the strut-and-tie model must coincide with the centroid of the actual tension tie reinforcement selected (ACI Code, Section 23.8.1). The tension tie reinforcement must be anchored by mechanical devices such as end plates, headed bars, or with standard hooks in or beyond the nodal zones (ACI Code, Section R23.2.1). For pile caps, banded tension tie reinforcement along the pile lines is preferred and commonly used in practice rather than uniformly distributed reinforcement.

Additional Reinforcement Requirements for Deep Beams

To control the width and growth of inclined cracks in deep beams, distributed vertical and horizontal reinforcement are required in deep beams per ACI Code, Section 9.9.3.1. The distributed *vertical* reinforcement, A_v, on the face of the beam, must satisfy the following relationship (see ACI Code, Section 9.9.3.1):

$$A_v \geq 0.0025 b_w s_1$$

where s_1 is the horizontal spacing of the distributed vertical or transverse reinforcement in the beam, where b_w is the width of the beam.

The distributed *horizontal* reinforcement, A_h, on the face of the beam should satisfy the following relationship:

$$A_h \geq 0.0025 b_w s_2$$

where s_2 is the vertical spacing of the distributed horizontal reinforcement in the beam.

The maximum spacing of the distributed reinforcement is *d*/5 but not greater than 12 in.

Terminate the horizontal and vertical reinforcement on the face of the deep beam with standard hooks.

Design Procedure for Pile Caps and Deep Beams Using the Strut-and-Tie Method

The design of pile caps and deep beams by the strut-and-tie method is an iterative process, and the procedure is as follows:

1. Determine the total load, *P*, on the beam or pile cap.
2. Determine the shear span, a, and establish that the beam is a deep beam and hence can be designed using the strut-and-tie method (see Figure 10.40).
3. Check the bearing stresses at the load and reaction points using the supported column or pier dimensions for calculating the bearing strength at the load points, and the supporting column or pile dimensions for calculating the bearing strength at the reaction points (see Section 10.5). These dimensions are used as the starting dimensions for the nodal zones, and they are combined with the height of the tension tie zone determined below and used to establish the geometry of the strut-and-tie model.
4. Check the pile cap or deep beam for one-way shear to establish an estimate of the minimum

   beam depth (see ACI Code, Sections 9.9.2.1, 9.9.3.1, and 9.9.4). The size of pile caps and deep beams should be selected so that the maximum shear in the pile cap or deep beam satisfies the following relationship—to control cracking at service loads and guard against diagonal compression failures in deep beams (ACI Code, Section R9.9.2.1):

   $$V_u \leq \phi 10 \sqrt{f_c'}\, b_w d$$

   where b_w is the width of the deep beam or pile cap.

   The minimum depth of the pile cap should be selected to ensure that the effective depth of the bottom reinforcement in the cap is at least 12 inches (ACI Code Section 13.4.6.1). This equation is used to establish the minimum dimension for the pile cap or deep beam that helps to control cracking under service loads and prevents diagonal compression failures in deep beams (ACI Code, Section R9.9.2.1).
5. For pile caps, check punching shear at the supported column and at the supporting piles.

6. Lay out the interior and boundary struts, tension ties, and nodal zones in the pile cap or deep beam as the idealized strut-and-tie model (see Figures 10.40 and 10.41). Ensure that the angle of inclination of the interior strut to the horizontal, θ, is not less than 25°.

7. The out-of-plane thickness of the strut and nodal zones is taken as the width, b_w, of the deep beam and the width of the pile cap for 2D pile cap problems. For pile caps with 3D strut-and-tie models (e.g., pile caps with three or more piles, as shown in Figure 10.41), the widths of the struts and nodal zones are not readily apparent because of the complex geometries. Initial estimates of the widths can be selected so that the ultimate strength of the struts and the bearing strength at the nodal zones exceed the factored forces in these elements. The struts and nodal zones with the assumed widths must also fit within the geometry of the pile cap. For 3D pile caps, in lieu of checking the strength of each individual strut and node, the strengths can be deemed adequate if the bearing strength at the supporting piles and at the supported column or pier are adequate [21].

8. For tension ties that are developed with hooks or straight bars beyond the nodal zone, the height of the tension tie (i.e., the in-plane dimension) can be determined by the number of layers of tension tie bars and the size of the bars. If a single layer of tension tie reinforcement is used, the minimum height, w_t, of the tension tie will be $d_b + 2c_c$ where d_b is the diameter of the tension tie bars and c_c is the clear concrete cover to the bars. If two layers of reinforcement are used and assuming a maximum clear distance of $2d_b$ between the bottom and upper layer reinforcement, the minimum height of the tension zone is $4d_b + 2c_c$. When the tension tie bars are developed by plate anchors at each end of the deep member, the height of the tension tie zone will be taken as the height of the anchor plates.

9. Calculate the forces in the struts and ties, and nodal zones.

10. Determine the dimensions of the struts, tension ties, and the nodal zones.

11. Using the assumed dimensions above, check the struts, tension tie, and nodal zones to determine if they have adequate strength to resist the forces calculated in the previous steps. If the strengths are inadequate, revise the strut-and-tie geometry and/or the sizes of the strut, tension tie zone, and nodal zones, and repeat the process until the strengths are adequate.

12. Check the minimum reinforcement required in the tension tie. For deep beams and pile caps (banded reinforcement), the minimum tension tie reinforcement is

$$A_{s,\min} = \frac{3\sqrt{f_c'}b_w d}{f_y} \geq \frac{200 b_w d}{f_y}$$

13. Detail the reinforcement, ensuring adequate development or anchorage in or beyond the nodal zones. For 2D pile caps, provide transverse reinforcement at the bottom of the pile cap, perpendicular to the main reinforcement, similar to temperature and shrinkage reinforcement in one-way slabs (see Chapter 2). Note that the exception to lateral confinement includes pile caps and continuous beam ledges, so distributed reinforcement on the vertical surfaces is not required in these members [25], though some engineers provide skin reinforcement to control crack widths in deep pile caps (see Figure 10.33). Distributed

vertical surface reinforcement or skin reinforcement as discussed in Chapter 7 are required in deep beams.

For piles that are subjected to net uplift forces, top reinforcement will be required in the pile cap, in addition to the bottom reinforcement; the dowels from the piles and the column must have adequate development length into the pile cap to resist the uplift tension force (see Figure 10.33).

EXAMPLE 10.9

Deep Beam Design Using the Strut-and-Tie Method

Analyze the 18 in. × 48 in. deep simply supported beam shown in Figure 10.42 for the unfactored loads shown. The beam spans 13.5 ft. center to center of supports with a clear span of 12 ft. (see also Figure 10.39). Ignore the self-weight of the beam. $f_c' = 4000$ psi and $f_y = 60{,}000$ psi. Determine the tension tie reinforcement and the additional reinforcement required in the beam.

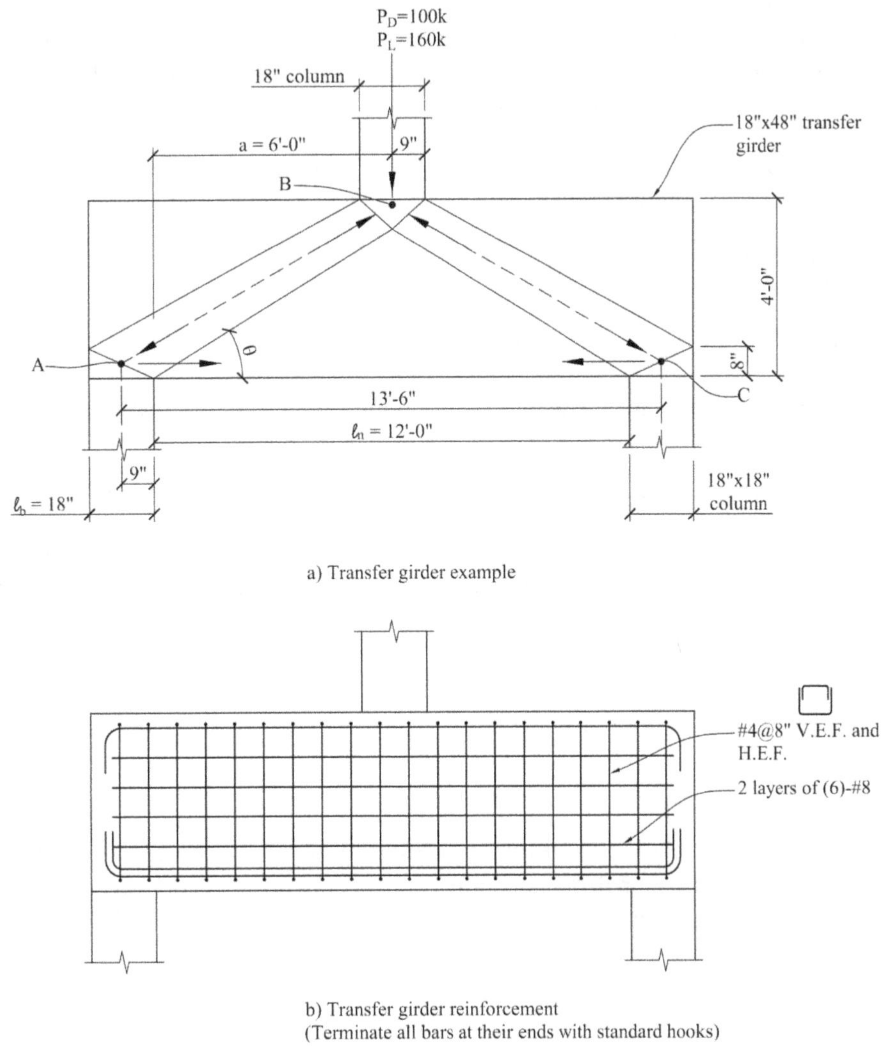

FIGURE 10.42 Transfer Girder for Example 10.9

Solution:

a. Analyze the beam under factored loads:

$P_D = 100$ kip and $P_L = 160$ kip

Therefore, the factored load, $P_u = 1.2(100) + 1.6(160) = 376$ kip

$$M_u = \frac{PL}{4} = \frac{(376)(13.5)}{4} = 1269 \text{ ft-kip}$$

$$V_u = \frac{P}{2} = \frac{376}{2} = 188 \text{ kip}$$

The factored reaction, $R_u = 188$ kip

Check to make sure that the beam is adequate in bearing at the reaction points (A and C) for the factored reaction, R_u, and at the load point (B) for the load, P_u, using the principles for determining the design bearing strength that was presented in Section 10.5 or ACI Code Section 22.8.3.2. The reader should verify that the load and reaction points (A or C and B) are adequate in bearing.

b. Check if this is a deep beam; otherwise, analyze using the flexural methods presented previously in Chapters 2 and 3.

Given the clear span of 12 ft., the shear span, a = 6 ft (see Figure 10.39) < $2h$ = 2(4 ft) = 8 ft O.K.

The clear span, $\ell_n = 12 \text{ ft} < 4h = 4(4 \text{ ft}) = 16 \text{ ft}$, O.K.

Therefore, the beam qualifies as a deep beam that can be designed or analyzed using the strut-and-tie method.

c. Check the maximum shear strength of the deep beam (ACI Code Section 9.9.2.1):

The width of the beam, $b_w = 18$ in. Assume the effective depth of the beam, $d = 0.8h$ (this will be checked later. If the actual d is greater than assumed, then the initial estimate will be conservative)

$$V_u = \phi 10\sqrt{f_c'}b_w d = (0.75)(10)\sqrt{4000}(18 \text{ in.})(0.8)(48 \text{ in.})$$
$$= 328 \text{ kip} > V_u = 188 \text{ kip, O.K.}$$

d. Assume the width of the tension zone, w_t (see Figure 10.43b):

Assuming the tension tie reinforcement will be placed in *two layers* and assuming 2 in. clear concrete cover to the tension tie reinforcement, and assuming #8 bars with 2 bar-diameters clear between the two layers of reinforcement, therefore, the width of the tension tie zone is

$$w_t = 4d_b + 2c_c = 4(1 \text{ in.}) + 2(2 \text{ in.}) = 8 \text{ in.}$$

Assume the depth of the nodal zone, w_{nz}, at point B just below the column load is also 8 in, therefore, from geometry of the deep beam (see Figure 10.42(a)), the angle of inclination of the strut, θ, is calculated as follows:

$$\tan\theta = \frac{h - \frac{w_t}{2} - \frac{w_{nz}}{2}}{\frac{\ell}{2}} = \frac{48\text{ in.} - \frac{8\text{ in.}}{2} - \frac{8\text{ in.}}{2}}{\frac{(13.5\text{ ft.})(12)}{2}} = 0.493$$

Therefore θ = 26.3° > 25° (see ACI Code, Section 23.2.7) O.K.

e. Calculate the factored or ultimate force in the strut, F_{us}, and check the capacity of the strut:

Equilibrium of forces at joint A requires that

$F_{us}\sin(26.3°)$ = maximum reaction, $R_{uA} = 188$ kip.

Therefore, the factored or ultimate compression force in the strut, $\mathbf{F_{us} = 424}$ **kip**.

From geometry and the selected tension tie zone depth (see Figure 10.42(a)), the in-plane width of the interior strut, w_s, (see Figure 10.43) is

$w_s = w_t \cos\theta + \ell_b \sin\theta = (8\text{ in.})\cos(26.3°) + (18\text{ in.})\sin(26.3°)$

= 15.2 in.

Assuming the strut is crossed by skin reinforcement satisfying ACI Code, Section 23.5, $\beta_s = 0.75$ (see Table 10.3). The strut and node confinement modification factor, β_c, can conservatively be taken as 1.0 (see Table 10.4). Therefore, the strut compression capacity is,

$$\begin{aligned}\phi F_{ns} &= \phi A_{cs}(0.85\beta_c\beta_s f_c')\\ &= (0.75)(18\text{ in.})(15.2\text{ in.})[(0.85)(1.0)(0.75)(4\text{ ksi})]\\ 523\text{ kip} &> F_{us} = 424\text{ kip},\quad \text{O.K.}\end{aligned}$$

f. Check the actual effective depth, d, and calculate the force in the tension tie and the area of the tension tie reinforcement:

Actual d = 48 in.– 8 in. /2 = 44 in. > 0.8h = 0.8(48) = 38.4 in. O.K.

Equilibrium of forces at joint A requires that $F_{us}\cos(26.3°)$ = Ultimate or factored Tension tie force, F_{ut}

Therefore, F_{ut} = 424 kip cos (26.3°) = 380 kip

Area of tension tie reinforcement required = $F_{ut}/\phi f_y$ = 380 kip/[(0.75)(60 ksi)] = 8.44 in.^{2}

The minimum reinforcement required is $0.0033b_w d$ or 2.64 in^2 which is much less than the required tension tie reinforcement, therefore, O.K.

Use 2 layers of 6 #8 bars for the tension tie (A_s provided = 9.48 in.^{2} > As required, O.K.)

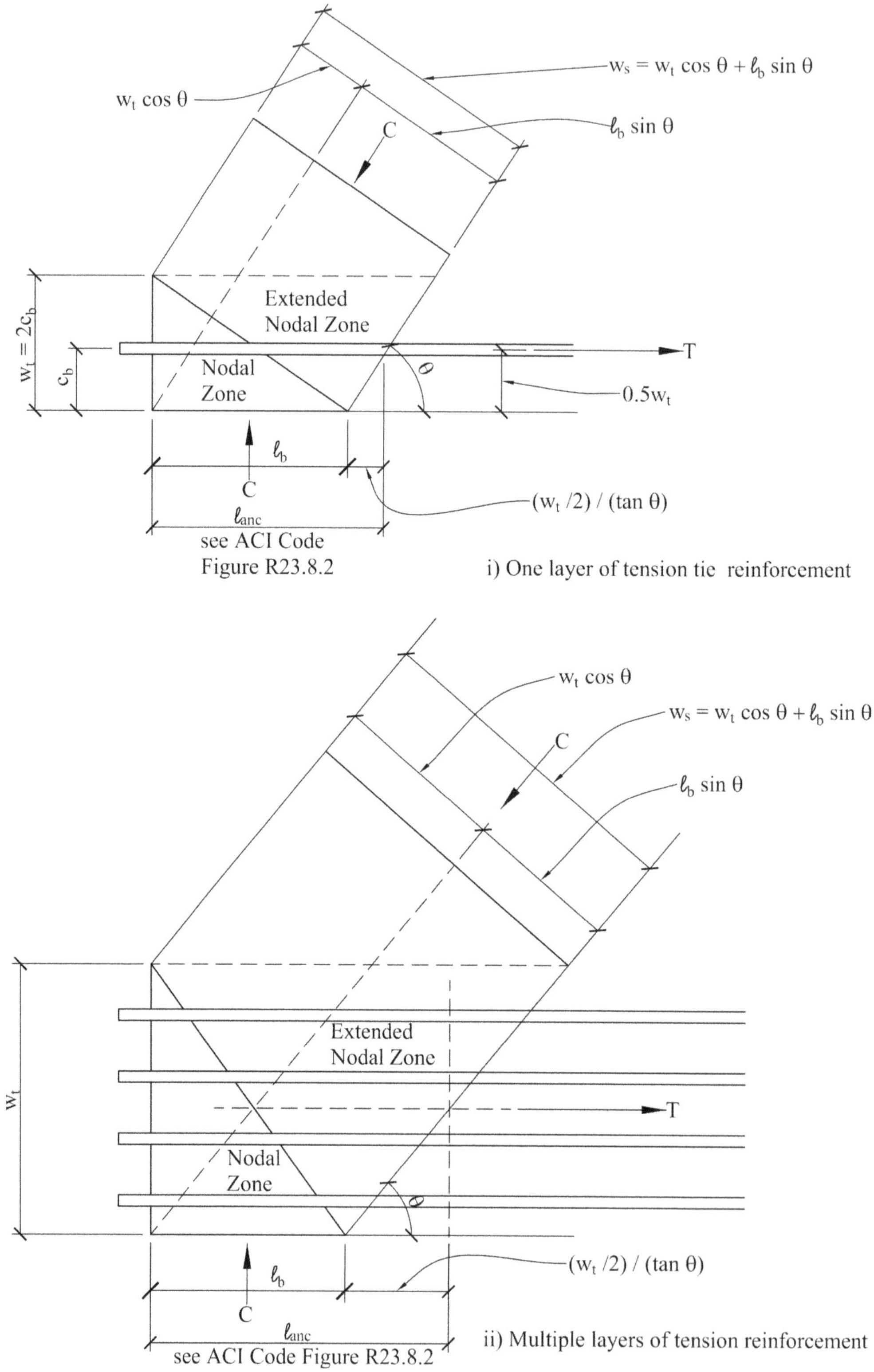

FIGURE 10.43 Size of Nodal Zones (Adapted from ACI 318-19 Figures R23.2.6a, b, and c).

Minimum beam width to accommodate 6 #8 bars in one layer (see Table A.3) = 15 in. < 18 in. O.K

Available length to develop hooked tension tie in two layers (see Figure 10.43) is

$$\ell_{anc} = \ell_b + \frac{\frac{w_t}{2}}{\tan\theta} - \text{concrete cover} = 18\text{ in.} + \frac{\frac{8\text{ in.}}{2}}{\tan 26.3} - 2\text{ in.} = 24.1\text{ in}$$

Required hooked bar development length, ℓ_{dh}, for #8 bars = 17.2 in. < available length, (see Table A.13) O.K.

g. Check the capacity of the nodal zones:

At nodal zone A:

Nodal zone A is a C-C-T node (see Figure 10.42), therefore, from Table 10.6, $\beta_n = 0.8$

From the geometry of the strut and the assumed depth of the tension tie zone depth (see Figure 10.42(a)), the in-plane width of the strut, w_s (see Figure 10.43) is

$$w_s = w_t \cos\theta + \ell_b \sin\theta = (8\text{ in.})\cos(26.3°) + (18\text{ in.})\sin(26.3°) = 15.2\text{ in.}$$

The compression capacity of nodal zone A is

$$\begin{aligned}\phi F_{nn} &= \phi A_{nz} f_{ce} = \phi A_{nz}(0.85\beta_c\beta_n f_c') \\ &= (0.75)(18\text{ in.})(15.2\text{ in.})[(0.85)(1.0)(0.8)(4\text{ ksi})] \\ &558\text{ kip} > F_{un} = 424\text{ kip,} \quad \text{O.K.}\end{aligned}$$

At nodal zone B:

Nodal zone B is a C-C-C node (see Figure 10.40), therefore, from Table 10.6, $\beta_n = 1.0$. From the geometry of the strut and the assumed nodal zone depth (see Figure 10.42(a)), the in-plane width of the strut at node B where it intersects with the interior strut is,

$$w_s = \sqrt{(8\text{ in.})^2 + (9\text{ in.})^2} = 12.04\text{ in.}$$

The compression capacity of nodal zone B is

$$\begin{aligned}\phi F_{nz} &= \phi A_{nz} f_{ce} = \phi A_{nz}(0.85\beta_c\beta_n f_c') \\ &= (0.75)(18\text{ in.})(12.04\text{ in.})[(0.85)(1.0)(1.0)(4\text{ ksi})] \\ &553\text{ kip} > F_{un} = 424\text{ kip,} \quad \text{O.K.}\end{aligned}$$

h. Provide additional distributed horizontal and vertical reinforcement on the vertical surfaces of the deep beam:

The maximum spacing of the distributed reinforcement is

$d/5 = 44\text{ in.}/5 = 8.8$ in. Use 8 in.

≤ 12 in.

Assuming #4 @ 8 in. o.c. vertical each face (VEF) and horizontal each face (HEF), therefore A_v provided = (2)(0.2 in.^{2}) = 0.4 in.^{2}

The required area of the minimum reinforcement on the vertical and horizontal surfaces of the beam are as follows:

$$A_v \geq 0.0025 b_w s_1 = 0.0025(18 \text{ in.})(8 \text{ in.}) = 0.36 \text{ in.}^2$$
$$< A_v \, provided, \text{ O.K.}$$

$$A_h \geq 0.0025 b_w s_2 = 0.0025(18 \text{ in.})(8 \text{ in.}) = 0.36 \text{ in.}^2$$
$$< A_h \, provided, \text{ O.K.}$$

where s_1 is the horizontal spacing of the vertical reinforcement, s_2 is the vertical spacing of the horizontal reinforcement, and b_w = width of the beam.

i. The reinforcement for the deep beam is shown in Figure 10.42(b).

NOTE *Though the self-weight of the deep beam was neglected in this example, we could conservatively have added the total factored self-weight of the beam to the factored concentrated load, P_u.*

EXAMPLE 10.10

Pile Cap Design Using the Strut-and-Tie Method

Determine the tension tie reinforcement required for a two-pile cap layout shown in Figure 10.44 that supports an 18 in. × 18 in. column with a factored load of 750 kip. The pile cap is 4 ft wide × 7 ft long × 3.5 ft deep and supported on two symmetrically located 12 in. diameter concrete-filled steel piles. The piles are spaced at 3 ft on centers with a 2 ft edge distance and an embedment of the pile in the pile cap of 6 in. Ignore the self-weight of the pile cap. f'_c = 5500 psi and f_y = 60,000 psi.

Solution:

a. Analyze the pile cap under factored loads:

$$P_u = 750 \text{ kip}$$
$$V_u = \frac{P_u}{2} = \frac{750}{2} = 375 \text{ kip}$$

The factored reaction in each pile, R_u = 375 kip

b. Check if this pile cap qualifies as a deep beam; otherwise, analyze and design the cap using the flexural design methods previously presented in Chapters 2 and 3.

The shear span, a = 18 in. – (12 in./2) = 12 in. (see Figures 10.39 and 10.44)

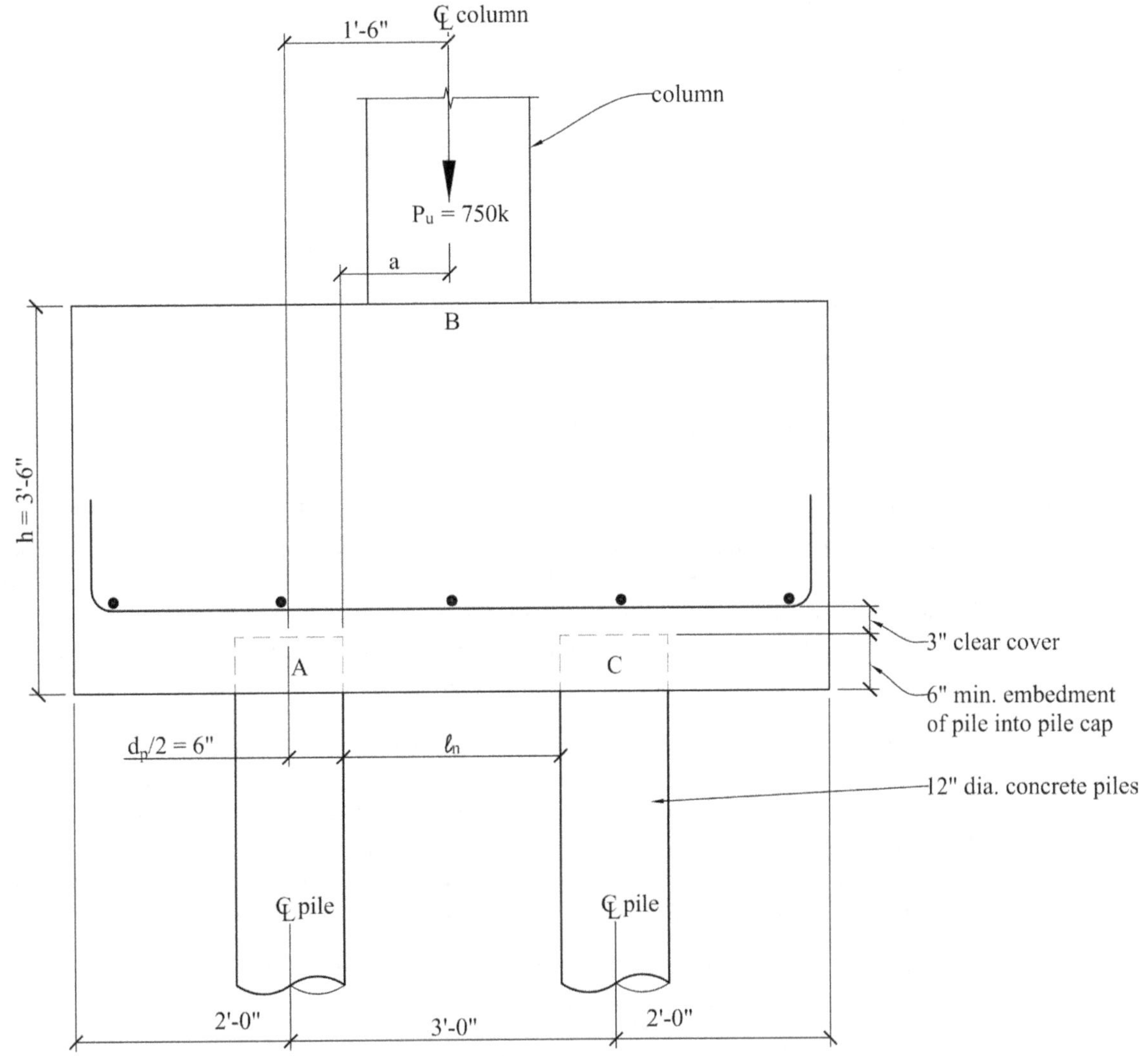

FIGURE 10.44 Pile cap for Example 10.10

$< 2h = 2(3.5\text{ ft}) = 7\text{ ft}$ O.K., this pile cap meets the first requirement to qualify as a deep beam.

The clear span,

$$\ell_n = 3\text{ ft.} - \left(\frac{1}{2}\right)\frac{12\text{ in.}}{(12)} - \left(\frac{1}{2}\right)\frac{12\text{ in.}}{(12)} = 2\text{ ft.}$$
$$< 4h = (4)(3.5\text{ ft}) = 14\text{ft.}$$

O.K., the pile cap meets the second requirement to qualify as a deep beam.

Therefore, since both requirements are satisfied, the pile cap qualifies as a deep beam and should be designed or analyzed using the strut-and-tie method.

c. Check maximum shear strength of the pile cap:

The width of the pile cap, $b_w = 48$ in.

Assume the effective depth of the pile cap, $d = 0.8h$ (this assumption can be checked later. If the actual d is greater than the assumed value, then the initial estimate will be conservative)

$$V_u = \phi 10\sqrt{f_c'}b_w d$$
$$= (0.75)(10)\sqrt{5500}(48)(0.8)(42)$$
$$= 897 \text{ kip} > V_u = 375 \text{ kip, O.K.}$$

d. Check the bearing capacity of the pile cap at the piles (the reaction points) and at the load point (i.e., at the column):

The bearing capacity can be obtained from Section 10.5 (ACI 22.8.3.2). The reader can verify that the bearing capacity for the 12 in. diameter concrete-filled steel piles is 687 kip (i.e., $\phi(0.85f_c'A_1)(2)$ = (0.65)(0.85)(5.5 ksi)(113.1 in.^{2})(2) = 687 kip) which far exceeds the factored pile reaction of 375 kip, and the bearing capacity at the 18 in. × 18 in. concrete column to pile cap interface ($\phi(0.85f_c'A_1)(2)$ = (0.65)(0.85)(5.5 ksi)(18)(18)(2) = 1969 kip) also far exceeds the factored column load of 750 kip.

e. Check the pile cap for punching shear at the supporting piles and at the supported column. Using the principles discussed in Section 10.12 under "Critical Sections for Shear in Pile Caps," we will now check the punching shear capacity at each pile in this 2-pile pile cap. From Figure 10.45, it is obvious that the critical perimeters of the two piles overlap to form the modified critical perimeter shown, thus, the punching shear capacity will be checked against factored load of the pile group (i.e., 750 kip).

The modified critical perimeter, b_o, for punching shear at both piles in the 2-pile pile cap shown in Figure 10.45 is calculated as follows:

The actual effective depth of the pile cap (assuming #7 bottom bars in the pile cap) is,

$$d = 42 \text{ in.} - 6 \text{ in. embedment} - 3 \text{ in. cover} - \left(\frac{1}{2}\right)\left(\frac{7}{8} \text{ in.}\right) = 32.6 \text{ in.}$$

The combined critical perimeter is,

$$b_o = (2)\ell + \pi\left(\frac{d_p}{2} + \frac{d}{2}\right) + \pi\left(\frac{d_p}{2} + \frac{d}{2}\right) = 2\ell + \pi(d_p + d)$$
$$= 2(3 \text{ ft.})(12) + \pi(12 \text{ in.} + 32.6 \text{ in.}) = 212 \text{ in.}$$

$$\phi V_c = \phi 4\sqrt{f_c'}b_0 d \ge V_u$$
$$= (0.75)4\left(\sqrt{5500}\right)(212 \text{ in.})(32.6 \text{ in.}) = 1538 \text{ kip} \ge 2V_u = 750 \text{ kip, O.K.}$$

Therefore, the pile cap is adequate for punching shear at the piles.

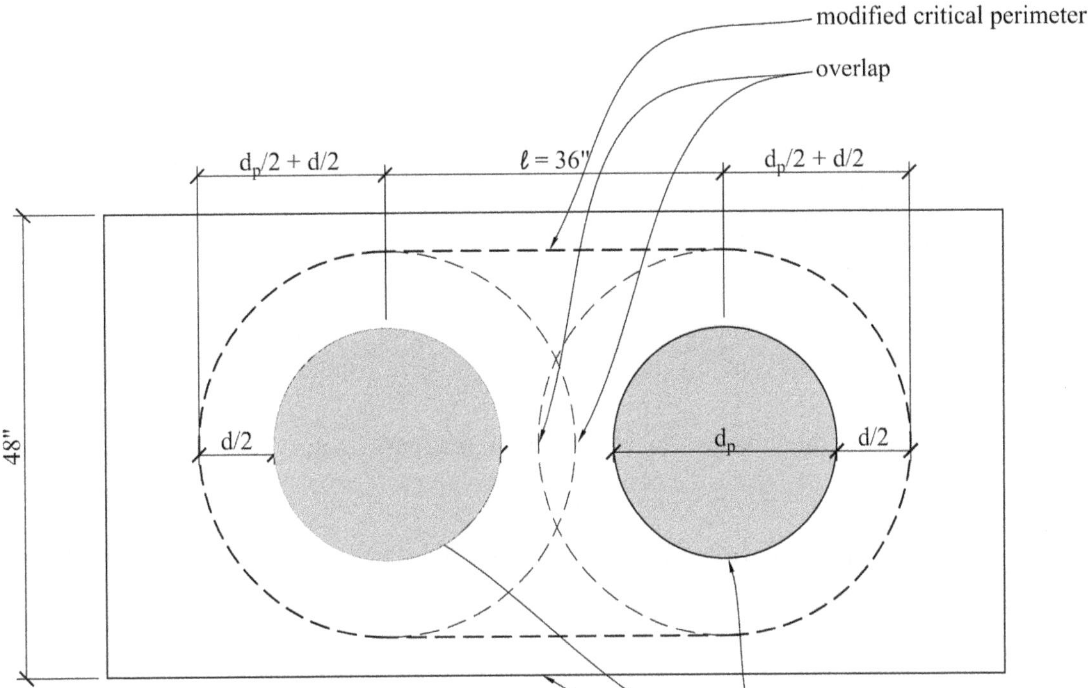

FIGURE 10.45 Modified critical perimeter for punching shear with overlapping critical perimeters

f. Assume the width of the tension zone, w_t (see Figure 10.43a). The piles are embedded 6 in. into the pile cap. Assuming the tension tie reinforcement will be placed in *one layer* with a 3 in. clear concrete cover to the tension tie reinforcement above the pile cut-off elevation, and assuming #7 bars.

Therefore, the width of the tension zone will be equal to the sum of two time the clear cover + and the bar diameter; that is,

$$w_t = d_b + 2c_c = 7/8 \text{ in.} + 2(3 \text{ in.}) = 6.88 \text{ in., say } 7 \text{ in.}$$

Assume the depth of the nodal zone, w_{nz}, at the column, below the load at point B is also 7 in.

Therefore, from geometry of the pile cap (see Figure 10.40), the angle of inclination of the interior strut will be calculated as follows:

$$\tan\theta = \frac{h - 6\text{ in.} - \frac{w_t}{2} - \frac{w_{nz}}{2}}{\frac{\ell}{2}}$$

$$= \frac{42\text{ in.} - 6\text{ in.} - \frac{7\text{ in.}}{2} - \frac{7\text{ in.}}{2}}{\frac{(3\text{ ft.})(12)}{2}} = 1.611$$

Therefore $\theta = 58.2° > 25°$ (see ACI Code Section 23.2.7) O.K.

g. Calculate the factored or ultimate force in the strut, F_{us}, and check the capacity of the strut.

Equilibrium of forces at joint A requires that $F_{us}\sin(58.2°)$ = pile reaction, $R_{uA} = 375$ kip,

Therefore, $F_{us} = 441$ kip.

From the geometry of the strut and the assumed depth of the tension tie zone depth (see Figure 10.43), the in-plane width of the strut, w_s, is

$w_s = w_t \cos\theta + \ell_b \sin\theta = (7\text{ in.}) \cos(58.2°) + (12\text{ in.}) \sin(58.2°)$

$= 13.89$ in.

Therefore, the cross-sectional area of the diagonal strut,

$A_{cs} = (b_w)(w_s) = (48\text{ in.})(13.89\text{ in.})$

For pile caps, the strut is unreinforced, therefore, from ACI Code Section 13.4.6.4, $\beta_s = 0.6$.

The strut and node confinement modification factor, β_c, is conservatively assumed as 1.0 (see Table 10.4).

Therefore, the strut compression capacity is,

$\phi F_{ns} = \phi A_{cs}\,(0.85\beta_c\beta_s f_c')$

$= 0.75\,(48\text{ in.})(13.89\text{ in.})[(0.85)\,(1.0)(0.6)(5.5\text{ ksi})]$

$= 1403\text{ kip} > F_{us} = 441$ kip O.K.

h. Calculate the force in the tension tie and the area of the tension tie reinforcement:

Equilibrium of forces at joint A requires that

$F_{us}\cos(58.2°) = F_{ut}$

Therefore, $F_{ut} = 441\text{ kip}\cos(58.2°) = 232$ kip

Area of tension tie rebar $= F_{ut}/\phi f_y = 232\text{ kip}/[(0.75)(60\text{ ksi})] = 5.2\text{ in.}^2$

The minimum reinforcement required is $0.0033 b_w d$, therefore,

$A_{s,min} = 0.0033(48\text{ in.})(42\text{ in.} - 6\text{ in.} - 3\text{ in. cover} - 0.875\text{ in.}/2)$

$= 5.16\text{ in}^2 < A_s$ required

Use one layer of 9 #7 bars for the tension tie (A_s provided $= 5.4\text{ in.}^2 > A_s$ required, O.K.).

Minimum width to accommodate 9 #7 bars in one layer.

(see Table A.3) $= 20\text{ in.} < b_w = 48$ in. O.K.

Available length to develop hooked tension tie in a single layer (see Figure 10.43) is

$$\ell_{anc} = 2d_p + \frac{d_p}{2} + \frac{\frac{w_t}{2}}{\tan\theta} - 3\text{" side cover}$$

$$= 2(12\text{ in.}) + \frac{12\text{ in.}}{2} + \frac{\frac{7\text{ in.}}{2}}{\tan 58.2} - 3\text{ in.} = 28.8\text{ in.}$$

Required hooked bar development length in 5500 psi concrete, ℓ_{dh}, for #7 bars = 12 in. < available length, O.K.

i. Check the capacity of the nodal zones:

At nodal zone A:

Nodal zone A is a C-C-T node (see Figure 10.40), therefore, from Table 10.6, $b_n = 0.8$

From the geometry of the strut and the assumed depth of the tension tie zone depth, and noting the 12 in. pile diameter, the in-plane width of the strut, w_s (see Figure 10.43) is

$w_s = w_t \cos\theta + \ell_b \sin\theta = (7\text{ in.})\cos(58.2°) + (12\text{ in.})\sin(58.2°)$

$= 13.89$ in.

At nodes A and C, assume the out-of-plane thickness of the nodal zone is equal to the width of the pile cap = 48 in.

The compression capacity of nodal zone A is

$\phi F_{nn} = \phi A_{cs}(0.85\beta_c\beta_n f_c')$

$= 0.75(48\text{ in.})(13.89\text{ in.})[(0.85)(1.0)(0.8)(5.5\text{ ksi})]$

$= 1870\text{ kip} > F_{un} = 441\text{ kip(O.K.)}$

At nodal zone B:

Nodal zone B is a C-C-C node (see Figure 10.40); therefore, from Table 10.6, $\beta_n = 1.0$.

From the geometry of the strut and the assumed nodal zone depth of 7 inches at node B (see Figure 10.44), the in-plane width of the strut at node B,

$$w_s = \sqrt{(7\text{in.})^2 + (9\text{ in.})^2} = 11.4\text{ in}$$

At node B, assume the out-of-plane thickness of the nodal zone is equal to the width of the pile cap = 48 in. The capacity of nodal zone B is

$$\phi F_{nn} = \phi A_{nz} f_{ce} = \phi A_{nz}(0.85\beta_n f_c') = 0.75(48\text{ in.})(11.4\text{ in.})(0.85)(1.0)(5.5\text{ ksi})$$

$$= 1919\text{ kip} > F_{un} = 441\text{ kip O.K.}$$

j. The minimum bottom reinforcement in the transverse direction (on top of the pile cap main bottom reinforcement) = 0.0018bh = 0.0018(12 in.)(42 in.) = 0.907 in.^{2} Use #7 @ 7½ in. on centers in the transverse direction, on top of and perpendicular to the main tension tie reinforcement.

k. The reinforcement in the pile cap will be laid out similar to the reinforcement shown in Figure 10.33. The design of this pile cap can be made more efficient by using a lower concrete strength. The reader should verify the lowest concrete strength that would be adequate for this pile cap.

An introduction to the design of pile caps has been provided in this chapter. For a more detailed design of pile caps under various loading conditions, the reader is referred to Ref. [20].

10.14 POLE FOOTINGS AND SONOTUBE FOUNDATIONS

A sonotube foundation is a simpler foundation type that is used in many lightly loaded structures. The foundations are typically cylindrical in shape and the term "sonotube" is used to describe the cylindrical tube-shaped formwork used for the concrete pour. The sonotubes vary in diameter from 8 inches to 48 inches. These are typically constructed by augering a hole into the ground down to the required design depth, and then the round form is placed. A formwork does not always have to be used, but the form is useful in preventing soil from falling into the hole during construction. The form also allows for a more aesthetically pleasing finish at the top for conditions that are left exposed. Sonotube foundations are used for agricultural buildings and for pole barns or pole buildings, as well as accessory structures, such as garages or storage buildings. Other common structures with posts embedded into the soil or into a concrete footing are residential decks, open structures such as pergolas, light poles, and smaller signs and flagpoles (see Figure 10.46).

These structures are typically not very sensitive to lateral deflections, and thus a simplified foundation and higher allowable lateral deflections can be justified. The lateral stability of the structure can be achieved by using a post buried into the concrete foundation with the post acting as a cantilever and the soil effectively providing a 'fixed' support through the lateral resistance of the soil. Empirical equations to select the required concrete foundation embedment depth based on the given loading and lateral resistance of the soil are found in Chapter 18 of the IBC. A key factor in the design of posts embedded into soil is the presence of restraint at the ground level. When restraint is provided at the ground level, which typically implies the use of a pavement or a concrete slab, the required embedment of the concrete foundation into the soil can be less than if a pavement or concrete slab is not present. It is conservative to assume that this restraint is not present or may not be there for the life of the structure.

In the case of a light structure such as a pergola or small frame that might support mechanical equipment exposed to the exterior, the lateral stability of the structure is achieved through a combination of vertical bracing or moment frames.In many of these cases where there might be more than a single sonotube (e.g. – a simple 4-post frame), the foundations would only support a gravity load and a horizontal shear force to avoid large moments in the foundation.

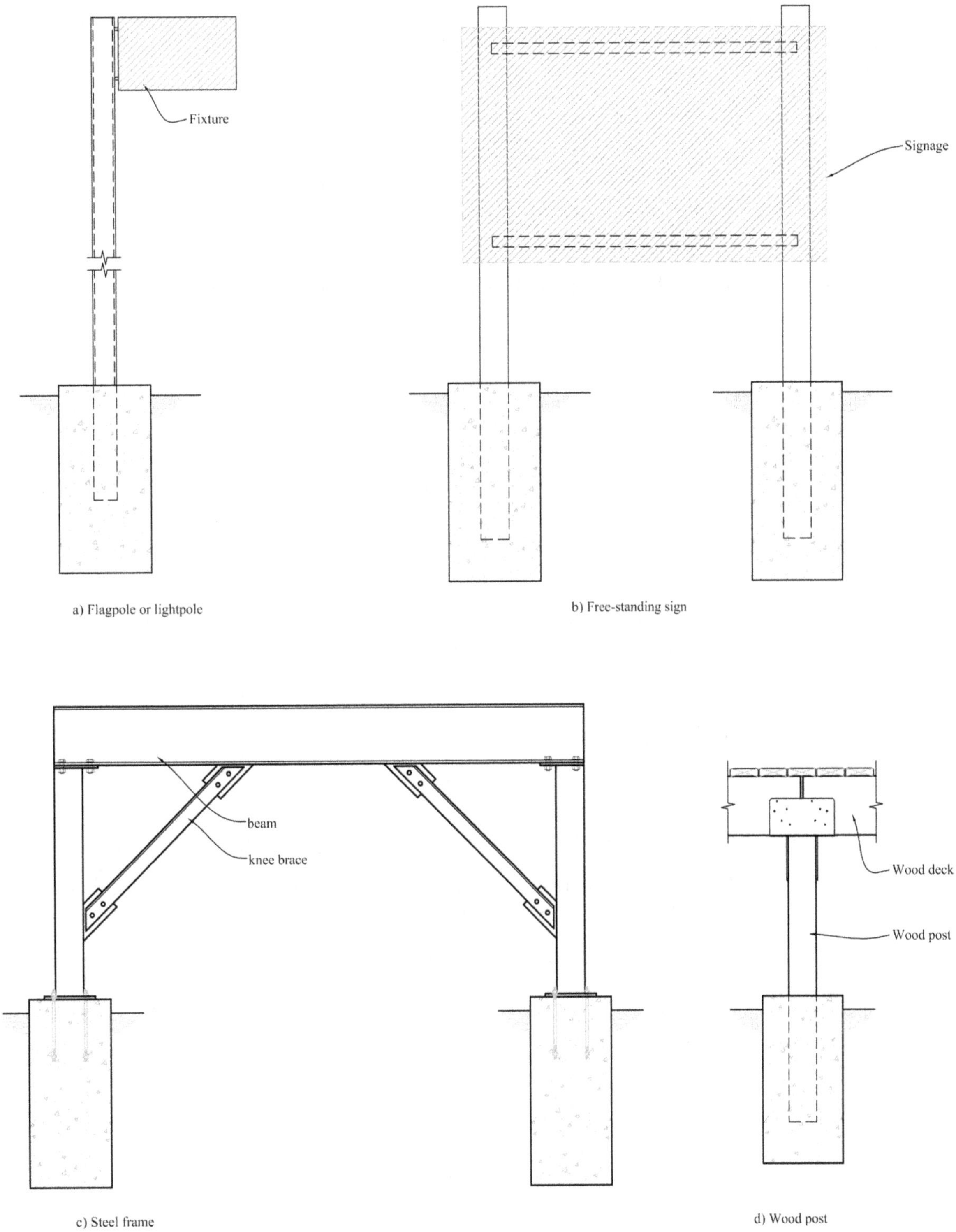

FIGURE 10.46 Pole footing types.

Using Chapter 18 of the International Building Code (IBC), the required embedment depth of a laterally loaded post embedded within a cylindrical concrete foundation into the soil is calculated as follows:

Nonconstrained at the ground level:

$$d = 0.5A[1 + [1 + (4.36h / A)]^{\frac{1}{2}}]$$

Constrained at the ground level:

(i.e. – pavement of concrete slab is present around the post)

$$d = \sqrt{\frac{4.25Ph}{S_3 b}} \quad or \quad \sqrt{\frac{4.25M_g}{S_3 b}}$$

Where

$$A = \frac{2.34P}{S_1 b}$$

b = diameter or width of the footing (in feet)

d = depth of embedment into the soil but not greater than 12 feet for the purposes of calculating the lateral soil pressure (in feet)

h = distance from the ground surface to the lateral force 'P' (in feet)

P = applied lateral force (in pounds)

M_g = Applied moment at the ground surface

S_1 = allowable lateral soil bearing pressure at a depth *equal to one-third of the depth of embedment*, (in psf), see Table 10.1 or IBC Table 1806.2. see IBC Table 1806.2 (in psf)

S_3 = allowable lateral soil bearing pressure at a depth *equal to the depth of embedment*, (in psf), see Table 10.1 or IBC Table 1806.2.

The footing itself will also need to be checked for bearing pressure at the bottom, axial load, bending, and shear. Reinforcing is often provided but only for temperature and shrinkage, but structures like these are typically lightly loaded and are often designed as unreinforced or plain concrete members (see Section 10.3).

To check bearing pressure:

$$p_a = \frac{P_V}{A} \leq P_{brg}$$

where:

p_a = actual soil bearing pressure

lte

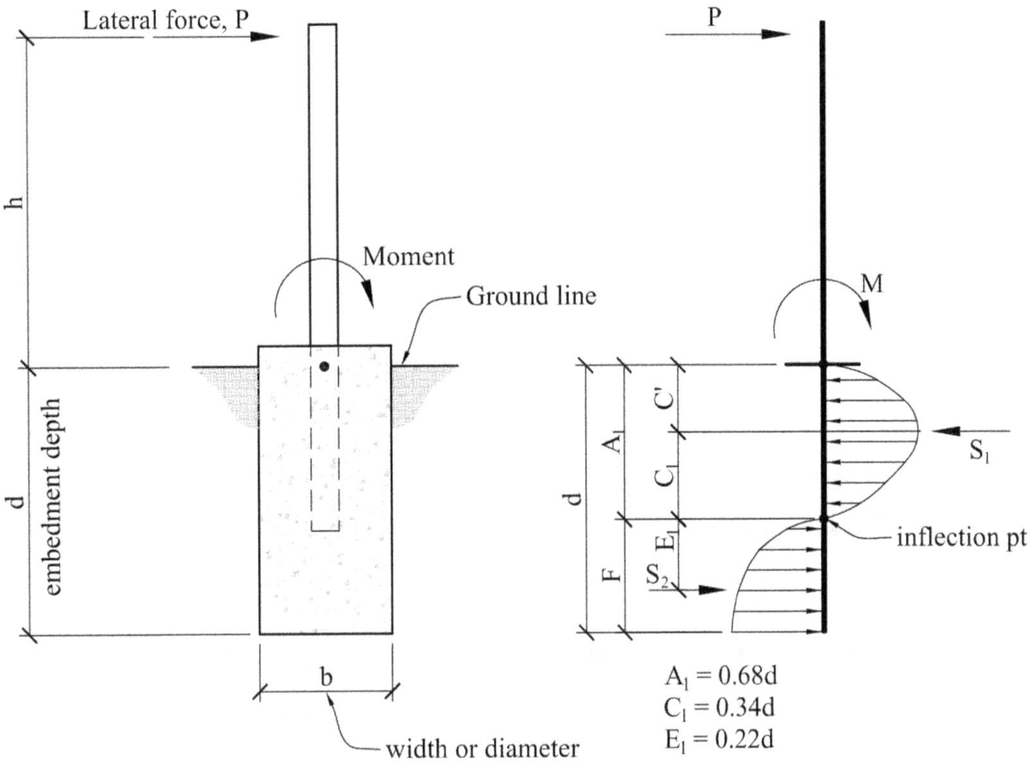

a) loading parameters

b) soil pressure [27]

FIGURE 10.47 Pole footing design parameters

P_v = Vertical load

P_{brg} = allow soil bearing pressure

The ultimate strengths of plain concrete members are calculated as follows:

For *axial load:*

$$\phi P_n = \phi 0.6 f_c' A_g \left[1 - \left(\frac{L_c}{32h}\right)^2\right]$$

Where

$\phi = 0.6$ (plain concrete, ACI 318-19, Section 22.1)

f_c' = Concrete strength, psi

A_g = gross area of the footing

L_c = effective height of the footing

h = least horizontal dimension of the footing

For *bending:*

$$\phi M_n = \phi 5\sqrt{f'_c}S_m < \phi 0.85 f'_c S_m$$

where:

$\phi = 0.6$ (plain concrete, ACI Code Section 22.1)

f'_c = Concrete strength, psi

S_m = section modulus of the footing

For *shear:*

$$\phi V_n = \phi \frac{4}{3}\sqrt{f'_c}A_g$$

where:

$\phi = 0.6$ (plain concrete, ACI Code Section 22.1)

f'_c = Concrete strength, psi

A_g = cross sectional area of the footing

EXAMPLE 10.11: POST-EMBEDDED CONCRETE FOOTING DESIGN

Determine the required embedment depth into soil of a concrete footing with the following parameters:

Lateral load: 1200 lb located 8 feet above the ground

Assume an 18 inch diameter footing

Soil type: sandy gravel

Concrete strength: f'_c = 4,500 psi

Check the nonconstrained and constrained cases.

Solution:

From IBC Table 1806.2 for sandy gravel, the allowable lateral soil bearing pressure is 200 psf per foot of embedment.

Assume an initial embedment depth of 8 feet

The lateral pressure is 0 psf at the top of the footing at the ground surface and the lateral pressure is (200 psf/ft)(8 ft) = 1600 psf at the bottom. Therefore, the pressure at one-third of the embedment is S_1 = (200 psf/ft)(8 ft/3) = 533 psf.

The parameter "A" is:

$$A = \frac{2.34P}{S_1 b} = \frac{2.34(1200)}{(533)(1.5)} = 3.51$$

Check the *non-constrained solution:*

Where b = 18 in. or 1.5 ft.

$$d = 0.5A\left[1 + \left[1 + (4.36h / A)\right]^{\frac{1}{2}}\right] = 0.5(3.51)\left[1 + \left[1 + (4.36(8) / 3.51)\right]^{\frac{1}{2}}\right] = 7.55 \text{ ft}$$

The required embedment depth was calculated to be less than the assumed embedment depth of 8 ft., so the design is considered adequate at a depth of 8 ft. Further iterations would yield an exact solution of d = 7.73 ft. For practical reasons, a depth of 8 ft. should be selected for this design.

Check the *constrained solution*:

Try an embedment depth of 5 feet.

S_3 = (200 psf/ft)(5) = 1000 psf

$$d = \sqrt{\frac{4.25Ph}{S_3 b}} = \sqrt{\frac{4.25(1200)(8)}{(1000)(1.5)}} = 5.21 \text{ ft}$$

The required embedment depth is larger than the assumed embedment depth, so a further iteration is required.

Try an embedment depth of 5'-3"

S_3 = (200 psf/ft)(5.25) = 1050 psf

$$d = \sqrt{\frac{4.25Ph}{S_3 b}} = \sqrt{\frac{4.25(1200)(8)}{(1050)(1.5)}} = 5.09 \text{ ft}$$

This is less than the assumed depth of 5'-3", so this could be chosen for the depth of the footing.

Again, it is conservative to assume a non-constrained condition if the actual conditions are not well defined.

Since the footing is subjected to both bending and shear, the plain concrete footing will also need to be checked. For bending, the factored moment and shear in the footing can conservatively be taken as:

M_u = (1.6)(1200 lb)(8 ft) = 15,360 ft-lb

V_u = (1.6)(1200 lb) = 1,920 lb

The section modulus of the footing is:

$$S_m = \frac{\pi b^3}{32} = \frac{\pi(18)^3}{32} = 572 \text{ in}^3$$

The bending strength for plain concrete is the smaller of:

$$\phi M_n = \phi 5\sqrt{f_c'}S_m = \frac{(0.6)5\sqrt{4,500}(572)}{12} = 9,592 \text{ ft-lb} < M_u = 15,360 \text{ ft-lb, N.G.}$$

and,

$$\phi M_n = \phi 0.85 f_c' S_m = \frac{(0.6)(0.85)(4{,}500)(572)}{12} = 109{,}395 \text{ ft-lb} (\textit{does not control})$$

The footing is not adequate for bending, so the diameter will have to be increased. Try a 22" diameter footing:

$$S_m = \frac{\pi b^3}{32} = \frac{\pi(22)^3}{32} = 1045 \text{ in}^3$$

$$\phi M_n = \phi 5\sqrt{f_c'} S_m = \frac{(0.6)5\sqrt{4{,}500}(1045)}{12} = 17{,}525 \text{ ft-lb} > 15{,}360 \text{ ft-lb, O.K.}$$

Check shear:

$$A_g = \frac{\pi b^2}{4} = \frac{\pi(22^2)}{4} = 380 \text{ in}^2$$

$$\phi V_n = \phi \frac{4}{3}\sqrt{f_c'} A_g = (0.6)\frac{4}{3}\sqrt{4500}(380) = 20{,}392 \text{ lb} > 1{,}920 \text{ lb, O.K.}$$

REFERENCES

[1] David G. Kittridge. "High Rise Shearwall Analysis, Design & Constructibility Guidelines," Proceedings of the *ASCE Structures Congress*, September 2010, pp. 381–382.

[2] Joseph A Amon. "The Geotechnical Report: What the Designer Needs to Know," *Concrete International*, September 2010, pp. 48–49.

[3] Joseph E. Bowles. *Foundation Analysis and Design, 5th ed.McGraw-Hill, New York 2001.*

[4] David Rogowsky and James K. Wight. "Load Factors are Load Factors," *Concrete International*, Vol 32, No. 7 July 2010, p. 75.

[5] Harry G. Poulos, "Tall Building Foundations: Design Methods and Applications," *Innovative Infrastructure Solutions* Vol 1, No. 1, 2016. Springer

[6] Braja M. Das. "Principles of Foundation Engineering," 5th Edition, Thomson, 2004.

[7] Steven C. Ball and James S. Notch. "Computer Analysis/Design of Large Mat Foundations," Journal of Structural Engineering, Vol. 110, No. 5, May 1984

[8] Edward J. Ulrich, Jr. "Subgrade Reaction in Mat Foundation Design," Concrete International, April 1991.

[9] ACI Committee 336. "Suggested Design Procedures for Combined Footings and Mats," Journal of the American Concrete Institute, Vol. 63, No. 10, pp. 1041–1077, 1988.

[10] Cheng Liu and Jack B. Evett. "Soils and Foundations," 8th Edition, Pearson, 2014.

[11] E. Alfred Picardi. "Structural System—Standard Oil of Indiana Building," Engineering Journal, American Institute of Steel Construction, Second Quarter, 1975.

[12] S. V. DeSimone. "Distribution of Wind Load to Soil," Technical Committee No. 11, Foundation Design, State of the Art Report No. 2, International Conference on the Planning and Design of Tall Buildings, Lehigh University, Bethlehem, PA, USA, August 21–26, 1972.
[13] CPCA. "Concrete Design Handbook," Third Edition, Canadian Portland Cement Association, 2012, pp. 9-3-9-49.
[14] R. Park and W.L. Gamble. Reinforced Concrete Slabs, pp. 513–514, John Wiley & Sons, New York.
[15] Alan Williams. "Design of Reinforced Concrete Structures," Engineering Press, 1997.
[16] Dennis C. K. Poon and Torsten G. Gottlebe. "Sky High in Shenzhen," ASCE Civil Engineering, pp. 48–53, December 2017.
[17] Dominic A. Webber and Howard A. Wells. "Grant Street Pier - Drilled Shaft Repair," Structure Magazine, pp. 22–25, December 2017.
[18] Gautier Chantelot and Alexandre Mathern. "Strut-and-tie modelling of reinforced concrete pile caps," Department of Civil Engineering, Chalmers University of Technology, Goteborg, Sweden, 2010.
[19] Perry Adebar and Luke (Zongyu) Zhou. "Design of Deep Pile Caps by Strut-and-Tie-Models," ACI Structural Journal, Vol. 94, No. 4, July-August 1996.
[20] Timothy W. Mays. "CRSI Design Guide for Pile Caps," Concrete Reinforcing Steel Institute (CRSI), 1st Edition, Shaumburg, IL, 2015.
[21] Chris Willams, Dean Deschenes, and Oguzhan Bayrak. "Strut-and-Tie Model Design Examples for Bridges: Final Report," Center for Transportation Research, Report No. FHWA/TX-12/5-5253-01-1, The University of Texas, Austin, Texas, June 2012.
[22] Ronald O. Hamburger, Stephen K. Harris, and Lachezar V. Handzhiyski. "Stabilizing San Francisco's Leaning Tower," Structure Magazine, pp. 8–10, June 2021.
[23] Tadeusz Granosik and Guy Michel Chovelon. "Insights on Pile Cap Construction," *Concrete International*, June 2018, pp. 38–39.
[24] CRSI. "CRSI Design Handbook," Concrete Reinforcing Steel Institute, 2002, pg. 13–18.
[25] Gary J. Klein, Nazanin Rezaei, David Garber, and A. Koray Tureyen. "Shear in Discontinuity Regions," *Concrete International,* May 2019, pp. 36–41.
[26] American Society of Civil Engineers, ASCE 7, Minimum design loads for buildings and other structures (Provisions and Commentary), Reston, VA, 2022.
[27] Donald Patterson, "Pole Building Design", American Wood Preservers Institute, 1969.
[28] Fernando Castilla, Phillippe Martin, and John Link, "Fixity of Members Embedded in Concrete," US Army Corp of Engineers Report No. CERL TR M-399, February 1984.
[29] Jason C. Varney, Jake E. Hughes, and Dominic J. Kelly, "Maintaining Stability on Unstable Ground," STRUCTURE Magazine, December 2022, pp. 26–28.

EXERCISES

NOTE *Assume that all steel is uncoated and the soil pressures given are gross allowable soil pressures.*

10.1 Design a plain concrete wall footing to carry a 12-in.-thick reinforced concrete wall. Service loads are 2.3 kip/ft dead load (includes the weight of the wall) and 2.3 kip/ft live load. Use $f_c' = 3000$ psi and an allowable soil pressure of 4000 psf. Assume that the bottom of the footing is to be 4 ft below grade. The weight of earth $w_e = 100$ lb/ft^3.

10.2 Redesign the footing for the wall of Exercise 10.1. Service loads are 8.0 kip/ft dead load and 8.0 kip/ft live load.

10.3 Design a reinforced concrete footing for the wall of Exercise 10.1 if the service loads are 6 kip/ft dead load (includes wall weight) and 15 kip/ft live load. Use $f_y = 60{,}000$ psi.

10.4 Design a square individual column footing (reinforced concrete) to support an 18-in.-square reinforced concrete tied interior column. Service loads are 200 kip dead load and 350 kip live load. Use $f'_c = 3000$ psi and $f_y = 60{,}000$ psi. The allowable soil pressure is 3500 psf, and $w_e = 100$ lb/ft^3. The column, reinforced with eight No. 8 bars, has $f'_c = 5000$ psi and $f_y = 60{,}000$ psi. The bottom of the footing is to be 4 ft below grade.

10.5 Design a square individual column footing (reinforced concrete) to support a 16-in.-square reinforced concrete tied interior column. Service loads are 200 kip dead load and 160 kip live load. Both column and footing have $f'_c = 4000$ psi and $f_y = 60{,}000$ psi. The allowable soil pressure is 5000 psf, and $w_e = 100$ lb/ft^3. The bottom of the footing is to be 4 ft below grade. The column is reinforced with eight No. 7 bars.

10.6 Design for load transfer from a 14-in.-square tied column to a 13-ft-0-in.-square reinforced concrete footing. Use $P_u = 650$ kip, footing $f'_c = 3000$ psi, column $f'_c = 5000$ psi, and $f_y = 60{,}000$ psi (all steel). The column is reinforced with eight No. 8 bars.

10.7 Redesign the footing of Exercise 10.5 if there is a 7-ft-0-in. restriction on the width of the footing.

10.8 Footings are to be designed for two columns A and B spaced with $D = 16$ ft as shown. There are no dimensional limitations. The service load from column A is 100 kip and from column B is 150 kip. The allowable soil pressure is 4000 psf. Determine the appropriate footing(s) size, type, and layout. Assume thickness and disregard reinforcing.

10.9 Same as Exercise 10.8, except service loads are 700 kip from column A and 900 kip from column B, and $D = 14$ ft.

10.10 For the column layout shown, there is a width restriction W of 16 ft. Here $D = 14$ ft. Determine an appropriate footing size and layout if the service load from column A is 700 kip and from column B is 800 kip. Assume thickness and disregard reinforcing. The allowable soil pressure is 4000 psf.

10.11 For the column layout shown, $D = 24$ ft. The footing for the 18-in.-square column A must be placed flush with the left side of the column. Service loads are 200 kip for column A and 300 kip for column B. The allowable soil pressure is 4500 psf. Assume a footing thickness of 2 ft-6 in. and determine an appropriate size, type, and layout for the footing.

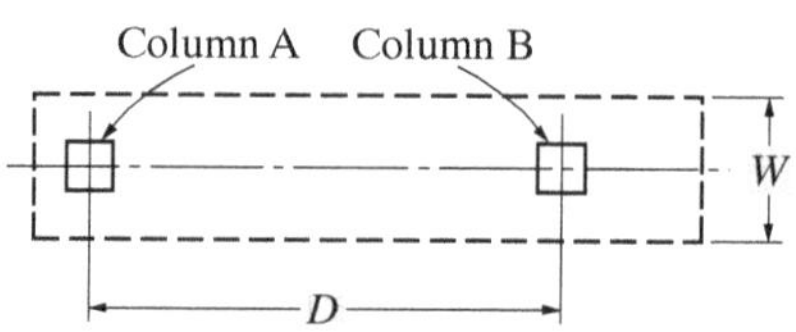

DETAILS FOR EXERCISES 10.8 TO 10.11

10.12 Refer to the Footing detail shown below where the Service dead load equals the service live load. Use $f_c' = 3000$ psi and $f_y = 60{,}000$ psi. The Net allowable soil pressure is 2,000 psf (q_{net})

a. Determine the service load 'P' that can be applied based on the net allowable soil pressure (service load)

b. Determine the service load 'P' that could be applied to the footing considering one-way shear.

c. Determine the service load 'P' that could be applied to the footing considering two-way shear.

d. Determine the service load 'P' that could be applied to the footing considering bending.

e. Based on the above considerations, what service load could this footing be rated for?

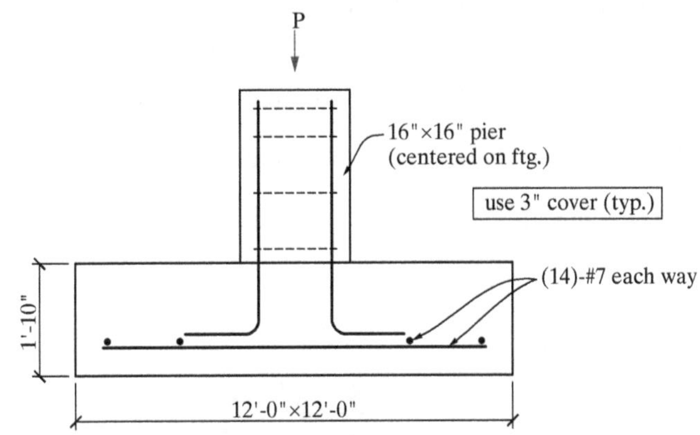

DETAIL FOR EXERCISE 10.12

10.13 Refer to the footing detail shown below. The Net allowable soil pressure is 6000 psf (q_{net}).

a. Determine the maximum column load, P, that can be applied based on the net allowable soil pressure (service load).

Use the result from part (a) to complete the rest of this problem assuming the service dead load equals the service live load.

b. Determine if the footing is adequate for one-way shear.

c. Determine if the footing is adequate for two-way shear.

d. Determine if the footing is adequate for bending.

e. Determine of the development length Ld1 is adequate.

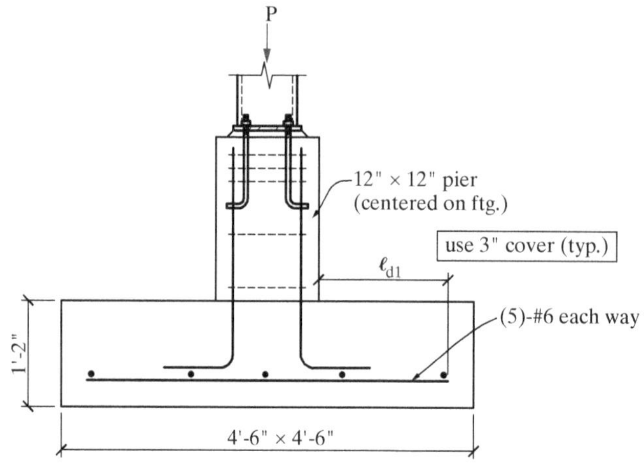

DETAIL FOR EXERCISE 10.13

10.14 Refer to the footing detail shown below. The Column service loads are: Dead = 40 kip, Live = 72 kip. The Net allowable soil pressure is 2000 psf (q_{net}).

a. Determine if the footing is adequate for one-way shear.

b. Determine if the footing is adequate for two-way shear.

c. Determine if the footing is adequate for bending.

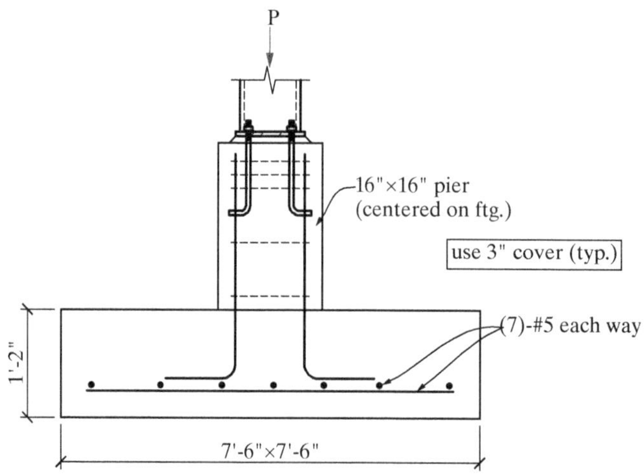

DETAIL FOR EXERCISE 10.14

CHAPTER 11

Prestressed Concrete Fundamentals

11.1 INTRODUCTION

According to the ACI definition, prestressed concrete is a material that has had internal stresses induced to balance out, to a desired degree, tensile stresses due to externally applied loads. Because tensile stresses are undesirable in concrete members, the object of prestressing is to create compressive stresses (prestress) at the same locations as the tensile stresses within the member so that the tensile stresses will be diminished or will disappear altogether. The diminishing or elimination of tensile stresses within the concrete will result in members that have fewer cracks or are crack-free at service load levels. This is one of the advantages of prestressed concrete over reinforced concrete, particularly in corrosive environments. Prestressed concrete offers other advantages. Because prestressed concrete beam cross sections are primarily in compression, diagonal tension stresses are reduced, and the beams are stiffer at service loads. In addition, sections can be smaller, resulting in less dead weight, reduced structural height, and savings on foundation costs.

Despite the advantages, we must consider the higher unit cost of stronger materials, the need for expensive accessories, the necessity for close inspection and quality control, and, in the case of precasting, a higher initial investment in plant equipment. Figure 11.1 shows precast bridge girder components ready for post-tensioning.

11.2 DESIGN APPROACH AND BASIC CONCEPTS

As most of the advantages of prestressed concrete are at service load levels and as permissible stresses in the "green" concrete often control the amount of prestress force to be used, a major part of analysis and design calculations is made using service loads, permissible stresses, and basic assumptions as outlined in Sections 20.3.2.5.1 and 24.5 of the ACI Code. The strength requirements of the code must also be met, however. Therefore, at some point, the design must be checked using appropriate load factors and strength-reduction factors. Three design checks are required for a *prestressed concrete* member: 1) service load stresses, 2) serviceability limit state such as deflections, and 3) the ultimate limit state or strength design check. In contrast, the design of *reinforced concrete* members only requires the latter two design checks.

FIGURE 11.1 Precast bridge girder components for post-tensioning. (George Limbrunner)

The normal method for applying prestress force to a concrete member is through the use of steel tendons. There are two basic methods of arriving at the final prestressed member: pretensioning and post-tensioning.

Pretensioning may be defined as a method of prestressing concrete in which the tendons are tensioned before the concrete is placed. This operation, which may be performed in a casting yard, is basically a five-step process:

1. The tendons are placed in a prescribed pattern on the casting bed between two anchorages. The tendons are then tensioned to a value not to exceed 94% of the specified yield strength, f_{py}, but not greater than the lesser of 80% of the specified tensile strength, f_{pu}, of the tendons and the maximum value recommended by the manufacturer of the prestressing tendons or anchorages (ACI Code, Section 20.3.2.5.1). The tendons are then anchored so that the load in them is maintained.

2. If the concrete forms are not already in place, they may then be assembled around the tendons.

3. The concrete is then placed in the forms and allowed to cure. Proper quality control must be exercised, and curing may be accelerated with the use of steam or other methods. The concrete will bond to the tendons.

4. When the concrete attains a prescribed strength, normally within 24 hours or less, the tendons are cut at the anchorages. Because the tendons are now bonded to the concrete, as they are cut from their anchorages the high prestress force must be transferred to

the concrete. As the high tensile force of the tendon creates a compressive force on the concrete section, the concrete will tend to shorten slightly. The stresses that exist once the tendons have been cut are often called the stresses at *transfer.* Because there is no external load at this stage, the stresses at transfer include only those due to prestressing forces and those due to the weight of the member.

5. The prestressed member is then removed from the forms and moved to a storage area so that the casting bed can be prepared for further use.

Pretensioned members are usually manufactured at a casting yard or plant that is somewhat removed from the job site where the members will eventually be used. In this case, they are usually delivered to the job site ready to be set in place. Where a project is of sufficient magnitude to warrant it economically, a casting yard may be built on the job site, thus decreasing transportation costs and allowing larger members to be precast without the associated transportation problems.

Figure 11.2 depicts the various stages in the manufacture of a precast, pretensioned member.

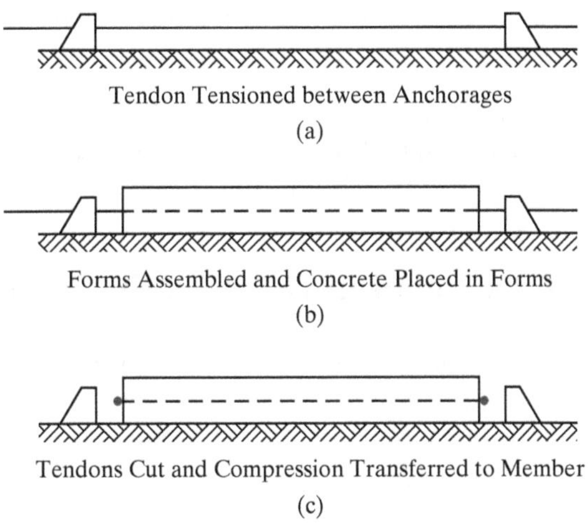

FIGURE 11.2 Pretensioned member.

Post-tensioning may be defined as a method of prestressing concrete in which the tendons are tensioned *after* the concrete has cured. (Refer to Figure 11.3.) The operation is commonly a six-step process:

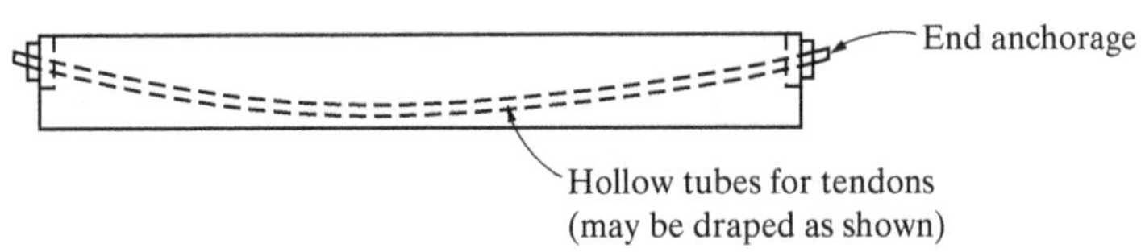

FIGURE 11.3 Post-tensioned member.

1. Concrete forms are assembled with flexible hollow tubes (i.e., metal or plastic ducts) placed in the forms and held at specified locations.
2. In most cases, the tendons are inserted or placed into the metal or plastic ducts prior to the concrete placement.
3. Concrete is then placed in the forms and allowed to cure to a prescribed strength.
4. The tendons are tensioned by jacking against an anchorage device or end plate that, in some cases, has been previously embedded in the end of the member. The anchorage device will incorporate some method for gripping the tendon and holding the tension force in the tendon.
5. If the tendons are to be bonded, the space in the tubes or ducts around the tendons may be grouted using a pumped grout. Some post-tensioned members use unbonded tendons and the tendons are usually covered with grease to reduce the friction between the tendons and the inside of the ducts. Unbonded tendons are commonly used in buildings, while bonded tendons are common in bridges and parking garages to reduce the likelihood of corrosion of the tendons.
6. The end anchorages may be covered with a protective coating to reduce the likelihood of corrosion.

Post-tensioning, in contrast to pre-tensioning, is done at the job site. The tensioning of the tendons is normally done with hydraulic jacks. Many patented devices are available to accomplish the anchoring of the tendon ends to the concrete.

11.3 STRESS PATTERNS IN PRESTRESSED CONCRETE BEAMS

The stress pattern existing on the cross-section of a prestressed concrete beam may be determined by superimposing the stresses due to the loads and forces acting on the beam at any particular time. For our purposes, the following sign convention will be adopted:

Tensile stresses are positive (+).

Compressive stresses are negative (−).

Because we will assume a crack-free cross-section at service load level, the entire cross-section will remain effective in carrying stress. Also, the entire concrete cross-section will be used in the calculation of the centroid and moment of inertia.

For purposes of explanation, we will consider a rectangular shape with tendons placed at the centroid of the section, and we will investigate the induced stresses. Although the rectangular shape is used for some applications, it is often less economical than the more complex shapes.

EXAMPLE 11.1

For the section shown in Figure 11.4, determine the stresses due to prestress immediately after transfer and the stresses at midspan when the member is placed on a 20-ft simple span.

Use $f'_c = 5000$ psi and assume that the concrete has attained a strength of 4000 psi at the time of transfer. Use a central prestressing force of 100 kip.

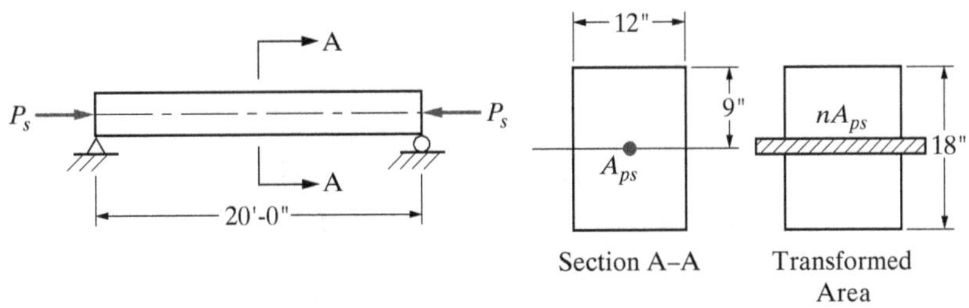

FIGURE 11.4 Sketches for Example 11.1.

Solution:

1. Compute the stress in the concrete at the time of initial prestress. With the prestressing force P_s applied at the centroid of the section and assumed acting on the gross section A_c, the concrete stress will be uniform over the entire section. Thus

$$f = \frac{P_s}{A_c} = \frac{-100}{12(18)} = -0.463 \text{ ksi}$$

2. Compute the stresses due to the beam dead load:

$$\text{Weight of beam: } w_{\text{DL}} = \frac{12(18)}{144}(0.150)$$
$$= 0.225 \text{ kip/ft}$$

$$\text{Moment due to dead load: } M_{\text{DL}} = \frac{w_{\text{DL}}\ell^2}{8} = \frac{0.225(20)^2}{8}$$
$$= 11.25 \text{ ft-kip}$$

Compute the moment of inertia I of the beam using the gross cross-section and neglecting the transformed area of the tendons (as the steel is at the centroid, it will have no effect on the moment of inertia):

$$I_g = \frac{bh^3}{12} = \frac{12(18)^3}{12} = 5832 \text{ in.}^4$$

$$\text{Dead load stresses: } f = \frac{Mc}{I} = \frac{11.25(12)(9)}{5832}$$
$$= \pm 0.208 \text{ ksi}$$

3. Compute the stresses due to prestress plus dead load:

$$f_{(\text{initial prestress + DL})} = -0.463 \pm 0.208$$
$$= -0.671 \text{ ksi} \quad (\text{compression, top})$$
$$= -0.255 \text{ ksi} \quad (\text{compression, bottom})$$

These results are shown on the stress summation diagram of Figure 11.5.

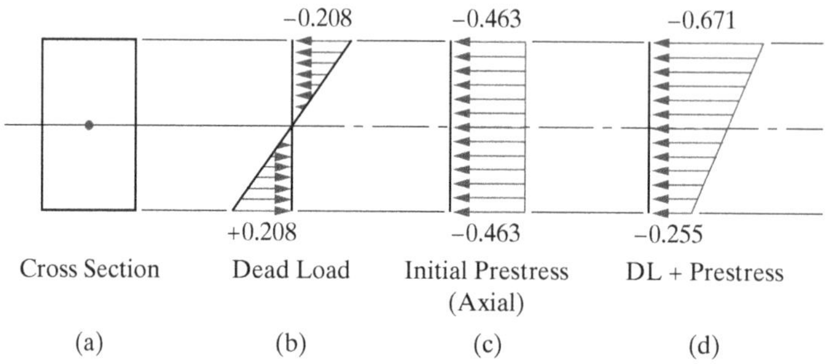

FIGURE 11.5 Midspan stresses for Example 11.1.

The tensile stresses due to the dead load (DL) moment in the bottom of the beam have been completely canceled out and compression exists on the entire cross-section. A limited additional positive moment may be carried by the beam without resulting in a net tensile stress in the bottom of the beam. This situation may be further improved by *lowering the location of the tendon* to induce additional compressive stresses in the bottom of the beam.

Example 11.1 reflects two stages of the prestress process. The transfer stage occurs in a pretensioned member when the tendons are cut at the ends of the member and the prestress force has been transferred to the beam, as shown in Figure 11.2. When the beam in this problem is removed from the forms (picked up by its end points), DL stresses are introduced, and in this second stage, both the beam weight (DL) and the prestress force are contributors to the stress pattern within the beam. This stage is important because it occurs early in the life of the beam (sometimes within 24 hours of casting), and the concrete stresses must be held within permissible values as specified in the ACI Code, Section 24.5.

If the prestress force were placed below the neutral axis in Example 11.1, negative bending moment would occur in the member at transfer, causing the beam to curve upward and pick up its dead load. Hence, for a simple beam such as this, where the prestress force is eccentric, the stresses due to the initial prestress would never exist alone without the counteracting stresses from the dead load moment.

11.4 PRESTRESSED CONCRETE MATERIALS

The application of the prestress force both strains or "stretches" the tendons and at the same time induces tensile stresses in them. If the tendon strain is reduced for some reason, the stress will also be reduced. In prestressed concrete, this is known as a *loss of prestress*. This loss occurs after the prestress force has been introduced. Contributory factors to this loss are creep and shrinkage of the concrete, elastic shortening of the concrete member due to compression from the prestress, relaxation of the tendon stress, anchorage seating loss, and friction losses due to intended or unintended curvature in post-tensioning tendons. Estimates on the magnitudes of these losses vary. It is generally known, however, that an ordinary steel bar tensioned to its yield strength (40,000 psi) would lose its entire prestress by the time all stress losses had taken place. Therefore, for prestressed concrete applications, it is necessary to use very high-strength steels, where the previously mentioned strain losses will result in a much smaller percentage of change in the original prestress force.

The most commonly used steel for pretensioned prestressed concrete is in the form of a seven-wire, uncoated, stress-relieved strand having a minimum tensile strength (f_{pu}) of 250,000 psi or 270,000 psi, depending on grade. The seven-wire strand is made up of seven cold-drawn wires. The center wire is straight, and the six outside wires are laid helically around it. All six outside wires are the same diameter, and the center wire is slightly larger. This in effect guarantees that each of the outside wires will bear on and grip the center wire.

Prestressing steel does not exhibit the definite yield point characteristic found in the normal ductile steel used in reinforcing steel (see Figure 11.6). The yield strength for prestressing wire and strand is a "specified yield strength" that is obtained from the stress-strain diagram at 1% strain, according to the American Society for Testing and Materials (ASTM). Nevertheless, the specified yield point is not as important in prestressing steel as is the yield point in the ductile steels. It is a consideration when determining the ultimate strength of a beam. For more information on prestressing wire and strand, the reader is referred to ASTM Standards A416 and A421 [1].

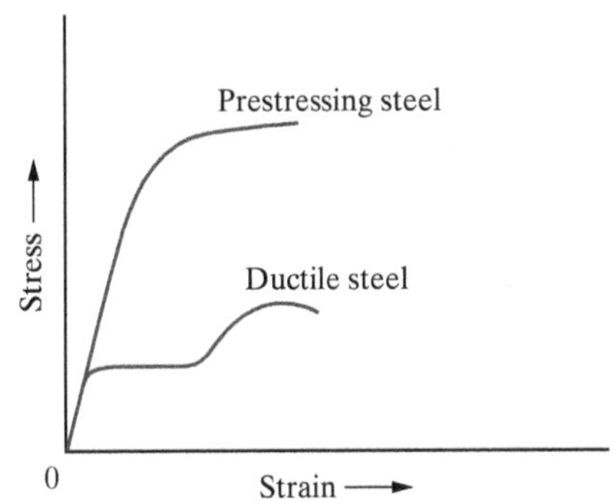

FIGURE 11.6 Comparative stress-strain curves.

In normal reinforced concrete members designed to ensure tension failures, the strength of the concrete is secondary in importance to the strength of the steel in the determination of the flexural strength of the member. In prestressed applications, concrete in the range from 4000 to 6000 psi is commonly used. Some of the reasons for this are as follows: (1) volumetric changes for higher-strength concrete are smaller, which will result in smaller prestress losses; (2) bearing and development stresses are higher; and (3) higher-strength concrete is more easily obtained in precast work than in cast-in-place work because of better quality control. In addition, high early strength cement (Type III) is normally used to obtain as rapid a turnover time as possible for optimum use of forms.

11.5 ANALYSIS OF RECTANGULAR PRESTRESSED CONCRETE BEAMS

The analysis of flexural stresses in a prestressed member should be performed for different stages of loading—that is, the initial service load stage, which includes dead load plus prestress before losses; the final service load stage, which includes dead load plus prestress plus live load after losses; and finally, the factored load stage, which involves load and strength-reduction factors. Generally, checking of prestressed members is accomplished at the service load level based on unfactored loads. The nominal strength of a member should be checked, however, using the same strength principles as for non-prestressed reinforced concrete members. Additionally, serviceability requirements such as deflections should be checked, similar to what is done for reinforced concrete members.

EXAMPLE 11.2

For the beam cross-section shown in Figure 11.7, analyze the flexural stresses at midspan at transfer and in service. Neglect prestress losses. Use a prestressed steel area A_{ps} of 2.0 in.^{2}, use $f_c' = 6000$ psi, and assume that the concrete has attained a strength of 5000 psi at the time of transfer. The initial prestress force = 250 kip. The service dead load = 0.25 kip/ft, which does not include the weight of the beam. The service live load = 1.0 kip/ft. Use $n = 7$. Assume that the entire cross-section is effective and use the transformed area (neglecting displaced concrete) for the moment of inertia.

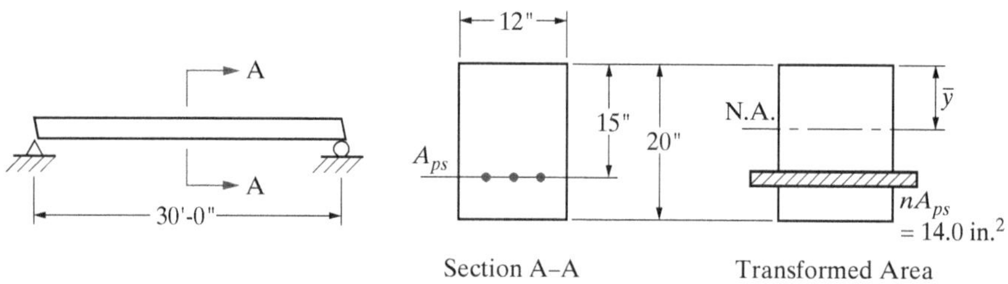

FIGURE 11.7 Sketches for Example 11.2.

Solution:

1. The beam weight is

$$w_{DL} = \frac{20(12)}{144}(0.150) = 0.25 \text{ kip/ft}$$

The moment due to the beam weight is

$$M_{DL} = \frac{w_{DL}\ell^2}{8} = \frac{0.25(30)^2}{8} = 28.1 \text{ ft-kip}$$

The moment due to superimposed loads (DL + LL) is

$$M_{DL+LL} = \frac{w_{(DL+LL)}\ell^2}{8} = \frac{1.25(30)^2}{8} = 140.6 \text{ ft-kip}$$

2. The location of the neutral axis is

$$\bar{y} = \frac{\Sigma(Ay)}{\Sigma A}$$

Using the top of the section as the reference axis,

$$\bar{y} = \frac{12(20)(10) + 14(15)}{12(20) + 14} = 10.28 \text{ in.}$$

The eccentricity e of the strands from the neutral axis is

$$15 - 10.28 = 4.72 \text{ in.}$$

The moment of inertia about the neutral axis is

$$I = \frac{12(20)^3}{12} + 12(20)\left(10.28 - \frac{20}{2}\right)^2 + 14(4.72)^2$$
$$= 8331 \text{ in.}^4$$

3. The stresses may now be calculated. These are summarized in Figure 11.8.

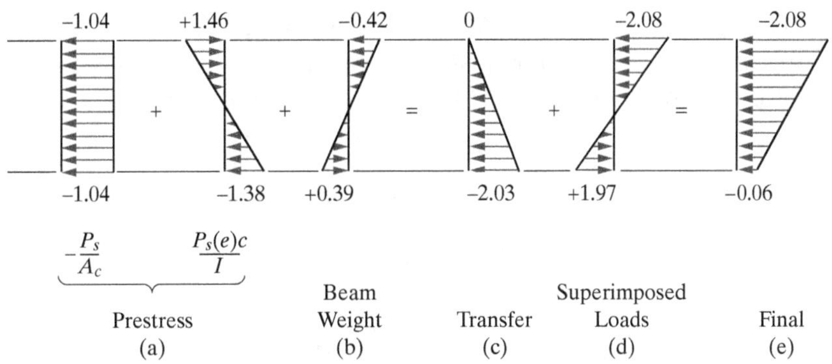

FIGURE 11.8 Midspan stress summary for Example 11.2.

a. Initial prestress: As a result of the eccentric prestressing force, the induced stress at the initial prestress stage will *not* be uniform but may be computed from

$$f = -\frac{P_s}{A_c} \pm \frac{Mc}{I}$$
$$= -\frac{P_s}{A_c} \pm \frac{P_s(e)c}{I}$$

where P_s/A_c represents the axial effect of the prestress force and $\frac{P_s(e)c}{I}$ is the eccentric, or moment, effect. Then

$$-\frac{P_s}{A_c} = -\frac{250}{12(20)} = -1.04 \text{ ksi}$$

(compression top and bottom)

$$+\frac{P_s(e)c}{I} = \frac{250(4.72)(10.28)}{8331} = +1.46 \text{ ksi (tension in top)}$$

$$-\frac{P_s(e)c}{I} = -\frac{250(4.72)(9.72)}{8331} = -1.38 \text{ ksi (compression in bottom)}$$

b. The stresses due to the beam self-weight are

$$f = \pm\frac{Mc}{I}$$

$$= +\frac{28.1(12)(9.72)}{8331} = +0.39 \text{ ksi (tension in bottom)}$$

$$= -\frac{28.1(12)(10.28)}{8331} = -0.42 \text{ ksi (compression in top)}$$

Summarizing the prestress transfer stage [i.e., prestress plus beam self-weight (DL)]:

Top of beam: $-1.04 + 1.46 - 0.42 = 0$

Bottom of beam: $-1.04 - 1.38 + 0.39 = -2.03$ ksi (compression)

The permissible stresses (ACI Code, Sections 24.5.3.1 and 24.5.3.2) immediately after prestress transfer (before losses) are in terms of f'_{ci}, which is the specified compressive strength (psi) of concrete at the time of initial prestress:

$$\text{compression} = 0.60 f'_{ci}$$
$$= 0.60(5000) = 3000 \text{ psi} = 3.0 \text{ ksi}$$
$$\text{tension} = 3\sqrt{f'_{ci}}$$
$$= 3\sqrt{5000} = 212 \text{ psi} = 0.212 \text{ ksi}$$

Because 2.03 ksi < 3.0 ksi and 0 < 0.212 ksi, the beam is satisfactory at this stage. Should the tensile stress exceed $3\sqrt{f'_{ci}}$, additional bonded reinforcement shall be provided.

Note in Figure 11.8c that no tensile stress exists at transfer.

c. At the service load stage, the stresses due to the superimposed loads (DL + LL) are

$$f = \pm\frac{Mc}{I}$$

$$= +\frac{140.6(12)(9.72)}{8331} = +1.97 \text{ ksi (tension in bottom)}$$

$$= -\frac{140.6(12)(10.28)}{8331} = -2.08 \text{ ksi (compression in top)}$$

Summarizing the service load stage when the beam has had service loads applied (prestress plus beam dead load plus superimposed loads) gives us

$$\text{Top of beam: } 0 - 2.08 = -2.08 \text{ ksi (compression)}$$

$$\text{Bottom of beam: } -2.03 + 1.97 = -0.06 \text{ ksi (compression)}$$

The permissible compressive stress in the concrete at the service load level (ACI Code, Section 24.5.4) due to prestress plus total load is 0.60 f'_c:

$$0.60(6000) = 3600 \text{ psi} = 3.60 \text{ ksi}$$

The permissible stress just given assumes that losses have been accounted for. Our analysis has neglected losses. Aside from this, the stresses in the beam would be satisfactory because 2.08 ksi < 3.60 ksi, and no tensile stress exists.

The final stress distribution (Figure 11.8e) in the beam of Example 11.2 shows that the entire beam cross-section is under compression. This is the stress pattern that exists at *midspan* (as the applied moments calculated were midspan moments). It is evident that moments will decrease toward the supports of a simply supported beam and that the stress pattern will change drastically. For example, if the eccentricity of the tendon were constant in the beam in question, the net prestress (Figure 11.8a) would exist at the beam ends, where moment due to beam weight and applied loads is zero. Because the tensile stress is undesirable, the location of the tendon is changed in the area of the end of the beam, so that eccentricity is decreased. This results in a *curved or draped* tendon within the member, as shown in Figure 11.9. Post-tensioned members may have curved tendons accurately placed to satisfy design requirements, but in pretensioned members, because of the nature of the fabrication process, only approximate curves are formed by forcing the tendon up or down at a few points.

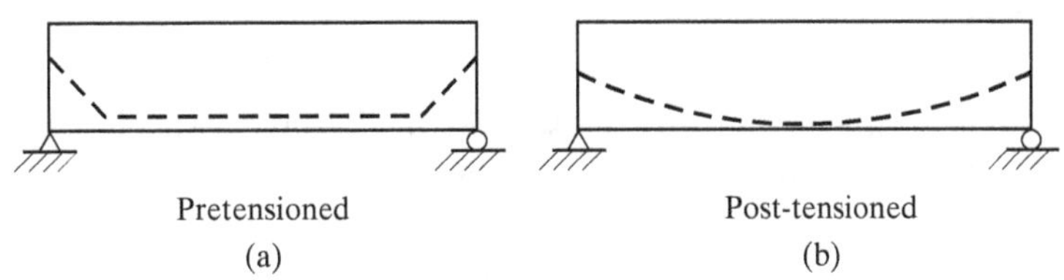

FIGURE 11.9 Draped tendons.

Although shear was not included in the beam analysis, the reader should be aware that the draping of the tendons, in addition to affecting the flexural stresses at the beam ends, produces a force acting vertically upward, which has the effect of reducing the shear force due to dead and live loads. Precast prestressed concrete flexural members are available in numerous shapes suitable for various applications. A few of the more common shapes are shown in Figure 11.10.

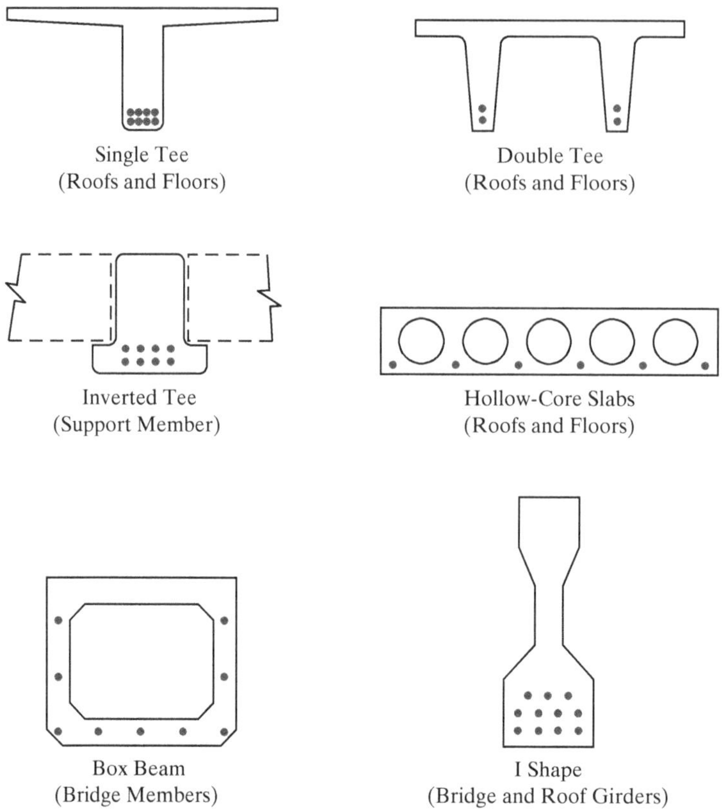

FIGURE 11.10 Typical precast, prestressed flexural members.

11.6 ALTERNATIVE METHODS OF ELASTIC ANALYSIS: LOAD BALANCING METHOD

The ACI Code requires that prestressed beams be analyzed elastically to establish whether the stresses developed by the combined action of service loads and prestressing force are within specified allowable stress limits. In effect, this limit on the magnitude of stress, specifically at transfer or with service loads in place, controls cracking and subsequently prevents crushing of the concrete.

The analysis method used in Section 11.5 is often designated as the *method of superposition* or the *combined loading concept*. To explain briefly, it is based on superimposing the stresses created by the applied loads with those created by the prestressing force. As previously shown, all these stresses are computed independently and then combined. This method provides the designer a complete picture of stress variation under various loading conditions.

A second method of elastic analysis is called the *load balancing method.* According to the Post-Tensioning Institute, this method is by far the most widely used method for analysis and design of post-tensioned structures. It is a technique of balancing the external load by selecting a prestressing force and tendon profile that creates a transverse load acting opposite to the external load. This transverse load may be equal to either the full external applied load or only part of it.

For example, to balance a uniformly distributed load (w) acting on a simply supported beam, a parabolic tendon profile with zero end eccentricities would be selected. The prestressing force P_s needed would be a function of the load to be balanced as well as the acceptable sag of the tendon. If the transverse load created by the tendons exactly balanced the external load, a uniform compressive stress distribution P_s/A_g will develop over the beam cross-section, and the beam will remain essentially level with no deflection or camber. To balance the load, the end eccentricities should be zero; otherwise, an end moment will be developed that disturbs the uniform stress distribution.

If only a portion of the external load is balanced, a net moment in the beam at any point will develop from that portion of the load that is *not* balanced by the prestressing. It is only this net moment that must be considered in computing the bending stress. Therefore, the stress acting at any point on a cross-section may be expressed as

$$f = \frac{P_s}{A_g} \pm \frac{M_{\text{net}}y}{I_g}$$

where

P_s = prestressing force applied to the member

M_{net} = net unbalanced moment on the section

and A_g, y, and I_g are as previously defined. The computed stress due to the uniform compression and the unbalanced moment must then (as in the other methods) be compared with code-allowable stresses.

The externally applied load that will be balanced need not necessarily be uniformly distributed. It can also be a concentrated load or be a combination of both types. For a uniformly distributed load, a shallow parabolic tendon profile is generally selected, whereas a linear profile with a sharp directional change is used for a concentrated load. These are illustrated in Figure 11.11.

With respect to the balancing of a concentrated load, note the sharp directional change of the tendon under the external load at midspan. This creates an upward force component

$$P_p = 2P_s \sin\theta$$

In the case shown, P_p exactly balances the applied load P. Therefore, if we neglect its own weight, the beam is not subject to net *transverse* load. At the ends of the beam, the horizontal components of P_s, which are shown as $P_s \cos\theta$ (and which are collinear with the centroidal axis of the beam), create a uniform compressive stress along the entire length of the beam. Therefore, the stress in the beam at any section may be expressed as

$$f = \frac{P_s \cos\theta}{A_g}$$

and for small values of θ,

$$f = \frac{P_s}{A_g}$$

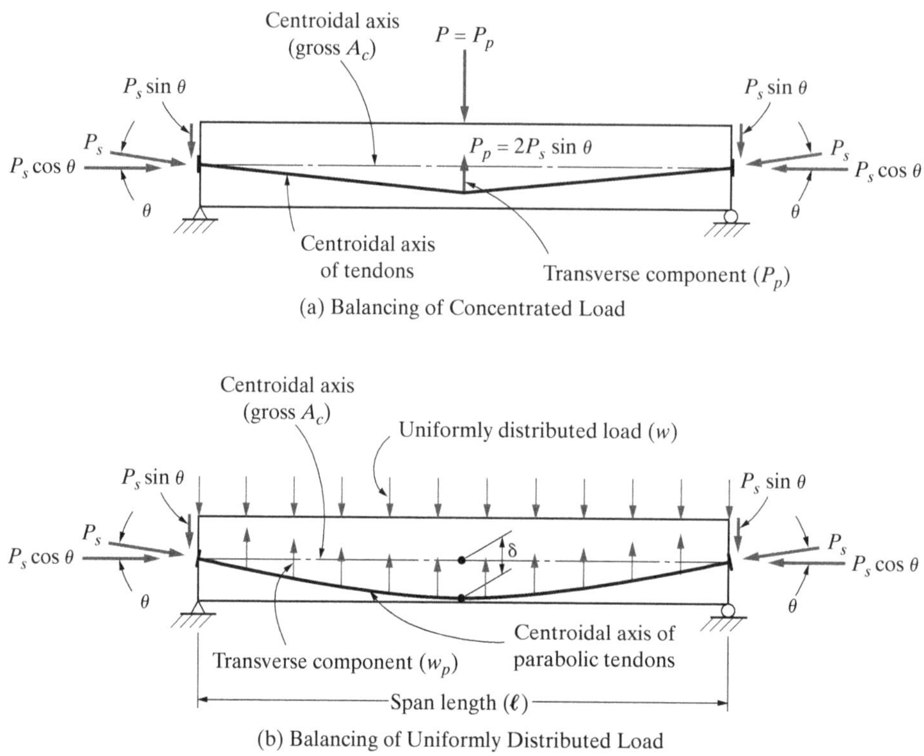

FIGURE 11.11 Load balancing: (a) balancing of concentrated load; (b) balancing of uniformly distributed load.

If any additional external load is applied, the beam will act as an elastic, homogeneous concrete beam (up to the point of cracking), and a bending stress will develop, which can be evaluated by

$$f = \frac{M_{\text{net}}c}{I_g}$$

where M_{net} is the moment developed by any load applied less the balanced load, in addition to P.

With respect to the balancing of the uniformly distributed load, note that the tendon profile is that of a parabolic curve whose upward transverse component or balanced load, w_p in lb/ft can be obtained from Figure 11.11b by considering the equilibrium of forces of a free body diagram of a portion of the beam between the midspan and the end of the beam, and is given by

$$w_p = \frac{8P_s\delta}{\ell^2} \qquad \textbf{(11.1)}$$

Assuming that the externally applied load w, including the weight of the beam, is exactly balanced by the transverse component w_p, there is no bending in the beam, and the beam is subject to uniform compressive stress calculated from

$$f = \frac{P_s}{A_g}$$

If the balanced load is different from the applied external load, the bending moment developed from the net loading will induce bending stress, which can be evaluated by

$$f = \frac{M_{\text{net}} c}{I_g}$$

where M_{net} is the moment developed by the net load = the applied load minus w_p.

It is questionable as to what portion of the external load should be balanced by the prestress. If too much of the applied load, such as DL plus $\frac{1}{2}$LL is to be balanced, excessive prestress may be required, which may result in undesirable tension stresses at the top of the beam. The designer must exercise judgment in determining the proper amount of loading to be balanced by prestressing. Some designers balance out 80% of the dead load for interior beams and 100% of the dead load for spandrel beams and girders supporting cladding loads.

EXAMPLE 11.3

The rectangular prestressed beam shown in Figure 11.12 carries uniformly distributed service loads of 1.0 kip/ft LL and 1.0 kip/ft DL (which includes the weight of the beam). A parabolic tendon, configured as shown, will be used. The tendon is to furnish a uniformly distributed upward balancing load of 1.5 kip/ft $\left(\text{DL} + \frac{1}{2}\text{LL}\right)$. Calculate the required prestressing force and determine the net moment at midspan.

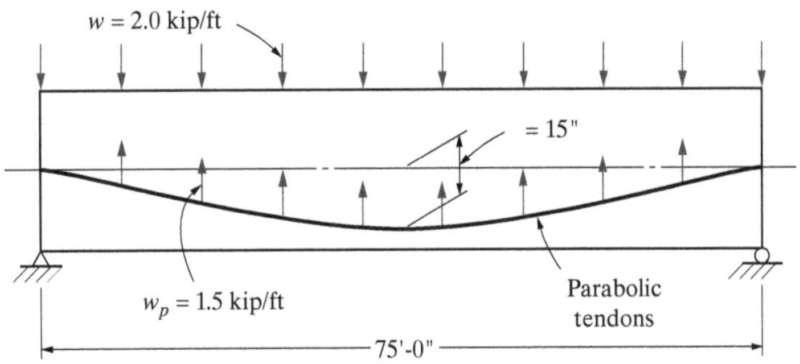

FIGURE 11.12 Sketch for Example 11.3.

Solution:

Solving Equation (11.1) for required P_s:

$$\text{required } P_s = \frac{w_p \ell^2}{8\delta} = \frac{1.5(75)^2}{8(15/12)} = 844 \text{ kip}$$

The net moment at midspan is calculated from

$$M_{net} = \frac{w_{net} \ell^2}{8} = \frac{(2.0 - 1.5)(75)^2}{8} = 352 \text{ ft-kip}$$

Example 11.4 illustrates the two methods of elastic analysis. In this example, we will neglect the transformed area of the steel as well as any displaced concrete, and we will use the gross properties of the cross-section to compute stresses. Note that this varies from the solution of Example 11.2.

EXAMPLE 11.4

A rectangular prestressed concrete beam shown in Figure 11.13 is simply supported and has a span length of 25 ft-0 in. The beam carries a superimposed service LL of 2.0 kip/ft. The only service DL is the weight of the beam. The parabolic prestressing tendon is located as shown and is subjected to an effective prestress force of 300 kip (neglect any prestress losses). Determine the outer fiber flexural stresses at midspan at (i) transfer (i.e., prestress + beam weight), and (ii) when the member is under full-service load conditions (i.e., prestress + beam weight + service LL).

Solution:

a. Using the **method of superposition -**

1. Calculate moments after determining the beam weight:

$$w_{\mathrm{DL}} = \frac{28(18)}{144}(0.150) = 0.525 \text{ kip/ft}$$

The moment due to beam weight is

$$M_{\mathrm{DL}} = \frac{w_{\mathrm{DL}}\ell^2}{8} = \frac{0.525(25)^2}{8} = 41.0 \text{ ft-kip}$$

The moment due to superimposed LL is

$$M_{\mathrm{LL}} = \frac{w_{\mathrm{LL}}\ell^2}{8} = \frac{2.0(25)^2}{8} = 156.3 \text{ ft-kip}$$

2. Using the gross concrete section and neglecting the transformed steel area, the neutral axis coincides with the centroidal axis. Therefore, the distance from the neutral axis to the outer fiber is 14 in. The eccentricity e of the tendon is given as 6 in. The moment of inertia about the neutral axis is found from

$$I_g = \frac{18(28)^3}{12} = 32{,}928 \text{ in.}^4$$

3. Calculate the stresses.

a. At transfer, the stress due to prestressing alone is determined from

$$\begin{aligned} f &= -\frac{P_s}{A_g} \pm \frac{P_s ec}{I_g} \\ &= -\frac{300}{18(28)} \pm \frac{300(6)(14)}{32{,}928} \\ &= -0.595 \pm 0.765 \end{aligned}$$

from which

$$\text{stress at top} = -0.595 + 0.765 = 0.170 \text{ ksi} \ \ (\text{tension})$$

$$\text{stress at bottom} = -0.595 - 0.765 = -1.36 \text{ ksi} \ \ (\text{compression})$$

The stress due to the service DL (beam weight) is calculated from

$$f = \pm\frac{Mc}{I} = \pm\frac{41.0(12)(14)}{32,928} = \pm 0.209 \text{ ksi}$$

Combining the preceding stresses for the transfer condition gives us

$$\text{stress at top} = +0.170 - 0.209 = -0.039 \text{ ksi} \ \ (\text{compression})$$

$$\text{stress at bottom} = -1.36 + 0.209 = -1.15 \text{ ksi} \ \ (\text{compression})$$

b. Under full-service load conditions, stress due to superimposed LL is calculated from

$$f = \pm\frac{Mc}{I} = \pm\frac{156.3(12)(14)}{32,928} = \pm 0.797 \text{ ksi}$$

That is, the stresses due to the superimposed LL are -0.797 ksi at the top, and +0.797 ksi at the bottom of the beam. Combining the preceding stresses for the full-service load condition gives us

$$\text{stress at top} = -0.039 - 0.797 = -0.836 \text{ ksi} \ \ (\text{compression})$$

$$\text{stress at bottom} = -1.15 + 0.797 = -0.353 \text{ ksi} \ \ (\text{compression})$$

b. Using the **load balancing** method -

1. Let $e = \delta = 6$ in. The parabolic tendons produce an upward balancing load of

$$w_p = \frac{8P_s\delta}{\ell^2} = \frac{8(300)(0.5)}{25^2} = 1.92 \text{ kip/ft}$$

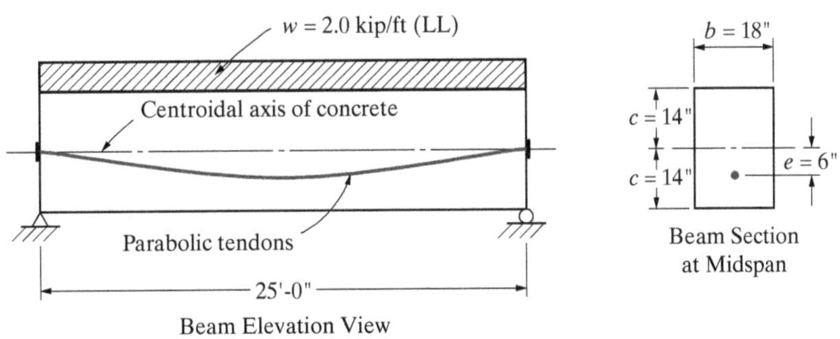

FIGURE 11.13 Sketches for Example 11.4.

a. At transfer, the net load on the beam is equal to the balanced load minus the dead load or self-weight of the beam, and is given as

$$w_{net} = 1.92 - 0.525 = 1.395 \text{ kip/ft} \qquad (\text{upward})$$

The net moment at midspan is

$$M_{net} = \frac{1.395(25)^2}{8} = 109.0 \text{ ft-kip} \quad \text{(tension on top)}$$

The bending stress due to the net moment is

$$f = \pm\frac{M_{net}c}{I_g} = \pm\frac{109(12)(14)}{32,928} = \pm 0.556 \text{ ksi}$$

The uniform compressive stress from the prestress force is

$$-\frac{P_s}{A_g} = -\frac{300}{18(28)} = -0.595 \text{ ksi}$$

$$\text{stress at top} = -0.595 + 0.556 = -0.039 \text{ ksi (compression)}$$

$$\text{stress at bottom} = -0.595 - 0.556 = -1.15 \text{ ksi (compression)}$$

b. Under full-service load, the net load on the beam is equal to the live load plus dead load minus the balanced load, and is given as

$$w_{net} = 2.0 + 0.525 - 1.92 = 0.605 \text{ kip/ft (downward)}$$

The net moment at midspan is

$$M_{net} = \frac{0.605(25)^2}{8} = 47.3 \text{ ft-kip (tension on bottom)}$$

The bending stress due to the net moment is

$$f = \pm\frac{M_{net}c}{I_g} = \pm\frac{47.3(12)(14)}{32,928} = \pm 0.241 \text{ ksi}$$

The uniform compressive stress from the prestress force is

$$-\frac{P_s}{A_g} = -\frac{300}{18(28)} = -0.595 \text{ ksi}$$

$$\text{stress at top} = -0.595 - 0.241 = -0.836 \text{ ksi (compression)}$$

$$\text{stress at bottom} = -0.595 + 0.241 = -0.354 \text{ ksi (compression)}$$

Note that the results from the method of superposition and the load balancing method agree (with some slight round-off error).

The load balancing method offers the advantages of simplicity and clarity, especially with continuous beams and slabs. It is the recommended method for those situations, particularly for preliminary designs. For simple spans, neither of the two methods has any particular advantage over the other; both are equally applicable.

11.7 FLEXURAL STRENGTH ANALYSIS

As part of the design and analysis procedure of a prestressed concrete beam, the ACI Code requires that the moment, due to factored service loads, M_u, not exceed the flexural design strength ϕM_n of the member. The design strength of prestressed beams may be computed using strength equations similar to those for reinforced concrete members, discussed in Chapter 2. The checking of the flexural (nominal) strength of a prestressed beam ensures that the beam is designed with an adequate factor of safety against failure.

The expression for the nominal bending strength of a rectangular-shaped prestressed beam is developed from the internal couple of an underreinforced beam at failure, as shown in Figure 11.14.

Assuming failure is initiated by the steel yielding, the magnitude of the internal couple is

$$M_n = A_{ps} f_{ps}\left(d_p - \frac{a}{2}\right)$$

where

A_{ps} = area of prestressed reinforcement in the tension zone
d_p = distance from extreme compression fiber to the centroid of the prestressed reinforcement
a = depth of stress block, determined by equating T and C and calculated from

$$a = \frac{A_{ps} f_{ps}}{0.85 f'_c b}$$

f_{ps} = stress in the prestressed reinforcement at nominal strength

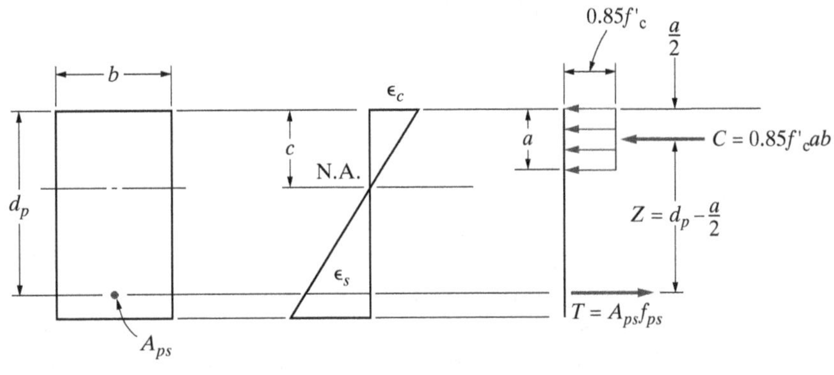

FIGURE 11.14 Equivalent stress block for strength analysis.

For members with **bonded** prestressing tendons and no non-prestressed tension or compression reinforcement $(i.e.\ \rho_s = \rho'_s = 0)$ and for $f_{se} \geq 0.5 f_{pu}$, the value of f_{ps} can be obtained from

$$f_{ps} = f_{pu}\left(1 - \frac{\gamma_p}{\beta_1}\rho_p \frac{f_{pu}}{f'_c}\right) \qquad \textbf{[Mod. ACE Eq. (20.3.2.3.1)]}$$

where

f_{pu} = ultimate tensile strength of the prestressing steel

γ_p = factor based on the type of prestressing steel: 0.55 when f_{py}/f_{pu} is not less than 0.80, 0.40 when f_{py}/f_{pu} is not less than 0.85, and 0.28 when f_{py}/f_{pu} is not less than 0.90, where f_{py} is the specified yield stress of the prestressing tendons (psi)

ρ_p = reinforcement ratio A_{ps}/bd_p

and β_1 is as defined in Chapter 2, Section 2.6. The expression for f_{ps} is valid when $f_{se} \geq 0.5f_{pu}$, where f_{se} is the effective stress in the prestressing steel after all losses. Here f_{ps} represents the average stress in the prestressing steel at failure. It is analogous to f_y, but as the high-strength prestressing steels do not have a well-defined yield point, it can be predicted using ACI Code Equation (20.3.2.3.1).

For members with **unbonded** prestressing tendons, and no non-prestressed tension or compression reinforcement (*i.e.* $\rho_s = \rho_s' = 0$), the value of f_{ps} can be obtained for $f_{se} \geq 0.5f_{pu}$ from ACI 318 Table 20.3.2.4.1:

TABLE 11.1 Approximate values of f_{ps} at nominal flexural strength for unbonded tendons.

$\dfrac{\ell_n}{h}$		f_{ps} (psi)
≤ 35 (i.e., beams and girders)	The least of:	$f_{se} + 10{,}000 + \dfrac{f_c'}{100\rho_p}$
		$f_{se} + 60{,}000$
		f_{py}
> 35 (i.e., slabs)	The least of:	$f_{se} + 10{,}000 + \dfrac{f_c'}{300\rho_p}$
		$f_{se} + 30{,}000$
		f_{py}

f_{se} = the effective stress in the prestressing steel after all losses
$= 0.7f_{pu} - \textit{prestress losses}$

Adapted from ACI 318-19 Table 20.3.2.4.1

Note that for prestressed one-way slabs and beams with unbonded tendons, ACI Code, Sections 7.6.2.3 require the prestressed member to have a minimum area of bonded non-prestressed longitudinal reinforcement equal to $0.004A_{ct}$, where A_{ct} is the area of the cross-section between the flexural tension face and the centroid of the gross section. Refer to ACI Code Section 8.6.2.3 for the bonded non-prestressed reinforcement requirement for two-way slabs.

EXAMPLE 11.5

Calculate the flexural strength ϕM_n of the prestressed beam of Example 11.4 and compare with M_u. Use $f_c' = 5000$ psi (normal-weight concrete) and seven-wire-strand grade 270 with f_{pu} = 270,000 psi (ordinary strand with $f_{py} = 0.85f_{pu}$). Use *bonded* prestressing tendons and neglect all prestress losses. Therefore, for the purpose of this example, $f_{se} = 0.70 f_{pu}$. Use A_{ps} = 1.224 in.^{2}. The beam has no non-prestressed tension or compression steel.

Solution:

Because prestress losses are neglected, $f_{se} = 0.70 f_{pu} > 0.5 f_{pu}$. Therefore, ACI Equation 20.3.2.3.1 is applicable. For the given conditions, $\gamma_p = 0.40$ and $\beta_1 = 0.80$. The reinforcement ratio and f_{ps} are calculated as follows:

$$\rho_p = \frac{A_{ps}}{bd_p} = \frac{1.224}{18(20)} = 0.00340$$

$$f_{ps} = f_{pu}\left(1 - \frac{\gamma_p}{\beta_1}\rho_p \frac{f_{pu}}{f_c'}\right)$$

$$= 270\left[1 - \frac{0.40}{0.80}(0.00340)\left(\frac{270}{5}\right)\right]$$

$$= 245 \text{ ksi}$$

Prestressed concrete sections are subject to the same conditions of being tension-controlled, compression-controlled, or transition sections as regular reinforced concrete sections, as discussed in Section 2.8. The appropriate values of ϕ, as previously discussed, apply. It is convenient to check the net tensile strain limit of 0.005 (for tension-controlled sections) through the use of a *reinforcement index* ω_p, where

$$\omega_p = \frac{\rho_p f_{ps}}{f_c'}$$

A reinforcement index of $0.32\beta_1$ corresponds to a net tensile strain of 0.005, the lower limit for a tension-controlled section. Therefore, checking ω_p:

$$\omega_p = \frac{\rho_p f_{ps}}{f_c'} = \frac{0.00340(245{,}000)}{5000} = 0.167$$

$$0.32\beta_1 = 0.32(0.80) = 0.256$$

Because $\omega_p < 0.256$, this is a tension-controlled section and $\phi = 0.90$.

Alternatively, as was done in Chapter 2, we can calculate the depth of the stress block,

$$a = \frac{A_{ps} f_{ps}}{0.85 f_c' b} = \frac{1.224(245)}{0.85(5)(18)} = 3.92 \text{ in}$$

With β_1 = 0.8 for f_c' = 5000 psi, the depth of the neutral axis, c = a/β_1 = 3.92 in./0.8 = 4.9 in. Therefore, c/d_p = 4.9/20 = 0.25 < 0.375, therefore the section is tension controlled.

Compute the nominal moment strength:

$$M_n = A_{ps} f_{ps}\left(d_p - \frac{a}{2}\right)$$
$$= 1.224(245)\left(20 - \frac{3.92}{2}\right)$$
$$= 5410 \text{ in.-kip}$$
$$= 451 \text{ ft-kip}$$
$$\phi M_n = 0.90(451) = 406 \text{ ft-kip}$$

Compute the factored service load moment:

$$M_u = 1.2M_{DL} + 1.6M_{LL}$$
$$= 1.2(41.0) + 1.6(156.3) = 299 \text{ ft-kip}$$

Thus $M_u < \phi M_n$ (299 ft-kip < 406 ft-kip). Therefore, O.K.

If a prestressed beam is satisfactorily designed based on service loads and then, when checked by the strength equations, is found to have insufficient strength to resist the factored loads, non-prestressed reinforcement may be added to increase the factor of safety. In addition, the ACI Code, Sections 7.6.2.1 (one-way slabs), 8.6.2.2 (two-way slabs), and 9.6.2.1 (beams), require that the total amount of reinforcement (prestressed and non-prestressed) be sufficient to develop a design moment capacity equal to at least 1.2 times the cracking moment calculated from the modulus of rupture of the concrete (i.e., $\phi M_n \geq 1.20M_{cr}$). This is necessary to ensure a ductile failure mode. It is permissible to waive this requirement when a flexural member has shear and flexural strength at least twice that required, that is, $\phi M_n \geq 2M_u$ and $\phi V_n \geq 2V_u$ (See ACI 318 Sections 7.6.2.2, 8.6.2.2.1, and 9.6.2.2).

11.8 NOTES ON PRESTRESSED CONCRETE DESIGN

Because the shape and dimensions of a prestressed concrete member may be established by a trial procedure or even assumed based on physical limitations, the design may be reduced to finding the prestress force, tendon profile, and the amount of the prestress steel area. The design problem then is further reduced to an analysis-type problem whereby service load stresses are checked at various stages and the flexural strength of the member is checked against the moment due to the factored service loads.

The intent of this chapter, however, is to present a conceptual approach to prestressed concrete members. Therefore, some significant topics normally considered in the analysis or design of prestressed concrete flexural members have been omitted. These include secondary moments and restraint reactions in continuous prestressed flexural member, shear design, deflection checks, prestress losses, and anchorage stresses. These topics are beyond the scope of our text. References [3] through [6] are among many other texts and publications available for further reading.

REFERENCES

[1] "ASTM Standards." American Society for Testing and Materials, 100 Barr Harbor Drive, West Conshohocken, PA 19428.

[2] George F. Limbrunner and Leonard Spiegel. *Applied Statics and Strength of Materials*, 5th ed. Upper Saddle River, NJ: Prentice Hall, 2009, pp. 470–473.

[3] T. Y. Lin and Ned H. Burns. *Design of Prestressed Concrete Structures*, 3rd ed. New York: John Wiley & Sons, Inc., 1981.

[4] Post-Tensioning Institute. *Post-Tensioning Manual*, 6th ed. Phoenix, AZ: Post-Tensioning Institute, 2006.

[5] Edward G. Nawy. *Prestressed Concrete: A Fundamental Approach*. Upper Saddle River, NJ: Prentice Hall, 2009.

[6] Arthur H. Nilson. *Design of Prestressed Concrete*, 2nd ed. New York: John Wiley & Sons, Inc., 1987.

EXERCISES

11.1 The *plain concrete* beam shown having a rectangular cross-section 10-in. wide and 18-in. deep is simply supported on a single span of 20 ft. Assuming no loads other than the dead load of the beam itself, find the bending stresses that exist at midspan.

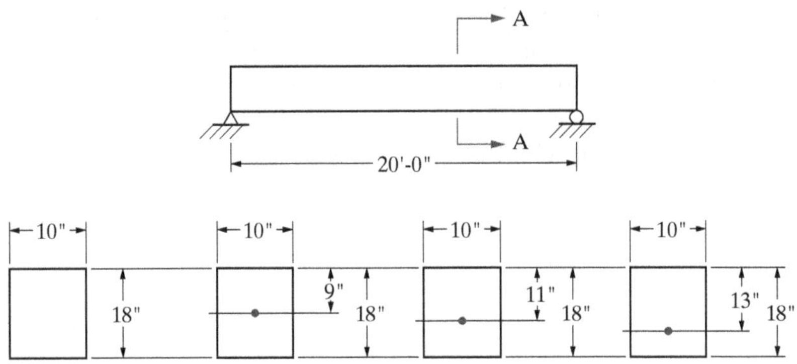

EXERCISES 11.1 TO 11.4

11.2 A beam identical to that of Exercise 11.1 is prestressed with a force P_s of 185 kip. The tendons are located at the center of the beam. Use $A_{ps} = 1.23\text{ in.}^2$. Find the stresses in the beam at midspan at transfer. There is no load other than the weight of the beam. Draw complete stress diagrams.

11.3 The tendons in the beam of Exercise 11.2 are placed 11 in. below the top of the beam at midspan. Assuming that no tension is allowed, determine the maximum service uniform load that the beam can carry in addition to its own weight as governed by the stresses at midspan. Assume that $f'_{ci} = 3500$ psi and $f'_c = 5000$ psi.

11.4 Same as Exercise 11.3, except that the tendons are 13 in. below the top of the beam at midspan and permissible stresses are as defined in the ACI Code, Section 24.5.

11.5 For the beam shown, calculate the outer fiber stresses at midspan and at the end supports at the following stages: (a) transfer (prestress + beam weight) and (b) under full-service load. The beam is simply supported with a span length of 20 ft-0 in. and carries a superimposed live load of 2.7 kip/ft. The tendon is straight and located 9 in. below the neutral axis of the beam. The prestress force is 250 kip. Neglect prestress losses and use the gross concrete section properties. Use the method of superposition for the solution.

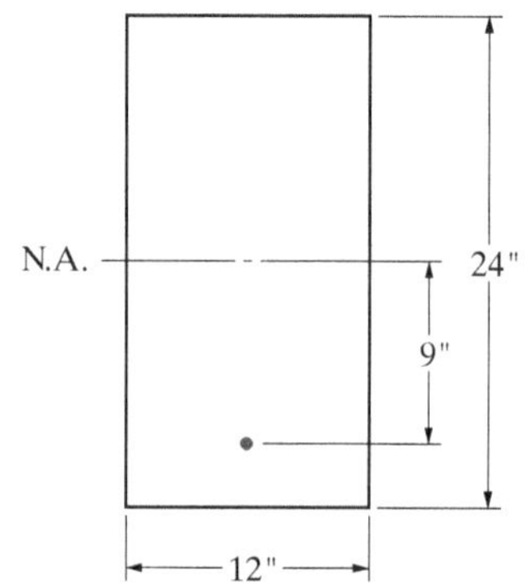

EXERCISE 11.5

11.6 The simply supported beam having a midspan cross-section as shown is prestressed by a parabolic tendon with an effective prestress force of 233 kip. The beam has a span length of 30 ft and carries a total service load of 1.20 kip/ft, which includes the weight of the beam of 0.21 kip/ft. Calculate the outer fiber stresses at midspan under full-service load. Use the gross properties of the cross-section. Use the load balancing method.

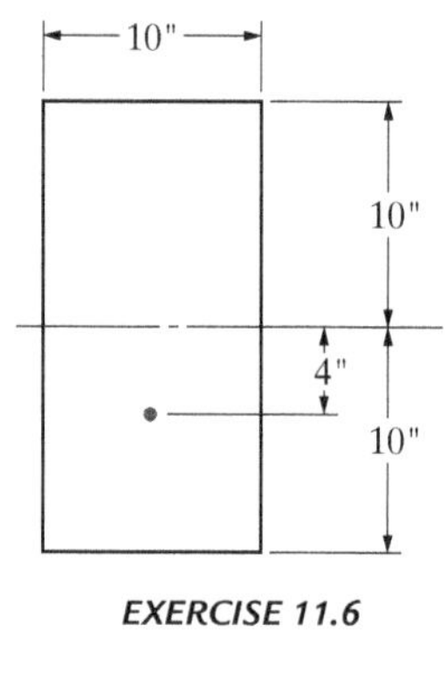

EXERCISE 11.6

11.7 For the beam of Exercise 11.5, calculate the outer fiber stresses at midspan for the following stages: (a) transfer and (b) under full service load. Assume the prestress tendons are parabolically draped with a midspan eccentricity of 9 in. as shown, and zero eccentricity at the end supports. Neglect prestress losses and use the gross concrete section properties. The prestress force is still 250 kip. Use the load balancing method.

11.8 Calculate the flexural strength ϕM_n of the prestressed beam shown and compare it with the factored design moment M_u. The service load moments are 125 ft-kip for dead load and 383 ft-kip for live load. Use $f'_c = 5000$ psi (normal-weight concrete) and grade 270 seven-wire strand with f_{pu} = 270,000 psi ($f_{py} = 0.85f_{pu}$). The tendons are bonded. Neglect all prestress losses.

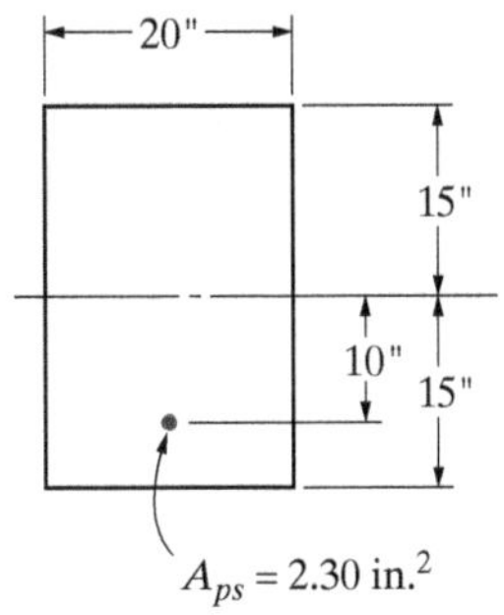

EXERCISE 11.8

CHAPTER 12

CONCRETE FORMWORK

12.1 INTRODUCTION

Forms are temporary engineered structures whose purpose is to provide containment for the freshly placed concrete and to support it until it can support itself. Forms usually must be *engineered* structures as they are required to support loads composed of the fresh concrete, construction materials, equipment, workers, impact of various kinds, and sometimes wind. The forms—which include the horizontal members and the vertical shoring and lateral bracings—must support all the applied loads without collapse or excessive deflection. In addition, the forms must provide for the molding of the concrete to the desired dimensions, shape, and surface finish.

Formwork failures are a major cause of collapses in concrete structures during construction, where instability failures of the formwork and its supporting elements are the most common culprits. The formwork must be designed to support its own self-weight, the weight of the wet concrete, the construction live loads (i.e., weight of construction personnel and equipment) as well as lateral loads and any loads induced by the post-tensioning for post-tensioned concrete structures.

The cost of formwork is significant, generally amounting to anywhere from 40% to 60% of the cost of a concrete structure, and this underscores why the contractor needs to pay proper attention to formwork design and construction. Economy in formwork design depends partly on the ingenuity and experience of the form designer, whether a contractor or an engineer. Judgment with respect to the development of a forming system could both expedite a project and reduce costs. Although forms may be job built (design examples are presented later in this chapter), many proprietary forming systems are available. Specialty companies will design and construct forms using patented systems. In multiple-reuse situations, steel formwork has replaced wood formwork to some degree, although the use of wood is still substantial because of its availability and ease of fabrication. Per ACI 301 Section 2.2.2.1 and ACI 347R-14(21) Section 4.1.3, the contractor and their formwork design engineer, not the structural design engineer of record, are responsible for the design and construction of the formwork. Since formwork are not part of the final structure and the contractor is responsible for the means and methods of construction, project specifications do not usually require approval of the formwork design by the structural engineer of record for the project. The formwork shop drawing submittals to the structural engineer of record are typically for information purposes only.

Whatever the eventual type of forming may be, economy in formwork begins when the architect/engineer, during the design of the structure, examines each member and considers ease of forming and reuse of forms. The American Concrete Institute (ACI) publishes the following standards that are necessary for the proper design of formwork for concrete structures:

1. ACI-PRC-347R-14(21), "Guide to Formwork for Concrete."
2. ACI 347.2R-17, "Guide for Shoring/Reshoring of Concrete Multi-story Buildings."
3. SP-4(14), Formwork for Concrete, 8th Edition

The goals of these documents are to achieve safety, quality control, and economy in concrete construction.

12.2 FORMWORK REQUIREMENTS

The essential requirements of good formwork may be categorized as quality, safety, and economy.

Quality in formwork requires that the forms be:

1. *Accurate:* The size, shape, position, and alignment of structural elements will depend on the accuracy with which the forms are built. The designer (engineer or architect) should specify the allowable tolerances in form dimensions for form design and construction but not make these tolerances finer than necessary, as this could add unnecessarily to formwork cost as well as delay the project.
2. *Rigid:* Forms must be sufficiently rigid to prevent movement, bulging, or sagging during the placing of concrete. Formwork must therefore be adequately shored, braced, and tied. Special consideration may have to be given to such items as corner details and the effect of any uplift pressures.
3. *Tight-jointed:* Joints that are insufficiently tight will leak cement paste. The surface of the concrete will thus be disfigured by fins of the cement paste, and honeycombing may result adjacent to the leaking joint.
4. *Properly finished:* The formwork in contact with the concrete should be so arranged and jointed as to produce a concrete surface of good appearance. Wires, nails, screws, and form surface flaws must not be allowed to disfigure the concrete surface. In some cases a provision of special form lining may be necessary to achieve the desired surface finish.

Safety in formwork requires that the forms be:

1. *Strong:* To ensure the safety of the structure and the protection of the workers, it is essential that formwork be designed to support the full load and lateral pressures from freshly placed concrete, together with the load from construction traffic and any necessary equipment. On large jobs, the design of formwork is usually left to a specialist, but

on smaller or routine jobs, it may be left to the carpenter supervisor using rule-of-thumb methods.

2. *Sound:* The materials used to construct the forms must be of the correct size and quantity, of good quality, and sufficiently durable for the job.

Economy in formwork requires that the forms be:

1. *Simple:* For formwork construction to be economical, it must be designed to be simple to erect and dismantle. Modular dimensions should be used.
2. *Easily handled:* The sizes of form panels or units should be such that they are not too heavy to handle.
3. *Standardized:* Comparative ease of assembly and the possibility of reuse will lower formwork costs when sizes are standardized.
4. *Reusable:* Formwork intended for reuse should be designed for easy removal. If formwork panels have to be ripped down, the concrete may be damaged, materials will be wasted, time will be lost, and expense will be incurred in repairing and replacing damaged panels.

12.3 FORMWORK MATERIALS AND ACCESSORIES

Materials used for forms for concrete structures include lumber, plywood, hardboard, fiberglass, plastic and rubber liners, steel, paper and cardboard, aluminum, fiber forms, and plaster of Paris. Additional materials include nails, bolts, screws, form ties, form clamps, anchors, various types of inserts, and various types of form oils and compounds as well as their accessories. Forms frequently involve the use of two or more materials, such as plywood facing attached to steel frames for wall or slab panels (see Figure 12.1a).

Form *lumber* generally consists of the softwoods with the species used being the type available in the local area. Various species are usually grouped together for grading and marketing purposes. Some of these groups and their included species commonly used for formwork are:

1. Douglas fir–larch (Douglas fir, western larch)
2. Douglas fir–south (only Douglas fir from southern growth areas)
3. Hem–fir (Western hemlock, California red fir, grand fir, noble fir, white fir, Pacific silver fir)
4. Spruce–pine–fir (alpine fir, balsam fir, black spruce, Englemann spruce, jack pine, lodgepole pine, red spruce, white spruce)
5. Southern pine (loblolly pine, longleaf pine, shortleaf pine, slash pine)

Form lumber is made up of standard sizes, either rough or dressed, with the grading quality of *construction grade* or *No. 2 grade* usually specified. Shoring or falsework requiring greater capacity should be of *select structural grade* lumber. Partially seasoned wood should be used because it has been found to be the most stable. Green lumber will warp and crack, whereas kiln-dried lumber will swell excessively when it becomes wet. Boards of various thicknesses may be used for sheathing when the imprint of the boards on the concrete surface is desired for architectural reasons.

Plywood is available as flat panels (4 ft × 8 ft sheets) made of thin sheets of wood (called plies) with a total thickness ranging from $\frac{1}{4}$ in. to $1\frac{1}{8}$-in. It is commonly used for sheathing or lining forms because it gives smooth concrete surfaces that require a minimum of hand finishing and because the relatively large sheets are economical and easy to use. U.S. plywood is built up of an odd number of plies with the grain of adjacent layers perpendicular. The plies are dried and joined under pressure with glues that make the joints as strong or stronger than the wood itself. The alternating direction of the grains of adjoining layers equalizes strains and thus minimizes shrinkage and warping of the plywood panels. Generally, the grain direction of the outer layers is parallel to the long dimension of the panel. For maximum strength, the direction of the grain in the face layers should be placed parallel to the span.

The plywood should be the exterior grade because of its waterproof glue. It should also be factory-treated with a form oil or parting compound. With proper care, plywood can be reused many times.

Coated plywood is occasionally used. Called *overlaid* or *plastic coated*, it is ordinary exterior plywood with a resin-impregnated fiber-facing material fused to one or both sides of the plywood sheet. The overlay covers the grain of the wood, resulting in a smoother and more durable forming surface. The resins used in overlay production are hard and resist water, chemicals, and abrasion. Two grades, high density and medium density, are available, and their difference is in the density of the surfacing material. Coated plywood is generally reused considerably more than the plain plywood form.

Nearly any exterior type of plywood can be used for concrete forming. A special panel called *plyform* is manufactured specifically for that purpose, however. Plyform is exterior-type plywood limited to certain wood species and veneer grades to ensure high performance and is manufactured in two classes, class I and class II. Class I is the strongest, stiffest, and most widely available type. Both classes have smooth, solid surfaces and can have many reuses. They are mill-oiled unless otherwise specified. Plyform is manufactured in thicknesses ranging from $\frac{15}{32}$ in. to $1\frac{1}{8}$ in., and the most commonly used thicknesses are $\frac{5}{8}$ in. and $\frac{3}{4}$ in.

Plyform can also be manufactured with a *high-density overlaid (HDO) surface* (on one side or both sides). HDO plyform has an exceptionally smooth, hard surface for the smoothest possible concrete finishes and the maximum number of reuses. The HDO surface is a hard, semiopaque resin-fiber overlay heat-fused to panel faces. It can have as many as 200 reuses, and light oiling between pours is recommended.

Structural I plyform is stronger and stiffer than plyform class I or II and is sometimes used when the face grain must be parallel to supports and there is a heavy loading against the forms. It is also available with HDO faces.

Hardboard (fiberboard) is made up of wood particles that have been impregnated, pressed, and baked. When used as a form liner, which is generally the case, hardboard must be backed with sheathing of some kind because it does not have the strength that plywood has. Tempered hardboard, which is preferred for formwork, has been impregnated with drying oil or other material that makes the board less absorbent and improves its strength. Hardboard has limited reuse capability.

Fiberglass is a glass-fiber-reinforced plastic. It is especially suitable for the repetitive production of complicated shapes, particularly precast elements. The initial cost is high, but the extensive reuse of one mold without evidence of wear contributes to economy. Suitable reinforcement and backing must be provided so that the mold will not be deformed when it is handled. Molds can be used with or without form oil. Wear of the mold surface will be slightly less if an oil is used.

Plastic and rubber form liners generally come in sheets, either flexible or rigid, and are attached to the solid form sheathing. The term *form liners* includes any sheet, plate, or layer of material attached directly to the inside face of forms to improve or alter the surface texture and quality of the finished concrete. Substances that are applied by brushing, dipping, mopping, or spraying to preserve the form material and to make form stripping easier are referred to as *form coatings*. Some of these coatings are so effective as to approximate the form liner in function. Repetitive use, a great variety of textures and patterns, and ease of stripping are among the advantages of these form liners.

Various types of rigid insulation boards are used as form liners and attached to the form sheathing. The boards are generally left in place, either bonded to the concrete or held in place by form plank clips.

Steel forms have been widely used as special-purpose forms. Steel panel systems have been fabricated and used, and steel framing and bracing are important in the construction of many wood and plywood panel systems. Patented steel pans and dome components form the underside of waffle slabs and pan joists. Corrugated steel sheets serve as permanent bottom forms for decks—that is, they remain in place after the concrete has been placed. Fabricated steel form systems are extensively used in the precast and prestressed concrete industries due to their suitability for being reused extensively.

Earth as a form is used in subsurface construction where the soil is stable enough to retain the desired shape of the concrete structure. With the resulting rough surface finish, the use of earth as a form is generally limited to footings and foundations.

Fiber forms are composed of multiple layers of heavy paper bonded together and impregnated with waxes and resins to create a water-repellent cardboard. These are generally single-use, cylindrical-shape molds for columns or other applications where preformed shapes are desirable. "Sonotube" fiber forms are patented lightweight units that can be adapted to various applications such as foundation forms.

Form accessories consist of a multitude of items necessary for form planning and construction. Types, availability, and detailed information can usually be obtained from manufacturers' catalogs as well as local concrete construction specialties distributors.

Some of the more common types of accessories are fasteners, spreaders, ties, anchors, hangers, clamps, and inserts. The most common *fastener* used is the double-headed nail, which may be withdrawn with a minimum of effort and little damage to the forms. *Spreaders* are braces that are inserted in forms to keep the faces a proper distance apart until the concrete is placed.

Preferably, they are removed during or after the placement of the concrete so that they are not cast in the concrete. They may be wood blocks or proprietary metal pieces. *Ties* are tensile units adapted to holding concrete forms secure against the lateral pressure of the fresh concrete, with or without provision for spacing the forms a definite distance apart. They pass through the concrete and are fastened on each side. Various types of ties are available, many of which are patented items. *Anchors* are devices used to secure formwork to previously placed concrete of adequate strength and are normally embedded in concrete during placement. There are two basic parts, the embedded anchoring device and the external fastener, which is removed after use. *Hangers* are devices used for suspending one object from another, such as the hardware attached to a building frame to support forms. *Inserts* are of many designs and are attached to the forms in such a way that they remain in the concrete when the forms are stripped. They provide for anchorage of brick or stone veneers, pipe hangers, suspended ceilings, duct work, and any building hardware or components that must be firmly attached to the concrete. *Clamps* consist of many varied devices but serve a similar purpose to that of the tie. Beam form clamps firmly hold the beam sides and bottom together with a minimum of nailing required. Column clamps or yokes encircle column forms and hold them together securely, withstanding the lateral pressure of the freshly placed concrete. Accessories used in formwork are required to have a minimum factor of safety of 2.0 to 3.0 based on the component (see Chapter 4 of reference [5]).

12.4 LOADS AND PRESSURES ON FORMS

The concrete form is a structure. Like all structures, it must be designed and constructed with due regard to the effects of the imposed loads. A well-designed form must have adequate strength to resist failure; in addition, it must have sufficient stiffness so that deflection will not be excessive. Efficient and economical use of the material of which the forms are built is also an integral part of the design process.

The first consideration in the design of a form for concrete is the load to be supported. The loads may be considered in two categories: vertical loads and lateral loads. The lateral loads may be further categorized as externally applied load (such as wind) and internally applied pressure due to the contained fresh concrete. These loads must be carried to the ground by the formwork system or resisted by in-place construction that has adequate strength for the purpose.

Vertical loads consist of dead load and live load. The dead load consists of the weight of the formwork and the freshly placed concrete. Live load consists of the weight of workers and equipment, stored materials, and impact due to moving loads. ACI-PRC 347R-14(21). *Guide to Formwork for Concrete* [1] recommends that the formwork be designed for the following:

1. Minimum live load of 50 psf of horizontal projection (75 psf if motorized carts are used for delivery of the concrete).
2. Minimum combined dead and live load of 100 psf (125 psf if motorized carts are used for delivery of the concrete).

Note that these minimum design loads would be applicable to the design of structural floor and roof slabs (not to the design of slabs on ground).

Lateral pressure is exerted on all nonhorizontal surfaces that contain the fresh concrete. The pressures are imposed on wall and column forms, and in essence are hydraulic loadings. The amount of lateral pressure is influenced by the weight, chemistry, and temperature of the concrete, the vertical rate of placement, size and shape of the form, height of the form, and method of consolidation (hand-spaded or mechanically vibrated). The amount and location of reinforcing also affect lateral pressure, although their effect is small and is usually neglected. ACI 347R-14 provides formulas that can be used to reasonably predict concrete pressure on wall forms and column forms. In the formulas, the following notation is used:

p = *concrete* lateral pressure (psf)
R = rate of placement (ft/h)
T = temperature of concrete in the forms (°F)
h = height of fluid or plastic concrete above point considered (ft)
w = unit weight of fresh concrete (pcf)
C_C = Chemistry coefficient
C_W = unit weight coefficient

A liquid head formula is the basis for the concrete lateral pressure calculations:

$$p = wh \qquad \textbf{[ACI 347R-14 Eq. (4.2.2.1a(a))]}$$

This formula will apply where a form is filled rapidly (i.e., when the slump is *greater* than 7 inches) before any hardening of the concrete occurs, and in situations where the conditions of Equation (4.2.2.1a(b)) and Equation (4.2.2.1a(c)) are not met. The pressure distribution in Equation (4.2.2.1a(a)) is triangular in shape, varying linearly from zero at the top of the form to wh at the base of the form. The limits stated for Equation (4.2.2.1a(b)) and Equation (4.2.2.1a(c)) do not apply to Equation (4.2.2.1a(a)).

For the special conditions of

a. concrete having a slump of 7 in. or less and

b. placed with normal internal vibration to a depth of 4 ft or less, formwork can be designed for a lateral pressure as follows:

1. For *columns* (note that a column, for purposes of applying the pressure formulas, is defined as a vertical element with no plan dimension exceeding 6.5 ft),

$$p_{max} = C_W C_C \left[150 + 9000\frac{R}{T}\right] \qquad \textbf{[ACI 347R-14 Eq. (4.2.2.1a(b))]}$$

   with a minimum of 600 C_W lb/ft^2, but in no case greater than wh.

2. For *walls*, with a rate of placement of less than 7 ft/hr and a placement height not exceeding 14 ft,

$$p_{max} = C_W C_C \left[150 + \frac{9000R}{T}\right] \qquad \textbf{[ACI 347R-14 Eq. (4.2.2.1a(b))]}$$

   with a minimum of 600 C_W lb/ft^2, but in no case greater than wh.

3. For walls with a placement rate of **less than 7 ft/h**, where placement **height exceeds 14 ft**, and for all walls with a **placement rate of 7 to 15 ft/h**,

$$p_{\max} = C_W C_C \left[150 + \frac{43,400}{T} + \frac{2800R}{T} \right] \qquad \textbf{[ACI 347R-14 Eq. (4.2.2.1a(c))]}$$

with a minimum of 600 C_W lb/ft ^{2}, but in no case greater than wh.

4. For all walls with a **placement rate greater than 15 ft/h (or with a slump greater than 7 inches)**,

$$p = wh \qquad \textbf{[ACI 347R-14 Eq. (4.2.2.1a(a))]}$$

For Cases 1 through 3 above (i.e., Equations 4.2.2.1a(b) and 4.2.2.1a(c)), the lateral pressure distribution from the wet concrete is trapezoidal in shape; it varies from zero at the top of the form to a maximum value of $p_{\max}$ at a height of $p_{\max}/w$ from the top of the form, and then remains constant at $p_{\max}$ to the base of the form, where w is the unit weight of fresh concrete [5].

The unit weight coefficient C_W has the value of 1.0 for concretes having a unit weight in the range of 140 to 150 lb/ft^3. For concrete unit weights outside this range, see Table 4.2.2.1a(c) in ACI 347R-14. The chemistry coefficient C_C has the value of 1.0 for Types I, II, and III cements without retarders. For concrete that uses flyash combined with slag cement as cement replacement, C_C is 1.5, and C_C is 1.4 for concrete with either slag cement or flyash [5]. For other types of cements or blends containing slag or fly ash, see Table 4.2.2.1a(b) in ACI 347R-14. For the examples in this text, we will assume both C_W and C_C equal to 1.0. Formwork should be designed for these lateral pressures to prevent blowout or collapse of the formwork.

The reader should note that there is discontinuity in the maximum lateral pressures for walls that are less than or equal to 14 ft. in height calculated using ACI 347R-14 Eq. 4.2.2.1a(b) versus the maximum lateral pressures in walls greater than 14 ft. in height calculated using ACI 347R-14 Eq. 4.2.2.1a(c). Consider two walls—one 14 ft. high and the other 14.5 ft. high, and assume a placement rate, R = 4 ft/h and the temperature of concrete in the forms, T = 90°F; assuming the unit weight of concrete is 150 pcf, and the chemistry coefficients are both 1.0, the maximum lateral pressures on the form for the 14 ft. tall wall using ACI 347R-14 Eq. 4.2.2.1a(b) will be 600 psf while the maximum lateral pressure on the form for the 14.5 ft. tall wall using ACI 347R-14 Eq. 4.2.2.1a(c) will be 757 psf, an almost 30% increase in lateral pressure for a very small increase in wall height beyond 14 ft.

Formwork should also be designed to resist all foreseeable *lateral loads*, such as seismic forces, wind, cable tensions, inclined supports, dumping of concrete, and impact due to equipment. ACI 347R-14 recommends for slabs a minimum horizontal design load of 100 lb per lineal foot of floor edge or 2% of the total dead load of the floor, whichever is greater. Wall forms exposed to the elements should be designed for a minimum wind load of 15 psf, and bracing for wall forms should be designed for a lateral load of at least 100 lb per lineal foot of wall applied at the top.

In addition, formwork should be designed for any special loads likely to occur, such as unsymmetrical placement of concrete, and uplift. Form designers must be alert to provide for any and all special loading conditions.

12.5 THE DESIGN APPROACH

The design of job-built forms may be considered largely as beam and post design. The bending members usually span several supports and are therefore indeterminate. Assumptions and approximations are made that simplify the calculations and facilitate the design process.

After establishing the appropriate design loads, the sheathing and supporting members are analyzed or designed in sequence. Bending members (sheathing, joists, studs, stringers, or wales) are considered uniformly loaded and supported on (1) a single span, (2) two spans, or (3) three or more spans. The uniform load assumption is common practice unless the spacing of point loads exceeds one-third to one-half of the span between supports, in which case the worst loading condition is investigated.

Each bending member should be analyzed or designed for bending moment, shear, and deflection. In addition, vertical supports (shoring) and lateral bracing (if applicable) must be analyzed or designed for either compressive or tensile loads. Bearing stresses at supports must be investigated (except for the sheathing).

In the analysis or design of the component parts, the traditional stress equations are used (see any strength of materials text). The expressions for the applied stresses are as follows:

For bending stress:

$$f_b = \frac{M}{S}$$

For shear stress:

$$f_v = \frac{1.5V}{A} \quad \text{(for rectangular wood members)}$$

$$f_v = \frac{VQ}{Ib} \quad \text{or} \quad \frac{V}{Ib/Q} \quad \text{(for plywood)}$$

For compression stress (both parallel and perpendicular to the grain):

$$f_{\text{comp}} = \frac{P}{A}$$

For tension stress:

$$f_{\text{tens}} = \frac{P}{A}$$

where

f_b = calculated unit stress in bending (psi)

f_v = calculated unit stress in shear (psi)

f_{comp} = calculated unit stress in compression (either parallel or perpendicular to the grain) (psi)

f_{tens} = calculated unit stress in tension (psi)

M = maximum bending moment (in.-lb)

S = section modulus (in.^{3})

P = concentrated load (lb)

V = maximum shear (lb)
A = cross-sectional area (in.^{2})
Ib/Q = rolling shear constant (in.^{2}/ft)

If we equate allowable unit stresses to the maximum unit stresses developed in a beam subjected to a uniformly distributed load w (lb/ft), expressions can be derived for the maximum allowable span length. Therefore, knowing the design loads and member section properties, a maximum allowable span length can be computed.

Table 12.1 contains expressions for maximum allowable span length as governed by moment, shear, and deflection. The reader may wish to derive these expressions from basic principles. The moment expressions are based on the maximum positive or negative moment. The maximum shear considered is the shear that exists at d distance from the support, where d is the depth of the member. Shear considerations for plywood vary from those accorded sawn lumber because of the cross-directional way in which the plies of the plywood are assembled. When plywood is loaded in flexure, the plies with the grain perpendicular to the direction of the span are the weakest aspect of the plywood with respect to shear. The wood fibers in these plies roll at stresses below the fiber shear strength parallel to the grain; hence the name *rolling shear*. For a discussion of the properties of plywood with respect to shear, see reference [2]. Deflection of forms must be limited to minimize unsightly bulges in the resulting concrete surface. The deflection limit may be specified as a fraction of span (i.e., $\ell/240$), as a limit ($\frac{1}{8}$ in.), or as the smaller of the two. The architect/engineer must decide on the deflection limits for the formwork based on bulging and sagging that can be tolerated in the surfaces of the finished structure. The casting of test panels may be warranted in some cases. Limiting deflection to $\frac{1}{360}$ of the span is acceptable in many cases where surfaces are coarse-textured and there is little reflection of light. Three

TABLE 12.1 Concrete Form Design Equations (to Determine Allowable Span Length).

	One span	**Two spans**	**Three or more spans**
Bending moment	$\ell = 9.8\sqrt{\frac{F_b' S}{w}}$	$\ell = 9.8\sqrt{\frac{F_b' S}{w}}$	$\ell = 10.95\sqrt{\frac{F_b' S}{w}}$
Shear	$\ell = \frac{16F_v' A}{w} + 2d$	$\ell = \frac{12.8F_v' A}{w} + 2d$	$\ell = \frac{13.3F_v' A}{w} + 2d$
	$\ell^a = \frac{24F_v'(Ib/Q)}{w} + 2d$	$\ell^a = \frac{19.2F_v'(Ib/Q)}{w} + 2d$	$\ell^a = \frac{20F_v'(Ib/Q)}{w} + 2d$
Deflection $\Delta_{all.} = \frac{\ell}{240}$	$\ell = 1.57\sqrt[3]{\frac{E'I}{w}}$	$\ell = 2.10\sqrt[3]{\frac{E'I}{w}}$	$\ell = 1.94\sqrt[3]{\frac{E'I}{w}}$
Deflection $\Delta_{all.} = \frac{\ell}{360}$	$\ell = 1.37\sqrt[3]{\frac{E'I}{w}}$	$\ell = 1.83\sqrt[3]{\frac{E'I}{w}}$	$\ell = 1.69\sqrt[3]{\frac{E'I}{w}}$
When deflection Δ (in.) is specified	$\ell = 5.51\sqrt[4]{\frac{\Delta E'I}{w}}$	$\ell = 6.86\sqrt[4]{\frac{\Delta E'I}{w}}$	$\ell = 6.46\sqrt[4]{\frac{\Delta E'I}{w}}$

Notes: ℓ, span length (center to center of supports) (in.); F_b', allowable bending stress (psi); S, section modulus (in.^{3}); w, uniform load (lb/ft); F_v', allowable shear stress (psi); A, cross-sectional area (in.^{2}); d, depth of member (in.); E', adjusted modulus of elasticity (psi); I, moment of inertia (in.^{4}); Ib/Q, rolling shear constant for plywood (in.^{2}/ft); Δ, deflection (in.).

[a]For plywood only.

design equations for deflection are given in Table 12.1. Alternatively, the designer may wish to compute the required size of the members when the design loads and span lengths are known. The same basic principles apply.

Section properties for selected thicknesses of plyform are given in Table 12.2. These section properties reflect that various species of wood used in manufacturing plywood have different stiffness and strength properties. Those species with similar properties are assigned to a species group. To simplify plywood design, the effects of using different species groups in a given panel as well as the effects of the cross-banded construction of the panel have been taken into consideration in establishing the section properties.

TABLE 12.2 Section Properties and Design Values for Plyform.

		Properties for stress applied parallel with face grain			Properties for stress applied perpendicular to face grain		
Thickness *t* (in.)	Approx. weight (psf)	Moment of inertia *I* ($in.^4/ft$)	Effective section modulus *KS* ($in.^3/ft$)	Rolling shear const. *Ib/Q* ($in.^2/ft$)	Moment of inertia *I* ($in.^4/ft$)	Effective section modulus *KS* ($in.^3/ft$)	Rolling shear const. *Ib/Q* ($in.^2/ft$)
Class I							
1/2	1.5	0.077	0.268	5.153	0.024	0.130	2.739
5/8	1.8	0.130	0.358	5.717	0.038	0.175	3.094
3/4	2.2	0.199	0.455	7.187	0.092	0.306	4.063
7/8	2.6	0.296	0.584	8.555	0.151	0.422	6.028
1	3.0	0.427	0.737	9.374	0.270	0.634	7.014
Class II							
1/2	1.5	0.075	0.267	4.891	0.020	0.167	2.727
5/8	1.8	0.130	0.357	5.593	0.032	0.225	3.074
3/4	2.2	0.198	0.454	6.631	0.075	0.392	4.049
7/8	2.6	0.300	0.591	7.990	0.123	0.542	5.997
1	3.0	0.421	0.754	8.614	0.220	0.812	6.987
Structural I							
1/2	1.5	0.078	0.271	4.908	0.029	0.178	2.725
5/8	1.8	0.131	0.361	5.258	0.045	0.238	3.073
3/4	2.2	0.202	0.464	6.189	0.108	0.418	4.047
7/8	2.6	0.317	0.626	7.539	0.179	0.579	5.991
1	3.0	0.479	0.827	7.978	0.321	0.870	6.981

	Design values		
	Plyform class I	Plyform class II	Structural I plyform
Modulus of elasticity (psi)	1,650,000	1,430,000	1,650,000
Bending stress (psi)	1930	1330	1930
Rolling shear stress (psi)	72	72	102

Source: APA—The Engineered Wood Association [3].

Note: All properties adjusted to account for reduced effectiveness of plies with grain perpendicular to applied stress.

In calculating these section properties, all plies were transformed to properties of the face ply. As a result, the designer need not be concerned with the actual panel layup but only with the allowable stresses for the face ply and the given section properties of Table 12.2. The section properties of Table 12.2 are generally the minimums that can be expected. Hence the actual panel obtained in the marketplace will usually have a section property greater than that represented in the table.

The plyform design values presented in Table 12.2 are based on wet strength and 7-day load duration, so no further adjustment in these values is required except for the modulus of elasticity. The modulus shown is an adjusted value based on the assumption that shear deflection is computed separately from bending deflection. These values should be used for bending deflection calculations (which is the usual case). To calculate shear deflection, the modulus should be reduced to 1,500,000 psi for class I and structural plyform and to 1,300,000 for class II plyform.

Section properties for selected members of standard dressed (S4S) sawn lumber are given in Table 12.3. Typical base design values for visually graded dimension lumber are furnished in Table 12.4. This table is simplified and brief and is primarily intended as a resource to accompany the examples and end-of-chapter exercises in this text. Those who require more detailed information with respect to section properties and design values should obtain reference [4].

TABLE 12.3 Properties of Structural Lumber.

Nominal size (in.)	Standard dressed size (S4S) (in.)	Area of section, A (in.^{2})	Moment of inertia,[a] I (in.^{4})	Section Modulus,[a] S (in.^{3})	Weight[b] (lb/ft)
2 × 4	$1\frac{1}{2}\times3\frac{1}{2}$	5.25	5.36	3.06	1.28
2 × 6	$1\frac{1}{2}\times5\frac{1}{2}$	8.25	20.80	7.56	2.01
2 × 8	$1\frac{1}{2}\times7\frac{1}{4}$	10.88	47.63	13.14	2.64
2 × 10	$1\frac{1}{2}\times9\frac{1}{4}$	13.88	98.93	21.39	3.37
2 × 12	$1\frac{1}{2}\times11\frac{1}{4}$	16.88	178.0	31.64	4.10
3 × 6	$2\frac{1}{2}\times5\frac{1}{2}$	13.75	34.66	12.60	3.34
3 × 8	$2\frac{1}{2}\times7\frac{1}{4}$	18.13	79.39	21.90	4.41
3 × 10	$2\frac{1}{2}\times9\frac{1}{4}$	23.13	164.9	35.65	5.62
3 × 12	$2\frac{1}{2}\times11\frac{1}{4}$	28.13	296.6	52.73	6.84
4 × 4	$3\frac{1}{2}\times3\frac{1}{2}$	12.25	12.51	7.15	2.98
4 × 6	$3\frac{1}{2}\times5\frac{1}{2}$	19.25	48.53	17.65	4.68
4 × 8	$3\frac{1}{2}\times7\frac{1}{4}$	25.38	111.1	30.66	6.17
4 × 10	$3\frac{1}{2}\times9\frac{1}{4}$	32.38	230.8	49.91	7.87
4 × 12	$3\frac{1}{2}\times11\frac{1}{4}$	39.38	415.3	73.83	9.57
6 × 6	$5\frac{1}{2}\times5\frac{1}{2}$	30.25	76.26	27.73	7.35
6 × 8	$5\frac{1}{2}\times7\frac{1}{2}$	41.25	193.4	51.56	10.03
6 × 10	$5\frac{1}{2}\times9\frac{1}{2}$	52.25	393.0	82.73	12.70
6 × 12	$5\frac{1}{2}\times11\frac{1}{2}$	63.25	697.1	121.2	15.37

[a] I and S are about the strong axis.

[b] Weight in lb/ft when the unit weight of wood is 35 lb/ft^3.

The base design values in Table 12.4 must be modified by applicable adjustment factors that are appropriate for the conditions under which the wood is used. Among the adjustment factors itemized in [4] are:

C_D = load duration factor
C_M = wet service factor
C_t = temperature factor
CF = size factor (not applicable to southern pine)
C_{fu} = flat use factor
C_r = repetitive member factor
C_L = beam stability factor
C_P = column stability factor
C_i = incising factor
C_b = bearing area factor

The product of the base design value (F_b for moment or F_v for shear) and any applicable adjustment factors provide an allowable unit stress for use in the design and analysis of wood members. The allowable unit stress is denoted as a primed value (e.g., F_b'), which indicates that it is an adjusted design value calculated for specific conditions. For instance, for allowable shear stress,

$$F_v' = F_v C_D C_M C_t C_i$$

If no adjustment factors apply, then the allowable unit stress will be equal to the tabulated base design value (e.g., $F_v' = F_v$). One required value that is not an allowable stress is modulus of elasticity E. It will be determined in a similar fashion as the product of a base design value and applicable adjustment factors.

Determination of Allowable Bending Stress F_b'

Applicable adjustment factors for allowable bending stress are load duration, wet service, temperature, size, flat use, repetitive member, and beam stability.

The load duration factor, C_D, for 7 days or less duration of load (common for form design) is given in Table 12.4.

The wet service factor, C_M, is applicable when the moisture content of the wood is more than 19% (which would occur in a wet service situation such as when the wood is in contact with fresh concrete) and the wood is at ordinary temperature. Table 12.4 shows the C_M factors for the different stresses, and for the modulus of elasticity.

The temperature factor, C_t, is generally not applicable in form design and may be disregarded.

The size factor, C_F, is based on member size and is shown in Table 12.5. Construction-grade-material allowable stresses are not adjusted for the size of the member. Southern pine base design values shown in Table 12.4 have size adjustments already included. The size factor C_F applies only to visually graded sawn lumber members.

The flat use factor, C_{fu}, is also shown in Table 12.5 and is applicable when dimension lumber 2 in. to 4 in. thick is loaded on the wide face.

TABLE 12.4 Base Design Values for Visually Graded Dimension Lumber (Normal Load Duration and Dry Service Condition).

No. 2 Grade (2″–4″ thick, 2″ and wider)						
	Design values (psi)					
Species	F_b	F_t	F_v	F_c	F_c	E
Douglas fir–larch	900	575	180	625	1350	1,600,000
Douglas fir–south	850	525	180	520	1350	1,200,000
Hem–fir	850	525	150	405	1300	1,300,000
Spruce–pine–fir	875	450	135	425	1150	1,400,000
Construction grade (2″–4″ thick, 2″–4″ wide)						
	Design values (psi)					
Species	F_b	F_t	F_v	F_c	F_c	E
Douglas fir–larch	1000	650	180	625	1650	1,500,000
Douglas fir–south	975	600	180	520	1650	1,200,000
Hem–fir	975	600	150	405	1550	1,300,000
Spruce–pine–fir	1000	500	135	425	1400	1,300,000
Southern pine (2″–4″ thick)						
	Design values (psi)					
No. 2 Grade	F_b	F_t	F_v	F_c	F_c	E
2″–4″ wide	1500	825	175	565	1650	1,600,000
5″–6″ wide	1250	725	175	565	1600	1,600,000
8″ wide	1200	650	175	565	1550	1,600,000
10″ wide	1050	575	175	565	1500	1,600,000
12″ wide	975	550	175	565	1450	1,600,000
Construction grade 4″ wide	1100	625	175	565	1800	1,500,000
Adjustment factors	F_b	F_t	F_v	F_c	F_c	E
Load duration C_D*	1.25	1.25	1.25	—	1.25	—
Wet service C_M	0.85**	1.0	0.97	0.67	0.80***	0.90

Source: Courtesy, American Wood Council, Leesburg, VA [4].

*Values shown are typical for forming applications.

**When $F_b(C_F) \le 1150$ psi, $C_M = 1.0$.

***When $F_c(C_F) \le 750$ psi, $C_M = 1.0$; for southern pine, when $F_c \le 750$ psi, $C_M = 1.0$

The repetitive member factor, C_r, applies only to F_b and to members 2 in. to 4 in. thick and must meet the following requirements per reference [4]: There are at least 3 members connected by continuous sheathing, and the spacing must not be greater than 24 inches.

The beam stability factor, C_L, is essentially a reduction factor for F_b where insufficient lateral restraint is furnished for the bending member. Because formwork assemblies are usually designed to meet lateral support criteria as prescribed by the *National Design Specification for Wood Construction* [4], the use of C_L is generally not applicable. For the purpose of this text, all formwork bending members are assumed to have adequate lateral support.

TABLE 12.5 Size and Flat Use Factors for Bending Stress F_b

	Size factor C_F^a		Flat use factor C_{fu}^b	
Width of lumber	2″–3″ thick	4″ thick	2″–3″ thick	4″ thick
2″–3″	1.5	1.5	1.0	—
4″	1.5	1.5	1.1	1.0
5″	1.4	1.4	1.1	1.05
6″	1.3	1.3	1.15	1.05
8″	1.2	1.3	1.15	1.05
10″	1.1	1.2	1.2	1.1
12″	1.0	1.1	1.2	1.1

Source: Courtesy, American Wood Council, Leesburg, VA [4].

[a]Applicable to No. 2 grade.

[b]Applicable to No. 2 grade and construction grade.

Determination of Allowable Shear Stress F'_v

Applicable adjustment factors for allowable shear stress are wet service, load duration, temperature, and incising factor.

The wet service factor C_M and the load duration factor C_D are furnished in Table 12.4. The temperature factor is generally not applicable in form design.

The incising adjustment factor C_i is a measure of the incisions that have to be made in a wood member to enhance its ability to receive pressure treatment. Thus, incising is used to increase the depth of penetration of wood preservatives in a wood member. The C_i factor is taken as 1.0 for non-incised wood, such as in the examples and problems of this chapter.

Determination of Allowable Stress for Compression Perpendicular to the Grain $F'_{c\perp}$

Applicable adjustment factors for allowable stress perpendicular to the grain are wet service, temperature, and bearing area. Note that the stress increase for short-term loading does not apply to $F_{c\perp}$.

The wet service factor C_M is furnished in Table 12.4. Again, the temperature factor is generally not applicable in form design.

The bearing area factor, C_b, is applicable where the bearing length is less than 6 in. long and at least 3 in. from the end of the member. It may be calculated from

$$C_b = \frac{\ell_b + 0.375}{\ell_b}$$

where ℓ_b is the length of the bearing (in.) measured parallel to the grain. For bearing at the end of a member and lengths of bearing equal to 6 in. or more, use $C_b = 1.0$. Note that C_b will never be less than 1.0. Therefore, it is conservative to omit it.

Determination of Modulus of Elasticity E′

The only applicable adjustment factor for modulus of elasticity in form design is the wet service C_M. It is shown in Table 12.4.

12.6 DESIGN OF FORMWORK FOR SLABS

Figure 12.1 shows a typical structural system for job-built forms for elevated slabs. The sequence of design is first to consider a strip of sheathing of the specified thickness and 12 in. in width. The maximum allowable span may then be determined based on the allowable values of bending stress, shear stress, and deflection for the sheathing. The lower of the computed values will determine the maximum spacing of the joists. This span value, usually rounded down to some lower modular value, becomes the spacing of the joists.

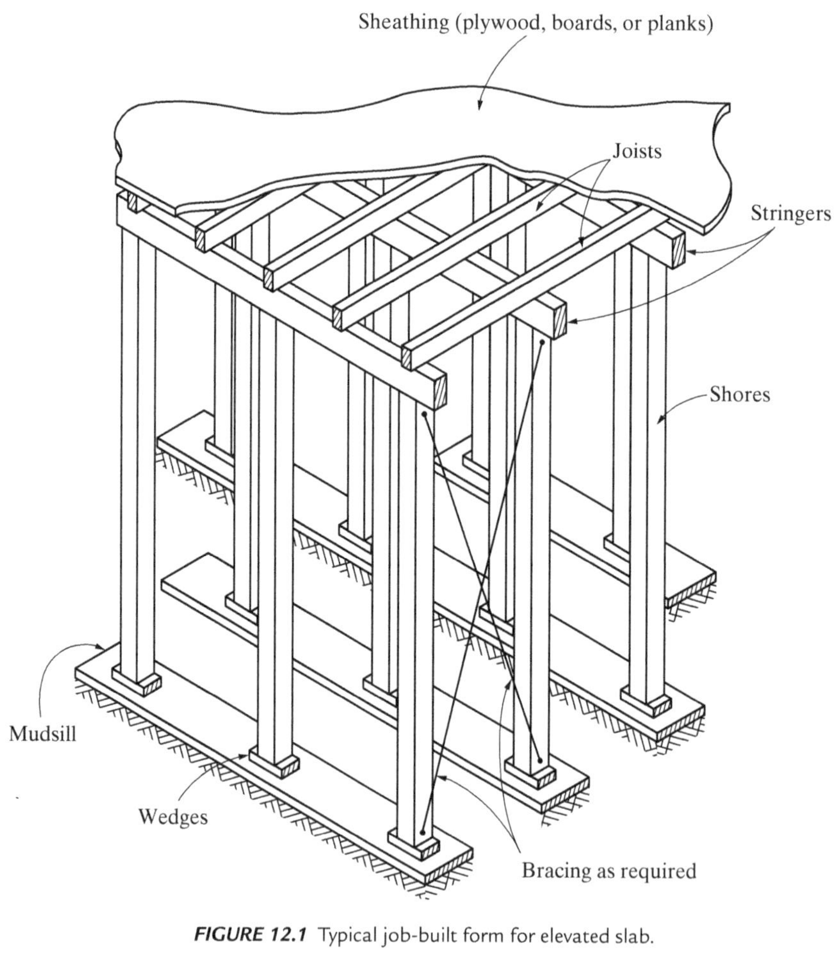

FIGURE 12.1 Typical job-built form for elevated slab.

Based on the joist spacing used, the joist itself is analyzed to determine its maximum allowable span. Each joist must support the load from the sheathing halfway over to the adjacent joist on either side. Therefore, the *width* of the load area carried by the joist is equal to the spacing of the joists. The joist span selected becomes the spacing of the stringers. Again, a modular value is selected for stringer spacing.

Based on the selected stringer spacing, the process is repeated to determine the maximum stringer span (distance between vertical supports or shores). Notice in the design of the stringers that the joist loads are actually applied to the stringer as a series of concentrated loads at the points where the joists rest on the stringer. It is simpler and sufficiently accurate to treat the load on the stringer as a uniformly distributed load, however. Again, the width of the uniform design load applied to the stringer is equal to the stringer spacing. The calculated stringer span must next be checked against the capacity of the shores used to support the stringers. The load on each shore is equal to the shore spacing multiplied by the load per foot of stringer. Thus, the maximum shore spacing (or stringer span) is limited to the lower span length as governed by stringer strength or shore strength. In addition, it is necessary to check the bearing at the point where each joist rests on the stringer. This is done by dividing the load at this point by the bearing area and comparing the resulting stress with the allowable unit stress in compression *perpendicular* to the grain. A similar procedure is applied at the point where each stringer rests on a vertical support.

The stringers shown in Figure 12.1 are supported by solid, rectangular wood shores that are columns (which we will assume are axially loaded). As with all axially loaded columns, the allowable load is a function of the slenderness ratio ℓ/d, the ratio of the unbraced length of the member to its least lateral dimension (not to exceed 50). The slenderness ratio ℓ/d is further modified and expressed as ℓ_e/d, where ℓ_e represents an effective unbraced column length and is defined as

$$\ell_e = K_e \ell$$

where K_e is an effective length factor based on column end conditions that affect rotation and translation. Factor K_e can be obtained from Table 12.6.

The following design approach applies for the determination of the allowable stress for compression parallel to the grain for simple, solid-sawn lumber columns, as recommended in [4].

Determination of Allowable Stress for Compression Parallel to the Grain F_c'

Applicable adjustment factors for allowable stress for compression parallel to the grain are load duration, wet service, temperature, size, and column stability. This may be expressed as

$$F_c' = F_c C_D C_M C_t C_F C_P$$

where

F_c = base design value for compression parallel to the grain (psi)

C_p = column stability factor

and all other quantities have been previously defined. This allowable stress is applicable to all values of the slenderness ratio (≤ 50) and replaces the previously used short-, intermediate-, and long-column equations of earlier design specifications. Note that although the slenderness ratio ℓ_e/d for solid columns shall not exceed 50, during construction this limit is increased to 75.

TABLE 12.6 Effective Length Factor K_e

Buckling modes	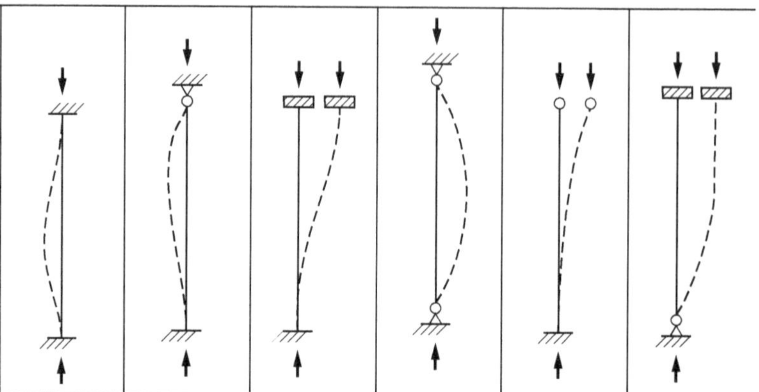					
Theoretical K_e value	0.5	0.7	1.0	1.0	2.0	2.0
Recommended design K_e when ideal conditions approximated	0.65	0.80	1.2	1.0	2.10	2.4
End condition code	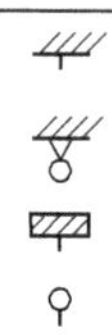	Rotation fixed, translation fixed Rotation free, translation fixed Rotation fixed, translation free Rotation free, translation free				

Source: Courtesy, American Wood Council, Leesburg, VA [4].

Because wood shores (columns) are generally reused repeatedly, ACI Committee 347 does not recommend the use of any adjustment factor that provides increased allowable stresses for short load duration [5]. In addition, temperature and wet service adjustments for wood shores are generally not required or considered. Therefore, only the column size factor and the column stability factor are normally considered in shore design.

The size factor, C_F, for compression parallel to the grain for the two grades indicated can be taken from Table 12.7.

TABLE 12.7 Size Factor C_F for Compression Parallel to the Grain

	Size factor C_F	
Width of lumber	**No. 2 grade**	**Const. grade**
2″–4″	1.15	1.0
5″	1.1	—
6″	1.1	—
8″	1.05	—
10″	1.0	—
12″	1.0	—

Source: Courtesy, American Wood Council, Leesburg, VA [4].

The column stability factor, C_p, is a function of the effective slenderness ratio ℓ_e/d of the shore, the adjusted modulus of elasticity E', and the adjusted base value of compression parallel to the grain before C_p is applied.

The unadjusted modulus of elasticity is normally used for wood shore design as shores are rarely in the wet service condition. Therefore, $E' = E$.

The column stability factor C_P for solid-sawn lumber can be calculated from Equation (3.7-1) in [4]. For convenience, and to use with the limited properties of Table 12.4, this equation is modified and written as

$$C_P = \frac{1+\alpha}{1.6} - \sqrt{\left(\frac{1+\alpha}{1.6}\right)^2 - \frac{\alpha}{0.8}}$$

Where

$$\alpha = \frac{F_{cE}}{F_c^*} = \frac{0.822E'_{min}}{\left(\frac{\ell_e}{d}\right)^2 F_c^*}$$

Since the tabulated values of the moduli of elasticity (E and E_{min}) are such that

$$E_{min} \approx 0.36E$$

Therefore, the equation for α can be rewritten as

$$\alpha = \frac{(0.822)(0.36E')}{\left(\frac{\ell_e}{d}\right)^2 F_c^*}$$

$$= \frac{0.3E'}{\left(\frac{\ell_e}{d}\right)^2 F_c^*}$$

and F_c^* is the adjusted base value of compression stress parallel to the grain before application of C_P.

Various tables are available to simplify the design approach and to expedite the selection of the formwork structural members [5]. These should be used with caution because of job-specific conditions and possible tabular limitations.

EXAMPLE 12.1

Design the formwork for a 6-in. structural concrete floor slab. The floor system is of the type shown in Figure 12.1. Use $\frac{3}{4}$-in. class I plyform for the sheathing and No. 2 Douglas fir–larch for the rest of the lumber. The maximum deflection for the sheathing may be taken as $\frac{1}{240}$ of the span. The maximum deflection for other bending members is taken as $\frac{1}{360}$ of the span. Based on end conditions, the effective unbraced length of the shores may be taken as 10 ft. The following conditions for design have been established:

1. Joists and stringers will be designed based on adequate lateral support.
2. Joists, stringers, and shores are used under dry service conditions.

Solution:

1. **Sheathing design** (find the joist spacing): Consider a 12-in.-wide strip of sheathing perpendicular to the supporting joists. The sheathing acts as a beam and is continuous over three or more supports. Determine the maximum allowable span for the sheathing. This becomes the maximum spacing for the joists (which support the sheathing).

   a. The design values for $\frac{3}{4}$-in. class I plyform (see Table 12.2) are

$$E = 1{,}650{,}000 \text{ psi} \quad (\text{modulus of elasticity})$$
$$F_b = 1930 \text{ psi} \quad (\text{bending stress})$$
$$F_v = 72 \text{ psi} \quad (\text{rolling shear stress})$$

   The plyform properties with face grain parallel to span (perpendicular to the joists) are

$$I = 0.199 \text{ in.}^4/\text{ft}$$
$$S = 0.455 \text{ in.}^3/\text{ft}$$
$$Ib/Q = 7.187 \text{ in.}^2/\text{ft} \quad (\text{rolling shear constant})$$
$$w_s = 2.2 \text{ psf} \quad (\text{weight of sheathing: for a 12 in. strip, this will be lb/ft})$$

   b. The loading on the sheathing is

$$\begin{aligned} &\text{DL (slab): } (6/12)(150) && = 75 \text{ psf} \\ &\text{LL (min.):} && = 50 \text{ psf} \\ &\text{Neglect sheathing weight} && \overline{125 \text{ psf}} = w \end{aligned}$$

   c. The maximum joist spacing based on the bending moment formula (from Table 12.1) is

$$\ell = 10.95\sqrt{\frac{F_b S}{w}}$$
$$= 10.95\sqrt{\frac{1930(0.455)}{125}} = 29 \text{ in.}$$

   d. The maximum joist spacing based on shear is

$$\ell = \frac{20F_v(Ib/Q)}{w} + 2d$$
$$= \frac{20(72)(7.187)}{125} + 2(0.75) = 84.3 \text{ in.}$$

   e. The maximum joist spacing based on deflection (maximum deflection is $\frac{1}{240}$ of the span) is

$$\ell = 1.94\sqrt[3]{\frac{EI}{w}}$$
$$= 1.94\sqrt[3]{\frac{1{,}650{,}000(0.199)}{125}}$$
$$= 26.8 \text{ in.}$$

Deflection controls. The maximum spacing of supporting joists should not exceed 26.8 in. Use joists at 24 in. o.c.

2. **Joist design** (find the stringer spacing): Assume 2 × 8 (S4S) joists and a 7-day maximum duration of load. Consider the joists as uniformly loaded beams continuous over three or more spans (the supports are the stringers).

   a. Obtain allowable stresses using base design values from Table 12.4 and appropriate adjustment factors.

      1. Bending: F_b = 900 psi. Adjustment factors:

         a. Load duration factor from Table 12.4: C_D = 1.25.

         b. Size factor from Table 12.5: C_F = 1.2.

         Therefore,

$$F'_b = F_b C_D C_F$$
$$= 900(1.25)(1.2) = 1350 \text{ psi}$$

      2. Shear: F_v = 180 psi. Adjustment factor:

         Load duration factor from Table 12.4: C_D = 1.25.

         Therefore,

$$F'_v = F_v C_D = 180(1.25) = 225 \text{ psi}$$

      3. Modulus of elasticity: E = 1,600,000 psi. No adjustment factors apply. Thus,

$$E' = E = 1{,}600{,}000 \text{ psi}$$

         Properties for the 2 × 8 (S4S) from Table 12.3 are

$$S = 13.14 \text{ in.}^3$$
$$I = 47.63 \text{ in.}^4$$
$$A = 10.88 \text{ in.}^2$$

   b. Loading: Because joists support a 2-ft width of sheathing, the loading on the joist is

$$w = 125 \text{ psf } (2) = 250 \text{ lb/ft}$$

      Assume the weight of the sheathing and joists to be 5 psf. Then

$$w = 250 + 5(2) = 260 \text{ lb/ft}$$

c. The maximum stringer spacing based on bending moment is

$$\ell = 10.95\sqrt{\frac{F'_b S}{w}}$$

$$= 10.95\sqrt{\frac{1350(13.14)}{260}} = 90.4 \text{ in.}$$

d. The maximum stringer spacing based on shear is

$$\ell = \frac{13.3F'_v A}{w} + 2d$$

$$= \frac{13.3(225)(10.88)}{260} + 2(7.25)$$

$$= 139.7 \text{ in.}$$

e. The maximum stringer spacing based on deflection (maximum deflection is $\frac{1}{360}$ of the span) is

$$\ell = 1.69\sqrt[3]{\frac{E'I}{w}}$$

$$= 1.69\sqrt[3]{\frac{1{,}600{,}000(47.63)}{260}}$$

$$= 112.3 \text{ in.}$$

Bending governs. Therefore, the maximum spacing of supporting stringers cannot exceed 90.4 in. Use a stringer spacing of 7 ft-0 in. (84 in.).

3. **Stringer design** (find the shore spacing): Use 4 in. × 8 in. (S4S) stringers and a 7-day maximum duration of load. Consider the stringers to be uniformly loaded beams continuous over three or more supports. The supports are the shores. The loads from the joists are concentrated loads, but for simplicity, we will assume a uniformly distributed load.

a. Obtain allowable stresses using base design values from Table 12.4 and appropriate adjustment factors.

1. Bending: F_b = 900 psi. Adjustment factors:
    a. Load duration factor from Table 12.4: C_D = 1.25.
    b. Size factor from Table 12.5: C_F = 1.3.

Therefore,

$$F'_b = F_b C_D C_F$$

$$= 900(1.25)(1.3) = 1463 \text{ psi}$$

2. Shear: F_v = 180 psi. Adjustment factor:
    Load duration factor from Table 12.4: C_D = 1.25.
    Therefore,

$$F'_v = F_v C_D = 180(1.25) = 225 \text{ psi}$$

3. Modulus of elasticity: E = 1,600,000 psi. No adjustment factors apply. Thus

$$E' = E = 1{,}600{,}000 \text{ psi}$$

Properties for the 4 × 8 (S4S) from Table 12.3 are

$$S = 30.66 \text{ in.}^3$$
$$I = 111.1 \text{ in.}^4$$
$$A = 25.38 \text{ in.}^2$$

b. Loading: Each stringer supports a 7 ft-0 in.-wide strip of design load. Assuming a formwork weight of 5 psf, the uniformly distributed load on a stringer is calculated from

$$w = 7.0(125 + 5) = 910 \text{ lb/ft}$$

c. The maximum shore spacing based on bending (of the stringers) is

$$\ell = 10.95\sqrt{\frac{F_b'S}{w}}$$
$$= 10.95\sqrt{\frac{1463(30.66)}{910}} = 76.9 \text{ in.}$$

d. The maximum shore spacing based on shear is

$$\ell = \frac{13.3F_v'A}{w} + 2d$$
$$= \frac{13.3(225)(25.38)}{910} + 2(7.25) = 98.0 \text{ in.}$$

e. The maximum shore spacing based on deflection is

$$\ell = 1.69\sqrt[3]{\frac{EI}{w}}$$
$$= 1.69\sqrt[3]{\frac{1{,}600{,}000(111.1)}{910}} = 98.1 \text{ in.}$$

Bending governs. Therefore, the maximum spacing of supporting shores cannot exceed 76.9 in. Use a post shore spacing of 6 ft-0 in. (72.0 in.).

4. **Design of post shores:** The stringers are spaced at 7 ft on center and are supported by shores at 6 ft-0 in. on center. Therefore, each shore must support a floor area of

$$7.0(6.0) = 42 \text{ ft}^2$$

Again, assuming formwork weight of 5 psf, the load per shore is calculated as

$$42(125 + 5) = 5460 \text{ lb}$$

Although commercial shores are usually readily available to support this load, we will design 4 × 4 wood shores. The effective unbraced length of the shore, ℓ_e, is 10 ft-0 in. in each direction. The capacity of the 4 × 4 (S4S) shore is calculated using the recommendations of Reference [4], as discussed previously.

a. The base design value for compression parallel to the grain from Table 12.4 is F_c = 1350 psi.

b. Adjustment factors:

1. Size factor from Table 12.7: C_F = 1.15.
2. For the column stability factor C_P, initially the following items must be established:

   a. For modulus of elasticity, there is no adjustment factor:

$$E' = E = 1{,}600{,}000 \text{ psi}$$

   b. Find F_c^*:

$$F_c^* = F_c C_F = 1350(1.15) = 1553 \text{ psi}$$

   c. $\dfrac{\ell_e}{d} = \dfrac{10(12)}{3.5} = 34.3 < 50$ (O.K.)

   d. Solve for α:

$$\alpha = \frac{0.3E'}{\left(\dfrac{\ell_e}{d}\right)^2 F_c^*} = \frac{0.3(1{,}600{,}000)}{34.3^2(1553)} = 0.263$$

Solve for C_P:

$$C_P = \frac{1+\alpha}{1.6} - \sqrt{\left(\frac{1+\alpha}{1.6}\right)^2 - \frac{\alpha}{0.8}} = \frac{1+0.263}{1.6} - \sqrt{\left(\frac{1+0.263}{1.6}\right)^2 - \frac{0.263}{0.8}} = 0.247$$

c. Compute the allowable stress F_c':

$$F_c' = F_c C_F C_P = 1350(1.15)(0.247) = 383 \text{ psi}$$

Therefore, the allowable load is

$$P = F_c' A = 383(3.5)^2 = 4690 \text{ lb} \quad \text{(N.G.)}$$

$$4690 \text{ lb} < 5460 \text{ lb}$$

There are several possible solutions. The shore size could be increased (try a 4 × 6 [S4S]), use lateral bracing (horizontal lacing) at midheight to reduce the effective length of the shore, or reduce the shore spacing. The latter choice is the simplest approach.

$$\text{new required spacing} = \frac{4690}{5460}(72 \text{ in.}) = 61.8 \text{ in.}$$

If a shore spacing of 5 ft-0 in. is used,

$$\text{load per shore} = 7.0(5.0)(125+5) = 4550 \text{ lb}$$

$$4550 \text{ lb} < 4690 \text{ lb} \qquad \text{(O.K.)}$$

Use 4 × 4(S4S) shores at 5 ft-0 in. on center.

5. **Bearing stresses:**
   a. Where stringers bear on shores (4 × 8 stringers on 4 × 4 shores),

$$\text{contact area} = 3.5(3.5) = 12.25 \text{ in.}^2$$
$$\text{total load on shore} = 4550 \text{ lb}$$

The actual bearing stress perpendicular to the grain of the stringer is

$$\frac{4550}{12.25} = 371 \text{ psi}$$

Determine the allowable bearing stress perpendicular to the grain of the stringer. From Table 12.4, $F_{c\perp} = 625$ psi. The adjustment factor applicable in this case is C_b, since the length of bearing ℓ_b is 3.5 in.:

$$C_b = \frac{\ell_b + 0.375}{\ell_b} = \frac{3.88}{3.5} = 1.107$$

Therefore,

$$F'_{c\perp} = 625(1.107) = 692 \text{ psi} \qquad \text{(O.K.)}$$
$$371 \text{ psi} < 692 \text{ psi}$$

   b. Where joists bear on stringers (2 × 8 joists on 4 × 8 stringers),

$$\text{contact area} = 1.5(3.5) = 5.25 \text{ in.}^2$$

Recalling from the design of the joists that the loading on each joist was 260 lb/ft, the load on the stringer from each joist is

$$260(7.0) = 1820 \text{ lb}$$

The actual bearing stress perpendicular to the grain is

$$\frac{1820}{5.25} = 347 \text{ psi}$$

Determine the allowable bearing stress perpendicular to the grain. Because the calculated bearing stress is low relative to the 625 psi base design value for $F'_{c\perp}$, the increase in $F'_{c\perp}$ due to C_b may be disregarded. Thus,

$$347 \text{ psi} < 625 \text{ psi} \qquad \text{(O.K.)}$$

6. **Lateral bracing:** For floor systems, the minimum load to be used in designing lateral bracing is the greater of 100 lb per lineal foot of floor edge or 2% of the total dead load of the floor. We will assume the slab to be 80 ft × 100 ft and placed in one operation. Guy wire bracing capable of carrying a load of 4000 lb each will be used on all four sides of the slab area attached at slab elevation and making a 45° angle with the ground. Guy wires can resist only tensile forces.

Calculating lateral load H as 2% of the dead load of the floor, again assuming the formwork weighs 5 psf, and recalling that the self-weight of the 6-in slab is 75 psf, yields

$$H = 0.02(75+5)(80)(100) = 12{,}800 \text{ lb}$$

Distributing this load along the long side yields

$$\frac{12{,}800}{100} = 128 \text{ lb/ft} > 100 \text{ lb/ft}$$

and along the short side yields

$$\frac{12{,}800}{80} = 160 \text{ lb/ft} > 100 \text{ lb/ft}$$

These results are shown in Figure 12.2a. For determination of the guy wire spacing, the 160 lb/ft lateral load will be used. From Figure 12.2b, the tension in the guy wire T is calculated as

$$\frac{T}{1.414} = \frac{160}{1}$$

$$T = 226.2 \text{ lb per ft of slab being braced}$$

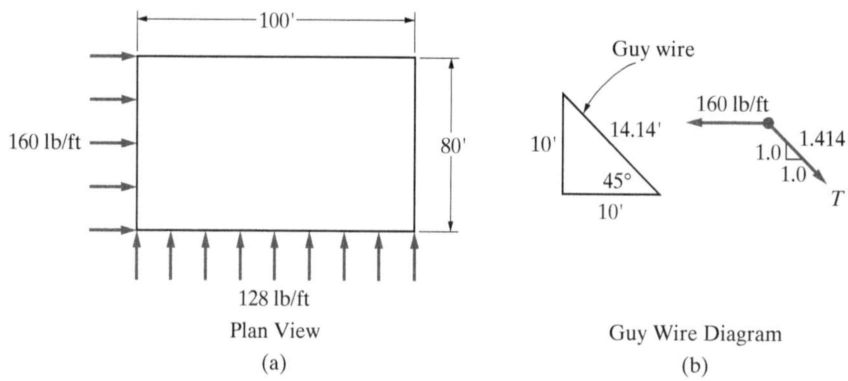

FIGURE 12.2 Floor slab lateral bracing design

The maximum spacing for the guy wires (for the given allowable load of 4000 Ib) is

$$\frac{4000}{226.2} = 17.7 \text{ ft}$$

Use guy wires spaced at 15 ft (max.) on center on all sides.

12.7 DESIGN OF FORMWORK FOR BEAMS

Figure 12.3 shows one of several common types of beam forms. The usual design procedure involves consideration of the vertical loads, with the following components to be designed: the beam bottom, the ledger that supports joists, and the supporting shores. Bearing stresses must also be checked.

For the deeper beams (24 in. and more), consideration should also be given to the lateral pressure produced by the fresh concrete, which must be resisted by the beam sides. The beam sides would be designed in much the same way as the sheathing in a wall form. Also of importance in Figure 12.3 are the *kickers*, which hold the beam sides in place against the pressure of the concrete, and *blocking*, which serves to transmit the slab load from the ledgers to the T-head shores.

Beam bottoms (or soffits) are usually made to the exact width of the beam. They may be composed of one or more 2-in. planks, or they may be of plywood backed by 2 × 4s. In the following example, the soffit is made of a 2 × 12, which is finished on two sides (S2S), giving it final dimensions of $1\frac{1}{2}$ in.×12 in.

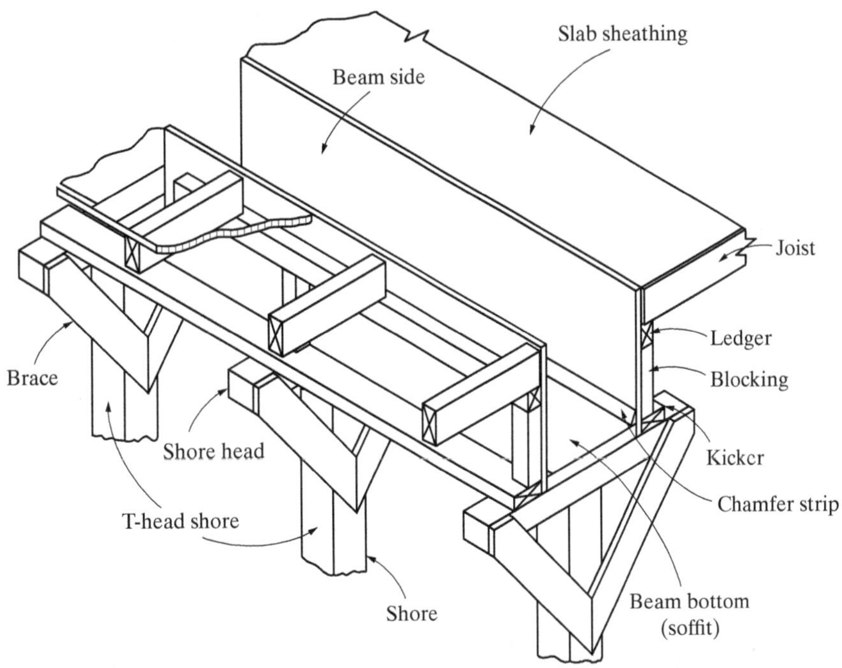

FIGURE 12.3 Typical beam formwork.

EXAMPLE 12.2

Design forms to support the 12 in. × 20 in. beam shown in Figure 12.4. The beam is to support a 4-in.-thick reinforced concrete slab. Use Douglas fir–larch No. 2 grade. The maximum allowable deflection is to be $\frac{1}{360}$ of the span for bending members. The unsupported shore height will be based on an assumed floor-to-floor height of 10 ft, from which the depth of the beam will be subtracted. All bending members are to be designed based on adequate lateral support.

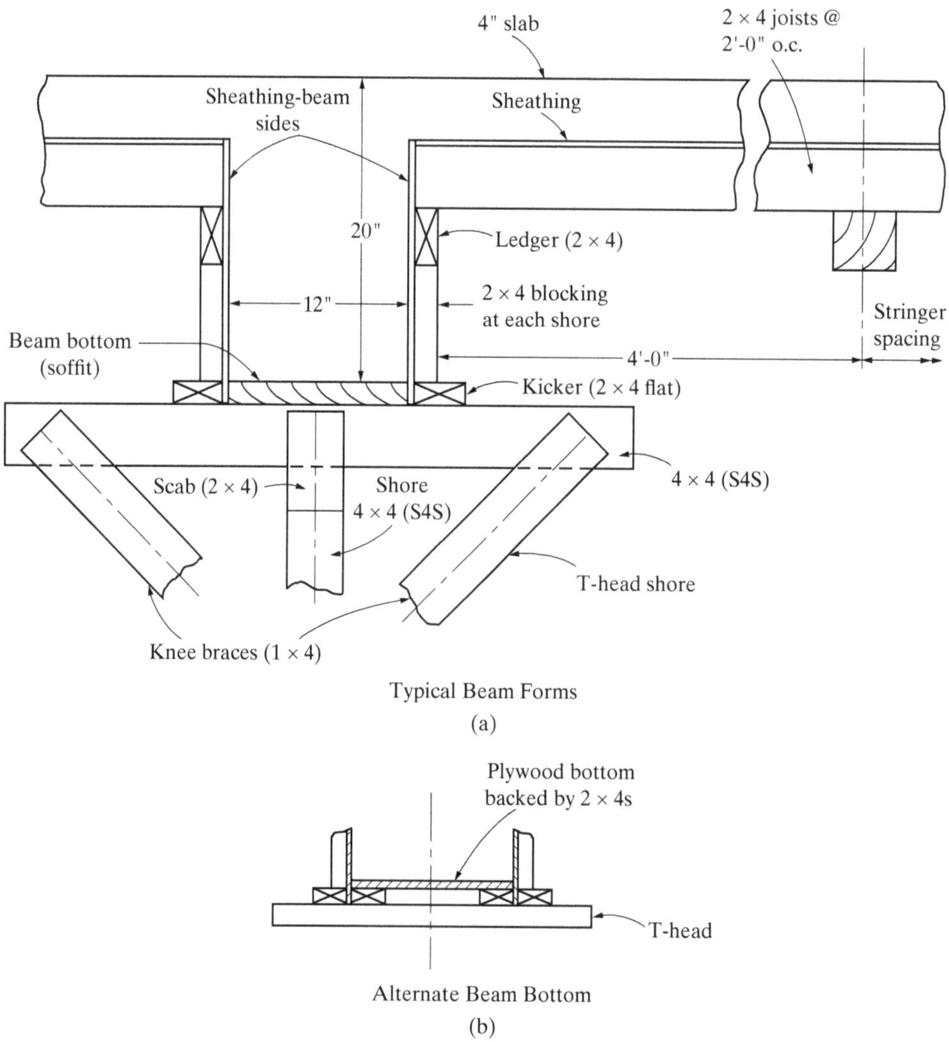

FIGURE 12.4 Beam form design.

Solution:

1. **Beam bottom design** (compute the maximum spacing between shores): Assume a 7-day maximum duration of load. Assume the plank to be continuous over three or more supports.

a. Obtain allowable stresses using base design values from Table 12.4 and appropriate adjustment factors from Tables 12.4 and 12.5.

1. Bending: $F_b = 900$ psi. Adjustment factors: $C_D = 1.25$, $C_F = 1.0$, $C_{fu} = 1.2$. Because $F_bC_F < 1150$ psi, $C_M = 1.0$ (see footnote, Table 12.4). Therefore,

$$\begin{aligned} F_b' &= F_b C_D C_F C_M C_{fu} \\ &= 900(1.25)(1.0)(1.0)(1.2) \\ &= 1350 \text{ psi} \end{aligned}$$

2. Shear: $F_v = 180$ psi. Adjustment factors: $C_D = 1.25$ and $C_M = 0.97$. Therefore,

$$F_v' = F_v C_D C_M = 180(1.25)(0.97) = 218.3 \text{ psi}$$

3. Modulus of elasticity: $E = 1{,}600{,}000$ psi. Adjustment factor: $C_M = 0.9$. Therefore,

$$E' = EC_M = 1{,}600{,}000(0.9) = 1{,}440{,}000 \text{ psi}$$

Properties for the 2 × 12 (S2S) are

$$S = \frac{bh^2}{6} = \frac{12(1.5)^2}{6} = 4.5 \text{ in.}^3$$

$$I = \frac{bh^3}{12} = \frac{12(1.5)^3}{12} = 3.38 \text{ in.}^4$$

b. The loading on the beam soffit is calculated as

$$\text{DL (reinforced concrete beam):}$$

$$\frac{12(20)}{144}(150) = 250 \text{ lb/ft}$$

$$\text{LL (use 50 psf)}: \frac{12}{12}(50) = 50 \text{ lb/ft}$$

$$\text{total load} = 250 + 50 = 300 \text{ lb/ft}$$

c. The maximum shore spacing based on bending is

$$\begin{aligned} \ell &= 10.95\sqrt{\frac{F_b' S}{w}} \\ &= 10.95\sqrt{\frac{1350(4.5)}{300}} \\ &= 49.3 \text{ in.} \end{aligned}$$

d. The maximum shore spacing based on shear is

$$\ell = \frac{13.3F_v'A}{w} + 2d$$
$$= \frac{13.3(218.3)(12)(1.5)}{300} + 2(1.5)$$
$$= 177.2 \text{ in.}$$

e. The maximum shore spacing based on deflection is

$$\ell = 1.69\sqrt[3]{\frac{E'I}{w}}$$
$$= 1.69\sqrt[3]{\frac{1{,}440{,}000(3.38)}{300}}$$
$$= 42.8 \text{ in.}$$

Deflection governs. Try shore spacing at 42 in. o.c.

2. **Ledger design:** Use $\frac{3}{4}$-in. plyform sheathing (vertically) for the beam sides and 2 × 4 kickers as shown. The ledger is supported at each shore by a blocking piece. Because the shores are to be 42 in. o.c., the ledger will be continuous over three or more spans. Use 2 × 4s (S4S) for the ledger as shown. Compute the required spacing for the ledger supports and compare with the 42-in. spacing previously determined. Neglect the connection of the ledger to the vertical sheathing.

   a. Obtain allowable stresses using base design values from Table 12.4 and appropriate adjustment factors from Tables 12.4 and 12.5.

      1. Bending: F_b = 900 psi. Adjustment factors: C_D = 1.25 and C_F = 1.5. Therefore,

$$F_b' = F_b C_D C_F$$
$$= 900(1.25)(1.5) = 1688 \text{ psi}$$

      2. Shear: F_v = 180 psi. Adjustment factor: C_D = 1.25. Therefore,

$$F_v' = F_v C_D = 180(1.25) = 225 \text{ psi}$$

      3. Modulus of elasticity: E = 1,600,000 psi. No adjustment factors apply. Thus,

$$E' = E = 1{,}600{,}000 \text{ psi}$$

   Properties for the 2 × 4 (S4S) from Table 12.3 are

$$A = 5.25 \text{ in.}^2$$
$$I = 5.36 \text{ in.}^4$$
$$S = 3.06 \text{ in.}^3$$

b. Loading: The 2 × 4 joists are supported by the ledger, as shown in Figure 12.4. Although the ledger is loaded with point loads, a uniform load will be assumed for simplicity. The loading on the slab sheathing is

$$\begin{aligned} \text{DL (slab)}: \left(\frac{4}{12}\right)(150) &= 50 \text{ psf} \\ \text{LL (min.)} &= 50 \text{ psf} \\ \text{assume sheathing weight} &= \underline{\ \ 5 \text{ psf}} \\ & \quad 105 \text{ psf} \end{aligned}$$

Because the span of the joists from the ledger to the adjacent stringer is 4 ft-0 in. (see Figure 12.4), the load to the ledger is calculated as

$$w = \tfrac{1}{2}(4)(105) = 210 \text{ lb/ft}$$

c. The maximum blocking spacing based on bending is

$$\begin{aligned} \ell &= 10.95\sqrt{\frac{F_b'S}{w}} \\ &= 10.95\sqrt{\frac{1688(3.06)}{210}} \\ &= 54.3 \text{ in.} \end{aligned}$$

d. The maximum blocking spacing based on shear is

$$\begin{aligned} \ell &= \frac{13.3F_v'A}{w} + 2d \\ &= \frac{13.3(225)(5.25)}{210} + 2(3.5) \\ &= 81.8 \text{ in.} \end{aligned}$$

e. The maximum blocking spacing based on deflection is

$$\begin{aligned} \ell &= 1.69\sqrt[3]{\frac{E'I}{w}} \\ &= 1.69\sqrt[3]{\frac{1,600,000(5.36)}{210}} \\ &= 58.2 \text{ in.} \end{aligned}$$

Because all the three foregoing spacings *exceed* 42 in., the 2 × 4 ledgers supported by blocking at 42 in. on center are satisfactory.

3. **Design of the post shores:** The shores are spaced 42 in. (or 3.5 ft) on center, and each must support a loading of

$$\text{from beam bottom}: 300(3.5) = 1050 \text{ lb}$$
$$\text{from slab forms (two sides)}: 210(3.5)(2) = 1470 \text{ lb}$$
$$\text{total load per shore} = 1050 + 1470 = 2520 \text{ lb}$$

Assume 4 × 4 (S4S) wood shores. The unsupported shore height will be based on an assumed floor-to-floor height of 10 ft-0 in., from which the depth of the beam will be subtracted. The unsupported height is

$$\ell = 10(12) - 20 = 100 \text{ in.} = 8.33 \text{ ft}$$

We will assume the shores are pin-connected. From Table 12.6, K_e = 1.0. Therefore, the effective unbraced length is

$$\ell_e = K_e\ell = 1.0(8.33) = 8.33 \text{ ft}$$

a. The base design value for compression parallel to the grain from Table 12.4 is F_c = 1350 psi.

b. Adjustment factors:

1. Size factor from Table 12.7: C_F = 1.15.
2. For the column stability factor C_P, initially the following items must be established:

   a. For modulus of elasticity, there is no adjustment factor:

   $$E' = E = 1{,}600{,}000 \text{ psi}$$

   b. Find F_c^*:

   $$F_c^* = F_c C_F = 1350(1.15) = 1553 \text{ psi}$$

   c. $\dfrac{\ell_e}{d} = \dfrac{8.33(12)}{3.5} = 28.6 < 50 \qquad \text{(O.K.)}$

   d. Solve for α:

   $$\alpha = \frac{0.3E'}{\left(\frac{\ell_e}{d}\right)^2 F_c^*} = \frac{0.3(1{,}600{,}000)}{28.6^2(1553)} = 0.378$$

   Solve for C_p:

   $$C_p = \frac{1+\alpha}{1.6} - \sqrt{\left(\frac{1+\alpha}{1.6}\right)^2 - \frac{\alpha}{0.8}} = \frac{1+0.378}{1.6} - \sqrt{\left(\frac{1+0.378}{1.6}\right)^2 - \frac{0.378}{0.8}} = 0.342$$

c. Compute the allowable stress F_c':

$$F_c' = F_c C_F C_P = 1350(1.15)(0.342) = 531 \text{ psi}$$

Therefore, the allowable load is

$$P = F_c' A = 531(3.5)^2 = 6500 \text{ lb} \qquad \text{(O.K.)}$$

$$6500 \text{ lb} > 2520 \text{ lb}$$

Use 4 × 4 (S4S) shores spaced 42 in. on center.

4. Bearing stresses:

a. Assume 4 × 4 (S4S) T-heads on the 4 × 4 (S4S) shores. Actual bearing stress perpendicular to the grain of the T-head is calculated as

$$\frac{\text{shore load}}{\text{contact area}} = \frac{2520}{3.5^2} = 206 \text{ psi}$$

From Table 12.4, the base design value for $F_{c\perp}$ is 625 psi. We will neglect the adjustment factor for bearing area (due to C_b > 1.0 because $3\frac{1}{2}$ in. < 6 in.). Therefore, use $F_{c\perp}' = 625$ psi. Thus,

$$206 \text{ psi} < 625 \text{ psi} \qquad \text{(O.K.)}$$

b. Check bearing stress between the 2 × 4 ledger and the 2 × 4 blocking. The load from the ledger to the blocking is 210 (3.5) = 735 lb. The actual bearing stress perpendicular to the grain of the ledger is

$$\frac{\text{load}}{\text{contact area}} = \frac{735}{1.5(3.5)} = 140 \text{ psi}$$

The allowable bearing stress, from part a, is 625 psi. Thus,

$$140 \text{ psi} < 625 \text{ psi} \qquad \text{(O.K.)}$$

c. Check bearing stress between the 2 × 4 joists and the 2 × 4 ledger. The joist loading is (105 psf)(2 ft) = 210 lb/ft, and the span of the joist is 4 ft. Therefore, the load from the joist to the ledger is (210 lb/ft) (2 ft) = 420 lb. The actual bearing stress perpendicular to the grain of the ledger (and the joists) is

$$\frac{\text{load}}{\text{contact area}} = \frac{420}{1.5(1.5)} = 186.7 \text{ psi}$$

The allowable bearing stress is determined as in part a (the bearing area adjustment factor C_b is not applicable because this is end bearing):

$$186.7 \text{ psi} < 625 \text{ psi} \qquad \text{(O.K.)}$$

d. Check the bearing of the beam soffit on the 4 × 4 (S4S) T-heads. The load from the beam soffit is (300 lb/ft)(3.5 ft) = 1050 lb. The actual bearing stress perpendicular to the grain of the T-head is

$$\frac{\text{load}}{\text{contact area}} = \frac{1050}{12(3.5)} = 25 \text{ psi}$$

For the allowable bearing stress, we will neglect the bearing area adjustment factor. The contact surface may be subjected to a wet condition, so this adjustment factor (C_M = 0.67 from Table 12.4) will be used:

$$F'_{c\perp} = F_{c\perp} C_M = 625(0.67) = 419 \text{ psi}$$

$$25 \text{ psi} < 419 \text{ psi} \qquad \text{(O.K.)}$$

12.8 WALL FORM DESIGN

The design procedure for wall forms is similar to that used for slab forms, substituting studs for joists, wales for stringers, and ties for shores. See Figure 12.5 for locations of these members.

The maximum lateral pressure against the sheathing must be determined first. We will assume conditions such that C_C and C_W are both 1.0. With the sheathing thickness specified, the maximum allowable span for the sheathing is computed based on bending, shear, and deflection. This will be the maximum stud spacing. (An alternative approach would be to establish the stud spacing and then calculate the required thickness of the sheathing.)

Next, the maximum allowable stud span is calculated based on stud size and loading, considering bending, shear, and deflection. This will be the maximum wale spacing. (An alternative approach would be to establish the wale spacing and then calculate the required size of the studs.)

The next step is to determine the maximum allowable spacing of wale supports (tie spacing). This is calculated based on wale size and loading. (An alternative approach would be to preselect the tie spacing and then calculate the wale size.) Double wales are commonly used to avoid the necessity of drilling wales for tie insertion.

The load supported by each tie must be computed and compared with the tie capacity. The load on each tie is calculated as the design load (psf) multiplied by the tie spacing (ft) and wale spacing (ft). If the load exceeds the tie strength, a stronger tie must be used or the tie spacing must be reduced.

Bearing stresses must also be checked where the studs bear on the wales and where the tie ends bear on the wales. Maximum bearing stress must not exceed the allowable compression stress perpendicular to the grain or crushing will result.

Finally, lateral bracing must be designed to resist any expected lateral loads, such as wind loads.

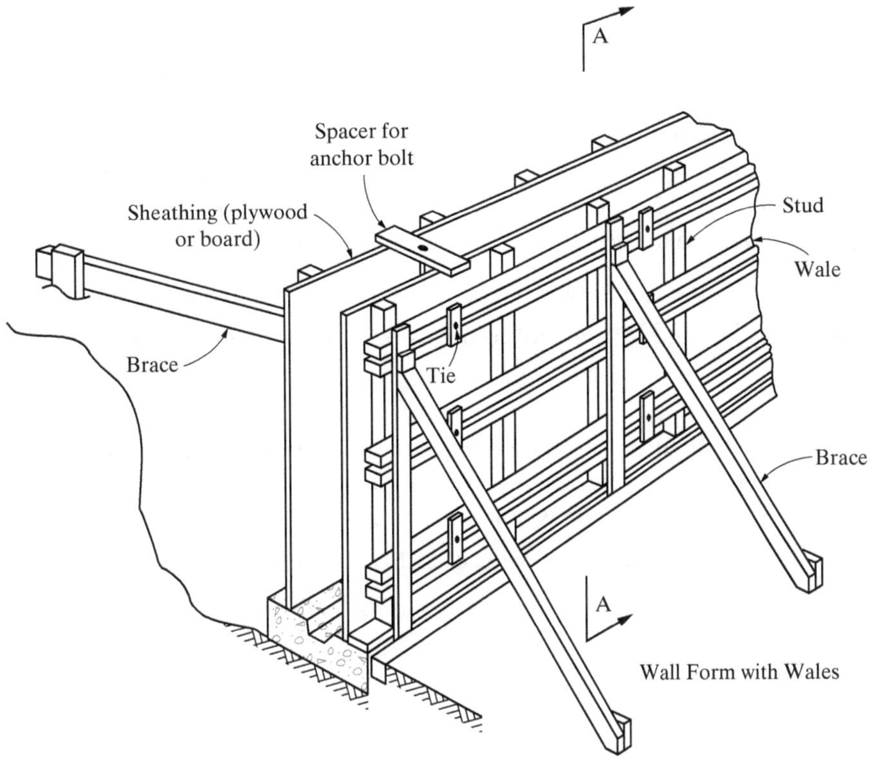

(a)

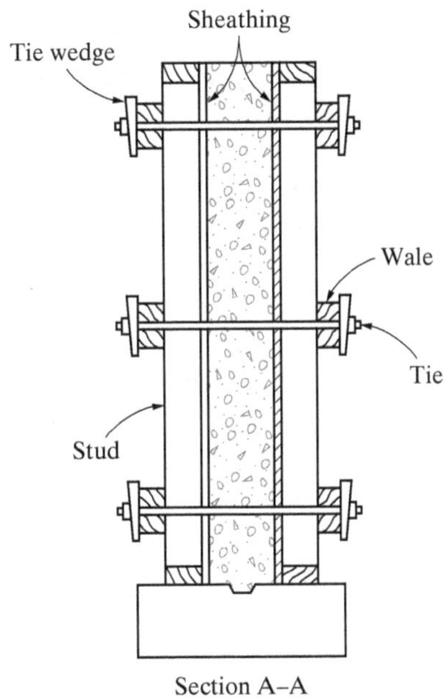

(b)

FIGURE 12.5 Typical wall forms.

EXAMPLE 12.3

Design formwork for an 8-ft-high wall. Refer to Figure 12.5. The concrete is to be placed at a rate of 4 ft/h and will be internally vibrated. Concrete temperature is expected to be 90°F. Assume normal weight concrete with a unit weight of 150 pcf. The maximum allowable deflection of bending members is to be $\frac{1}{360}$ of the span. Use $\frac{3}{4}$-in. class I plyform for the sheathing and No. 2 Douglas fir–larch for the rest of the lumber. Assume all bending members to be supported on three or more spans. The following conditions for design have been established:

1. Studs and wales are to be designed based on adequate lateral support.
2. Studs, wales, and bracing are to be used under dry service conditions.

Solution:

1. **Sheathing design** (find the stud spacing): Consider a 12-in.-wide strip of sheathing perpendicular to the supporting studs and acting as a beam continuous over three or more spans. The sheathing spans horizontally. Place the face grain perpendicular to studs.

   a. The design values for $\frac{3}{4}$-in. class I plyform (see Table 12.2) are

$$E = 1{,}650{,}000 \text{ psi}$$
$$F_b = 1930 \text{ psi}$$
$$F_v = 72 \text{ psi}$$

   Plyform properties for face grain parallel to the span are

$$I = 0.199 \text{ in.}^4$$
$$S = 0.455 \text{ in.}^3$$
$$Ib/Q = 7.187 \text{ in.}^2/\text{ft}$$

   b. Loading: The sheathing will be designed for concrete pressure, which is the lesser of $150h$ (where h will be taken as 8 ft), 2000 psf, or as determined by formula (noting that $R < 7$ ft/h and $h < 14$ ft):

$$p = 150 + \frac{9000R}{T} = 150 + \frac{9000(4)}{90}$$
$$= 550 \text{ psf} < 600 \text{ psf}$$
$$150(h) = 150(8) = 1200 \text{ psf}$$

   Use the ACI-recommended minimum of 600 psf for the sheathing design load, although the pressure could be decreased near the top of the form where $150(h)$ controls. The length in which the decreased pressure could be used may be calculated as p_{max}/w

$$\frac{600}{150} = 4.0 \text{ ft} \quad \text{(from the top of the form)}$$

   Thus, the actual lateral pressure distribution due to wet concrete is trapezoidal in shape, varying from zero at the top of the wall form to 600 psf at 4 ft from the top of the form, and then remains constant at 600 psf to the bottom of the form. It is conservative to design the full height of the wall formwork for a uniform lateral pressure of 600 psf, however.

c. The maximum stud spacing based on bending is

$$\ell = 10.95\sqrt{\frac{F_b'S}{w}}$$

$$= 10.95\sqrt{\frac{1930(0.455)}{600}} = 13.25 \text{ in.}$$

d. The maximum stud spacing based on shear is

$$\ell = \frac{20F_v'(Ib/Q)}{w} + 2d$$

$$= \frac{20(72)(7.187)}{600} + 2(0.75) = 18.75 \text{ in.}$$

e. The maximum stud spacing based on deflection is

$$\ell = 1.69\sqrt[3]{\frac{E'I}{w}}$$

$$= 1.69\sqrt[3]{\frac{1{,}650{,}000(0.199)}{600}} = 13.82 \text{ in.}$$

Bending is critical. Use a stud spacing of 12 in. o.c.

2. **Stud design** (compute the wale spacing): Assume 2 × 4 (S4S) studs and a 7-day maximum duration of load.

a. Obtain allowable stresses using base design values from Table 12.4 and appropriate adjustment factors from Tables 12.4 and 12.5.

1. Bending: F_b = 900 psi. Adjustment factors: C_D = 1.25 and C_F = 1.5. Therefore,

$$F_b' = F_b C_D C_F$$

$$= 900(1.25)(1.5) = 1688 \text{ psi}$$

2. Shear: F_v = 180 psi. Adjustment factor: C_D = 1.25. Therefore,

$$F_v' = F_v C_D = 180(1.25) = 225 \text{ psi}$$

3. Modulus of elasticity: E = 1,600,000 psi. No adjustment factors apply. Thus,

$$E' = E = 1{,}600{,}000 \text{ psi}$$

For the 2 × 4 lumber, from Table 12.3,

$$A = 5.25 \text{ in.}^2$$

$$I = 5.36 \text{ in.}^4$$

$$S = 3.06 \text{ in.}^3$$

b. Loading: Since the stud spacing is 12 in. o.c., the load w will be 600 lb/ft (see step 1, part [b]).

c. The maximum wale spacing based on bending is

$$\ell = 10.95\sqrt{\frac{F_b'S}{w}}$$

$$= 10.95\sqrt{\frac{1688(3.06)}{600}} = 32.1 \text{ in.}$$

d. The maximum wale spacing based on shear is

$$\ell = \frac{13.3F_v'A}{w} + 2d$$

$$= \frac{13.3(225)(5.25)}{600} + 2(3.5) = 33.2 \text{ in.}$$

e. The maximum wale spacing based on deflection is

$$\ell = 1.69\sqrt[3]{\frac{E'I}{w}}$$

$$= 1.69\sqrt[3]{\frac{1,600,000(5.36)}{600}} = 41.0 \text{ in.}$$

Therefore, bending governs. Use a wale spacing of 24 in. o.c. (maximum).

3. **Wale design** (compute the tie spacing): Double 2 × 4 (S4S) wales will be assumed, and the allowable stresses will be adjusted for a 7-day maximum duration of load.

a. Design values: Allowable stresses and E' will be the same as for the studs. Properties for the wales will be twice those for the studs because the wales are doubled. Thus, the properties for (2) 2 x 4 are:

$$A = 10.5 \text{ in.}^2$$

$$I = 10.72 \text{ in.}^4$$

$$S = 6.12 \text{ in.}^3$$

b. Loading: Each wale will support a strip of wall form that has a height equal to the spacing of the wales:

$$w = \frac{24}{12}(600) = 1200 \text{ lb/ft}$$

c. The maximum tie spacing based on bending is

$$\ell = 10.95\sqrt{\frac{F_b'S}{w}}$$

$$= 10.95\sqrt{\frac{1688(6.12)}{1200}} = 32.1 \text{ in.}$$

d. The maximum tie spacing based on shear is

$$\ell = \frac{13.3F'_v A}{w} + 2d$$

$$= \frac{13.3(225)(10.5)}{1200} + 2(3.5) = 33.2 \text{ in.}$$

e. The maximum tie spacing based on deflection is

$$\ell = 1.69\sqrt[3]{\frac{E'I}{w}}$$

$$= 1.69\sqrt[3]{\frac{1{,}600{,}000(10.72)}{1200}} = 41.0 \text{ in.}$$

Bending governs. A modular spacing of 24 in. would be desirable. Use tie spacing of 24 in. o.c.

FIGURE 12.6 Wall forms and bracing. (George Limbrunner)

4. **Ties.** Check the load on the ties (P_{tie}) with the capacity of the ties. Assume the tie capacity to be 3000 lb (ties of various capacities are widely available). Also assume that the ties have $1\frac{1}{2}$-in. wedges bearing on the wales. Then

$$P_{tie} = (\text{wale spacing}) \times (\text{tie spacing}) \times (\text{pressure})$$

$$= \frac{24}{12}\left(\frac{24}{12}\right)(600) = 2400 \text{ lb}$$

$$2400 \text{ lb} < 3000 \text{ lb}$$

Therefore, the capacity of the tie is satisfactory.

5. Check bearing stresses.

a. Where tie wedges bear on wales (wedges are $1\frac{1}{2}$ in. wide),

$$P_{\text{tie}} = 2400 \text{ lb}$$

$$\text{bearing contact area} = (2)(1.5)(1.5) = 4.5 \text{ in.}^2$$

$$\text{bearing stress (actual)} = \frac{2400}{4.5} = 533 \text{ psi}$$

The allowable compressive stress perpendicular to the grain is $F'_{c\perp} = 625$ psi (neglect bearing area adjustment factor). Thus,

$$533 \text{ psi} < 625 \text{ psi} \qquad \text{(O.K.)}$$

b. Where studs bear on wales (double wales),

$$\text{bearing contact area} = (2)(1.5)(1.5) = 4.5 \text{ in.}^2$$

The load on the wale from the stud is

$$P = (\text{load/ft on stud}) \times (\text{wale spacing})$$

$$= 600\left(\frac{24}{12}\right) = 1200 \text{ lb}$$

$$\text{actual bearing stress} = \frac{1200}{4.5} = 267 \text{ psi}$$

As in part (a), $F'_{c\perp} = 625$ psi. Thus,

$$267 \text{ psi} < 625 \text{ psi} \qquad \text{(O.K.)}$$

6. Lateral bracing: These should be designed for wall forms based on the greater of wind load (using 15 psf as a minimum) or 100 lb/ft applied at the top of the wall. Calculate wind load on wall forms using the minimum 15 psf:

$$\text{wind load} = (15 \text{ psf})(1 \text{ ft})(8 \text{ ft}) = 120 \text{ lb per ft of wall}$$

This load would be considered to act at midheight of the wall, 4 ft above the base, and would create an overturning moment about the base of

$$M_{OT} = 120 \text{ lb/ft}(4 \text{ ft}) = 480 \text{ ft-lb} \quad \text{(per ft of wall)}$$

The equivalent force, acting at the top of the wall, that would create the same overturning moment is

$$\text{force} = \frac{480 \text{ ft-lb}}{8 \text{ ft}} = 60 \text{ lb} \quad \text{(per ft of wall)}$$

$$60 \text{ lb/ft} < 100 \text{ lb/ft}$$

Therefore, use 100 lb/ft. This load, assumed to act at the top of the wall, can act in either direction. If *guy wires* are used, they must be placed on *both* sides of the wall. If *wooden strut bracing* is used, it can resist tension or compression and therefore *single-side* bracing may be used.

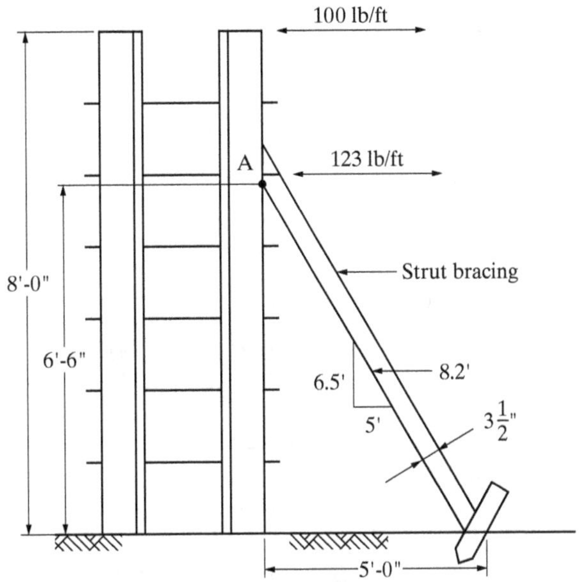

FIGURE 12.7 Lateral bracing for wall form.

In this example, use single-side strut bracing as shown in Figure 12.7, and design for compression. The horizontal load H on the strut at point A is calculated by considering moment taken at the base of the wall:

$$H(6.5) = 100(8)$$
$$H = 123 \text{ lb}$$

The force F in the strut, using the slope triangle shown, is determined from

$$\frac{F}{8.2} = \frac{123}{5}$$

Therefore, F = 202 lb (per foot of wall).

Use double 2 × 4 (S4S) lumber for the strut and compute the capacity as a compression member. (This will be adequate for tension also.) For No. 2 grade Douglas fir–larch:

a. The base design value for compression parallel to the grain from Table 12.4 is F_c = 1350 psi.

b. Adjustment factors:
   1. Size factor from Table 12.7: C_F = 1.15.
   2. For the column stability factor C_P, initially the following items must be established:
      a. For modulus of elasticity, there is no adjustment factor:

$$E' = E = 1{,}600{,}000 \text{ psi}$$

      b. Find F_c^*:

$$F_c^* = F_c C_F = 1350(1.15) = 1553 \text{ psi}$$

c. Assume that the ends are pin connected. Therefore, $K_e = 1.0$ and $\ell_e = 8.2$ ft. Then

$$\frac{\ell_e}{d} = \frac{8.2(12)}{3.0} = 32.8 < 50 \qquad \text{(O.K.)}$$

d. Solve for α:

$$\alpha = \frac{0.3E'}{\left(\frac{\ell_e}{d}\right)^2 F_c^*}$$

$$= \frac{0.3(1{,}600{,}000)}{32.8^2(1553)}$$

$$= 0.287$$

Solve for C_P:

$$C_P = \frac{1+\alpha}{1.6} - \sqrt{\left(\frac{1+\alpha}{1.6}\right)^2 - \frac{\alpha}{0.8}}$$

$$= \frac{1+0.287}{1.6} - \sqrt{\left(\frac{1+0.287}{1.6}\right)^2 - \frac{0.287}{0.8}}$$

$$= 0.267$$

c. Compute the allowable stress F_c':

$$F_c' = F_c C_F C_P = 1350(1.15)(0.267) = 415 \text{ psi}$$

Therefore, the allowable load is

$$P = F_c' A = 415(2)(5.25) = 4360 \text{ lb}$$

The maximum allowable strut spacing is calculated from

$$\frac{4360 \text{ lb}}{202 \text{ lb/ft}} = 21.6 \text{ ft}$$

Use struts at 21 ft-0 in. on center.

Single 2×4 struts could have been used if an intermediate brace were used to reduce the unbraced length.

12.9 FORMS FOR COLUMNS

Concrete columns are usually one of five shapes: square, rectangular, L-shaped, octagonal, or round. Forms for the first four shapes are generally made of sheathing, consisting of vertical planks or plywood, with wood yokes and steel bolts, patented steel clamps, or steel bands used to resist the concrete pressure acting on the sheathing. Forms for round columns may be wood, steel plate, or patented fiber tubes.

Because forms for columns are usually filled rapidly, frequently in less than 60 min, the pressure on the sheathing will be high, especially for tall columns. The ACI recommendations for lateral concrete pressure in column forms were discussed in Section 12.4. ACI 347R-14 (Equation [4.2.2.1a(b)]) is applicable for concrete having a slump of 7 in. or less and placed with normal internal vibration to a depth of 4 ft or less. However, assuming normal-weight concrete with a unit weight of 150 pcf, the pressure should not be taken as greater than $150h$ (psf), where h is the depth in feet below the upper surface of the freshly placed concrete. Thus, the *maximum* pressure at the bottom of a form 10-ft high should be taken as $150(10) = 1500$ psf regardless of the rate of filling the form or concrete temperature. It is suggested that the pressure be conservatively calculated using the equation

$$p = 150h$$

Figure 12.8a illustrates typical construction of a column form using plywood sheathing backed by vertical stiffening members and clamped with adjustable metal column clamps. The sheathing must be selected to span between the vertical stiffening members using the concrete pressure that exists at the bottom of the column form. The vertical stiffening members must span between the column clamps, the spacing of which can be increased as the pressure decreases toward the top of the form.

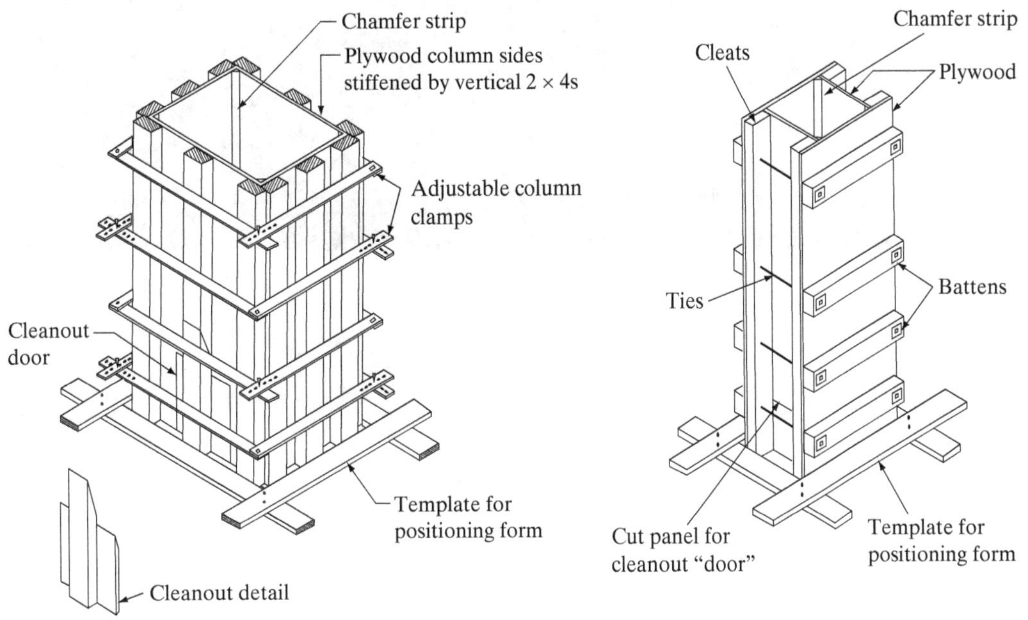

Typical Construction for Larger Column Forms
(a)

Typical Construction for Smaller Column Forms
(b)

FIGURE 12.8 Typical column forms. (*Source:* ACI 347R-14, courtesy of the American Concrete Institute)

Figure 12.8b illustrates a method suitable for forming smaller columns where no vertical stiffening members are required, and the plywood sheathing is backed directly by battens that are part of a wood and bolt column yoke. Column clamps can be used in this situation as well. If the thickness of the sheathing is selected, the design consists of determining the maximum safe spacing of the column clamps considering the pressure from the concrete as well as the permissible deflection, allowable bending stress, and allowable shearing stress.

Figure 12.9 shows the formwork for columns and walls in a multi-story building. Note the proprietary slab formwork support system in Figure 12.9 that is supported by the already cast concrete columns below, thus obviating the need for post shores.

FIGURE 12.9 Formwork for concrete columns in a multi-story building (Abi Aghayere)

TABLE 12.8 Safe Span in Inches for Class I Plyform, Continuous over Four or More Supports

	Stress parallel to gain				Stress perpendicular to gain			
Pressure (psf)	**1/2 in.**	**5/8 in.**	**3/4 in.**	**1 in.**	**1/2 in.**	**5/8 in.**	**3/4 in.**	**1 in.**
75	20	24	26	32	14	16	21	28
100	18	22	24	30	12	14	19	26
125	17	20	23	28	12	13	18	25
150	16	19	22	27	11	13	17	24
175	15	18	21	26	10	12	16	23
200	15	17	20	25	10	11	15	22
300	13	15	17	22	9	10	13	19
400	12	14	16	20	8	9	12	18
500	11	13	15	18	7	8	11	16
600	10	12	13	17	7	8	11	15
700	9	11	12	16	6	8	10	14
800	9	10	11	15	6	7	9	14
900	8	10	11	14	5	6	8	13
1000	8	9	10	13	5	6	7	12
1100	8	9	10	12	5	5	7	11
1200	7	8	9	12	4	5	6	10
1300	7	8	9	11	4	5	6	10
1400	6	7	9	11	4	4	6	9
1500	6	7	8	11	4	4	5	9
1600	6	6	8	10	—	4	5	8
1700	5	6	8	10	—	4	5	8
1800	5	6	7	9	—	4	5	8
1900	5	6	7	9	—	4	5	7
2000	5	5	7	9	—	—	4	7
2200	4	5	6	8	—	—	4	7
2400	4	5	6	8	—	—	4	6
2600	4	4	5	7	—	—	4	6
2800	4	4	5	7	—	—	4	6
3000	—	4	5	6	—	—	—	5

Notes: F_b = 1930 psi; F_v = 72 psi; E = 1,650,000 psi; allowable deflection = span/360 but not greater than $\frac{1}{16}$ in. Safe spans less than 4 in. are not shown. Tabulated spans are rounded to the nearest inch.

The sheathing span length ℓ may be calculated for moment, shear, and deflection, with the shortest of these span lengths being the controlling value.

Because the pressure against the forms varies with height, however, the determination of the optimum clamp spacing becomes laborious. As a result, tables have been developed that aid in the quick determination of support (clamp) spacing. Table 12.8 is an example of one such table specifically set up for plywood sheathing. The tabular values are based on the assumption that the lateral pressure is uniform between clamps and of an intensity equal to that at the lower clamp.

Clamps must also be investigated to determine if they can resist the applied loads. The manufacturer usually has recommended capacities for steel column clamps. Wood-yoke-type clamps with tie rods must be designed.

Generally, the type of forming system is a function of column size and height. As column size increases, either the thickness of the sheathing must be increased, or vertical stiffeners must be added to prevent sheathing deflection. If vertical supports or stiffeners are used (see Figure 12.8) in combination with plywood sheathing, the sheathing should span between the vertical supports, and the plywood face grain should be horizontal (in the direction of the span) for maximum strength. The clamp spacing is then a function of the vertical support member strength. If plywood sheathing spans between clamps (without vertical supports), the face grain should be vertical (in the direction of the span) for maximum strength.

EXAMPLE 12.4

Determine a clamp spacing pattern for column form sheathing made up of $\frac{3}{4}$-in.-thick plywood. The column height is to be 12 ft-0 in., and nomal weight concrete with a unit weight of 150 pcf is used. Assume the sheathing continuous over four or more supports and its face grain parallel to the span (vertical). Use class I plyform design values of

$$F_b = 1930 \text{ psi}$$
$$F_v = 72 \text{ psi}$$
$$E = 1{,}650{,}000 \text{ psi}$$

with allowable deflection of span/360 but not greater than $\frac{1}{16}$ in.

Solution:

1. Table 12.8 is used to determine the maximum span of the plywood between clamps. This depends on the pressure on the form, which is determined from

$$p = wh$$

where

w = unit weight of concrete (pcf)

h = depth of fresh concrete (ft)

2. Denoting the vertical distance from the bottom of the form as y (ft), the pressure is determined from

$$p = wh = 150(12 - y)$$

For some arbitrary values of y, the calculated pressures and the maximum spans (clamp spacings) from Table 12.8 are shown in Table 12.9.

3. A plot of maximum clamp spacing as a function of distance above top-of-footing is shown in Figure 12.10. The final clamp layout, also shown in Figure 12.10, is determined by trial and error. This procedure is similar to stirrup design (Chapter 4). One should attempt to minimize the number of clamps without having too many different-size spacings.

TABLE 12.9 Clamp Spacing Determination

y[a] (ft)	Pressure (psf)	Maximum clamp spacing (in.)[b]
0	1800	7
3	1350	9
6	900	11
9	450	15
11.5	75	26

[a]Quantity y is measured upward from the bottom of the form.
[b]From Table 12.8.

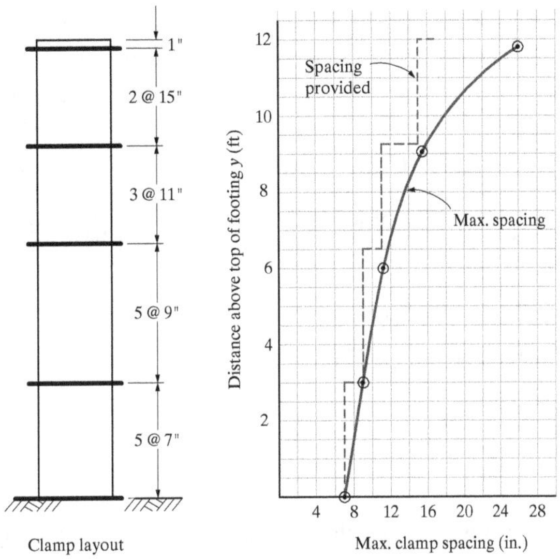

FIGURE 12.10 Sketch for Example 12.4.

EXAMPLE 12.5

A bridge column under construction collapsed during a concrete pour [6]. You have been asked to calculate the maximum pressures caused by the wet concrete and to investigate the possible cause(s) of the collapse of the concrete column steel forms given the following parameters:

- Column size: 12 ft x 6 ft x 39 ft high
- The height of placement of the concrete at the time of collapse was 35 ft.
- Normal-weight concrete with a density of 145 pcf
- Slump = 8 inches
- Concrete temperature, T = 90°F
- Rate of concrete placement, R = 11.33 ft per hour
- Fly ash and retarder was used in the concrete mix, so Cc = 1.4
- Cw = 1.0
- The maximum allowable hydrostatic pressure in the proprietary steel formwork for the column = 1000 psf, with a factor of safety of 2.7 [6].

Solution:

h = depth of fresh concrete (ft) at time of collapse = 35 ft

w = unit weight of concrete (pcf) = 145 pcf

(a) With the slump greater than 7 inches, the equations from Section 12.4 based on ACI 347 requires the concrete pressure to be equal to

$$wh = (145 \text{ pcf})(35 \text{ ft}) = 5075 \text{ psf}$$

This pressure is more than five times the steel form manufacturer's prescribed allowable pressure of 1000 psf.

(b) Let's consider the scenario where the slump effect is neglected, and determine if the pressures would have been adequate: This column does not fully qualify as a column since the 12 ft dimension is greater than 6.5 ft. If the column is considered as a wall because of the 12 ft dimension, the concrete pressure equation for walls with a rate of concrete placement of 7 to 15 ft/h (i.e., Case 3 from Section 12.4), we obtain the concrete pressures as

$$p_{\max} = C_W C_C\left[150+\frac{43,400}{T}+\frac{2800R}{T}\right]=(1.0)(1.4)\left[150+\frac{43,400}{90}+\frac{2800(11.33)}{90}\right]=1379\,\text{psf}$$

$$\geq 600C_W = 600(1) = 600\,\text{psf}$$

$$\leq wh = (145\,\text{pcf})(35\,\text{ft}) = 5,075\,\text{psf}$$

Therefore, $p_{\max}$ = 1379 psf

Conclusions:

All the lateral pressures calculated considering slump (5,075 psf) and excluding slump (1379 psf) exceed the steel form manufacturer's prescribed allowable pressure of 1000 psf, therefore, the wet concrete lateral pressures was definitely a contributing factor to the collapse of the steel forms. It was reported that there "was very little concrete adhering to the reinforcing bars, indicating the highly fluid state of concrete up to the time of the failure" [6], in which case the maximum pressure would have been more likely to be closer to the upper limit of 5,075 psf. This pressure was more than five times the manufacturer's prescribed allowable pressure for the concrete column steel forms, and more than twice the allowable ultimate hydrostatic pressure of 2.7 x 1000 psf or 2700 psf. Therefore, the excessive lateral pressures in the steel forms was the cause of the steel formwork collapse, which unfortunately led to the loss of two lives [6].

Proprietary Formwork Support and Shoring Systems.

For multi-story concrete buildings, the joists, stringers and shores supporting the plywood forms are often proprietary systems (see Figure 12.11), and the manufacturers of these systems typically provide load tables to aid in their design and selection. Figure 12.12 shows proprietary post re-shores supporting already hardened concrete slab that in turn supports the weight of wet concrete from the concrete pours of the upper level slabs in a multi-story building.

FIGURE 12.11 Plywood forms supported by proprietary shoring systems (Abi Aghayere)

FIGURE 12.12 Proprietary post re-shores (Abi Aghayere)

REFERENCES

[1] *Guide to Formwork for Concrete* (ACI PRC-347R-14(21)). American Concrete Institute, 38800 Country Club Drive, Farmington Hills, MI 48331.

[2] *Plywood Design Specification*. APA—The Engineered Wood Association, P.O. Box 11700, Tacoma, WA 98411-0700, January 1997.

[3] *Concrete Forming*. APA—The Engineered Wood Association, P.O. Box 11700, Tacoma, WA 98411-0700, 2012.

[4] *National Design Specification for Wood Construction* (ANSI/AF&PA NDS-2018), with supplements and commentary. American Wood Council, Leesburg, VA, 2018.

[5] David W. Johnston and ACI Committee 347, *Formwork for Concrete*, SP-4, 8th ed. American Concrete Institute, 38800 Country Club Drive, Farmington Hills, MI 48331, 2014

[6] M. Ayub, "Investigation of the June 14, 2007, Incident at U.S. Highway 90 across St. Louis Bay, Pass Christian, MS," U.S. Department of Labor, Occupational Safety and Health Administration, Directorate of Construction, December 2007

EXERCISES

Where applicable in these exercises, and unless otherwise noted, *assume:*

1. The vertical live load is to be 75 psf (motorized carts).
2. Bending members are continuous over three or more spans and are assumed to have adequate lateral support.
3. There is a 7-day duration of load.
4. Lumber is No. 2 grade hem–fir.
5. The forming weight is 5 psf; neglect sheathing weight.
6. The maximum deflection is $\frac{1}{360}$ of the span.
7. All lumber is (S4S).
8. Plywood is placed with the face grain perpendicular to the supports.
9. No available information exists with respect to wood splits, checks, and shakes. Assume $C_i = 1.0$, $C_C = 1.0$, and $C_W = 1.0$.

12.1 Using basic principles, derive the equations of Table 12.1 for allowable span length ℓ as governed by moment, shear, and deflection. Handbooks (such as the AISC *Manual*) may be helpful for moment, shear, and deflection equations. Show load diagrams.

a. One span

b. Two spans

c. Three spans

12.2 A slab form is to be built for an 8-in.-thick concrete slab. The plywood sheathing face grain is perpendicular to the joists. The maximum allowed deflection is to be the smaller of $\frac{1}{360}$ of the span or $\frac{1}{8}$ in. Compare the maximum allowable joist spacing for class II plyform $\frac{1}{2}$ in. thick and 1 in. thick. Draw a sectional view showing slab, sheathing, and joists.

12.3 A 6-in.-thick concrete slab is to be formed using plywood supported on 2 × 8 (S4S) joists that are spaced 1 ft-6 in. on center. Find the maximum allowable stringer spacing if maximum allowable deflection is 1/240 of the span. Draw a sketch of the formwork.

12.4 For the slab of Exercise 12.3, assume that double 2 × 10 stringers spaced 5 ft-0 in. o.c. will support the joists. Determine the maximum shore spacing. Draw a sketch of the formwork. (Neglect shore capacity.)

12.5 In Exercise 12.4, if the shores were to be spaced 6 ft-0 in. o.c., select a new stringer size. The stringer may be either doubled 2-in.-thick planks or a single wood beam.

12.6 Compare the capacities of 6 ft-0 in. and 12 ft-0 in. 4 × 4 (S4S) wooden shores. What are the capacities if the shores are full nominal size (4 in. × 4 in.)?

12.7 Design a soffit (beam bottom) for an $11\frac{1}{4}$ in.×24 in. reinforced concrete beam form. Use a 2 × 12. Determine the maximum shore spacing. Draw a sectional view through the beam.

12.8 Design a plywood beam bottom form shown as an alternative in Figure 12.4. The beam will be 14 in. × 24 in. First design the class I plyform to span the clear span between 2 × 4s (single span, maximum deflection of $\frac{1}{16}$ in.). Then determine maximum shore spacing assuming that the weight of the beam is supported by the 2 × 4s, which back the beam bottom.

12.9 In Exercise 12.8, replace the 2 × 4s with 2 × 6s and determine the new maximum shore spacing.

12.10 A 12-ft-high concrete wall is to be placed at a rate of 6 ft/h, and the temperature is expected to be 90°F. What is the *maximum* lateral pressure due to fresh concrete for which the wall forms must be designed? Draw a diagram showing lateral pressure versus distance from the bottom of the wall for the full 12-ft height of wall.

12.11 Design the formwork for an 8-in.-thick reinforced concrete floor slab. Use $\frac{3}{4}$ in. class II plyform and No. 2 grade Douglas fir–larch for the other lumber. Use 2 × 8 joists and double 2 × 8 stringers. Use 4 × 4 shores with an unsupported height of 10 ft. Assume that the slab will be 120 ft × 120 ft in plan. Use guy wire bracing (at 45° to the 10-ft height) that has a tensile capacity of 4400 lb.

12.12 Design forms to support reinforced concrete beams as shown in Figure 12.4a with the following changes. The beam will be $11\frac{1}{4}$ in.×22 in., joists are 1 ft-8 in. o.c., and stringers are spaced 7 ft-0 in. o.c. Shores will be full nominal size 4 × 4s with an unsupported height of 10 ft. All lumber will be No. 2 grade Southern pine.

12.13 Design the formwork for a 12-ft-high reinforced concrete wall. Concrete will be placed at a rate not to exceed 5 ft/h and will be internally vibrated. Temperature is expected to be 80°F. Use $\frac{3}{4}$-in. class I plyform for the sheathing and No. 2 grade Douglas fir–larch for the rest of the lumber. Use 2 × 4s for studs and doubled 2 × 6s for wales. Ties will have a capacity of 5000 lb (2-in.-wide wedges). For lateral bracing design, assume the wind to be 15 psf and use a guy wire (at 45° to the 10-ft height) that has a tensile capacity of 4400 lb.

12.14 Determine a clamp spacing pattern for a column form that has sheathing of 1-in.-thick plywood. The column height is to be 14 ft-0 in. Use Table 12.8. Assume that F_b = 1930 psi, F_v = 72 psi, and E = 1,650,000 and that the face grain is parallel to the span between clamps.

12.15 Assuming normal-weight concrete, calculate the maximum lateral pressure in psf on the formwork for a 24 in. x 24 in. x 15 ft high concrete column due to the wet concrete assuming the following parameters:

- The rate of concrete placement is 11 ft./h
- The concrete slump is less than 7 inches
- The concrete temperature is 80°F
- Assume the unit weight coefficient, C_W = 1.0
- Assume the chemistry coefficient, C_C = 1.4

CHAPTER 13

Detailing of Reinforced Concrete Structures

13.1 INTRODUCTION

The contract documents package for a typical building as developed by an architect/engineer's office commonly includes both construction drawings and specifications. The construction drawings typically cover the following areas: site, architectural, structural, mechanical, electrical, and plumbing. The structural drawings, in particular, include the general notes, foundation plan, structural floor and roof plans, schedules, sections, and typical details. The specifications supplement and amplify the drawings. The contract documents package is the product that results from what may be categorized as the planning and design phase of a project.

The next sequential phase may be categorized as the construction phase. It includes many subcategories for reinforced concrete structures, two of which are detailing and fabricating of the reinforcing steel. As described in the *ACI Detailing Manual-2020* [MNL-66(20)] [1], detailing consists of the preparation of placing drawings, reinforcing bar details, and bar lists that are used for the fabrication and placement of the reinforcement in a structure. Fabricating consists of the actual shopwork required for the reinforcing steel, such as cutting, bending, bundling, and tagging.

Most bar fabricators not only supply the reinforcing steel but also prepare the placing drawings and bar lists, fabricate the bars, and deliver them to the project site. In some cases, the bar fabricator may also act as the placing subcontractor.

It is general practice in the United States for all reinforced concrete used in building projects to be designed, detailed, and fabricated in accordance with the latest ACI Code. In addition, the Concrete Reinforcing Steel Institute (CRSI) regularly publishes its *Manual of Standard Practice* [2], which contains the latest recommendations of the reinforcing steel industry for standardization of materials and practices.

Techniques have also been developed that make use of electronic computers and other data-processing equipment to facilitate the generation of bar lists and other components of the detailing process. This not only aids in standardization and accuracy of the documents produced but also can be readily incorporated into the stock control system and the shopwork planning of the reinforcing steel fabricator.

FIGURE 13.1 Reactor containment foundation mat. Seabrook Station, New Hampshire. (George Limbrunner).

13.2 PLACING OR SHOP DRAWINGS

The placing drawing (commonly called a shop drawing) consists of a plan view with sufficient sections to clarify and define bar placement. As such, it is the guide that the ironworkers will use as they place the reinforcing steel on the job. In addition, the placing drawing will contain typical views of beams, girders, joists, columns, and other members as necessary. Frequently, tabulations called "schedules" are used to list similar members, which vary in size, shape, and reinforcement details. A bar list, bending details, or both may or may not be shown on the placing drawing, as some fabricators not only have their own preferred format but prefer the list to be prepared as a separate entity.

The preparation of the placing drawing is based on the complete set of contract documents and generally contains only the information necessary for bar fabrication and placing. Building dimensions are not shown unless they are necessary to locate the steel properly.

The structural drawing Figure 13.2, which is a part of the structural plans of the contract documents, is the drawing on which the placing drawing is based. Figure 13.3 shows a placing drawing of the same system shown in Figure 13.2. This building is an example of a framing system that uses girders between columns to support beams, which in turn support one-way slabs. The girders support only two-thirds of the beams; the remainder frame directly into the columns.

The structural drawing, as may be observed, shows the floor plan view locating and identifying the structural elements, along with views of typical beams, girders, and slabs and

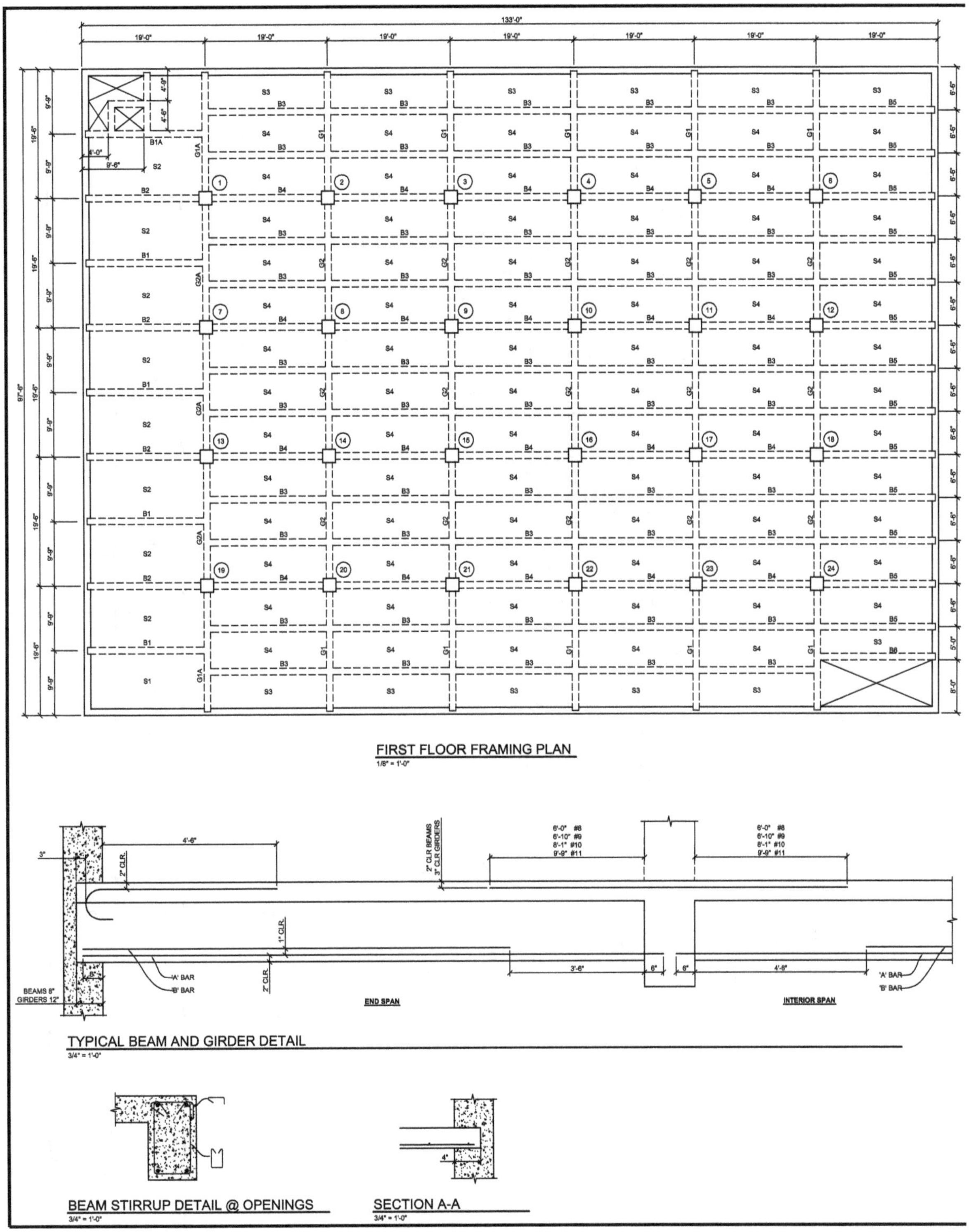

FIGURE 13.2 Structural drawing for beam and girder framing. (ACI Detailing Manual (SP66-04), Courtesy of the American Concrete Institute)

BEAM AND GIRDER SCHEDULE

MARK	SIZE		BOTTOM		TOP	STIRRUPS	
	B	D	'A' BARS	'B' BARS		NO. - SIZE	SPACING FROM FACE OF SUPPORT
G1	14	32	3 - #9	3 - #8 ★	3 - #9 @ NON CONT. ENDS	22 - #3	1@2, 3@6, 3@9, 2@10, 2@12
G1A	SAME AS G1 U.O.N., ADD 3 - #4 E.F.				3 - #9, 3 - #8 } @ COLS 1 TO 6 & 19 TO 24 (2 LAYERS, #9 BARS IN TOP LAYER	26 - #4	1@2, 3@6, 3@5, BAL @10
G2	14	32	3 - #8	2 - #6 ★	3 - #11 AT COLS 7 TO 18	16 - #3	1@2, 7@12
G2A	SAME AS G2 U.O.N., ADD 3 - #4 E.F.					22 - #4	1@2, 10@10
B1	12	22	3 - #9	3 - #7 ★	2 - #8 NON CONT. END 4 - #9 CONT. END	16 - #3	1@2, 2@5, 5@6
B2	12	22	3 - #7	2 - #6 ★	2 - #8 NON CONT. END 4 - #9 CONT. END	16 - #3	1@2, 2@5, 5@6
B3	10	22	2 - #7	2 - #6	2 - #7 EACH END	10 - #3	1@2, 4@6
B4	10	22	2 - #6	2 - #6	2 - #10 EACH END EXCEPT @ END CONT. WITH B2, B3, B5 (SEE B2, B3, B5 FOR THIS REINF.)	10 - #3	1@2, 4@6
B5	10	22	2 - #7	2 - #7 ★	2 - #11 CONT. END 2 - #8 NON-CONT. END	10 - #3	1@2, 4@6
B6	10	16	2 - #7	2 - #7	2 - #6 CONT	10 - #3	1@2, 4@6
B7	10	18	2 - #5		2 - #6 CONT	10 - #3	1@2, 4@7
B8	8	8	2 - #5			NONE	
L1	8	12	2 - #5			NONE	

★ UPPER LEVEL

SLAB SCHEDULE

MARK	DEPTH	REINFORCING		
		BOTTOM	TOP	TEMP.
S1	5"	#4@ 6 1/2"	#4@12" NON CONT., END #4@7" CONT., END	#3@11"
S2	5"	#4@ 9 1/2"	#4@7" CONT., END	#3@11"
S3	4"	#3@12"	#3@12" NON CONT., END #4@12" CONT., END	#3@14"
S4	4"	#3@12"	#4@12" CONT., END	#3@14"

NOTES:

1. ALL CONCRETE WORK SHALL CONFORM TO ACI 318-11.
2. REINFORCING STEEL SHALL CONFORM TO ASTM A615, GRADE 60.
3. f'c = 4000psi AT 28-DAYS MAXIMUM AGGREGATE SIZE IS 3/4".
4. PLACE MAIN REINFORCING STEEL SO THAT BOTTOM OF STEEL IS 2" ABOVE FORMS IN BEAMS AND 3/4" IN SLABS.
5. WHERE BEAM OR GIRDER IS PARALLEL TO MAIN SLAB REINFORCING, PLACE #4x5'-0"@12" IN TOP OF SLAB OVER GIRDER AND AT RIGHT ANGLES TO SAID MEMBER.
6. STIRRUPS TO HAVE 2 - #3 SUPPORT BARS AS REQUIRED.
7. LINTELS TO BEAR 8" ON EACH SIDE OF OPENING.
8. PROVIDE 2 - #3 BARS TOP AND BOTTOM AT ALL OPENINGS AND EXTEND 1'-9" BEYOND OPENING.
9. PROVIDE #3 BARS AT RIGHT ANGLES TO MAIN REINFORCING STEEL AT OPENINGS IN SLAB AS SHOWN ON PLAN.
10. LAP ALL TEMPERATURE BARS 16".
11. PLACE E/W TOP GIRDER BARS AT COLUMNS FIRST BELOW N/S BEAM TOP BARS.
12. BAR SUPPORTS TO BE CLASS 3 (NO PROTECTION).
13. PROVIDE DETAILS IN ACCORDANCE WITH ACI-315 DETAILING MANUAL.

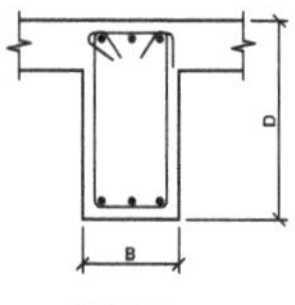

BEAM SECTION

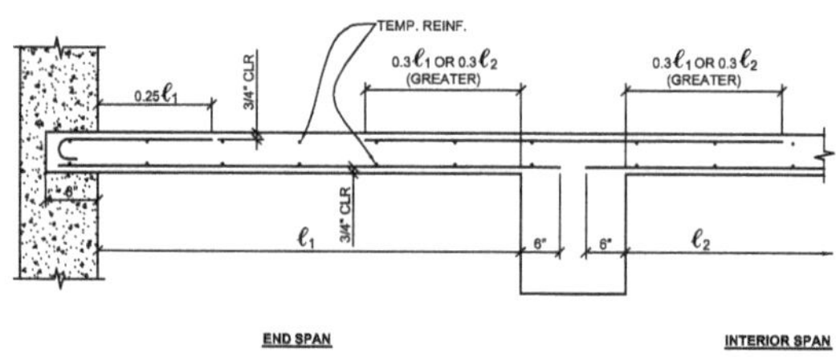

TYPICAL SLAB DETAIL

3/4" = 1'-0"

PROJECT: NEW OFFICE BUILDING

LOCATION:

TITLE: FIRST FLOOR FRAMING PLAN

ENGINEER:

REVISIONS

BY	DATE

DRAWN BY: CAD

DATE: 01-01-12

JOB. NO.: ##

DWG. NO.: S6

FIGURE 13.2 Continued

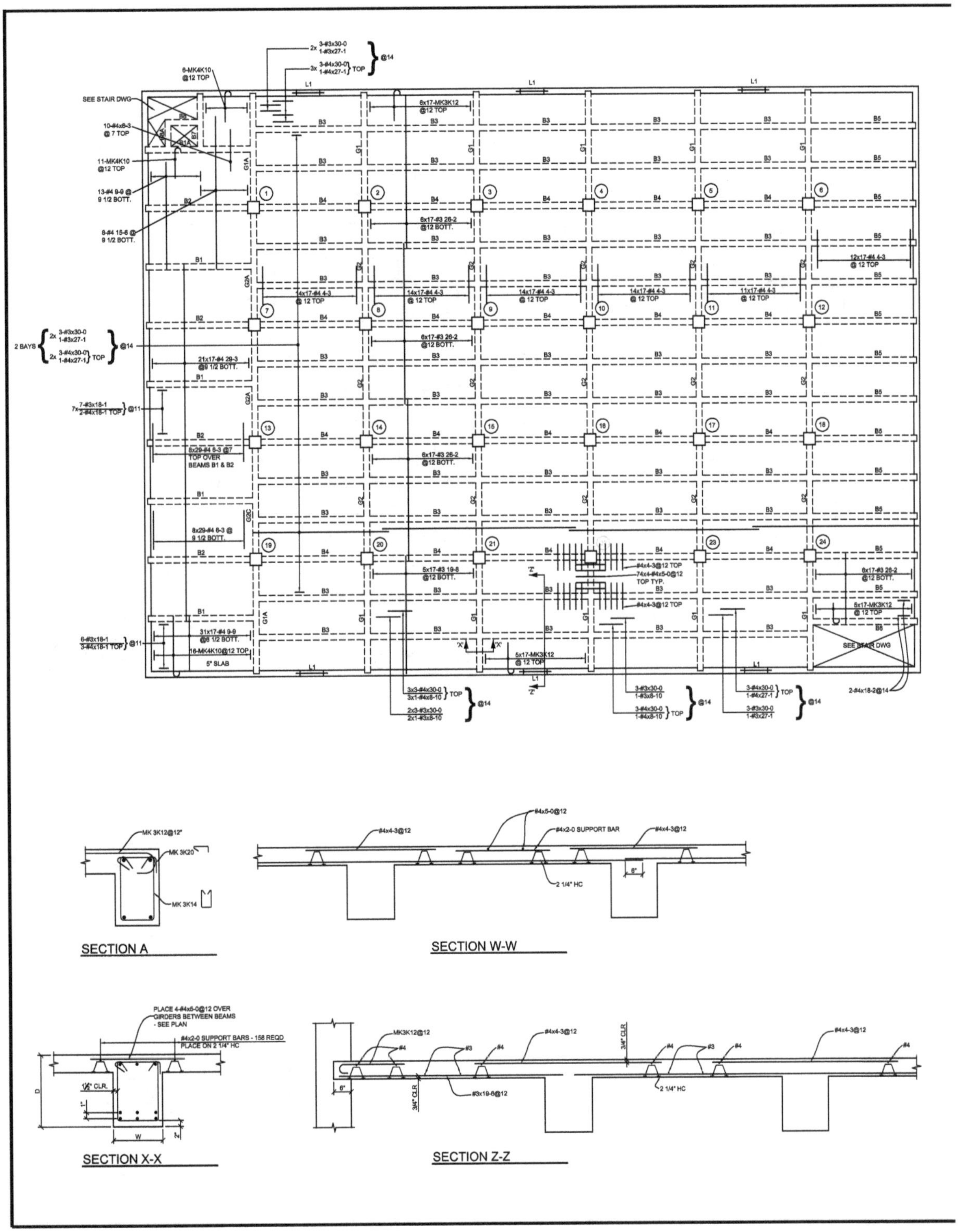

FIGURE 13.3 Placing drawing for beam and girder framing. (ACI Detailing Manual (SP66-04), Courtesy of the American Concrete Institute)

BENDING DETAILS														
MARK	SIZE	LGTH	TYPE	A	B	C	D	E	F	G	J		H	K
9K1	9	6-7	1	1-3	5-4						11 3/4			
3K2	3	6-3	S3	4	2-4	11	2-4			4				
3K3	3	1-7	T9	4	11					4				
10K4	10	10-9	3		8-1	6-3							4-5	4-5
8K6	8	5-11	1	11	5-0						8			
3K8	3	4-5	S3	4	1-7	7	1-7			4				
5K9	5	13-6	1	7	12-11						5			
4K10	4	3-2	1	6	2-8						4			
4K11	4	4-0	3		2-0	2-0							1-5	1-5
3K12	3	2-4	1	3	1-11						3			
3K13	3	3-5	S3	4	1-1	7	1-1			4				
3K14	3	3-9	S3	4	1-3	7	1-3			4				
3K15	3	4-7	S3	4	1-7	9	1-7			4				
4K16	4	4-0	1	6	3-6						4			
4K17	4	4-5	1	6	3-11						4			
4K18	4	2-7	1	6	2-1						4			
8K19	8	22-5	1	11	21-6						8			
3K20	3	1-3	T9	4	7					4				
4K21	4	6-4	S3	4 1/2	2-4	11	2-4			4 1/2				
4K22	4	1-8	T9	4 1/2	11					4 1/2				
3K23	3	1-5	T9	4	9					4				

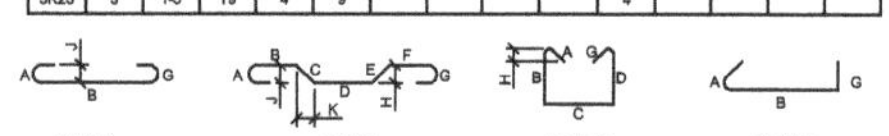

TYPE 1 TYPE 3 TYPE S3 TYPE T9

BAR SUPPORTS (CLASS 3)

490 PCS. 2" BEAM BOLSTERS x 5'-0" (BB)
63 PCS. 1" BEAM BOLSTERS UPPER x 5'-0" (BBU)
402 PCS. 3/4" SLAB BOLSTERS x 10'-0" (SB)
149 - 3 1/4" HC SPACE @ 3'-0" FOR 5" SLAB
1522 - 2 1/4" HC SPACE @ 3'-0" FOR 4" SLAB

NOTES:
ALL SLABS ARE 4" THICK UNLESS OTHERWISE SPECIFIED
MAIN SLAB BARS IN BOTTOM TO BE SUPPORTED IN EACH INTERIOR BAY BY TWO ROWS OF 3/4" SLAB BOLSTERS
MAIN SLAB BARS IN BOTTOM TO BE SUPPORTED IN EACH EXTERIOR BAY & 5" SLABS BY THREE ROWS OF 3/4" SLAB BOLSTERS
FOR BEAM & GIRDER DETAIL SEE ARCH'S DWG. 3 OF 12
SLAB BARS TO BE CUT IN FIELD AT OPENINGS
REINFORCING - ASTM A615 GRADE 60
SEE ACI 315 DETAILING MANUAL FOR TYPICAL BAR BENDS

BEAM AND GIRDER SCHEDULE									
NO.	MARK	SIZE		REINFORCING				STIRRUPS SPACE EACH END	STIRRUP SPACERS
				BOTTOM		TOP *			
		W	D	'A' BARS	'B' BARS	NON-CONT. END	CONT. END		
9	G1	14	32	3 - #9x19-0	3 - #8 x 16-3 * EXTEND INTO EXTERIOR SUPP.	3 - MK 9K1	3-#9x17-8 & 3-#8x12-0 COLS. 2 TO 6 & 20 TO 23 2 LAYERS, #9 BARS INTO TOP LAYER	22 MK 3K2 22 MK 3K3 (1@2", 3@6" 3@9", 2@10", 2@12")	2-#3x7-10 LAP 12" WITH TOP BARS
2	G1A	14	32	3 - #9x19-0	3 - #8 x 16-3 * EXTEND INTO EXTERIOR SUPP.	3 - MK 9K1	3-#9x17-8 & 3-#8x12-0 COLS. 1 & 19 2 LAYERS, #9 BARS INTO TOP LAYER	25 MK 4K21 25 MK 4K22 (1@2", 3@6", 3@9", BAL@10")	2-#4x7-10 TOP LAP 12" 4-#4x19-0 2 EF
1	G1B	14	32	3 - #9x19-0	3 - #8 x 16-3 EXTEND INTO EXTERIOR SUPP.	3 - MK 9K1 SEE PLAN	3-#9x17-8 & 3-#8x12-0 COL. 24 #9 BAR IN TOP LAYER FOR DOWELS TO STAIR	25 MK 4K21 25 MK 4K22 (1@2", 3@6", 3@9", BAL@10")	2-#4x7-10 TOP LAP 12" WITH TOP BARS 4-#4x19-0 2 EF
10	G2	14	32	3 - #8x19-0	2 - #6 x 13-6 * PLACE SYM.		3-#11x19-6 COLS. 8-12, 14-18	16 MK 3K2 16 MK 3K3 (1@2", 7@12")	
2	G2A	14	32	3 - #8x19-0	2 - #6 x 13-6 * PLACE SYM.		3-#11x19-6 COLS. 7 & 13	22 MK 4K21 22 MK 4K22 (1@2", 10@10")	2-#4x3-0 TOP 4-#4x19-0 2 EF
5	G2B	14	32	3 - #8x19-0	2 - #6 x 13-6 PLACE SYM.		SEE G1, G1B & G2 FOR TOP STEEL	16 MK 3K2 16 MK 3K3 (1@2", 7@12")	
1	G2C	14	32	3 - #8x19-0	2 - #6 x 13-6 PLACE SYM.		SEE G1A & G2A FOR TOP STEEL	22 MK 4K21 22 MK 4K22 (1@2", 10@10")	2-#4x3-0 TOP 4-#4x19-0 2 EF
5	B1	12	22	3 - #8x18-0	3 - #7 x 15-5 * EXTEND INTO EXTERIOR SUPP.	2 - MK 8K6	4 - #9x15-0	16 MK 3K15 16 MK 3K23 (1@2", 2@5", 5@8")	
4	B2	12	22	3 - #7x18-4	2 - #6 x 15-5 * EXTEND INTO EXTERIOR SUPP.	2 - MK 8K6	4 - #9x17-6	16 MK 3K15 16 MK 3K23 (1@2", 2@5", 5@6")	
10	B3	10	22	3 - #7x18-10	2 - #6 x 13-0 PLACE SYM.	2 - #10x14-4	2 - #10x16-2	10 MK 3K8 10 MK 3K20 (1@2", 4@6")	
1	B3A	10	22	2 - #7x18-10	2 - #6 x 15-10 EXTEND INTO G1B	2 - MK 10K4 @ STAIR	SEE B4	10 MK 3K8 10 MK 3K20 (1@2", 4@6")	
46	B4	10	22	2 - #6x18-10	2 - #6 x 13-0 PLACE SYM.		2 - #10x16-2 SEE B2, B3, B5	10 MK 3K8 10 MK 3K20 (1@2", 4@6")	
13	B4A	10	22	2 - #6x18-10	2 - #6 x 13-0 PLACE SYM.	SEE B3, B4, B5 FOR TOP STEEL		10 MK 3K8 10 MK 3K20 (1@2", 4@6")	
13	B5	10	22	2 - #7x18-4	2 - #7 x 15-5 * EXTEND INTO EXTERIOR SUPP.	2 - MK 8K6	2 - #11x19-6	10 MK 3K8 10 MK 3K20 (1@2", 4@6")	
1	B6	10	16	2 - #6x18-4	2 - #8 x 15-5 * EXTEND INTO EXTERIOR SUPP.	2 - MK 8K19 EXT. 3-8 PAST @ G1B		10 MK 3K8 10 MK 3K20 (1@2", 4@6")	
1	B7	10	18	2 - #5x9-9		2 - MK 5K9 CONT. EXT. 2-8 @ B1		10 MK 3K14 10 MK 3K20 (1@2", 4@7")	
1	B8	8	8	2 - #5 x 7-0					
1	B8A	8	8	2 - #5 x 5-4					
6	L1	8	12	2 - #5 x 5-0					

* NOTE - PLACE GIRDER TOP BARS FIRST UPPER LAYER

REINFORCING STEEL PLACING DRAWING

USE THIS DRAWING IN CONJUNCTION WITH THE ARCHITECTURAL & STRUCTURAL DRAWINGS

ELEVATIONS & DIMENSIONS SHOWN ON THIS DRAWING ARE FOR PURPOSES OF PLACING REINFORCING BARS ONLY AND ARE NOT TO BE USED FOR CONSTRUCTION UNLESS VERIFIED BY THE CONTRACTOR.

REV. NO.	DATE	DESCRIPTION

DWG.: FIRST FLOOR FRAMING PLAN		
JOB: NEW OFFICE BUILDING		
LOCATION:		
ENGINEER:		
CONTRACTOR:		
DATE:	DWG. BY:	FILE NO.:
	CHK. BY:	DWG NO.: P-6

FIGURE 13.3 Continued

their accompanying schedules. The placing drawing supplements the structural drawing by furnishing all the information necessary for bar fabrication and placement. The precise size, shape, dimensions, and location of each bar are furnished, using a marking system discussed later in this chapter. Information relative to bar supports may also be included. Figure 13.3 illustrates a placing drawing that includes placing data for bar supports, as well as a bending details schedule.

Placing drawings, in addition to controlling the placement of the steel in the forms, serves as the basis for ordering the steel. Therefore, a proper interpretation of the contract documents by the fabricator is absolutely essential. Generally, all placing drawings are submitted to the architect or engineer for checking and reviewing for conformance to the specifications and contract documents before shop fabrication begins.

13.3 MARKING SYSTEMS AND BAR MARKS

With respect to buildings, two identification systems are required. The first involves the identification of the various structural members, and the second involves the identification of the individual bars within the members. The marking system for the structural members may consist of an alphabetical-numerical identification for each beam, girder, and slab, with the columns designated numerically, as shown in Figures 13.2 and 13.3. Also used is a system of alphabetical and numerical coordinates in which the centerlines of columns are numbered consecutively in one direction and lettered consecutively in the other. A coordinate system may be observed in the foundation-engineering drawing of Figure 13.4, where a column may have a coordinate designation such as B2 or C3. The system is generally established on the architectural and structural drawings and adopted by the detailer unless the detailer requires a more precise identification system.

Footings, as may be observed in Figure 13.4, are generally designated with an F prefix followed by a number, such as F1 and F3, without regard to a coordinate system. Footing piers or pedestals may be identified using the coordinate system, such as B2 or D4, or may be designated with a P prefix followed by a number, such as P1 or P3. Beams, joists, girders, lintels, slabs, and walls are generally given designations that indicate the specific floor in the building, the type of member, and an identifying number. For example, 1G2 indicates a first-floor girder numbered 2, and RB4 indicates a roof beam numbered 4.

In some situations, the floor designation is omitted, as shown in Figures 13.2 and 13.3. The beams and girders are then designated with the prefix B or G, respectively, followed by a number (e.g., B4 or G2). In some cases, suffixes have been added, such as G2A, indicating that there is a difference in the member.

Along with a marking system established for the structural members, a system of identifying and marking the reinforcing bars must be established. In buildings, only bent bars are furnished with a mark number or designation. The straight bar has its own identification by virtue of its size and length. Numerous systems are in use throughout the industry, with the system choice generally a function of the building type, size, and complexity as well as the standards of each fabricator.

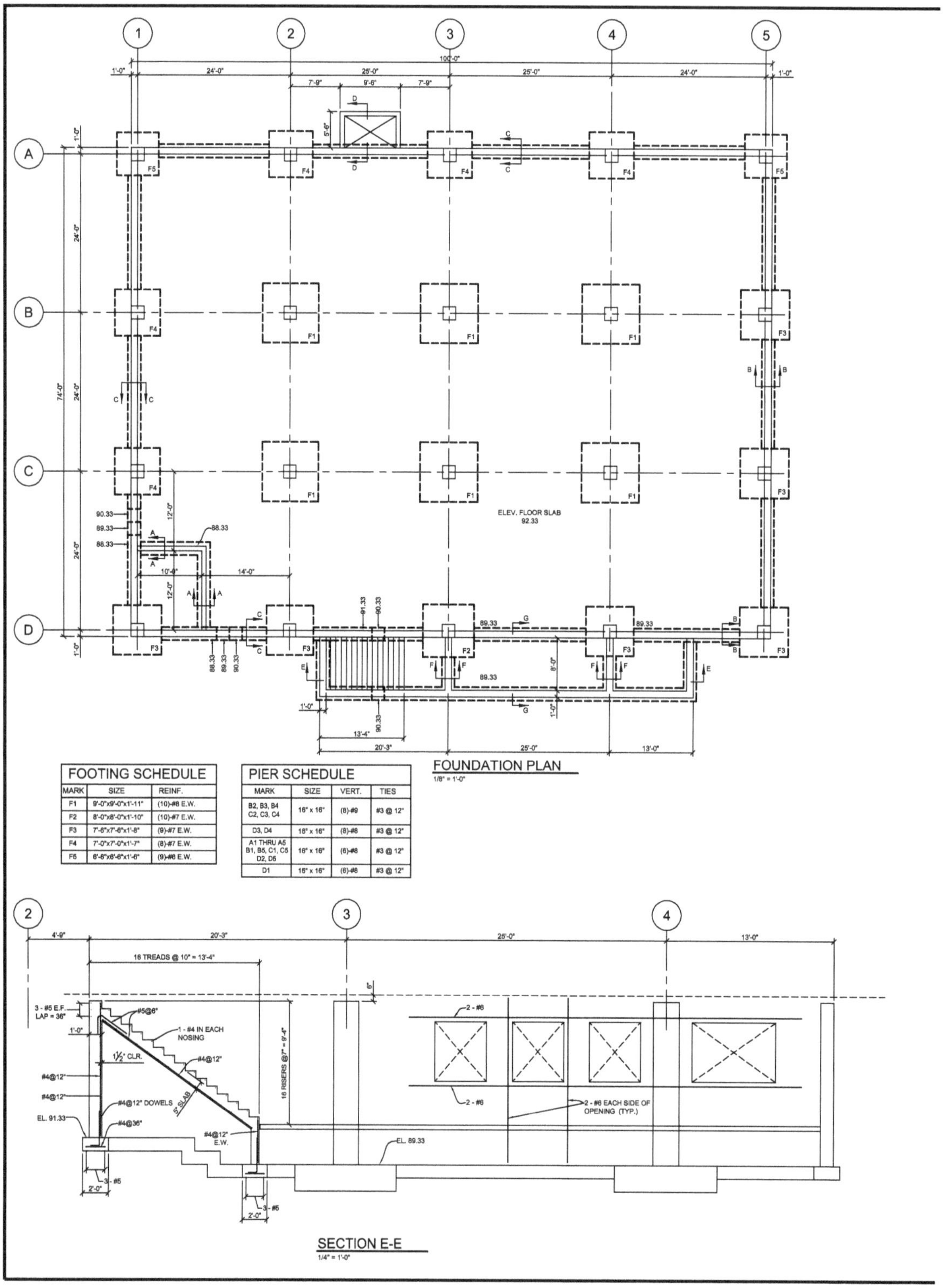

FOOTING SCHEDULE		
MARK	SIZE	REINF.
F1	9'-0"x9'-0"x1'-11"	(10)-#8 E.W.
F2	8'-0"x8'-0"x1'-10"	(10)-#7 E.W.
F3	7'-6"x7'-6"x1'-8"	(9)-#7 E.W.
F4	7'-0"x7'-0"x1'-7"	(8)-#7 E.W.
F5	6'-6"x6'-6"x1'-6"	(9)-#6 E.W.

PIER SCHEDULE			
MARK	SIZE	VERT.	TIES
B2, B3, B4 C2, C3, C4	16" x 16"	(8)-#9	#3 @ 12"
D3, D4	16" x 16"	(8)-#8	#3 @ 12"
A1 THRU A5 B1, B5, C1, C5 D2, D5	16" x 16"	(6)-#8	#3 @ 12"
D1	16" x 16"	(6)-#8	#3 @ 12"

FIGURE 13.4 Structural engineering drawing for foundations. (ACI Detailing Manual (SP66-04), Courtesy of the American Concrete Institute)

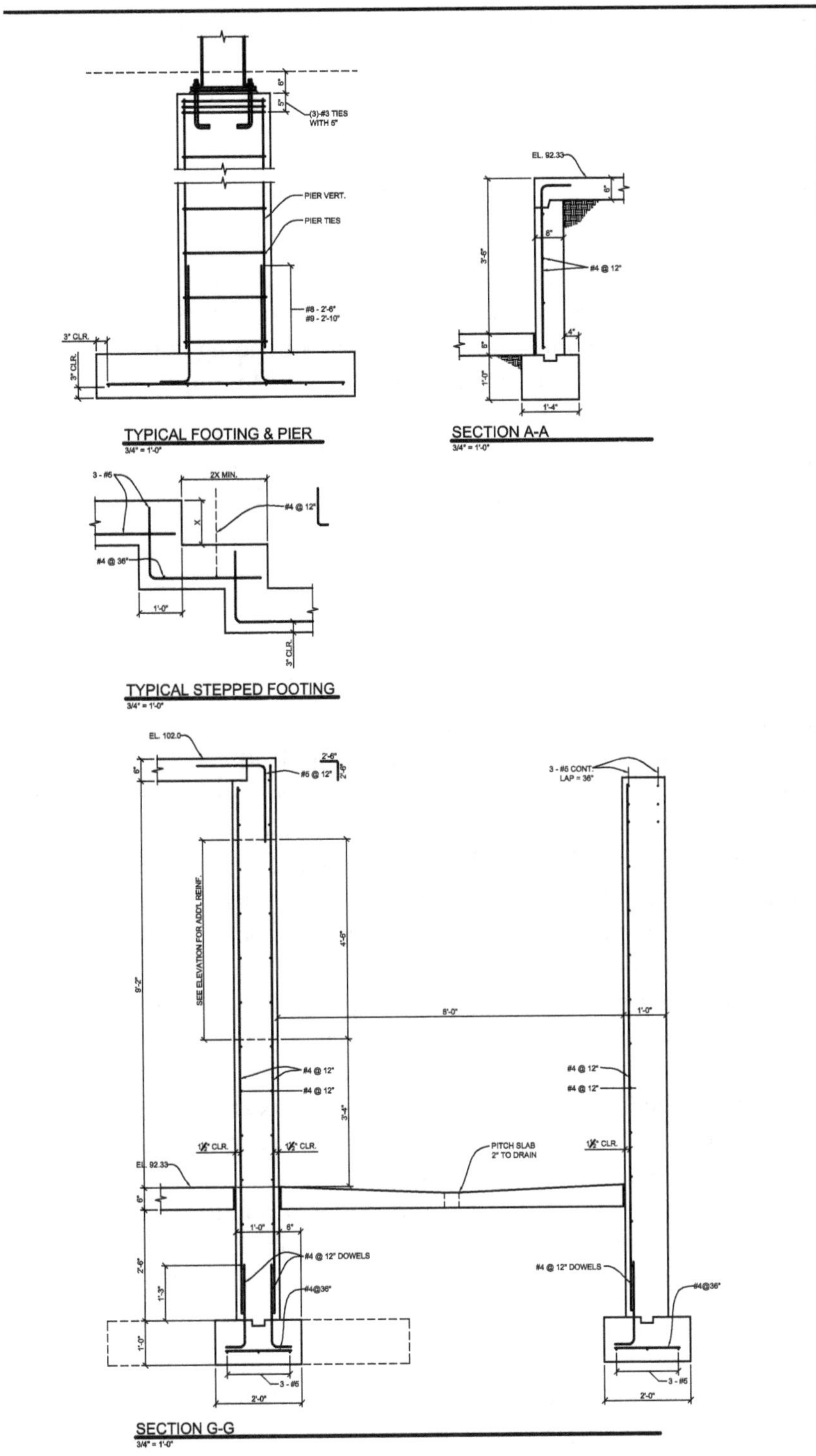

NOTES

1. ALL CONCRETE WORK SHALL CONFORM TO ACI 318-11.
2. REINFORCING STEEL SHALL CONFORM TO ASTM A615, GRADE 60.
3. F'c = 4000psi AT 28-DAYS FOR FOOTINGS AND WALLS.
   F'c = 5000psi AT 28-DAYS FOR PIERS.
4. MAXIMUM AGGREGATE SIZE IS 3/4".
5. ALL REINFORCING BAR SPLICES ARE TO BE CLASS B TENSION SPLICES PER LAP SPLICE SCHEDULE UNLESS OTHERWISE SHOWN. WALL HORIZONTALS ARE TO BE CONSIDERED 'TOP BARS'.
6. EXTEND #5 LONGITUDINAL BARS IN WALL FOOTINGS 1'-0" INTO COLUMN FOOTINGS.
7. E.W. = EACH WAY
   E.F. = EACH FACE
8. ALL ELEVATIONS SHOWN ON PLAN ARE TO TOP OF FOOTING.
9. UNLESS OTHERWISE NOTED, TOP OF ALL FOOTINGS AT ELEV. 91.33.

LAP SPLICE SCHEDULE

BAR SIZE	TOP BARS	OTHER BARS
#3	24"	19"
#4	32"	25"
#5	40"	31"
#6	48"	37"
#7	70"	54"
#8	80"	62"
#9	91"	70"

REVISIONS

MARK	DESCRIPTION	DATE

NEW OFFICE BUILDING

DRAWING TITLE:
FOUNDATION AND PIERS

ARCHITECT:

CONTRACTOR:

DATE: 01-01-12	DRAWN BY: CAD	SHEET: S1
JOB NO. 12-345	CHECKED BY: PM	

FIGURE 13.4 Continued

One common system is the use throughout the project of an arbitrary letter such as K followed by consecutive numbers, without regard to bar location or shape. This system may be observed in the placing drawing of Figure 13.3. The letter is prefixed with the size of the bar. For example, 8K19 represents a No. 8 bar whose shape and dimensions may be observed in the bending details schedule and whose location, in this case, may be established by the beam and girder schedule as being the top reinforcing at the noncontinuous end of the B6 members.

An alternative system is to use many letters rather than one arbitrary letter. Column bars may be designated with a C and footing bars with an F. For example, a 7F5 would be a No. 7 footing bar, whose shape and dimensions would be established in a bending details schedule and whose location would be observed in a footing schedule, typical footing details, or the foundation plan.

Other acceptable systems are currently being used. Of primary importance in any system that is chosen is that it should be simple, logical, easy to understand, and not lead to ambiguity or confusion.

13.4 SCHEDULES

Schedules generally appear both on engineering drawings and placing drawings. Typical schedules may be observed in Figures 13.2, 13.3, and 13.4. On the engineering drawings (Figure 13.2), the schedules are presented in a tabular format indicating a member mark number, concrete dimensions, and the member reinforcing steel size and location. Specific design details pertinent to the member reinforcing may also be furnished in the schedule. On the engineering drawing, the schedule must be correlated with a typical section and plan view to be meaningful; for example, the member marks must be shown on the plans. Schedules are generally used for the typical members, among which are slabs, beams, girders, joists, columns, footings, piers, shear walls, boundary zones or end zones of shear walls, and shear wall link or coupled beams. The shear wall schedule provides, in a tabular format, the wall thickness, concrete strength, and the required distributed vertical and horizontal reinforcement in the walls for each story of the building. The shear wall boundary or end zone schedule provides the required lengths/dimensions of the boundary or end zones, the additional concentrated vertical reinforcement, and horizontal ties for each story of the building. The shear wall link beam or coupled beam schedule provides the depth of the beam connecting the shear wall segments, the stirrup size and spacings, and the continuous top and bottom reinforcement and any additional diagonal reinforcement. Typical footing and pier schedules may be observed in Figure 13.4, and typical beam, girder, and slab schedules may be observed in Figure 13.2. A typical concrete column schedule was presented previously in Section 9.13 of Chapter 9.

Similar schedules are used for placing drawings, but additional information is furnished. The placing drawing schedule is more detailed and generally indicates the number of bars, member mark number, and physical dimensions of member reinforcing steel. In addition, it indicates size and length of straight bars, mark numbers (if bent bars), location of bars, and spacing, along with all specific notes and comments relative to the reinforcing bars. This schedule must also be coordinated with the typical sections and plan views. The placing drawing schedules are generally accompanied by a bar list, indicating bending details that may or may not be presented on the placing drawing. There is no standard format for either engineering drawing or placing drawing schedules. They are merely a convenient technique of presenting information for a group of similar items, such as groups of beams, girders, columns, shear walls, and footings. The schedule format will vary somewhat to conform to the requirements of a particular job.

A schedule for all structural components may not be necessary on a placing drawing. In Figure 13.3, the slab reinforcement is shown directly on the plan, so there is no need for a slab schedule.

Another system used to specify slab reinforcement that obviates the need for a slab schedule include providing uniform top and bottom grid reinforcement in both orthogonal directions in the slab, and then specifying additional bottom bars on the plan between the column/beam/wall supports, and additional top bars in the slab at the column, beam or wall supports. As an example, a designer may specify #4 @ 10 in. o.c bottom each way (BEW) and top each way (TEW) in the slab. Additional bars in the slab (which might vary from location to location) might be indicated on the plan at a particular location as follows: B-3-6, indicating additional 3#6 bottom bars, or T-7-5, indicating additional 7#5 top bars.

13.5 FABRICATING STANDARDS

The fabrication process consists of cutting, bending, bundling, and tagging the reinforcing steel. Our discussion will primarily be limited to cutting and bending because of their effect on a member's structural capacity. Bending, which includes the making of standard hooks, is generally accomplished in accordance with the requirements of the ACI Code, the provisions of which have been discussed in Chapter 5.

In the fabricating shop, bars to be bent are first cut to length as stipulated and then sent to a special bending department, where they are bent as designated in the bending details schedule or bar lists. The common types of bent bars have been standardized throughout the industry, and applicable configurations are generally incorporated in the placing drawing or bar lists in conjunction with the bending details (see Figure 13.3). Each configuration, sometimes called a bar type, has a designation such as 7, 8, or 9 and S1 or T1, with each dimension designated by a letter. Typical bar bends are shown in Figure 13.5. In addition, standards have been established with respect to the details of the hooks and bends. The ACI Code (Sections 25.3 and 26.6) establishes minimum requirements and is graphically portrayed in Figure 13.6. The tables in Figure 13.6 also show the extra length of the bar needed for the hook (A or G), which must be added to the sum of all other detailed dimensions to arrive at the total length of the bar. It is common practice to show all bar dimensions as out to out (meaning outside to outside) of the bar. The ACI Code also stipulates that bars must be bent cold unless indicated otherwise by the engineer. Field bending of bars partially embedded in concrete is not allowed unless specifically permitted by the engineer (ACI Code, Section 26.6.3.2b).

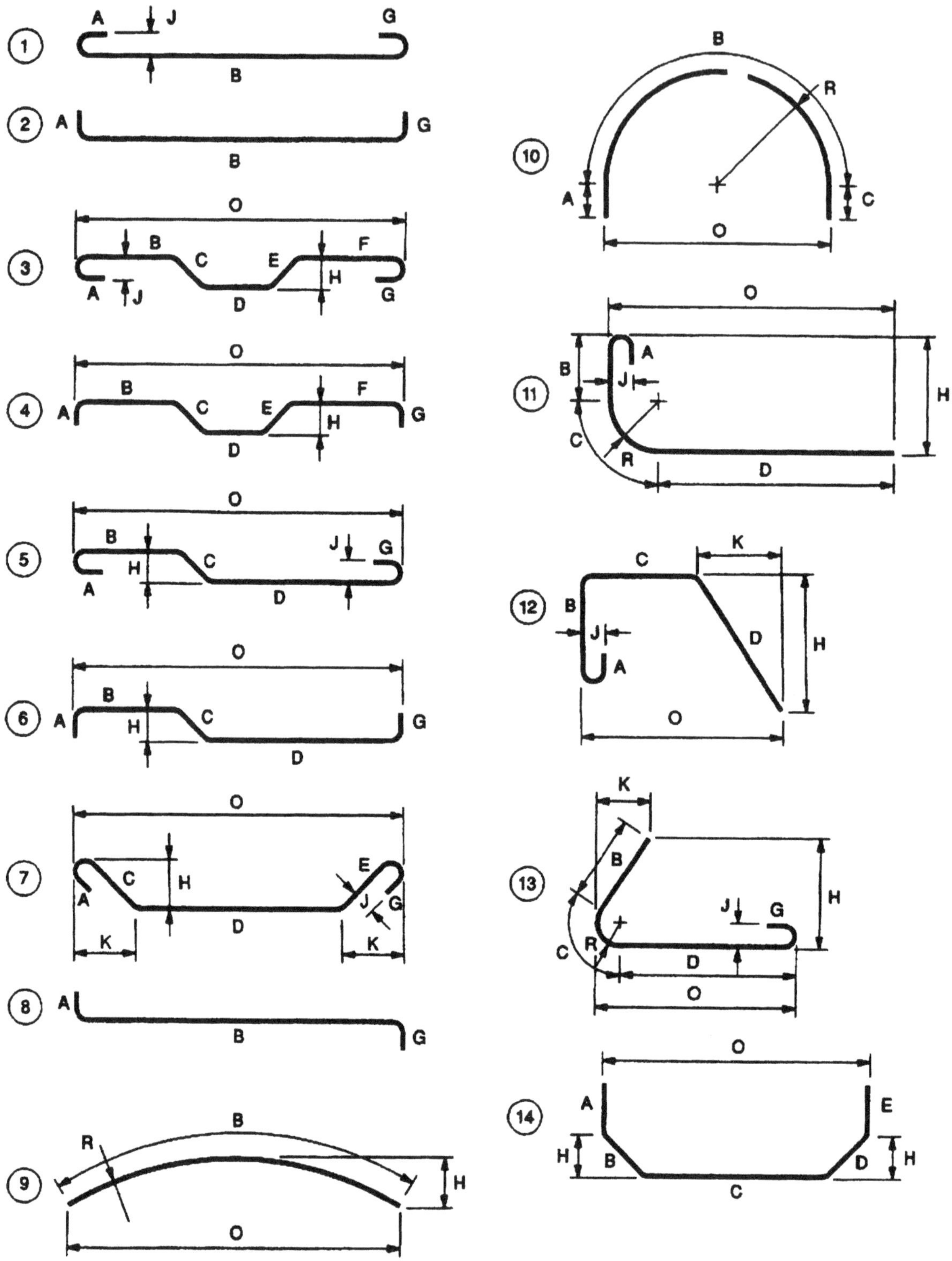

FIGURE 13.5 Typical bar bends. (ACI Detailing Manual (SP66-04), Courtesy of the American Concrete Institute)

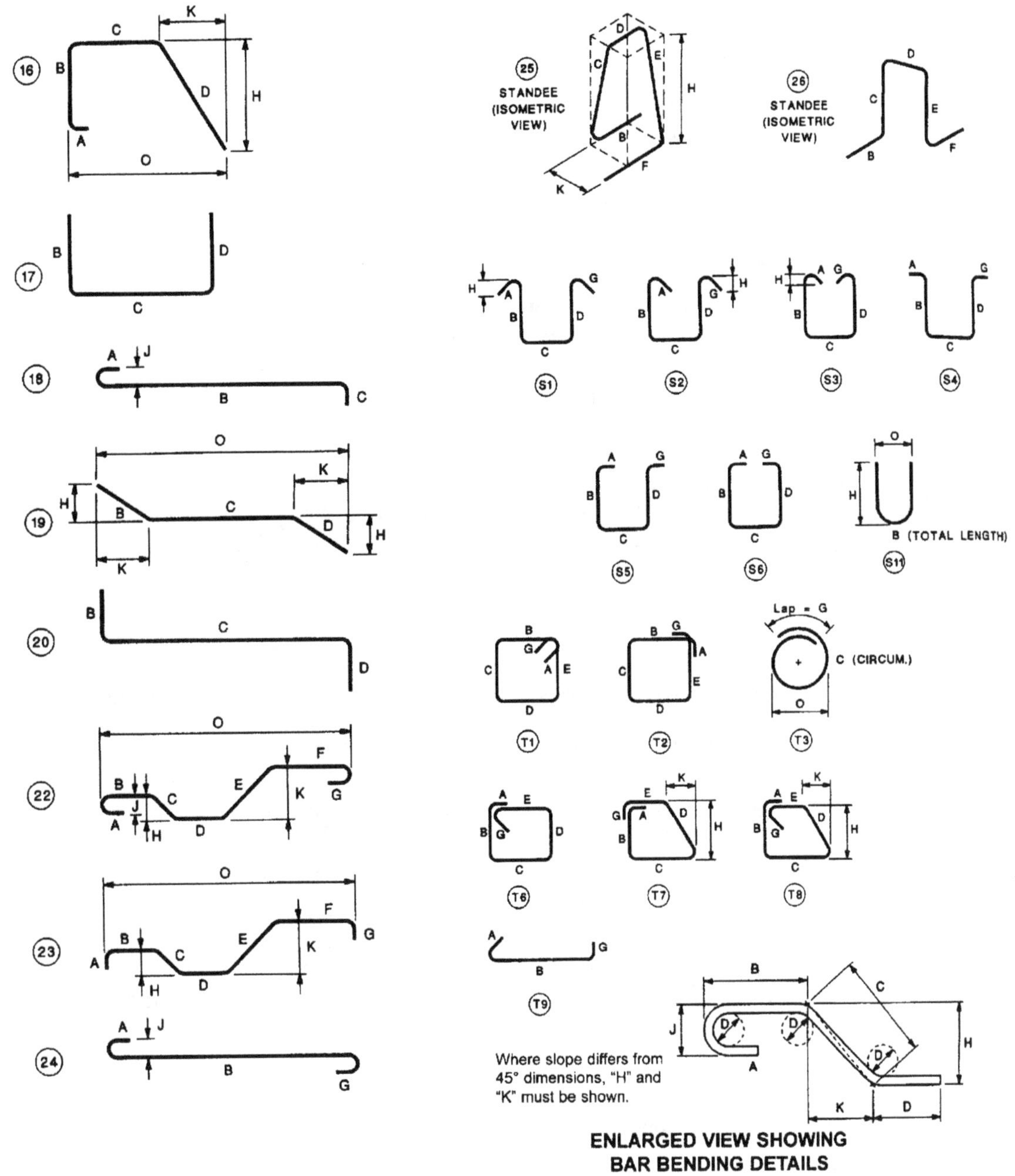

Notes:

1. All dimensions are out-to-out of bar except "A" and "G" on standard 180 and 135 degree hooks.
2. "J" dimensions on 180 degree hooks to be shown only where necessary to restrict hook size, otherwise ACI standard hooks are to be used.
3. Where "J" is not shown, "J" will be kept equal or less than "H" on Types 3, 5, and 22. Where "J" can exceed "H," it should be shown.
4. "H" dimension stirrups to be shown where necessary to fit within concrete.
5. Where bars are to be bent more accurately than standard fabricating tolerances, bending dimensions that require closer fabrication should have limits indicated.
6. Figures in circles show types.
7. For recommended diameter "D" of bends and hooks, see Section 3.7.1; for recommended hook dimensions, see Table 1.
8. Type S1 through S6, S11, T1 through T3, T6 through T9: apply to bar sizes No. 3 through 8 (No. 10 through 25).
9. Unless otherwise noted, diameter "D" is the same for all bends and hooks on a bar (except for Types 11 and 13).

FIGURE 13.5 Continued

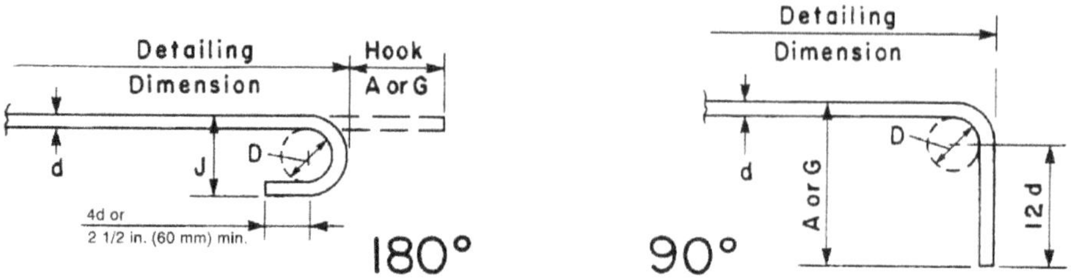

RECOMMENDED END HOOKS
All grades
D = Finished bend diameters

Bar size, No.	D,* in (mm)	180 degree hook		90 degree hook
		A or G, ft-in (mm)	J, ft-in, (mm)	A or G, ft-in, (mm)
3 (10)	2 1/4 (60)	5 (125)	3 (80)	6 (155)
4 (13)	3 (80)	6 (155)	4 (105)	8 (200)
5 (16)	3 3/4 (95)	7 (180)	5 (130)	10 (250)
6 (19)	4 1/2 (115)	8 (205)	6 (155)	1-0 (300)
7 (22)	5 1/4 (135)	10 (250)	7 (175)	1-2 (375)
8 (25)	6 (155)	11 (275)	8 (205)	1-4 (425)
9 (29)	9 1/2 (240)	1-3 (375)	11 3/4 (300)	1-7 (475)
10 (32)	10 3/4 (275)	1-5 (425)	1-1 1/4 (335)	1-10 (550)
11 (36)	12 (305)	1-7 (475)	1-2 3/4 (375)	2-0 (600)
14 (43)	18 1/4 (465)	2-3 (675)	1-9 3/4 (550)	2-7 (775)
18 (57)	24 (610)	3-0 (925)	2-4 1/2 (725)	3-5 (1050)

Table 1(cont.)—Standard hooks: All specific sizes recommended meet minimum requirements of ACI 318

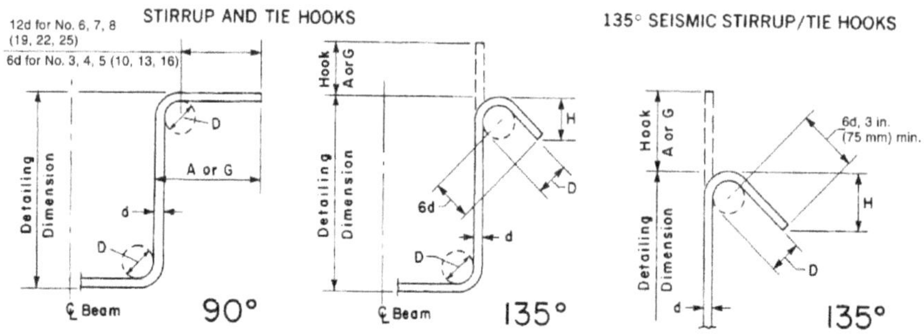

STIRRUP
(TIES SIMILAR)
STIRRUP AND TIE HOOK DIMENSIONS
ALL GRADES

Bar size, No.	D,* in. (mm)	90 degree hook	135 degree hook	
		Hook A or G, ft-in. (mm)	Hook A or G, ft-in. (mm)	H approx., ft-in. (mm)
3 (10)	1 1/2 (40)	4 (105)	4 (105)	2 1/2 (65)
4 (13)	2 (50)	4 1/2 (115)	4 1/2 (115)	3 (80)
5 (16)	2 1/2 (65)	6 (155)	5 1/2 (140)	3 3/4 (95)
6 (19)	4 1/2 (115)	1-0 (305)	8 (205)	4 1/2 (115)
7 (22)	5 1/4 (135)	1-2 (355)	9 (230)	5 1/4 (135)
8 (25)	6 (155)	1-4 (410)	10 1/2 (270)	6 (155)

135 DEGREE SEISMIC STIRRUP/TIE
HOOK DIMENSIONS
ALL GRADES

Bar size, No.	D,* in. (mm)	135 degree hook	
		Hook A or G, ft-in. (mm)	H approx., ft-in. (mm)
3 (10)	1 1/2 (40)	4 1/4 (110)	3 (80)
4 (13)	2 (50)	4 1/2 (115)	3 (80)
5 (16)	2 1/2 (65)	5 1/2 (140)	3 3/4 (95)
6 (19)	4 1/2 (115)	8 (205)	4 1/2 (115)
7 (23)	5 1/4 (135)	9 (230)	5 1/4 (135)
8 (25)	6 (155)	10 1/2 (270)	6 (155)

*Finished bend diameters include "spring back" effect when bars straighten out slightly after being bent and are slightly larger than minimum bend diameters in 3.7.2.

FIGURE 13.6 Standard hook details. (ACI Detailing Manual (SP66-04), Courtesy of the American Concrete Institute)

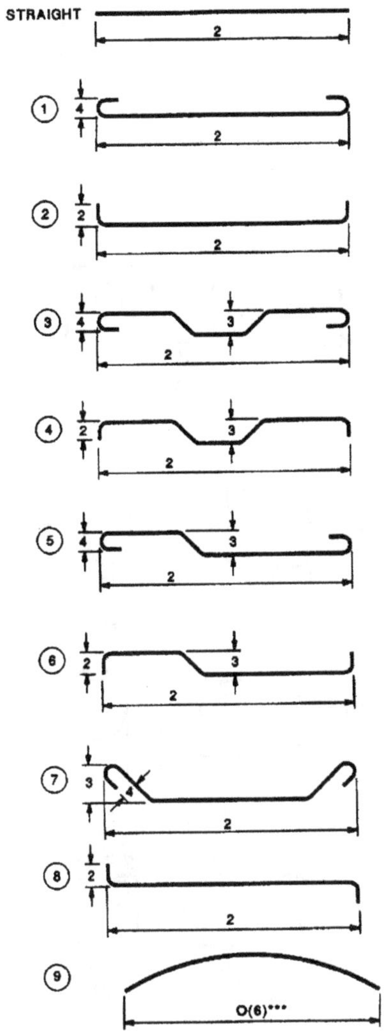

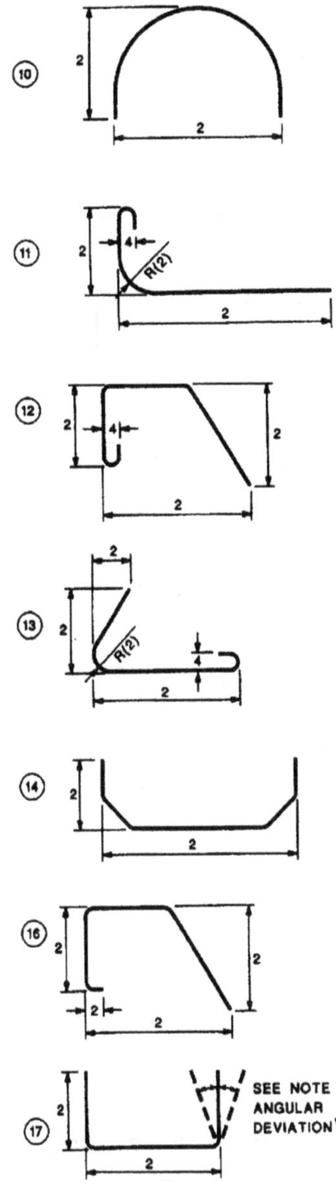

TOLERANCE SYMBOLS

1	=	±1/2 in. (15 mm) for bar size No. 3, 4, and 5 (No. 10, 13, and 16) (gross length < 12 ft. 0 in. (3650 mm))
1	=	±1 in. (25 mm) for bar size No. 3, 4, and 5 (No. 10, 13, and 16) (gross length ≥ 12 ft. 0 in. (3650 mm))
1	=	±1 in. (25 mm) for bar size No. 6, 7, and 8 (No. 19, 22, and 25)
2	=	± 1 in. (25 mm)
3	=	+ 0, -1/2 in. (15 mm)
4	=	±1/2 in. (15 mm)
5	=	±1/2 in. (15 mm) for diameter ≤ 30 in. (760 mm)
5	=	±1 in. (25 mm) for diameter > 30 in. (760 mm)
6	=	± 1.5% × "O" dimension, ≥ ± 2 in. (50 mm) minimum

Note: All tolerances single plane and as shown.
*Dimensions on this line are to be within tolerance shown but are not to differ from the opposite parallel dimension more than 1/2 in. (15 mm).
**Angular deviation—maximum ± 2-1/2 degrees or ± 1/2 in./ft (40 mm/m), but not less than 1/2 in. (15 mm) on all 90 degree hooks and bends.
***If application of positive tolerance to Type 9 results in a chord length ≥ the arc or bar length, the bar may be shipped straight.
Tolerances for Types S1-S6, S11, T1-T3, T6-T9 apply to bar size No. 3 through 8 (No. 10 through 25) inclusive only.

FIGURE 13.7 Standard fabricating tolerances for bar sizes No. 3 through No. 11 (No. 10 through No. 36).
(ACI Detailing Manual (SP66-04), Courtesy of the American Concrete Institute)

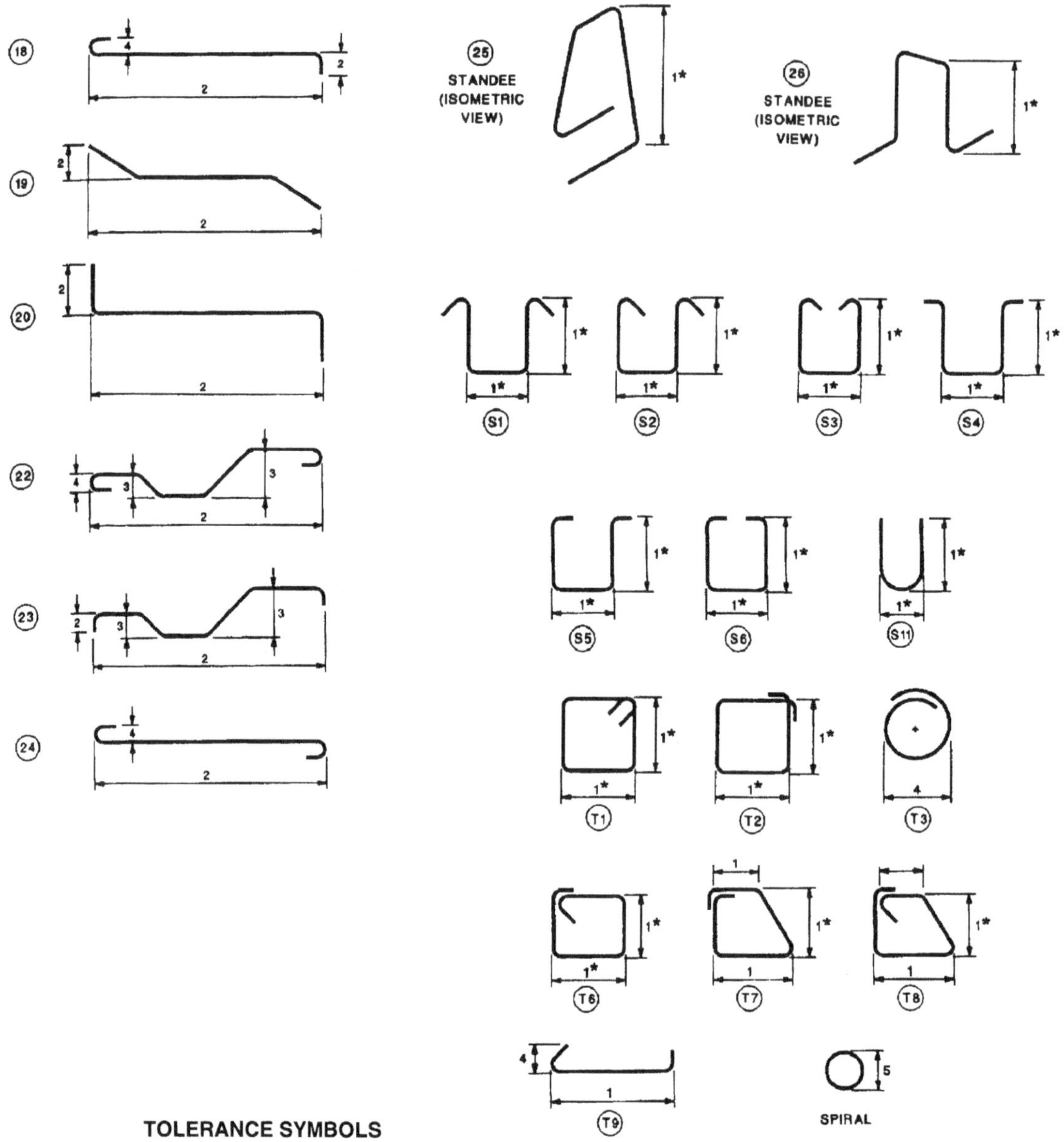

TOLERANCE SYMBOLS

1	=	±1/2 in. (15 mm) for bar size No. 3, 4, and 5 (No. 10, 13, and 16) (gross length < 12 ft. 0 in. (3650 mm))
1	=	±1 in. (25 mm) for bar size No. 3, 4, and 5 (No. 10, 13, and 16) (gross length ≥ 12 ft. 0 in. (3650 mm))
1	=	±1 in. (25 mm) for bar size No. 6, 7, and 8 (No. 19, 22, and 25)
2	=	± 1 in. (25 mm)
3	=	+ 0, -1/2 in. (15 mm)
4	=	±1/2 in. (15 mm)
5	=	±1/2 in. (15 mm) for diameter ≤ 30 in. (760 mm)
5	=	±1 in. (25 mm) for diameter > 30 in. (760 mm)
6	=	± 1.5% × "O" dimension, ≥ ± 2 in. (50 mm) minimum

Note: All tolerances single plane and as shown.

*Dimensions on this line are to be within tolerance shown but are not to differ from the opposite parallel dimension more than 1/2 in. (15 mm).

**Angular deviation—maximum ± 2-1/2 degrees or ± 1/2 in./ft (40 mm/m), but not less than 1/2 in. (15 mm) on all 90 degree hooks and bends.

***If application of positive tolerance to Type 9 results in a chord length ≥ the arc or bar length, the bar may be shipped straight.

Tolerances for Types S1-S6, S11, T1-T3, T6-T9 apply to bar size No. 3 through 8 (No. 10 through 25) inclusive only.

FIGURE 13.7 Continued

Straight bars are cut to the prescribed length from longer stock-length bars, which are received in the fabricating shop from the mills. Tolerances in the fabrication of reinforcing steel are generally standardized and are given in Figure 13.7. For instance, the cutting tolerance for straight bars is the specified length ±1 in., unless special tolerances are called for. Due consideration for these tolerances must be made by both the engineer and contractor in the design and construction phases.

13.6 BAR LISTS

The bar list serves several purposes. It is used for fabrication, including cutting, bending, and shipping, as well as placement and inspection. It represents a bill of materials indicating complete descriptions of the various bar items. The information in a bar list is obtained from the placing drawing or while the placing drawing is being prepared. A typical bar list form is shown in Figure 13.8; its similarity to the bending details schedule of Figure 13.3 is apparent.

The bar list generally includes both straight and bent bars and indicates all bar dimensions and bends as well as the grade of steel and the number of pieces. A bar list of this type may be used in addition to a bending detail schedule on the placing drawing. Some fabricators, however, prefer not to use a bending detail schedule. With this system, the detailers will use sketches until the drawing is complete and then transfer the information to separate bar lists. Two bar lists are prepared, one for straight bars and one for bent bars. Dimensions less than 12 in. are given in inches; over 12 in. they are given in feet and inches. In some cases, when the dimension is less than one inch, a digit is added at the end to indicate the fractional length in increments of one-eighth of an inch (see Figure 13.9).

13.7 EXTRAS

Bars are sold on the basis of weight. To a base price are added various extra charges (extras) dependent principally on the amount of effort required to produce the final product. Among these extras are bending extras and special fabrication extras.

Bent bars are generally classified as heavy bending or light bending. Extra charges are made for all shop bending, with the charge a function of the classification. Due to the increased amount of handling and number of bends per pound of steel, light bending charges per pound are appreciably more than the charges for heavy bending. According to the ACI, heavy bending is defined as bar sizes No. 4 through No. 18 that are bent at not more than six points, with a radius bent to one radius, and bending not otherwise defined. Light bending includes all No. 3 bars, all stirrups and column ties, and all bars No. 4 through No. 18 that are bent at more than six points, bent in more than one plane, and radius bent with more than one radius in any one bar, or a combination of radius and other bending. Special fabrication includes the fabrication of bars especially suited to conditions for a given project. This may include special tolerances and variations from minimum standards as well as unusual bends and spirals.

X Y Z PRODUCTS COMPANY
CHICAGO, ILLINOIS

PROJECT NO. 27693
CUSTOMER: JONES BROS. CONST. CO. DRAWING NO. Figs.18-5a,18-6a
PROJECT: FIELDCREST APT. BLDG. SHEET 1 of 2
LOCATION: SMITHVILLE, N.C. DATE 9/15/97 REVISED 9/19/97
MATERIAL FOR: PARTIAL BASEMENT COLUMNS DRAWN BY H.N.H.

ITEM	NO. PCS.	SZ	LENGTH	BAR MARK	TYPE	A	B	C	D	E	F	G	H	J	K	O	R
1	STRAIGHT																
2	4	57	23-11														
3	4	57	18-11														
4	12	57	8-11														
5																	
6	8	43	23-11														
7	4	43	8-11														
8																	
9	12	29	12-8														
10	6	29	10-8														
11	6	29	4-1														
12																	
13	STRAIGHT (SAW CUT BOTH ENDS)																
14	8	57	23-11	57W1													
15	8	57	11-5	57W2													
16																	
17	HEAVY BENDING																
18	4	36	20-0	36BC5	3		16-0	1-0	3-0				0-3	1-0			
19	64	36	13-6	36C16	3		3-0	1-3	9-3				0-4	1-3			
20																	
21	18	29	12-8	29C4	3		1-11	1-8	9-1				0-4½	1-8			
22																	
23	LIGHT BENDING																
24	22	10	8-4	10T6	T2	0-4	2-1	1-9	2-1	1-9		0-4					
25	22	10	7-8	10T9	T2	0-4	1-11	1-7	1-11	1-7		0-4					
26	50	10	6-3	10BT1	T2	0-4	1-4¾	1-4¾	1-4¾	1-4¾		0-4					
27	26	10	6-3	10BT3	T2	0-4	1-0¾	1-8¾	1-0¾	1-8¾		0-4					
28	44	10	3-4	10T10	S10		1-3½	0-9	1-3½								
29	22	10	2-10	10T23	T5	0-5	2-1	0-4									
30	52	10	2-10	10BT4	S10		1-0¼	0-9¼	1-0¼								
31	22	10	2-8	10T8	T5	0-5	1-11	0-4									
32	22	10	2-6	10T20	T5	0-5	1-9	0-4									
33																	
34			SPIRALS														
35			Height		Dia.	Pitch	Turns	Spcrs									
36	4	13	8-9	SP5	21"	3"	38	3									

ALL DIMENSIONS ARE OUT TO OUT
ALL BARS ASTM A615M GRADE 420

FOR STANDARD BEND TYPES REFER TO
CRSI MANUAL OF STANDARD PRACTICE

FIGURE 13.8 Typical bar list for buildings. (ACI Detailing Manual (SP66-04), Courtesy of the American Concrete Institute)

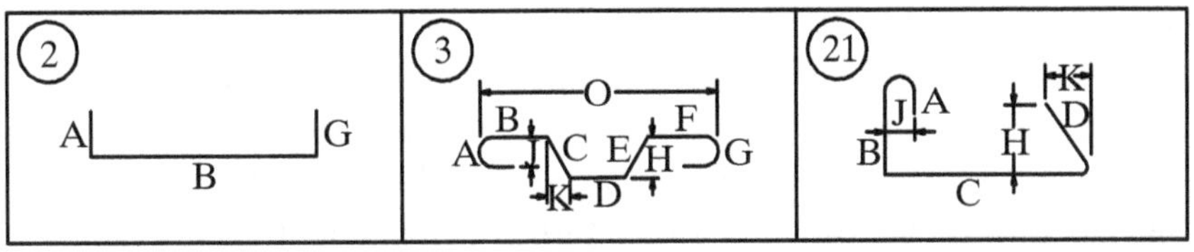

QTY	SIZE	THEO LENGTH	EP	BAR MARK	V	BND TYP	CD	A	B	C	D	E	F/R	G	H		K	O
2	5	10–08		5A9		3			2–06	5–073	2–06				4–00		3–113	
2	5	8–04		5A10		3				5–102	2–06				4–013		4–02	
1	5	4–04		5A11		2		0–10	3–062									
7	4	1–01		4A4		21				0–063	0–062				0–061		0–05	

$3'\text{-}6\frac{1}{4}''$ (the "2" is $\frac{2}{8} = \frac{1}{4}$)

$6\frac{1}{8}''$

$3'\text{-}11\frac{3}{8}''$

FIGURE 13.9 Rebar bend details.

13.8 BAR SUPPORTS AND BAR PLACEMENT

Bar supports are used to hold the bars firmly at their designated locations before and during the placing of concrete. The ACI Code Section 26.6.2.2 requires that all reinforcing be in place and secured prior to pouring concrete, which precludes the placing or positioning of reinforcing while the concrete is wet (i.e., "wet-setting" is precluded). These supports may be of metal, plastic, precast concrete, or other approved materials. The most commonly used bar supports are factory-made wire bar supports, which are available in various sizes and types and may provide some corrosion resistance by having exposed parts covered or capped with plastic or being made wholly or in part of galvanized or stainless steel. CRSI publishes information intended to serve as a guide for the selection and utilization of steel wire bar supports used to position reinforcing steel. It is a general practice that unless the engineers' drawings or specifications show otherwise, bar supports will be furnished in accordance with CRSI standards. More detailed information relative to bar supports may be found in the CRSI publications *Placing Reinforcing Bars* [3] and *Manual of Standard Practice* [2].

13.9 DETAILING SOFTWARE

The use of software for the detailing and production of shop drawings is now fairly common for many construction projects. During the design and production of the contract drawings, it is common to use 3-D modeling software for the quick and accurate generation of rebar quantities and details. However, because the contracts for the shop drawings and the contract drawings are typically separate and independent, the sharing of the electronic data for the contract drawings - which would help reduce the time and effort to create the shop drawings - is often times not guaranteed. As such, the fabricator creates the shop drawings completely independently from the electronic data that exists for the contract drawings. However, on some projects, the project delivery model increasingly involves the use of integrated 3-D models like Building Information Modeling (BIM) software to share information among the project stakeholders and to produce the shop drawings, resulting in the added advantage of helping to identify any potential conflicts and errors in the contract drawings. Several commercially available detailing software perform at various levels of sophistication, and bar fabricators sometimes create their own in-house software to perform unique functions geared to their own needs. Each package and each in-house detailing software is different, and the person entering the field can expect to receive training on the particular equipment and software that the company is using. The development and availability of increasingly sophisticated software can be expected, specifically software that could be used in conjunction with the software that is used to create the design or contract drawings.

The functions most widely performed by computers in the bar fabricating business involve the generation and printing of notes and labels, schedules (beam, column, slab, pier, footing, joist, and the like), bending details, bar tags, and weight summaries. The software will also optimize the cutting schedule to minimize waste (which is very important in the high-volume, low-margin rebar fabricating business). In addition, information can be generated that will limit bar bundles based on maximum weight and configuration.

The process begins with the preparation of the placing drawing (see Figure 13.3). For ease of use, it is preferable to furnish the labels on the placing drawing. If the drawing is complex, the detailer will label a reinforcing bar or a group of bars with a single letter or number, which is then keyed to a label list and becomes part of the placing drawing.

The data becomes part of a database, which then provides information for the various other operations. A bill of material, or bar list, is commonly generated on which the bars are segregated by grade and size in descending order. Some detailing software allow the merging of rebar requirements for several jobs; lower-level (and less expensive) detailing software may not have this capability. It is not difficult to imagine the complexities involved. The bar shop must keep orderly track of bars that vary in grade, type (uncoated, epoxy-coated, galvanized), shape (see Figure 13.5), length, and size.

Several illustrations provide examples of the types of documents that are commonly used in the bar fabricating business. Figure 13.3 shows a placing drawing and has been previously discussed. The detailing software can also generate a rebar shearing schedule and a rebar shearing schedule summary (weight summary). The cutting of the 60-ft-long bars is optimized to minimize the scrap. The software also aids in the sequencing, phasing, and delivery of rebar for projects.

REFERENCES

[1] *ACI Detailing Manual* [MNL-66(20)], includes "Details and Detailing of Concrete Reinforcement" (ACI 315R-18), "Guide to Presenting Reinforcing Steel Design Details." American Concrete Institute, 38800 Country Club Drive, Farmington Hills, MI 48331, 2018.

[2] *Manual of Standard Practice*, 29th ed. Concrete Reinforcing Steel Institute, 933 North Plum Grove Road, Schaumburg, IL 60173, 2018.

[3] *Placing Reinforcing Bars*, 10th ed. Concrete Reinforcing Steel Institute, 933 North Plum Grove Road, Schaumburg, IL 60173, 2019.

CHAPTER 14

Practical Considerations in the Design of Reinforced Concrete Buildings

14.1 INTRODUCTION

In this chapter, several additional topics of importance to the structural engineer for the proper design and construction of reinforced concrete structures are covered. The topics covered range from the rules of thumb and practical means for designing reinforced concrete structures, to the retrofitting of existing structures and the load testing of structures. The topic of diaphragms, drag struts, and chords, important elements in the lateral load path of structures, is introduced –a topic often neglected, and obscure to the typical student in a concrete design course, but important enough that the ACI Code devotes an entire chapter to it (see ACI 318-19, Chapter 12). The reader is also introduced to concrete slabs subjected to concentrated loads, and construction pour gaps that are needed in the casting of large concrete floor areas, fire resistance of concrete elements, edge-supported two-way slabs on stiff supports, and concrete specifications—another topic that is often obscure to the typical civil engineering student and is not often touched upon in the concrete design courses or textbooks. The chapter concludes with the introduction of two real-world type reinforced concrete building design projects that will help the reader not only to hone their design skills of the individual structural elements presented in previous chapters, but more importantly enable them to learn how each of these individual structural elements fit together in a holistic fashion within the context of an entire building structure—an invaluable skill in the architecture, engineering, and construction (AEC) industry.

14.2 RULES OF THUMB AND PRACTICAL CONSIDERATIONS FOR REINFORCED CONCRETE DESIGN

In reinforced concrete construction, simplicity and repetitiveness are the keys to economy [1,2,3] since formwork is a major cost component in concrete construction. In beam and girder framing schemes with one-way slabs, the following should be considered:

1. When laying out reinforced concrete beams and girders in one-way slab systems, the girders should span in the shorter direction along the column lines and the beams should

span in the longer direction. See Figure 3.2 for typical one-way slab framing schemes. For greater economy, bay sizes are typically 20–35 ft maximum for beam and girder framing schemes with one-way slabs in residential and office buildings [3,4]. The beam spacing determines the span of the one-way slabs and typically ranges from 10 ft to 16 ft maximum in order to keep the slab depth to a reasonable size.

2. The depths of reinforced concrete beams and one-way slabs should be determined using the minimum depths specified in Table 6.1 (see ACI 318 Code Table 7.3.1.1 and 9.3.1.1)

   The depths of beams or girders should be varied in 1 in. increments and the depths of slabs should be varied in ½-in. increments. *Even* sizes for the beam width and depth are more commonly specified. The width of the beam or girder should be no less than (1/3)h to (½)h, where h is the overall beam depth, but for economy and simplicity of formwork, a beam or girder width equal to or greater than the column width is recommended [29]. For beam and girder design, the beam stem width, b_w, and effective depth, d, should also satisfy the following relationship [29]:

   $b_w d^2 \geq 20\, M_u$,

   where b_w and d are in inches and M_u is in ft-kip.

   The overall depth of the beam, $h \approx d + 2.5''$ for beams and girders with single layer rebar and, $h \approx d + 3.5''$ for beams and girders with double layer rebar

   For slabs, the width b_w is always 12 inches, so the required minimum effective depth of the slab can be found from the equation above as $d^2 \geq 3.83M_u$

   Knowing the minimum slab effective depth, the minimum depth or thickness of the slab to resist the applied factored moment, M_u is: $h = d$ + clear cover + $d_b/2$, where d_b is the diameter of the main reinforcement in the slab.

3. The concrete columns should be laid out in a regular orthogonal grid, as much as possible, preferably with square bays, where possible. In one-way slab construction, the columns are usually spaced a maximum of approximately 35 ft apart for economy.

4. Since concrete formwork constitutes a major component of the cost of reinforced concrete, it is more cost-effective to use the same depth for the beams and girders throughout the height of the building, where feasible, and vary the beam width and/or reinforcement where necessary. Similarly, it is more economical to keep the size of continuous beams constant from span to span, and vary the reinforcement as required [29].

5. Where possible, column sizes should be kept constant for multiple stories of the building, varying the concrete strength and/or the amount of reinforcing steel where higher capacity is required. Where changes in column sizes need to be made, it is cost-effective to change only one dimension at a time and to limit the size change to increments of 2 in. [5]. Consider using rectangular columns instead of square or circular columns to help in hiding large portions of the columns within walls.

6. Using the charts in Reference [5], the preliminary column size (square column) in inches for normal-height columns (less than or equal to 15 ft story height) can be calculated as a function of the sum of the tributary areas supported by the column as follows:

$$h = 0.0025 \sum A_t + 5$$

where,

h = square column cross-sectional dimension in inches

ΣA_t = sum of the total tributary floor/roof area in ft^2 supported by the column.

7. A concrete building structure is considered "braced" if the stiffness of the shear walls is at least equal to or greater than *12* times the sum of the gross stiffness of the columns in a given direction within a story (ACI Code 6.2.5). Since in reinforced concrete buildings the elevator or stair walls typically function as shear walls, most small to moderate height buildings are usually considered "braced" by the shear walls.

   For columns in "braced" frames bent in *double curvature* with approximately equal end moments (i.e. $M_1/M_2 = +1$) which would apply to columns above the second floor level, the maximum clear story heights for which slenderness can be neglected in the design of the columns is approximately $12h$; for ground floor columns supported on spread footings which provide minimal restraint to bending at the bottom of the column, the maximum clear story heights is approximately $10h$; the values for maximum clear story height for different column sizes in braced frame buildings are given in Table 14.1 [3]:

TABLE 14.1 Maximum Clear Story Heights for Columns in Braced Frames for which Slenderness Can Be Neglected

Column size, h	Ground floor columns $\left(\frac{\ell_u}{h} \approx 10\right)$	Columns on the second floor and above $\left(\frac{\ell_u}{h} \approx 12\right)$
	Maximum clear story height, ℓ_u (ft)	Maximum clear story height, ℓ_u (ft)
12	10	12
14	11.67	14
16	13.33	16
18	15	18
20	16.67	20
22	18.33	22
24	20	24

8. Visualize the construction sequence for the structure and its elements and how the reinforcement will be placed. This will help ensure constructability.

9. Provide adequate spacing between adjacent reinforcing bars to ensure free passage of concrete between the rebars and to prevent voids or honeycombs from forming.

10. Draw to scale the detail of reinforcing in slabs, beams, girders, and columns in order to catch any potential congestion of the reinforcing steel.

11. Preferable to use 90° instead of 180° hooks in concrete slabs because of the difficulty in placing bars with 180° hook around the edge reinforcement in the slab in the perpendicular direction. When using 90° hooks in reinforced concrete slabs, using reinforcement with a small-enough diameter that results in a hook length no greater than 80% of the slab thickness will ensure that the hook will fit within the thickness of the slab. If the length of the hook is greater than the slab thickness, the 90° hook would need to be oriented at an angle in order to fit within the slab thickness.

12. Though the ACI Code allows the maximum reinforcement in columns of up to 8% of in the gross area of the columns (which amounts to 4% maximum when lap splices are used), use between 1% and 2% of vertical reinforcing steel in reinforced concrete columns for economy and to avoid congestion of reinforcement, especially at beam-column connections [3, 4]. Instead of using more than 2% reinforcement in a column, consider increasing the concrete strength and/or the cross-sectional size of the column.
13. Instead of using bundled vertical bars in columns, increase the size of the column to obviate the need for bundled bars.
14. Coordinate placement of mechanical ducts and piping to avoid interference with location of reinforcement in slabs, beams, girders, and columns.
15. For typical concrete buildings, a 7-day floor-to-floor cycle (i.e., 7 days between casting the concrete floors) will ensure adequate strength development, but the use of admixtures can reduce this time to as low as three days.
16. As previously mentioned, the most economical formwork is achieved when the beam or girder has the same width as the column [3]. Care should be taken when the beam or girder width is much larger than the column width because only a portion of the girder top reinforcement will be developed within the column width. If, in the modeling and analysis of such a girder-column connection, a full moment connection was assumed, this assumption would be incorrect because the connection will have less than a full moment capacity since all the girder top reinforcement cannot be fully developed within the column width.

    Avoid the use of spandrel beams or girders that are much wider than their supporting columns because of the aforementioned issue, and the undesirable eccentricity of loading and the ensuing torsional moments on the spandrel beam or girder.

EXAMPLE 14.1

Preliminary Sizing of Columns

Determine the preliminary size of the first-floor typical interior column for a 10-story building with 30-ft square bay dimensions, considering axial loads only. The concrete and steel strengths are 4000 psi and 60,000 psi, respectively. Assume a dead load of 100 psf and a live load of 50 psf. Ignore floor live load reduction.

Solution:

The factored floor load, w_u = 1.2 (100 psf) + 1.6 (50 psf) = 200 psf

The tributary floor area of a typical interior column = (30 ft)(30 ft) = 900 ft^2

Sum of tributary areas supported by the first floor interior column = (10 floors)(900 ft^2) = 9,000 ft^2

The preliminary square column cross-sectional size, h = (0.0025)(9,000) + 5 = 27.5 in.

Therefore, use 28″ × 28″ column at the ground floor level.

Total factored axial load on the typical interior ground floor column, P_u = (9,000 ft^2)(200 psf) = 1,800 kip

It should be noted that the self-weight of the column has been neglected in this example, and the tributary area method for calculating the column load above neglects to account for the self-weight of the beam and girder stems below the soffit of the slab. The self-weight of the beam and girder stems will add, approximately, an additional 5% to 10% to the factored load calculated above. Furthermore, the first interior columns will also support a higher load than the typical interior columns. An additional 7.5% (i.e., $[1.15 + 1]/2 - 1] \times 100\%$) should be added to the factored load for the first interior columns to account for the higher shears in the end span beams and girders at the first interior columns (see the ACI shear coefficients for beams in Figure 6.1b).

In Example 14.1, the load-reducing effect of live load reduction permitted by the building codes has not been taken into account. Taking advantage of the live load reduction will result in a smaller column size.

14.3 APPROXIMATE MOMENTS AND SHEARS IN CONTINUOUS GIRDERS

Girders usually frame into columns and support the concentrated reactions from the infill beams, in addition to any other uniformly distributed loads from the self-weight of the girder stem and cladding loads. Consequently, the ACI coefficients for continuous beams and one-way slabs, introduced in Chapter 6, are not applicable to continuous girders. The approximate moments and shears in continuous girders can be obtained from Tables 14.2a and 14.2b, respectively [6].

TABLE 14.2a Approximate Factored Moment for Continuous Girders [6].

Positive Moment
At end spans:
$M_u^+ = \dfrac{w_u \ell_n^2}{14} + \dfrac{\ell_n}{6} \sum P_u$
At interior spans:
$M_u^+ = \dfrac{w_u \ell_n^2}{16} + \dfrac{\ell_n}{7} \sum P_u$
Negative Moment at Supports
At interior face of external column or perpendicular structural wall:
$M_u^+ = \dfrac{w_u \ell_n^2}{16} + \dfrac{\ell_n}{10} \sum P_u$
At exterior face of first internal column or perpendicular structural wall, only two spans:
$M_u^+ = \dfrac{w_u \ell_n^2}{9} + \dfrac{\ell_n}{6} \sum P_u$
At faces of internal columns or perpendicular structural walls, more than two spans:
$M_u^+ = \dfrac{w_u \ell_n^2}{10} + \dfrac{\ell_n}{6.5} \sum P_u$
At faces of structural walls parallel to the plane of the frame:
$M_u^+ = \dfrac{w_u \ell_n^2}{12} + \dfrac{\ell_n}{7} \sum P_u$
At support of girder cantilevers:
$M_u^+ = \dfrac{3 w_u \ell_n^2}{4} + \ell_n \sum P_u$

(Adapted from ACI 314-16 Table 8.7.4.1)

The moment for which the columns are designed, in addition to the axial loads, can be obtained by calculating the unbalanced moment in the column due to the factored gravity loads on the girders. The factored live loads on the girder should be alternated to create the maximum value for the unbalanced moments in the column at the column–girder connection. The two cases to be considered are as follows: in Case 1, the *odd* spans of the girder (i.e., the exterior spans in Figure 14.1) should be loaded with the factored live loads (i.e., the factored live load reactions from the infill beams), in addition to the factored uniform dead load and the factored dead load reactions from the infill beams on *all* girder spans

TABLE 14.2b Approximate Factored Shear for Continuous Girders [6]

At exterior face of first interior column:

$$V_u = 1.15\frac{w_u \ell_n}{2} + 0.80\,\Sigma P_u$$

At faces of all other columns:

$$V_u = \frac{w_u \ell_n}{2} + 0.75\,\Sigma P_u$$

At supports of girder cantilevers:

$$V_u = w_u \ell_n + \Sigma P_u$$

where,

w_u = factored uniformly distributed load on girder (kip/ft)

ΣP_u = sum of all the factored concentrated loads acting on the girder span (kip)

ℓ_n = clear span of the girder (ft)

(Adapted from ACI 314-16 Table 8.7.4.1)

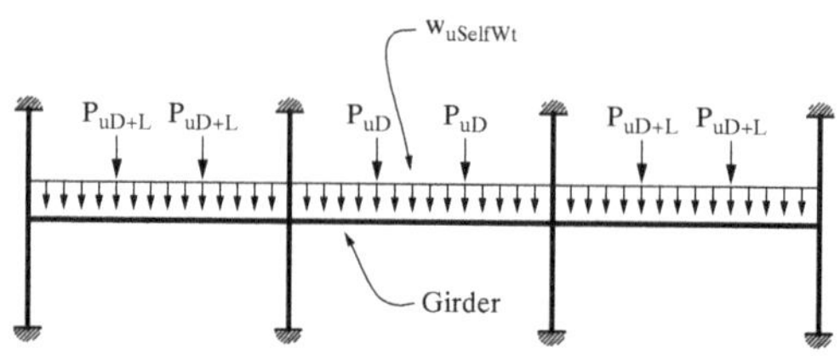

a) Case 1

$P_{uD+L} = 1.2P_D + 1.6P_L$

$P_{uD} = 1.2P_D$

$w_{uSelfWt}$ = 1.2 x (weight of girder stem and any wall load directly supported by the girder)

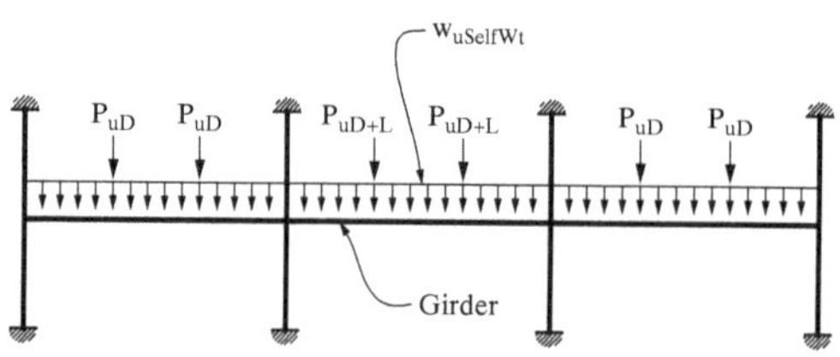

b) Case 2

FIGURE 14.1 Pattern Live Load Analysis for Girders.

(see Figure 14.1); in Case 2, the *even* spans of the girder (i.e., the interior span in Figure 14.1) should be loaded with the factored live loads (i.e., the factored live load reactions from the infill beams), in addition to the factored uniform dead load and the factored dead load reactions from the infill beams on *all* girder spans (see Figure 14.1). The unbalanced moment, ΔM_u, in the column at a girder–column connection is the maximum difference between the girder-factored moments on both sides of the column. This unbalanced moment is distributed to the column above and below the floor level under consideration in proportion to the ratio of the flexural stiffness of the columns above and below the floor level (i.e., $I_{c,\text{above}}/h_{\text{above}}: I_{c,\text{below}}/h_{\text{below}}$), where I_c is the moment of inertia of the column for bending in the plane of the girder and h is the floor-to-floor height of the column. For roof girders, the top story columns resist all the unbalanced moment, ΔM_u.

14.4 STRUCTURAL CONDITION ASSESSMENT, AND STRENGTHENING AND REHABILITATION OF EXISTING REINFORCED CONCRETE STRUCTURES

The occupancy of a reinforced concrete structure may change during the life of the structure, necessitating modifications to the structure in cases where the change results in increased loads. Converting a building from office use to a library or storage occupancy, for example, will result in a significant increase in the applied loads on the structure. A floor opening may need to be cut in the slabs of an existing building to provide natural lighting, which may reduce the capacity of the existing slab, beams, or girders. Also, the structural members in a building may have deteriorated or been degraded due to exposure to a corrosive environment (e.g., concrete exposed to moisture, oxygen, and chlorides from seawater or deicing salts) or the members may have been damaged due to a seismic or fire event, or they may need to be upgraded to meet new code requirements.

In the visual structural condition assessment of existing structures, some important signs to look out for in a concrete structure that impact structural integrity include water leaks, water ponding – leading to water intrusion into the concrete, noticeable water stains or damage (e.g., efflorescence on the concrete surface, peeling paint, etc.), scaling, cracks, spalling and delamination of concrete, honeycombs, and sagging of structural and non-structural elements (e.g., sagging floors and sagging ceilings). The results of the visual condition assessment phase may lead to a more thorough evaluation phase where nondestructive testing methods like ground-penetrating radar (GPR), half-cell potential corrosion survey, ultrasonic pulse velocity, or impact echo methods, etc., are used, and/or destructive evaluation techniques are used - such as obtaining concrete cores and conducting compressive strength tests and petrographic analysis on the concrete cores. See Ref. [39] for more detailed discussion of structural condition assessment. The importance of routine annual maintenance inspections, periodic condition assessments, and milestone condition assessment and evaluation of existing building structures has taken on greater importance since the tragic collapse of the Champlain Towers South condominium in Surfside Florida on June 24, 2021. One important information needed from these condition assessments includes whether life safety issues are present, and whether or not the structural elements should immediately be temporarily shored pending repairs and/or if the building should be evacuated.

The exposure of concrete to moisture, oxygen, and chlorides from seawater or deicing salts results in the corrosion of the steel reinforcement when water reaches the reinforcement through cracks in the concrete. This corrosion causes an increase in the size of the reinforcement due to the surrounding rust layer. The increase in the surface area of the steel reinforcement causes tension forces to be exerted on the surrounding concrete which leads to cracking and spalling of the concrete, and debonding of the reinforcement from the concrete. The corrosion also causes a loss or reduction in the cross-sectional area of the reinforcement. These factors—combined together—lead to a reduction in the strength of the concrete member. Consequently, either as a result of a structural condition assessment or because of some of the other reasons previously discussed, the existing slabs, beams, girders, and columns in a reinforced concrete building may need to be repaired and/or strengthened.

There are several options available for increasing the strength of reinforced concrete structural elements, but the common goal is to achieve composite action between the existing concrete member and the added reinforcement:

1. For reinforced concrete slabs, Carbon Fiber Reinforced Polymer (CFRP) strips could be applied to the top or bottom surface of the slab to add flexural capacity to an existing slab [7]. Steel plates with drilled epoxy anchors installed at the bottom and top surfaces of a reinforced concrete slab can also be used to increase the moment capacity of the slab [8]. For slabs with large openings, steel beam framing at the underside of the slab at the edges of the floor opening could be used to increase the load-carrying capacity of the slab. In addition to reinforcing and strengthening the edges of the slab opening, the adjacent slabs should be evaluated for redistribution of moments, and the adjacent reinforced concrete beams should also be evaluated for torsion and a reduced moment capacity due to the floor opening.

   The punching shear capacity of two-way slabs can be increased by providing a steel angle collar around the top of the column at the underside of the flat plate or flat slab. The angles are through-bolted to the existing column and epoxy anchored into the existing slab. The horizontal leg of the angle increases the effective perimeter of the critical section for punching shear, b_o, thus leading to a reduction in the punching shear stress.

   The punching shear capacity of a flat plate can also be enhanced by casting a concrete collar or jacket around the top of a column at the underside of the existing slab. The concrete jacket or collar is attached to the existing column with through-bolts or dowels and connected to the underside of the slab with epoxy dowels or anchors.
2. For the retrofit and strengthening of existing reinforced concrete beams and girders, the following options are possible:

   a. Add a steel plate bonded with epoxy to the underside of an existing concrete beam or girder. Epoxy-anchor the steel plate into the existing concrete beam to hold the steel plate in place. Note that the steel plate will need to be fireproofed [14]. Through-bolted or epoxy-anchored steel side plates have also been used to strengthen existing beams in shear and flexure.

   b. Provide CFRP reinforcement on the sides, top, or bottom of existing concrete beams or girders to resist bending and shear. For shear, the CFRP strips are placed at an angle on the two vertical sides of the beam. CFRPs are preferred to bonded steel plates because no corrosion protection is required and they possess good fatigue properties. CFRPs are also lighter than steel plates and therefore more easily handled on site.

   c. Externally post-tension the concrete beam or girder using parabolic cables to relieve the beam/girder of some of the existing dead loads.
   d. Enlarge the size of the beam or girder by casting additional concrete around the existing member. The surfaces of the existing member would need to be roughened and dowels drilled and epoxied into the existing concrete member in order to bond the new concrete to the existing concrete.

3. For reinforced concrete columns, the methods for strengthening or repair include the use of CFRP wraps around existing columns. Fiber Reinforced Polymer (FRP) laminates are also used where the FRP acts like a jacket around the existing column and the annular gap between the laminate and the face of the existing column is filled with resin or grout [9]. Concrete columns can also be strengthened using steel jackets around the existing column. The annular space between the steel jacket and the existing column is filled with grout or epoxy resin. These wraparounds help to confine the concrete, thus increasing the compressive strength of the column.

   Reinforced concrete columns can also be strengthened or repaired by concrete jacketing or encasement of the existing column. To ensure composite action between an existing reinforced concrete member and a new concrete jacket, it is recommended that the surface of the existing concrete be roughened, and with a bonding agent applied.

4. Other considerations include ensuring the strength of the retrofitted member is maintained during a fire. CFRP reinforcements are typically exposed and will lose their strength in a fire. Before deciding which methodology to use, the engineer should determine the cause of the deficiency and choose a repair method accordingly that will correct the problems and minimize the impact on building operations. Other factors to consider include the available amount of space for the retrofit, the cost of the retrofit, aesthetics, fire protection, and the local contractors' experience with the retrofit methods.

When an opening needs to be cut in an existing floor slab, the location of existing reinforcement can be determined using ultrasonic pulse velocity (i.e., X-ray) or ground penetrating radar or other proprietary equipment to avoid cutting existing reinforcement. Examples of structural details for the repair and strengthening of damaged or degraded reinforced concrete columns, beams, and girders [10, 11] are shown in Figures 14.2 and 14.3.

The reinforcement methods described above are passive methods which means that they only help to resist the additional loads on the structure above and beyond the existing loads. Active reinforcement methods (see ACI PRC-546 Guide to Concrete Repair) help to resist both existing and additional loads. Examples of such methods include external post-tensioning. For additional information, the reader should refer to the International Concrete Reinforcing Institute's ICRI 330.1: Guideline for the Selection of Strengthening Systems for Concrete Structures and ACI PRC-440.2: Guide for the Design and Construction of Externally Bonded FRP Systems for Strengthening Concrete Structures.

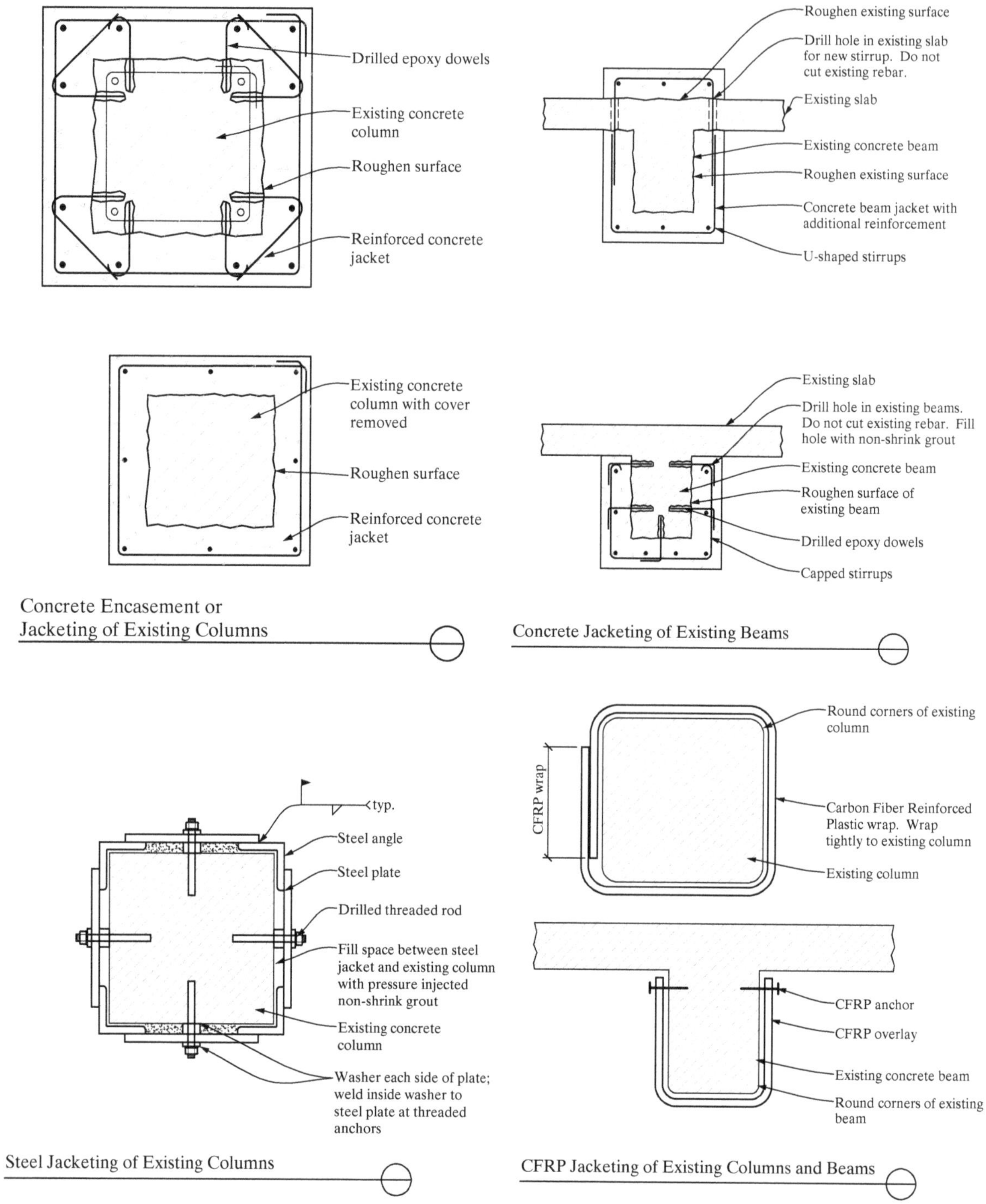

FIGURE 14.2 Concrete Repair and Strengthening Details (see Ref. [10]).

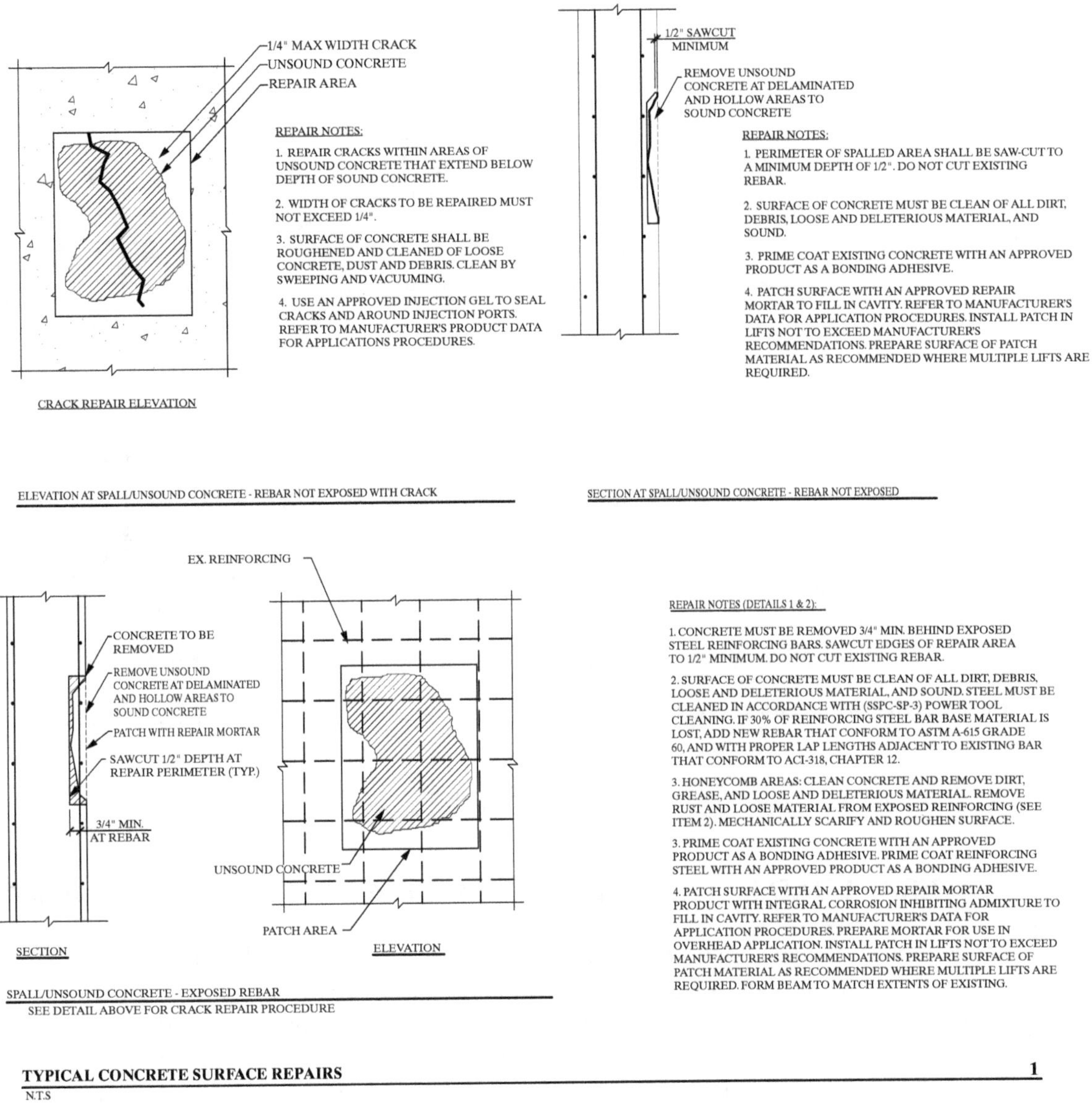

FIGURE 14.3 Concrete Repair Details.

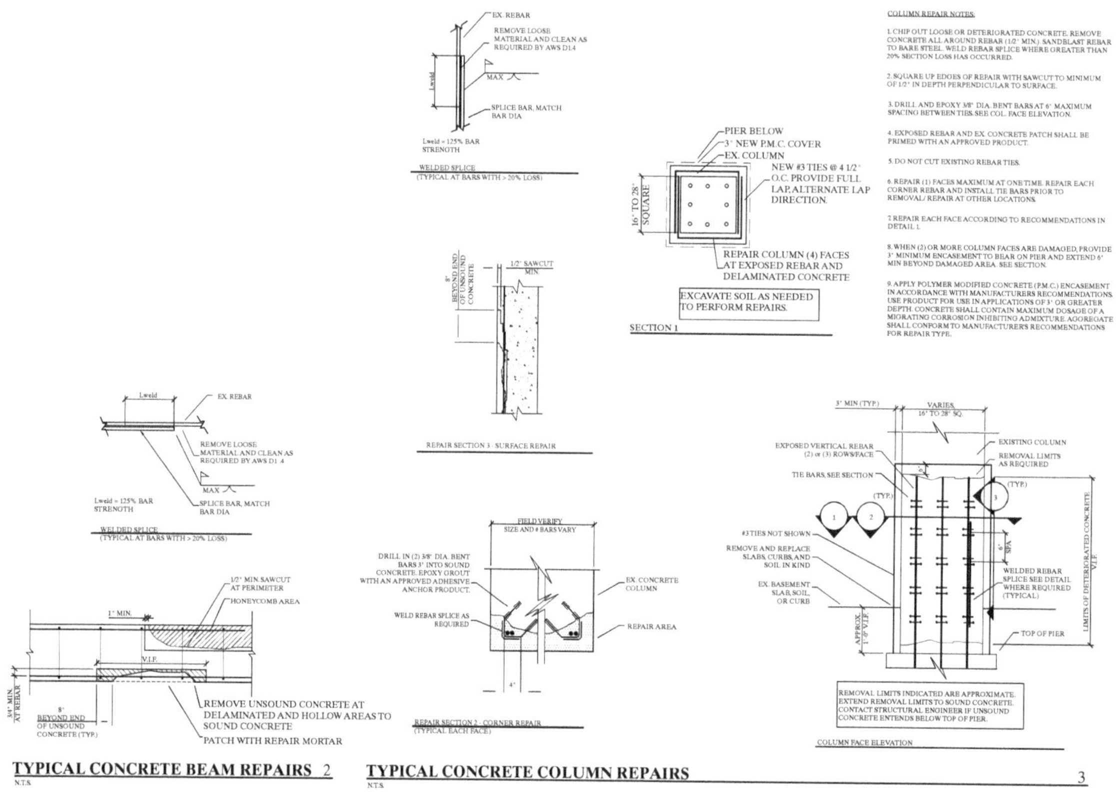

FIGURE 14.3 Continued.

14.5 LATERAL LOAD PATH, DIAPHRAGMS, DRAG STRUTS, AND CHORDS

The lateral wind load applied to a building structure causes vertical bending of the exterior wall cladding that is perpendicular to the wind direction. The exterior cladding then transfers the wind load to the floors and roof slab or the horizontal diaphragms of the building. The horizontal diaphragm then transfers the lateral loads to the vertical lateral force resisting systems (i.e., shear walls and/or moment frames) that are parallel to the lateral load direction. This is the lateral load path for wind loads. For seismic loads, the lateral load path is as follows: the ground shaking from the earthquake induces a lateral or horizontal base shear on the building that is then distributed to the various floor and roof diaphragms as a function of their heights from grade and the period of vibration of the building (see Chapter 3 of Ref. [37] for calculation of lateral wind and seismic loads). Similar to the wind load path, the horizontal diaphragms transmits the seismic forces to the vertical lateral force resisting systems (i.e., shear walls and/or moment frames) that are parallel to the seismic lateral load direction. In building structures, the floors and the roof act as deep horizontal plates or deep beams in transferring the in-plane lateral shear forces (from wind and seismic loads) to the vertical lateral force resisting systems (i.e., shear walls or moment frames). When floor and roof slabs, in addition to supporting gravity loads, resist the in-plane moments and shear forces from lateral wind, seismic or unbalanced soil pressure loads, they are referred to as diaphragms. The horizontal diaphragm consists of tension and compression chords at the diaphragm edges that are perpendicular to the direction of the lateral force, and drag struts or collectors that are parallel to the direction of the lateral force and located along the same vertical planes as the shear walls or moment frames. Diaphragms are covered in Chapter 12 of ACI 318-19. In the authors' experience, diaphragms, chords, and drag struts or collectors are some of the very few structural elements that are either ignored or taken for granted in the structural design of buildings, so it is instructive to note that a chapter is dedicated to this all-important structural element in the ACI Code. In high-rise buildings, the ground floor or podium concrete slab may be designed to act as a "super" diaphragm to transfer all or a portion of the lateral shears in the shear walls into the perimeter basement or slurry walls. In this and all other similar situations, the necessary reinforcement required in the floor diaphragm (including the chords and drag struts or collectors) to facilitate this transfer of lateral shears must be provided, in addition to the reinforcement required in the slab to support the gravity loads. See Figures 14.4 through 14.7b.

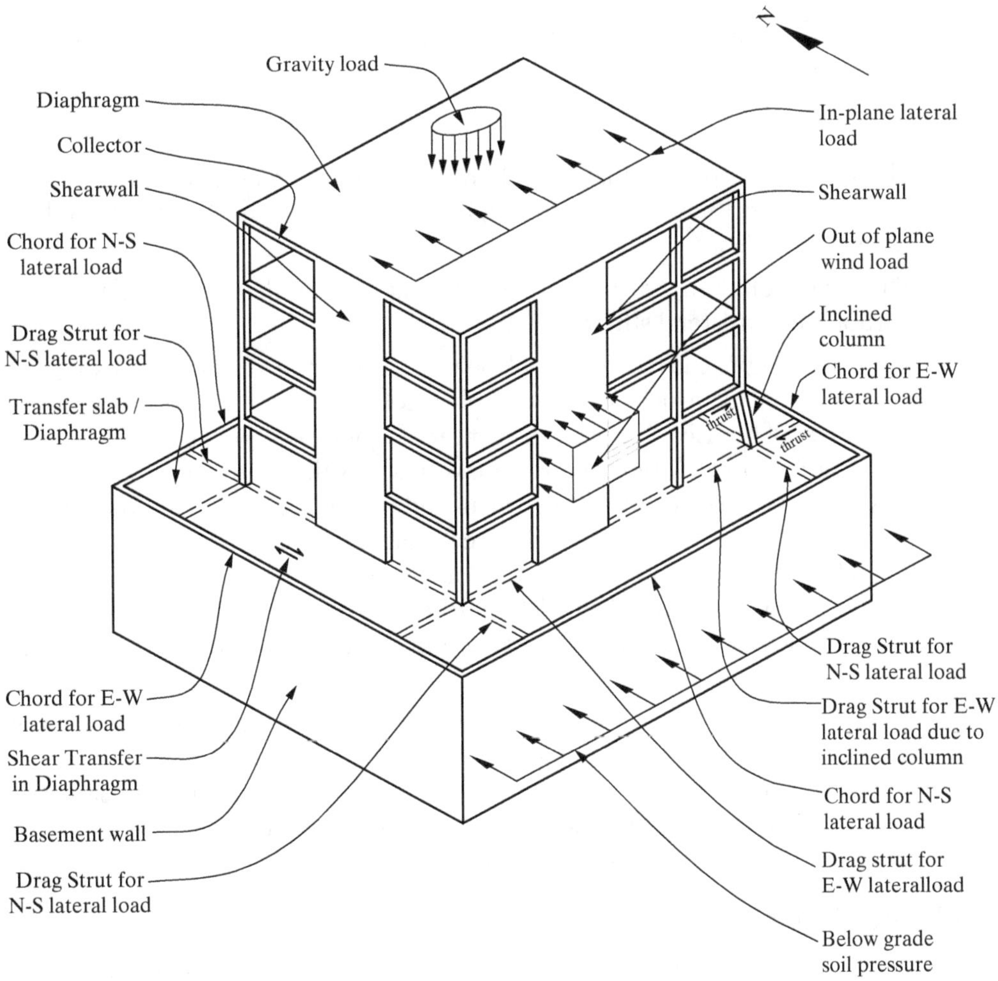

FIGURE 14.4 Diaphragms, drag struts, and chords: Multistory building. (Adapted from ACI 318-19 Figure R12.1.1 [18])

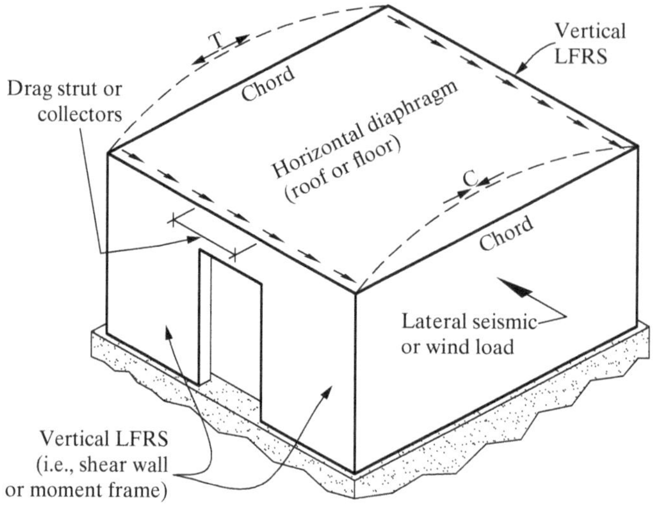

FIGURE 14.5 Diaphragms, drag struts, and chords: One-story building.

There are three types of horizontal diaphragms—rigid diaphragms, flexible diaphragms, and semi-rigid diaphragms:

Rigid diaphragms:

- These are floors or roof that are relatively stiffer in the horizontal plane when compared to the lateral stiffness of the shear walls or moment frames (i.e., the lateral force resisting systems (LFRS)).
- The lateral forces in structures with rigid diaphragms are distributed to the LFRS in proportion to their lateral stiffnesses.
- Rigid diaphragms can transmit in-plane rotational or torsional forces arising from the lateral force acting on the structure. These in-plane rotational forces are due to the non-symmetrical placement of the LFRS; that is, the center of rigidity (CR) of all the LFRS in the structure does not line up with the center of mass (CM) of the structure. The CM is the center of gravity at each floor or roof level, calculated for each orthogonal direction. The CR is the center of lateral stiffness of the LFRS calculated for each orthogonal direction. The calculation for CR only includes the LFRS that are parallel to the direction under consideration. The resultant wind and seismic lateral forces act through the CM at each floor/roof level. A method for calculating the CM and CR in diaphragms is covered in detail in Chapter 14 of Ref. [37].
- In structures with no horizontal structural irregularities, the concrete floor and roof slab can be considered rigid if the ratio of the maximum diaphragm span, S (i.e., the distance between adjacent LFRS that are parallel to the lateral force), to the diaphragm width, D_e (i.e., the diaphragm plan dimension in the direction parallel to the lateral force), is 3 or less (i.e., $S/D_e \le 3$). See ASCE 7, Section 12.3.1.2 [41].

Flexible diaphragms:

- These are floors or roofs with in-plane stiffness much smaller than the lateral stiffness of the shear walls or moment frames (i.e., the LFRS).

- The lateral forces in structures with flexible diaphragms are distributed to the LFRS in proportion to the tributary floor or roof area of each LFRS.
- Flexible diaphragms cannot transmit in-plane rotational or torsional forces; they can only transmit the direct lateral force.
- In structures with no horizontal structural irregularities, the concrete floor and roof slab can be considered flexible if the maximum diaphragm span-to-diaphragm width ratio, $S/D_e > 3$ (See ASCE 7, Section 12.3.1.3) [41].

Diaphragms that do not qualify as rigid or flexible diaphragms are termed semi-rigid diaphragms. The roof or floor diaphragm resists the in-plane moment and in-plane shear stresses caused by the lateral loads. It is not absolutely necessary to determine the type of diaphragm because ACI 318 Section 12.4.2.4 allows the envelope method to be used for the diaphragm analyses where the design forces are taken as the larger of the forces obtained from two separate analyses assuming rigid and flexible diaphragms. According to ACI 318 Sections 12.5.3.3 and 12.5.3.4, the in-plane shear strength of a reinforced concrete diaphragm is

$$\phi V_n = \phi A_{cv}\left(2\lambda\sqrt{f_c'} + \rho_t f_y\right) \leq \phi 8 A_{cv}\sqrt{f_c'}$$

where

$\phi = 0.75$

A_{cv} = "Cross-sectional area of the effective deep beam that forms the diaphragm" = diaphragm thickness x length of diaphragm parallel to the lateral force

ρ_t = Area of distributed diaphragm reinforcement oriented parallel to the in-plane shear divided by the gross concrete area of the diaphragm perpendicular to that reinforcement. Note that ρ_t represents the additional slab reinforcement used solely to resist in-plane lateral shear forces and can be conservatively assumed to be zero.

$\lambda = 1.0$ for normal weight concrete

$\sqrt{f_c'} \leq 100$ psi

As previously indicated, the floor or roof diaphragms consist of *drag struts* and *chords*. The chords are the edge members in the diaphragm perpendicular to the direction of the lateral force, and they resist the in-plane moment on the diaphragm caused by the lateral force through a force couple, with one chord in tension and the opposite chord in compression. In a concrete structure, the location of the additional top and bottom reinforcement in the slab (perpendicular to the direction of the lateral force) resulting from the axial forces in the chords are shown in Figures 14.4, 14.5, and 14.7a.

The drag struts or collectors are structural elements that lie along the same lines or vertical planes as the LFRS (i.e., shear walls and moment frames) and parallel to the direction of the lateral force. They help to transfer or collect or drag the lateral force from the roof/floor diaphragm into the shear walls or moment frames. The locations of drag struts are shown in Figures 14.4 through 14.6b, and 14.7b.

The transfer of the lateral force from the diaphragm into the shear wall can take place either directly through the in-plane shear resistance provided by shear friction rebar within the diaphragm that is oriented perpendicular to the lateral force, or through the drag struts. The drag strut in a reinforced concrete structure might consist of (1) additional reinforcement within the slab diaphragm parallel to the lateral force and developed into the shear wall, or (2) drag strut reinforcement parallel to the lateral force placed within a thickened concrete slab or collector beam to ensure sufficient slab-to-shear wall connectivity and to provide additional space to accommodate the reinforcement that will be embedded deep into the shear wall or moment frame (see Figure 14.7b).

The location of the additional top and bottom reinforcement in the slab (parallel to the direction of the lateral force) resulting from the axial forces in the drag struts or collectors can be seen in Figures 14.6a and 14.6b. In some instances, the drag strut may need to be a beam or thickened slab to accommodate the required amount of longitudinal reinforcement needed to transfer or drag the in-plane force into the shear wall or moment frame (see Figure 14.7b).

The tension-compression couple in a **chord member** is calculated as

$$T_u = \frac{M_u}{d}$$

where

M_u = in-plane moment in the diaphragm
d = perpendicular distance between the tension and compression chords
≈ 0.95x length of diaphragm parallel to the in-plane shear [38].

If the lateral shear from the slab diaphragm cannot be transferred directly into a shear wall or moment frame through shear friction, the excess lateral force that cannot be transferred by shear friction will have to be "dragged" into the shear wall by the drag struts. The maximum tension force, T_u, in the drag struts is therefore the difference between the lateral force resisted by a shear wall and the lateral force that can be transferred directly from the slab diaphragm into the shear wall by shear friction and dowels.

The required area of the additional reinforcement due to a chord or drag strut force can be calculated using the relationship,

$$A_s = \frac{T_u}{\phi f_y}$$

For drag struts and chords subjected to seismic loads, the required area of additional reinforcement is

$$A_S = \frac{\Omega_O T_u}{\phi f_y}$$

where

T_u is the factored axial force in a chord or drag strut member
ϕ is the strength reduction factor = 0.9, and
f_y is the yield strength of the reinforcement
Ω_O = overstrength factor for seismic analysis from ASCE-7 Section 12.10.2.1 [41].

The roof and floor diaphragms need to be reinforced to transfer the lateral shear to the drag struts or collectors, and from the collectors to the shear walls and moment frames. The transfer of the drag strut force into the shear walls and moment frames is accomplished through shear friction (See ACI Code Sections R12.5.3.7 and R12.5.4.1) and/or through longitudinal reinforcement in the drag strut that resists direct in-plane tension or compression force. The shear friction reinforcement is placed perpendicular to the drag strut tension/compression reinforcement, as shown in Figure 14.6b.

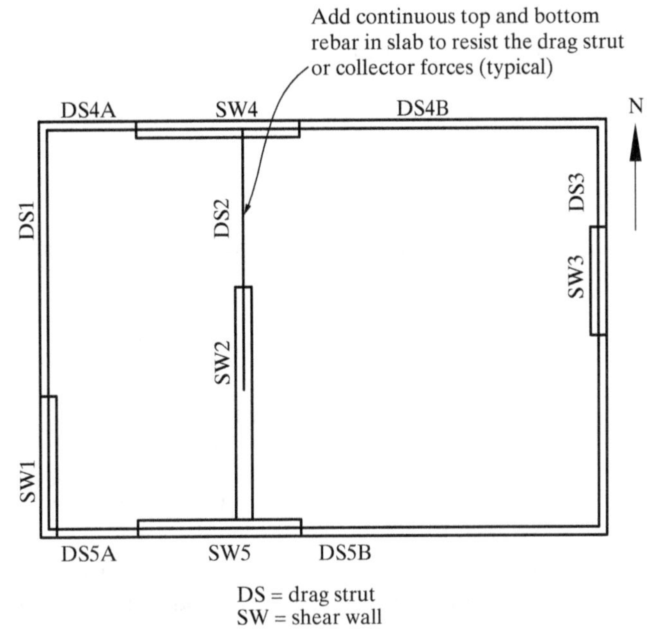

FIGURE 14.6a Plan view of a typical floor or roof diaphragm (slab) showing **drag struts** and additional reinforcement.

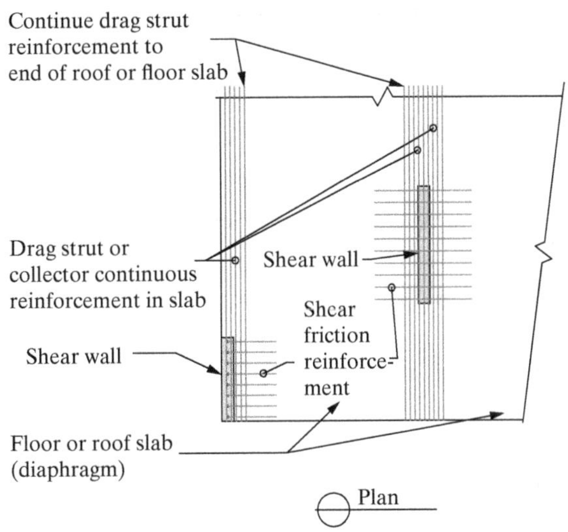

FIGURE 14.6b Drag strut and shear friction reinforcement in a typical floor or roof diaphragm.

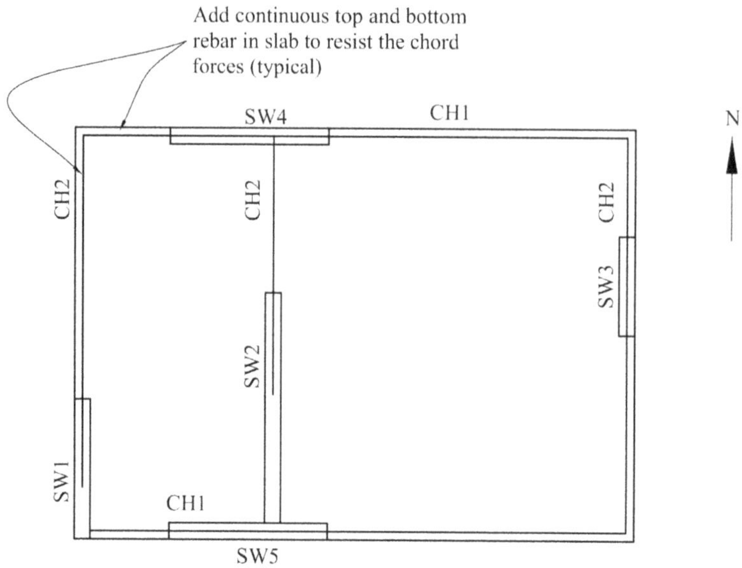

CH1 = Chord and additional reinforcement for N-S lateral force

CH2 = Chord and additional reinforcement for E-W lateral force

FIGURE 14.7a Plan view of a typical floor or roof diaphragm (slab) showing **chord**s and additional reinforcement.

Below are the plan and section detail of a drag strut and the forces for which the different elements have to be designed to ensure adequate lateral load transfer in a floor diaphragm with openings.

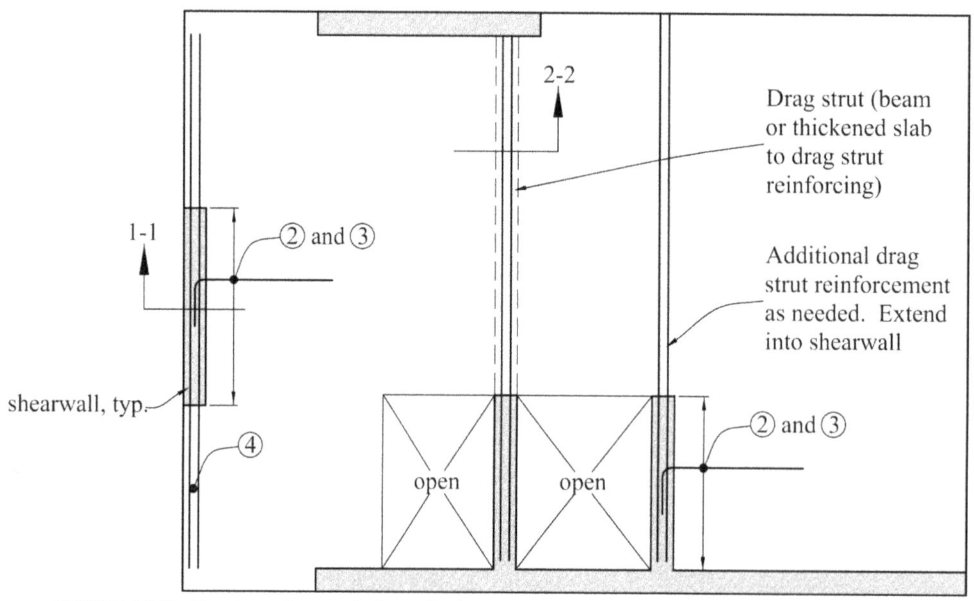

FIGURE 14.7b Plan view of a sample floor diaphragm with openings, and sections showing **drag struts** and additional reinforcement for N-S lateral force only.

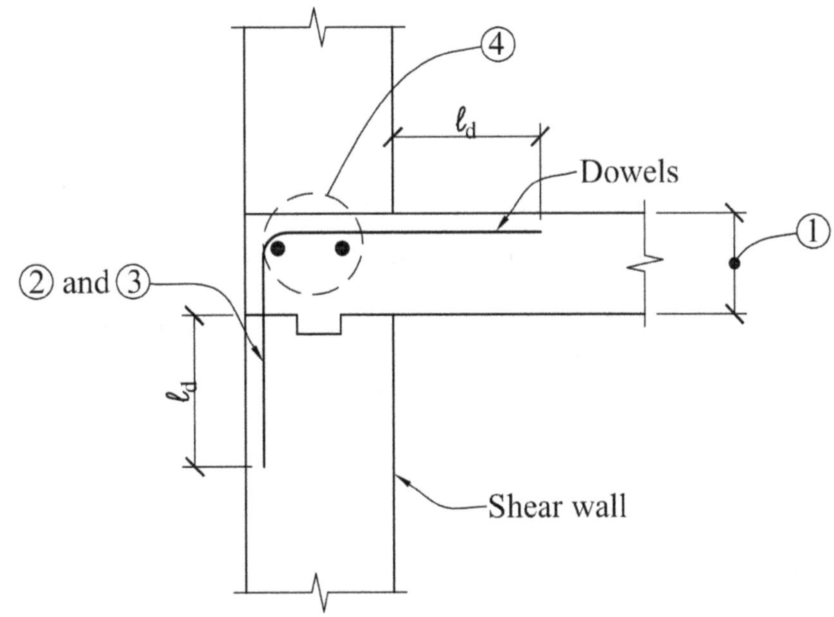

Note: Slab and shear wall reinforcement not shown

Section 1-1

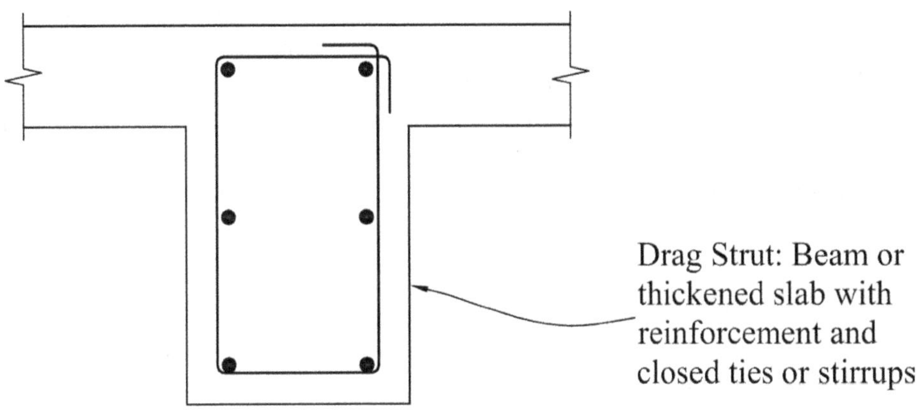

Note: Slab reinforcement not shown

Section 2-2

FIGURE 14.7b contd. Plan view of a sample floor diaphragm with openings, and sections showing **drag struts** and additional reinforcement for N-S lateral force only.

Legend for Figure 14.7b:

1. Check that the slab has adequate thickness to resist the in-plane lateral shear parallel to the shear wall (ignoring the slab reinforcement parallel to the lateral load). That is, check that $\phi V_n\ (slab) = \phi A_{cv}\left(2\lambda\sqrt{f'_c}\right) \geq V_u$ along the shear wall or moment frame lines, where all the terms in the equation are as previously defined.
2. Check if the slab can transfer the lateral shear, V_u along the shear wall or moment frame lines from the slab diaphragm directly into the shear wall through the construction joint via shear friction, otherwise drag strut reinforcement or reinforced drag strut members are required. The required area of dowels, A_{vf}, is calculated as follows using the requirement that $\phi V_n = \mu\phi A_{vf} f_y \geq V_u$ along the shear wall or moment frame lines (ACI 318 Section 22.9.4.2):

   $$A_{vf} = \frac{V_u}{\mu\phi f_y}$$

   where

   μ = coefficient of friction = 0.6, when the slab-to-shear wall connection is not roughened intentionally.

   ϕ = 0.75

   f_y = yield strength of steel reinforcement
3. These dowels within the length of the shear wall (i.e., A_{vf}) must be sufficient to transfer all the in-plane lateral force or shear from the diaphragm into the shear wall, otherwise drag strut reinforcement or reinforced drag strut members are required.

   The shear friction or dowel reinforcement, A_{vf}, calculated in item #2 above is assumed to be uniformly distributed over the full depth of the diaphragm (parallel to the lateral load) because of the presence of the drag struts. If no drag struts are provided, then A_{vf} will have to be distributed over the length of the shear wall only.
4. Beyond the shear wall footprint, the reinforced drag strut member or drag strut reinforcement must resist the drag strut tension force, T_u (refer to Example 14.2 for a sample calculation of drag strut axial forces). The required tension reinforcement in the drag strut or collector is $A_s = \dfrac{T_u}{\phi f_y}$ or $A_s = \dfrac{\Omega_o T_u}{\phi f_y}$.

   where ϕ = 0.9

The presence of large floor openings due to elevators, stairs, or shafts adjacent to a shear wall, especially when the openings are on both sides of the shear wall, requires that collectors or drag struts be located at such a shear wall to ensure an adequate lateral load path from the floor diaphragm into the shear wall, otherwise the lateral load transfer will not occur. That is, because of these floor openings, the shear wall will not receive the lateral load that the designer may have assigned to it, and this collector or drag strut deficiency could be catastrophic for the building. A retrofit for a hospital building with this type of collector or drag strut deficiency using "externally bonded fiber-reinforced polymer (FRP) collectors" was recently completed in California [40].

EXAMPLE 14.2

Diaphragm, drag strut, and chord forces

The roof diaphragm shown in Figure 14.7c is subjected to the lateral force of 110 kip acting in the N-S direction. Assume a 6 in. thick slab, an 8 in. thick shear wall W1, a height of 15 ft, and a concrete strength at 28 days of 4000 psi.

a. Calculate the maximum shear in the diaphragm along line 1 and check the adequacy of the diaphragm to resist this lateral shear. Ignore any inplane torsional moments caused by actual or accidental eccentricities. See Ref. [37] for how to account for the impact of these eccentricities on the lateral forces in lateral force resisting systems.

b. The maximum force in the drag strut along line 1 and the required reinforcement. Neglect the gravity loads on the drag strut.

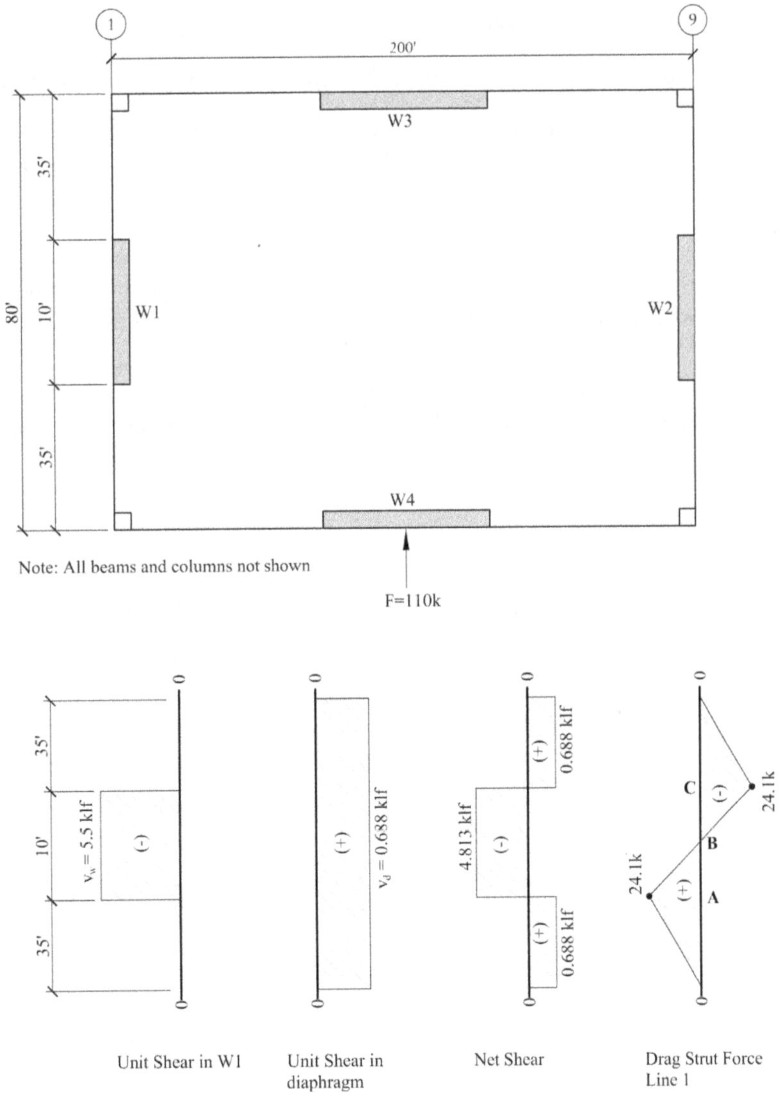

FIGURE 14.7c Plan view of roof diaphragm for Example 14.2

c. Calculate the maximum shear and overturning moment in shear wall W1.

d. Determine the distributed horizontal and vertical reinforcement in the shear wall.

e. Neglecting the gravity load on the shear wall, determine the end zone vertical reinforcement in the shear wall.

f. Design the diaphragm chord reinforcement.

g. Determine the dowel reinforcement required between the slab and shear wall W1 assuming shear friction.

Solution:

a. Diaphragm:

For the N-S lateral force, the distance, S, between adjacent shear walls that are parallel to the lateral force (i.e., shear walls W1 and W2) is S = 200 ft (see Figure 14.7c).

The depth of the diaphragm parallel to the lateral force, D_e = 80 ft

S/D_e = 200 ft/80 ft = 2.5 < 3, therefore, this qualifies as a rigid diaphragm

Therefore, the lateral force of 110 kip will be distributed to shear wall W1 and W2 in proportion to their stiffness. Since walls W1 and W2 have the same length and thickness, their stiffness will be the same, and therefore, the factored lateral force on wall W1 along line 1 = 110 kip/2 = 55 kip.

The lateral force calculated above ignores any accidental inplane eccentricities and the resulting inplane torsional moments. The full depth of the diaphragm of 80 ft is available to resist this shear because of the drag struts that will be provided along line 1 from the shear wall W1 to the North and South edges of the diaphragm. If no drag struts were provided, then the depth of the diaphragm that would have been available to resist the inplane shear would have been only 10 ft or the length of the shear wall.

The shear strength of the diaphragm (ignoring any rebar in the slab parallel to the lateral force) is

$$\phi V_n\ (\text{slab}) = \phi A_{cv}\left(2\lambda\sqrt{f_c'}\right) = (0.75)(6\,\text{in.})(80\,\text{ft.})(12\,\text{in./ft.})(2)(1)\left(\sqrt{4000}\right)$$
$$= 546\,\text{kip} > V_u = 55\,\text{kip} \quad (\text{O.K.})$$

b. Drag strut and reinforcement:

The unit shears in the shear wall W1 and the unit shears in the horizontal diaphragm along grid line 1, and the net unit shears and drag strut forces along line 1 are shown in Figure 14.7c. The drag strut forces at points A, B, and C are calculated as follows:

Point A (South end of wall W1): $(0.688\ \text{kip/ft.})(35\,\text{ft.}) = 24.06\ \text{kip}$

Point B (Mid-length of wall W1): $(0.688\ \text{kip/ft.})(35\,\text{ft.}) - (4.813\ \text{kip/ft.})\left(\dfrac{10\,\text{ft.}}{2}\right) = 0\ \text{kip}$

Point C (North end of wall W1): $(0.688\ \text{kip/ft.})(35\,\text{ft.}) - (4.813\ \text{kip/ft.})(10\,\text{ft.}) = -24.06\ \text{kip}$

Therefore, the maximum force in the drag strut along Grid line 1 = 24.06 kip

The longitudinal reinforcement in the drag strut is $A_s = \frac{T_u}{\phi f_y} = \frac{24.06 \text{ kip}}{(0.9)(60 \text{ ksi})} = 0.45 \text{ in.}^2$

Try 2#5 bars in the drag strut.

A_s provided = 0.62 in.^{2} > As required (O.K.)

This reinforcement will be extended an adequate length into shear wall W1 so that the reinforcement is fully developed in tension (see Chapter 5).

c. Maximum Shear and Overturning Moment in Shear Wall W1:

$V_u = 55 \text{ kip}$

$\ell_{W1} = (10\text{ft.})(12 \text{ in./ft}) = 120 \text{ in.}$

$\frac{h_{W1}}{\ell_{W1}} = \frac{15\text{ft.}}{10\text{ft.}} = 1.5, \alpha_c = 3$ (See Chapter 8)

$0.5\phi\alpha_c\lambda A_{cv}\sqrt{f_c'} = 0.5(0.75)(3)(1)(8 \text{ in.})(120 \text{ in.})\sqrt{4000} = 68 \text{ kip} > V_u$ (O.K.)

Therefore, only minimum distributed vertical and horizontal reinforcement is required in shear wall W1 (see Chapter 8).

Maximum overturning moment, M_u = (55 kip)(15 ft.) = 825 ft.-kip

d. Distributed vertical or longitudinal reinforcement in the shear wall W1:

Using the methods for the design of shear walls presented in Chapter 8, minimum distributed reinforcement is required in the shear wall W1 (see ACI 318, Table 11.6.1). The required distributed minimum reinforcement is as follows:

Maximum spacing (ACI 318 Section 11.7.2.1) = 3h or 18 in

That is, 3(8 in.) = 24 in. or 18 in. Use 18 in. maximum spacing.

Assuming No. 4 bars: $A_{s,min}$ (vertical) = (0.0012)(8")(12") = 0.12 in.^{2}/ft

Use #4 at 18 in. o.c. VEF

A_s provided = 2 faces x 0.13 in.^{2}/ft = 0.26 in.^{2}/ft > $A_{s,min}$ required O.K.

Distributed horizontal or transverse reinforcement in shear wall W1:

Maximum spacing (ACI 318 Section 11.7.2.1) = 3h or 18 in

That is, 3(8 in.) = 24 in. or 18 in. Use 18 in. maximum spacing.

Assuming No. 4 bars: $A_{s,min}$ (horizontal) = (0.002)(8")(12") = 0.192 in.^{2}/ft

Use #4 at 18 in. o.c. VEF

A_s provided = 2 faces x 0.13 in.^{2}/ft = 0.26 in.^{2}/ft > $A_{s,min}$ required O.K.

e. End zone vertical reinforcement in Shear Wall W1:

The maximum factored shear and bending moment, respectively, at the base of the wall due to the factored lateral load are

$V_u = 55$ kip

$M_u = (55\,\text{kip})(15\,\text{ft}) = 825\,\text{ft.-kip}$

The limit states design equation for flexure requires that $\phi M_n = \phi h d^2 \bar{k} \geq M_u$. Initially, assume ϕ = 0.9 (this will be checked later after the net tensile strain in the extreme tension steel, ε_t, is determined) and then calculate the required $\bar{k}$ as follows:

Assume the effective flexural depth, $d = 0.8\ell_w = (0.8)(120\text{ in.}) = 96\text{ in.}$

$$\bar{k} = \frac{M_u}{\phi h d^2} = \frac{825(12)}{(0.9)(8\text{ in.})(96\text{ in.})^2} = 0.15$$

From Table A.10, we obtain $\rho = 0.0026$ and $\varepsilon_t \gg 0.005$; therefore, $\phi = 0.9$, as initially assumed. The concentrated vertical reinforcement required at each end zone of the shear wall is

$$A_s\text{ required} = \rho h d = (0.0026)(8\text{ in.})(96\text{ in.}) = 2.0\text{ in.}^2$$

The minimum area of concentrated vertical steel required for bending at the ends of the shear wall is

$$A_{s,\min} = \frac{3\sqrt{f'_c}}{f_y} hd \geq \frac{200}{f_y} hd$$

$$= \frac{3\sqrt{4000}}{60{,}000}(8)(96) = 2.42\text{ in.}^2$$

$$\geq \frac{200}{60{,}000}(8)(96) = 2.56\text{ in.}^2$$

Because A_s required < $A_{s,\min}$ = 2.56 in.^{2} Use $A_{s,\min}$ = 2.56 in.^{2}

Try 4 No. 8 vertical reinforcement at each end of the wall (i.e., 2 No. 8 VEF)

Total area of steel provided at each end of the wall = 3.16 in.^{2} > A_s required (O.K.)

Provide additional ties of No. 4 @ 8 in. on centers to the end zone reinforcement.

f. Diaphragm chord reinforcement:

The chords of the diaphragm for the lateral force shown is located at the slab edges that are perpendicular to the lateral force (i.e., the slab edges where the shear walls W3 and W4 are located). The diaphragm in-plane moment due to the lateral force, F = 110 kip uniformly distributed over the 200 ft length of the diaphragm is calculated as

$$M_u = \frac{w\ell^2}{8} = \frac{\frac{110\text{ kip}}{200\text{ ft.}}(200\text{ ft})^2}{8} = 2750\text{ ft.-kip}$$

The diaphragm chord forces are calculated as

$$T_{chord} = C_{chord} = \frac{M}{d} = \frac{2750 \text{ ft.-kip}}{(0.95)(80\,\text{ft})} = 36.2\,\text{kip}$$

The area of reinforcement required in the diaphragm chords is calculated as

$$A_s = \frac{T_u}{\phi f_y} = \frac{36.2\,\text{kip}}{(0.9)(60\,\text{ksi})} = 0.67 \text{ in.}^2$$

Provide 2#6 (i.e., 1#6 Top and Bottom) continuous chord reinforcement at the edge of the slab located parallel to shear walls W3 and W4. A_s provided = 0.88 in.^{2} > 0.67 in.^{2} O.K.

g. Shear friction reinforcement between diaphragm and shear wall W1 (i.e., Dowel reinforcement):

We will provide adequate dowel reinforcement between the slab diaphragm and the shear wall W1 to ensure that the full lateral shear of 55 kip along line 1 is transferred.

The coefficient of friction μ is 0.6 for hardened concrete connected to hardened concrete, and 1.4 for monolithic pour. In this case, $\mu = 0.6$

Since drag struts are provided, the full 80 ft depth of the diaphragm is available for the transfer of lateral shear from the diaphragm into the shear wall.

The shear capacity of the dowels in shear friction is calculated as

$$\phi V_n = \phi \mu A_{vf} f_y \geq V_u$$
$$= (0.75)(0.6) A_{vf} (60) = 55 \text{ kip}$$

$$A_{vf} = \frac{55 \text{ kip}}{(0.75)(0.6)(60)} = 2.04 \text{ in.}^2$$

Therefore, the area of dowel reinforcement required per foot depth of diaphragm = 2.04 in.^{2}/80 ft. = 0.026 in.^{2}/ft

Therefore, provide No. 4 dowels at 18 in. o.c. Top in the slab to connect the slab to shear wall W1 and to the drag struts, The area of steel furnished is 0.13 in.^{2}/ft > A_{vf} required = 0.026 in.^{2}/ft (O.K.)

14.6 ONE-WAY SLABS SUBJECTED TO CONCENTRATED LOADS

There are a number of situations in practice where one-way slabs may be subjected to concentrated loads. Examples include slabs that support wheel loads due to trucks or forklift or heavy safes, or slabs that support concentrated loads from hung equipment. The question of what effective width to use in the design of one-way slabs in these situations is not readily obvious.

For hollowcore planks in bending, the Spancrete Manufacturers Association [20–22] recommends for bending moment calculations an effective width at the midspan of 0.50ℓ for an interior concentrated load or line load, and 0.25ℓ at the midspan when the concentrated or line load is close to an unsupported edge of the slab, where ℓ is the span of the hollowcore plank. The load near an unsupported slab edge must be a minimum of 1 ft from the slab edge. For shear design, the effective width varies from 4 ft to a maximum of 0.50ℓ at a distance of 0.25ℓ from the support face for interior loads. For loads close to an unsupported edge, the effective width for shear design varies from 1.0 ft to a maximum of 0.25ℓ at a distance of 0.25 ℓ from the support face. In summary, the effective widths are as follows:

Bending Moment:

Interior Concentrated Loads at slab midspan –

Effective width, $b_{eff} = 0.50\ell$

Concentrated Loads at slab midspan Close to Unsupported Slab Edge –

Effective width, $b_{eff} = 0.25\ell$

Shear:

Interior Concentrated Loads –

Effective width at a distance x(ft) from the face of support is:

$$b_{eff} = 4.0\,\text{ft.} + \left[\frac{0.50\ell - 4.0\,\text{ft}}{0.25\ell}\right]x \le 0.50\ell$$

Concentrated Loads Close to Unsupported Slab Edge –

Effective width at a distance x(ft) from the face of support is:

$$b_{eff} = 1.0\,\text{ft} + \left[\frac{0.25\ell - 1.0\,\text{ft}}{0.25\ell}\right]x \le 0.25\ell$$

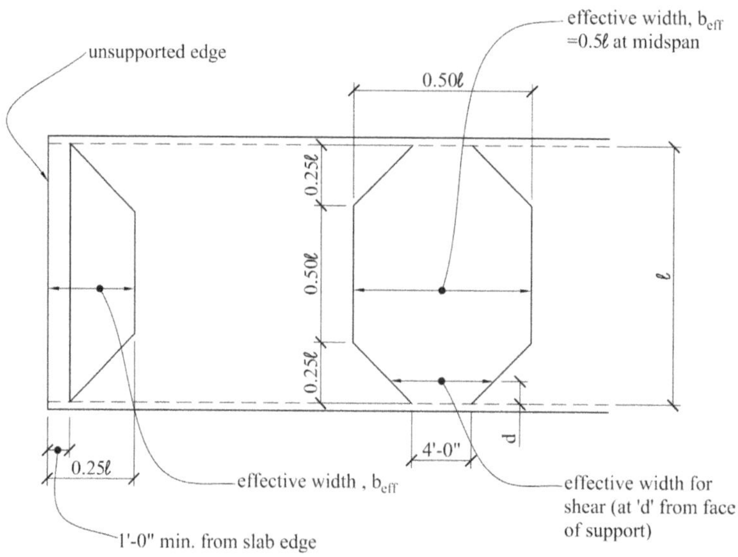

FIGURE 14.8 Effective widths for one-way slabs subject to concentrated loads (adapted from Ref. 21 and 22).

Effective width equations similar to the above have been recommended by others [23, 24]. Note that one-way slabs subjected to concentrated loads (e.g., forklift or truck loads) should also be checked for two-way or punching shear (see Chapter 6).

EXAMPLE 14.3

One-way slabs under concentrated loads

A one-way slab spanning 12 ft. between simple rigid supports is subjected to a uniformly distributed dead load of 100 psf (includes its self-weight and superimposed dead load) and a concentrated live load of 12 kip at midspan applied in the interior of the slab. Calculate the maximum design moment in ft.-kip/ft. width of the slab and the maximum shear in kip/ft. width of slab.

Solution:

Bending Moment:

For the concentrated load, the effective width at midspan = $0.5\ell = 0.5(12 \text{ ft.}) = 6$ ft.

Therefore, the concentrated load is 12 kip/6 ft. = 2 kip/ft width of slab

$P_u = 1.6(2 \text{ kip/ft}) = 3.2$ kip/ft.

$w_u = 1.2(100 \text{ psf})(1 \text{ ft width of slab}) = 120$ Ib/ft = 0.12 kip/ft width of slab

The maximum design moment at the slab midspan due to the uniformly distributed dead load and the concentrated live load at the midspan is calculated as

$$M_u = \frac{P_u \ell}{4} + \frac{w_u \ell^2}{8} = \frac{(2 \text{ kip/ft})(12 \text{ ft.})}{4} + \frac{(0.12 \text{ kip/ft.})(12 \text{ ft.})^2}{8}$$
$$= 8.16 \text{ ft.-kip/ft. width of slab}$$

Shear Force:

For an interior concentrated load (i.e., not close to an unsupported slab edge), the effective width at the face of the support (i.e., at x = 0) = 4 ft.

Therefore, the concentrated load for calculating the shear at the slab support is 12 kip/4 ft. = 3 kip/ft width of slab

$P_u = 1.6(3 \text{ kip/ft}) = 4.8$ kip/ft.

$w_u = 1.2(100 \text{ psf})(1 \text{ ft width of slab}) = 120$ Ib/ft = 0.12 kip/ft width of slab

The maximum design shear at the slab support due to the uniformly distributed dead load and the concentrated live load at the midspan is calculated as

$$V_u = \frac{P_u}{2} + \frac{w_u \ell}{2} = \frac{4.8 \text{ kip/ft.}}{2} + \frac{(0.12 \text{ kip/ft.})(12 \text{ ft.})}{2}$$
$$= 3.12 \text{ kip/ft. width of slab}$$

14.7 LOAD TESTING OF STRUCTURES

The load testing of an existing structure may become necessary for a number of reasons: (1) there is an increase in the occupancy load and the load capacity of an existing concrete structure cannot be determined analytically because there are no available structural drawings and the size and location of the existing reinforcement cannot be determined; or (2) the existing structure has been damaged due to corrosion, fire, or other causes, and the in situ load capacity is desired; or (3) it is necessary to determine the load capacity of a repaired or retrofitted structure; or (4) it is necessary to determine the effects of construction or design errors such as misplacement or omission of reinforcement or the effect of lower-than-specified concrete strength on the load-carrying capacity of a structure. There are two available methods for conducting these load tests on existing structures [12–19]:

1. The static and monotonic 24-hour uniformly distributed load test; this is the most commonly used load test.
2. The cyclic load test (this is a more rapid load test). For a more detailed discussion of this load test, the reader should refer to Ref. [12].

The static and monotonic load tests have been in use since the 1,800s, and ACI Code Chapter 27 provides the procedure for evaluating the strength of existing structures using the static and monotonic load test. Other standards for load testing include Section 1708 of the IBC—In situ Load Tests and ASTM E196-06—Standard Practice for Gravity Load Testing of Floors and Low Slope Roofs. The loading materials that can be used for the load test include non-hydraulic loads such as sandbags, steel plates, CMU blocks, and water containers (with a water meter). These materials can be stacked to achieve the desired load. Care should be taken if these materials are placed on the structure using forklifts or other equipment whose own weights could negatively affect the safety of the structure [16, 17]. These non-hydraulic loading methods can only be used for static and monotonic load tests. The lack of ability to control or promptly reduce the non-hydraulic loads in case of imminent failure of the structure is one of the disadvantages of these types of loads. On the other hand, hydraulic loading can be used for both static and cyclic load tests, and it has the obvious advantage that the applied load can be controlled, thus enhancing safety during a load test.

Safety is of paramount importance during a load test and shoring towers below the structural element to be tested must be provided. The top of the shoring tower should be located below the tested structure, low enough to allow the structural element to deflect freely but high enough to prevent collapse of the structural element.

Static and Monotonic Uniformly Distributed Load Test

According to Section 27.4 of the ACI 318 Code, the requirements and specifications for the strength evaluation of an existing structure using a static and monotonic load test are as follows (see Figure 14.9):

1. The minimum age of the structure at the time of the load test is 56 days. This requirement can be waived and the structure tested at a younger age if all the stakeholders involved with the structure including the owner, the contractor, and the structural engineer are in agreement.
2. The duration of time over which the total test load is held constant on the structure is 24 hours.

3. The total test load is applied in a minimum of four equal increments (ACI 318, Section 27.4.3.1).
4. Survey the structure and record the maximum deflections and other load effects at each load increment, allowing about one hour between each load increment to accomplish these tasks. Within 1 hour before the start of the load test, the initial deflection of the structural member, $\Delta_{initial}$, should be taken (ACI 318, Section 27.4.4.2).

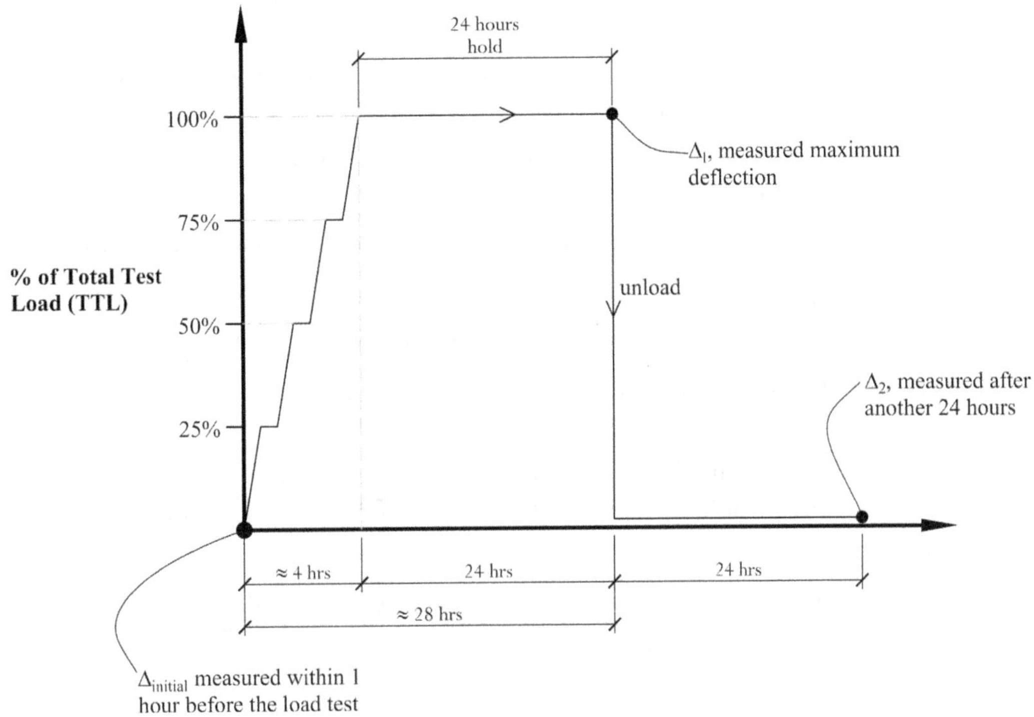

FIGURE 14.9 Load-time graph for initial load test.

5. Measure and record the deflection of the structural element at each loading stage. After the total test load has remained on the structure for 24 hours, measure the deflection of the structural element, Δ_1. The load is left on the structure for the 24-hour duration to account for the effect of creep on the deflections. The test load is incrementally removed from the structure, again allowing about one hour between each unloading stage, and the deflections are measured at each unloading increment, and measured again 24 hours after the full test load has been removed from the structure, and this final deflection is denoted Δ_2. This will give an idea as to how much deflection rebound occurred in the structure.
6. Check for cracking or spalling or crushing of concrete at each load stage.
7. The test load arrangement and location should be selected so as to maximize the deflection and load effects on the member that is being evaluated (ACI 318, Section 27.4.2.1). The load combinations for the load test (ACI 318, Section 27.4.2.2) are as follows:
   a. $T_t = 1.0D_w + 1.1D_s + 1.6L + 0.5(L_r \text{ or } S \text{ or } R)$
   b. $T_t = 1.0D_w + 1.1D_s + 1.0L + 1.6(L_r \text{ or } S \text{ or } R$
   c. $T_t = 1.3(D_w + D_s)$

where

D_w = self-weight dead load
D_s = superimposed dead load
L = live load
L_r = roof live load
S = snow load
R = rain load
T_t = total load per ACI Code including the self-weight of the member
TTL = total test load = $T_t - D_w$

The total test load, TTL, is applied in four equal increments, as shown in Figure 14.9 (ACI 318, Section 27.4.3)

8. The structure is deemed capable of supporting the applied test load if Criterion (a) *or* (b), *and* Criterion (c) are satisfied:

   a. The maximum deflection, $\Delta_t \leq$ larger of 0.05-in. and $\dfrac{\ell_t}{2000}$

   b. The residual deflection, $\Delta_r = \Delta_2 - \Delta_{initial} \leq \dfrac{\Delta_1 - \Delta_{initial}}{4}$ (that, is 75% or more deflection recovery after 48 hours)

   c. There should be no crushing or spalling of concrete, or cracks indicating imminent shear failure (ACI 318, Sections 27.4.5.1 and 27.4.5.2).

where,

ℓ_t = span of the concrete member in inches (for two-way slabs, use the shorter span).
Δ_t = maximum deflection in inches
h = overall depth of the member in inches

The equations above were derived based on simple-span members.

Second or Repeated Load Test

If the structure does not satisfy the first two criteria of the first load test for the load rating considered, a second load test can be carried out after at least 72 hours after the removal of the test load from the first load test (see Figure 14.10). The structure shall be deemed acceptable for the second load test if there are no concrete cracks or spalling or crushing, and if the following acceptance criterion (ACI 318, Section 27.4.5.7) is satisfied:

$$\Delta_{r(repeated)} = \Delta_{2(repeated)} - \Delta_{initial(repeated))} \leq \frac{\Delta_{1(repeated)}}{5}$$

If a structure fails to satisfy the acceptance criterion for the second or repeated load test or the first two acceptance criteria for the initial load test, a reduced maximum load rating can be calculated for the structure or structural member using these acceptance criteria.

A Safety plan is essential for each load test to prevent injury or death of personnel, and to prevent structural damage or collapse. In establishing a safety plan, the following should be considered:

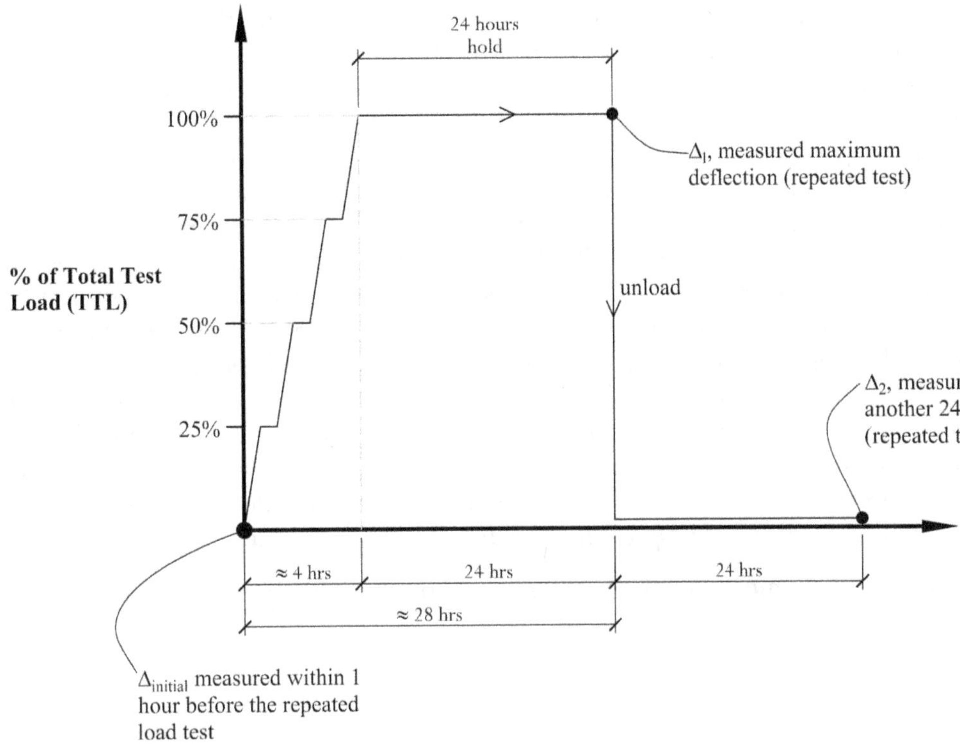

FIGURE 14.10 Load-time graph for second or repeated load test.

- The safety plan should include a shoring tower or platform below the structural member that is being load tested, to catch the member or structure in case of sudden failure, and to prevent progressive collapse.
- The amount of gap needed between the underside of the structural member being tested and the shoring tower that allows enough clearance for the structural member to deflect freely and without restraint is obtained from the structural analysis of the member.
- The deflection gap should not be too large to prevent large deformations of the structure that is being load tested.
- The shoring tower should be designed and detailed to be capable of supporting the self-weight of the structural member plus the test load in case of sudden failure.

14.8 CLOSURE OR POUR STRIPS IN REINFORCED CONCRETE FLOORS

As concrete cures, it shrinks and this movement, if restrained, will lead to shrinkage cracks. The larger the concrete pour, the greater the shrinkage movement, unless strategies are adopted to reduce the movement due to shrinkage as well as minimize the restraint to shrinkage, and thus control the distribution and size of the shrinkage cracks. For reinforced concrete floor or roof plans larger than approximately 200 ft to 250 ft, the movement due to shrinkage can be mitigated through the use of expansion joints or through the use of closure or pour strips. Shrinkage-compensating cement can also be used in the concrete mix. Closure strips are also called shrinkage strips or construction or pour gaps. They are often used in multistory building construction and involve pouring adjacent portions of the concrete floor, leaving a gap between two adjacent concrete pours to allow the two portions of the concrete floor to shrink independently of each other without restraint from each portion. The pour strip is then filled in with concrete at a later date. Closure strips can be provided on floors in two orthogonal directions so that no concrete pour is larger than 200 ft × 200 ft in plan. The width of the closure strip is typically 3 ft to 5 ft and should be large enough to accommodate a tension lap splice for the slab or beam reinforcement, since the slab or beam reinforcement on either side of the pour gap must be continuous through the closure strip, but "physically independent of each other" until the pour gap is filled in. The closure strip or pour gap is subsequently filled in after 14 days to 90 days [29]. In some structures where shrinkage cracks must be greatly minimized, the length of time it takes before filling in the closure strip might be much longer and could be as long as 120 days. Before the closure strip is filled in, the primary shoring for the floor slab and beams near the closure strip and the adjacent floor bays must be kept in place, and fall prevention measures at the location of the closure strip should be installed by the contractor. The use of closure strips and the length of time that the strip is unfilled can greatly impact the construction cost and construction schedule of a project. Recently, "lockable dowels" have been used to *obviate* the need for shoring near the closure strips [30].

Typical reinforcement details for closure strips in slabs, beams, and walls are shown in Figure 14.11.

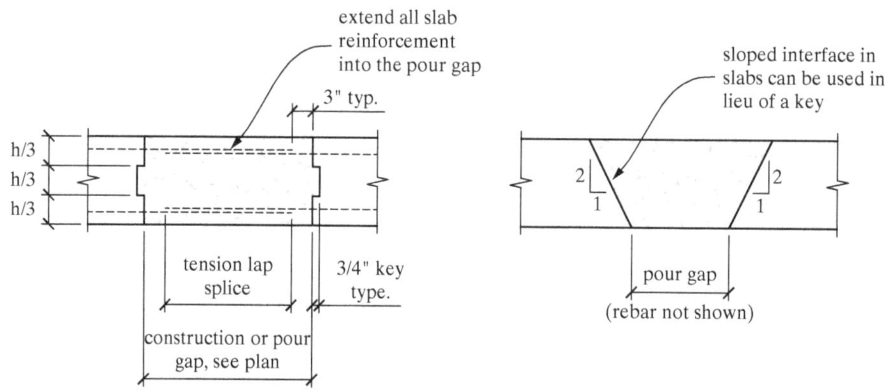

(a) Construction or Pour Gap in Reinforced Concrete Slabs

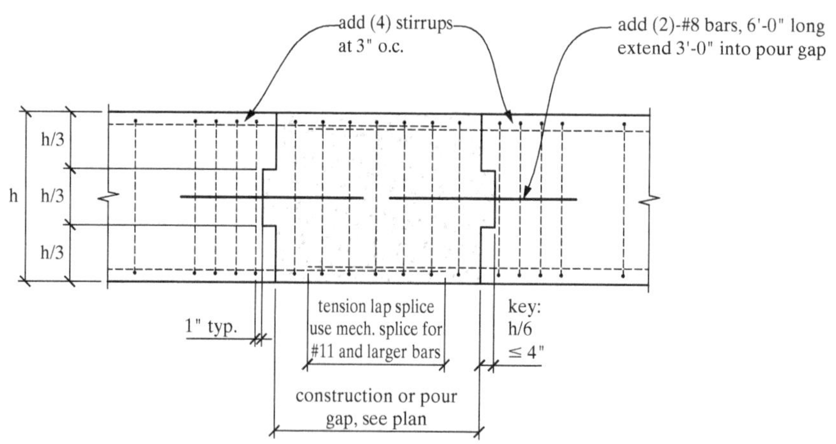

(b) Construction or Pour Gap in Reinforced Concrete Beams/Girders

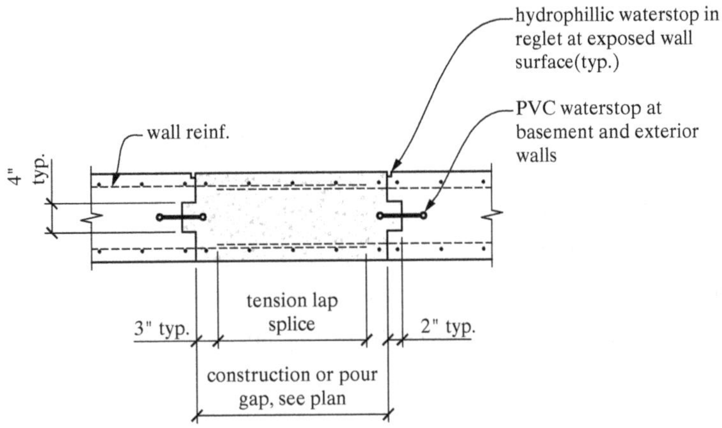

(c) Construction or Pour Gap in Reinforced Concrete Walls

FIGURE 14.11 Closure or pour strip details.

Similar to the pour gap detail in Figure 14.11 that is specifically used to control shrinkage, temporary openings are sometimes required in floor/roof slabs and beams to accommodate construction cranes that have to be located within the building footprint because of existing adjacent properties. The detail used for such cases will be similar to Figure 14.11, except that the reinforcement within the opening (that will be filled in later) will be temporarily bent upwards to accommodate the crane. After the crane is removed, the bars are bent back to the straight position before the opening is filled in. Also, additional shear reinforcement may be required for the beams and girders near the interface between the concrete that is already poured and the concrete in the opening that will be filled in at a later date.

14.9 FIRE RESISTANCE OF CONCRETE STRUCTURAL ELEMENTS

Since the attack on the World Trade Center on September 11, 2001, there has been renewed interest in ensuring that the structural elements in buildings have adequate fire resistance [25]. Concrete has excellent inherent fire-resistive properties and does not burn; this is one of the major advantages of concrete over other building materials such as wood or steel which have to be protected from fire. The required clear cover for reinforcing steel in concrete and the required minimum thickness of a structural concrete member are a function of the following:

1. The type of aggregate used:

   That is, whether siliceous, carbonate, or sand-lightweight aggregate. Examples of siliceous aggregates include aggregates from granite and sandstone, while examples of carbonate aggregates include aggregates from limestone and dolomite.

2. Whether the structural member is *restrained* or *unrestrained*:

   According to Section 703.2.3 of the International Building Code (IBC), "fire-resistance-rated assemblies tested under ASTM E119 or UL 263 shall not be considered to be restrained unless evidence satisfactory to the building official is furnished by the registered design professional showing that the construction qualifies for a restrained classification in accordance with ASTM E119 or UL 263." [27]. Examples of a *restrained* element include the following:

a. The interior span of a cast-in-place continuous concrete slab.

b. Beams securely fastened to the framing members (i.e., columns).

c. All types of cast-in-place concrete floor or roof systems (such as beam-and-slab, flat slabs, and waffle slabs) where the floor or roof system is cast with the framing members (i.e., columns).

Examples of *unrestrained* structural element include the following:

a. A single-span slab or beam with simply supported ends.

b. The simply supported end span of a continuous slab or beam.

3. The fire rating:

The fire rating is defined as "the duration of time that an assembly (roof, floor, beam, wall, or column) can endure a "standard fire" as defined in ASTM E 119" [25, 28]. There are four criteria defined in ASTM E119 that act as a measure of the fire endurance of the structural element. The fire endurance is reached when any of those four criteria is reached first. One of those conditions is the time in hours it takes for the structural element or member to lose its strength and collapse under the applied service load when subjected to a fire.

Concrete with siliceous aggregates is more susceptible to the deleterious effects of fire and requires larger clear covers and thicknesses to achieve the same fire rating as concrete with carbonate or sand-lightweight aggregates. Also, a restrained structural member (e.g., the interior span of a continuous slab or beam) requires a smaller clear cover and minimum thickness than an unrestrained structural member (e.g., the end span of a continuous slab or beam, or a simply supported slab or beam). The minimum concrete covers specified in ACI 318, Section 20.6.1.3.1 sometimes exceeds or is equal to the prescriptive values specified in ACI 216 - Code Requirement for Determining Fire Resistance of Concrete and Masonry Construction Assemblies or Section 721 of the IBC [25–28], but for fire ratings greater than or equal to 2 hours, this is not necessarily the case. Therefore, it is imperative that the minimum cover and minimum thickness requirements needed to satisfy the fire resistance ratings in ACI 216 or the IBC not be overlooked during the design phase because ignoring these requirements could lead to costly repairs. For the higher fire ratings (e.g., 3 hours or 4 hours), the values of the minimum concrete cover tabulated in ACI 216 are more stringent than the minimum covers prescribed in ACI 318. Since most ready-mix concrete contractors generally use aggregates that are indigenous to the area where the project is located, the minimum concrete covers and minimum thicknesses given below from ACI 216 and the IBC [25–27] are for the worst-case scenario; that is, for concrete with *siliceous* aggregates and assuming *unrestrained* structural members:

- Minimum cover to main rebar in floor and roof slabs:
  - 1-hour and 1.5-hour fire rating: Use the concrete covers specified in ACI 318, Table 20.6.1.3.1
  - 2-hour fire rating: Minimum cover for floor and roof slabs = 1 in.
  - 3-hour fire rating: Minimum cover for floor and roof slabs = 1¼ in.
  - 4-hour fire rating: Minimum cover for floor and roof slabs = 1-5/8 in.

- Minimum cover to main rebar in beams and girders:
    - Less than 4-hour fire rating: Minimum cover to main rebar = values from ACI 318, Table 20.6.1.3.1
    - 4-hour fire rating: For beams that are 10 in wide or greater, minimum clear cover to main rebar = 1¾ in
- Minimum cover to main rebar in columns:
    - 1-hour and 1.5-hour fire rating, minimum clear cover = values from ACI 318, Table 20.6.1.3.1
    - 2-hour, 3-hour, or 4-hour fire rating, minimum clear cover = 2 in.
- Minimum thickness of floor and roof slab (practical minimum slab thickness = 4 in.):
    - 1.5-hour fire rating, minimum slab thickness = 4.3 in.
    - 2-hour fire rating, minimum slab thickness = 5.0 in.
    - 3-hour fire rating, minimum slab thickness = 6.2 in.
    - 4-hour fire rating, minimum slab thickness = 7.0 in.
- Minimum concrete column dimensions (practical minimum column size = 12 in.):
    - 3-hour fire rating, minimum column size = 12 in
    - 4-hour fire rating, minimum column size = 14 in.

14.10 ANALYSIS AND DESIGN OF EDGE-SUPPORTED TWO-WAY SLABS ON STIFF SUPPORTS

The analysis of two-way slabs is an indeterminate and complex problem. In this section, the analysis of rectangular slab panels supported on stiff supports (e.g., walls and stiff beams) along all four edges of the panel is presented. The slab bends in two directions with curvature in the two orthogonal directions, and a portion of the load is carried by bending in the short direction while the remaining portion is carried by bending in the long direction of the slab. An approximate equation for distributing the uniform load on the slab can be derived by assuming equal deflection of simply supported slab strips at the center of the slab panel in both orthogonal directions and neglecting the effects of the twisting moments at the corners of the slab.

The maximum deflection of the center strip in the long (ℓ_1) direction is

$$\delta_1 = \frac{5w_1\ell_1^4}{384EI}$$

The maximum deflection of the center strip in the short (ℓ_2) direction is

$$\delta_2 = \frac{5w_2\ell_2^4}{384EI}$$

For deflection compatibility, $\delta_1 = \delta_2$

Therefore,

$$\frac{5w_1\ell_1^4}{384EI} = \frac{5w_2\ell_2^4}{384EI}$$

Which yields $w_1 = w_2\left(\frac{\ell_2}{\ell_1}\right)^4$

Noting that $w_1 + w_2 = w$ results in the following expressions for the distribution of the uniform load in the two orthogonal directions of the slab:

$$w_1 = w\left(\frac{\ell_2^4}{\ell_1^4 + \ell_2^4}\right)$$

$$w_2 = w\left(\frac{\ell_1^4}{\ell_1^4 + \ell_2^4}\right)$$

where

w_1 = the portion of the total uniformly distributed load carried by bending in the ℓ_1 direction

w_2 = the portion of the total uniformly distributed load carried by bending in the ℓ_2 direction

w = total uniformly distributed load supported by the slab

Note that in deriving the above equations, though the corners are usually restrained, the torsional moments at the corners of the slab have been neglected for simplicity. The load distribution above is conservative when compared to the elastic moment distribution in a two-way edge-supported slab [34].

In the preceding approximate method, the complex behavior of the two-way edge-supported concrete slab is reduced to the behavior of only the two intersecting center strips, and the inelastic moment redistribution that occurs in the concrete slab is neglected. In actuality, other strips in the slab will participate in supporting the load through bending and twisting. The actual behavior of the 2-way edge-supported slab will involve both bending and twisting moments in both orthogonal directions, and it also involves the redistribution of moments, resulting in a higher load-carrying capacity for the slab, and hence, reduced bending moments compared to the approximate moment values calculated above.

A more accurate method for calculating the moments in edge-supported concrete two-way slabs, which has been in use for many decades, is the moment coefficient method presented in Tables E.1 and E.2 in Appendix E of Ref [35].

The moment coefficients in both orthogonal directions are tabulated for various slab aspect ratios, $\frac{\ell_2}{\ell_1}$, and edge support conditions—simply supported or continuous. The minimum slab thickness for edge-supported two-way slabs can be calculated as follows [35]:

- h_{min} = (perimeter of slab panel)/140 ≥ 4 in. for simply supported two-way slab panels
- h_{min} = (perimeter of slab panel)/160 ≥ 4 in. for continuous two-way slab panels

Ref. [6] also provides moment coefficients that can be used for calculating the design moments for continuous edge-supported two-way slabs on stiff supports.

EXAMPLE 14.4

A slab is simply supported on walls on all four sides. The clear span of the long side of the slab is 27 ft, and the short side is 21 ft. The slab supports a total unfactored dead load of 150 psf (includes the self-weight of the 10 in. thick slab, 15 psf partition load, and 10 psf for floor finishes, ceiling, and mechanical and electrical equipment) and a live load of 50 psf. Calculate the factored moment in ft.-kip/ft in the long direction (i.e., M_1) and in the short direction (i.e., M_2) using the moment coefficient method. Calculate the required area of steel using the methods presented in Chapter 2. Assume f_c' is 4000 psi and f_y is 60,000 psi.

Solution:

w_u = 1.2 (150 psf) + 1.6(50 psf) = 260 psf = 0.26 ksf

ℓ_1 = 27 ft

ℓ_2 = 21 ft

$\frac{\ell_1}{\ell_2} = \frac{27\,\text{ft}}{21\,\text{ft}} = 1.29 < 2.0$ Therefore, this is a two-way slab.

The slab panel aspect ratio, $\frac{\ell_2}{\ell_1} = \frac{21\,\text{ft}}{27\,\text{ft}} = 0.78$

h_{min} = 2(21 ft. + 27 ft.)(12 in./ft)/140 = 8.2 in.

Also, use Table 6.4 to check the minimum slab thickness as follows:

$$h\min = \frac{\ell_n}{33} = \frac{\ell_{n1}}{33} = \frac{(27\text{ ft})(12)}{33} = 9.8 \text{ in. say 10 in. slab. O.K.}$$

For the slab aspect ratio of 0.78, the maximum design moment coefficients can be obtained by linear interpolation from the table of moment coefficients in Ref. [35] as follows:

Long direction:

$C_a = 0.021;\ M_1 = 0.021 w_u \ell_1^2 = 0.021(0.26)(27)^2 = 4$ ft.-kip/ft. width

Short direction:

$C_a = 0.059;\ M_2 = 0.059 w_u \ell_2^2 = 0.059(0.26)(21)^2 = 6.8$ ft.-kip/ft. width

Note that these are the moments in the middle strips of the slab panel (i.e., the middle 10.5 ft. for long direction bending, and the middle 13.5 ft. for short direction bending). Though the moments can be assumed to vary from a maximum at the edge of the middle strips to 1/3 of the maximum value at the non-deflecting edge support, for convenience and ease of rebar placement, some engineers for practical reasons assume a uniform maximum moment across the entire slab width, resulting in a uniform layout of the slab reinforcement.

Using the methods presented in Chapter 2, the slab reinforcement can be determined. For the M_2 moment, use an effective depth, d = 9" (i.e., 10" – ¾" cover – 0.5"/2) in the short direction assuming #4 reinforcement in the bottom lower layer, BLL; for the moment, M_1, use an effective depth, d = 8.5" (i.e., 10" – ¾" cover – 0.5" - 0.5"/2) in the long direction (i.e., bottom upper layer or BUL). Note that the bottom reinforcement will be uniformly distributed in the slab.

In the short direction,

$$\bar{k} = \frac{M_{u2}}{\phi bd^2} = \frac{6.8(12)}{(0.9)(12\,\text{in.})(9\,\text{in.})^2} = 0.093$$

From Table A.10, we get $\rho = 0.00157 = \dfrac{A_s}{bd}$

Therefore, the required $A_s = 0.00157(12\,\text{in.})(9\,\text{in.}) = 0.17\,\text{in.}^2 \geq A_{s,\min}$

$A_{s,\min} = 0.0018bh = 0.0018(12\,\text{in.})(10\,\text{in.}) = 0.22\,\text{in.}^2$

Therefore, use $A_{s,\min}$.

Use #4 @ 11" o.c. BLL (see Table A.4)

In the long direction:

$$\bar{k} = \frac{M_{u1}}{\phi bd^2} = \frac{4(12)}{(0.9)(12\,\text{in.})(8.5\,\text{in.})^2} = 0.0615$$

From Table A.10, we get $\rho = 0.0011 = \dfrac{A_s}{bd}$

Therefore, the required $A_s = 0.0011(12\,\text{in.})(8.5\,\text{in.}) = 0.11\,\text{in.}^2 \geq A_{s,\min}$

$A_{s,\min} = 0.0018bh = 0.0018(12\,\text{in.})(10\,\text{in.}) = 0.22\,\text{in.}^2$

Therefore, use $A_{s,\min}$.

Use #4 @ 11" o.c. BUL (see Table A.4)

Maximum spacing of main reinforcement in the slab = $2h$ = 2(10 in.) = 20 in. > 11 in. O.K.

To resist the twisting moments at the corners of the slab, provide additional reinforcement at the top of the slab at the four corners in accordance with Section 6.4.11 and Figure 6.36. Since the slab is assumed to be simply supported at all four edges, minimum top reinforcement (hooked at the wall supports) should also be provided in the slab at the edges in the two orthogonal directions. For continuous two-way slabs supported on concrete or masonry walls or stiff beams/girders, the positive and negative moment coefficients, and the shear coefficients can be obtained from Tables 7.9.2a through 7.9.2d in Ref. [6].

14.11 CAST-IN PLACE CONCRETE SPECIFICATIONS

The contract documents for building structures include the contract drawings (i.e., the structural, civil/site, architectural, mechanical, electrical, and plumbing or MEP drawings) and the written specifications for the materials used in the structure. The contract specifications contain directions for the contractor. There is so much information regarding the performance requirements for the concrete materials, products, and accessories used in the structure that cannot all be included in the structural or architectural drawings. Therefore, more in-depth performance requirements are usually provided in the contract specifications. The structural drawings are the graphic representation, and the contract specifications are the written representation, of the design intent of the structural engineer of record (SER). Both documents are meant to be complementary to each other and there is sometimes some overlap between the drawings and the contract specifications, but conflicts between both documents should be avoided. In practice, it is common to find notes on the drawings or in the specifications stating that the more stringent specification requirements shall govern, as determined by the engineer of record, when there is a conflict between the specifications and the drawings. The concrete specifications, and indeed all material specifications, follow the standard master format prescribed by the Construction Specifications Institute (CSI). The most used master specification is the AIA Master Spec. There are fifty divisions in the CSI master specifications: the concrete specifications are in division 030000, and the cast-in-place concrete specifications are in section 033000 of division 030000; the masonry specification is in division 040000, while the metals specification is in division 050000 [31, 32].

The typical cast-in-place concrete specification includes three parts with many subsections: Part 1—General, Part 2—Products, and Part 3—Execution [32]. The subsections for each of the three parts include specifications for items that are pertinent to the proper functioning of the concrete structure. Some pertinent areas that are covered in the concrete specifications are listed below [32]:

Part 1 - General - includes the following sections:

- Summary—this section lists the applicable structural members for which this specification is applicable (e.g., “Section includes cast-in-place concrete, including formwork, reinforcement, concrete materials, mixture design, placement procedures, and finishes, for the following: footings, foundation walls, slabs-on-grade, suspended slabs, concrete toppings, building frame members, and building walls”).
- Related sections—these are the related sections in the master specification (e.g., Architectural concrete and Concrete topping)
- References (i.e., the applicable codes and standards)
- Submittals—these include design mixtures, shop drawings for reinforcement, shop drawings for formwork, construction joint layout, welding certificates, material certificates, material test reports for aggregates, floor surface flatness and levelness measurements, field quality control report
- Quality assurance - installer, manufacturer, and testing agency qualifications, concrete testing, mockups
- Delivery, storage, and handling of materials

Part 2 - Products - includes the following subsections [32]:

- Form-facing materials - plywood, metal, tubes, void forms, form ties, etc
- Steel reinforcing—Grades of steel used (e.g., ASTM A 615, ASTM A 706, etc), recycled content of steel reinforcement, etc.
- Reinforcement accessories such as chairs or bar supports, bolsters, and spacers

- Concrete materials—For example, type of cementitious material used (e.g. Portland cement ASTM C150 Type I or II, III, V, flyash-ASTM C618, silica fume - ASTM C1240), maximum size of coarse aggregate, and water
- Cementitious material—Portland Cement and flyash.
- Aggregates—"Normal-Weight Aggregates: ASTM C 33, coarse aggregate (gravel) and fine aggregate (sand)
- Admixtures—The admixtures used are listed and includes a note prohibiting the use of calcium chloride or admixtures containing calcium chloride because of its corrosive effect on steel reinforcement. Examples of permitted admixtures include Water-Reducing and Retarding Admixture: High-Range, Water-Reducing Admixture: Plasticizing and Retarding Admixture: Set-Accelerating Corrosion-Inhibiting Admixture; Non-Set-Accelerating Corrosion-Inhibiting Admixture
- Water stops—Permitted types and manufacturers
- Vapor barriers—Permitted types and manufacturers
- Curing materials—For example, "Absorptive cover: AASHTO M 182, Class 2, burlap cloth made from jute or kenaf, weighing approximately 9 oz./sq. yd. when dry; Moisture-Retaining Cover: ASTM C 171, polyethylene film or white burlap-polyethylene sheet"
- Related materials—For example, bonding agent, epoxy bonding adhesive "for bonding hardened or freshly mixed concrete to hardened concrete"
- Concrete mixtures—General: In this section, the limits in percentage by weight of cementitious materials other than Portland cement are listed (e.g., fly ash is limited to 25% and silica fume is limited to 10%, and ground granulated blast-furnace slag is limited to 50%)
- Concrete mixtures for each of the different structural elements are listed in this section (e.g., footings, foundation walls, suspended slabs, building frame members, building walls, concrete toppings, slabs on grade, etc). A sample proportion of the elements in normal weight concrete are as follows:
    - Minimum Compressive Strength: 4500 psi at 28 days
    - Maximum Water-Cementitious Materials Ratio: 0.45
    - Slump Limit: 4 inches plus or minus 1 inch. If admixtures are used to improve workability, the maximum slump limits may be relaxed with Engineer's approval (i.e., a higher slump could be allowed)
    - Air Content: 6% at point of delivery
    - Coarse Aggregate: 1-in. nominal maximum aggregate size
- Fabricating reinforcement—reference is made to the "Manual of Standard Practice" of the Concrete Reinforcing Steel Institute (CRSI)
- Concrete mixing—"Ready-Mixed Concrete: Measure, batch, mix, and deliver concrete according to ASTM C 94/C 94M and ASTM C 1116/C 1116M, and furnish batch ticket information"

Part 3 - Execution - includes the following subsections [32]:

- Formwork - "Design, erect, shore, brace, and maintain formwork according to ACI 301.."
- Vapor Retarders - (see ASTM E1643)

- Steel reinforcement—In this section, instructions and performance requirements are given for the placement of steel reinforcement which must conform to CRSI's "Manual of Standard Practice" for placing reinforcement.
- Construction joints—The locations of construction joints are specified (e.g., "Place joints perpendicular to main reinforcement. Continue reinforcement across construction joints unless otherwise indicated. Do not continue reinforcement through sides of strip placements of floors and slabs; Locate joints for beams, slabs, joists, and girders in the middle third of spans. Offset joints in girders a minimum distance of twice the beam width from a beam-girder intersection; Locate horizontal joints in walls and columns at underside of floors, slabs, beams, and girders and at the top of footings or floor slabs")
- Contraction or Control joints in slab-on-grade
- Isolation joints in slab-on-grade (i.e., slab joints around vertical surfaces such as columns, foundation walls, piers, and grade beams)
- Water stops (installed at construction joints using the longest practical lengths). These are used to prevent water infiltration through the joints.
- Concrete placement—In this section, the performance requirements for placing the wet concrete are specified (e.g., "Do not add water to concrete during delivery, at Project site, or during placement unless approved by Architect; Deposit concrete continuously in one layer or in horizontal layers of such thickness that no new concrete will be placed on concrete that has hardened enough to cause seams or planes of weakness. If a section cannot be placed continuously, provide construction joints as indicated. Deposit concrete to avoid segregation")
- Cold-weather placement—In this section, the performance requirements for concrete placement during cold weather are covered. The specification requirements include the following instructions: "Comply with ACI 306.1 and as follows. Protect concrete work from physical damage or reduced strength that could be caused by frost, freezing actions, or low temperatures. When average high and low temperature is expected to fall below 40°F for three successive days, maintain delivered concrete mixture temperature within the temperature range required by ACI 301. Do not use frozen materials or materials containing ice or snow. Do not place concrete on frozen subgrade or on subgrade containing frozen materials. Do not use calcium chloride, salt, or other materials containing antifreeze agents or chemical accelerators unless otherwise specified and approved in mixture designs."
- Hot weather placement—In this section, the performance requirements for concrete placement during hot weather are covered. The specification requirements include the following instructions: "Comply with ACI 301 and as follows: Maintain concrete temperature below 90°F at time of placement. Chilled mixing water or chopped ice may be used to control temperature, provided water equivalent of ice is calculated to total amount of mixing water. Using liquid nitrogen to cool concrete is Contractor's option. Fog-spray forms, steel reinforcement, and subgrade just before placing concrete. Keep subgrade uniformly moist without standing water, soft spots, or dry areas."
- Concrete repairs and fill
- Finishing formed surfaces

- Finishing floors and slabs—The overall flatness and levelness values are specified for slab-on-grade and floor and roof slabs (e.g., "Finish slab-on-grade and elevated slab surfaces to the following tolerances, according to ASTM E 1155 and ACI 117, ACI 302, and ACI 302.1, for a randomly trafficked floor surface: Specified overall values of flatness, F(F) 25; and of levelness, F(L) 20; with minimum local values of flatness, F(F) 17; and of levelness, F(L) 15")
- Miscellaneous concrete items—The performance requirements for items such as curbs, equipment bases and foundations, steel pan stairs, and filling-in holes and openings in concrete surfaces are covered in this section (e.g., "Fill in holes and openings left in concrete structures after work of other trades is in place unless otherwise indicated. Mix, place, and cure concrete, as specified, to blend with in-place construction. Provide other miscellaneous concrete filling indicated or required to complete the Work")
- Concrete protection and curing—Instructions are given in this section for protecting freshly poured concrete from premature drying, and the curing of formed and unformed surfaces per ACI 308.1. One or a combination of the several curing methods listed—moisture curing, moisture-retaining-cover curing, or curing compound—can be used. Slab-on-grade are moist cured for a minimum of 10 days. Examples of specification items in this section include the following:
    - "Moisture curing—Keep surfaces continuously moist for not less than seven days with the following materials: Water, Continuous water-fog spray, Absorptive cover, water saturated, and kept continuously wet. Cover concrete surfaces and edges with 12-in. lap over adjacent absorptive covers; *Moisture-Retaining-Cover Curing*: Cover concrete surfaces with moisture-retaining cover for curing concrete, placed in widest practicable width, with sides and ends lapped at least 12 in., and sealed by waterproof tape or adhesive. Cure for not less than seven days. Immediately repair any holes or tears during curing period using cover material and waterproof tape. Moisture cure or use moisture-retaining covers to cure concrete surfaces to receive floor coverings. Moisture cure or use moisture-retaining covers to cure concrete surfaces to receive penetrating liquid floor treatments. Cure concrete surfaces to receive floor coverings with either a moisture-retaining cover or a curing compound that the manufacturer certifies will not interfere with bonding of floor covering used on Project."
- Floor treatments—Such as polished concrete floor treatment and the application of a seal coat to the concrete surface are covered in this section.
- Joint filling—In this section, the requirements for the installation of joint fillers are covered.
- Concrete surface repairs—That is, the repair of defective or damaged concrete surfaces using patching mortar and epoxy bonding agent or adhesives is covered in this section.
- Field quality control—This includes specifications on testing and inspection of materials and the work. The owner engages the services of an inspector and a testing agency to test the materials such as the concrete. Some of the areas of inspection include: "Steel reinforcement placement; Steel reinforcement welding; Headed bolts and studs; Verification of use of required design mixture; Concrete placement, including conveying and depositing; Curing procedures and maintenance of curing temperature; Verification of concrete strength before removal of shores and forms from beams and slabs."). The testing of the fresh concrete is to

be in accordance with ASTM C 172, and the testing frequency for concrete properties such as compressive strength, slump, concrete temperature, unit weight, air content, etc. are specified in this section; the number of concrete cylinders to be tested are also specified.

The compression test specimens are obtained in accordance with ASTM C31/C31M. Two sets of two standard 6 in. diameter by 12 in. long cylinder specimens are to be laboratory cured, and two sets of two standard cylinder specimens are to be field cured. Collect at least one sample per 100 cubic yards of concrete mixture with at least 5 compressive tests for each concrete mixture. Some of the pertinent requirements specified in this field quality control section for concrete testing include the following items:

- "Cylinders for compression test shall be made in accordance with ASTM C 31/C 31M
- Cylinders for compression strength test shall be tested in accordance with ASTM C 39/C 39M:
    - A compressive-strength test shall be the average compressive strength from a set of two specimens obtained from same composite sample and tested at the age indicated. One set of two field-cured and one set of two laboratory-cured specimens are tested at 7 days, and one set of two specimens are tested at 28 days for the field-cured and laboratory-cured specimens.
- "When strength of *field-cured* cylinders is less than 85% of companion *laboratory-cured* cylinders, Contractor shall evaluate operations and provide corrective procedures for protecting and curing in-place concrete."
- "Strength of each concrete mixture will be satisfactory if every average of any three consecutive compressive-strength tests equals or exceeds specified compressive strength, and no compressive-strength test value falls below specified compressive strength by more than 500 psi."
- "Test results shall be reported in writing to Architect, concrete manufacturer, and Contractor within 48 hours of testing. Reports of compressive-strength tests shall contain Project identification name and number, date of concrete placement, name of concrete testing and inspecting agency, location of concrete batch in Work, design compressive strength at 28 days, concrete mixture proportions and materials, compressive breaking strength, and type of break for both 7- and 28-day tests."
- "Nondestructive Testing: Impact hammer, sonoscope, or other nondestructive device may be permitted by Architect but will not be used as sole basis for approval or rejection of concrete."
- Additional Tests: At Contractor's expense, "testing and inspecting agency shall make additional tests of concrete when test results indicate that slump, air entrainment, compressive strengths, or other requirements have not been met, as directed by Architect. Testing and inspecting agencies may conduct tests to determine adequacy of concrete by cored cylinders complying with ASTM C 42/C 42M or by other methods as directed by Architect."
- At Contractor's expense, "additional testing and inspecting will be performed to determine compliance of replaced or additional work with specified requirements."
- At Contractor's expense, "correct deficiencies in the Work that test reports and inspections indicate do not comply with the Contract Documents."

Additional Specifications for Thick Concrete Slabs

For thick concrete members such as mat or raft foundations that could be classified as mass concrete, the heat of hydration generated during the concrete curing process could result in excessive shrinkage which leads to excessive cracking of the mass concrete. Temperatures of up to 170–200°F have been measured in the cores of mass concrete foundations [33]. To mitigate the effects of high temperatures on thick concrete elements, additional specification items are required. The reader should refer to ACI 301 and ACI 207—Mass Concrete for further guidance. In one of the author's projects—a high-rise building supported on a 6 ft thick mat foundation, the 6,000 cubic yards (or 162,000 ft^3) of concrete was poured continuously (i.e., a monolithic pour) over 28 hours using several pumps. This was done to avoid any construction joints in such a thick slab, especially since the area where the high-rise building is located has a high-water table. In fact, the water table was so high on the site that, in order to prevent uplift of the mat foundation, the water table was lowered and monitored until the tenth-floor slabs and beams were cast, thus ensuring enough dead load to resist the buoyancy forces from the hydrostatic uplift pressures at the base of the mat foundation. To further control the heat of hydration, the 28-day strength of the concrete was specified as 3000 psi, with a 120-day strength of 4500 psi. Prequalification trial mixtures submitted by three concrete suppliers were used to cast 6 ft thick trial blocks of concrete that were tested by the owner's materials consultant prior to the mat foundation concrete pour; and the concrete supplier with the trial mix that best met the performance specifications was selected. Additional specification items for the concrete mat foundation that were successfully used on the project are as follows:

- Cementitious material: Moderate heat of hydration Portland cement (Type II) with flyash (Type F) and cementitious hydraulic slag (Type H).
- A portion of the water in the concrete mix should be in the form of crushed ice with temperatures below 32°F to achieve the specified placing temperature for the concrete.
- To further reduce the shrinkage in thick mat foundations, a maximum aggregate thickness of 1½ inches was specified.
- The concrete mix for the mass concrete was selected to achieve the specified placing temperature and to ensure that the temperature in the concrete and the maximum temperature differential did not exceed the specified limits.
- To measure the temperatures, thermocouples were placed in the concrete in three layers and at different locations within the mat foundation. The delivery temperature of the mat foundation concrete was specified to be between 41°F and 50°F, and the maximum temperature of the concrete was limited to 130°F.

The preceding requirements were included in Part 2 of the cast-in-place concrete specifications for the project. In Part 3 of the cast-in-place concrete specifications, the following additional items were added to the specification:

- Provide a 3 in thick mud slab or skim coat with f_c' of 750 psi at the underside of the mat foundation.
- Ensure the stability of the mat top reinforcement at all times and during the concrete pour.

- The construction of the concrete mat foundation should be accomplished in one continuous pour to ensure that no cold joints develop between successive lifts or sections of concrete placement.
- The maximum temperature differential should not exceed 68°F. To measure the maximum temperature and the maximum temperature differential, thermocouples should be inserted in the concrete at 2 inches from the bottom and top surfaces, and at the mid-depth of the mat foundation. The thermocouples should be spaced longitudinally and transversely a maximum distance of 100 ft. Protective covers and insulation placed over the mat should be kept in place until the temperature differential is within limits, and the temperatures should be regularly monitored to check compliance.

14.12 STUDENT DESIGN PROJECTS

In this section, two reinforced concrete building projects are presented, and any one of them could be assigned as a term project in a reinforced concrete design course. These design projects are very similar to building projects found in consulting structural engineering practice. In fact, low-rise buildings like the ones presented here constitute approximately 90% of the floor area built in the United States in any given year [36]. These design projects will help the student tie together the design of the individual structural concrete members covered in previous chapters within the context of an entire building design, and this will engender "a solid understanding of how those members are integrated within a complete structural system" [36]. This holistic approach to structural concrete design which bridges the gap between academia and engineering practice has been the underlying philosophy in every chapter of this text. The design brief for each project is as follows:

1. **Building type:**

   a. **Residential Building:** A two-story reinforced concrete residential building with plan dimensions of 135 ft by 57 ft. The floor-to-floor height is 15 ft (see Figures 14.12 and 14.13).

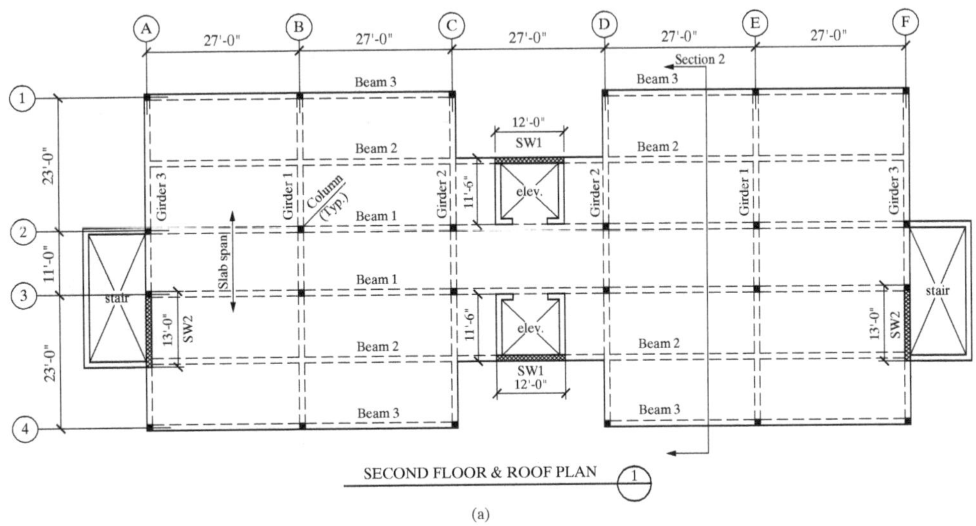

Figure 14.12 Residential building - Typical floor and roof plan.

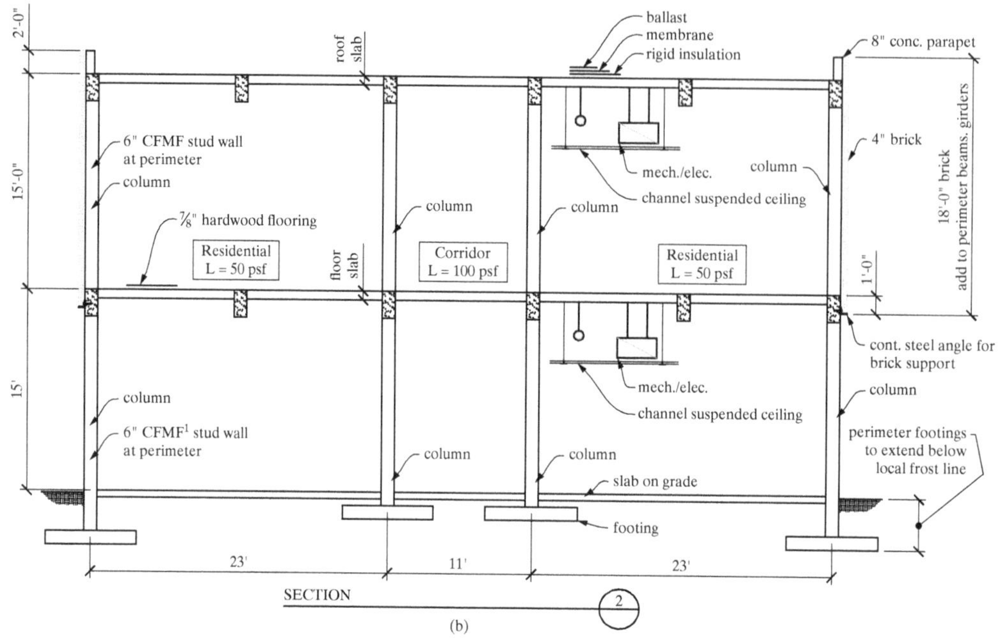

FIGURE 14.13 Residential building - Section
[1]CFMF is cold-formed metal framing

b. **Hospital Building:** A four-story reinforced concrete hospital building with plan dimensions of 150 ft by 100 ft. The floor-to-floor height varies (see Figures 14.14 and 14.15).

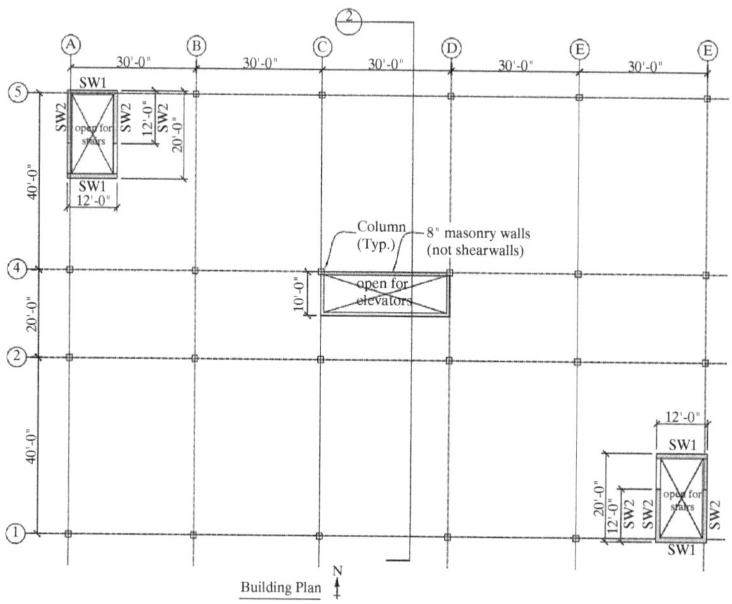

FIGURE 14.14 Hospital building - Typical floor and roof plan.

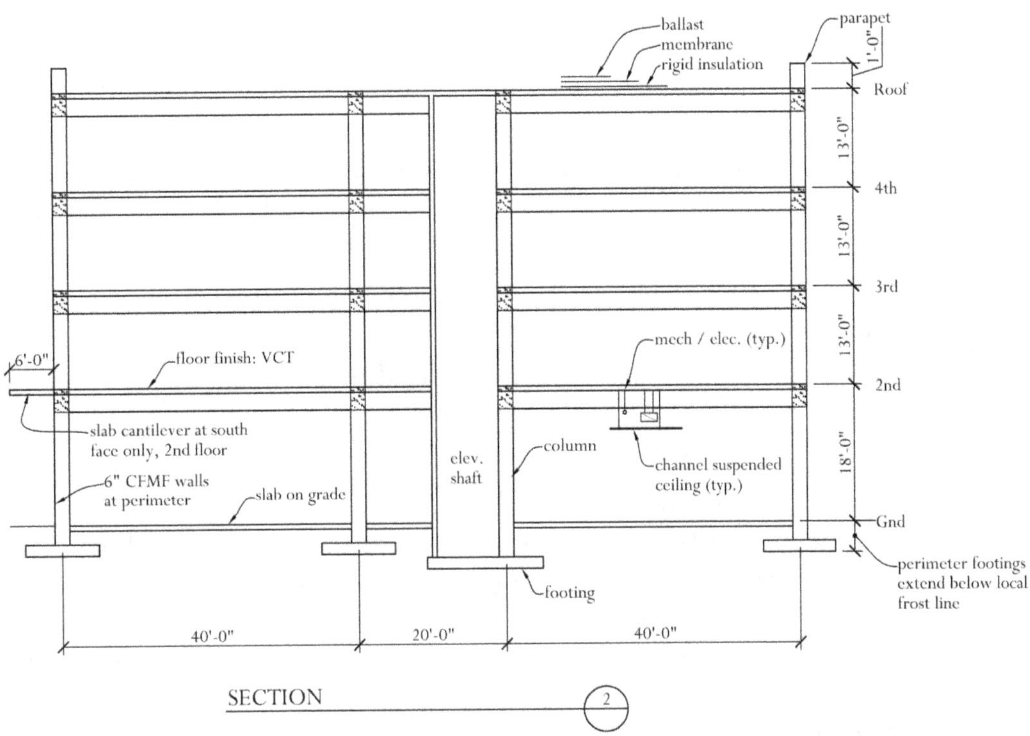

FIGURE 14.15 Hospital building - Section.

- Each building is located in Philadelphia, PA. or in any other City of the instructor's choice.
- The elevated slabs construction supported on beams and girders.
- Lateral loads are to be resisted by shear walls that are symmetrically located as indicated.
- The ACI 318 specification and the ASCE-7 load standard should be used.
- Assume that the stairs will be designed by others.
- Determine the critical lateral loading—seismic or wind—and use this to design the LFRS or shear walls.
- *Member Properties:* Use $f_c' = 4000$ psi and $f_y = 60$ ksi.

2. **Drawings and details:** May be either large size, foldout size, or on 8-1/2″ × 11″ and should include the following:

- *Framing Floor and Roof Plans* that show the floor and roof slabs, floor and roof beams and girders, columns, and shear walls.
- Provide appropriate marks for all of the framing members (e.g., B-1, B-2, G-1, etc.)
- *Foundation Plans* that show the slab-on-grade and foundations or footings.
- Show control joint locations in the slab-on-grade.
- *Schedules*: Slab, beam, girder, column, and footing schedules.
- *Sections, Details, and Elevations* that show typical beams, slabs, columns, footings, and shear walls with reinforcing sizes and placement, as well as cover requirements and lap splice details.

3. **Loads:**

- Calculate the dead and live loads (floor live loads and roof live loads).
- Live loads:
  - **a.** Residential Building—Use a design live load of 50 psf for the second floor except at the corridor between grids 2 and 3 where the live load of 100 psf is required.
  - **b.** Hospital Building—Use a design live load of 80 psf for all floors.
- Use at least 15 psf for partitions (if applicable) and 10 psf for mechanical and electrical fixtures.
- Follow the ASCE-7 load standard to calculate the wind load, snow load, and seismic load.
- Assume a net allowable soil bearing capacity of 3000 psf.
- Perimeter Cladding:
  - **a.** Residential Building—Assume the perimeter cladding (4″ brick wall with steel stud backing), which weighs 40 psf, will be supported by the perimeter (spandrel) beams and girders only.
  - **b.** Hospital Building—Perimeter is curtain wall and weighs 20 psf supported at each floor.

4. **Checklist of Design Items:**

- *Provide the following at the front of the project submission*: cover page, an executive summary, design assumptions and a table of contents.
- *Gravity and Lateral Loads*: Provide a load summation table for the floor and roof loads (gravity) and calculate the lateral wind and seismic loads at each level of the building. Provide a column load summation table.
- *Floor and Roof Slabs*: Continuous one-way slabs, check deflections against ACI requirements on the floors; framing plans should show typical slab reinforcing.
- *Floor Beams*: Design the interior "T" beams and girders, and the exterior "L" beams and girders. Draw the shear force diagrams for each and check deflections against Code requirements; show final beam and girder sizes in a beam schedule.
- *Columns*: Show the column loads in a load summation table and show final column sizes in a column schedule.
- *Foundations*: Design column footings, wall footings for shear walls, and exterior wall footings and show final footing sizes in a footing schedule.
- *Shear walls*: Design for lateral loads in both orthogonal directions and show reinforcing details - distributed vertical and horizontal reinforcement, and the end zone reinforcement.
- Check the development lengths for all flexural reinforcements.
- Provide quantities of rebar and concrete and installed cost estimates of each.

5. **Analysis and Design considerations:**

- For simplicity, the box-type behavior of the stair and elevator core walls is ignored. That is, each wall segment is assumed to act independently.

- For the analysis of the continuous slabs, beams, and girders, use a computer analysis to more accurately determine the shears, moments, and deflections, and compare the results with the approximate moments and shears obtained from the ACI coefficients.
- Provide line diagrams of the slabs, beams, and girders showing the loads and other input data used for either hand calculations or the computer analysis.
- The following deflection limits for long-term deflections accounting for creep are recommended: Total Loads: L/480, Live Load: L/360, Brick plus Live Load: L/600 or 0.3 inch (use the smaller value) for members supporting brick cladding.
- Shear walls should be modeled with lateral and gravity loads to design the footings. The footings are designed as a continuous wall footing and should be checked for overturning from the lateral loads.

A tabular project checklist is presented in Table 14.3.

TABLE 14.3 Project Checklist.

Design Item	Description
Project Description	Title page, Table of contents, Executive summary, Design assumptions.
Design Loads	Dead, Live, Snow, Wind.
Slab Design	Continuous span design, shear/moment diagrams, Bending, shear, main steel, temperature steel.
Beams	Line diagrams, Continuous span design, shear/moment diagrams, Bending, strain, shear, deflection.
Girders	Line diagrams, Continuous span design, shear/moment diagrams, Bending, strain, shear, deflection.
Columns	Load summation tables, compression check, ties and vertical bars.
Footings (spread and wall ftgs.)	1- and 2-way shear, bending.
Shear walls	Load diagrams (lateral and gravity), shear/moment diagrams, shear, bending.
Slab-on-Grade	Joints, reinforcing.
Bond/Anchorage	Lap splices and dowels: columns, walls; dev length: footings; slabs and beams.
Schedules	Slab, beams, girders, columns, footings.
Drawings and Details	**Notes:** Material properties and assumptions **Plans:** marks for beams, columns, footings **Slabs:** rebar layout in plan and section **Beams:** rebar and stirrup layout in an elevation and section **Shear walls:** plans, sections, detail **Columns:** section; **Slab-on-grade:** Joints and reinforcing **Footings:** sections **Schedules:** as needed for slabs, beams, columns, and footings All details are to be drawn to scale and include all appropriate dimensions and callouts
Spreadsheets/computer analysis	Line diagrams for analysis input, shear/moment diagrams for output; show beam marks where appropriate; clearly identify variables
Quantities	Cubic yards of concrete, tons of reinforcing

REFERENCES

[1] Jim Delahay and Brad Christopher. "Current Trends in Economical Concrete Construction," *STRUCTURE Magazine*, October 2007, pp. 20–21.

[2] Clifford W. Schwinger. "Tips for Designing Constructible Concrete Structures." *STRUCTURE Magazine*, February 2011, pp. 42–43.

[3] David B. Fanella and S. K. Ghosh. *Simplified Design—Reinforced Concrete Buildings of Moderate Size and Height*." Skokie, IL: Portland Cement Association, 1993.

[4] R. E. Shaeffer. *Reinforced Concrete—Preliminary Design for Architects and Builders*. Blacklick, OH: McGraw Hill, 1992, 196 pp.

[5] Edward Allen and Joseph Iano. *The Architect's Studio Companion—Rules of Thumb for Preliminary Design*, 5th ed. John Wiley, Hoboken, New Jersey, 2011.

[6] ACI 314-16. "Guide to Simplified Design for Reinforced Concrete Buildings," American Concrete Institute, P.O. Box 9094, Farmington Hills, MI 48333-9094, 2016.

[7] B. Keith Brenner. "Innovative Reinforcing Gives Old Structure New 'Light.'" *STRUCTURE Magazine*, January 2012, pp. 26–27.

[8] T. Alkhrdaji and J. Thomas. "Keys to Success: Structural Repair and Strengthening Techniques for Concrete Facilities." Structural Engineer, V. 5, No. 4, May 2004, pp. 24–27.

[9] Mo Ehsani, Majid Farahani, and Eric Raatz. "Repair of Columns with FRP Laminates." *STRUCTURE Magazine*, January 2012, pp. 35–37.

[10] FEMA-547. "Techniques for the Seismic Rehabilitation of Existing Buildings." Federal Emergency Management Agency, Washington, D.C., 2006.

[11] Institute for Research in Construction. "Guidelines for Seismic Upgrading of Building Structures." National Research Council of Canada, 1995, pp. 16.

[12] M. K. ElBatanouny, Antonio Nanni, Paul H. Ziel, and Fabio Matta. "Condition Assessment of Prestressed Concrete Beams Using Cyclic and Monotonic Load Tests. "*ACI Structural Journal*, V. 112, No. 1, January-February 2015, pp. 81–89.

[13] Subhash Kulkarni. "Forensic Study of a Partial Collapse." *ACI Concrete International*, October 2000, pp. 45–50.

[14] Alexander Newman. "Structural Renovation of Buildings. McGraw Hill, September 2000.

[15] Dov Kaminestzky. "Construction and Design Failures: Lessons from Forensic Investigations." McGraw-Hill, January 1991.

[16] Gustavo Tumialan, Nestore Galati and Antonio Nanni. "In Situ Load Testing of Concrete Structures: Part 1 – Rational, Objectives and Execution. "*STRUCTURE Magazine*," April 2014, pp. 10–12.

[17] Gustavo Tumialan, Nestore Galati and Antonio Nanni. "Load Testing of Concrete Structures: Part 2 – Test Protocols and Case Studies. "*STRUCTURE Magazine*," June 2014, pp. 16–18.

[18] ACI Committee 318, "Building Code Requirements for Structural Concrete (ACI 318R19(22))," American Concrete Institute, Farmington Hills, MI, 2019, 623 pp. (Reapproved 2022)

[19] ACI 437-12. "Code Requirements for Load Testing of Existing Concrete Structures and Commentary." American Concrete Institute (ACI), Farmington Hills, Michigan, 2012.

[20] *PCI Design Handbook – Precast and Prestressed Concrete*, 7th Edition, Precast/Prestressed Institute, Chicago, IL 2010.

[21] PCI Design Manual for the Design of Hollow Core Slabs, Precast/Prestressed Institute, Chicago, IL 1998.

[22] Alex Aswad and Francis J. Jacques, "Behavior of Hollow-Core Slabs Subject to Edge Loads," PCI Journal, March-April 1992.

[23] AS 3600-2009, Australian Standard, Concrete Structures, Section 9.6, Council of Standards Australia, Sydney, Australia, 2009.

[24] Eva O. L. Lantsought, Cor van der Veen, and Joost C. Walraven, "Shear in One-Way Slabs under Concentrated Load Close to Support," *ACI Structural Journal*, V 110, No. 2, March-April 2013.

[25] David N Bilow and Mahmoud E. Kamara. *Fire and Concrete Structures*, 2008 ASCE Structures Congress, Vancouver, B.C., Canada. April 24–26, 2008.

[26] ACI. Standard Method for Determining Fire Resistance of Concrete and Masonry Construction Assemblies, ACI 216.1, American Concrete Institute, Farmington Hills, MI, 2007.

[27] IBC. International Building Code, International Code Council, Falls Church, VA, 2016.

[28] ASTM E 119-00a, "Standard Test Methods for Fire Tests of Building Construction and Materials." ASTM International, West Conshohocken, PA, 2000, p. 21.

[29] David A. Fanella "Time-saving Design Aids for Reinforced Concrete." *Structural Engineer, pp. 39–41, August 2001.*

[30] Andre Brault, Neil Hoult, Tom Greenough, Ian Trudeau, and Barry Charnish. "Closure Strip Strategies," *ACI Concrete International*, pp. 29–35, July 2017.

[31] Madan Mehta, Walter Scarborough, and Diane Armpriest. "Building Construction: Principles, Materials, and Systems," Pearson, 2010.

[32] CSI MasterFormat®, "Master List of Numbers and Titles for the Construction Industry." Construction Specifications Institute (CSI), 2012.

[33] Abdol R. Chini, Larry C. Muszynski, Lucky Acquaye, and Sophia Tarkhan, "Determination of the maximum placement and curing temperature of mass concrete to avoid durability problems and DEF." Final Report to Florida Department of Transportation (Contract No. BC 354–29), February 2003.

[34] Stephen P. Timoshenko and Woinowsky-Krieger. "Theory of Plates and Shells," Second Edition, McGraw-Hill, 1959, New York.

[35] CAC "Concrete Design Handbook." Cement Association of Canada (CAC), 3rd Edition, 2006, Ottawa.

[36] David A. Fanella and Michael Mota. "A Capstone Course for Reinforced Concrete Buildings." Concrete International, June 2018, pp. 24–30.

[37] Abi Aghayere and Jason Vigil. "Structural Steel Design," 3rd Edition, Mercury Learning, Boston MA 2020.

[38] David A. Fanella and Michael Mota. "Design of Reinforced Concrete Diaphragms for Wind." STRUCTURE Magazine, April 2018, pp. 9–11.

[39] Robert T. Ratay (Editor). "Structural Condition Assessment – Evaluation Techniques for Strength, Serviceability, Restoration, Adaptive Reuse, Code Compliance, and Vulnerability," John Wiley & Sons, Inc., Hoboken, New Jersey, 2005.

[40] Erik Moore, Aniket Borwankar, Sarah Outzen, and Jay Love. "FRP Collector Strengthening in a California Hospital." STRUCTURE Magazine, March 2023, pp. 46–49.

[41] American Society of Civil Engineers, ASCE 7-22, Minimum Design Loads for Buildings and Other Structures (Provisions and Commentary), Reston, VA 2022.

APPENDIX

A

TABLES AND DIAGRAMS

TABLE A.1 Reinforcing Steel.

Type of steel and ASTM specification number	Bar sizes	Grade	Minimum tensile strength (psi)	Minimum yield strength f_y (psi)	Yield strain ε_y
Billet Steel A615	Nos. 3–6	40	60,000	40,000	0.00138
	Nos. 3–18	60	80,000	60,000	0.00207
	Nos. 6–18	80	100,000	80,000	0.00276
	Nos. 6–18	100	115,000	100,000	0.00345
Low-Alloy Steel A706	Nos. 3–18	60	80,000	60,000	0.00207
	Nos. 3–18	80	100,000	80,000	0.00276

Bar number	3	4	5	6	7	8	9	10	11	14	18
Unit weight per foot (lb)	0.376	0.668	1.043	1.502	2.044	2.670	3.400	4.303	5.313	7.65	13.60
Diameter[a] (in.)	0.375	0.500	0.625	0.750	0.875	1.000	1.128	1.270	1.410	1.693	2.257
Area (in.^{2})	0.11	0.20	0.31	0.44	0.60	0.79	1.00	1.27	1.56	2.25	4.00

[a]The nominal dimensions of a deformed bar (diameter and area) are equivalent to those of a plain round bar having the same weight per foot as the deformed bar.

TABLE A.2 Areas of Multiples of Reinforcing Bars (in.^{2}).

Number of bars	Bar number								
	#3	#4	#5	#6	#7	#8	#9	#10	#11
1	0.11	0.20	0.31	0.44	0.60	0.79	1.00	1.27	1.56
2	0.22	0.40	0.62	0.88	1.20	1.58	2.00	2.54	3.12
3	0.33	0.60	0.93	1.32	1.80	2.37	3.00	3.81	4.68
4	0.44	0.80	1.24	1.76	2.40	3.16	4.00	5.08	6.24
5	0.55	1.00	1.55	2.20	3.00	3.93	5.00	6.35	7.80
6	0.66	1.20	1.86	2.64	3.60	4.74	6.00	7.62	9.36
7	0.77	1.40	2.17	3.08	4.20	5.53	7.00	8.89	10.9
8	0.88	1.60	2.48	3.52	4.80	6.32	8.00	10.2	12.5
9	0.99	1.80	2.79	3.96	5.40	7.11	9.00	11.4	14.0
10	1.10	2.00	3.10	4.40	6.00	7.90	10.0	12.7	15.6
11	1.21	2.20	3.41	4.84	6.60	8.69	11.0	14.0	17.2

Number of bars	Bar number								
	#3	#4	#5	#6	#7	#8	#9	#10	#11
12	1.32	2.40	3.72	5.28	7.20	9.48	12.0	15.2	18.7
13	1.43	2.60	4.03	5.72	7.80	10.3	13.0	16.5	20.3
14	1.54	2.80	4.34	6.16	8.40	11.1	14.0	17.8	21.8
15	1.65	3.00	4.65	6.60	9.00	11.8	15.0	19.0	23.4
16	1.76	3.20	4.96	7.04	9.60	12.6	16.0	20.3	25.0
17	1.87	3.40	5.27	7.48	10.2	13.4	17.0	21.6	26.5
18	1.98	3.60	5.58	7.92	10.8	14.2	18.0	22.9	28.1
19	2.09	3.80	5.89	8.36	11.4	15.0	19.0	24.1	29.6
20	2.20	4.00	6.20	8.80	12.0	15.8	20.0	25.4	31.2

TABLE A.3 Minimum Required Beam Widths (in.)

Number of bars in one layer	Bar number							
	#3 and #4	#5	#6	#7	#8	#9	#10	#11
2	6.0	6.0	6.5	6.5	7.0	7.5	8.0	8.0
3	7.5	8.0	8.0	8.5	9.0	9.5	10.5	11.0
4	9.0	9.5	10.0	10.5	11.0	12.0	13.0	14.0
5	10.5	11.0	11.5	12.5	13.0	14.0	15.5	16.5
6	12.0	12.5	13.5	14.0	15.0	16.5	18.0	19.5
7	13.5	14.5	15.0	16.0	17.0	18.5	20.5	22.5
8	15.0	16.0	17.0	18.0	19.0	21.0	23.0	25.0
9	16.5	17.5	18.5	20.0	21.0	23.0	25.5	28.0
10	18.0	19.0	20.5	21.5	23.0	25.5	28.0	31.0

Note: Tabulated values based on No. 3 stirrups, minimum clear distance of 1 in., and a .. in. cover.

TABLE A.4 Areas of Reinforcing Bars per Foot Width of Slab (in.^{2})

Bar spacing (in.)	Bar number								
	#3	#4	#5	#6	#7	#8	#9	#10	#11
2	0.66	1.20	1.86						
$2\frac{1}{2}$	0.53	0.96	1.49	2.11					
3	0.44	0.80	1.24	1.76	2.40	3.16	4.00		
$3\frac{1}{2}$	0.38	0.69	1.06	1.51	2.06	2.71	3.43	4.35	
4	0.33	0.60	0.93	1.32	1.80	2.37	3.00	3.81	4.68
$4\frac{1}{2}$	0.29	0.53	0.83	1.17	1.60	2.11	2.67	3.39	4.16
5	0.26	0.48	0.74	1.06	1.44	1.90	2.40	3.05	3.74
$5\frac{1}{2}$	0.24	0.44	0.68	0.96	1.31	1.72	2.18	2.77	3.40
6	0.22	0.40	0.62	0.88	1.20	1.58	2.00	2.54	3.12
$6\frac{1}{2}$	0.20	0.37	0.57	0.81	1.11	1.46	1.85	2.34	2.88
7	0.19	0.34	0.53	0.75	1.03	1.35	1.71	2.18	2.67

Bar spacing (in.)	Bar number								
	#3	#4	#5	#6	#7	#8	#9	#10	#11
$7\frac{1}{2}$	0.18	0.32	0.50	0.70	0.96	1.26	1.60	2.03	2.50
8	0.16	0.30	0.46	0.66	0.90	1.18	1.50	1.90	2.34
9	0.15	0.27	0.41	0.59	0.80	1.05	1.33	1.69	2.08
10	0.13	0.24	0.37	0.53	0.72	0.95	1.20	1.52	1.87
11	0.12	0.22	0.34	0.48	0.65	0.86	1.09	1.39	1.70
12	0.11	0.20	0.31	0.44	0.60	0.79	1.00	1.27	1.56
13	0.10	0.18	0.29	0.41	0.55	0.73	0.92	1.17	1.44
14	0.09	0.17	0.27	0.38	0.51	0.68	0.86	1.09	1.34
15	0.09	0.16	0.25	0.35	0.48	0.64	0.80	1.02	1.25
16	0.08	0.15	0.23	0.33	0.45	0.59	0.75	0.95	1.17
17	0.08	0.14	0.22	0.31	0.42	0.56	0.71	0.90	1.10
18	0.07	0.13	0.21	0.29	0.40	0.53	0.67	0.85	1.04

TABLE A.5 Design Constants

f'_c (psi)	$\frac{3\sqrt{f'_c}}{f_y} \geq \frac{200}{f_y}$	Recommended design values	
		ρ	$\bar{k}$ (ksi)
	$f_y = 40,000$ psi		
3000	0.0050	0.0135	0.4828
4000	0.0050	0.0180	0.6438
5000	0.0053	0.0225	0.8047
6000	0.0058	0.0270	0.9657
	$f_y = 50,000$ psi		
3000	0.0040	0.0108	0.4828
4000	0.0040	0.0144	0.6438
5000	0.0042	0.0180	0.8047
6000	0.0046	0.0216	0.9657
	$f_y = 60,000$ psi		
3000	0.0033	0.0090	0.4828
4000	0.0033	0.0120	0.6438
5000	0.0035	0.0150	0.8047
6000	0.0039	0.0180	0.9657
	$f_y = 75,000$ psi		
3000	0.0027	0.0072	0.4828
4000	0.0027	0.0096	0.6438
5000	0.0028	0.0120	0.8047
6000	0.0031	0.0144	0.9657

[a]Does not apply to T-beams with flanges in tension (see Section 3-2). To compute $A_{s,min}$, see Section 2-8.

TABLE A.6 Properties and Constants for Normal-Weight Concrete

	f_c (psi)			
	3000	3500	4000	5000
E_c (psi)[a]	3,120,000	3,370,000	3,605,000	4,030,000
n[b]	9	9	8	7
$7.5\sqrt{f'_c}$ (ksi)[c]	0.411	0.444	0.474	0.530

[a] E_c for normal-weight concrete $= 57{,}000\sqrt{f'_c}$.

[b] Nearest whole number.

[c] Modulus of rupture (f_r).

TABLE A.7 Coefficient of Resistance ($\bar{k}$) versus Reinforcement Ratio (ρ) ($f'_c = 3000$ psi; $f_y = 40{,}000$ psi; units of $\bar{k}$ are ksi)

ρ	$\bar{k}$	ρ	$\bar{k}$	ρ	$\bar{k}$
0.0010	0.0397	0.0054	0.2069	0.0098	0.3619
0.0011	0.0436	0.0055	0.2105	0.0099	0.3653
0.0012	0.0476	0.0056	0.2142	0.0100	0.3686
0.0013	0.0515	0.0057	0.2178	0.0101	0.3720
0.0014	0.0554	0.0058	0.2214	0.0102	0.3754
0.0015	0.0593	0.0059	0.2251	0.0103	0.3787
0.0016	0.0632	0.0060	0.2287	0.0104	0.3821
0.0017	0.0671	0.0061	0.2323	0.0105	0.3854
0.0018	0.0710	0.0062	0.2359	0.0106	0.3887
0.0019	0.0749	0.0063	0.2395	0.0107	0.3921
0.0020	0.0788	0.0064	0.2431	0.0108	0.3954
0.0021	0.0826	0.0065	0.2467	0.0109	0.3987
0.0022	0.0865	0.0066	0.2503	0.0110	0.4020
0.0023	0.0903	0.0067	0.2539	0.0111	0.4053
0.0024	0.0942	0.0068	0.2575	0.0112	0.4086
0.0025	0.0980	0.0069	0.2611	0.0113	0.4119
0.0026	0.1019	0.0070	0.2646	0.0114	0.4152
0.0027	0.1057	0.0071	0.2682	0.0115	0.4185
0.0028	0.1095	0.0072	0.2717	0.0116	0.4218
0.0029	0.1134	0.0073	0.2753	0.0117	0.4251
0.0030	0.1172	0.0074	0.2788	0.0118	0.4283
0.0031	0.1210	0.0075	0.2824	0.0119	0.4316
0.0032	0.1248	0.0076	0.2859	0.0120	0.4348
0.0033	0.1286	0.0077	0.2894	0.0121	0.4381
0.0034	0.1324	0.0078	0.2929	0.0122	0.4413

ρ	$\bar{k}$	ρ	$\bar{k}$	ρ	$\bar{k}$
0.0035	0.1362	0.0079	0.2964	0.0123	0.4445
0.0036	0.1399	0.0080	0.2999	0.0124	0.4478
0.0037	0.1437	0.0081	0.3034	0.0125	0.4510
0.0038	0.1475	0.0082	0.3069	0.0126	0.4542
0.0039	0.1512	0.0083	0.3104	0.0127	0.4574
0.0040	0.1550	0.0084	0.3139	0.0128	0.4606
0.0041	0.1587	0.0085	0.3173	0.0129	0.4638
0.0042	0.1625	0.0086	0.3208	0.0130	0.4670
0.0043	0.1662	0.0087	0.3243	0.0131	0.4702
0.0044	0.1699	0.0088	0.3277	0.0132	0.4733
0.0045	0.1736	0.0089	0.3311	0.0133	0.4765
0.0046	0.1774	0.0090	0.3346	0.0134	0.4797
0.0047	0.1811	0.0091	0.3380	0.0135	0.4828
0.0048	0.1848	0.0092	0.3414	0.0136	0.4860
0.0049	0.1885	0.0093	0.3449	0.0137	0.4891
0.0050	0.1922	0.0094	0.3483	0.0138	0.4923
0.0051	0.1958	0.0095	0.3517	0.0139	0.4954
0.0052	0.1995	0.0096	0.3551	0.0140	0.4985
0.0053	0.2032	0.0097	0.3585	0.0141	0.5016

ρ	$\bar{k}$	ρ	$\bar{k}$	ρ	$\bar{k}$	ε_t
0.0142	0.5047	0.0173	0.5981	**0.02033**	**0.6836**	**0.00500**
0.0143	0.5078	0.0174	0.6011	0.0204	0.6855	0.00497
0.0144	0.5109	0.0175	0.6040	0.0205	0.6882	0.00493
0.0145	0.5140	0.0176	0.6069	0.0206	0.6909	0.00489
0.0146	0.5171	0.0177	0.6098	0.0207	0.6936	0.00485
0.0147	0.5202	0.0178	0.6126	0.0208	0.6963	0.00482
0.0148	0.5233	0.0179	0.6155	0.0209	0.6990	0.00478
0.0149	0.5264	0.0180	0.6184	0.0210	0.7017	0.00474
0.0150	0.5294	0.0181	0.6213	0.0211	0.7044	0.00470
0.0151	0.5325	0.0182	0.6241	0.0212	0.7071	0.00467
0.0152	0.5355	0.0183	0.6270	0.0213	0.7097	0.00463
0.0153	0.5386	0.0184	0.6298	0.0214	0.7124	0.00460
0.0154	0.5416	0.0185	0.6327	0.0215	0.7150	0.00456
0.0155	0.5447	0.0186	0.6355	0.0216	0.7177	0.00453
0.0156	0.5477	0.0187	0.6383	0.0217	0.7203	0.00449
0.0157	0.5507	0.0188	0.6412	0.0218	0.7230	0.00446
0.0158	0.5537	0.0189	0.6440	0.0219	0.7256	0.00442
0.0159	0.5567	0.0190	0.6468	0.0220	0.7282	0.00439
0.0160	0.5597	0.0191	0.6496	0.0221	0.7308	0.00436

ρ	$\bar{k}$	ρ	$\bar{k}$	ρ	$\bar{k}$	ε_t
0.0161	0.5627	0.0192	0.6524	0.0222	0.7334	0.00432
0.0162	0.5657	0.0193	0.6552	0.0223	0.7360	0.00429
0.0163	0.5687	0.0194	0.6580	0.0224	0.7386	0.00426
0.0164	0.5717	0.0195	0.6608	0.0225	0.7412	0.00423
0.0165	0.5746	0.0196	0.6635	0.0226	0.7438	0.00419
0.0166	0.5776	0.0197	0.6663	0.0227	0.7464	0.00416
0.0167	0.5805	0.0198	0.6691	0.0228	0.7490	0.00413
0.0168	0.5835	0.0199	0.6718	0.0229	0.7515	0.00410
0.0169	0.5864	0.0200	0.6746	0.0230	0.7541	0.00407
0.0170	0.5894	0.0201	0.6773	0.0231	0.7567	0.00404
0.0171	0.5923	0.0202	0.6800	0.0232	0.7592	0.00401
0.0172	0.5952	0.0203	0.6828	0.02323	0.7600	0.00400

$*d = d_t$.

TABLE A.8 Coefficient of Resistance ($\bar{k}$) versus Reinforcement Ratio (ρ) ($f'_c = 3000$ psi; $f_y = 60{,}000$ psi; units of $\bar{k}$ are ksi)

ρ	$\bar{k}$	ρ	$\bar{k}$	ρ	$\bar{k}$	ε_t^*
0.0010	0.0593	0.0059	0.3294	0.0108	0.5657	
0.0011	0.0651	0.0060	0.3346	0.0109	0.5702	
0.0012	0.0710	0.0061	0.3397	0.0110	0.5746	
0.0013	0.0768	0.0062	0.3449	0.0111	0.5791	
0.0014	0.0826	0.0063	0.3500	0.0112	0.5835	
0.0015	0.0884	0.0064	0.3551	0.0113	0.5879	
0.0016	0.0942	0.0065	0.3602	0.0114	0.5923	
0.0017	0.1000	0.0066	0.3653	0.0115	0.5967	
0.0018	0.1057	0.0067	0.3703	0.0116	0.6011	
0.0019	0.1115	0.0068	0.3754	0.0117	0.6054	
0.0020	0.1172	0.0069	0.3804	0.0118	0.6098	
0.0021	0.1229	0.0070	0.3854	0.0119	0.6141	
0.0022	0.1286	0.0071	0.3904	0.0120	0.6184	
0.0023	0.1343	0.0072	0.3954	0.0121	0.6227	
0.0024	0.1399	0.0073	0.4004	0.0122	0.6270	
0.0025	0.1456	0.0074	0.4054	0.0123	0.6312	
0.0026	0.1512	0.0075	0.4103	0.0124	0.6355	
0.0027	0.1569	0.0076	0.4152	0.0125	0.6398	
0.0028	0.1625	0.0077	0.4202	0.0126	0.6440	
0.0029	0.1681	0.0078	0.4251	0.0127	0.6482	
0.0030	0.1736	0.0079	0.4300	0.0128	0.6524	
0.0031	0.1792	0.0080	0.4348	0.0129	0.6566	
0.0032	0.1848	0.0081	0.4397	0.0130	0.6608	

ρ	$\bar{k}$	ρ	$\bar{k}$	ρ	$\bar{k}$	ε_t^*
0.0033	0.1903	0.0082	0.4446	0.0131	0.6649	
0.0034	0.1958	0.0083	0.4494	0.0132	0.6691	
0.0035	0.2014	0.0084	0.4542	0.0133	0.6732	
0.0036	0.2069	0.0085	0.4590	0.0134	0.6773	
0.0037	0.2123	0.0086	0.4638	0.0135	0.6814	
0.0038	0.2178	0.0087	0.4686	**0.01355**	**0.6835**	**0.00500**
0.0039	0.2233	0.0088	0.4734	0.0136	0.6855	0.00497
0.0040	0.2287	0.0089	0.4781	0.0137	0.6896	0.00491
0.0041	0.2341	0.0090	0.4828	0.0138	0.6936	0.00485
0.0042	0.2396	0.0091	0.4876	0.0139	0.6977	0.00480
0.0043	0.2450	0.0092	0.4923	0.0140	0.7017	0.00474
0.0044	0.2503	0.0093	0.4970	0.0141	0.7057	0.00469
0.0045	0.2557	0.0094	0.5017	0.0142	0.7097	0.00463
0.0046	0.2611	0.0095	0.5063	0.0143	0.7137	0.00458
0.0047	0.2664	0.0096	0.5110	0.0144	0.7177	0.00453
0.0048	0.2717	0.0097	0.5156	0.0145	0.7216	0.00447
0.0049	0.2771	0.0098	0.5202	0.0146	0.7256	0.00442
0.0050	0.2824	0.0099	0.5248	0.0147	0.7295	0.00437
0.0051	0.2876	0.0100	0.5294	0.0148	0.7334	0.00432
0.0052	0.2929	0.0101	0.5340	0.0149	0.7373	0.00427
0.0053	0.2982	0.0102	0.5386	0.0150	0.7412	0.00423
0.0054	0.3034	0.0103	0.5431	0.0151	0.7451	0.00418
0.0055	0.3087	0.0104	0.5477	0.0152	0.7490	0.00413
0.0056	0.3139	0.0105	0.5522	0.0153	0.7528	0.00408
0.0057	0.3191	0.0106	0.5567	0.0154	0.7567	0.00404
0.0058	0.3243	0.0107	0.5612	0.01548	0.7597	0.00400

$*d = d_t$.

TABLE A.9 Coefficient of Resistance ($\bar{k}$) versus Reinforcement Ratio (ρ) ($f'_c = 4000$ psi; $f_y = 40{,}000$ psi; units of ($\bar{k}$) are ksi)

ρ	$\bar{k}$	ρ	$\bar{k}$	ρ	$\bar{k}$	ρ	$\bar{k}$
0.0010	0.0398	0.0054	0.2091	0.0098	0.3694	0.0142	0.5206
0.0011	0.0437	0.0055	0.2129	0.0099	0.3729	0.0143	0.5239
0.0012	0.0477	0.0056	0.2166	0.0100	0.3765	0.0144	0.5272
0.0013	0.0516	0.0057	0.2204	0.0101	0.3800	0.0145	0.5305
0.0014	0.0555	0.0058	0.2241	0.0102	0.3835	0.0146	0.5338
0.0015	0.0595	0.0059	0.2278	0.0103	0.3870	0.0147	0.5372
0.0016	0.0634	0.0060	0.2315	0.0104	0.3906	0.0148	0.5405
0.0017	0.0673	0.0061	0.2352	0.0105	0.3941	0.0149	0.5438
0.0018	0.0712	0.0062	0.2390	0.0106	0.3976	0.0150	0.5471

ρ	$\bar{k}$	ρ	$\bar{k}$	ρ	$\bar{k}$	ρ	$\bar{k}$
0.0019	0.0752	0.0063	0.2427	0.0107	0.4011	0.0151	0.5504
0.0020	0.0791	0.0064	0.2464	0.0108	0.4046	0.0152	0.5536
0.0021	0.0830	0.0065	0.2501	0.0109	0.4080	0.0153	0.5569
0.0022	0.0869	0.0066	0.2538	0.0110	0.4115	0.0154	0.5602
0.0023	0.0908	0.0067	0.2574	0.0111	0.4150	0.0155	0.5635
0.0024	0.0946	0.0068	0.2611	0.0112	0.4185	0.0156	0.5667
0.0025	0.0985	0.0069	0.2648	0.0113	0.4220	0.0157	0.5700
0.0026	0.1024	0.0070	0.2685	0.0114	0.4254	0.0158	0.5733
0.0027	0.1063	0.0071	0.2721	0.0115	0.4289	0.0159	0.5765
0.0028	0.1102	0.0072	0.2758	0.0116	0.4323	0.0160	0.5798
0.0029	0.1140	0.0073	0.2795	0.0117	0.4358	0.0161	0.5830
0.0030	0.1179	0.0074	0.2831	0.0118	0.4392	0.0162	0.5863
0.0031	0.1217	0.0075	0.2868	0.0119	0.4427	0.0163	0.5895
0.0032	0.1256	0.0076	0.2904	0.0120	0.4461	0.0164	0.5927
0.0033	0.1294	0.0077	0.2941	0.0121	0.4495	0.0165	0.5959
0.0034	0.1333	0.0078	0.2977	0.0122	0.4530	0.0166	0.5992
0.0035	0.1371	0.0079	0.3013	0.0123	0.4564	0.0167	0.6024
0.0036	0.1410	0.0080	0.3049	0.0124	0.4598	0.0168	0.6056
0.0037	0.1448	0.0081	0.3086	0.0125	0.4632	0.0169	0.6088
0.0038	0.1486	0.0082	0.3122	0.0126	0.4666	0.0170	0.6120
0.0039	0.1524	0.0083	0.3158	0.0127	0.4701	0.0171	0.6152
0.0040	0.1562	0.0084	0.3194	0.0128	0.4735	0.0172	0.6184
0.0041	0.1600	0.0085	0.3230	0.0129	0.4768	0.0173	0.6216
0.0042	0.1638	0.0086	0.3266	0.0130	0.4802	0.0174	0.6248
0.0043	0.1676	0.0087	0.3302	0.0131	0.4836	0.0175	0.6279
0.0044	0.1714	0.0088	0.3338	0.0132	0.4870	0.0176	0.6311
0.0045	0.1752	0.0089	0.3374	0.0133	0.4904	0.0177	0.6343
0.0046	0.1790	0.0090	0.3409	0.0134	0.4938	0.0178	0.6375
0.0047	0.1828	0.0091	0.3445	0.0135	0.4971	0.0179	0.6406
0.0048	0.1866	0.0092	0.3481	0.0136	0.5005	0.0180	0.6438
0.0049	0.1904	0.0093	0.3517	0.0137	0.5038	0.0181	0.6469
0.0050	0.1941	0.0094	0.3552	0.0138	0.5072	0.0182	0.6501
0.0051	0.1979	0.0095	0.3588	0.0139	0.5105	0.0183	0.6532
0.0052	0.2016	0.0096	0.3623	0.0140	0.5139	0.0184	0.6563
0.0053	0.2054	0.0097	0.3659	0.0141	0.5172	0.0185	0.6595

ρ	$\bar{k}$	ρ	$\bar{k}$	ρ	$\bar{k}$	ε_t^*
0.0186	0.6626	0.0229	0.7927	**0.0271**	**0.9113**	**0.00500**
0.0187	0.6657	0.0230	0.7956	0.0272	0.9140	0.00497
0.0188	0.6688	0.0231	0.7985	0.0273	0.9167	0.00494

ρ	$\bar{k}$	ρ	$\bar{k}$	ρ	$\bar{k}$	ε_t^*
0.0189	0.6720	0.0232	0.8014	0.0274	0.9194	0.00491
0.0190	0.6751	0.0233	0.8043	0.0275	0.9221	0.00488
0.0191	0.6782	0.0234	0.8072	0.0276	0.9248	0.00485
0.0192	0.6813	0.0235	0.8101	0.0277	0.9275	0.00482
0.0193	0.6844	0.0236	0.8130	0.0278	0.9302	0.00480
0.0194	0.6875	0.0237	0.8159	0.0279	0.9329	0.00477
0.0195	0.6905	0.0238	0.8188	0.0280	0.9356	0.00474
0.0196	0.6936	0.0239	0.8217	0.0281	0.9383	0.00471
0.0197	0.6967	0.0240	0.8245	0.0282	0.9410	0.00469
0.0198	0.6998	0.0241	0.8274	0.0283	0.9436	0.00466
0.0199	0.7029	0.0242	0.8303	0.0284	0.9463	0.00463
0.0200	0.7059	0.0243	0.8331	0.0285	0.9490	0.00461
0.0201	0.7090	0.0244	0.8360	0.0286	0.9516	0.00458
0.0202	0.7120	0.0245	0.8388	0.0287	0.9543	0.00455
0.0203	0.7151	0.0246	0.8417	0.0288	0.9569	0.00453
0.0204	0.7181	0.0247	0.8445	0.0289	0.9596	0.00450
0.0205	0.7212	0.0248	0.8473	0.0290	0.9622	0.00447
0.0206	0.7242	0.0249	0.8502	0.0291	0.9648	0.00445
0.0207	0.7272	0.0250	0.8530	0.0292	0.9675	0.00442
0.0208	0.7302	0.0251	0.8558	0.0293	0.9701	0.00440
0.0209	0.7333	0.0252	0.8586	0.0294	0.9727	0.00437
0.0210	0.7363	0.0253	0.8615	0.0295	0.9753	0.00435
0.0211	0.7393	0.0254	0.8643	0.0296	0.9779	0.00432
0.0212	0.7423	0.0255	0.8671	0.0297	0.9805	0.00430
0.0213	0.7453	0.0256	0.8699	0.0298	0.9831	0.00427
0.0214	0.7483	0.0257	0.8727	0.0299	0.9857	0.00425
0.0215	0.7513	0.0258	0.8754	0.0300	0.9883	0.00423
0.0216	0.7543	0.0259	0.8782	0.0301	0.9909	0.00420
0.0217	0.7572	0.0260	0.8810	0.0302	0.9935	0.00418
0.0218	0.7602	0.0261	0.8838	0.0303	0.9961	0.00415
0.0219	0.7632	0.0262	0.8865	0.0304	0.9986	0.00413
0.0220	0.7662	0.0263	0.8893	0.0305	1.0012	0.00411
0.0221	0.7691	0.0264	0.8921	0.0306	1.0038	0.00408
0.0222	0.7721	0.0265	0.8948	0.0307	1.0063	0.00406
0.0223	0.7750	0.0266	0.8976	0.0308	1.0089	0.00404
0.0224	0.7780	0.0267	0.9003	0.0309	1.0114	0.00401
0.0225	0.7809	0.0268	0.9031	0.03096	1.0130	0.00400
0.0226	0.7839	0.0269	0.9058			
0.0227	0.7868	0.0270	0.9085			
0.0228	0.7897					

$*d = d_t$.

TABLE A.10 Coefficient of Resistance ($\bar{k}$) versus Reinforcement Ratio (ρ) ($f'_c = 4000$ psi; $f_y = 60{,}000$ psi; units of $\bar{k}$ are ksi)

ρ	$\bar{k}$	ρ	$\bar{k}$	ρ	$\bar{k}$	ρ	$\bar{k}$
0.0010	0.0595	0.0039	0.2259	0.0068	0.3835	0.0097	0.5322
0.0011	0.0654	0.0040	0.2315	0.0069	0.3888	0.0098	0.5372
0.0012	0.0712	0.0041	0.2371	0.0070	0.3941	0.0099	0.5421
0.0013	0.0771	0.0042	0.2427	0.0071	0.3993	0.0100	0.5471
0.0014	0.0830	0.0043	0.2482	0.0072	0.4046	0.0101	0.5520
0.0015	0.0889	0.0044	0.2538	0.0073	0.4098	0.0102	0.5569
0.0016	0.0946	0.0045	0.2593	0.0074	0.4150	0.0103	0.5618
0.0017	0.1005	0.0046	0.2648	0.0075	0.4202	0.0104	0.5667
0.0018	0.1063	0.0047	0.2703	0.0076	0.4254	0.0105	0.5716
0.0019	0.1121	0.0048	0.2758	0.0077	0.4306	0.0106	0.5765
0.0020	0.1179	0.0049	0.2813	0.0078	0.4358	0.0107	0.5814
0.0021	0.1237	0.0050	0.2868	0.0079	0.4410	0.0108	0.5862
0.0022	0.1294	0.0051	0.2922	0.0080	0.4461	0.0109	0.5911
0.0023	0.1352	0.0052	0.2977	0.0081	0.4513	0.0110	0.5959
0.0024	0.1410	0.0053	0.3031	0.0082	0.4564	0.0111	0.6008
0.0025	0.1467	0.0054	0.3086	0.0083	0.4615	0.0112	0.6056
0.0026	0.1524	0.0055	0.3140	0.0084	0.4666	0.0113	0.6104
0.0027	0.1581	0.0056	0.3194	0.0085	0.4718	0.0114	0.6152
0.0028	0.1638	0.0057	0.3248	0.0086	0.4768	0.0115	0.6200
0.0029	0.1695	0.0058	0.3302	0.0087	0.4819	0.0116	0.6248
0.0030	0.1752	0.0059	0.3356	0.0088	0.4870	0.0117	0.6296
0.0031	0.1809	0.0060	0.3409	0.0089	0.4921	0.0118	0.6343
0.0032	0.1866	0.0061	0.3463	0.0090	0.4971	0.0119	0.6391
0.0033	0.1922	0.0062	0.3516	0.0091	0.5022	0.0120	0.6438
0.0034	0.1979	0.0063	0.3570	0.0092	0.5072	0.0121	0.6485
0.0035	0.2035	0.0064	0.3623	0.0093	0.5122	0.0122	0.6532
0.0036	0.2091	0.0065	0.3676	0.0094	0.5172	0.0123	0.6579
0.0037	0.2148	0.0066	0.3729	0.0095	0.5222	0.0124	0.6626
0.0038	0.2204	0.0067	0.3782	0.0096	0.5272	0.0125	0.6673

ρ	$\bar{k}$	ρ	$\bar{k}$	ρ	$\bar{k}$	ε_t^*
0.0126	0.6720	0.0154	0.7985	**0.01806**	**0.9110**	**0.00500**
0.0127	0.6766	0.0155	0.8029	0.0181	0.9126	0.00498
0.0128	0.6813	0.0156	0.8072	0.0182	0.9167	0.00494
0.0129	0.6859	0.0157	0.8116	0.0183	0.9208	0.00490
0.0130	0.6906	0.0158	0.8159	0.0184	0.9248	0.00485
0.0131	0.6952	0.0159	0.8202	0.0185	0.9289	0.00481
0.0132	0.6998	0.0160	0.8245	0.0186	0.9329	0.00477
0.0133	0.7044	0.0161	0.8288	0.0187	0.9369	0.00473

ρ	$\bar{k}$	ρ	$\bar{k}$	ρ	$\bar{k}$	ε_t^*
0.0134	0.7090	0.0162	0.8331	0.0188	0.9410	0.00469
0.0135	0.7136	0.0163	0.8374	0.0189	0.9450	0.00465
0.0136	0.7181	0.0164	0.8417	0.0190	0.9490	0.00461
0.0137	0.7227	0.0165	0.8459	0.0191	0.9529	0.00457
0.0138	0.7272	0.0166	0.8502	0.0192	0.9569	0.00453
0.0139	0.7318	0.0167	0.8544	0.0193	0.9609	0.00449
0.0140	0.7363	0.0168	0.8586	0.0194	0.9648	0.00445
0.0141	0.7408	0.0169	0.8629	0.0195	0.9688	0.00441
0.0142	0.7453	0.0170	0.8671	0.0196	0.9727	0.00437
0.0143	0.7498	0.0171	0.8713	0.0197	0.9766	0.00434
0.0144	0.7543	0.0172	0.8754	0.0198	0.9805	0.00430
0.0145	0.7587	0.0173	0.8796	0.0199	0.9844	0.00426
0.0146	0.7632	0.0174	0.8838	0.0200	0.9883	0.00422
0.0147	0.7676	0.0175	0.8879	0.0201	0.9922	0.00419
0.0148	0.7721	0.0176	0.8921	0.0202	0.9961	0.00415
0.0149	0.7765	0.0177	0.8962	0.0203	0.9999	0.00412
0.0150	0.7809	0.0178	0.9003	0.0204	1.0038	0.00408
0.0151	0.7853	0.0179	0.9044	0.0205	1.0076	0.00405
0.0152	0.7897	0.0180	0.9085	0.0206	1.0114	0.00401
0.0153	0.7941			0.02063	1.0126	0.00400

$*d = d_t$.

TABLE A.11 Coefficient of Resistance ($\bar{k}$) versus Reinforcement Ratio (ρ) ($f'_c = 5000$ psi; $f_y = 60{,}000$ psi; units of $\bar{k}$ are ksi)

ρ	$\bar{k}$	ρ	$\bar{k}$	ρ	$\bar{k}$	ρ	$\bar{k}$
0.0010	0.0596	0.0048	0.2782	0.0086	0.4847	0.0124	0.6789
0.0011	0.0655	0.0049	0.2838	0.0087	0.4899	0.0125	0.6838
0.0012	0.0714	0.0050	0.2894	0.0088	0.4952	0.0126	0.6888
0.0013	0.0773	0.0051	0.2950	0.0089	0.5005	0.0127	0.6937
0.0014	0.0832	0.0052	0.3005	0.0090	0.5057	0.0128	0.6986
0.0015	0.0890	0.0053	0.3061	0.0091	0.5109	0.0129	0.7035
0.0016	0.0949	0.0054	0.3117	0.0092	0.5162	0.0130	0.7084
0.0017	0.1008	0.0055	0.3172	0.0093	0.5214	0.0131	0.7133
0.0018	0.1066	0.0056	0.3227	0.0094	0.5266	0.0132	0.7182
0.0019	0.1125	0.0057	0.3282	0.0095	0.5318	0.0133	0.7231
0.0020	0.1183	0.0058	0.3338	0.0096	0.5370	0.0134	0.7280
0.0021	0.1241	0.0059	0.3393	0.0097	0.5422	0.0135	0.7328
0.0022	0.1300	0.0060	0.3448	0.0098	0.5473	0.0136	0.7377
0.0023	0.1358	0.0061	0.3502	0.0099	0.5525	0.0137	0.7425
0.0024	0.1416	0.0062	0.3557	0.0100	0.5576	0.0138	0.7473

ρ	$\bar{k}$	ρ	$\bar{k}$	ρ	$\bar{k}$	ρ	$\bar{k}$
0.0025	0.1474	0.0063	0.3612	0.0101	0.5628	0.0139	0.7522
0.0026	0.1531	0.0064	0.3667	0.0102	0.5679	0.0140	0.7570
0.0027	0.1589	0.0065	0.3721	0.0103	0.5731	0.0141	0.7618
0.0028	0.1647	0.0066	0.3776	0.0104	0.5782	0.0142	0.7666
0.0029	0.1704	0.0067	0.3830	0.0105	0.5833	0.0143	0.7714
0.0030	0.1762	0.0068	0.3884	0.0106	0.5884	0.0144	0.7762
0.0031	0.1819	0.0069	0.3938	0.0107	0.5935	0.0145	0.7810
0.0032	0.1877	0.0070	0.3992	0.0108	0.5986	0.0146	0.7857
0.0033	0.1934	0.0071	0.4047	0.0109	0.6037	0.0147	0.7905
0.0034	0.1991	0.0072	0.4100	0.0110	0.6088	0.0148	0.7952
0.0035	0.2048	0.0073	0.4154	0.0111	0.6138	0.0149	0.8000
0.0036	0.2105	0.0074	0.4208	0.0112	0.6189	0.0150	0.8047
0.0037	0.2162	0.0075	0.4262	0.0113	0.6239	0.0151	0.8094
0.0038	0.2219	0.0076	0.4315	0.0114	0.6290	0.0152	0.8142
0.0039	0.2276	0.0077	0.4369	0.0115	0.6340	0.0153	0.8189
0.0040	0.2332	0.0078	0.4422	0.0116	0.6390	0.0154	0.8236
0.0041	0.2389	0.0079	0.4476	0.0117	0.6440	0.0155	0.8283
0.0042	0.2445	0.0080	0.4529	0.0118	0.6490	0.0156	0.8329
0.0043	0.2502	0.0081	0.4582	0.0119	0.6540	0.0157	0.8376
0.0044	0.2558	0.0082	0.4635	0.0120	0.6590	0.0158	0.8423
0.0045	0.2614	0.0083	0.4688	0.0121	0.6640	0.0159	0.8469
0.0046	0.2670	0.0084	0.4741	0.0122	0.6690	0.0160	0.8516
0.0047	0.2726	0.0085	0.4794	0.0123	0.6739	0.0161	0.8562

ρ	$\bar{k}$	ρ	$\bar{k}$	ρ	$\bar{k}$	ε_t
0.0162	0.8609	0.0194	1.0047	**0.02125**	**1.0838**	**0.00500**
0.0163	0.8655	0.0195	1.0090	0.0213	1.0859	0.00498
0.0164	0.8701	0.0196	1.0134	0.0214	1.0901	0.00494
0.0165	0.8747	0.0197	1.0177	0.0215	1.0943	0.00491
0.0166	0.8793	0.0198	1.0220	0.0216	1.0985	0.00487
0.0167	0.8839	0.0199	1.0263	0.0217	1.1026	0.00483
0.0168	0.8885	0.0200	1.0307	0.0218	1.1068	0.00480
0.0169	0.8930	0.0201	1.0350	0.0219	1.1110	0.00476
0.0170	0.8976	0.0202	1.0393	0.0220	1.1151	0.00473
0.0171	0.9022	0.0203	1.0435	0.0221	1.1192	0.00469
0.0172	0.9067	0.0204	1.0478	0.0222	1.1234	0.00466
0.0173	0.9112	0.0205	1.0521	0.0223	1.1275	0.00462
0.0174	0.9158	0.0206	1.0563	0.0224	1.1316	0.00459
0.0175	0.9203	0.0207	1.0606	0.0225	1.1357	0.00456
0.0176	0.9248	0.0208	1.0648	0.0226	1.1398	0.00452

ρ	$\bar{k}$	ρ	$\bar{k}$	ρ	$\bar{k}$	ε_t
0.0177	0.9293	0.0209	1.0691	0.0227	1.1438	0.00449
0.0178	0.9338	0.0210	1.0733	0.0228	1.1479	0.00446
0.0179	0.9383	0.0211	1.0775	0.0229	1.1520	0.00442
0.0180	0.9428	0.0212	1.0817	0.0230	1.1560	0.00439
0.0181	0.9473			0.0231	1.1601	0.00436
0.0182	0.9517			0.0232	1.1641	0.00433
0.0183	0.9562			0.0233	1.1682	0.00430
0.0184	0.9606			0.0234	1.1722	0.00426
0.0185	0.9651			0.0235	1.1762	0.00423
0.0186	0.9695			0.0236	1.1802	0.00420
0.0187	0.9739			0.0237	1.1842	0.00417
0.0188	0.9783			0.0238	1.882	0.00414
0.0189	0.9827			0.0239	1.1922	0.00411
0.0190	0.9872			0.0240	1.1961	0.00408
0.0191	0.9916			0.0241	1.2001	0.00405
0.0192	0.9959			0.0242	1.2041	0.00402
0.0193	1.0003			0.02429	1.2076	0.00400

$*d = d_t$.

TABLE A.12 Development Length (ℓ_{dc}) for Compression Bars with $f_y = 60{,}000$ psi (in.)

Bar Size	f'_c (normal-weight concrete), psi			
	3000	4000	5000	6000
3	8.2	7.1	6.8	6.8
4	11.0	9.5	9.0	9.0
5	13.7	11.9	11.3	11.3
6	16.4	14.2	13.5	13.5
7	19.2	16.6	15.8	15.8
8	21.9	19.0	18.0	18.0
9	24.7	21.4	20.3	20.3
10	27.8	24.1	22.9	22.9
11	30.9	26.8	25.4	25.4
14	37.1	32.1	30.5	30.5
18	49.4	42.8	40.6	40.6

Note: See Chapter 5 for calculation of development length for compression bars.

TABLE A.13 Development Length (ℓ_{dh}) for Hooked Bars with $f_y = 60{,}000$ psi (in.)

Bar Number	f'_c **(psi)**			
	3000	**4000**	**5000**	**6000**
3	6.0	6.0	6.0	6.0
4	7.0	6.1	6.0	6.0
5	9.8	8.5	7.6	7.0
6	12.9	11.2	10.0	9.1
7	16.3	14.1	12.6	11.5
8	19.9	17.2	15.4	14.1
9	23.9	20.7	18.5	16.9
10	28.5	24.7	22.1	20.2
11	33.3	28.9	25.8	23.6

Note: Modification factors may apply.

ℓ_{dh} × applicable modification factors ≥ larger of $8d_b$ or 6 in. See Chapter 5 for calculation of development length for hooked bars.

TABLE A.14 Preferred Maximum Total Number of Column Bars

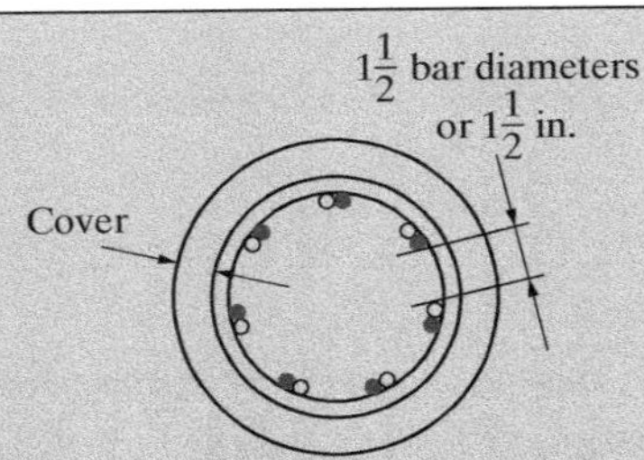

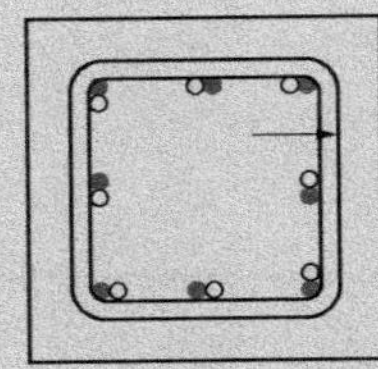

Recommended spiral or tie bar number	Core size (in.) = column size − 2 × cover	Circular area (in.^{2})	Bar number							Square area (in.^{2})	Bar number						
			#5	#6	#7	#8	#9	#10	#11		#5	#6	#7	#8	#9	#10	#11[a]
3[a]	9	63.6	8	7	7	6	—	—	—	81	8	8	8	8	4	4	4
	10	78.5	10	9	8	7	6	—	—	100	12	8	8	8	8	4	4
	11	95.0	11	10	9	8	7	6	—	121	12	12	8	8	8	8	4
	12	113.1	12	11	10	9	8	7	6	144	12	12	12	8	8	8	8
	13	132.7	13	12	11	10	8	7	6	169	16	12	12	12	8	8	8
	14	153.9	14	13	12	11	9	8	7	196	16	16	12	12	12	8	8
	15	176.7	15	14	13	12	10	9	8	225	16	16	16	12	12	12	8
4	16	201.1	16	15	14	12	11	9	8	256	20	16	16	16	12	12	8
	17	227.0	18	16	15	13	12	10	9	289	20	20	16	16	12	12	8
	18	254.5	19	17	15	14	12	11	10	324	20	20	16	16	16	12	12
	19	283.5	20	18	16	15	13	11	10	361	24	20	20	16	16	12	12
	20	314.2	21	19	17	16	14	12	11	400	24	24	20	20	16	12	12
	21	346.4	22	20	18	17	15	13	11	441	28	24	20	20	16	16	12
	22	380.1	23	21	19	18	15	14	12	484	28	24	24	20	20	16	12
5	23	415.5	24	22	21	19	16	14	13	529	28	28	24	24	20	16	16
	24	452.4	25	23	21	20	17	15	13	576	32	28	24	24	20	16	16
	25	490.9	26	24	22	20	18	16	14	625	32	28	28	24	20	20	16
	26	530.9	28	25	23	21	19	16	14	676	32	32	28	24	24	20	16
	27	572.6	29	26	24	22	19	17	15	729	36	32	28	28	24	20	16

[a] No. 4 tie required for No. 11 or larger longitudinal reinforcement (ACI Section 7.10.5.1).

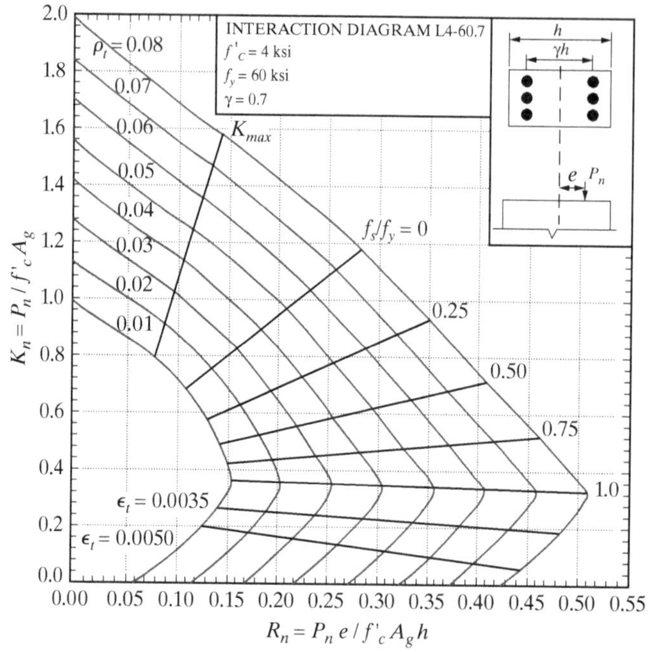

FIGURE A.15 Column Interaction Diagram

Source: Diagrams A.15 through A.22 are from the ACI Reinforced Concrete Design Handbook Design Aid—Analysis Tables (ACI SP-17DA(14)) and are reprinted here with the permission of the American Concrete Institute.

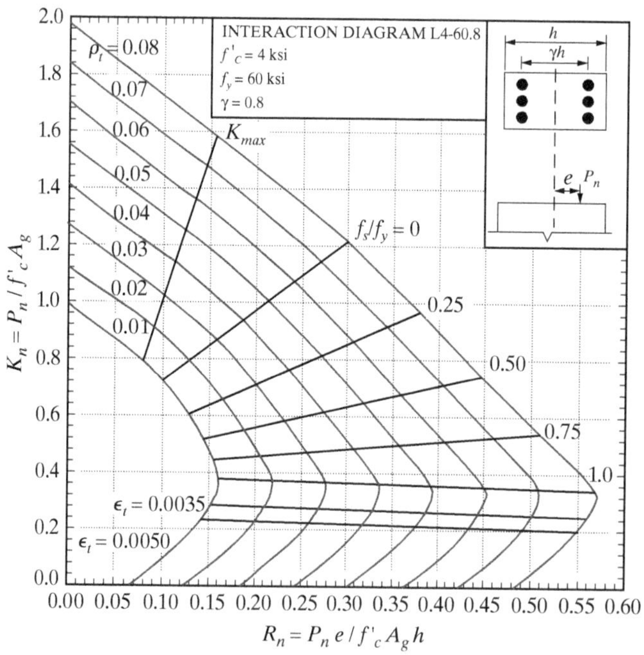

FIGURE A.16 Column Interaction Diagram

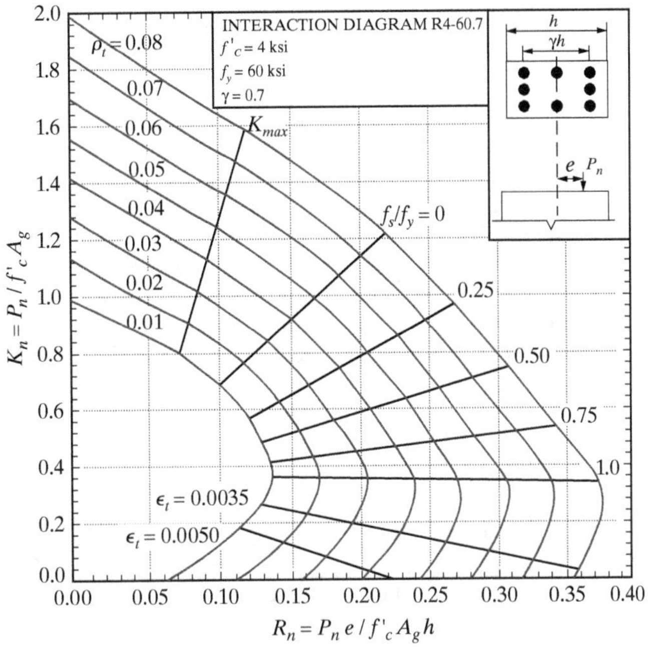

FIGURE A.17 Column Interaction Diagram

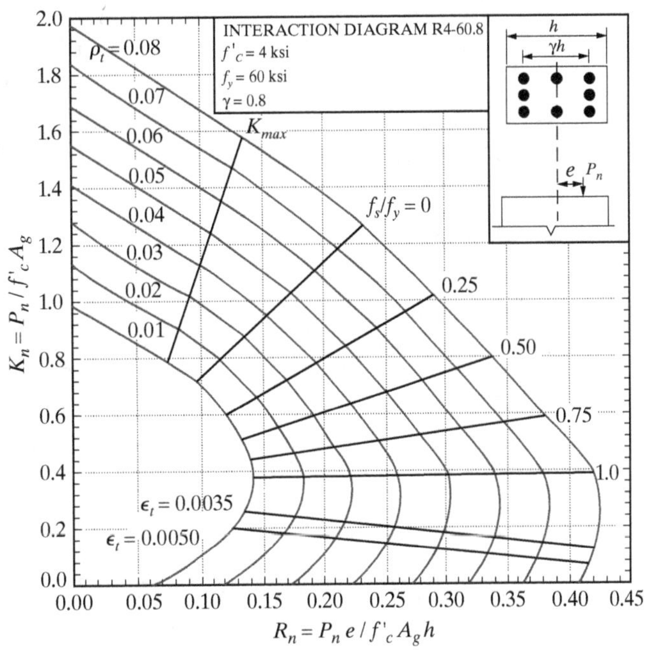

FIGURE A.18 Column Interaction Diagram

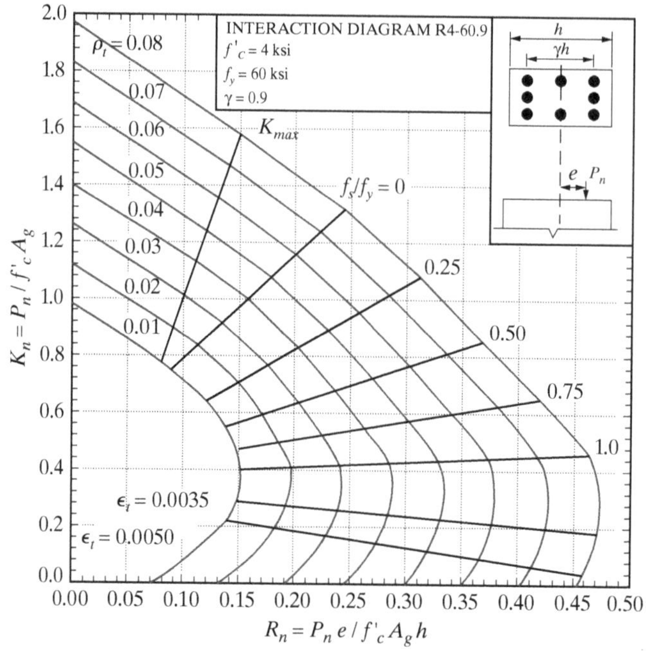

FIGURE A.19 Column Interaction Diagram

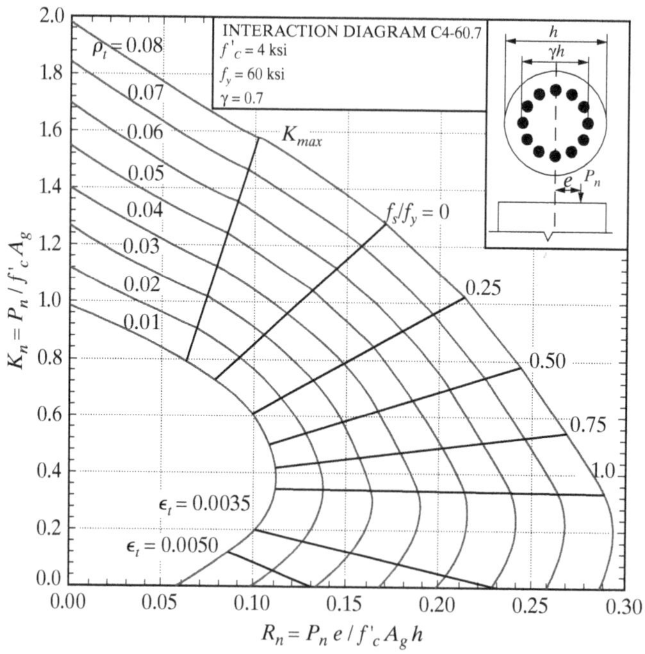

FIGURE A.20 Column Interaction Diagram

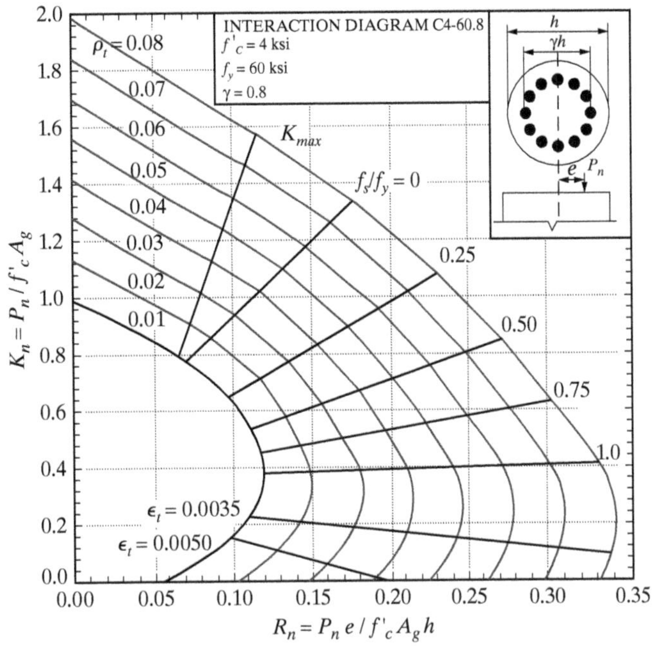

FIGURE A.21 Column Interaction Diagram

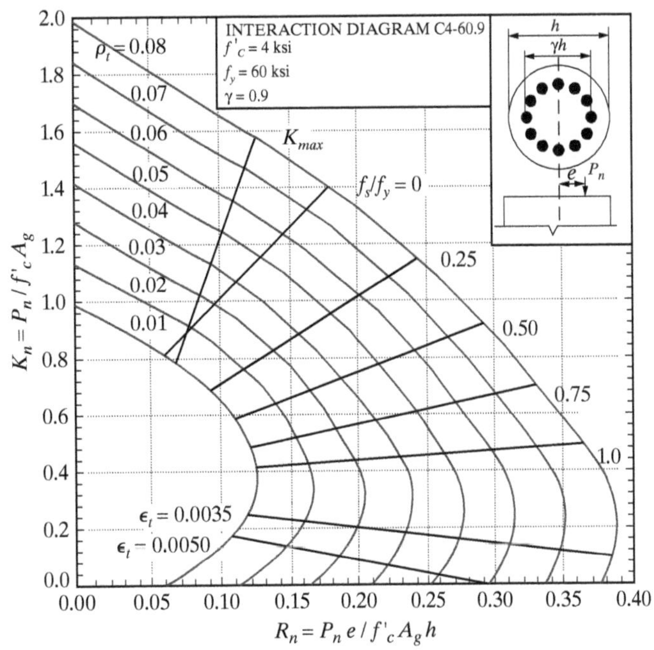

FIGURE A.22 Column Interaction Diagram

APPENDIX B

SUPPLEMENTARY AIDS AND GUIDELINES

B.1 ACCURACY FOR COMPUTATIONS FOR REINFORCED CONCRETE

The widespread availability and use of electronic calculators and computers for even the simplest of calculations have led to the use of numbers that represent a very high order of accuracy. For instance, a calculator with an eight-digit display will yield the following:

$$\frac{8}{0.7} = 11.428571$$

It should be recognized that the numerator and denominator, each of one-figure accuracy, resulted in a number that indicates eight-figure accuracy. Because the quotient cannot be expected to be more accurate than the numbers that produced it, the result as represented may lead one into a false sense of security associated with numbers of very high accuracy. For instance, it is illogical to calculate a required steel area to four-figure accuracy when the loads were to two-figure accuracy, and the bars to be chosen have areas tabulated to two- (sometimes three-) figure accuracy. Likewise, the mathematical expressions developed in this book and those presented by the ACI Code should be thought of in a similar light. They deal with a material, concrete, that

1. Is made on site or plant-made, is subject to varying amounts of quality control, and will vary from the design strength.
2. Is placed in forms that may or may not produce the design dimensions.
3. Contains reinforcing steel of a specified minimum strength, but that may vary above that strength.

In addition, the reinforced concrete member has reinforcing steel that may or may not be placed at the design location, and the design itself is generally based on loads that may be only "best estimates."

With the foregoing in mind, the following has been suggested by the Concrete Reinforcing Steel Institute as a rough guide for numerical accuracy in reinforced concrete calculations:

1. Loads to the nearest 1 psf; 10 lb/ft; 100 lb concentration
2. Span lengths to about 0.1 ft
3. Total loads and reactions to 0.1 kip or three-figure accuracy
4. Moments to the nearest 0.1 ft-kip or three-figure accuracy
5. Individual bar areas to 0.01 in.^{2}
6. Concrete sizes to $\frac{1}{2}$ in.
7. Effective beam depth of 0.1 in.
8. Column loads to the nearest 1.0 kip

In general (admittedly, not always), the reader will find that in this text, we have represented numbers used in calculations to an accuracy of three significant digits. If the number begins with 1, then four significant digits are shown. We round intermediate and final numerical solutions in accordance with this rule of thumb. When working on a calculator, however, one will normally maintain all digits and round only the final answer. For this reason, the reader may frequently obtain numerical results that are slightly different from those printed in the text. This should not cause undue concern.

B.2 FLOW DIAGRAMS

The step-by-step procedures for the analysis and design of reinforced concrete members may be presented in the form of flow diagrams (see Figures B.1 to B.4). To aid the reader in grasping the overall calculation approach, which may sometimes include cycling steps, flow diagrams for the analysis and design of rectangular beams and T-beams are presented here. These flow diagrams represent, on an elementary level, the type of organization required to develop computer programs to aid in analysis and design calculations.

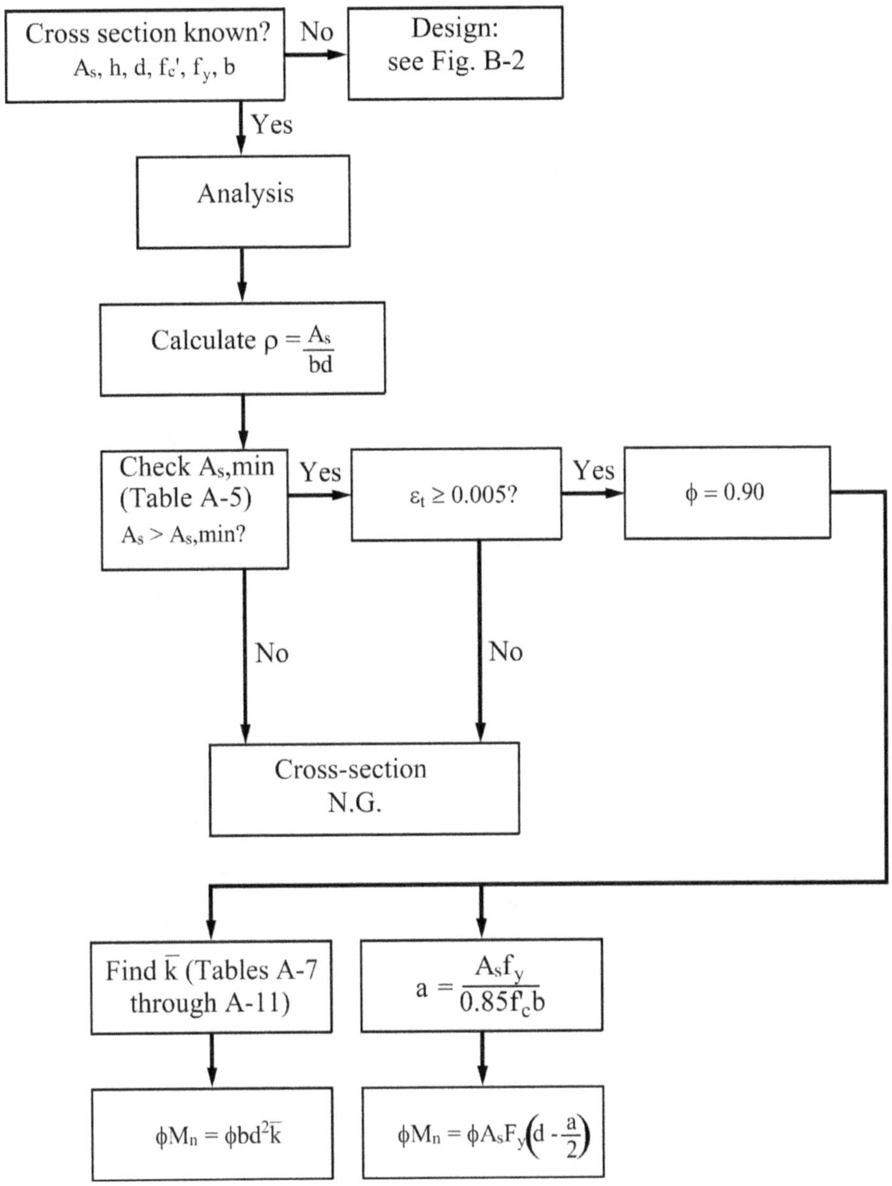

FIGURE B.1 Rectangular beam analysis for moment (tension steel only).

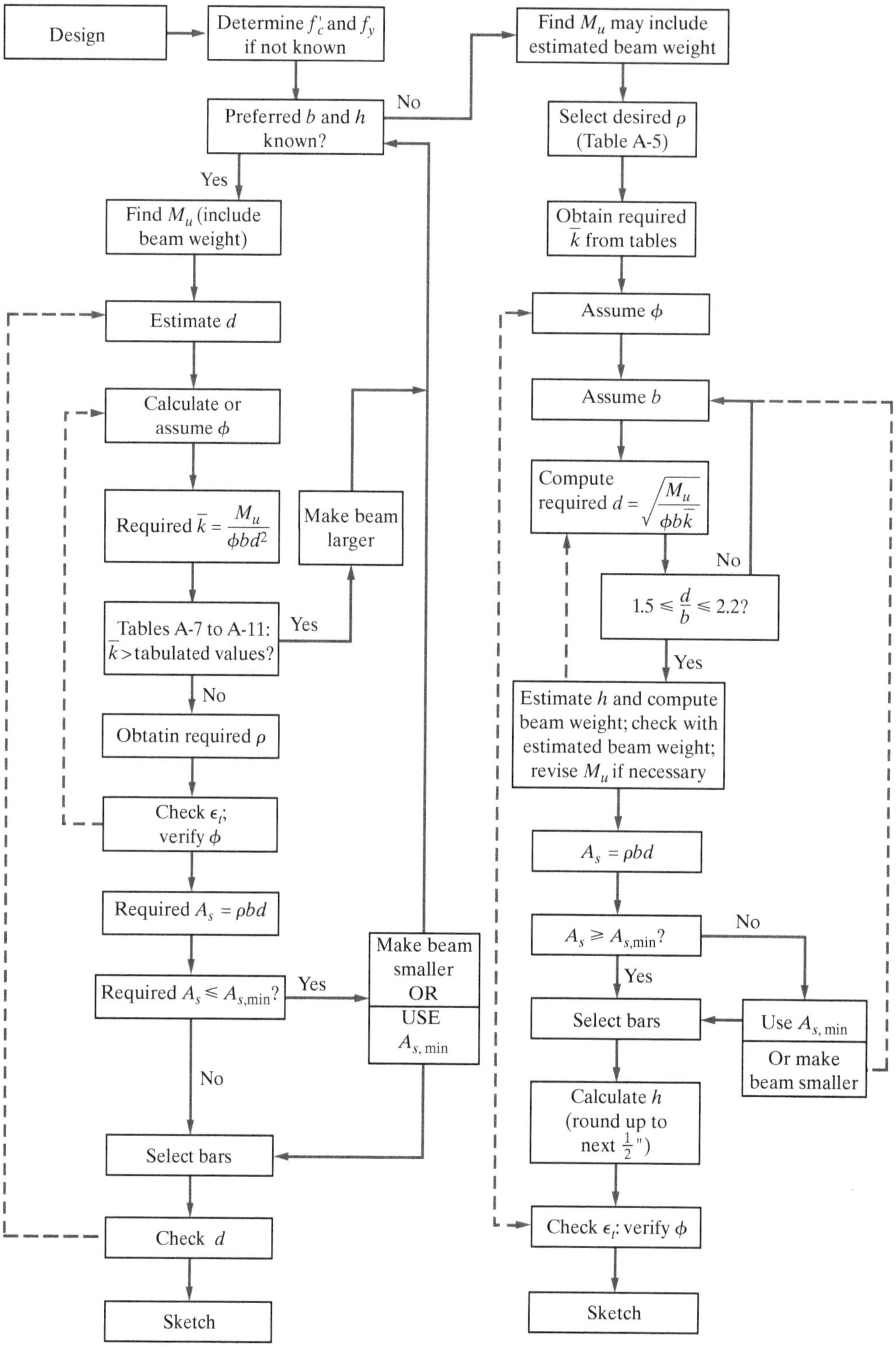

FIGURE B.2 Rectangular beam design for moment (tension steel only).

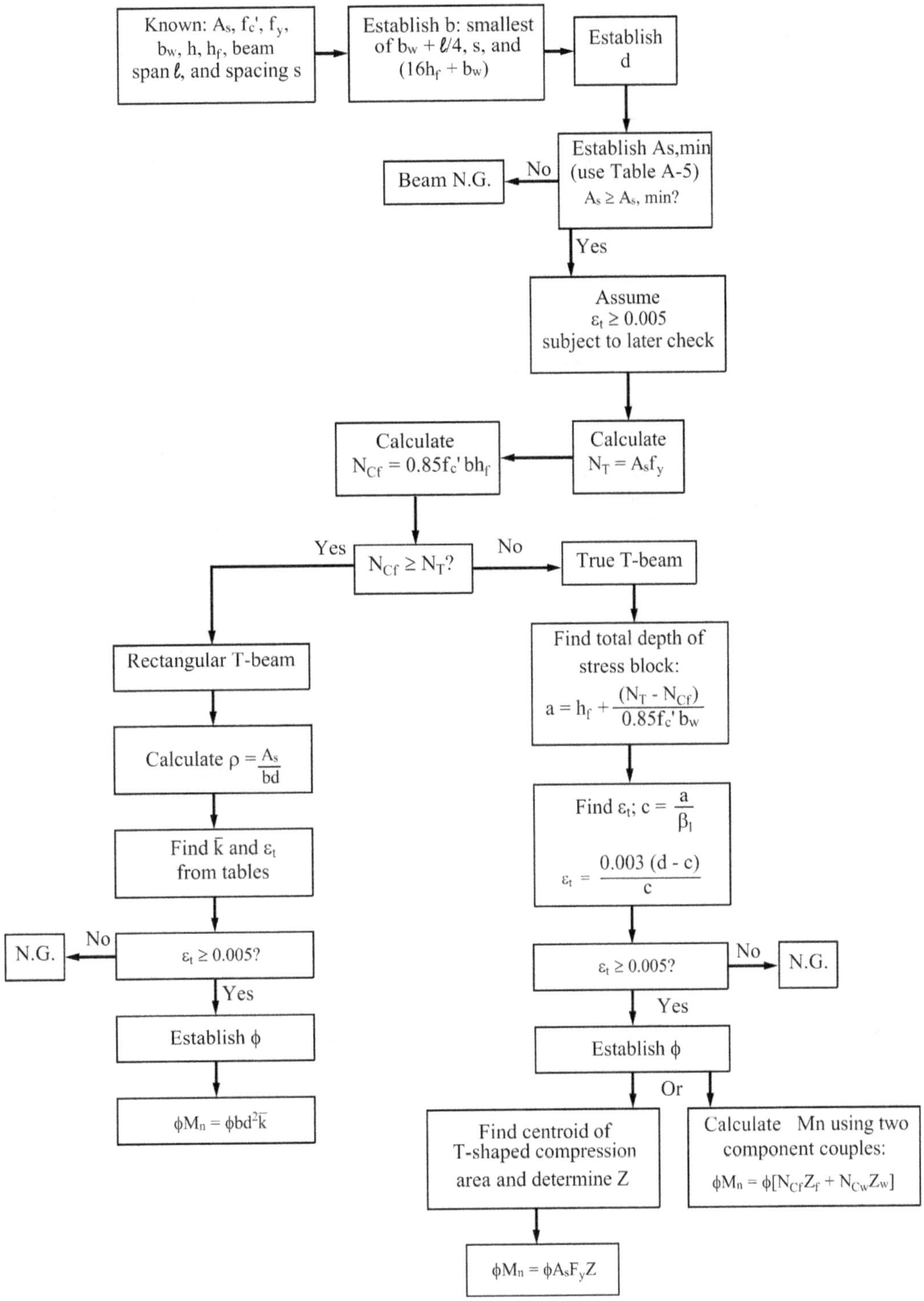

FIGURE B.3 T-beam analysis for moment.

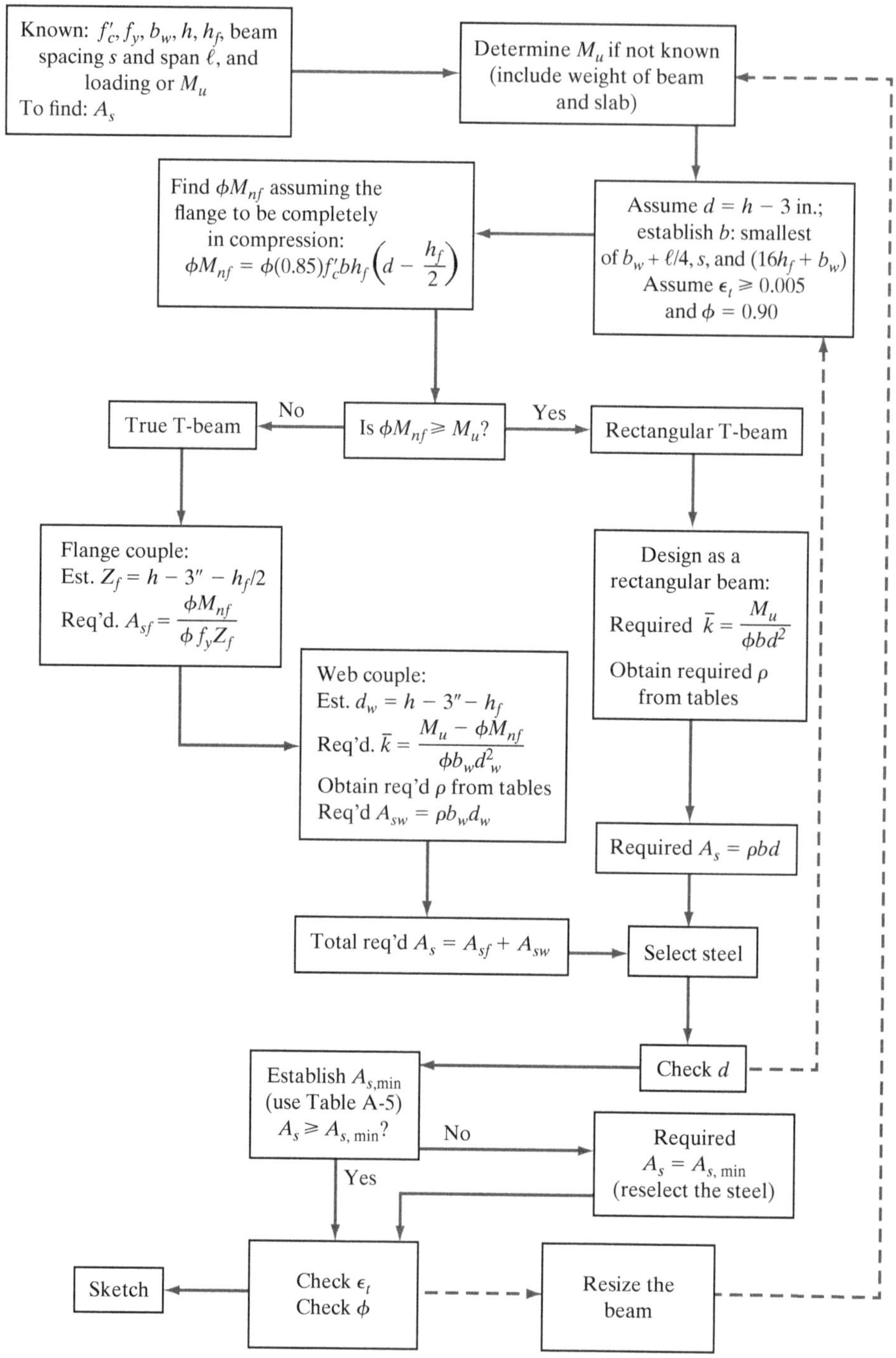

FIGURE B.4 T-beam design for moment.

APPENDIX C

METRICATION

C.1 THE INTERNATIONAL SYSTEM OF UNITS (SI)

The U.S. Customary System (or "inch-pound system") of weights and measures has been used as the primary unit system in this book. This system developed from the English system (British), which had been introduced in the original 13 colonies when they were under British rule. Even though the English system was spread to many parts of the world during the past three centuries, it was widely recognized that there was a need for a single international coordinated measurement system. As a result, a second system of weights and measures, known as the metric system, was developed by a commission of French scientists and was adopted by France as the legal system of weights and measures in 1799.

Although the metric system was not accepted with enthusiasm at first, adoption by other nations occurred steadily after France made its use compulsory in 1840. In the United States, an Act of Congress in 1866 made it lawful throughout the land to employ the weights and measures of the metric system in all contracts, dealings, or court proceedings.

By 1900, a total of 35 nations, including the major nations of continental Europe and most of South America, had officially accepted the metric system. In 1971, the secretary of commerce, in transmitting to Congress the results of a 3-year study authorized by the Metric Study Act of 1968, recommended that the United States change to the predominant use of the metric system through a coordinated national program. Congress responded by enacting the Metric Conversion Act of 1975 which established a U.S. Metric Board to carry out the planning, coordination, and public education that would facilitate a voluntary conversion to a modern metric system. Today, with the exception of a few small countries, the entire world is using the metric system or is changing to its use. Because of the many versions of the metric system that developed, an International General Conference on Weights and Measures in 1960 adopted an extensive revision and simplification of the system. The name *Le Système International d'Unités* (International System of Units), with the international abbreviation SI, was adopted for this modernized metric system.

The American Society of Civil Engineers (ASCE) resolved in 1970 to actively support conversion to SI and to adopt the revised edition of the American Society for Testing and Materials (ASTM) *Metric Practice* guide. In 1988, Congress passed the Omnibus Trade and

Competitiveness Act that, among other things, mandated that by the end of fiscal 1992, the federal government would require metric specifications on all the goods it purchased. In 1991, President George H. W. Bush signed Executive Order 12770, *Metric Usage in Federal Government Programs*, which required federal agencies to develop specific timetables and milestones for the transition to metric. A deadline of the year 2000 was set by the Federal Highway Administration for state implementation of the metric system for the design and construction of federally funded highway projects. Under the influence of vigorous lobbying, Congress cancelled this deadline in 1998. Currently, in state departments of transportation (DOTs), there is no uniformity of systems. Some use the U.S. Customary System, some use SI, and some allow either system. Most federal building construction is metricated, while very few private construction projects are being built using metric units [1].

Metric conversion involves two distinct aspects. On the one hand, the new units will affect design calculations and detailing practices. Many publications such as specifications, building codes, and design handbooks (and their associated software) are in various stages of production based on SI. This is essentially a paper (and electronic) change and is called a "soft" conversion. On the other hand, the physical sizes of some products will be affected (for instance, plywood will change from a 4-ft width to a slightly smaller 1200-mm width); this is called a "hard" conversion.

Metric reinforcing steel sizes have been standardized based on a soft conversion rather than a hard conversion. That is, the physical sizes of the 11 (Nos. 3–11, 14, and 18) inch-pound bars remain the same, but the designations change and the dimensions (diameter, area, and so on) are specified in metric units. The metric bar number is the nominal diameter rounded to the nearest mm. Virtually all new reinforcing steel is labeled in metric units (see Figure C.1). Table C.1 provides in inch-pound bar sizes the soft metric bar sizes and soft metric bar data.

TABLE C.1 Inch-Pound Bar Designation versus Soft Metric Bar Designation (with Soft Metric Bar Data)

	Soft metric		
Inch-pound bar designation	**Bar designation**	**Nominal diameter (mm)**	**Nominal area (mm^2)**
#3	#10	9.5	71
#4	#13	12.7	129
#5	#16	15.9	199
#6	#19	19.1	284
#7	#22	22.2	387
#8	#25	25.4	510
#9	#29	28.7	645
#10	#32	32.3	819
#11	#36	35.8	1006
#14	#43	43.0	1452
#18	#57	57.3	2581

Minimum yield strengths have been established as 420 MPa (grade 420) and 520 MPa (grade 520). These values are intended to be equivalent to inch-pound grade 60 and grade 75 materials, respectively. ASTM A615/A615M-04b also includes a grade 280 (280 MPa), which is equivalent to inch-pound grade 40. These minimum yield strengths are summarized in Table C.2.

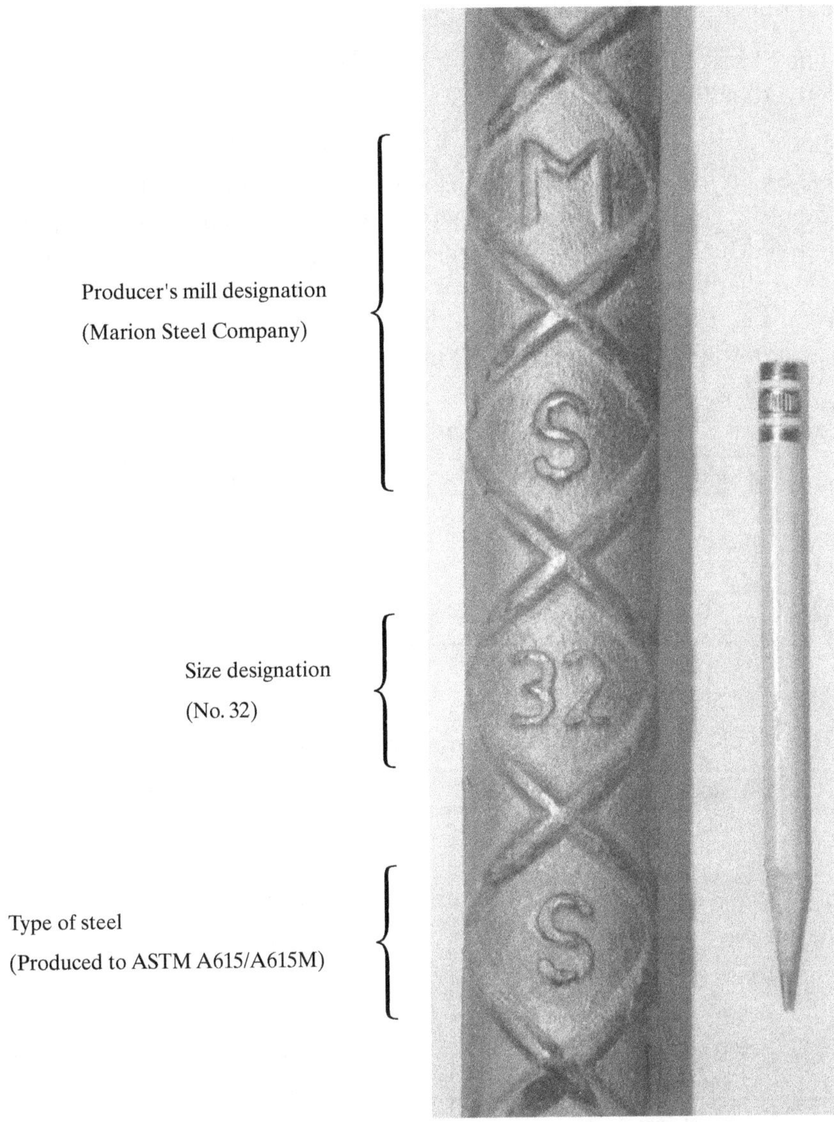

FIGURE C.1 Identification marks. This bar has a single longitudinal grade line (indicating grade 420 steel) on its opposite side (hidden in this photo).

TABLE C.2 Minimum Yield Strengths (ASTM A615/A615M-04b)

Metric (MPa)	Inch-pound equivalent (psi)
520 (grade 520)	75,400 (grade 75)
420 (grade 420)	60,900 (grade 60)
280 (grade 280)	40,600 (grade 40)

C.2 SI STYLE AND USAGE

The SI consists of a limited number of *base* units that establish fundamental quantities and a large number of *derived* units, which come from the base units, to describe other quantities.

The SI base units pertinent to reinforced concrete design are listed in Table C.3, and the SI-derived units pertinent to reinforced concrete design are listed in Table C.4.

Because the orders of magnitude of many quantities cover wide ranges of numerical values, SI *prefixes* have been established to deal with decimal-point placement. SI prefixes representing steps of 1000 are recommended to indicate orders of magnitude. Those recommended for use in reinforced concrete design are listed in Table C.5. From Table C.5, it is clear that there is a choice of ways to present numbers. It is preferable to use numbers between 1 and 1000, whenever possible, by selecting the appropriate prefix. For example, 18 m is preferred to 0.018 km or 18,000 mm.

TABLE C.3 SI Units and Symbols

Quantity	Unit	SI symbol
Length	meter	m
Mass	kilogram	kg
Time	second	s
Angle[a]	radian	rad

[a] It is also permissible to use the arc degree *and its decimal submultiples* when the radian is not convenient.

TABLE C.4 SI-Derived Units

Quantity	Unit	SI symbol	Formula
Acceleration	meter per second squared	—	m/s^2
Area	square meter	—	m^2
Density (mass per unit volume)	kilogram per cubic meter	—	kg/m^3
Force	Newton	N	$kg{\cdot}m/s^2$
Pressure or stress	pascal	Pa	N/m^2
Volume	cubic meter	—	m^3
Section modulus	meter to third power	—	m^3
Moment of inertia	meter to fourth power	—	m^4
Moment of force, torque	newton meter	—	N·m
Force per unit length	newton per meter	—	N/m
Mass per unit length	kilogram per meter	—	kg/m
Mass per unit area	kilogram per square meter	—	kg/m^2

A brief note must be made on this point concerning the presentation of numbers with many digits. It is common practice in the United States to separate digits into groups of three by means of commas. To avoid confusion with the widespread European practice of using a comma on the line as the decimal marker, the method of setting off groups of three digits with a gap, as shown in Table C.5, is recommended international practice. Note that this method is used on both the left and right sides of the decimal marker for any string of five or more digits. A group of four digits on either side of the decimal marker need not be separated.

TABLE C.5 SI Prefixes

Prefix	SI symbol	Factor
mega	M	$1{,}000{,}000 = 10^6$
kilo	K	$1000 = 10^3$
milli	M	$0.001 = 10^{-3}$
micro	*m*	$0.000\ 001 = 10^{-6}$

A significant difference between SI and other measurement systems is the use of explicit and distinctly separate units for mass and force. The SI base unit kilogram (kg) denotes the base unit of mass, which is the quantity of matter of an object. This is a constant quantity that is independent of gravitational attraction. The derived SI unit newton (N) denotes the absolute derived unit of force (mass times acceleration: kg·m/s^2). The term *weight* should be avoided since it is confused with mass and because it describes a particular force that is related solely to gravitational acceleration, which varies on the surface of the earth. For the conversion of mass to force, the recommended value for the acceleration of gravity in the United States may be taken as $g = 9.81$ m/s^2.

As an example, we will consider the mass of reinforced concrete, which, in the design of bending members, must be considered as a load or force per unit length of span. In the U.S. Customary System, reinforced concrete weighs 150 lb/ft^3. This is equivalent in the SI to

$$150\,\frac{\text{lb}}{\text{ft}^3}\times\left(\frac{3.2808\text{ ft}}{1\text{ m}}\right)^3\times\frac{1\text{ kg}}{2.2046\text{ lb}}\approx 2400\,\frac{\text{kg}}{\text{m}^3}$$

This represents a mass per unit volume (density), where the unit volume is 1 cubic meter, m^3. To use the density of the concrete to obtain a force per cubic meter, Newton's law must be applied:

$$\begin{aligned}F &= \text{mass times acceleration of gravity}\\ &= mg\\ &= 2400(9.81)\\ &= 23{,}500\,\frac{\text{kg}\cdot\text{m}}{\text{s}^2\text{m}^3}\\ &= 23{,}500\text{ N/m}^3\\ &= 23.5\text{ kN/m}^3\end{aligned}$$

The dead load per unit length of a beam of dimensions $b = 500$ mm and $h = 1000$ mm can then be determined:

$$\left(\frac{500}{1000}\right)\left(\frac{1000}{1000}\right)(23.5) = 11.75\text{ kN/m}$$

Thus, the load or force per unit length (1 meter) equals 11.75 kN.

C.3 CONVERSION FACTORS

Table C.6 contains conversion factors for the conversion of the U.S. Customary System units to SI units for quantities frequently used in reinforced concrete design.

TABLE C.6 Conversion Factors: U.S. Customary to SI Units

	Multiply		By		To Obtain
Length	inches	×	25.4	=	millimeters
	feet	×	0.3048	=	meters
	yards	×	0.9144	=	meters
	miles (statute)	×	1.609	=	kilometers
Area	square inches	×	645.2	=	square millimeters
	square feet	×	0.0929	=	square meters
	square yards	×	0.8361	=	square meters
Volume	cubic inches	×	16,387.	=	cubic millimeters
	cubic feet	×	0.028 32	=	cubic meters
	cubic yards	×	0.7646	=	cubic meters
	gallons (U.S. liquid)	×	0.003 785	=	cubic meters
Force	pounds	×	4.448	=	newtons
	kip	×	4,448.	=	newtons
Force per unit length	pounds per foot	×	14.594	=	newtons per meter
	kip per foot	×	14,594.	=	newtons per meter
Load per unit volume	pounds per	×	0.157 14	=	kilonewtons per
	cubic foot	×	0.0283		cubic meter
Bending moment or torque	inch-pounds	×	0.1130	=	newton meters
	foot-pounds	×	1.356	=	newton meters
	inch-kip	×	113.0	=	newton meters
	foot-kip	×	1356.	=	newton meters
	inch-kip	×	0.1130	=	kilonewton meters
	foot-kip	×	1.356	=	kilonewton meters
Stress, pressure, loading (force per unit area)	pounds per square inch	×	6895.	=	pascals
	pounds per square inch	×	6.895	=	kilopascals
	pounds per square inch	×	0.006 895	=	megapascals
	kip per square inch	×	6.895	=	megapascals
	pounds per square foot	×	47.88	=	pascals
	pounds per square foot	×	0.047 88	=	kilopascals
	kip per square foot	×	47.88	=	kilopascals
	kip per square foot	×	0.047 88	=	megapascals
Mass	pounds	×	0.454	=	kilograms
Mass per unit volume (density)	pounds per cubic foot	×	16.02	=	kilograms per cubic meter
	pounds per cubic yard	×	0.5933	=	kilograms per cubic meter
Moment of inertia	inches^4	×	416,231.	=	millimeters^4
Mass per unit length	pounds per foot	×	1.488	=	kilograms per meter
Mass per unit area	pounds per square foot	×	4.882	=	kilograms per square meter

Although specified in the SI, the pascal is not universally accepted as the unit of stress. Because section dimensions and properties are generally in millimeters, it is more convenient to express stress in newtons per square millimeter (1 N/mm^2 = 1 MPa).

Reference is made to the metric version of the code, ACI 318M-19 [2]. The metric version of the code furnishes equivalents for equations and data necessary for use in the SI. Other reference sources that contain treatment of the many aspects of metrication in the design and construction field are listed at the end of this appendix [3,4].

EXAMPLE C.1

Find ϕM_n for the beam of cross section shown in Figure C.2. The steel is grade 420, and $f_c' = 20\ \text{N/mm}^2$.

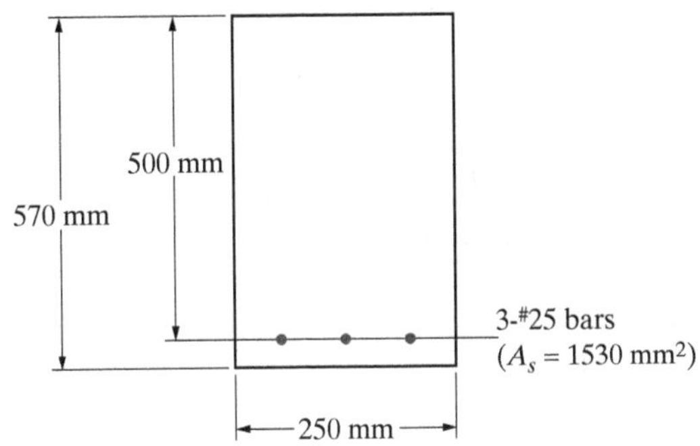

FIGURE C.2 Cross section for Example C.1.

Solution:

1. $f_y = 420\ \text{N/mm}^2$

   $f_c = 20\ \text{N/mm}^2$

   $b = 250\ \text{mm}, \quad d = 500\ \text{mm}$

   $h = 570\ \text{mm}, \quad A_s = 1530\ \text{mm}^2$

2. To be found: ϕM_n.

3. $\rho = \dfrac{A_s}{bd} = \dfrac{1530}{250(500)} = 0.0122$

4. Check $A_{s,\min}$ (ACI 318M-19, Section 9.6):

$$A_{s,\min} = \frac{0.25\sqrt{f_c'}}{f_y} b_w d \geq \frac{1.4}{f_y} b_w d$$

$$\frac{0.25\sqrt{f_c'}}{f_y} = \frac{0.25\sqrt{20}}{(420)} = 0.00266$$

$$\frac{1.4}{f_y} = \frac{1.4}{420} = 0.00333$$

Therefore, use 0.00333. Then

$$A_{s,\min} = 0.00333(250 \text{ mm})(500 \text{ mm}) = 416 \text{ mm}^2$$

$$1530 \text{ mm}^2 > 416 \text{ mm}^2 \qquad \text{(O.K.)}$$

5. Determine ε_t from Equation (2.2) (Chapter 2):

$$\begin{aligned}\varepsilon_t &= \frac{0.00255\, f'_c \beta_1}{\rho f_y} - 0.003 \\ &= \frac{0.00255(20 \text{ N/mm}^2)(0.85)}{0.0122(420 \text{ N/mm}^2)} - 0.003 \\ &= 0.00546 > 0.005\end{aligned}$$

Therefore, $\phi = 0.90$

$$a = \frac{A_s f_y}{0.85 f'_c b} = \frac{1530(420)}{0.85(20.0)(250)} = 151.2 \text{ mm}$$

6. $Z = d - \dfrac{a}{2} = 500 - \dfrac{151.2}{2} = 424 \text{ mm}$

$M_n = A_s f_y Z$

$\phi M_n = 0.9(A_s f_y Z)$

All the quantities needed for the ϕM_n calculation have been determined, but some conversion is required to make prefixes compatible. Rather than set up prefix conversion for each lengthy calculation, it is suggested for situations such as this that quantities be substituted in units of meters and newtons. The results will be in the same units. This method also lends itself to the use of numerical values expressed in powers-of-10 notation. For the ϕM_n calculation,

$$A_s = 1.530 \times 10^{-3} \text{ m}^2,\ f_y = 420 \times 10^6 \text{ N/m}^2,$$

$$Z = 0.424 \text{ m}$$

from which

$$\begin{aligned}\phi M_n &= 0.9(1.530 \times 10^{-3})(420 \times 10^6)(0.424) \\ &= 245 \times 10^3 \text{ N} \cdot \text{m} \\ &= 245 \text{ kN} \cdot \text{m}\end{aligned}$$

Note that the final ϕM_n is changed to kN·m. The kilo prefix is the most appropriate prefix for the majority of flexural problems that are presented in this booktext using the U.S. Customary System of units (see "Bending moment or torque" in Table C.6 of this text for comparison with ft-kip).

EXAMPLE C.2

Design a rectangular reinforced concrete beam for a simple span of 10 m to carry service loads of 17.1 kN/m dead load (does not include the dead load of the beam) and 31.0 kN/m live load.

The maximum width of beam desired is 400 mm. Use $f_c' = 20.0\ \text{N/mm}^2$ and $f_y = 420\ \text{N/mm}^2$. Assume a No. 10 stirrup and sketch the design. (See Chapter 2 for design procedure.)

Solution

1.
$$\begin{aligned} w_u &= 1.2\, w_{\text{DL}} + 1.6 w_{\text{LL}} \\ &= 1.2(17.1) + 1.6(31.0) \\ &= 70.1\ \text{kN/m} \end{aligned}$$

$$M_u = \frac{w_u \ell^2}{8} = \frac{70.1(10)^2}{8} = 876\ \text{kN}\cdot\text{m}$$

2. Assume that $\rho = 0.0090$ (Table A.5). Note that the given f_c' and f_y correspond approximately to 3000 psi and 60,000 psi, respectively.

3. From Table A.5(or A.8) We know that

$$\text{required } \bar{k} = 0.4828\ \text{ksi}$$

Converting to SI (see Table C.6),

$$\bar{k} = 0.4828(6.895) = 3.329\ \text{N/mm}^2$$

4. Assume that $b = 400$ mm:

$$\text{required } d = \sqrt{\frac{M_u}{\phi b \bar{k}}}$$

Substituting quantities in terms of meters and newtons yields

$$\begin{aligned} \text{required } d &= \sqrt{\frac{876 \times 10^3}{0.9(0.400)(3.329 \times 10^6)}} = 0.855\ \text{m} \\ &= 855\ \text{mm} \end{aligned}$$

$$\frac{d}{b}\ \text{ratio} = \frac{855}{400} = 2.14 \qquad \text{(O.K.)}$$

5. Estimate the total beam depth for purposes of determining the beam dead load. Assume a No. 36 main bar, a No. 10 stirrup, and a minimum cover of 40 mm. Then

$$h = 855 + 35.8/2 + 9.5 + 40 = 922\ \text{mm}$$

Use a beam depth of 950 mm. The density of reinforced concrete is 23.5 kN/m^3; therefore, the dead load of the beam per meter length is

$$0.400(0.950)(23.5) = 8.93\ \text{kN/m}$$

6. The additional M_u due to the beam dead load is

$$M_u = \frac{1.4w_{DL}\ell^2}{8} = \frac{1.4(8.93)(10)^2}{8}$$
$$= 156.3 \text{ kN}\cdot\text{m}$$
$$\text{total } M_u = 876 + 156.3 = 1032 \text{ kN}\cdot\text{m}$$

7. Using ρ, $\bar{k}$, and b as before, the effective depth required is

$$\text{required d} = \sqrt{\frac{1.032\times10^6}{0.9(0.400)(3.329\times10^6)}} = 0.928 \text{ m}$$
$$= 928 \text{ mm}$$
$$\frac{d}{b} \text{ ratio} = \frac{928}{400} = 2.32 \qquad \text{(O.K.)}$$

8. required $A_s = \rho bd$

$$= 0.0090(400)(928)$$
$$= 3340 \text{ mm}^2$$

Check $A_{s,min}$ (use Table A.5 because the given values of f'_c and f_y correspond approximately to 3000 psi and 60,000 psi):

$$A_{s,min} = 0.0033b_w d$$
$$= 0.0033(400 \text{ mm})(928 \text{ mm})$$
$$= 1225 \text{ mm}^2$$
$$3340 \text{ mm}^2 > 1225 \text{ mm}^2 \qquad \text{(O.K.)}$$

9. Use four No. 36 bars:

$$A_s = 1006(4) = 4024 \text{ mm}^2$$

The minimum beam width for four No. 36 bars may be determined (closely) from Table A.3, noting that a No. 36 bar is approximately equivalent to a No. 11 bar. Thus,

$$\text{minimum } b = 14(25.4) = 356 \text{ mm} < 400 \text{ mm} \quad \text{(O.K.)}$$

10. The total beam depth h may be taken as the effective depth required, plus minimum concrete cover, plus stirrup diameter, plus one-half the main steel diameter:

$$\text{required } h = 928 + 40 + 9.5 + 35.8/2 = 995 \text{ mm}$$

Use $h = 1000$ mm.

11. Check ε_t by calculation. Due to rounding of h, $d = 928 + 5 = 933$ mm. The final ρ is

$$\rho = \frac{A_s}{bd} = \frac{4024 \text{ mm}^2}{400 \text{ mm}(933 \text{ mm})} = 0.01078$$

From Equation (2.2) (Chapter 2) We know that:

$$\varepsilon_t = \frac{0.00255 f_c' \beta_t}{\rho f_y} - 0.003$$

$$= \frac{0.00255(20 \text{ N/mm}^2)(0.85)}{0.01078(420 \text{ N/mm})^2} - 0.003$$

$$= 0.0066 > 0.005$$

Therefore, $\phi = 0.90$ as assumed.

The design sketch is shown in Figure C.3

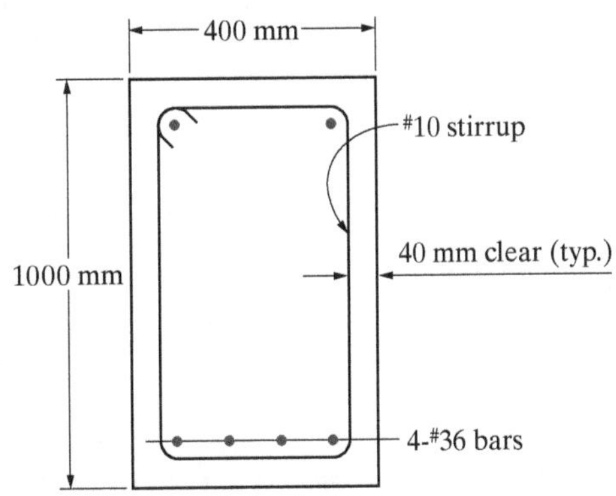

FIGURE C.3 Design sketch for Example C.2.

EXAMPLE C.3

A simply supported, rectangular, reinforced concrete beam 300 mm wide and having an effective depth of 500 mm carries a total factored load w_u of 70 kN/m on a 9.0-m clear span. (The given load includes the dead load of the beam.) Design the web reinforcement (stirrups). The steel is grade 280, and $f_c' = 20 \text{ N/mm}^2$.

Solution

1. Draw the shear force V_u diagram (see Figure C.4):

$$V_u = \frac{w_u \ell}{2} = \frac{70(9)}{2} = 315 \text{ kN}$$

At the critical section

$$V_u^* = 315 - \frac{500}{1000}(70) = 280 \text{ kN}$$

2. Determine if stirrups are required:

$$V_c = 0.17\sqrt{f'_c}b_w d$$
$$= 0.17\sqrt{20}(300)(500) = 114 \times 10^3 \text{ N} = 114 \text{ kN}$$
$$\phi V_c = 0.75(114) = 85.5 \text{ kN}$$

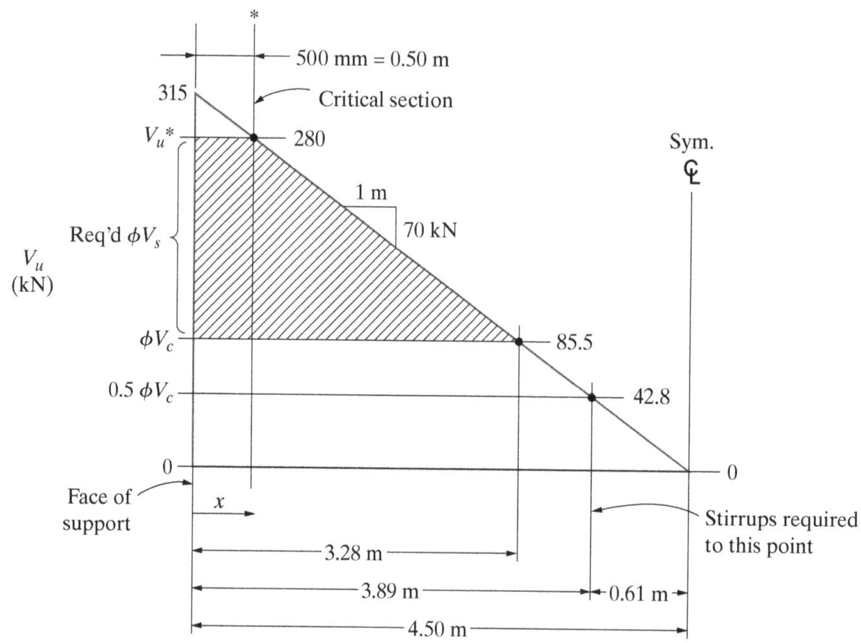

FIGURE C.4 V_u diagram.

Stirrups must be provided if $V_u > 0.5\phi V_c$:

$$0.5\,\phi V_c = 0.5(85.5) = 42.8 \text{ kN}$$

Stirrups *are* required, as $280 > 42.8$.

3. Find the length of span over which stirrups are required, referencing from the face-of-support:

$$\frac{315 - 42.8}{70} = 3.89 \text{ m}$$

Note this location on the V_u diagram as well as the location where $V_u = \phi V_c = 85.5$ kN. This location is obtained from

$$\frac{315 - 85.5}{70} = 3.28 \text{ m}$$

4. Designate "Req'd ϕV_s" on the V_u diagram:

$$\text{required } \phi V_s = \text{max.}V_u - \phi V_c - mx$$
$$= 315 - 85.5 - 70x$$
$$= 230 - 70x$$

This applies in the range

$$500\text{ mm} \le x \le 3280\text{ mm}$$

5. Assume a No. 10 stirrup $\left(A_v = 2A_s = 142\text{ mm}^2\right)$ and compute the spacing requirement at the critical section based on the required ϕV_s^*. At this location, the stirrups will be most closely spaced. From ACI Equation (22.5.10.5.3),

$$\text{required } s* = \frac{A_v f_{yt} d}{V_s^*} = \frac{\phi A_v f_{yt} d}{\text{required } \phi V_s^*}$$

where the denominator is determined with reference to Figure C.4:

$$\text{required } \phi V_s^* = V_u^* - \phi V_c$$

Using basic units of meters and newtons,

$$\begin{aligned}\text{required } s* &= \frac{0.75\left(142\times10^{-6}\right)\left(280\times10^{6}\right)(0.500)}{(280-85.5)\times10^{3}}\\ &= 0.077\text{ m} = 77\text{ mm}\end{aligned}$$

This is less than our 100-mm (4-in.) minimum spacing rule of thumb. Therefore, increase the stirrup size to a No. 13 bar $\left(A_v = 258\text{ mm}^2\right)$. Then

$$\begin{aligned}\text{required } s* &= \frac{0.75\left(258\times10^{-6}\right)\left(280\times10^{6}\right)(0.500)}{(280-85.5)\times10^{3}}\\ &= 0.139\text{ m} = 139\text{ mm}\end{aligned}$$

We will use a 130-mm spacing as the stirrup spacing between the face-of-support and the critical section, subject to further checks.

6. Establish the ACI Code maximum spacing requirements. From the ACI Code (318M-19), Section 9.7.6.2.2, if V_s is less than $0.33\sqrt{f_c'}b_w d$, the maximum spacing is $d/2$ or 600 mm, whichever is smaller; otherwise, the maximum spacing will be the smaller of $d/4$ or 300 mm.

$$\begin{aligned}0.33\sqrt{f_c'}b_w d &= 0.33\sqrt{20.0}(300)(500)\\ &= 221\times10^{3}\text{ N}\\ &= 221\text{ kN}\end{aligned}$$

At the critical section, the required V_s is

$$\begin{aligned}V_s^* &= \frac{\phi V_s}{\phi} = \frac{V_u^* - \phi V_c}{\phi}\\ &= \frac{280-85.5}{0.75} = 259\text{ kN}\end{aligned}$$

Because 259 kN > 221 kN, the maximum spacing will be $d/4$ or 300 mm, whichever is smaller, from the face-of-support out to where the required V_s drops below 221 kN. This maximum spacing is

$$\frac{d}{4} = \frac{500}{4} = 125 \text{ mm}$$

125 mm is less than 300; therefore, use 125 mm. Next, determine where $V_s = 221$ kN, which is where the maximum spacing can be increased to the smaller of $d/2$ or 600 mm.

$$\frac{d}{2} = \frac{500}{2} = 250 \text{ mm}$$

250 mm < 600 mm; therefore, use 250 mm.

$$V_s = \frac{\phi V_s}{\phi} = \frac{V_u - \phi V_c}{\phi} = \frac{(315 - 70x) - \phi V_c}{\phi} = 221 \text{ kN}$$

from which

$$x = \frac{\phi 221 + \phi V_c - 315}{-70} = 0.911 \text{ m}$$

Therefore, the maximum spacing allowed increases to 250 mm at 0.911 m from the face-of-support.

A second criterion for maximum spacing is based on the code minimum area requirement (ACI 318M-19, Section 9.6.3.3). The governing equation may be rewritten in the form

$$s_{max} \leq \frac{A_v f_{yt}}{0.062\sqrt{f'_c}\, b_w} = \frac{258(280)}{0.062\sqrt{20}\,(300)} = 868 \text{ mm}$$

Check the upper limit:

$$s_{max} = \frac{A_v f_{yt}}{0.35 b_w} = \frac{258(280)}{0.35(300)} = 688 \text{ mm}$$

Of the foregoing maximum spacing criteria, the smallest value will be controlled. Maximum spacing requirements are summarized in Figure C.5.

7. Next, determine the spacing requirements based on shear strength. At the critical section, the required spacing is 130 mm. The maximum spacing is 125 mm to 0.911 m from the face-of-support and 250 mm thereafter. At other points along the span (x meters from the face-of-support), the required spacing may be determined as follows:

$$\text{required } s = \frac{A_v f_{yt} d}{\text{required } V_s} = \frac{\phi A_v f_{yt} d}{\text{required } \phi V_s}$$

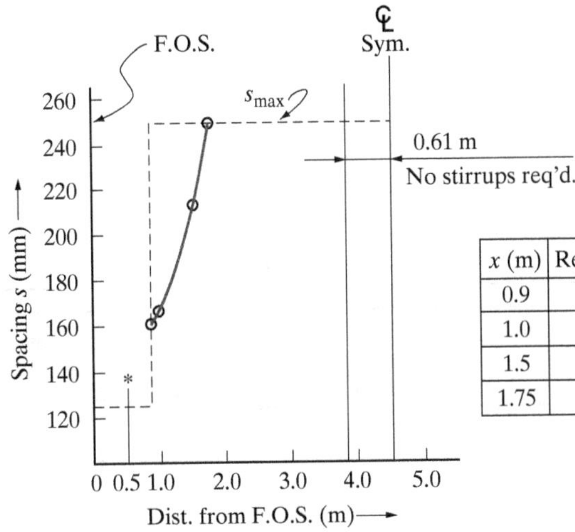

x (m)	Req'd s (mm)
0.9	162
1.0	169
1.5	217
1.75	252

FIGURE C.5 Stirrup spacing requirements for Example C.3.

where the denominator can be determined from the expression given in step 4, where

$$\text{required } \phi V_s = 230 - 70x$$

Using basic units of meters and newtons, the calculation for required spacing results in

$$\text{required } s = \frac{0.75(258 \times 10^{-6})(280 \times 10^{6})(0.500)}{230 \times 10^{3} - \left(70 \times 10^{3}\right)x}$$

$$= \frac{27.1 \times 10^{3}}{230 \times 10^{3} - (70 \times 10^{3})x}$$

$$= \frac{27.1}{230 - 70x}$$

where the resulting spacing is in meters.

The results for several arbitrary values of x are shown tabulated and plotted in Figure C.5. As an example, compute the required stirrup spacing at a distance of 1 m from the face-of-support (x = 1 m):

$$\text{required } s = \frac{27.1}{230 - 70(1)} = 0.169 \text{ m} = 169 \text{ mm}$$

Similarly, the required spacing may be found at other points along the beam (see Figure C.5).

8. Using Figure C.5, the stirrup pattern shown in Figure C.6 may be developed. Stirrups have been placed the full length of the span, which is a common, conservative practice.

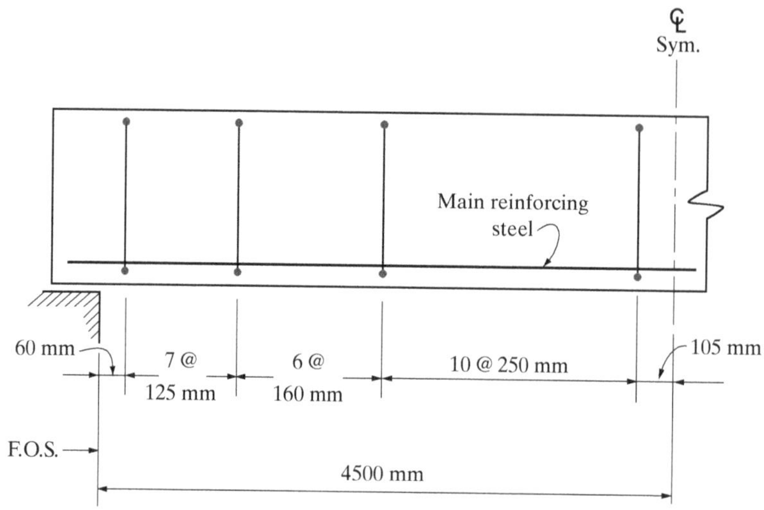

FIGURE C.6 Stirrup spacing, Example C.3.

REFERENCES

[1] *Construction Metrication*. "Construction Metrication Council of the National Institute of Building Sciences," 1090 Vermont Ave. NW, Washington, D.C. 20005-4905, V. 9, No. 4 (4th Qtr.), 2000.

[2] *Building Code Requirements for Structural Concrete* (ACI 318M-19). American Concrete Institute. 38800 Country Club Drive, Farmington Hills, MI 48331, 2019.

[3] ISO 1000/AMD1:1998. *SI Units and Recommendations for the Use of Their Multiples and of Certain Other Units,* 3rd ed., American National Standards Institute, Inc., 1819 L Street, NW, Washington, D.C. 20036, 1992 (amended 1998).

[4] *Using Soft-Metric Reinforcing Bars in Non-Metric Construction Projects*, Engineering Data Report No. 42. Concrete Reinforcing Steel Institute, 933 N. Plum Grove Road, Schaumburg, IL 60173-4758, 1997.

APPENDIX D

ANSWERS TO SELECTED EXERCISES

Chapter 1

1.1 **a.** w = 467 lb/ft

b. w = 488 lb/ft

1.3 f_r = 0.356 ksi; ACI f_r = 0.411 ksi

1.5 f_{top} = 0.396 ksi

1.9 M_{cr} = 531 in.-kip

Chapter 2

2.1 **a.** M_n = 421 ft-kip

b. M_n = 515 ft-kip (+22%); (A_s: +27%)

c. M_n = 501 ft-kip (+19%); (d: +16.7%)

d. M_n = 436 ft-kip (+3.6%); $\left(f_c': +33.3\%\right)$

2.3 **a.** ϕM_n = 213 ft-kip

b. ϕM_n = 310 ft-kip (+45.5%); (f_y: +50%)

2.5 M_u = 939 ft-kip

a. ϕM_n = 894 ft-kip (N.G.)

b. ϕM_n = 1072 ft-kip (O.K.)

2.7 M_u = 164.5 ft-kip

ϕM_n = 169 ft-kip (O.K.)

2.9 ϕM_n = 350 ft-kip > 304 ft-kip (O.K.)

2.11 ϕM_n = 41.4 ft-kip > 35.5 ft-kip (O.K.)

2.13 As designed, ϕM_n = 19.32 ft-kip; as built, ϕM_n = 11.96 ft-kip (−38.1%)

2.15 4 No. 9

2.17 3 No. 11 ϕM_n = 417 ft-kip

2.19 6 No. 9 (two layers, 1-in. clear.) ϕM_n = 540 ft-kip

2.21 b = 16 in., h = 32 in., 5 No. 9

2.23 b = 18 in., h = 35 in., 6 No. 9

2.25 b = 16 in., h = 34 in., 5 No. 10

2.27 b = 12 in., h = 27 in., 3 No. 11

2.29 (+Moment): b = 11 in., h = 24 in., and 2 No. 9 (−Moment): same b and h, and 2 No. 6

2.31 **a.** 6 in. slab, No. 5 @ 15 in. o.c. main steel, No. 4 @ 18 in. o.c. shrinkage and temperature steel

b. 4 in. slab, No. 6 @ 12 in. o.c. main steel, No. 3 @ 14 in. o.c. shrinkage and temperature steel

Chapter 3

3.1 ϕM_n = 303 ft-kip

3.3 ϕM_n = 439 ft-kip

3.5 **a.** ϕM_n = 1527 ft-kip

b. A_s = 12.43 in.^{2}

3.7 ϕM_n = 856 ft-kip

3.9 ϕM_n = 1120 ft-kip

3.11 ϕM_n = 422 ft-kip

3.13 6 No. 10

3.15 **a.** 4 No. 9 bars (two layers)

b. 4 No. 8 bars (one layer)

3.17 **a.** $w_{LL} = w_{DL}$ = 1.87 kip/ft

b. **1.** ϕM_n = 341 ft-kip

2. ϕM_n = 344 ft-kip

3.19 **a.** ϕM_n = 582 ft-kip

b. with 4 No. 8 compression bars, ϕM_n = 696 ft-kip

3.21 3 No. 8 (compression steel), 5 No. 11 (tension steel 2 layers, 2 up, 3 down)

3.23 2 No. 9 bars (compression steel), 4 No. 9 bars (two layers, tension steel)

3.25 2 No. 10 bars (compression steel), 5 No. 9 bars (tension steel, in one layer)

Chapter 4

4.1 7150 lb < 9000 lb (N.G. in shear)

4.3 Max. V_u = 35.6 kip

4.5 s = 7.76-in. spacing

4.7 No. 3 stirrups (from F.O.S.): 3 in., 8 sp @ 6 in., 11 sp @ 10 in.

4.9 Double-loop No. 3 stirrups (from F.O.S.): 4 in., 11 sp @ 9 in., 6 sp @ 20 in.

4.11 No. 3 stirrups (from F.O.S.): 2 in., 7 sp @ 9 in., 7 sp @ 16 in.

4.13 No. 3 stirrups (from F.O.S.): 2 in., 6 sp @ 6 in., 5 sp @ 8 in., 3 sp @ 12 in.

4.15 No. 3 stirrups (from F.O.S.): 4 in., 14 sp @ 13 in.

4.17 T_{cr} = 24.3 ft-k: torsion may not be neglected

Chapter 5

5.1 55.4 in.

5.3 32.7 in. Use 2-in. (min.) side cover.

5.5 No. 4

5.7 $\ell_d = 69.6$ in. > 49.5 in. (N.G.); 180° hook: $\ell_{dh} = 16.9$ in.

5.9 At A, req'd lap = 32.2 in.; at B: req'd lap = 54.3 in.

5.11 Req'd lap = 25.4 in.

5.13 **a.** Cut 2 No. 9 @ 13 ft-0 in. from centerline.

b. No. 3 stirrups (from F.O.S.): 3 in., 3 sp @ 15 in., 6 sp @ 5 in., 8 sp @ 15 in.

Chapter 6

6.1 M_u = –3.03 ft-kip, +5.19 ft-kip, –7.26 ft-kip, +4.54 ft-kip, –6.60 ft-kip; V_u = 3.30 kip, 3.8 kip

6.3 b = 12 in., h = 24 in.; end span: 2 No. 6 bars for –M @ end support (with 180° hook), 2 No. 8 bars for +M: interior span: 3 No. 8 bars for –M @ interior support, 2 No. 8 bars for +M

Chapter 7

7.1 **a.** $\bar{y} = 11.61$ in., $I_{cr} = 20{,}350$ in.^{4}

b. $\bar{y} = 1.68$ in., $I_{cr} = 72.7$ in.^{4}

c. $\bar{y} = 7.63$ in., $I_{cr} = 6159$ in.^{4}

7.3 **a.** Δ = 0.36 in.

b. Δ = 0.43 in.

7.5 Δ = 0.34 in. < 0.40 in. (O.K.)

7.7 **a.** s = 3.3 in.; max. s = 10.31 in. (O.K.)

b. s = 8.0 in.; max. s = 12 in. (O.K.)

Chapter 8

8.1 **a.** Total H = 7.01 kip/ft

b. Total H = 6.9 kip/ft

c. Total H = 8.0 kip/ft

d. Total H = 12.5 kip/ft

8.3 Overturning F.S. = 3.29, sliding F.S. = 1.35, p_{max} = 0.80 ksf, p_{min} = 0.55 ksf

8.5 No. 7 @ 8 in. o.c.; use a 90° standard hook

8.9 No. 4 HEF @ 18 in. o.c., No. 4 VEF @18 in. o.c., 6 No. 9 vertical bars each end of wall spread over an end zone length of 12 in. (required minimum end zone length is 6 in.)

Chapter 9

9.1 **a.** $\phi P_{n(max)}$ = 530 kip, required ties are No. 4 @ 14 in. o.c.

b. $\phi P_{n(max)}$ = 1156 kip, required ties are No. 3 @ 18 in. o.c. (three per set)

c. $\phi P_{n(max)}$ = 759 kip, required ties are No. 3 @ 16 in o.c. (two per set)

9.3 $\phi P_{n(max)}$ = 636 kip, required ties = No. 4 @ 16 in. o.c., $P_{DL} = P_{LL}$ = 227 kip

9.5 $\phi P_{n(max)}$ = 840 kip, required spiral = $\frac{3}{8}$ in. diam. @ 2 in. o.c.

9.7 Use 12 No. 9 bars (four per face), required ties are No. 3 @ 18 in. o.c. (three per set)

9.9 Use a column 14 in. × 14 in., 6 No. 9 bars, No. 3 ties @ 14 in. o.c.

9.11 e_b = 13.29 in., ϕP_n = 341 kip

9.13 Use a column 24 in. diameter; 10 No. 10 bars, $\frac{3}{8}$ in. diameter spiral at $2\frac{1}{4}$ in. o.c.

Chapter 10

10.1 Width = 24 in., depth = 12 in., longitudinal steel: 3 No. 4 bars

10.3 Width = 6 ft-0 in depth = 1′-3″, transverse steel: No. 6 @ 11 in. o.c.; longitudinal steel: 7 No. 5 bars

10.5 9 ft-0 in. square, depth = 2 ft-0 in., 11 No. 7 bars each way

10.7 Rectangular footing, 7 ft-0 in × 11 ft.-6 in., depth = 2 ft-0 in., 9 No. 8 bars (long direction), 15 No. 6 bars (short direction)

10.9 Rectangular footing, 19 ft-6 in. × 23 ft- 9 in.

10.11 Footing for column A: 7 ft-6 in. square; footing for column B: 8 ft-2 in. square

Chapter 11

11.1 $f = \pm 0.209$ ksi

11.3 w = 1.34 kip/ft

11.5 Midspan (transfer): f_{top} = 0.929 ksi (tens), f_{bott} = 2.66 ksi (comp); midspan (full load): f_{top} = 0.477 ksi (comp), f_{bott} = 1.259 ksi (comp); at end supports (only prestressing stresses exist): f_{top} = 1.085 ksi (tension), f_{bott} = 2.82 ksi (comp)

11.7 f_{top} = 2.20 ksi (comp), f_{bott} = 0.132 ksi (comp)

11.9 ϕM_n = 889 ft-kip, M_u = 763 ft-kip (O.K.)

Chapter 12

12.3 Max. stringer spacing = 92.9 in.

12.5 Use 4 × 10(S4S)

12.7 Max. shore spacing = 37.0 in.

12.9 Max. shore spacing = 35.4 in.

12.11 Joists @ 16 in. o.c., stringers @ 5 ft-9 in. o.c., 4 × 4 (S4S) shores @ 4 ft-6 in. o.c., guy wire bracing @ 12 ft (max.) o.c. on all sides

12.13 Studs @ 12 in. o.c., wales @ 28 in. o.c., max. tie spacing = 32 in. o.c., guy wire bracing @ 25 ft (max.) o.c. on each side of the wall

INDEX

D

T

www.ingramcontent.com/pod-product-compliance
Lightning Source LLC
LaVergne TN
LVHW062125200926
848484LV00007B/1769
* 9 7 8 1 6 8 3 9 2 9 0 7 9 *